181371 10.95

Rieger/Michaelis/Green

Glossary of Genetics and Cytogenetics

ALLE ZEIT WACH 1842

SPRINGER STUDY EDITION

R. Rieger A. Michaelis M. M. Green

Glossary of Genetics and Cytogenetics

Classical and Molecular

Fourth Completely Revised Edition

Springer-Verlag
Berlin Heidelberg New York 1976

With 100 Figures and 8 Tables.

ISBN 3-540-07668-9 (4th edition) Springer-Verlag Berlin · Heidelberg · New York
ISBN 0-387-07668-9 (4th edition) Springer-Verlag New York · Heidelberg · Berlin

ISBN 3-540-04316-0 (3rd edition) Springer-Verlag Berlin · Heidelberg · New York
ISBN 0-387-04316-0 (3rd edition) Springer-Verlag New York · Heidelberg · Berlin

The first and second edition of this book were published in German in 1954 and 1958 by Springer-Verlag under the title: „Genetisches und Cytogenetisches Wörterbuch".
The third edition was published in English by Springer-Verlag in 1968.

Library of Congress Cataloging in Publication Data. Rieger, Rigomar. A glossary of genetics and cytogenetics.
"The first and second edition of this book were published in German ... under the title: Genetisches und cytogenetisches Wörterbuch". Bibliography: p. Includes index. 1. Genetics—Dictionaries. 2. Cytogenetics—Dictionaries. I. Michaelis, Arnd, joint author. II. Green, Melvin M., joint author. III. Title.
QH427.R54 1976 575.1'03 76-16183

Printing: VEB Druckhaus Köthen
Bookbinding: Konrad Triltsch, Graphischer Betrieb, FRG.

Authors

Professor Dr. Rigomar Rieger
Zentralinstitut für Genetik und Kulturpflanzenforschung
der Akademie der Wissenschaften der DDR, Gatersleben

Dr. Arnd Michaelis
Zentralinstitut für Genetik und Kulturpflanzenforschung
der Akademie der Wissenschaften der DDR, Gatersleben

Professor Dr. Melvin M. Green
Department of Genetics, University of California,
Davis/Calif., USA

Preface

In preparing the new completely revised edition of this glossary, which in the meantime has been translated into Russian and Polish, we have attempted to include the most important new terms and to revise the text in those cases where new data demanded it*. As a result about fifty percent of the text is completely rewritten. Once more we have tried to provide material suitable and usable both for students and research workers. Accordingly, depending upon our evaluation, some terms have been simply defined, others have been described at some length even to the extent of providing experimental data. Wherever possible, synonymy and redundancy have been pointed out, and in the interest of historical accuracy the individual responsible for introducing a particular term or concept listed with the specific paper included in the literature citations. Cross references between related terms are designated by an arrow (→) before each relevant term.

To keep the book, as far as possible, to a reasonable size the terms carried over from the earlier edition have once more been critically selected and, where necessary, revised. In spite of these efforts a certain increase in volume was unavoidable.

We hope that the new edition will once more prove useful to a wide audience and enjoy the same cordial reception as the earlier ones. Comments and suggestions from the reviewers and users of the earlier editions have contributed significantly to the revision. We are grateful to all those colleagues who directed our attention to mistakes, omissions and ambiguities. Needless to say any comment of the users of the new edition will be much appreciated.

Many colleagues and friends, too numerous to identify here, have once more aided in the compilation of the contents of the book. To all our sincere thanks. We are, however, especially indebted to Mrs. Erdmuthe Rieger who participated directly in the preparation of the manuscript; without her untiring and unstinting help the revised text would, in all probability, never have been typed and prepared for the printer.

Gatersleben and Davis
June 1976

Rigomar Rieger *Arnd Michaelis*
Melvin M. Green

---

* Some terms and references have been added in proof (see pages 634 to 647).

A

AI, AII — abbreviations for the anaphase of the first and second meiotic division, resp. (→ meiosis).
abbreviation — the successive shortening of ontogenesis through cessation of individual stages.
aberration rate — → chromosome mutation.
abortive infection (*Lwoff* 1953) — infection of prokaryotic (bacterial) or eukaryotic cells by bacteriophages or viruses, respectively, which does not result in the formation of infectious virus although one or more viral components are synthesized (→ productive infection; reductive infection; lysogenization).
abortive transfer — any DNA transfer from bacterial donor to recipient cells that fails to establish the incoming DNA as part of the hereditary material of the recipient. A. t. has been observed following → transduction, → transformation, and → conjugation. In all cases the transferred fragment is diluted out as the culture grows. Failure of integration of transferred DNA into the hereditary material of the recipient cell may be due to: 1. the failure of incoming DNA to form circular molecules; 2. circularization takes place, but the circular molecule fails to take up its maintenance system. A. t. of extrachromosomal elements (→ plasmids), as opposed to chromosomal fragments, is relatively uncommon since plasmids are genetic elements of autonomous survival in a bacterial cell. It is only when a mutation in the recipient or a resident plasmid makes the host component of the plasmid maintenance system inactive that a. t. of a plasmid occurs. Genes carried on abortive pieces of DNA may be expressed in the recipient cells.
accessory chromosome (*McClung* 1900) — → sex and → B-chromosome.
accessory DNA — surplus DNA present in certain cells or cell stages due to, e.g., → gene amplification.
accessory nucleus — in the oocytes of some insects, any of the structures present in the peripheral cytoplasm during → vitellogenesis and probably originating from the oocyte nucleus. The electron-dense a. nuclei contain RNA, possibly of nuclear origin. A. nuclei seem to be involved in the control of the synthesis of albuminous yolk and afterwards perhaps with the formation of the vitelline membrane.
accessory plate (*Darlington* 1936) — a supplementary metaphase plate which owes its origin to bivalents which, during the metaphase either remain outside the equatorial plate (because of a lack of → centromere orientation) or have not yet attained this (noncongression) (→ congression).
accomodation (*Thach & Thach* 1971) — a GTP-dependent reaction in bacteria which occurs during → genetic translation. → Messenger RNA (mRNA) is moved a distance of about three nucleotides in the 5 direction relative to the → ribosome during the translocation of peptidyl-tRNA from the A to the P site of the ribosome. This reaction is catalyzed by the → G

factor (→ translocation factor) and is dependent on the hydrolysis of GTP. In contrast, mRNA is not moved during the IF 2-catalyzed (→ initiation factor) hydrolysis of GTP which is involved in the activation of ribosome-bound fMet-tRNA (→ initiator tRNA). This second type of GTP-dependent reaction is called a. of fMet-tRNA.

acentric — ref. to → chromosomes or chromosome segments without a → centromere.

achiasmate — of a → meiosis without → crossing over and chiasmata. Where a. meiosis occurs, it is usually confined to one of the two sexes. Generally, the other sex has a chiasmate type of meiosis. A common feature of a. meiosis is the absence of any opening-out of paired chromosomes (bivalents) at diplotene; the four chromatids remain parallel until the beginning of first metaphase. A. meiosis is a distinct and characteristic mechanism which has arisen independently in a large number of organisms. In higher plants it is rare and not yet observed in vertebrates (*White* 1974). (→ cryptochiasmate).

achromatic (*Flemming* 1879) — those parts of the → nucleus which may not be stained with dyes characteristic for chromosomes (→ chromatin).

achromatic figure — the → spindle during → mitosis and → meiosis (→ mitotic apparatus).

achromatic lesion — → gap.

A-chromosome (*Randolph* 1928) — any of the standard chromosomes of the → chromosome complements of all eukaryotic organisms. A-chromosomes represent a delicately balanced system and each of them usually has to be present to secure normal viability. In many species, the standard complement of A-chromosomes may be supplemented by varying numbers of supernumerary or → B-chromosomes. These are derived from the A.-c. set and have become functionally subordinate.

acquired character — a → modification impressed on an organism by environmental influences during development.

acrocentric (*White* 1945) — of chromosomes where the → centromere is very close to one end so that one → chromosome arm is small or minute and the other very much longer (→ metacentric).

acrosome (*Lenhossek* 1897) — a cap-like structure which invests the front part of the sperm head. It is limited externally by an outer membrane that is reflected forwards at the posterior boundary of the a. to form an inner membrane contiguous with the nuclear membrane of the spermatozoon (*Hancock* 1966).

The details of a. formation vary but in general follow one of two paths (*Nath* 1956): either the Golgi elements (→ dictyosomes) are transformed directly into the a. or they "secrete" the a. and are then shed with the residual cytoplasm.

In function the a. is concerned with the enzymatic penetration of the protective cover of the egg cell after formation of a surface layer. When sperm draw near to recently ovulated eggs, the outer acrosomal membrane and overlying plasma membrane fragment, creating openings to release enzymes of the a. This process is called "acrosomal reaction" (*Bedford* 1970). Genetically controlled defects of the a. may impair zygote formation.

acrosyndesis (*Percival* 1932) — incomplete end-to-end pairing of two chromosomes during → meiosis.

action system (*Hamburger*) — in embryology, a system consisting of the → organizer and its organization field, which, in the → reaction system, brings about the realization of one or more developmental potencies (→ induction system).

activation — in egg cells, a chain of responses made by an unfertilized egg to fertilization or penetration by a pipette. These responses include the rupture of cortical granules, the lifting off of a vitelline membrane, and the free rotation of the egg nucleus by the influence of gravity (*Gurdon* 1974).

activator (*Huxley* 1935) — a gene-dependent substance which stimulates the development of a certain embryonic tissue or organ, i.e., is morphogenetically active.

1. **Local activator:** active only in the cell or in the tissue in which it is produced.

a) intracellular activator: active only in the cell;

b) chemodifferentiator: active on the tissue to bring about determination of embryonic parts.

2. **Distance activator** or **hormone:** active beyond the region of its formation. The transport in the body may proceed by

a) diffusion (diffusion hormone),

b) body fluids (circulation hormone).

activator RNA (*Britton & Davidson* 1969) — a regulator molecule capable of recognizing specific sites on DNA. A. RNA is assumed to play a role in the regulation of → gene expression in eukaryotes. According to the *Britton-Davidson* model of → genetic regulation in eukaryotes, the genetic material is in its normal state repressed. Particular control systems function specifically to overcome this repression and to switch on appropriate genes by the functioning of a. RNA. A. RNA is assumed to be coded for by "integrator" genes which are active only when an adjacent segment of DNA (= "sensor" gene) interacts with a specific protein.

active transport — passage of a substance across the → cell membrane against a concentration gradient; energy must be supplied for a. t.

adaptation — any change in an organism's structure or function that allows it to better cope with conditions in the environment. Harmonious adjustment to environmental conditions is the result of a. Adaptation as a process or the result of a process taking a different course in different individuals, signifies the build-up or the possession of → characters which prove advantageous for the individual or the population under the environmental conditions in which it lives and through which the organism acquires → adaptive value or fitness in a certain environment.

A. can be achieved in two different ways, as pure phenotypic a. or as genotypic a. In the first case, the → reaction norm of the genotype can be attuned to those environmental conditions prevalent under natural circumstances. In the second case, genotypic specialization leads to a. in such a way that a change of genotype results in the formation of a new reaction norm which makes it possible for the genotype to harmonize where the original form fails to do so.

Adaptations consist of harmoniously working combinations of genes which are built up and preserved by the process of natural → selection. The adaptation presupposes in each case the possession of a genotype with a favorable reaction norm. The reaction is then favorable if the frequently recurrent environmental influences give rise to phenotypic modifications which permit the organism to remain viable and produce offspring (→ flexibility).

An adaptation process which furthers the group at the cost of the individual, i.e., the chance of the survival of the individual is lowered while the survival rate of the group increases, is designated as **"altruistic adaptation"** (*Haldane* 1932).

The possibility of certain genotypes adapting immediately to changed environmental conditions by means of prospective functions is so-called **"prospective adaptation"** (*Simpson* 1953) and means the formation of characters which are without adaptive value at the time of formation but which prove to be adaptive in a new environment (= **preadaptation**).

"Pseudoexogenous" (*Waddington* 1953) is an a. which appears to be occasioned directly by environmental influences but actually results without environmental stimulus or is independent thereof.

adaptedness — the state of being adapted. The a. of a genotype is a function of its → norm of reaction and of the range of environments in which it occurs.

adaptibility — the potentiality for → adaptation. Physiological a. depends upon a → norm of reaction yielding a certain adaptedness in the environments which individuals or populations meet. Genetic a. (= evolutionary → plasticity) occurs by changing the norm of reaction to produce an improved adaptedness in some or all, old or new environments. A genotype that gives rise to adaptive traits in some or all of the environments it meets may be said to possess high a. which may depend upon physiological → homeostasis or upon developmental homeostasis (*Dobzhansky* 1968).

adaptiogenesis — the formation of new → adaptations.

adaptive — of those changes of an organism which act to preserve life, i.e., tend to increase viability, survival rate, and reproductive rate. An a. trait or character is a structural or functional characteristic of the organism which enables or enhances the probability of survival and reproduction (*Dobzhansky* 1956).

adaptive norm (*Schmalhausen* 1949) — a well-adapted more or less stable complex of genetic diversity within the population.

adaptive peak (*Wright* 1932) — a symbolic representation (from an evolutionary point of view) of the relations between organism and environment in the form of a topographic map. Groups of related genotypes which make their carriers able to occupy certain ecological niches are said to occupy "adaptive peaks" situated in different parts of the map and separated from each other by "adaptive valleys" comprising unfit gene combinations. Any such peak represents a particular configuration of → gene (allele) frequencies and → genotype frequencies characteristic for a population equilibrium and (relatively) stabilized by a complex interaction of opposing forces. If one of these interacting factors becomes modified in its action, three reactions may occur (*Dobzhansky* 1951, *Lerner* 1958):

1. compensating adjustments in the responses of the population to the other forces;

2. movement of the population from one such equilibrium across a "valley" to another new a. p. which may be represented by a completely different array of gene frequencies and may require a far reaching rebuilding of the → gene pool. A shift from one to another a. p. is connected with the formation of stages intermediate between the two which may be more or less unbalanced;

3. disappearance of an original a. p. and extinction of the particular gene combinations in those cases in which no mutations occur creating new adaptive gene combinations or if the new combinations do not occur at the right time.

The construction of adaptive peaks is as follows: all genotypes are plotted as points on a plane — the more closely they lie the less they differ from each other — and as ordinate the prevailing → adaptive value of these genotypes within the appropriate environment is used. In this way a "mountain range" is set up with "peaks" separated by "valleys" and "saddles". Each of the various genotypes is then represented by a point on this "mountain surface" and each population occupies a region on the same.

adaptive radiation — the evolutionary diversification of a group of organisms (of a single phyletic line) which leads, often within a relatively short period of time, to the formation by natural → selection of a variety of types from a single ancestral species. These types are adapated to certain environmental conditions. A. r. results in a ramification of an adaptive type in a series of new adaptive zones (*Simpson* 1953).

adaptive value — the survival value and the reproductive capability of a → genotype in comparison with other genotypes of the population in a particular environment. The a. v. (= fitness) represents a property of the genotype as a whole and is more than the sum of the values of its constituent genes, since, e.g., gene A may interact unfavorably in combination with B, neutrally with C, and favorably with D. A higher a. v. of a genotype means that the bearer, on average, will produce more surviving offspring than the bearer of another genotype in the same environment. This superiority can be the result of a higher resistance towards environmental influences and a greater longevity, or one genotype may be sexually more active or more fertile than the other (*Dobzhansky* 1951). The a. v. of a particular genotype in a given environment can be increased or detrimentally affected by the presence of other genotypes in the same environment (*Weisbrot* 1966).

adaptive zone — the "way of life" of a taxonomic group of organisms in a broad sense. An a. z. may be subdivided into adaptive subzones.

adaptor hypothesis (*Crick* 1958; *Hoagland* 1959) — an hypothesis (now experimentally proved) which states that the amino acid sequence of a protein is determined, during the course of → genetic translation, by the alignment of aminoacyl-transfer RNA's at corresponding nucleotide →

codons in → messenger RNA. The specificity of the translation mechanism depends on base pairing between a nucleotide region of → transfer RNA (the → anticodon) and nucleotides in the → codon.

adaptor modification hypothesis (*Sueoka* & *Kano-Sueoka* 1964) — an hypothesis proposed to explain the regulation of protein synthesis by modification of tRNA molecules at a site which affects recognition of either the mRNA codon, the enzyme (→ aminoacyl-tRNA-synthetase), or the ribosome. When multiple species of tRNA correspond to the same amino acid, modification of one tRNA species may prevent the → genetic translation of mRNA with the corresponding codon(s).

adaptor molecules — → genetic translation.

additive genes — genes interacting and showing no → dominance if → alleles or showing no → epistasis if nonalleles (→ gene interaction).

additive theorem — the a. t. of → exchange percentages states: the exchange percentage occurring on → crossing-over between the loci A and C of a linkage group is equal to the sum of the values AB and BC, if the locus B lies between A and C, or equals the difference if locus B is localized outside the stretch AC (Fig. 1).

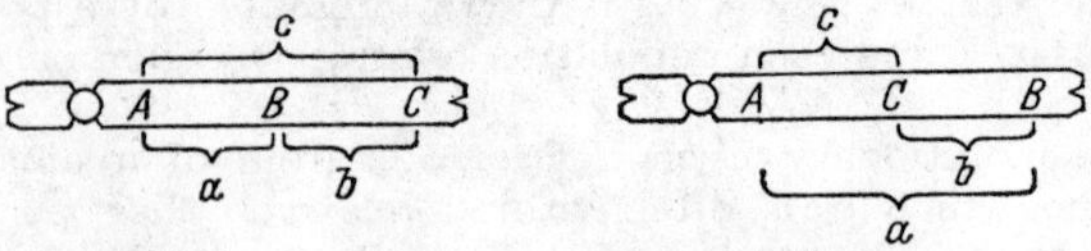

Fig. 1. Diagram showing the possible arrangements of three genes when only two of the distances (a and b) are known ("additive theorem of exchange percentages").

adelphogamy — sib pollination. Pollen and stigma belong to two different individuals which derive vegetatively from the same mother plant.

adjacent distribution (*McClintock* 1945) — the orientation and distribution of chromosomes lying adjacent in the ring or chain configuration of translocation heterozygotes (→ translocation) to the same pole in the first meiotic division (→ meiosis) in contrast to → alternative distribution, in the course of which alternating chromosomes of the configuration are distributed to the same pole. It is a consequence of a. d. that either all or a part of the products of meiosis (gones or gametes) are genetically unbalanced and contain → duplications or → deletions. The proportion of unbalanced meiotic products is determined by the position and number of the chiasmata (in the → pairing segments or → interstitial segments). A. d. is responsible for partial sterility of translocation heterozygotes (→ semisterility). Two types of a. d. may be distinguished (Fig. 2a, b):

1. **adjacent-1 distribution** (= nonhomologous adjacent distribution; nondisjunctional distribution): chromosomes adjacent in the configuration with nonhomologous centromeres are distributed to the same pole in meiosis I. In such a case the translocated segments and the homologous segments of the structurally unchanged chromosomes are not separated from one another. For them it is a case of "non-disjunction".

2. **adjacent-2 distribution** (= homologous adjacent distribution; disjunctional distribution): chromosomes adjacent in the configuration with homologous centromeres are distributed to the same pole in meiosis I. The translocated segments and the homologous segments of the structurally unchanged chromosomes are separated from one another. For them it is a case of "disjunction".

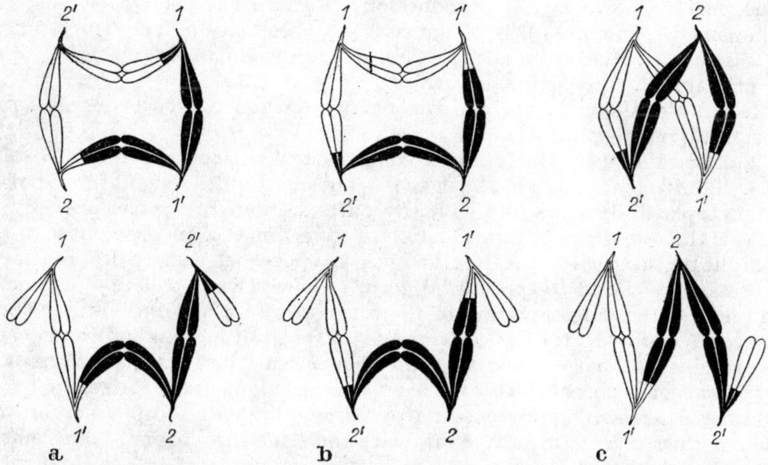

Fig. 2. The principal possible orientations in an individual heterozygous for a reciprocal translocation of a ring (upper row) or chain (lower row) of four chromosomes at metaphase I, assuming the spindle poles are towards the upper and lower margins of each row. The modes of distribution of the chromosomes resulting from these types of orientation are called adjacent —1 distribution (a), adjacent — 2 distribution (b), and alternative distribution (c).

adventitious embryony (*Strasburger* 1878) — a form of → apomixis (agamospermy); the production of seeds without a sexual process.

affinity — 1. nonrandom → assortment of unlinked markers by some sort of a. between nonhomologous chromosomes causing them to pass preferentially to the same pole of the → spindle in anaphase of the first or second meiotic division. A. of this type has been inferred from genetic studies in mice and yeast (*Michie & Wallace* 1953).

2. In the case of selective → fertilization, the genetically controlled mutual attraction of male and female gamete types differing as to their genetic constitution. A. is a measure of strength of attraction, while the velocity with which this attraction proceeds, is called "reaction velocity" (*Haustein* 1955).

3. → differential affinity.

4. → terminal affinity.

agameon (*Camp & Gilly* 1942) — a species reproducing exclusively by → apomixis.

agamete — any noncopulating → germ cell (= spore). Agametes are formed as products either of meiosis (meio-a., meiospores, tetraspores, gonia) or mitosis (mito.-a. gonidia). Reproduction accomplished by a. is designated as → agamogony, monocytogony, monogenic, or monocytogenic reproduction, agamic or agametic reproduction, agamogenesis, or sporogony (→ gamete).

agamic — reproducing (→ reproduction) asexually.

agamogenesis — asexual → reproduction (→ gamogenesis).

agamogony (*Hartmann* 1904) — asexual → reproduction. The development of a new individual from a single cell. A. in unicellular organisms proceeds in principle in three different ways:

1. Simple fission of the cell the result of which are two roughly equal fission (division) products.
2. "Budding", in which only a small part of the cell is constricted off.
3. "Multiple fission", in which the → nucleus divides several times before the cytoplasm disrupts into as many parts as there are nuclei present.

A. is the sole mode of reproduction of some amoeba and one phase of the complicated (sexual) reproduction cycle of sporozoa. Widely differentiated processes of a. lead to specialized germ cells which are called → agametes and lead without fertilization to the formation of a new individual. These agametes can be formed as the immediate result of meiosis (in Ectocarpus, the mosses, the basidiomycetes, and angiosperms) and give rise to haploid organisms or stages of the life cycle as constant elements of the sexual cycle. Agametes can also be produced mitotically by haploid or diploid organisms and in this case yield individuals with the same chromosome complement as the original form.

agamont (*Hartmann* 1904) — the asexual individual, or resp. the asexual generation, in whose agametangia the agametes are formed (→ gamont).

agamospecies (*Turesson* 1929) — an apomictic population whose members are of common origin (→ apomixis).

agamospermy — a form of → apomixis: the formation of seeds without sexual processes through → adventitious embryony, diplospory, or apospory. The process can begin autonomously or after pollination (→ pseudogamy). In pseudogamous apomixis the pollen does not make any genetic contribution to embryo formation, but is necessary to initiate the growth of the ovary or to fertilize the endosperm nucleus.

The various embryological pathways of a. are alike in bypassing both meiosis and fertilization in the cell leading to the new embryo. This results normally in the formation of seeds containing embryos which are genetically identical with the maternal parent.

In some plant groups, a. replaces sexual reproduction completely (*obligate a.*); in other plants some seeds form by agamospermous processes and some by sexual processes (*facultative a.*).

AG complex (*Correns* 1928) — the complete set of factors assumed to be responsible for the formation of sexual organs and expression of sexual differences which are not sex-determining. Those factors acting to produce male sex organs are designated by A, those acting to produce female sex organs by G. A and G are carried by the autosomes (or the cytoplasm).

In a diploid cell, A and G each occur twice as AAGG, in a haploid cell only once each. Thus each cell possesses the possibility of development in both directions. Which of the two directions is chosen depends on the specific → sex realizers which are designated M and F. The sex realizers act on the alternative → reaction norm evolved through the → bisexual potency and the AG-system and this is fixed in a male or female direction (→ sex determination).

agmatoploidy (*Malheiros-Gardé* 1950) — increase of chromosome number by → fragmentation of chromosomes with diffuse or multiple → centromere organization leading to → pseudopolyploidy or → pseudoaneuploidy.

agmato-pseudopolyploidy (*Battaglia* 1956) — → pseudopolyploidy.

akaryotic — without a → nucleus (→ eukaryotic; prokaryotic).

akinetic — = → acentric.

akinetoplastic — → kinetoplast.

albomaculatus (*Correns* 1904) — ref. to a → variegation or mottling ("status albomaculatus") in plants caused by genes or extrachromosomal hereditary determinants and consisting of an irregular distribution of white and green regions (white green variegation), (→ paralbomaculatus).

aleuroplast — a → leucoplast in which protein granules predominate as a storage product.

algeny (*Lederberg* 1966) — the (purposeful) altering of genes in the body cells or in germinal tissues, or the introduction of desired genes from outside (= "genetic engineering" or "genetic surgery").

alien addition line (*Leighty & Taylor* 1924; *O'Mara* 1940) — a line with an extra chromosome (monosomic addition) or an extra chromosome pair (disomic addition) from a related species (→ substitution line; alien substitution line).

alien substitution line (*Kattermann* 1938; *Unrau* et al. 1956) — a line in which an alien chromosome or a pair of alien chromosomes from a donor species replace one chromosome or a pair of chromosomes of the recipient species (→ alien addition line; substitution line). If a pair of alien chromosomes compensates for the missing chromosomes they are considered → homoeologous to the pair they replace.

allele (*Johannsen* 1909) — one of two or more alternate forms of a → gene occupying the same → locus on a particular chromosome or linkage structure and differing from other alleles of that locus at one or more → mutational sites whose number per gene is between 10^2 and 10^3. Members of a set of alleles are mutually exclusive → genetic markers and arise by → gene mutation. Their activity is concerned with the same biochemical and developmental process. A haploid organism or phase of the life cycle has a single representative of each a., a diploid two, a polyploid more than two from the total number existing in the population as a whole.

With respect to a given locus a diploid may be homo- (two identical alleles in the pair of homologous chromosomes) or heterozygous (two different alleles). If the phenotype of a heterozygous allele pair (A/a) resembles that of the homozygous pair (A/A), A is said to be dominant, a recessive. Any new allele may be characterized by its genotypic effect when it is homozygous and by its phenotypic effect when combined in heterozygotes with a sister allele, that is, they are known by their actions and interactions (→ gene interaction). On this basis the following classes of alleles may be distinguished:

1. **amorphs** (*Muller* 1932): inactive alleles which act as → genetic blocks to normal biosynthesis. In the broadest sense an amorph may be an a. that is inoperative, one that fails to produce a measurable effect, or even the absence (→ deletion) of a gene.

2. **hypomorphs** (*Muller* 1932): alleles that function imperfectly in comparison with → wild-type alleles; sometimes known as → leaky genes. They are prevalent among induced gene mutations.

3. **hypermorphs** (*Muller* 1932): alleles that produce an excess amount of a product (the antithesis of hypomorphs), whereby excess is defined in terms of the wild-type alleles.

4. **antimorphs** (*Muller* 1932): alleles with an action opposite to that of the wild-type; this type is rare and poorly defined.

5. **neomorphs** (*Muller* 1932): alleles whose action differs qualitatively from that of the wild-type allele. Heterozygotes of neomorphs commonly display the allelic products of both alleles.

6. **isoalleles** (*Stern & Schaeffer* 1943): alleles producing only very slight differences in phenotypic expression making their identification bound to the use of special methods.

The individual alleles of a group consisting of more than two are called a **series of "multiple alleles"** (*Morgan* 1914). A diploid individual or phase of the life cycle may be homozygous for any one allele or heterozygous for a combination of any two. → Segregation in meiosis results in gametes with only a single a. The dominance relations among multiple alleles vary from one group to another. For some groups of alleles every homo- and heterozygous → genotype produces a different → phenotype. In others, the alleles may be arranged in a descending series in which every a. is dominant over all alleles below it. The number of genotypes in a series of multiple alleles is $1/2\,[n \cdot 1\,(n + 1)]$ where n is the number of alleles of the group (for four alleles: $1/2\,(4 \cdot 5) = 10$). Multiple alleles of a gene affect similar parts or processes in an organism and are indicated by attaching distinguishing letters or numbers as superscripts to the base symbol (→ genetic nomenclature).

Different alleles may produce detectable effects upon the variability at any stage in the life history of an individual and may modify the expected phenotypic segregation ratio so that certain classes of offspring are in excess, in reduced frequency, or are totally absent (→ lethals).

Two recessive mutations of independent origin are allelic if the heterozygote of m/m′ has a mutant phenotype. If among the products of meiosis in heterozygotes — but not in homozygotes — an occasional revertant or an a. different from the two original ones arises, the two alleles are different. Different alleles are called "nonidentical" (*Demerec* 1956) (in contrast to identical alleles) when recombination has been obtained between them. The "nonidentity" refers to the fact that each represents a mutation at a different mutational site. Another term is "heteroallelic" (*Roman* 1956) for combinations of two alleles of different mutational origin which yield the wild type by recombination or other mechanisms (in contrast to "homoallelic" combinations). (→ genetic complementation; genetic recombination; pseudoalleles.)

allele center (*Reinig* 1938) — = → gene center.

allele shift — an alteration of allele frequency caused by selection which can lead in extreme cases to complete (± rapidly occurring) loss of an allele depending on its initial frequency. Similar effects may be a consequence of genetic drift.

allele trend — a directed change in allele frequency within a given time.

allelic (*Johannsen* 1909) — ref. to the relationships between → alleles of the same gene.

allelic complementation — the production of a non-mutant phenotype when two independent mutations at the same gene locus, but on different homologous chromosomes, are introduced into the same cell. A. c. can be accounted for in terms of the subunit structure of multimeric protein molecules and occurs in a wide variety of microorganisms and in some higher organisms (→ genetic complementation).

allelic interaction — → allele; gene interaction.

allelism (*Johannsen* 1909) — the relationship between → alleles.

allelobrachial — of → chromosome structural changes in which the arms of homologous chromosomes (→ heterobrachial; homobrachial) are involved.

allelogenous (*Vandel* 1938) — of females producing only males or only females in different broods (→ arrhenogeny; thelygeny).

allelomorph (*Bateson & Saunders* 1902) — = → allele.

allelosomal — of → chromosome structural changes in which both homologues of a chromosome pair are involved (→ heterosomal; homosomal).

allelotype (*Strandskov* 1950) — a term which corresponds on a population basis to the → genotype (which always relates to a single individual) and refers to the genetic composition — allele frequency — of a → breeding population. Changes in the a. of a population can come about by → mutation, → selection, sampling error (→ genetic drift), and → migration processes in case of incomplete → isolation of populations. With incomplete isolating mechanisms immigrant individuals may increase the frequency of particular alleles or introduce new ones, while emigrating individuals may decrease the frequency of particular alleles or, in extreme cases, produce total loss of a particular allele.

allocycly (*Darlington* 1941) — ref. to differences in → chromosome coiling which are shown by chromosome segments, whole chromosomes, or even whole chromosome complements and may be environmental, genotypic, or cellular in origin.

Regions most frequently subject to allocyclic behavior are the centromeres, telomeres and → nucleolus organizing regions, while the chromosomes most often allocyclic are the → sex chromosomes, → B-chromosomes, and → sex-limited chromosomes of certain Diptera.

The commonest type of a. is that in which the concerned segments appear overcondensed in comparison to regions which follow the standard cycle. This so-called **positive** → **heteropycnosis** may manifest itself during the interphase or during nuclear division with the exception of the meta-anaphase in which the regions following the standard cycle are also maximally condensed. The opposite allocyclic state, **negative heteropycnosis,** is less frequent and is detectable at all times during cell division. Different regions of the chromosomes can follow different types of allocyclic behavior and the same region can behave differently in different cycles of nuclear division.

allodiploid (*Serra* 1948) — of cells or individuals in which one or more chromosome pairs are exchanged for one or more pairs from another species. In the extreme case, two → genomes, one from either species, are combined in the zygote.

1. Allodiplomonosomes: allodiploids in which two foreign single chromosomes are present in the chromosome complement.

2. allomonodiplosomes: allodiploids which have exchanged one pair of chromosomes for a pair of foreign chromosomes. If there are two exchanged pairs, the individuals are called "allodidiploids", etc.

allogamy — = → cross-fertilization or exogamy.

allogenic — 1. of bacterial → transformation if the induced change produces a new character different from the characters of either the strain being transformed or of the strain from which the inducing DNA was isolated (*Ephrussi-Taylor* 1951). A. transformations are distinguished from "autogenic" transformations in which the induced change corresponds exactly with the source of the transforming DNA.

2. Of recombination by → crossing over within a → codon which may occur in heterozygotes bearing at the same site codons for different amino acids or different codons for the same amino acid and give rise to a codon for a third, different amino acid (*Zamenhof* 1967). Allogenic recombination closely resembles → gene mutation in its effects.

allograft — a tissue graft (= allogeneic graft; homograft) from a donor of one genotype to a host of another genotype, host and donor being members of the same species. Host and donor are called allogeneic.

allohaploid (*Ivanov* 1938) — → haploid (= allopolyhaploid).

alloheteroploid (*Sharp* 1934) — 1. of → heteroploid strains, individuals, or cells whose chromosomes derive from various chromosome sets (→ autoheteroploid);

2. of strains, individuals, or cells which contain additional chromosomes of another species (*Serra* 1948). Those individuals with one extra foreign chromosome are allomonoheteroploid (→ allodiploid).

alloiogenesis — → alternation of generations.

allolysogenic — → lysogenic.

allometric — ref. to growth, whereby the growth rate of one part of the organism differs from that of another part or of the rest of the body.

allomixis — = → cross-fertilization.

allopatric (*Poulton* 1903; *Mayr* 1942) — of populations or species which are mutually exclusive, but usually inhabit adjacent geographical regions. Genotypically distinct allopatric populations of a species are called geographic races, subspecies, or local varieties. Gene exchange between allopatric populations is either limited or completely absent (→ sympatric).

A. speciation is the term used to denote the differentiation of geographically isolated a. populations to the point where they are considered to be taxonomically discrete → species (→ evolution).

allophene (*Hadorn* 1955) — → allophenic.

allophenic (*Hadorn* 1955; *Mintz* 1962, 1967) — 1. of → characters (allophenes) which in a given cell system arise by intercellular gene action, i.e., through the action of genes manifest in other cell systems of the organism (*Hadorn*, 1955). In such cases, a gene has produced an → "autophene" in one cell system and the allophene characteristic of another system is the direct or indirect result of the gene which is manifest in the first cell system (→ gene activation). For allophenes a nonautonomous behavior on transplantation is characteristic. If an autophene secondarily gives rise to one or a series of allophenes, this is called "relational pleiotropy" (*Hadorn* 1954, 1955) and does not represent a true case of → pleiotropy.

2. Of individuals who display an orderly arrangement of two or more concurrent, allelically alternative cellular → phenotypes, or allophenes (*Mintz* 1962, 1967). A. individuals (mammals) have been artificially formed by aggregating cleavage-stage blastomeres *in vitro* from embryos of different genotypes and transfer of the composites to an "incubator" mother for further development. Such mosaic embryos may develop to form healthy and long-lived adults.

alloplasm (*Meyer* 1896) — those "cell organelles" which serve a special purpose and are not of regular occurrence such as the neuro- and myofibrils, the cilia, flagella, contractile vacuoles, and Nessel capsules (→ paraplasm).

alloploid (*Clausen, Keck & Hiesey* 1945) — of individuals (called "alloploids" or "amphiploids") which arise after natural or experimental ("synthetic alloploids") crossing of two (or more) species or genera and contain the (structurally and genetically different) chromosome sets of the particular parents, each set being present either once ("allodiploids") or in greater number ("allopolyploids").

Allopolyploidy or **amphiploidy** means the doubling (= allotetraploidy or amphidiploidy) or higher multiplication (allohexaploidy, allooctoploidy, etc.) of chromosome sets from different species or genera either spontaneously or induced experimentally by → spindle poisons. Polyploidization may occur following the critical cross through somatic or generative (formation and fusion of unreduced gametes) increase of the number of chromosome sets.

In allopolyploids or amphiploids, each chromosome of the original allodiploid or amphihaploid hybrid is represented twice and such forms generally display characteristics of both the contributing parents. By formation of allopolyploids, fertile derivatives may be obtained from hybrids in which sterility was caused by chromosomal imbalance and meiotic irregularity. Meiotic behavior of allopolyploids is dictated by the homology relationships of the chromosome sets they contain. Some form only → bivalents and no → multivalents at → meiosis. In this case every chromosome finds one identical partner with which it pairs (→ chromosome pairing). There is no segregation; the allopolyploid breeds true for its original genotype. Other allopolyploids are multivalent forming. This may result from

1. the presence of homologous chromosome sets in more than double dose (combination of, e.g., diploid and tetraploid species), or

2. the introduction of partly homologous (= homoeologous) chromosome sets.

Segregation of the allopolyploid is the consequence in both cases. Generally, extreme genetical imbalance occurs among segregants of allopolyploids with numerous or complex multivalents in meiosis resulting in reduced fertility.

In conformity with these variations in their meiotic behavior, allopolyploids are classified as follows (*Stebbins* 1945):

1. **Genome allopolyploids:** Chromosome pairing in meiosis is exclusively in the form of bivalents. Pairing restricted to exactly homologous chromosomes ("homogenetic pairing") may be the consequence of the striking structural dissimilarity between the chromosome sets of the parents or may result from the activity of specific genes which prevent the pairing of partially homologous chromosomes. The sterility of the diploid hybrid from which the allopolyploid arose is usually complete in this category; fertility is fully restored with the production of the genome allopolyploid derivative. Once formed this derivative is usually completely isolated by sterility barriers from its nearest relatives and may show all the criteria of a new species.

2. **Segmental allopolyploids:** Chromosome pairing in meiosis is characterized usually by bi- and multivalent formation. The parental chromosome sets combined in the diploid hybrid are in this case partially homologous (homoeologous) and correspond in a considerable number of segments permitting "heterogenetic pairing". The sterility of the diploid hybrid is thus not pronounced. In contrast to category I., segmental allopolyploids show more or less extensive segregation, the products of which are markedly imbalanced genetically and infertile. Hence segmental allopolyploids are usually unstable. Through segregation they may give rise to stable segregants, which in regard to their chromosome constitution are either → autoploid, genome alloploid, or stably segmental alloploid.

3. **Autoallopolyploids:** Chromosome pairing in meiosis is in form of bi- and multivalents as in 2. because homologous chromosome sets are present in more than double dose. Polyploids of this type are possible from the hexaploid level upwards and combine characteristics of auto- and allopolyploidy.

allopolyploid (*Kihara* & *Ono* (1926) — → alloploid.

allosome (*Montgomery* 1906) — a chromosome (= sex chromosome) deviating in size, form, or behavior from the other chromosomes (→ autosome).

allosteric (*Monod* & *Jacob* 1961) — of an enzyme whose activity is altered when its structure is distorted by an organic compound at a nonsubstrate site. A. enzymes possess at least two distinct non-overlapping receptor sites. One of these, the **active** (or catalytic) **site**, binds the substrate and is responsible for the biological activity of the protein. The other, the **allosteric site,** is complementary to the structure of another low-molecular molecule, the so-called **allosteric effector,** which binds specifically and reversibly. This bond modifies the catalytic activity of the active site (selective increase or decrease) without the allosteric effect relying on a direct chemical connection between the substrate and the allosteric effector, and operates through introduction of conformational changes in the enzyme molecule which in turn lead to changes in the catalytic site (allosteric transition).

Two classes of → a. effects are distinguished:

1. homotropic effects, i.e., interactions between identical ligands.
2. heterotropic effects, i.e., interactions between different ligands.

The general properties of a. systems may be stated as follows (*Monod* et al. 1969):

1. Most a. proteins are polymers rather than oligomers involving several identical subunits.

2. A. interactions frequently appear to be correlated with alterations of the quaternary structure of the proteins, i.e., alterations of the bonding between subunits.

3. While heterotropic effects may be either positive or negative (i.e., cooperative or antagonistic), homotropic effects appear to be always cooperative.

4. Few, if any, a. systems exhibiting exclusively heterotropic effects are known. Cooperative, homotropic effects are almost invariably observed with at least one of the two or more ligands of the system.

5. Conditions, treatments, or mutations which alter the heterotropic interactions simultaneously alter the homotropic interactions.

A. effects are assumed to play a role as elements of the negative feedback control and the end-product activation of enzyme activity (positive feed-back control). The allosteric sites of control point enzymes are genetically determined. Mutations can lead to breakdown of the allosteric site and this type of mutation conditions the collapse of the control of the metabolic path in question. The concept of the a. enzyme was developed to interpret a large number of cellular control systems including gene → repression (→ end-product inhibition).

allosteric effect — the effect of a modifier bound at a site of an → allosteric enzyme which is topologically distinct from the active site at which reaction is occurring and which influences the binding or activity at this site by indirect means.

allosteric effector — → allosterism, allasteric, autosteric effector.

allosteric inhibition — → isosteric inhibition.

allosteric preconditioning (*Alpers & Paulus* 1971) — → preconditioning.

allosteric site (*Monod* et al. 1963) — a topologically separate site of an enzyme which lacks catalytic activity but which can exert regulatory control over the active site.

allosterism — a mechanism which regulates enzyme activity by means of specific activators or inhibitors (allosteric effectors). These combine with the → allosteric enzyme at sites other than the substrate catalytic site. Such allosteric effectors are usually compounds of low molecular weight, and are believed to exert their influence by changing the conformation of the protein.

allosubstitution (*Karpetschenko* 1935) — → chromosome substitution.
allosynapsis (*Sharp* 1943) — = → allosyndesis.
allosyndesis (*Ljungdahl* 1924) — in → polyploids and → aneuploids, meiotic → chromosome pairing of completely or partially homologous (homoeologous) chromosomes which have been introduced into the zygote by different parental gametes at → fertilization (→ allosyndesis).

A. can be classified as:

1. complete a.: all chromosomes are paired allosyndetically;

2. one-sided a.: the chromosomes derived from one gamete pair allosyndetically, those from the other remain unpaired;

3. complete auto-allosyndesis; all chromosomes occur part auto-, part allosyndetically paired.

allotopic (*Krooth* 1969) — of a mutation which imparts to one tissue a characteristic metabolic feature normally found in another tissue.
allotypic nuclear division (*Strasburger* 1905) — = → meiosis.
allozygote (*McKusick* 1973) — a genetic compound, i.e., → heterozygous for different mutant alleles (→ allele).
alternation of generations (*Hofmeister* 1851) — the alternation of two or more → generations, reproducing themselves in different ways. By generation is understood that phase of the life cycle which lasts from one reproduction process to the next. Most often there is an alternation of sexual and asexual, in rare cases (Protozoa) of different asexual modes of reproduction. If the alternation between the modes of reproduction is not firmly established the term "facultative a. of g." is used; if the alternation proceeds according to an exactly determined sequence the phenomenon is spoken of as "obligatory a. of g." The following types are distinguished (*Hartmann* 1939):

1. **Primary a. of g,** i.e., the alternation between generations which produce → gametes and those which reproduce through → agametes.

a) Homophasic a. of g.: all generations are equal in respect to their degree of ploidy (number of chromosome sets per nucleus) (= monophasic a. of g., homogeneous a. of g., homologous a. of g.).

b) Heterophasic a. of g.: alternation between generations which produce gametes and those which produce agametes, exhibiting various degrees of ploidy. A. of g. and → alternation of nuclear phases are coupled with one another (= diphasic a. of g., heterogeneous a. of g., antithetic a. of g., botanic a. of g.).

2. **Secondary a. of g.,** i.e., the alternation between generations which produce gametes and those which reproduce by parthenogenesis or by vegetative reproduction.

a) metagenesis: generations producing gametes alternate with generations which reproduce vegetatively.

b) heterogony: generations producing gametes alternate with generations which reproduce by parthenogenesis (= cyclic parthenogenesis, alloiogenesis, heterogamy).

3. **Asexual a. of g.,** i.e., the alternation of two asexual modes of reproduction differing in the products of reproduction (*Grell* 1956).

alternation of nuclear phases — the regular change of chrosomome number connected in the case of sexual → reproduction with the fusion of the → gametes (zygote formation) and the subsequent → meiosis. After meiosis and before → fertilization of the ovum, the organism is in the haplophase (gametic chromosome number), after fertilization and before meiosis in the diplophase (zygotic chromosome number).

1. **Zygotic a. of n. p.:** meiosis takes place in the zygote ("zygotic meiosis"). This type of a. of n. p. is restricted to → haplontic organisms.

2. **Gametic a. of n. p.:** meiosis takes place at the time of → gametogenesis ("gametic meiosis"). This type occurs with → diplontic organisms.

3. **Intermediary a. of n. p.:** meiosis occurs during → sporogenesis of a diploid generation in the case of → diplohaplontic organisms.

alternative distribution — adjacent → centromeres of a ring or chain configuration (composed of four or more chromosomes) in → translocation heterozygotes are orientated towards opposite spindle poles and alternative chromosomes are distributed to the same cell pole in the first meiotic division (Fig. 2c). With ring formation, a. d. is more frequent than → adjacent distribution; in the case of chain formation the two types of distribution are of about the same frequency. (→ adjacent distribution).

alysogenic — ref. to prophage-free sensitive mutants which occur in → lysogenic bacteria.

amber mutation (*Epstein* et al. 1963) — a suppressible genetic change which results in the → terminator codon UAG in messenger RNA (→ ochre mutation; opal mutation). The corresponding polypeptide chain is terminated at the site of the amber codon if no phenotypic or genotypic suppression (→ amber suppressor) occurs.

A. mutations in bacteria and bacteriophages can be phenotypically suppressed by 5-fluorouracil (FU). This may be due to incorporation of FU into messenger RNA, where it induces coding errors. When incorporated into the amber triplet UAG in the mRNA, FU will occasionally cause misreading of the amber codon as CAG leading to insertion of glutamine at this site and completion of the peptide chain (FU rescue of amber mutants).

amber suppressor — a → suppressor mutation which allows amino acid insertion into the polypeptide at the site of an amber codon (UAG) in messenger RNA. The efficiency with which completion of the peptide chain occurs depends (in *E. coli*) on the particular a. s. Functional protein may result depending on the amino acid inserted in place of the amber codon. In *E. coli*, different a. s. have been shown to insert serine, glutamine, tyrosine, lysine, or leucine at the site of the amber codon. A. suppressors result from single base substitutions in the DNA sequences coding for transfer RNAs specific for these amino acids. The mutations result in changed anticodons of the tRNA molecules (→ suppressor tRNA), allowing them to pair with UAG at the ribosome (→ nonsense suppression).

ambiguity — the phenomenon in which a given → codon can be translated in more than one way (→ genetic code; coding ambiguity; translational ambiguity).

ambivalent (*Huxley* 1955) — ref. to genes with both advantageous and disadvantageous action for the organism. A distinction is made between genes with "full ambivalence" and genes with "dosage ambivalence". In the first case, the positive and negative effects occur both with heterozygotes (Aa) and with homozygotes (aa); in the second, the deleterious effects are confined to the homozygotes (aa) or are more strongly marked there than with the heterozygotes.

amDNA — abbreviation for → antimessenger DNA.

ameiosis — the failure of meiosis and its replacement by a mode of nuclear division which does not involve a reduction of the chromosome number.

ameiotic parthenogenesis (*White* 1945) — → parthenogenesis in which → meiosis has been entirely suppressed.

amino acid activation — the first step of protein synthesis (→ genetic translation), whereby an amino acid reacts with adenosine triphosphate. The reaction is catalyzed by a specific → aminoacyl tRNA synthetase. The first reaction product is an amino acid adenylate, in which the necessary energy is conserved for the second stage — the attachment of the amino acid to a specific → transfer RNA molcule. The adenylate remains bonded to the enzyme until transfer to the transfer RNA molecule. The same enzyme also catalyzes the transfer reaction.

amino acid attachment site — → transfer RNA.

amino acids — the building blocks of the → proteins. There are 20 common a. a. of natural occurrence, as L-stereoisomers. They all have the same basic structure but differ from one another according to the side group (R):

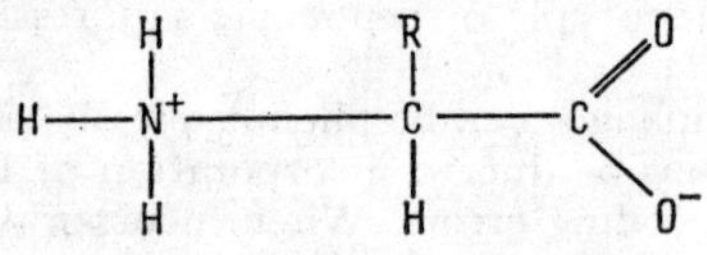

The linear order of the a. a. in a peptide or protein is called an a. a. sequence and is genetically determined (→ genetic translation).

aminoacyl adenylate — in protein synthesis an activated compound (AA ~ AMP) which is an intermediate in the formation of a covalent bond between an amino acid and its tRNA acceptor (→ aminoacyl-tRNA synthetase).

aminoacyl transferase — two soluble protein factors (T I and T II) which are involved in the transfer of amino acids from tRNA to ribosome-bound polypeptides in mammalian systems. **Aminoacyl transferase I** catalyzes the interaction between aminoacyl-tRNAs and 80 S ribosomes, placing the aminoacyl-tRNA at the A site of the ribosomes. **Aminoacyl transferase II** aids in the translocation of the peptidyl-tRNA at the ribosomal A site to the P site (→ peptidyl transferase).

aminoacyl-tRNA — the product of the reaction between aminoacyl adenylates and the appropriate → transfer RNA. The reaction is catalyzed by specific → aminoacyl-tRNA synthetases.

aminoacyl-tRNA synthetase — any of about 20 specific synthetases which activate the 20 amino acids to form bound aminoacyl-adenylates as the initial step in the synthesis of proteins (→ genetic translation). The activated amino acid is then transferred to a specific → transfer RNA (tRNA) to form the aminoacyl-tRNA. The specific aminoacyl-tRNAs are recognized by a triplet base sequence in the → messenger RNA and the activated amino acid is then incorporated into a polypeptide.

The reactions catalyzed by an aminoacyl tRNA synthetase are:

1. $AA(\text{Amino acid}) + ATP + E(\text{Enzyme}) \xrightleftharpoons{Mg^{++}} E \cdot AA - AMP + PP$
2. $E \cdot AA - AMP + tRNA \rightleftarrows E + AA - tRNA + AMP.$

In addition the aminoacyl-tRNA synthetases are important in the regulation of → RNA synthesis, enzyme → repression, and intergenic → suppression.

Aminoacyl-tRNA synthetases usually exhibit molecular weights between 90,000 and 180,000 daltons. They are either single polypeptide chains or oligomeric proteins of apparently identical subunits. There are indications that these enzymes are present in the microsomal or membrane fractions of the cell, i.e., do not float free in the cytoplasm.

All the enzymes differ from each other in the rate of reaction (PPi exchange as well as aminoacylation of tRNA), composition, or electrophoretic mobility, in addition to recognizing the cognate tRNA.

The majority of a.-tRNA synthetases recognize two or more tRNA species. The specificity of the various synthetases for their substrate tRNAs could be related to the base sequence of the tRNA, to its overall geometry, or to both factors.

amitosis (*Flemming* 1882) — nuclear division by a process other than → mitosis ("direct nuclear division") which in typical cases involves a dumbbell shaped cleavage of the cell nucleus, in the course of which no chromosomes are recognizable, nor is a → spindle formed. A. occurs *inter alia* in ciliates, certain protists, and in specialized animal tissues. Part of the phenomenon described by a. involves so-called → nuclear fragmentation.

amixis (*Burnett* 1956) — a mode of reproduction (→ heteromixis; homomixis) in haploid organisms where the essential events of sexual reproduction are absent but the pre-conjugation and post-meiotic events normally associated with sexual reproduction may occur (for the same situation in higher plants the term apomixis is applied).

amniocentesis — a procedure for diagnosing genetic abnormalities in utero. By puncturing the abdominal and uterine walls with a needle a little amniotic fluid, containing fetal cells, is taken from a pregnant woman. From these cells it is possible to definitely establish whether or not the developing fetus is affected with certain specific genetic and chromosomal defects. A. has made for a significant advance in → genetic counselling.

Three types of disorders are suitable for intrauterine diagnosis by a.: 1. chromosomal disorders (culturing of cells and determining their karyotype). 2. Autosomal or X-linked recessive diseases due to deficiencies of a specific enzyme that is normally present in cultured amniotic cells. 3. X-linked recessive conditions for which no biochemical lesion can be detected in amniotic fluid, but for which determination of fetal sex will distinguish between girls (who will be unaffected) and boys (who have a high risk of being affected).

If a defect is present, the pregnancy can be terminated. A. is generally not recommended in pregnancies with less than a one percent recurrence risk for a certain disease.

amorph (*Muller* 1932) — → allele.

amphiagamospecies (*Turesson* 1929) — a → species consisting of a group of facultative sexual biotypes whose reproduction is essentially apomictic (→ agamospecies; apomixis).

amphiapomictic (*Turesson* 1926) — of biotypes which propagate partly by sexual → reproduction and partly apomictically (→ apomixis).

amphidiploid (*Navashin* 1927) — of tetraploid (genome allotetraploid) species hybrids (amphidiploids) whose somatic cells contain the diploid chromosome complements of both parent species (syn. allotetraploid, didiploid, double diploid). The diploid hybrid from which the amphidiploids originate is generally sterile owing to the significant nonhomology between the chromosome sets and the consequent difficulty with → chromosome pairing during meiosis. With the doubling of the chromosome number the sterility barrier is removed. Amphidiploids commonly possess a better viability and competitive ability than the original species and the diploid hybrid. These properties frequently result in the establishment of amphidiploids as new species (→ alloploid).

Each genetic locus of an a. form which was represented in both parent species occurs four times in the amphidiploids and shows a corresponding → segregation of the → alleles. If a dominant allele (+) be present, a recessive mutation shows no phenotypic effect, provided a second does not occur, since in "amphiploidy" → multivalents are not formed in meiosis (see table 1). In order that a recessive gene manifest itself phenotypically, two independent mutations of the locus are necessary. No segregation results for genes which do not occur in both parent species.

amphigamy (*Renner* 1916) — 1. the fusion of two sex cells and the formation of conjugated pairs of nuclei (→ "dikaryophase"). If a. immediately follows a → karyogamy, the process is referred to as amphimixis (*Renner* 1916); 2. the normal fertilization process (*Battaglia* 1947).

Table 1. *Pattern of segregation in the case of amphidiploidy*

locus 1	locus 2	
+/+	+/+	without segregation
+/+	+/a_2	segregation for a_2 without phenotypic effect
+/+	a_2/a_2	without segregation
+/a_1	+/+	segregation for a_1 without phenotypic effect
+/a_1	+/a_2	segregation for a_1 a_2 with phenotypic segregation ratio of 15:1
+/a_1	a_2/a_2	segregation for a_1 with phenotypic segregation ratio of 3:1
a_1/a_1	+/+	without segregation
a_1/a_1	+/a_2	segregation for a_2 with phenotypic segregation ratio of 3:1
a_1/a_1	a_2/a_2	mutant-phenotype, without segregation

amphigony (*Haeckel*) — → reproduction by → fertilization, as opposed to → parthenogenesis

amphihaploid (*Zukov* 1941) — of "haploid" (allodiploid) types produced from amphidiploids (allotetraploids) (*Olsson* & *Hagberg* 1955).

amphikaryon (*Boveri* 1905) — the nucleus of the zygote produced in the course of → fertilization.

amphimixis (*Weismann* 1891) — the fusion of two gametes in sexual → reproduction in contrast to → apomixis. A. is cross-fertilization if the two gametes are produced by dioecious organisms (→ automixis).

amphiorientation — → centromere orientation.

amphiplasty (*Navashin* 1928) — a term proposed to denote morphological changes occurring in chromosomes following interspecific hybridization. Changes affecting individual chromosomes of the complement are called "differential a.".

amphiploid (*Clausen, Keck* & *Hiesey* 1945) — of individuals (called "amphiploids") which are either segmental allopolyploid, genome allopolyploid, autoallopolyploid or aneuploid (by addition or substitution of alien single chromosomes) in type (→ alloploid, allodiploid).

amphitelic (*Bauer, Dietz* & *Röbbelen* 1961) — → syntelic.

amphitene (*Janssens* 1905) — = → zygotene.

amphithallic (*Lange* 1952) — → heteromixis.

amphitoky (*Leuckart* 1857) — → parthenogenesis.

amphogenous (*Vandel* 1945) — female individuals which give rise to male and female offspring in the ratio of about 1:1 (→ monogenous, thelygenous).

amphoterotoky (*Taschenberg* 1892) = — amphitoky (→ parthenogenesis).

amplicon (*Lima-de-Faria* et al. 1973) — a chromosomal (DNA) region which takes part in → gene amplification and becomes amplified (→ repetitious DNA).

amyloplast (*Errera* 1882) — in plant cells, a → leucoplast specialized for the synthesis and storage of starch (→ plastid). Amyloplasts commonly have a less highly developed internal system of membranes than → chloroplasts. The starch is always stored in the stroma, either without or accompanied by only a few short → thylakoids. Internal membranes, together with the stroma, become progressively reduced as starch deposits increase in size.

anagenesis (*Rensch* 1947) — a mode of → evolution characterized by cumulative changes (production of new organs, structural types, and building plans) in an evolutionary lineage in the course of time without subdivision in the lineage (= phyletic evolution).

According to *Huxley* (1957) a. is a comprehensive term for all forms and degrees of biological improvements in the course of evolution from specific adaptation to general organizational advances (→ cladogenesis; stasigenesis).

anaphase (*Strasburger* 1884) — → mitosis, meiosis.

anaphase movement — → chromosome movement.

anaphase separation — during anaphase of → mitosis and → meiosis II, the disjunction of the chromatids of each → chromosome; during anaphase of meiosis I the separation of chromosomes paired as → bi- or multivalents. In all cases the disjunction results with causal participation of the → spindle.

anaphragmic — ref. to mutations which by removal of an inhibiting influence lead to an increased enzyme activity in the mutant.

anauxotrophic — = → prototrophic.

aneucentric — ref. to → chromosome mutations in the course of which chromosomes with more than one → centromere (dicentric, tricentric chromosomes) are formed from monocentric (= eucentric) chromosomes.

andro-autosome (*Yamamoto* 1938) — an → autosome carrying → sex realizers for male character (→ gyno-autosomes).

androdioecious (*Darwin* 1877) — → dioecious.

androecious (*v. Uexküll-Syllenband* 1901) — of plants having only male flowers (→ gynoecious).

androecium (*Roeper*) — the male reproductive organs of a plant; the stamens of a flower taken collectively.

androgenesis (*Verworn* 1891) — male → parthenogenesis, i.e., the development of a haploid embryo from a male nucleus (= ephebogenesis; merogony), as opposed to maternal haploidy (gynogenesis) originating from the egg by incomplete fertilization, from other nuclei of the embryo sac in plants, by degeneration of the fertilized or unfertilized egg, or from other nuclei or other embryo sacs in connection with → polyembryony.

In case of a. the maternal nucleus is eliminated or inactivated subsequent to fertilization of the egg cell and the haploid individual (referred to as "androgenetic") contains in its cells the chromosome set of the male gamete only. A. is easily detected by the use of suitable cytological or genetic markers.

androgenous — producing only male offspring (→ gynogenous).

androgynodioecious (*Darwin* 1877) — → dioecious.

androgynous — ref. to plants bearing both stamens and pistillate flowers in the same inflorescence (= hermaphroditic).

andromerogony — the development of an egg fragment with chromosomes exclusively from the father.

andromonoecious (*Darwin* 1877) — → monoecious.

androsome — any chromosome occurring exclusively in the nuclei of the male → germ line. Androsomes are present neither in the somatic cells of either sex nor in the female germ line (= male limited chromosomes).

androsporogenesis (*Battaglia* 1955) — = → microsporogenesis.

aneugamy (*Austin* 1960) — an abnormal fertilization process, in which a polpyloid ovum is fertilized by a normal sperm, or a normal ovum by a diploid sperm (→ polyandry, polygyny).

aneuhaploid (*Kimber & Riley* 1963) — → haploid.

aneuploid (*Täckholm* 1922) — 1. of cells, individuals, or populations having one or more whole chromosomes of a → euploid complement absent from or in addition to that complement. In the case of diploidy this type of chromosomal aneuploidy (*Dyer* et al. 1970) includes *inter alia* → nullisomics (loss of a chromosome pair, 2 n — 2), → monosomics (loss of a single chromosome, 2 n — 1), → trisomics (presence of a single extra chromosome of the standard set, 2 n + 1), and tetrasomics (occurrence of an extra chromosome pair, $2n + 2$.) If the loss or addition concerns more than one specific chromosome or chromosome pair, the situation may be characterized by the terms "doubly monosomic" ($2n - 1 - 1$) or "doubly trisomic" ($2n + 1 + 1$), etc. (→ pseudoaneuploid).

Chromosomal aneuploidy does not include chromosome complements where the numerical variation only involves → B-chromosomes. It represents a form of → heteroploidy and occurs in somatic or generative cells either spontaneously or following the action of physical and chemical factors by:

1. Loss of individual chromosomes in mitosis or meiosis leading to the formation of nuclei with hypoploid chromosome numbers.

2. Non-disjunction during mitosis and meiosis, whereby hypo- and hyperploid nuclei are formed.

3. Irregularities of the chromosome distribution in meiosis of polyploids, especially those with an uneven number of chromosome sets (triploids, pentaploids, etc.).

4. Multipolar mitoses with irregular chromosome distribution to the daughter cells; this can result in so-called "multiform aneuploidy" (*Böök* 1945) characterized by cells with various aneuploid chromosome numbers occurring in the same tissue.

The prefixes "hypo-" and "hyper-" applied to the haploid, diploid and polyploid number allow a classification of aneuploid chromosome numbers.

2. Of cells or individuals having one or more chromosome segments of a basic chromosome set absent from or in addition to one or more basic sets. This type of segmental aneuploidy (*Dyer* et al. 1970) includes heterozygosity for → deletions or → duplications where one or more acentric segments of a basic set have been gained or lost, or where centric fragments (sometimes forming new chromosome arm combinations, as in secondary and tertiary → trisomics), have been gained (→ chromosome mutation).

aneuploid reduction — the reduction of → genetic variability by decreasing the number of → linkage groups. A. r. is supposed to play a role in the evolution of plant species in which outcrossing prevails and which occupy pioneer habitats. Such conditions produce a high degree of heterozygosity on the one hand and require repeated colonization of new areas on the other hand (*Stebbins* 1966).

aneusomatic — ref. to individuals whose cells exhibit variable numbers of individual (mostly heterochromatic) chromosomes (→ B-chromosomes), or, generally, individuals in which both euploid and aneuploid cells occur.

anisogamete — → anisogamy; gamete.

anisogamy — the → gametes fusing during fertilization vary in respect to size, shape, and behavior. The distinction may include a) the behavior at fertilization only, b) size only, or c) both size and shape (the most extreme case: oogamy). A. can occur in the form of → gametogamy, → gametangiogamy, or → somatogamy.

anisomeric (*Sirks* 1933) — ref. to non-equivalent genes which by interacting (→ gene interaction) produce a particular phenotype (→ isomeric).

anisotrisomic (*Renner* 1949) — → trisomic.

anlage — 1. in genetics, originally a "hereditary factor" (now replaced by → "gene");
2. in embryology, the primordium or cell group from which a part of the organism develops.

annealing — the "incubation" of a mixture of DNAs in single-stranded form and the quantitative determination of the helical material formed. The method permits the detection of homologous regions in any two species of DNA, provided the genetic complement is relatively small as in bacteria and viruses.

annidation (*Ludwig* 1948) — adaptation of genotypes to different ecological niches. By a. a mutant capable of utilizing an ecological niche not available to the parental form may be maintained in the population even if it happens to be inferior to the parent in the ecological niche of the latter. A. is a factor in → evolution.

annulate lamella (*Swift* 1956) — a class of cytomembranes which occur as stacks of parallel, double-membraned elements each having the same dimensions as the → nuclear envelope and containing pore complexes identical to those of the nuclear envelope. The elements are arranged in highly ordered fashion with a regular spacing between them. A. lamellae occur in oocytes, spermatocytes, a variety of somatic cells and some cancer cells; generally, they are indicative of actively growing cells and may function in the transmission and storage of genetic information from the nucleus. It seems certain that a. lamellae are directly or indirectly derived from the nuclear envelope. Such membranes occur in both the cytoplasm and the nucleoplasm. Those of the nucleoplasm are small, irregular, and also short-lived. In some instances ribosomes have been observed as being directly attached to the membranes of a. lamellae indicating a role in protein synthesis.

annulus (*Callan* & *Tomlin* 1950) — → nuclear pore.

anormogenesis — a process of → development outside the realm of variability of the normal development, deviating from normal development (normogenesis) and resulting in malformation or lethality.

anorthogenesis — adaptive changes of evolutionary significance based on preadaptations (→ adaptation).

anorthoploid (*Winkler* 1916) — = → aneuploid.

anorthospiral — → chromosome coiling.

antephase (*Bullough* 1963) — → prophase.

anther — that part of the stamen in which the pollen grains are produced (→ microsporogenesis).

antheridium (*Bischoff* 1835) — the male gametangium (microgametangium), within which in ferns, mosses, fungi, and gymnosperms, the male gametes are formed (→ archegonium).

antibody — a protein (immunoglobulin) usually found in serum whose presence can be demonstrated by its specific reactivity with an → antigen or hapten.

anticipation — the apparent tendency of certain diseases to appear at earlier onset ages and with increasing severity in successive generations.

anticodon — a sequence of three nucleotides in → transfer RNA next in sequence to an invariant uridine in the a. loop of the molecule. The a. serves to recognize the appropriate nucleotide triplet that constitutes the → codon in → messenger RNA (→ genetic code). Mutations in the a. occur that change codon-anticodon pairing without affecting amino acid recognition by the tRNA molecule. Such mutations result in the "wrong" amino acid being placed in many polypeptide sequences during → genetic translation and may, therefore, be expected to be deleterious or lethal in many cases (→ suppressor tRNA).

→ Ambiguity in the pairing between the first base of the anticodons and the third base of codons is responsible for a definite pattern of the → genetic code (→ wobble). There are indications that replacement of uridine by 2-thiouridine in the a. may be an evolutionary device for preventing →mistranslation; 2-thiouridine is much less likely to form wobble base pairs.

antigen — any agent (substance) which, when introduced into a vertebrate, elicits the production of → antibodies which bind specifically to the inducing substance (→ immunogenetics).

anti-inducer — a compound that inhibits → operon induction caused by → inducers. Anti-inducers compete for inducer binding to free → repressors and also stabilize the repressor-operator complex. Both anti-inducers and inducers are able to bind either to free repressor or to operator-bound repressor.

anti-messenger DNA — a DNA strand (symbol: am DNA) copied from → messenger RNA by RNA-dependent → DNA polymerase (reverse transcriptase, reverse polymerase, revertase) by a process called → reverse transcription (= cDNA).

antimetabolite — a compound which is similar to an essential metabolite and binds to replace the metabolite.

antimitotic — of agents that lead to the cessation of → mitosis either by decreasing the number of cells entering mitosis **(mitodepressive effect),** by disturbing the → mitotic apparatus **(mitoclastic effect),** by stopping cell proliferation (**mitostatic effect**), or by causing immediate cell death (→ spindle poison).

antimorph (*Muller* 1932) — → allele.

antimutagenic — of agents that can reduce the rate, not only of induced, but also of spontaneous → mutation. Antimutagens act either as "protective agents" or promote the → repair of premutational DNA lesions. Generally, mutagens and antimutagens are counteracting factors; type, extent, and result of the interaction processes are dependent on the mutagen and the biological material. Some antimutagens are a. under some conditions, and → mutagenic under other conditions.

antimutator gene (*Drake & Allen* 1968) — any mutant gene which lowers the → mutation rate. A. alleles have been found which suppress both errors of replication and errors of incorporation (→ gene mutation). → Mutator and a. genes both affecting → DNA polymerase demonstrate the profound role of this enzyme in determining the spontaneous mutation rate.

antipolarity (*Ito* & *Crawford* 1965) — the lowering by → polar mutations of enzyme production of the genes located on the side of a mutated structural gene proximal to the operator of an → operon (→ polarity).

antirepressor (*Oppenheim* et al. 1970) — a gene product capable of abolishing repression due to a → repressor (in lambda phage). The a. may inactivate the repressor, compete for the repressor binding site on the DNA, or allow the transcription of the operons repressed by the repressor.

anti-sigma factor — a protein that prevents recognition of initiation sites by the → sigma factor of → RNA polymerase.

antisuppressor — a mutation which counteracts suppressor effects (→ nonsense suppression). A. mutations can occur in any of the following components affecting the → genetic translation system (*Cox* 1973):

1. A ribosomal component affecting codon-anticodon recognition or tRNA binding. 2. A termination factor altered in such a way that polypeptide release from the ribosome is enhanced thus decreasing the efficiency of suppression. 3. An enzyme affecting → suppressor tRNA (either an → aminoacyl-tRNA synthetase, a base modifying enzyme affecting the anticodon, an enzyme modifying a base or bases involved in ribosome binding, an enzyme affecting tRNA maturation, or a factor controlling total tRNA synthesis).

antitermination factor — a protein (specified by an infecting phage DNA) that prevents normal termination of RNA synthesis, perhaps by interfering with the action of the → rho factor. The a. f. allows the host → RNA polymerase to "read through" sites at which it would otherwise terminate on the phage genome.

antitermination signal — in some phages, a protein which allows transcription beyond → termination signals.

anucleolate — of mutants lacking → nucleoli.

apoamphimict (*Turesson* 1926) — a biotype which reproduces predominantly by → apomixis but also sexually.

apocyte — a multinucleate cell or mass of cytoplasm.

apoenzyme — the protein part of an → enzyme to which a coenzyme can attach.

apogamety (*Renner* 1916) — → apomixis.

apogamogony (*Fagerlind* 1944) — → apomixis.

apogamy (*de Bary* 1877; *Winkler* 1908). — 1. = → apomixis (*de Bary* 1877).

2. The production of a sporophyte, not from the ovum but from other cells of the gametophyte ("euapogamy" according to *Farmer* & *Digby* 1907) or from cells of the sporophyte. A distinction is made between somatic (diploid) and generative (haploid) a.

apomeiosis (*Renner* 1916) — sporogenesis without reduction of chromosome number during meiosis and giving rise to → apomixis.

apomixis (*Winkler* 1906) — in plants, the replacement of sexual → reproduction (= amphimixis) by various types of asexual reproduction which do not result in fusion of the → gametes. Owing to failure or suppression of → fertilization and → meiosis, no alternation of nuclear phases is coupled with a. provided it occurs regularly in successive generations as obligatory a. Obligate apomicts have a completely closed → recombination system and heterozygotic genotypes are preserved at the cost of evolutionary flexibility. In the case of facultative apomicts, apomictic and sexual modes of reproduction coexist. The mode of reproduction in animals equivalent to a. is → parthenogenesis.

A., like amphimixis, is a genetically controlled reproduction system and is brought about by mutations which modify the course of sexual reproduction to the point of non-function. This occurs *inter alia* by transformation of meiosis into an → apomeiosis, through degeneration of the megaspore mother cells, the spores or the embryo sacs, through hindrance of gamete fusion, the introduction of parthenogenetic egg cell development, or development of synergides.

Apomictic reproduction is divided into two main types (*Gustafsson* 1935):

1. **Agamospermy:** all forms of a., in which reproduction proceeds via asexually produced embryos and seeds. Fertilization is always by-passed, meiosis usually. The embryo generally has the same chromosome number and the same genotype as the mother individual provided there is no → autosegregation. Mechanisms of this type are termed "agamogony" (*Fagerlind* 1940, *Stebbins* 1941), "apogamogony" (*Fagerlind* 1944), "gametophytic apomixis" (*Stebbins* 1950), and "apospory and displospory with following parthenogenesis" (*Gustafsson* 1935). In **"apospory"** a diploid embryo sac is formed directly from a nucellus or integument cell; in **"diplospory"**, the embryo sac forms from an archespore cell, but meiosis is either omitted or does not lead to reduction of the chromosome number. From the diploid gametophyte produced by apo- or diplospory, embryos are produced either by division of the ovum (parthenogenesis) or another cell (apogamety). The simplest mode of agamospermy is adventitious embryony, whereby the embryos develop directly from the diploid sporophyte tissue and the gametophyte stage is omitted.

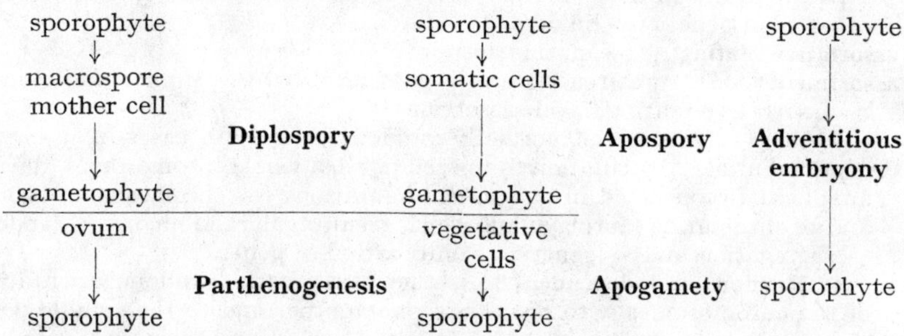

2. **Vegetative reproduction:** the new individual arises from a group of undifferentiated or differentiated cells, in which case neither embryo nor seed are produced.

apophase (*Bullough* 1965) — the period of postmitotic reconstruction, i.e., the time when the cell machinery is rebuilt and when the cell grows in mass to recover the size typical of its kind (→ prophase).

apospory (*Druery* 1886; *Bower* 1886) — → apomixis.

aptitude — a specific physiological state of → lysogenic bacterial strains in which, under the action of inducing agents (→ induction), they react to produce infectious bacteriophages.
archegonium — the female gametangium (macrogametangium) of ferns, mosses, fungi, and gymnosperms, within which the ovum is contained (→ antheridium).
archesporium (*Goebel* 1880) — a cell or group of cells formed by mitosis of the micro- and macrospore mother cells (→ microsporogenesis, macrosporogenesis).
arrhenotoky (*Leuckart* 1857) — → parthenogenesis.
artioploid — of even-order polyploids (4 n, 6 n, etc.), as distinguished from odd-order polyploids (3 n, 5 n, etc.) or perissoploids.
A-site — → ribosome.
ascertainment — in human genetics, the finding or selection of families for inclusion in a genetic study.
ascogonium — the female → gametangium (oogonium) of anisogamous fungi.
ascospore — one of the spores (regularly eight are formed pro → ascus) of ascomycetes, which are produced in the course of meiosis and a subsequent mitosis.
ascus — a membrane enclosed spore-sac of ascomycetes which usually contains eight ascospores. The a. is produced from an ascogenic hypha in which haploid nuclei, previously present as → dikaryons, fuse to form a diploid nucleus (karyogamy). Meiosis follows immediately and then, as a rule, one mitosis, so that from the 4 meiotic products, 8 haploid nuclei are formed, each of which gives rise to an ascospore.
asexual — → reproduction.
association — 1. = → chromosome pairing;

1. in human genetics, the non-random occurrence of two genetically separate traits in a population; association has its basis in mechanisms other than genetic → linkage.

assortative mating — → mating system.
assortment — 1. preference for → mating between similar individuals (= assortative mating; → disassortment);

2. in → meiosis, the normally random, in certain cases, nonrandom ("preferential") distribution to the cell poles of whole chromosomes (during anaphase I) contained in pairing configurations (→ chromosome pairing) and of chromatids (during anaphase II) resulting in random or non-random → segregation and → genetic recombination of genes.

1. **Random or independent a.** is the result of the random orientation of the chromosomes with respect to one another and to the spindle poles in anaphase (→ chromosome movement) I and II and their corresponding distribution to the cell poles and meiotic products. It is prerequisite for random → segregation of the gene pairs.

2. **Non-random a.** is the preferential distribution of certain chromosomes to one of the cell poles and leads to distorted segregation ratios. It can be the result of different processes and in all cases is caused by a preferential → centromere orientation (→ affinity, meiotic drive, automatic frequency response). Pairing disturbances, heterozygosity for certain

chromosome structural changes (→ complex heterozygosity), aneuploidy, and autopolyploidy are frequently associated with non-random a. of the chromosomes.

aster (*Fol* 1877) — a symmetrical star-shaped figure which surrounds the → centrosome during mitosis and meiosis.

asynapsis (*Beadle* 1931) — → desynapsis; asynaptic.

asynaptic (*Beadle* 1931) — ref. to chromosomes in the first meiotic division in which pairing (→ chromosome pairing) either fails completely or is incomplete. Asynapsis is recognizable by the presence of a number (varying from case to case) of → univalents and can be caused by incomplete homology of the chromosomes (e.g. in species hybrids), by gene mutations (asynaptic genes), or environmental influences. As a rule, asynapsis leads to the formation of hypo- and hyperploid meiotic products and frequently to fertility disturbances (→ desynaptic).

asyntenic (*Renwick* 1971) — → syntenic.

atelocentric (*Levan* et al. 1964) — of chromosomes whose → centromere is not localized at the end point (terminally) in contrast to telocentric chromosomes.

A + T/G + C ratio — = → base pair ratio.

ATP — adenosine triphosphate, a major carrier of phosphate and energy in biological systems which consists of the purine (adenine) the five-carbon sugar, ribose, and three phosphate groups.

attached X chromosome (*L. V. Morgan* 1922) — a metacentric → X chromosome in *Drosophila*, which is an isochromosome and consists of two, normally acrocentric X chromosomes with a single centromere (→ compound X chromosome; detached-X).

attachment site (*Campbell* 1962) — particular sites in both bacterial and phage DNA molecules (lambda phage). The phage DNA integrates into the bacterial DNA by recombination between these sites. Both host DNA and phage DNA molecules are broken in these regions; the genomes then fuse to form a continuous structure in which the phage genome (the → prophage) is stably integrated into the bacterial genome. Reversal of integration results in the excision of the phage genome. Integration and excision are cases of → genetic recombination in which → crossing over is restricted to special (attachment) sites. A special phage recombination systeme, called INT, promotes recombination only at these sites. The recombination system is site-specific and the INT-allele is required for its function. Site specificity is due to the capability of the INT system to recognize structural features, presumably specific nucleotide sequences, of the a. sites. If the a. s. is missing or altered, efficient integration of phage DNA does not occur.

autoallopolyploid (*Kostoff* 1939) — of cells or individuals whose → chromosome complement shows the characteristics of auto- and alloploidy (→ alloploid, autoploid). Generally, autoallopolyploids are hexaploid or of even higher ploidy and contain chromosome sets of different species: e.g. AAAABB or AAAABBBB, where each capital represents a chromosome set.

autobivalent (*Hakansson & Levan* 1957) — a → bivalent in first meiotic division, formed in each case from two structurally and genetically completely identical sister-chromosomes produced in the course of a supernumerary reduplication. Thus, autobivalents can be formed if a supplementary reduplication of the chromosomes takes place in a premeiotic interphase nucleus or in early meiotic prophase. The number of autobivalents then corresponds with the somatic chromosome number of the species. The chromosomes formed by supernumerary reduplication are positionally associated with one another from the outset and preferentially pair together. In this way, their pairing forces are completely neutralized, so that for the special case of a meiosis with autotetraploid chromosome number, → multivalents are not generally formed (→ chromosome pairing, autoploid).

autogamy — obligatory → self-fertilization (automixis) within one undivided cell (gamont) by pairwise fusion of haploid nuclei; cell fusion is entirely lacking (→ paedogamy; parthenogamy).

A. is a particular mode of nuclear reorganization in *Paramecium* which takes place in single, unpaired individuals. The two → micronuclei divide twice by mitosis and seven of the eight resulting nuclei disintegrate. The remaining nucleus divides again mitotically and gives rise to a pair of identical haploid nuclei. In the absence of → conjugation the members of the pair of nuclei fuse. Fusion results in homozygosity of each of the newly formed diploid nuclei (→ killer).

autogenic (*Ephrussi-Taylor* 1951) — of → transformations in which the induced change corresponds exactly with the source of the transforming DNA (→ allogenic).

autogenomatic (*Levan* 1937) — of → chromosome sets which are strictly homologous and pair normally in → meiosis (→ heterogenomatic).

autogenous regulation (*Goldberger* 1974) — = → autoregulation.

autograft — a tissue transplant from one site to another in the same individual (→ homograft).

autolysis — digestion of a cell or cell material by some of the cellular enzymes (→ lysosome).

automatic frequency response (*Brown* 1963) — the automatic increase or decrease in the frequency of genes which are responsible for a change in the life cycle (including meiosis). Selection and random fluctuations are without effect on the a. f. r. which is a phenomenon parallel to → meiotic drive. Both the latter and a. f. r. affect the → gene frequency, but usually in different ways. While, in the case of meiotic drive there is a direct gene influence on the behavior of chromosomes in meiosis, the influence proceeds indirectly in the case of a f. r. (*Brown* 1964).

automixis — obligatory self-fertilization by → autogamy, → paedogamy, or → parthenogamy, as opposed to amphimixis (cross-fertilization).

automutagen — any → mutagen, formed in the organism as a normal or abnormal product of metabolism, which may induce → gene and → chromosome mutations.

automutation (*Imai* 1936) — → exomutation.

auto-orientation (*Darlington* 1936) — → centromere orientation.

autophagic vacuole — a type of digestive → vacuole in which some of the cell's own organelles (mitochondria, endoplasmatic reticulum, etc.) are destroyed. An a. v. (= cytolysosome, cytosegrosome, composite body, autolysosome) is formed by the development of a membrane around a small area of cytoplasm containing one or more organelles. Like the → phagosomes containing material from outside the cell (→ endocytosis), the a. vacuoles fuse with primary → lysosomes and thus acquire hydrolases which digest the structure inside the a. v. Autophagy occurs in various types of cells under physiologic as well as pathologic conditions.

autophene (*Hadorn* 1955) — a genetically controlled character which is manifested by the cell's own genotype and which shows "autonomous" behavior in transplants and explants (→ allophene). Gene action leading to an a. is designated "intracellular"; an a. is the direct or indirect result of such action confined to the single cell. The term "mosaic pleiotropy" is used to describe the case of a gene influencing or controlling more than one a. (→ pleiotropy, gene action).

autoploid (*Clausen, Keck & Hiesey* 1945) — of cells or individuals with chromosome sets characteristic of the species in question; the chromosome sets are homologous to each other and thus pair fully in meiosis (→ alloploid).

The terms autodiploidy, autotriploidy, autotetraploidy, etc. are used when there are two, three, four, etc. homologous chromosome sets per cell. The presence of more than two (diploidy) of the monoploid chromosome sets characteristic of the species is generally referred to as **autopolyploidy** (*Kihara & Ono* 1926). It is associated with a corresponding increase in the number of → genomes and → genetic information content.

Owing to the structural identity of the chromosome sets of autopolyploid forms, the homologous chromosomes can pair during meiosis and participate in → crossing-over. Since each chromosome has more than one pairing partner, crossing-over produces → multivalents (depending on the degree of ploidy, chiasma frequency and → chiasma distribution), i.e. paired groups consisting of more than two chromosomes, referred to as tri- (three paired chromosomes), quadri- (four paired chromosomes), pentavalents (five paired chromosomes per multivalent), etc.

The polyploidization of a diploid species may occur spontaneously or be induced by spindle poisons; it represents a → genome mutation and, like allopolyploidy, it is the result of the formation of a → restitution nuleus in somatic (somatic polyploidisation) or generative (generative or gametic polyploidisation) tissue.

The autopolyploid derivative of a diploid species generally only differs quantitatively in → phenotype from its parent. If qualitative differences are present they can be due to the following causes:

1. an increased size of nuclei and cells can differentially influence the gene activity (*Goldschmidt* 1937);

2. the increase in the number of → alleles per locus associated with autopolyploidy can modify the → expressivity of individual genes;

3. if the diploid form was heterozygous for the allele pairs concerned, an increase in the number of chromosome sets may lead to modifications of the dominance relationships of the alleles.

As a rule, autopolyploids show fertility disturbances. for several reasons:

1. During meiosis, depending on the ploidy-level, the multivalent type

and the type of centromere orientation, there is an irregular distribution of the chromosomes, leading to the formation of unbalanced gametes and zygotes;

2. autopolyploidy is frequently associated with genetic-physiological disturbances which can adversely affect the functioning of the meiotic products independently of and in addition to the irregularities in chromosome distribution;

3. these physiological disturbances associated with autopolyploidy may of themselves cause irregularities in chromosome distribution.

The meiotic → **segregation of genes** of autopolyploid forms differs in many respects from the segregation of the allele pairs of diploid organisms. With autopolyploidy, the number of alleles of each gene is increased according to the ploidy-level and gametes are formed which, in contrast to diploid forms, contain more than one allele per gene (homo- or heterozygotic). According to the number of dominant and recessive alleles of the particular locus, the genotypes are distinguished as "quadruplex and monoallelic" (AAAA or A^4), "triplex and biallelic" (AAAa or A^3a), "duplex and biallelic" (AAaa or A^2a^2), "simplex and biallelic" (Aaaa or Aa^3), and "nulliplex and monoallelic" (aaaa or a^4), respectively, and in the case of autotetraploidy have the genetic constitution given in brackets. Autopolyploids show a so-called **"polysomic inheritance".** Further, the segregation of genes of autopolyploids is affected by factors which play no essential part in diploids. Amongst these factors are the number and position of chiasmata in the multivalents, the distance of the particular locus from the → centromere, the behavior in distribution of the homologues paired as multivalents and the presence of → univalents.

1. **Segregation in the case of autotriploidy:** in this case (as in all other ploidy levels with uneven numbers of chromosome sets) meiosis takes a very irregular course. During meiosis I, the chromosomes are present as tri-, bi-, and univalents. Since the distribution of each homologous group proceeds independently of other such groups, all possible gamete types between haploid to diploid chromosome numbers are formed. These two extreme types are the rarest, the unbalanced intermediate types most common. The actual segregation ratios correspond in principle with those of → trisomics, but as a rule, are strongly perturbed by irregularities of the chromosome distribution and by chromosome elimination; they are thus difficult to predict.

2. **Segregation in the case of autotetraploidy:** In this case, i.e. for the occurrence of four homologous chromosome sets, the prediction of segregation ratios is easier. The chromosomes of each group are present as → quadrivalents, or otherwise, as two bivalents, one tri- and one univalent, or four univalents. Assuming that the four particular homologous chromosomes are distributed to the poles in the ratio 2:2 in anaphase I, the theoretical segregation ratios for the various autotetraploid genotypes of a locus may be calculated. Theoretically calculated segregation ratios which do not take into consideration → crossing-over between the concerned locus and the centromere are referred to as "chromosome segregation", those which consider a regularly formed crossing-over in this region of the

chromosomes and the thereby modified segregation of heterozygous loci, as → "chromatid segregation" (= double reduction). In the first case the segregation expectations result from the summation of the chance distribution combinations of the four homologues into in each case two to the poles. Table 2 shows the relative frequencies of the gamete types and zygote types of autotetraploid genotypes after selfing or, resp., crossing identical genotypes, calculated for pure "chromosome segregation".

Table 2. *Segregation ratios (pure chromosome segregation) for the selfing of autotetraploid genotypes*

parent genotype	gametes				zygotes					
	AA	Aa	aa	divisor	A^4	A^3a	A^2a^2	Aa^3	a^4	divisor
quadruplex A^4	1	—	—	1	1	—	—	—	—	1
triplex A^3a	1	1	—	2	1	2	1	—	—	4
duplex A^2a^2	1	4	1	6	1	8	18	8	1	36
simplex Aa^3	—	1	1	2	—	—	1	2	1	4
nulliplex a^4	—	—	1	1	—	—	—	—	1	1

On the other hand, if pure "chromatid segregation" is assumed and two sister chromatid segments with identical alleles of one chromosome become distributed after crossing-over and corresponding chromatid distribution into the same meiotic product, modified expectations as to the gamete and zygote classes are realized (Table 3).

Table 3. *Segregation ratios (pure chromatid segregation) of autotetraploid genotypes*

parent genotype	gametes				zygotes					
	AA	Aa	aa	divisor	A^4	A^3a	A^2a^2	Aa^3	a^4	divisor
A^4	1	—	—	1	1	—	—	—	—	1
A^3a	15	12	1	28	225	360	174	24	1	784
A^2a^2	3	8	3	14	9	48	82	48	9	196
Aa^3	1	12	15	28	1	24	174	360	225	784
a^4	—	—	1	1	—	—	—	—	1	1

The presence of "chromatid segregation" is implicit in cases where an unexpected excess of recessive and homozygous genotypes relative to "chromosome segregation" expectancies occur. The extent of "chromatid segregation" actually observed varies from locus to locus between certain limits and depends *inter alia* upon the crossing-over frequency between the locus and the centromere, on the frequency of multivalent formation, and on the types of orientation and distribution of the multivalents.

Segregation ratios of autotetraploid forms generally lie between those theoretically expected for pure "chromosome segregation" and pure "chromatid segregation". The different phenotype ratios in the offspring produced by crossing autotetraploid genotypes with complete dominance of the allele A over a and pure "chromosome segregation", respectively pure "chromatid segregation" are summarized in table 4.

Table 4. *The phenotypic segregation ratios of autotetraploids (complete dominance of A and pure chromosome (I), or chromatid segregation (II))*

type of cross	phenotypic segregation ratio in the offspring I	II
$A^4 \times A^4$	all A	all A
$A^3a \times A^3a$	all A	783A:1a
$A^2a^2 \times A^2a^2$	35A:1a	20,8A:1a
$Aa^3 \times Aa^3$	3A:1a	2,5A:1a
$a^4 \times a^4$	all a	all a
$A^3a \times A^2a^2$	all A	130A:1a
$A^3a \times Aa^3$	all A	51,3A:1a
$A^3a \times a^4$	all A	27A:1a
$A^2a^2 \times Aa^3$	11A:1a	7,7A:1a
$A^2a^2 \times a^4$	5A:1a	3,7A:1a
$Aa^3 \times a^4$	1A:1a	0,87A:1a

The special features of the segregation of autopolyploids compared with the segregation of diploids may be summarized as follows:

1. The occurrence of "chromosome" and "chromatid segregation".
2. The predominance of heterozygous genotypes.
3. The occurrence of several types of heterozygotes in the offspring of biallelic genotypes (e.g., A^3a, A^2a^2 and Aa^3).
4. The occurrence of heterozygous genotypes (e.g., A^3a) which lead to phenotypic segregation only after two generations.

autopolyploid (*Kihara & Ono* 1926) — → autoploid.

autoreduplication — applies to biological systems (organisms, cells, or subcellular units such as chromosomes, plastids, viruses, or genes) which generate the template for their own reproduction and duplicate themselves or produce mutant systems which are autoreduplicative in the same manner (syn. autoduplication, autoreproduction, identical reduplication). According to the present state of knowledge, the basis of a. are the nucleic acids as carriers of → genetic information.

autoregulation (*Calhoun & Hatfield* 1973) — a regulatory mechanism of gene expression (= autogenous regulation; self-regulation) in prokaryotic and eukaryotic organisms in which the product (protein) of a → structural gene modulates its own expression or the expression of the operon to which that structural gene belongs. The regulatory gene product has several functions; it may act as a regulatory protein and also as an enzyme, structural protein, or antibody. Often the regulatory protein is the first enzyme in the pathway and participates in control by → feedback inhibitions; its regulatory function is not dependent upon any catalytic or other function this protein may have and it is not involved in the transport or metabolism of an → inducer or → corepressor.

A. may provide a mechanism for amplification of gene expression, for severe and prolonged inactivation of gene expression, for buffering the response of structural genes to changes in the environment, and for maintaining a constant intracellular concentration of a protein, independent of cell size or growth rate (*Goldberger* 1974). (→ closed regulatory loop).

autosegregation (*Gustafsson* 1935) — the occurrence of changes in the chromosome complement during the formation of the egg cell by plants with gametophytic → apomixis. A. can lead to phenotypically changed, usually → aneuploid individuals and is a result of → subsexual recombination in strongly heterozygous but normally true-breeding apomicts.

autosome (*Montgomery* 1904) — any chromosome with the exception of the → sex-chromosomes which are referred to as allosomes or heterochromosomes. The genes localized in the autosomes follow the mode of distribution of these chromosomes to the gametes in the course of meiosis; they show autosomal inheritance, those localized in the sex-chromosomes show "sex-linked" inheritance.

autosteric effector — an effector which is part of or immediately adjacent to the active site of an → enzyme, in contrast to an allosteric effector localized at a distant site (→ allosteric).

autosyndesis (*Ljungdahl* 1924) — in contrast to → allosyndesis, the pairing (→ chromsome pairing) of complete or partial homologues (→ homologous chromosomes) contained in the same gametes at → fertilization, during the prophase of the first meiotic division (*Stebbins* 1947). A. can occur as → homo- or heterogenetic pairing and is associated with the presence of polyploidy or aneuploidy. The types of a. can be classified as:

1. complete a.: all chromosomes are paired autosyndetically;

2. one-sided a.: the chromosomes derived from one gamete pair autosyndetically, those from the other remain unpaired;

3. complete auto-allosyndesis: all chromosomes occur part auto-, part allosyndetically paired.

autotrophic — of cells or organisms that synthesize their cell components from simple chemical substances (e.g., ammonia, carbon dioxide, etc.) (= prototrophic). Obligatory autotrophs are limited to this way of life; facultative autotrophs are adaptable. Obligatory autotrophs always require carbon dioxide as a carbon source and either light or reduced inorganic compounds as a source of energy (→ auxotrophic, heterotrophic, prototrophic).

auxocyte — 1. each cell in which → meiosis is incipient (= meiocyte or gonotocont);

2. the androcyte, sporocyte, or primary spermatocyte during the growth period.

auxotrophic (*Ryan & Lederberg* 1946) — of nutritionally dependent cells, individuals, or strains ("auxotrophs") whose growth depends on a supplement to the basic food medium (minimal synthetic medium) in which → prototrophs are able to grow freely. Auxotrophs arise from prototrophs by → mutations causing one ("monoauxotrophs") or more ("polyauxotrophs") → genetic blocks.

axial core — any of the linear electron-dense cores in leptotene chromosomes (→ meiosis) developed before the onset of → chromosome pairing. The a. cores later become the lateral element of the → synaptonemal complex.

axoneme — the central thread or bundle of fibrils in a → cilium or a flagellum.

axostyl — in flagellates, an intracellular bundle of several thousand cross-bridged → microtubules which propagates a wave of bending from its anterior to its posterior end. Motility of flagellates results from undulations in the a.

B

backcross — the crossing of a heterozygote (→ heterozygous) with one of its parents. The first, second, etc., b. generations are symbolized by B_1, B_2, etc. The second and following backcrosses are made by crossing of individuals of the B_1-, B_2-generations, etc., with the same parent.

background genotype — → genotype.

background radiation — that radiation dose to which the organism is subject under natural conditions. It is a mutagenic factor (→ mutagen) and consists of two components:

1. **Cosmic radiation:** this consists of elementary particles (protons, alpha-particles, and heavy particles) which come from outer space, meet the layers of the atmosphere and there give rise to photons and various types of elementary particles. The total dose per unit time is dependent upon the height above sea-level, the atmospheric pressure and the geomagnetic latitude.

2. **Terrestrial radiation** from the natural radioactivity of the ground and the environment including the radio-isotopes taken into the body. It is subject to great variation in different geographical zones and is essentially dependent upon the rock formation.

The radiation dose received by humans due to b. r. lies between 2–4 milliroentgens per week. At present, 300 milliroentgens per week is accepted as the maximum permissible dose.

back mutation — a heritable change in a mutant → gene (→ gene mutation) resulting in a → revertant which has regained the enzyme or function that was lost with a socalled "forward mutation". A true back mutation restores the original nucleotide sequence which had been changed by forward mutation. The observation of a wildtype phenotype in a mutant line does not necessarily mean that this reversion has occurred by an actual "reversal" of the original mutational event, that is, by true back mutation.

True back mutation can be simulated by (*Green* 1959):

1. Unequal → crossing-over in the presence of a tandem duplication or by crossing-over between → pseudoalleles with positive or negative interference;

2. → gene conversion;

3. → position effects in consequence of chromosomal structural changes;

4. → suppressor mutations.

bacterial sporulation — the conversion of a bacterial cell from the vegetative to the sporulating state as a result of cellular differentiation. Sporulating cells undergo significant changes in morphology and physiology. They synthesize enzymes and → messenger RNA not found in vegetative cells and many vegetative enzymes and mRNA molecules disappear during early

b. s. These changes are the result of expression of a new set of genes and the turn-off of certain genes active in the vegetative cell.

bacterial transformation — → genetic transformation.

bacteriocide — any agent which kills bacteria, usually within a specified time.

bacteriocin (*Grazia* 1925; *Jacob* et al. 1953) — any substance (protein) produced by bacterial strains which, after adsorption at specific receptors on the surface of sensitive bacterial cells, is capable of killing these cells.

Bacteriocins represent a heterogeneous group of substances ranging from simple protein molecules to high molecular weight particles resembling those of phages or parts of phages. In every case, however, the part responsible for bacteriocidal action is the protein. The resistance and the ability to synthesize bacteriocins are genetically controlled by an accessory genetic element called → bacteriocinogenic factor which replicates in the bacteria as an autonomous → plasmid without physical association with the bacterial chromosome. Transfer of "bacteriocinogenesis" between bacteria of different systematic units takes place via → conjugation and → transduction.

Bacteriocins produced by *Escherichia coli* are called → colicins, those of *Pseudomonas aeruginosa*, pyocins; *Pasteurella pestis* produces pesticins. The mechanism of killing action differs from one b. to another. All activities of bacteriocins are exerted by b. molecules attached externally to the bacterial surface at specific receptors on the bacterial cell wall. Phage receptors and b. receptors are often the same and a mutation can render a bacterium resistant to a phage and one or more bacteriocins.

bacteriocinogenic factor — covalently closed circular DNA molecules in bacteria which represent independent genetic units (→ plasmid) of varying complexity and determine the production of specific bacteriocidal antibiotics called → bacteriocins. B. factors (which are not essential for normal cell life) may contain, besides the genes for the production of bacteriocins, genes specifying properties such as regulation, immunity, fertility, resistance to UV-radiation, susceptibility to male-specific phages, repression of fertility and restriction of phages, depending on the complexity of the b. f. (→ colicinogenic factor).

bacteriophage — a virus whose host is a bacterial cell. Bacteriophages (= phages) may have a complex morphology. They consist essentially of a core of genetic material (→ DNA or RNA) carrying the particle's → genetic information which is surrounded by a protein coat. Phages exhibit differences in their life cycles and may be grouped as either **virulent** or **temperate.** The introduction by infection of the genetic material of virulent phages into a susceptible host results invariably in the death and dissolution of the cell with the release of (100—10000) new phage particles (**"lytic cycle"**). Temperate phages can become integrated into the bacterial chromosome at specific points and replicate with it as a unit, the viral functions then being indefinitely unexpressed **("lysogenic cycle").** In the integrated stage the viral material is called a → "prophage" and each bacterium of the "lysogenic" clone carries a copy of the prophage in question. Once integrated into the bacterial chromosome, the viral genes can be considered additional bacterial genes and studied by the usual methods of bacterial genetics. The lysogenic cycle can become converted into a lytic one spontaneously or by → induction. In that case the prophage (→ episome) is released from the bacterial chromosome, multiplies independently and the cycle ends with the liberation of infectious phages by lysis of the host cell.

During the lytic cycle one pool of "vegetative" nucleic acid and another pool of structural units (such as head, sheath and tail proteins) are produced. By withdrawing nucleic acid and proteins from both pools essentially at random (→ host induced modification) mature phage particles are assembled. Within the pool, nucleic acid molecules with different contents of genetic information can give rise by interaction and → genetic recombination to recombinant molecules. Mature phages may incorporate genetic material from the bacterial host cell and transfer it to another cell they infect after their release from the lysed cell. This process is called → transduction.

The genetic information of a b. is contained within one linkage structure or phage chromosome which may be partly → heterozygous in some cases. Infection of a host cell with phages carrying different → genetic markers at homologous loci of the linkage structure constitutes "phage crossing".

The susceptibility of a bacterium to phage infection is primarily dependent on whether or not the phage can attach to specific attachment sites, receptors, on the cell surface. Male-specific DNA phages attach to the tips of → sex pili (F- or I-pili). For male-specific RNA phages nothing is known about their receptors other than they attach to pili.

bacteriostasis — inhibition of bacterial multiplication without immediate killing.

balance (*Bridges* 1922) — → genetic balance.

balanced lethals (*Muller* 1917) — → lethal factor.

balanced load (*Muller* 1950) — → genetic load.

Balbiani ring (*Beermann* 1952) — a large RNA → puff of → giant chromosomes which is present during a greater portion of larval development in the chironomid family of Diptera. This structural modification of specific loci is remarkable for its large size and characteristic ring shape. Similar to RNA puffs observed in the multistranded polytene chromosomes of a large number of dipteran species, the Balbiani rings have a high content of RNA and show a rapid uptake and turnover of RNA precursors. The B. r. structure and its extreme degree of puffing can be understood as the result of the unraveling of a thick, folded cable into individual filamentous elements, probably in the form of extended loops (see Fig. 81).

Balbiani ring granule (*Beermann & Bahr* 1954) — any of the numerous ribonucleoprotein (RNP) particles in → Balbiani rings. These particles are assembled from nascent mRNA and nuclear proteins. Similar particles have been found in some → puffs. The finished B. r. particles apparently arise from filamentous precursor particles and are released into the nuclear sap.

Baldwin-effect (*Simpson* 1953) — the exchange of "non-hereditary" by "hereditary" → characters via → mutation and → selection (→ genetic assimilation).

ball metaphase (*Barber & Callan* 1943) — a form of → c-mitosis with characteristically clumped chromosomes. The b. m. is followed by either a complete degeneration of the cell or a state similar to interphase.

band (*Painter* 1939) — the specific association of a large number of → chromomeres in → polytene chromosomes of Diptera (the polytene equivalent of "ordinary" chromomeres). Chromosome bands usually have clearcut outlines and characteristic dimensions. Each b. is well defined in relation to neighboring bands and separated from these by regions called interbands (→ banding pattern).

Bands contain about 95% of the chromosomal DNA. They (or band + interband) are likely units of genetic function and replication and show a selective transcriptional activity which is morphologically manifested as → puffing.

One band contains DNA in the range between 3,000 and 100,000 nucleotide pairs (→ service unit; transcription).

banding pattern — in → polytene chromosomes of *Diptera*, the linear pattern of deeply staining → bands and weakly staining interbands which results from defined local differences in the degree of compaction along the nucleoprotein elements composing the polytene chromosomes. The b. p. is thought to reflect the structural organization of these elements in the diploid interphase nucleus and represents, for any particular species of Diptera, a constant character, disturbed only by chromosome structural changes or temporary modifications such as those in → puffs and → Balbiani rings. If distinct species show precisely the same b. p. they are called homosequential species, as opposed to heterosequential species.

Barr body — → sex chromatin.

basal body (*Fawcett* 1961) — a → centriolar pinwheel that bears a → flagellum or → cilium (= kinetosome, blepharoplast, basal granule, basal corpuscle, and proximal centriole). The b. b. is a specialized, approximately cylindrical part of flagella and cilia located within the cell. Its outer wall is made up of nine tubular fibrils evenly spaced around the circumference of a circle and somewhat skewed towards the center, essentially like the peripheral fibrils of the flagellum or cilium. However, in the b.b. each peripheral fibril contains three rather than two subfibrils (central elements are missing). The b.b. of most types of cilia bears fibrous structures extending into the cell and called ciliary roots or rootlets. → Microtubules may also be connected with the b.b.

In various species, the b. b. varies from 120 to 150 nm in diameter and from 300 to 2000 nm in length.

basal granule — = basal body.

base pair — two nitrogenous bases that pair in double-stranded → DNA or → RNA molecules. The relative amount of adenine-thymine to guanine-cytosine pairs, which are those to which pairing in double stranded DNA is restricted, is called the base pair ratio (A—T/G—C ratio). In RNA thymine is replaced by uracil.

base pair substitution — a class of "lesions" in DNA molecules which may give rise to → gene mutations. Base pair substitutions consist of transitions, which preserve the purine-pyrimidine axis (AT ↔ GC), and of transversions, which reverse it (AT ↔ TA ↔ GC ↔ CG ↔ AT). Operationally, b. p. substitutions are recognized either by the amino acid substitutions in the corresponding polypeptide (missense mutations) and suppressible chain terminating signals (nonsense mutations) which they frequently produce, or by their patterns of chemically induced → reversion. Only very rarely are b. p. substitutions measured directly, or within RNA transcribed from DNA.

basic chromosome set (*Dyer* et al. 1970) — → chromosome set.

basic defect — a uniform defect of a structural or chemical nature which is the basis of the manifold syndromes of a → genetic disease.

basic number — the number of → chromosomes in a basic → chromosome set. The b. n. represents the smallest → monoploid chromosome number (symbolized by x) of a polyploid series. All chromosome numbers being exact multiples of the b. n. are called → euploid. All numbers deviating by single chromosomes from x and from multiples of it are called → aneuploid. The specific → karyotype comprising the b. n. is the basikaryotype (*Sinoto & Sato* 1940).

basikaryotype (*Sinoto & Sato* 1940) — → basic number.

B-chromosome (*Randolph* 1928) — any chromosome of a heterogeneous category of → chromosomes (also called supernumerary, accessory or extra chromosomes) present in many species of both plants and animals which differ in their behavior from normal or A-chromosomes in the following characteristics (some of them may be missing in the special case under observation):

1. Their morphology (they are usually smaller in size than A-chromosomes, frequently but not exclusively heterochromatic and telocentric).
2. Their genetic effectiveness (they normally do not strongly influence the viability and phenotype of their carriers).
3. Their numerical variation (between different cells, tissues, individuals, and populations).
4. Their meiotic behavior (no chiasmate pairing with the A-chromosomes, lower degree of pairing among themselves, lagging and elimination or preferential non-disjunction during meiosis and post-meiotic divisions).
5. Their mitotic behavior (lagging and elimination, polymitosis or preferential distribution).

According to *White* (1945) two main groups of B-chromosomes may be distinguished: those which are mitotically stable, where all cells of the individual have the same number of B's, and those which are mitotically unstable, giving rise to cells with different numbers of B's within the same individual.

Accumulation systems of B-chromosomes in populations are various but normally fall into two categories: 1. those in which the B. undergoes distorted segregation at meiosis and is thereby preferentially transmitted to the gametes, and 2. those in which the B. undergoes mitotic → nondisjunction in some particular divisions or tissues.

Special types of accumulation systems result in a tendency for the B-chromosomes to increase in frequency in males and to decrease correspondingly in females.

B-chromosomes in both plants and animals may affect growth, development of the external phenotype, genetic recombination and chiasma formation during meiosis.

The idea is current that supernumerary chromosomes might have originated from small fragments caused by structural rearrangement and might eventually become as large as other chromosomes by repeated duplication.

It is generally held that B-chromosomes are of adaptive importance since their occurrence and frequency within the species varies between populations of different origin and habitat. A minority view is that B-chromosomes are parasitic and maintained more or less independently of any adaptive benefit or handicap they confer upon individuals or populations.

behavior flexibility (*Thoday* 1953) — all means of behavior permitting temporary → adaptation to environmental conditions such as habitat selection and measures for controlling the habitat or providing protection against environmental influences (→ phenotype).

behavior genetics — a branch of genetics which deels with the inheritance of different types or forms of behavior, such as intelligence, personality traits, etc.

bimitosis (*Gonzales* 1967) — the simultaneous occurrance of two → mitoses in binucleate cells which arise by interference with → cytokinesis.

binemic — of → chromosomes that contain or are assumed to contain two DNA helices per metaphase chromatid, as opposed to uninemic chromatids with one DNA helix, or polynemic chromatids with more than two DNA helices.

binom (*Grant* 1957) — = agamospecies, agameon.

binucleate — of cells with two nuclei.

bioassay — determination of the amount or effectiveness of a biologically active substance by measuring its effectivity on a living organism (→ mutagenicity screening).

biochemical genetics — a branch of genetics dealing with the chemical nature of hereditary determinants and the manner of their action in both development and function (→ gene, gene action).

biocoenosis (*Möbius* 1877) — a community of organisms which consists of a number of different kinds of plants and animals, living together in a particular habitat or biotope.

bioid (*Decker & Speidel* 1972) — any of the simplest chemical open systems capable of Darwinian → evolution by mutation from one steady state into a more stable one.

biological altruism — all forms of behavior by which some individuals benefit other individuals of the same species, without benefiting themselves. The benefits may be voluntary or involuntary, direct or indirect. B. a. is involved in the organization and evolution of social systems of insects and other animals including, of course, man (*P. I. Darlington* 1972).

biome — a community of plants and animals (→ ecosystem). Adjacent biomes usually blend into one another (no sharp demarcation line) through a fairly broad transition region called an ecotone.

biometrical genetics — a branch of genetics concerned with the study of quantitative → inheritance.

biometry — statistics applied to biological problems.

biopoesis (*Pirie* 1937) — the origin of the first living thing, including the chemical history that preceded it. The first "organism" is called an eobiont (→ eobiogenesis; neobiogenesis). B. can be divided arbitrarily into two phases (*Keosian* 1965): a) the phase that covers the period of physical and chemical processes in the abiotic synthesis of macromolecular systems; b) the phase (eobiogenesis) that includes the transformation of macromolecular systems into the first living things.

biosome (*Lehmann* 1947) — any autonomous cell constituent multiplying or presumed to multiply by autoreduplication.

biosynthesis — the synthesis of the chemical components of the → cell from simple precursors and their assembly into structures such as, e.g., → membrane systems, → mitochondria, → nuclei, and → ribosomes. The b. of the cell components and the cell's functional morphology are genetically programmed.

biotic potential — the inherent ability of a → population to increase in numbers when the age ratio is stable and all environmental conditions are optimal (= reproductive potential).

biotope (*Dahl* 1921) — → biocoenosis.

biotype (*Johannsen* 1903) — a group of genetically identical individuals.

biparental cross — → triparental cross.

bipolarity (*Margolin* 1965) — the phenomenon wherein both strands of a DNA helix (chromosome) serve in → genetic transcription, although at different regions.

bipolaron (*Piperno* et al. 1966, *Sermonti & Carere* 1968) — a nucleic acid segment whose polarity in → genetic recombination between pairs of mutations is opposite at the two ends of the → gene cluster and changes gradually as a function of the position of the selected → crossing over.

bisexual — 1. of organisms having both male and female reproductive organs and producing male and female → gametes within the same individual, which is either monoecious or hermaphroditic;

2. of populations or generations composed of male and female individuals as opposed to a unisexual parthenogenetic population or generation (→ parthenogenesis).

bit — the basic unit ("binary unit") of information content which represents the answer to a binary choice, i.e. describes the possible conditions of which the given condition represents one of two mutually exclusive possibilities (such as "yes" or "no").

bithallic (*Ahmad* 1954) — → heteromixis.

bivalent (*Haecker* 1892) — a pairing configuration (→ chromosome pairing) during the first meiotic division which consists of two completely or partially homologous chromosomes (→ pseudobivalent, quasibivalent). The number of bivalents per cell (meiocyte) normally corresponds to half the somatic chromosome number of diploid and genome-allopolyploid species. In chiasmate meiosis the pairing partners are held together after → crossing-over by → chiasmata, in achiasmate meiosis by functionally equivalent structures (→ collochores) until the beginning of anaphase I. The formation of bivalents is prerequisite for the reduction of the somatic chromosome number and the random → assortment of the pairing partners (→ segregation). The following types of b. are to be distinguished:

1. **Homomorphic bivalents** in which the two pairing partners are in complete structural agreement.

a) ring b.: a b. appearing in the form of a ring during meta-anaphase I following chiasma formation in both arms and complete or almost complete → chiasma terminalization.

b) rod b.: a b. appearing as a rod during meta-anaphase I following chiasma formation and chiasma terminalization in only one arm.

2. **Heteromorphic bivalents** in which (due to heterozygous → chromosome mutations) the two partners of the pair differ structurally and are only partially homologous.

a) unequal b.: the partners are of unequal size, as a consequence of → deletions or various types of → translocation.

b) asymmetrical b.: The partners are of equal size but the → centromeres are located differently.

3. **Precentric** and **precocious bivalents** in which the centromere regions of the partners are already separated in metaphase I (a departure from normal behavior).

4. **C-bivalents,** i.e. two → C-pairs, lying close together, whose chromatids are separated from one another, apart from the centromere region. C-bivalents are characteristic of → C-meiosis and are produced by the action of colchicine and other spindle poisons.

bivalent interlocking — → interlocking.

blastocyst — an early (pre-implantation) stage of mammalian development. The b. is a hollow ball-shaped structure consisting of 100 to 1000 cells (according to species); some of these cells will form the embryo.

blastocyte — any of the undifferentiated embryonic cells (→ cytodifferentiation)

blastomere — a single large (undifferentiated) cell of a very early multicellular embryo. Blastomeres originate from the first few cleavages of a → zygote in animal embryogenesis (→ cytodifferentiation).

blastulation — → embryonic development.

blending inheritance — → inheritance.

blepharoplast (*Webber* 1897) — = → basal body.

bordered pit — a pit in tracheids or other plant cells involved in water conduction having a distinct rim of the cell wall overarching the pit membrane (*Frey-Wyssling & Mühlethaler* 1965).

bouquet stage (*Eisen* 1900) — a meiotic prophase stage (leptotene to pachytene) in those species where the chromosomes are oriented by one or both ends towards one point of the → nuclear envelope. The polarization process responsible for this arrangement originates in the cytoplasm. Where → centrosomes occur, there is an orientation relationship between these and the chromosomes.

brachymeiosis — an abnormal type of → meiosis characterized by the omission of the second meiotic division.

bradytelic (*Simpson* 1944) — of rates of → evolution which are lower than average (→ horotelic, tachytelic).

break through (*Hadorn* 1945) — a genotype which in spite of the presence of a → lethal factor in effective dose, survives the lethal crisis and continues to develop despite its autolethal constitution. Having overcome a first critical life-period resulting from the → phase specificity of lethals, the b. t. can reach, under certain circumstances, a second sensitive stage in which it perishes.

breeding size — the number of individuals in a population that are actually involved in → reproduction in a given generation (symbolized by N).

Effective b. s. (symbolized by N_e) is the b. s. mathematically adjusted so that populations with varying → sex ratios, degrees of → inbreeding etc. may be compared. N_e is equivalent to N only in an "ideal population" of large size, with equal numbers of the two sexes, with random → mating and random fusion of the gametes. Normally the effective b. s. is smaller than the b. s., since uneven sex ratios, inbreeding, cyclic changes of b. s. as well as non-random sampling of the gametes all reduce the effective b. s.

breeding system — a term used to cover all those variables apart from → mutation which affect the genetic relations of the gametes that fuse in sexual → reproduction. The variable components of the b. s. are numerous and often not easy to estimate. Two main groups may be distinguished (*Lewis & John* 1964):

1. those which affect the ability of particular gametes to fuse or parents to mate, and

2. those which affect their probability within the limits set by the first. The variables under 1. comprise an aspect of the b. s. known as the → mating system.

The b. s. may be an essential factor controlling population structure and evolutionary divergence (→ evolutionary rate). In plants, it controls the extent of → outbreeding which may take various forms: exclusive or predominant outcrossing, predominant selfing, and a mixture of selfing and crossing.

breeding value. — the value of an individual as defined by the mean value of its progeny. The b. v. of an indvidual equals the sum of the average effects of the genes contained in its → genotype ("average effect of a gene" is the mean deviation from the population mean of those individuals who received that particular gene from one parent, whereas the other gene of the pair under consideration came at random from the population studied).

bridge-breakage-fusion-bridge cycle (*McClintock* 1951) — a process which may arise from the formation of → dicentric chromosomes (chromosome type of the BBFB-cycle) or chromatids (chromatid type of the BBFB-cycle), whereby daughter cells are formed (via breakage of the dicentric bridge between the cell poles during the anaphase of meiosis and/or mitosis) differing in their content of → genetic information due to → duplications and → deletions in the chromosomes. A prerequisite for the maintenance of such a cycle is that the daughter cells produced be viable and that the chromosome ends formed by breakage of the bridge should not be stabilized but should fuse (Fig. 3). In the course of a BBFB-cycle there may be a repeated regrouping of the genetic loci in the chromosomes or chromatids involved since the breakage of the bridge may occur in different positions. The cycle is initiated by primary → chromosomes mutations and results in secondary changes of the chromosomes taking part.

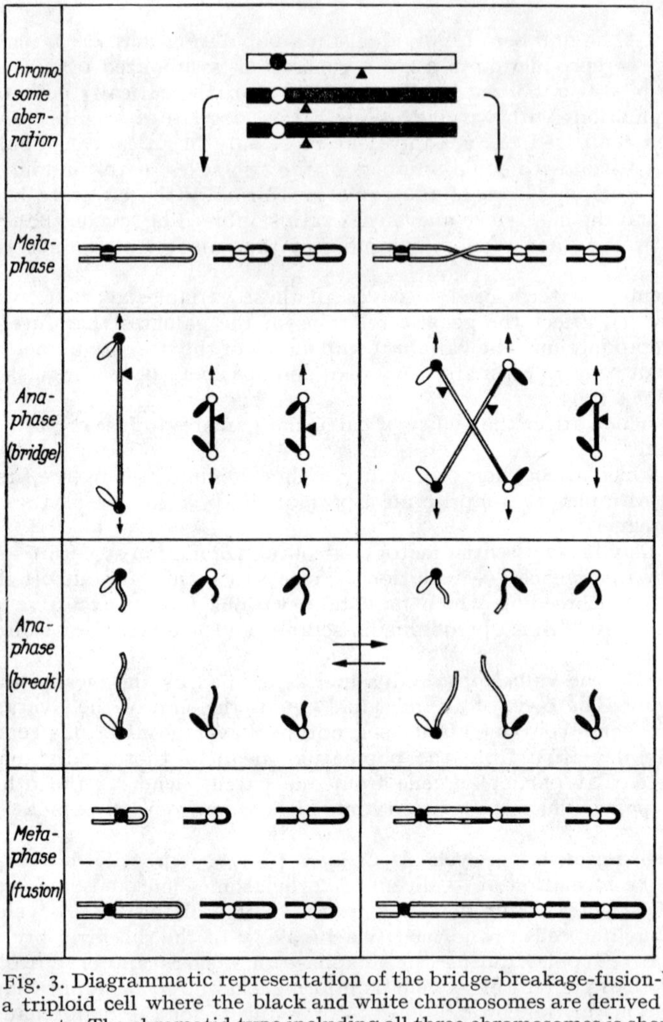

Fig. 3. Diagrammatic representation of the bridge-breakage-fusion-bridge cycle in a triploid cell where the black and white chromosomes are derived from separate parents. The chromatid type including all three chromosomes is shown on the left; the chromosome type on the right shows two of the three chromosomes joined to form a dicentric chromosome. Possible interconversion of the two types indicated by arrows (modified after *Faberge* 1958).

brights (*Preer, Siegel & Stark* 1953) — → killer.
budding — → cell budding.
bud sport — a somatic → mutation expressed in a bud or branch of plants = bud mutation).
bulbifery — in plants, a common form of asexual reproduction by small bulbils formed in the axils of leaves, in the floral axis, or in floral branches. In many cases bulbils fully supplant flower formation and the sexual process; in others bulbils are formed in addition to normal flower formation.
burst size — the mean number of → bacteriophage particles liberated per infected bacterium, or the ratio of phage titers before and after lysis of phage-infected bacterial cells. The interval preceeding the increase in titer is called the latent period; the interval of increasing titer is called the rise period or the release period. The b. s. varies widely with strain of phage and bacterium used and is dependent upon cultural conditions.

C

callus — a disorganized mass of undifferentiated (parenchyma) plant cells which may accumulate at a wound surface, or grow out from pieces of plant tissue placed in culture.

cAMP — → cyclic adenosine-3′, 5′-monophosphate.

canalization (*Waddington* 1940) — the property of developmental pathways of achieving a standard → phenotype in spite of genetic or environmental disturbances (→ homeostasis). C. is a developmental buffering system reducing potential variation that may act at the gene level or on variation due to either genotype or environment.

Phenotypic characters whose development is normally unaffected by environmental stresses or by underlying genetic variability are called canalized characters.

canalizing selection (*Waddington* 1940) — → selection.

CAP (*Zubay* et al. 1970) — abbreviation for → catabolite activator protein.

capsid (*Lwoff, Anderson & Jacob* 1959) — the protein coat of a virus particle or → virion which may take different forms (either helical or isometric), depending on the nature of the protein subunits from which the c. is formed. Components (protein subunits) of capsids observed in the electron microscope on axes of symmetry are referred to as capsomeres. Composite capsids may consist of more than one kind of capsomere.

The shape or design of the c. of a virion is determined by the specific bonding properties of its capsomeres, which may be composed of one or more proteins.

capsomere (*Lwoff, Anderson & Jacob* 1959) — → capsid.

capsule — a jelly-like polysaccharide or, less frequently, polypeptide coat surrounding cells of certain bacteria.

carrier — → genetic carrier.

carrier cell — any virus-infected cell which continuously divides and replicates virus.

caryonide (*Sonneborn* 1938) — any cell line in ciliates derived from the products of the first cell division after conjugation. Cells in such lineages possess → macronuclei descended from a common ancestral macronucleus. Two caryonides from one exconjugant form a clone, and the four caryonides of a pair represent a synclone. The synclone is the unit of Mendelian inheritance in ciliates.

cascade regulation (*Pontecorvo* 1963, *Scherrer & Marcaud* 1968) — 1. in prokaryotes, a form of → genetic regulation on the transcriptional level where one of several → operons codes for the production of an internal inducer which turns on one or more of the other prokaryotic operons.

2. A multistep model of genetic regulation in eukaryotic cells according to which regulation operates at a multiplicity of levels by mechanisms

interfering with messenger RNA formation (transcriptional control), mRNA transport, and mRNA translation (translational control).

catabolite activator protein (*Zubay* et al. 1970) — a protein (abbreviation CAP) which acts as a transcription-initiation factor necessary for the efficient transcription of bacterial → operons which are subject to → catabolite repression (bio-degradative operons). CAP is activated by → cyclic adenosine monophosphate (cyclic AMP) and the combination of cAMP and CAP is bound to the DNA enabling → RNA polymerase to bind to the → promoter of the operon and to initiate transcription. Possibly, the promoter of these operons is composed of two sites that interact with CAP and RNA polymerase holoenzyme. In the presence of the appropriate → repressor, the operon cannot synthesize mRNA, even in the presence of CAP and cyclic AMP.

Operons subject to catabolite repression show a reduced rate of expression in the presence of, e.g., glucose. Glucose apparently results in a decreased availability of cAMP to CAP. CAP is also called cyclic AMP receptor protein (CRP).

catabolite repression (*Magasanik* 1961) — the inhibition by glucose and some biochemically related compounds of the synthesis of certain inducible enzymes in bacteria. C. r. is a mechanism of genetic regulation at the DNA level (transcriptional control). The extent of inhibition varies from a few percent to more than 90% and is dependent on the bacterial strain, the source of the catabolite, and the growth conditions. Most genes classified as inducible show c. r.

C. r. is due to a lowering of the cyclic adenosine monophosphate (cAMP) level; expression of operons subject to c. r. requires cAMP and → catabolite activator protein (CAP).

catabolite sensitive (*Zubay* et al. 1970) — of genes whose activities are subject to → catabolite repression.

catalase reactivation (*Monod, Torriani & Jolit* 1949) — the increased survival of UV-irradiated bacteria by treatment of the cells with catalase or peroxidase (→ reactivation). C. r. is enhanced by small doses of visible light and in the presence of oxidizable substances (*Rupert & Harm* 1966).

catonated — of topologically interlinked, circular DNA molecules.

caveola — any of the cell surface invaginations which (like → microvilli) serve to increase the surface area. In a c. the → unit membrane forms a blind, cavelike or tubular pouch opening to the extracellular space. A large number of caveolae are transient membrane configurations associated with secretion (excytosis) and with uptake of molecules into the cell by encytosis.

C-bivalent — → C-meiosis; bivalent.

cDNA — DNA complementary to a purified messenger RNA and produced by RNA-dependent → DNA polymerase (reverse transcriptase, revertase). cDNA (= anti-messenger DNA) can be used as very sensitive and specific probes in molecular hybridization studies. These allow one to distinguish between nucleotide sequences that bear a high degree of similarity and also provide a practical means of assessing the purity of any RNA population containing the sequence reverse-transcribed.

cell (*Hooke* 1665) — the basic structural and functional unit of all living matter which represents a multimembraned system, compartmentalized into specific functional areas or organelles with discrete but complexly interrelated and coordinated roles.

Cells may exist as independent units of life; they may associate with one another to form cell colonies (facultative association) or tissues (obligatory association).

Cells are the building elements of all organisms. In the probionta, the cells possess specific organelles to accomplish all the important life functions; in the metabionta the cells are specialized, resulting in a division of labor whereby particular cells adopt specific functions (→ cytodifferentiation).

Each cell is a strictly delimited system which may be easily identified by its material nature, its structural organization and its behavioral expressions. Cells show manifold variation in size and form, but in all organisms they conform to a single fundamental plan: the protoplasm, i.e. the cellular substance as a whole, is differentiated into the → cytoplasm and the cell → nucleus (in eukaryotes) or → nucleoid (in prokaryotes). Each cell is surrounded by a → plasma membrane and in certain species by a → cell wall as well. The various cell organelles are situated in the → ground plasm of the cell (the "matrix"), i.e. the soluble phase composed of water, salts and free molecules (including many enzymes). In addition to the cell nucleus or the nucleoid, such cell organelles are the → mitochondria and → chondrioids, the → plastids, the → ribosomes, the → Golgi apparatus, the → centrioles, the → lysosomes, the → cortex, the → vacuoles, and the → endoplasmic reticulum.

Cells of eukaryotes (higher plants and animals, protozoa, fungi and most algae) and prokaryotes (bacteria, blue-green algae) are of different complexity.

1. **The eukaryotic cell** (*Chatton* 1925) (Fig. 4) may be understood as a three-phase system (*Morrison* 1966): one phase is represented by the nucleo-cytoplasmic matrix in which → nucleoplasm and cytoplasm have continuity via the pores of the → nuclear envelope. The second phase is formed by the membraneous organelles of the cell and the third consists of the contents of these membrane-enclosed structures. Membranes (→unit membrane) subdivide the cells of eukaryotes into the two major compartments, nucleus and cytoplasm, and the cytoplasm into a system of interconnected cavities and various separate compartments. The discrete, membrane-bounded compartments represent the mitochondria, plastids and lysosomes, while the system of continuous cavities forms the endoplasmic reticulum and the Golgi apparatus. The chemical constituents (enzymes) by which cells perform their functions are localized mostly in organized membrane structures.

Unlike animal cells, plant cells usually possess → plastids and vacuoles, and are covered by a → cell wall. Mitochondria, ribosomes and a nucleus are present as in animal cells. Plant cells have structures termed → dictyosomes which are similar to the → Golgi complex of animal cells.

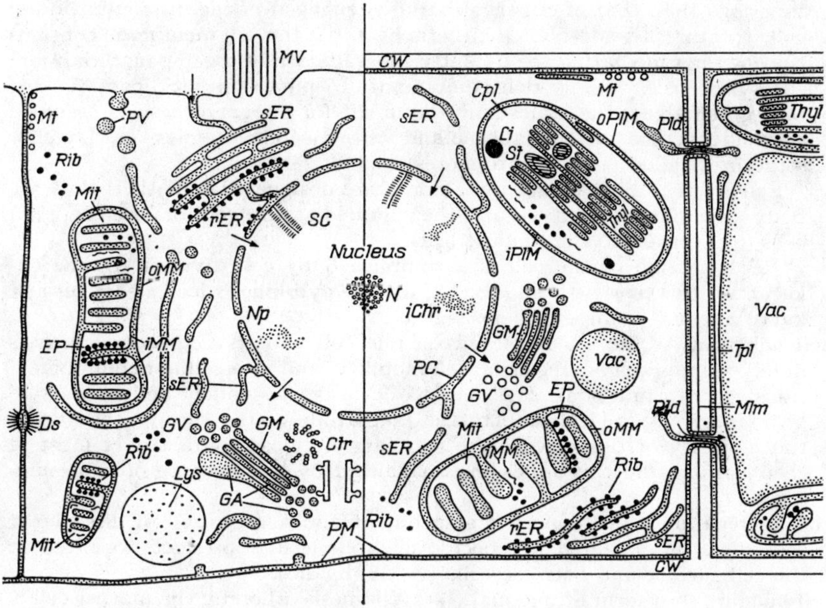

Fig. 4. Diagrammatic representation of the architecture and ultrastructure of an ideal animal (left) and plant cell (right). The mitochondria (Mit) are of the cristae type, the chloroplasts (Cpl) of the grana type. Arrows indicate membraneless connections between different cell compartments. Elementary membranes are symbolized by single continuous lines. Cpl: chloroplast; CW: cell wall; Ds: desmosome; EP: elementary particle; GA: Golgi apparatus; GM: Golgi membrane; GV: Golgi vesicle; iChr: interphasic chromatin; iMM: inner mitochondrial membrane; iPLM: inner plastid membrane; Li: Lipid droplet; Ly: lysosome; Mit: mitochondrion; Mlm: middle lamella; Mt: microtubulus; MV: microvillus; N: nucleolus; Np: nuclear pore; oMM: outer mitochondrial membrane; oPLM: outer plastid membrane; PC: perinuclear cisterna; Pld: plasmodesma; PM: plasma (cell) membrane; PV: pinocytotic vesicle; rER: rough endoplasmic reticulum; Rib: ribosome; SC: synaptonemal complex in meiotic prophase; sER: smooth endoplasmic reticulum; SG: starch grain; Thyl: thylakoid; Tpl: tonoplast; Vac: vacuole (slightly modified after *Klima* 1967).

2. The **prokaryotic cell** (*Chatton* 1925) is the unit of structure in bacteria and blue-green algae and shows less structural components classifiable as organelles than the eukaryotic cell. These organelles include the nuclear equivalent or nucleoid, the cell membrane and a few membraneous structures regularly present in the cytoplasm. Localized invaginations (→ mesosome) of the cell membrane give rise to tube-like structures. Particles equivalent to the ribosomes and mitochondria (→ chondrioid) of eukaryotes are present in bacterial cells.

The cell membrane of prokaryotes seems to perform more complex functions than that of eukaryotes and is engaged in the differentiation of cell structure. In non-photosynthetic bacteria, the cell membrane contains the enzymes mediating the oxidative degradation of organic materials and the synthesis of energy-rich compounds. In photosynthetic bacteria and blue-green algae, the molecular equipment for photosynthesis is assumed to reside in the cell membrane and its intrusions (vesicles and lamellar structures called → chromatophores).

Cells of the multicellular metabionta are designated somatic if they are body cells. The → gametes and their immediate progenitors are the germ cells, in contrast to the somatic cells.

Reproduction of the cell is accomplished by cell division. In the eukaryotes, this process is subdivided into → cytokinesis (cell cleavage) and karyokinesis (→ mitosis, meiosis).

cell adhesion — the contact between one cell and its neighbors which is involved in → cell aggregation, cell mobility, and intercellular communications (→ gap junction). The extent of c. a. varies widely and appears to bear an inverse relationship to the magnitude of the surface charge. In the majority of tissues cells are in relatively close contact over most of their surface and this adhesion is enhanced by the presence of → desmosomes.

cell aggregation — histogenetic attachment and aggregation of dissociated cells. C. a. is mediated by specific macromolecular products localized at the cell surface and between cells (→ cell ligand).

cell budding — a form of unequal → cytokinesis whereby the mother cell is not halved; instead, bud-like growths form at one or more narrowly restricted sites and one daughter nucleus migrates to each such growth, followed by the separation of the daughter cells from the mother cell.

cell center — the cell structure which determines the cell poles during the nuclear division (→ mitotic center).

cell cleavage — → cleavage.

cell colony — → cell.

cell communication — long-range and short-range interaction between cells in both excitable and nonexcitable tissues. Short-range c. c. requires direct physical contact between cells and is frequently accompanied by distinct intercellular membrane specializations (→ gap junction; tight junction; septate junction; synaptic junction; desmosome).

cell compartment — any of the membrane-bound units or parts within cells. Presumably each c. c. is able to regulate its contents by virtue of the selective permeability of its double or single → unit membrane and may serve as a regulatory control in the cell. Main compartments of eukaryotic cells are: 1. the cell → nucleus, 2. → mitochondria, 3. → plastids, → cytosol, 4. → endoplasmic reticulum, 5. → ribosomes, 6. → Golgi apparatus, 7. → lysosomes, 8. → peroxisomes, and 9. glycogen granules. (→ compartmentalization).

cell cortex (*Chambers* 1940) — → cortex.

cell culture — the growing of dispersed cells *in vitro* including the culture of single cells. The terms **tissue culture** and **organ culture** denote the maintenance or growth of tissues, organ primordia, or the whole or part of an organ *in vitro* in a way that may allow differentiation and preservation of their architecture and/or function (→ culture alteration). A culture started from cells, tissues, or organs taken directly from organisms is called a **primary culture.** The first subculture of a primary culture gives rise to a → cell line (*Feodoroff* 1967).

cell cycle — the life cycle of the individual cell. In proliferating somatic cells it consists of four phases: the mitotic phase (symbol M) and a three-stage interphase consisting of the pre-DNA synthesis or G1-phase, the DNA-synthesis phase (S) and the post-DNA synthesis or G2-phase. During G1 the chromosomes are unreduplicated; chromosome reduplication occurs in S (*Howard & Pelc* 1953). A similar subdivision may be applied to the interphase preceding meiosis. In this case the G2-phase is generally either short or completely absent, i.e. meiosis takes place shortly or immediately after the synthesis period (S) of the interphase nucleus. Besides the orderly sequence of replication of the → genome, the development of a cell from one division to the next includes the formation of cell → organelles and the morphological events of cell division. C. cycles reprogramming the synthetic repertoire of daughter cells are called "quantal cell cycles", those yielding replicas of the mother cell are termed "proliferative cell cycles" (*Ishikawa* et al. 1968).

Progress through the c. c. is based on and regulated by the temporal sequence of → genetic transcription (→ cell synchrony).

cell differentiation — = → cytodifferentiation.

cell division — the reproduction of cells by → karyokinesis and → cytokinesis.

cell division lag (*Ryan* 1954) — delayed manifestation of induced → mutations which may occur after heavy doses of some → mutagens and is the result of the delayed onset of renewed division (→ mutational lag, phenotypic lag).

cell envelope — the sum of all surface components of the cell that are located outside the cytoplasmic membrane.

cell fusion — → cell hybridization.

cell generation time — the interval between consecutive divisions of a cell.

cell heredity — heredity at the cellular level. In eukaryotes, the informational patterns of c. h. consist of nuclear DNA, organelle DNA, and priming configurations (*Sonneborn* 1967) whose presence is required to direct the assembly of molecular species into functional two- and three-dimensional structures (→ cortex).

cell hybridization (*Barski* et al. 1960) — in eukaryotes, the fusion of somatic cells in cell culture and the formation of viable hybrid cells. In animals the hybrid cells produced by co-cultivation of cell lines or inactive virus mediated cell fusion possess a nucleus containing chromosomes of both parents. (Virus-mediated cell fusion is about two orders of magnitude more efficient than cell fusion by co-cultivation). In interspecific somatic cell hybrids, segregation occurs by species-dominant loss of the chromosomes of one parental cell. This makes these cells particularly suitable for genetic analysis (→ synteny; synkaryon).

cell interaction — influences of one cell upon another which play an important role in embryonic development (location and sequence of tissue growth), wound healing, and tissue culture (e.g., → contact inhibition of division and movement).

cell junction (*Farquhar & Palade* 1963) — in animals, any type of specialization of the cell periphery related to → cell adhesion and attachment (→ desmosome). C. junctions represent a variety of morphological forms and all types may exist in various geometrical forms. They arise at characteristic times during ontogeny. The location and specificity of cell junctions play a decisive role in the specificity of many histological and cytological characteristics, such as tissue permeability, transmission of electrical signals, recognition of cells during embryogenesis, tissue differentiation and the shape, polarity and internal organization of cells. (→ zonula occludens; macula adherens).

cell lethality — the death of individual cells which contain "cell lethals". Cell lethals (*Demerec & Hoover* 1936) are → mutations (mostly → deletions) which normally bring about the death of the cells in which they occur (→ lethals). A distinction is made between primary and secondary cell lethality (*Hadorn* 1949):

1. **Primary c. l.:** cells with cell lethals perish even as mosaic spots or implants in association with genetically "normal" tissue.

2. **Secondary c. l.:** following appropriate transplantation into genetically "normal" tissue, the cells with cell lethals circumvent a lethal crisis and survive; their death results from the natural death of the carrier organism.

cell lethals (*Demerec & Hoover* 1936) — → cell lethality.

cell ligand (*Moscona* 1968) — any factor associated with the cell surface and intercellular spaces which mediates histogenetic attachment and aggregation of cells. The molecular features, differences or complementarities and distribution patterns of c. ligands on and between cells are assumed to be responsible for differential adhesiveness of the cells, cell recognition, affinities, and preferential association.

cell line — a population of cells which arises from a primary → cell culture at the time of the first subculture (→ cell strain). A **cloned line** is a c. l. descended directly from a clone. A **diploid c. l.** is one in which, arbitrarily, at least 75% of the cells have the same → karyotype as the normal cells of the species from which the line was derived. If the karyotype has changed, but the diploid number of chromosomes remains the same, such cells may be referred to as "pseudodiploid". A **heteroploid c. l.** is one having less than 75% of cells with diploid chromosome constitution (*Feodoroff* 1967).

An established c. l. consists of cultured cells of unique origin capable of stable growth for many generations.

cell lineage (*Wilson* 1882) — the derivation of a tissue or part of tissue from a definite → blastomere of the embryo.

cell membrane (*Nägeli & Cramer* 1855) — a component of the → cell surface (= plasma membrane). The c. m. is a tri-laminate structure (usually 8 to 10 nm thick) with two dense lines spaced 4 to 6 nm apart on each side of an intervening element of lesser densitiy in the electron microscope

(→ unit membrane). Differences in thickness, in symmetry, and in spacing of the two dense lamellae have been found. C. membranes possess pores with diameters greater than 0.4 nm but less than 0.8 nm. Size of pores and charge density within them are the main factors that control the fluxes of materials. Hormones, calcium and polybasic substances are assumed to regulate the size of membrane pores.

The c. m. acts as a passive diffusion barrier to charged and large molecules, and in addition carries out facile diffusion (passive transport), active transport, endocytosis (pino- and phagocytosis), and exocytosis. It is further involved in locomotion and chemotaxis, secretion and reception of hormones, conduction of nerve impulses, and direct cellular interactions such as → cell adhesion and → contact inhibition. Differentiated functional sites of the c. m. are involed in these various functions (→ mesosome).

The assembly of the c. m. is a complex process in which the → Golgi apparatus of eukaryotic cells might be involved (→ membrane system).

The c. m. and internal membranes (→ membrane system; mesosome; endoplasmic reticulum) of the cell are essentially alike. Both are composed of proteins and lipids; in mammalian cells, small amounts of carbohydrate are also present, associated either with protein (glycoproteins), or with lipid (glycolipids). There is likely to be an interrelation between the protein array and membrane function in cells with a variable distribution of proteins.

cell migration — in animal embryos, the selective movement of cells from one part of the embryo to another. Migration of cells, either individually or *en masse,* is critically timed and results in new juxtapositions of tissues, which in many cases appear to control the initiation of specific tissue-to-tissue → inductions.

cell nucleus (*Brown* 1831) — → nucleus.

cell plate (*Strasburger* 1882) — in plant cells, the structure formed between daughter nuclei (→ cytokinesis) after → karyokinesis. The c. p. complex or phragmoplast is composed both of fibrillar elements (microtubules), which arise at the plate and are oriented at right angles to its plane, and numerous cytoplasmic droplets which eventually accumulate and fuse in the mid region of the cell. Presumably the vesicles originate from → dictyosomes. The microtubules probably determine the channels along which vesicles move before they fuse.

cell population — a group of cells designated as either static, expanding, or renewing (*Leblond* 1964).

1. **Static c. populations** represent homogeneous groups of cells in which no mitotic activity can be detected; the total DNA content remains constant.

2. **Expanding c. populations** are homogeneous groups of cells showing scattered mitoses in numbers which account for the increase in total DNA content.

3. **Renewing c. populations** are homogeneous groups of cells in which mitosis is abundant and exceeds that required for the total increase in DNA content.

cell recognition — the mutual recognition of cells which is probably due to antigen-antibody type reactions or enzyme-substrate complexes, specifically involving glycosyltransferases.

cell sap — the interparticular fluid of the cell = hyaloplasm or soluble phase of the cell). The c. s. is the milieu in which cellular formed elements persist and function and the main processes of cellular metabolism are performed. The c. s. is defined operationally as the supernatant that fails to sediment in a centrifugal field strong enough to deposit all particular material (as e.g. → ribosomes, → mitochondria, → plastids, etc.).

The c. s. or soluble phase may consist of at least three general subfractions:

1. micromolecular subfraction, including water, inorganic ions, and dissolved gases;

2. mesomolecular subfraction, including all metabolic intermediates, lipids, sugars, nucleotides, nucleotide derivates, and other low-information content molecules which are not integrated into a preformed structural element;

3. free macromolecules, chiefly proteins and RNA (including → transfer RNA).

cell selection (*Darlington* 1937) — → selection within a population of genetically different cells. C. s. involves competition between genetically balanced or wild type cells and unbalanced or mutant cells with the result that the latter are usually eliminated.

cell strain — a population of cells derived either from a primary culture (started from cells, tissues, or organs taken directly from organisms), or a → cell line by the selection or cloning of cells having specific properties or markers. A c. s. derived by isolation of a single cell or groups of cells having properties or markers not shared by all cells of the c. s. is called a **substrain.** A **cloned strain** is a c. s. descended directly from a → clone, i.e. a cell population derived from a single cell by mitosis (*Feodoroff* 1967).

cell surface — a two-component structure surrounding the → cell. The inner component is the → cell membrane or plasma membrane. A second component can usually be demonstrated outside the cell membrane. This component is rich in saccharides and always contains other compounds as well. It is called → cell wall, glycocalyx, cell coat, mucous coat, extraneous coat or zona pellicula.

Specialized areas of the c. s. serving to bind that surface to another structure are termed → desmosomes.

cell synchrony → the situation where a population of cells proceeds through the sequences of the → cell cycle with synchrony, i.e., divides at one time.

C. s. is characteristic for some tissues and may be obtained experimentally by two general procedures: 1. Enforcement, i.e., by a metabolic block which causes a temporary arrest of a particular process (e.g., DNA replication or mitosis). Enforcement permits a high recovery of synchronous cells. 2. Selection, i.e., by a procedure which exploits properties of the cell system so that stage-specific cells may be physically separated from an exponentially growing population in *in vitro* culture (→ cell culture).

There is evidence that a high degree of c. s. exists in systems (multinucleate cells, syncytia, somatic cell hybrids) where one nucleus in the process of synthesizing DNA transmits inducers through the cytoplasm to stimulate DNA sythesis in an adjacent nucleus.

cell tetrad — → tetrad.

cell theory (*Schwann* 1838) — the theory, based upon unequivocal experiments, originally put forward by *Dutrochet* in France and by *Schleiden* and *Schwann* in Germany and refined by *Schwann* (1838/1839) according to which;

1. The → cell is the smallest building element of a multicellular organism and as a unit is itself an elementary organism.

2. Each cell in a multicellular organism has a specific task to accomplish and represents a working unit.

3. A cell can only be produced from another cell (or a → syncytium) by cell division.

cell transformation — a heritable change in the biochemical and morphological properties of eukaryotic cells due to infection by (oncogenic) RNA and DNA viruses. Viral transformation of mammalian cells is manifested by the loss of regulatory restraints on their growth potential most frequently accompanied by alteration of cell shape, karyotype and cell surface architecture, by loss of contact inhibition, changes in glycolytic rate, production of increased amounts of organic acids and acid mucopolysaccharides as well as acquisition of malignant potential (production of tumors by transformed cell when inoculated into host animals). These viral induced changes reflect alterations in → gene expression. Transformed cells do not contain or produce infectious virus or infectious virus nucleic acid; they do, however, harbor the information of the virus genome incorporated into their chromosomes (synthesis of infectious virus is induced by fusion with → permissive cells).

C. t. by the RNA tumor viruses involves the synthesis (by RNA-dependent → DNA polymerase) of a double-stranded DNA copy of the transforming viral genome and the integration of this viral DNA into one or more host chromosomes. Besides reverse transcriptase, RNA tumor viruses carry into the cells they infect DNA endonuclease, DNA exonuclease and DNA ligase.

cell wall — a rigid external coat which surrounds plant cells and some animal cells (one of the variants of the → glycocalyx). It is formed outside the → cell membrane and contributes to the mechanical strength of plants and to the ability of cells to withstand turgor and osmotic variations.

C. walls normally contain carbohydrates, lipids, proteins, and minerals; usually carbohydrates predominate in the form of polysaccharides (cellulose, chitin, xylans, mannans, and lignin). Ultrastructure of the c. w. nearly always comprises long, thin threads (microfibrils 10 to 20 nm wide) embedded in amorphous material. Lipid substances (suberin, cutin, waxes) are deposited either on the inside (suberin) or mainly on the outside (cutin, waxes) of the microfibrillar wall layers (adcrusting layers).

The c. w. of plants is perforated in many places by → bordered pits and plasmodesmata which provide direct contact between neighboring cells of a tissue.

cellular affinity — the tendency of cells to adhere specifically to cells of the same type, but not to different types. C. a. is lost in cancer cells.
center — → mitotic center.
centimorgan — the unit of → crossing over, equal to 1% crossing over (→ Morgan unit; map unit).
central body — = → centriole.
central spindle (*Hermann* 1891) — → spindle.
centric — of chromosomes or chromosome segments having a → centromere, as opposed to acentric.
centric fission — a chromosome structural change (→ chromosome mutation) converse to → centric fusion or Robertsonian translocation, i.e., the origin of two → acrocentric or → telocentric chromosomes from one metacentric chromosome (= dissociation). C. f. may be due to → centromere misdivision or may involve a donor chromosome providing one centromere and two telomeres. Evolutionary replacement of a metacentric chromosome by two telocentrics has occurred in some animal species.
centric fusion (*Robertson* 1916) — the fusion by reciprocal → translocation of two → acrocentric chromosomes to give a → metacentic chromosome and a small fragment which is usually lost. C. fusions are the main way by which chromosome numbers have been changed in (animal) → karyotype evolution. In the heterozygous condition, c. fusions show quite regular → meiosis with the two acrocentrics invariably passing to one pole and the metacentric to the other. Aneuploid gametes may result, if the three centromeres orientate in a linear manner (→ centromere orientation) on the meiotic spindle instead of in an alternate configuration.

In the case of c. f. the number of large chromosome arms is maintained (no change of the so-called "fundamental number of chromosome arms"). C. fusions may be due to:

1. Two breaks very close to the centromere of acrocentric chromosomes, one in the long arm, the other in the short arm. Reciprocal translocation may then give rise to a large metacentric and a very small chromosome which will usually be lost (usual mechanism of c. f.).

2. two breaks in the centromere regions of the acrocentrics or in their short arms giving rise to a metacentric with two centromeres so closely together that they function as one and an acentric which is immediately lost (→ tandem fusion).

centric region — = → centromere.
centrifugal selection — → selection.
centriolar pinwheel (*Fulton* 1971) — any structure which, in electron microscope, resembles a → centriole, i.e., a cylinder about 0.2 nm in diameter with a wall composed of nine equally spaced groups of → microtubules, usually triplets.
centriole (*Boveri* 1895) — a cylinder some 120 to 150 nm wide and of variable length in the cells of animals, some algae and fungi. The cylinder wall is composed of 27 → microtubules arranged in triplets parallel to the long axis of the c. (Fig. 5). The c. is a highly specialized organelle for → tubulin assembly and serves in intracytoplasmic anchoring for the outgrowth of

ciliar and flagellar microtubules (→ basal body; cilium) and as organization center of → spindle microtubules.

Centrioles involved in spindle organization are most often found associated in orthogonal pairs (diplosomes) next to the → nuclear envelope of interphase nuclei. At the beginning of karyokinesis these two centrioles separate to the poles. After karyokinesis, new centrioles are assembled orthogonal (at right angles) to the old ones (Fig. 6). Centriolar assembly appears to

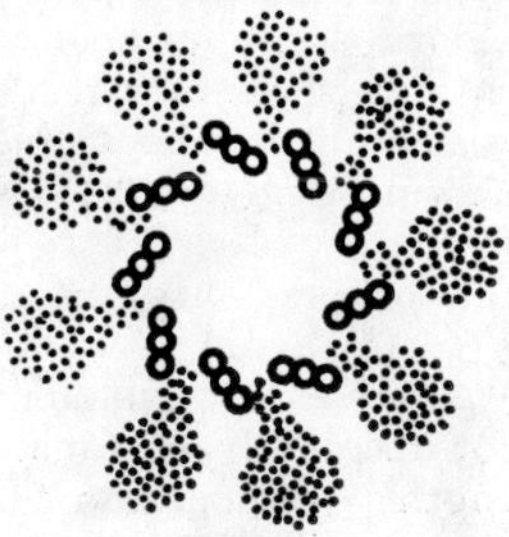

Fig. 5. Diagrammatic representation (cross section) of a centriole (the satellite areas are symbolized by punctuation) (from *Sitte* 1965).

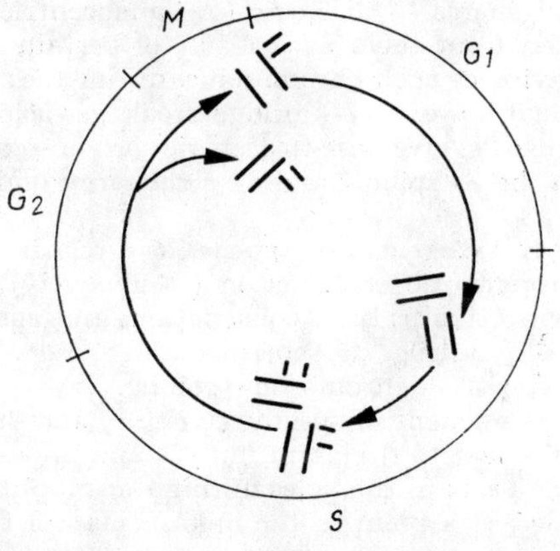

Fig. 6. The orthogonal assembly of new centrioles during the centriole cycle (from *Guttman* (1971).

be analogous to the spontaneous assembly of a virion and coincides with nuclear DNA replication (S period of interphase). There are indications that some centrioles may contain DNA (→ centriole satellite).

centriole satellite (*de Harven & Bernhard* 1956) — any of the appendages which may surround the nine centriolar triplets of microtubules. (Fig. 5).

centripetal selection — → selection.

centrodesmose (*Heidenhain* 1894) — a set of fibrous connections between the cell centers during the early stages of their separation (→ spindle).

centrolecithal (*Balfour* 1880) — of egg cells the yolk of which is accumulated towards the middle instead of at one end (→ telolecithal ova) or scattered more or less evenly throughout the cytoplasm (oligolecithal ova).

centromere (*Waldeyer* 1903) — the region (or regions) of each → chromosome with which the → spindle fibers become associated during → mitosis and → meiosis. This association is a prerequisite for ordered → chromosome movement in the course of karyokinesis. Recent studies support the conclusion that the c. is directly involved in the organization of microtubular subunits into chromosomal spindle fibers.

The c. is subject to its own cycle of functional reproduction: it divides, ordinarily, quite regularly at mitosis and at the second meiotic division (meiosis II). In mitosis and meiosis II the c. of each chromosome behaves as two units (one for each chromatid), in first meiotic division (meiosis I) as an undivided unit (→ centromere orientation). Transverse instead of longitudinal splitting of the c. at meiosis or mitosis is called → centromere misdivision and may give rise to telocentric chromosomes and → isochromosomes.

The following main types of centromeres (kinetochores) may be distinguished:

1. **Localized centromere:** the → chromosome possesses a permanently localized c. region which — provided the c. is not absolutely terminal — divides it into two arms of equal or unequal length (→ idiogram). Such chromosomes are called "monocentric" (= "eucentric"). Chromosomes with two (or more) localized centromeres — dicentric (or "aneucentric") chromosomes — may be produced from them as a result of certain → chromosome mutations. The behavior of such chromosomes during the nuclear division is generally abnormal however (→ bridge-breakage-fusion-bridge-cycle). The loss of the localized c. gives rise to acentric chromosomes which are incapable of moving in the → spindle and as a rule are rapidly eliminated.

During metaphase of mitosis, the c. generally appears as a relatively despiralized, negatively → heteropycnotic constriction (→ allocycly) of the chromosomes ("primary constriction"); these constrictions are generally not detectable ("isopycnosis") during the corresponding stage in meiosis I because of a different type of → chromosome coiling.

The localized c. is a chromosome segment of specific structure and has a special cycle of division; it is the last to divide in mitosis and appears to drag the remaining chromosome parts to the poles during the anaphase movement. Genetically it is defined as a point in the linkage map of the linkage group which normally segregates at the first division of meiosis provided it is not a postreductional meiosis.

According to light microscope studies the structure of the centromere region corresponds to a reversed → repeat composed of two groups of three zones (Fig. 7). The outermost zone is the least differentiated and contains → chromomeres of average size. The middle zone is the most strongly differentiated and is occupied by smaller chromomeres; the innermost zone of the tripartite structure is composed of fibrils with several small chromomeres. This structural pattern is repeated in the reverse order in monocentric chromosomes (→ centromere misdivision). In the case of a divided centromere region, the adherence of the → chromatids of each chromosome up to anaphase separation is guaranteed by the (proximal) chromosome segments lying to right and to left of the centromere region, which are composed largely of → heterochromatin (*Lima-de-Faria* 1949, 1956).

Ultrastructural studies of the localized c. generally fit into the following categories: 1. The c. appears morphologically undifferentiated from the rest of the chromosome; 2. The c. appears as a less electron-dense sphere located in a recess of the more electron-dense chromatin; 3. the c. appears as an electron-dense plate; 4. The c. appears to consist of several layers of varying electron density and complexity. During metaphase about 50 to 100 microtubules are attached to a single c.

2. **Neo-centromere** (*Rhoades* 1952): chromosome ends (→ telomere) which, under certain conditions, show movement on the spindle during mitosis and meiosis, just as do the localized centromeres of these chromosomes. They are secondary centromeres whose activity results in the chromosome ends moving first during anaphase movement.

3. **Non-localized centromere:** in this case the attachment of the spindle fibers is not confined to a strictly localized segment of the chromosome but takes place along the entire length of the chromosome. As a rule all parts of the chromosome have active spindle-mobility and acentric fragments are lacking. Distinction is made between two variants of this type:

a) The multiple or polycentric (polykinetic) condition: each chromosome is furnished with many c., separated by small non-centric segments. Chromosomes of this type are actually multiple or compound structures. The only certain example of this organizational type is that of the germline chromosomes of some ascarid nematodes (→ nuclear differentiation).

b) The diffuse or holocentric (holokinetic) condition (*Hughes-Schrader* & *Ris* 1941): every point along the entire chromosome shows active mobility on the spindle. This type of chromosome occurs *inter alia* in Hemiptera, Homoptera, a few Protista and, among higher plants, in the genus *Luzula*.

4. **Semilocalized centromere** (*Vaarama* 1954): the active spindle mobility is coupled to a localized primary c. in mitosis but in meiosis is displaced to another localized site (secondary, tertiary c.). Chromosomes with semilocalized c., as are found in *Pleurozium*, are chromosomes with potentially multiple centromeres and behave functionally like those with localized c. but the position of spindle activity undergoes a regular shift (→ centromere shift). Semilocalized centromeres are regarded as intermediate between the multiple and the strongly localized centromere types.

centromere distance — the distance of a gene from → centromere, measured in terms of → recombination frequency.

centromere interference — an inhibitory influence (positive → interference) exerted by the → centromere on → crossing-over and the distribution of → chiasmata in its vicinity. For each individual chromosome there is a characteristic distance between the centromere and the point at which the first (proximal) crossing-over takes place; this is the so-called "differential distance" (*Mather* 1936).

centromere misdivision (*Darlington* 1939) — an anomalous transverse — instead of lengthwise — division of the → centromere region in chromosomes with localized centromeres. It corresponds to a break in the centromere and can occur in each of the structurally distinguishable zones. If the break occurs in the innermost zone, → isochromosomes with identical arms may be produced. Breaks in other zones lead to chromosomes with terminally localized centromeres, known as telocentric chromosomes (Fig. 7).

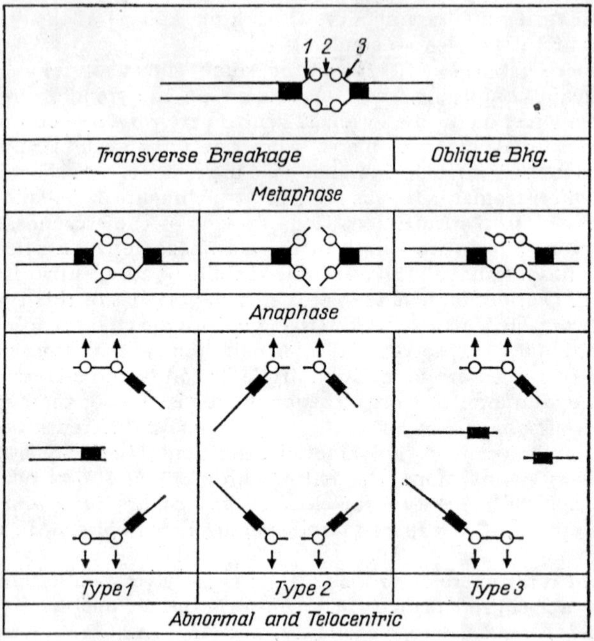

Fig. 7. The consequences of transverse and oblique breakage (centromere misdivision) of the centromere (modified after *Lewis & John* (1963).

centromere orientation — The process of orientation of → centromeres during prometaphase of mitosis and meiosis. C. o. is achieved as follows: each chromosome (mitosis) or → bivalent (meiosis I) exploits two independently acting centromeres, each of which is orientated towards one of the two spindle poles. A stable orientation (→ congression) of the centromeres of all the chromosomes along the spindle equator ("metaphase plate") results from the interaction between the spindle poles and centromeres via the spindle fibers.

Sister centromeres orient to opposite spindle poles (amphiorientation of *Bauer* et al. 1961) in mitosis, but to the same pole (synorientation) in one (usually the first) of the two meiotic divisions. Mitotic chromosomes with one oriented and one unoriented c. are called mono-oriented (monotelic orientation). If both centromeres are oriented towards the same pole in mitosis, the chromosome is maloriented. During metaphase an equilibrium position is established and is maintained until the beginning of the anaphase movement (→ chromosome movement).

Initial orientation established at the start of prometaphase is often inappropriate and, therefore, re-orientation is of critical importance for controlled chromosome distribution. **Re-orientation** means the loss of chromosomal spindle fiber connection to one pole followed by the formation of a new connection to the opposite pole. Without re-orientation abnormal chromosome complements would result.

The terms auto- and co-orientation of centromeres (*Darlington* 1936) are used in describing the orientation process.

1. **Auto-orientation of centromeres:** The orientation of the two sister centromeres of mitotic chromosomes and (in meiosis) of → univalents towards opposite cell poles ("amphiorientation"); the centromeres do not orientate themselves in any direct relation to one another but lie directly on the equator.

2. **Co-orientation of centromeres:** The process of mutual orientation of centromeres of → bi- and multivalents during one meiotic division (usually meiosis I), leading to the arrangement of pairing associations on the spindle equator (in this case sister centromeres orient to the same pole: "synorientation"). For bivalents the arrangement is such that one centromere is "above", the other "below" the spindle equator. In cases where the pairing associations are composed of more than two chromosomes ("multivalents") several types of co-orientation are distinguished (Fig. 8). The type of orientation to which a multivalent conforms depends on the location of the → chiasmata:

a) **linear co-orientation:** all the centromeres of the multivalents lie one behind the other in the spindle; as a rule the chromosomes in the middle of the chain give rise to"false univalents";

b) **convergent co-orientation:** alternate centromeres are aligned to the same pole;

c) **indifferent co-orientation:** individual centromeres of the multivalents appear to have no special bearing on the orientation of each other. This also generally leads to the formation of "false univalents";

d) **parallel co-orientation:** neighboring centromeres are aligned to the same pole. This type of orientation first becomes possible in quadrivalents which consist of four paired chromosomes.

The various types of orientation of multivalents give rise to various forms of chromosome distribution to the poles and influence the relative proportions of genetically balanced and unbalanced meiotic products.

centromere repulsion — the mutual repulsion of the centromeres of the paired chromosomes towards the end of the meiotic prophase (in diplotene and diakinesis). The extent of this repulsion is possibly wholly or partially responsible for the particular degree of → chiasma terminalization.

centromere shift — 1. the displacement of the centromere region by structural changes of chromosomes (→ chromosome mutation) such as → inver-

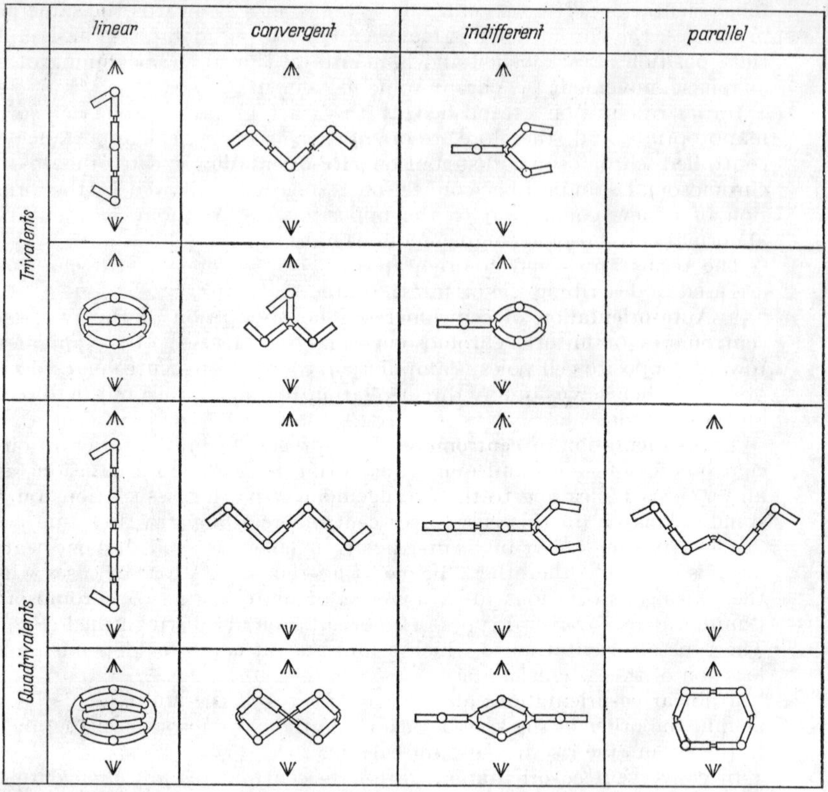

Fig. 8. Types of co-orientation of the centromeres in trivalents and quadrivalents (metaphase of the first meiotic division).

sions, → transpositions and → translocations (= true c. s. according to *White* 1949);

2. a change of centromere position on the chromosome without detectable structural changes during meiosis of certain species (*Coleman* 1948). It results in the occurrence of homologous chromosomes (within a cell or in different cells) with differently located centromeres; their behavior corresponds to that associated with semilocalized → centromeres.

centromeric chromomere (*Östergren* 1947) — → chromomere.

centromeric fission = → centric fission.

centromeric fusion — a type of → centric fusion.

centromeric granule — a specific part of the → centromere associated with the → spindle microtubules.

centromeric heterochromatin — → heterochromatin associated with the → centromere. There is evidence that c. h. contains DNA with a specific and relatively uniform base composition, socalled → satellite DNA or → repetitious DNA.

centronucleus (*Boveri* 1901) — a cell nucleus (→ nucleus) with an intranuclearly localized → centrosome.

centroplasm (*Erlanger* 1897) — 1. = → centrosome;

2. the central region of the protoplasts of blue algae which remains free of the lamellar packets of the → chromatoplasm and in which DNA-containing structures exist.

centrosome (*Boveri* 1888) — a spherical zone of cytoplasm adjacent to the interphase nucleus or at the spindle poles which is free from → ribosomes, → mitochondria, and → Golgi elements. If → centrioles are present, each c. contains a pair of → centrioles (a diplosome) and associated osmiophilic material. According to electron microscopy, the c. (= centrosphere) contains vacuole-like elements apparently derived from the → endoplasmic reticulum, radiating tubular "astral rays", and the termini of spindle filaments in irregular configuration.

centrosphere (*Strasburger* 1893) — the area around → centrioles (= → centrosome).

cephalobrachial (*Levitsky* 1931) — of → heterobrachial chromosomes whose short arm is represented by a small chromosome segment (= acrocentric).

certation (*Nilsson* 1915) — competition in growth rate between pollen tubes (= pollen tube competition) of different genotypes, resulting in unequal chances of accomplishing → fertilization (*Darlington* & *Mather* 1949). Competition takes place in the style of the female flower and results in a greater percentage of fertilization by pollen of the more compatible genotype, which reach the ovules first and leave insufficient egg cells for fertilization by genotypically less compatible pollen. As a result of c. segregation ratios may be distorted or the → sex ratio may shift in those cases where male and female determining pollen (male heterogamy: → sex determination) have a differential chance of fertilization.

Cf — = → colicinogenic factor.

C-factor (*Muller* 1926) — any gene or heterozygous structural change (in most cases → inversions) reducing the frequency of → crossing-over.

CG value — → GC value.

chain-initiating codon — → initiator codon.

chain-initiating mutation — a mutation which gives rise to an initiator codon.

chain-terminating codon — → terminator codon.

chain-terminating mutation (*Sarabhai* et al. 1964) — a → nonsense mutation which gives rise to a → terminator codon.

chain termination — termination of polypeptide synthesis during → genetic translation which requires both → terminator codons and → release factors.

chalone (*Bullough* 1967) — any of the tissue-specific mitotic inhibitors that control cell proliferation and tissue growth in vertebrates by negative feedback inhibition. Chalones are characterized as follows:

1. They inhibit mitosis *in vitro* and *in vivo*; 2. their action is reversible and they are not cytotoxic; 3. they are synthesized by mature cells of the

tissue upon which they exert influence; 4. they are released from cells and circulate in blood and humoral fluids (chemical messengers); 5. they are tissue-specific, but not species-specific.

Many chalones appear to be glycoproteins with a mass of about 30,000 to 50,000 daltons. A smaller group of such agents has a mass of only 1,000 to 3,000 daltons.

The mode of action of chalones is largely conjectural but their activity may depend on the state of cellular membranes which may serve to concentrate c. within the cell. Membrane-chalone interaction is the probable source of the tissue-specificity of a c. Stress hormones commonly act as cofactors.

One suggestion is that the "mitotic genes" are activated if the c. action is weak, while the genes for tissue function are activated if the c. action is strong.

character (*Bateson* 1907) — a c. (= trait, phene) is any observable (phenotypic) feature of the developing or fully developed individual: a biochemical property, a cellular form or process, an anatomical structure, an organ function or a mental characteristic (*Stern* 1960). Genetic characters are in most cases advanced stages, often end points, of developmental sequences, the individual steps of which may be governed by different genes. They are produced by → gene action, are removed from the genes by at least one and, often, numerous steps and may be mainly controlled by the → alleles of one or more gene loci in cooperation with the environmental conditions (→ expressivity, penetrance). It is not the character per se which is inherited, but the genetically determined → reaction norm of the → genotype. Genes merely provide the potentiality to produce or contribute to the production of a given c. Whether or not the c. is realized depends upon the influence of the external as well as the genetic environments (→ genetic background).

Characters may be controlled by autosomal or sex-linked genes; they may be either → dominant, → recessive, intermediate, → sex-linked, → sex-limited, or → sex controlled. They are determined autonomously (→ autophene) if the c. in question is produced as a consequence of the allele being in the cell or are dependent for their expressiou on the type of cell or organ (→ allophene) independent of the gene (intra- versus intercellular gene action).

1. **Discontinuous, qualitative, or oligogenic characters:** Characters controlled primarily by a few genes each of which has a large contribution to the process of character formation as compared with the non-hereditary fluctuations of the character concerned.

2. **Continuous, quantitative, or polygenic characters:** In contrast to 1. these characters are controlled by many genes with small individual contributions and show continuous variation. Their study depends on measurements rather than counts. They almost invariably show a large environmental component of variation (→ heritability).

3. **Acquired characters:** Phenotypic modifications arising purely by environmental influences during the developmental process of an organism.

Phenotypic variance (→ variation) of a genetically controlled c. may first be divided into its genetic and environmental components (symbols V_G and V_E). The genetic variance is further subdivisible into the additive, dominance, and nonallelic components, and into the phenotypic variance expressed as $V_p = V_A$ (additive) $+ V_D$ (dominance) $+ V_N$ (nonallelic) $+ V_E$ (environmental). In the absence of dominance and nonallelic interactions, the genetic variance of a c. is entirely additive. Any two alleles that show different quantitative effects contribute to V_A. Additive genetic variance is the most important cause of resemblance among parents and offspring as to a specific c., and commonly exceeds the dominance and nonallelic variances.

If correlated variations between two or more characters are evident and the basis of the correlation is genetic, these are referred to as "genetic character correlations" (*Simpson* 1953). Such correlations may be the result of genetically controlled gradients of growth (including allometries), → pleitropy, → linkage of the genes responsible for the formation of the characters, correlated or random incorporation of genes into a genetic system which is integrated on another basis.

character displacement (*Brown* & *Wilson* 1956) — in the case of two populations or species whose distributions geographically overlap the accentuation of differences between them in the zone of overlap (→ sympatric) and the weakening or entire loss of differences in the parts of their range outside this zone. The differences may involve morphological, physiological, ecological, or behavioral characters (= character divergence).

character divergence (*Darwin* 1859) — = → character displacement.

character pair — two more or less markedly differing → characters, the origins of which may be of variable complexity. In the simplest case two different → alleles of a genetic locus are responsible for the difference in character.

character progression — the geographic gradation of the expression of specific characters over the range of distribution of a race or species; the gradation may be considered from the center to the border or from one end of the distribution to the other.

chasmogamy — in plants, fertilization after opening of flower or floret as opposed to → cleistogamy.

check cross — the crossing of an unknown → genotype with a phenotypically similar individual of known genotype. The F_2 segregation serves to establish whether the same phenotypes result from the action of identical or non-identical alleles of the same genetic locus or from the action of non-allelic genes.

chemodifferentiation — the chemical changes which precede visible differentiation in the process of → cytodifferentiation during embryonic development.

chiasma (*Janssens* 1909) — that place in a → bivalent, → multivalent, or in the case of intrachromosomal (internal) pairing within a chromosome (→ isochromosome), from late prophase of meiosis up to first anaphase where a) pairs of homologous → chromatids remain in contact although they are separated in other parts, and b) an exchange of homologous parts between nonsister chromatids of the chromosomes has taken place by → crossing-over (Fig. 9) ("chiasmatype theory"). A site of contact, however, does not in all cases necessarily mean exchange of parts. In achiasmate meiosis so-called → collochores may take over function a) of chiasmata.

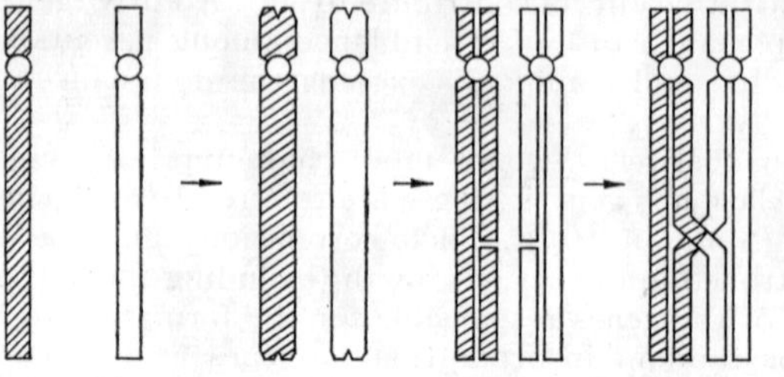

Fig. 9. Diagram illustrating the origin of a chiasma (preceded by a crossover) according to the chiasmatype theory.

With normal chiasmate meiosis at least one c. is necessary to hold the pairing partners in contact after repulsion ("chiasma theory of → chromosome pairing", *Darlington* 1929) between them begins in diplotene. Premature separation of the pairing partners is the consequence of failure of chiasma formation bringing → univalents into being (→ asynapsis; desynapsis). In organisms with large chromosomes two chromatids are seen to cross reciprocally from one chromosome to the other at each c.

Classes of single chiasmata (*Darlington* 1929) are:

1. **Interstitial chiasma:** on each side of the c. is a length of chromatid.
2. **Lateral chiasma:** the chiasma is terminal as to two chromatids and interstitial as to two others (two kinds, symmetrical and asymmetrical).
3. **Terminal chiasma:** the chiasma is localized at the ends of the pairing partners following → chiasma terminalization.
4. **Multiple chiasma:** a terminal chiasma in which three or four pairs of chromatids participate (as in → multivalents).
5. **Imperfect chiasma:** a chiasma in which one of the four cross-wise associations is broken prior to anaphase (= "incomplete chiasma").

If two chiasmata are present in one bivalent (its 4 chromatids represented by A, A′, B, B′) there may be three different relationships between them (→ crossing-over) (*Darlington* 1937):

1. **Reciprocal (comparate) chiasmata:** the chromatids A and B or A′ and B′ participate in both chiasmata (= "regressive chiasmata", "two-strand double crossing-over").
2. **Complementary (comparate) chiasmata:** chromatids A and B participate in one chiasma, chromatids A′ and B in the other ("digressive chiasmata", "four-strand double crossing-over").
3. **Diagonal (disparate) chiasmata:** the chromatids involved in one chiasma are, e.g., A and B and those in the other A and B′ or A′ and B (= "progressive chiasmata", "three-strand double crossing-over").

The formation of a c. is understood to be the consequence of a process of breakage and reunion at homologous sites of both chromatids during → crossing-over in which normally only the nonsister chromatids of the paired chromosomes are involved. The reunion process is directed such that no dicentric chromatids are produced by asymmetric reunion.

The formation of a c. represents the culmination of an interrelated sequence of processes each presumably with its own control mechanism. C. formation is temporally associated with a repair type replication (→ repair replication) and there is increasing evidence that both nucleic acid synthesis and protein synthesis are involved. The crossing-over frequency is one half of the chiasma frequency per pairing configuration since only two chromatids of the four are engaged in crossing over at one point. The "map length" of a chromosome is the average chiasma frequency of that pair of homologues, times 50. The average frequency of chiasmata per pairing configuration, per meiocyte, or per karyotype is controlled both by genetic and environmental factors.

Differential distance (*Mather* 1936) is the interval between the first chiasma and a fixed point (usually the → centomere or → telomere) in a bivalent relative to which this c. is believed to originate. The interval between the centromere and the nearest chiasma may be called the **residual centric distance** (*Southern* 1967), the distance between the telomeric end and the nearest chiasma may be called **residual telomeric distance** (*Southern* 1967). Both residual distances are determined by an interaction between the length of a chromosome arm and its chiasma frequency.

Interstitial distance (*Mather* 1936) is the distance between successive chiasmata in those cases where more than one is present in a bivalent.

chiasma centralization (*Resende* 1953) — the movement of the chiasmata to the chromosome center in chromosomes with nonlocalized → centromeres or in dicentric chromosomes (→ chiasma terminalization).

chiasma interference (*Mather* 1933) — the occurrence less often (positive c. i.) or more often (negative c. i.) than expected by chance of two or more crossovers and chiasmata in one segment of a chromosomal pairing configuration in individual cells at meiosis (→ interference). The c. i. normally does not but in some cases may extend across the centromere region (= chiasma position interference.)

chiasma localization (*Darlington* 1931) — → chiasma terminalization.

chiasma movement index (*Slizynski* 1955) — a measure of the degree of → chiasma terminalization derived from the average number of chiasmata per → bivalent in an early and in a later stage of prophase of first meiotic division. A high c. m. i. results from quick movement of chiasmata between the two prophase stages compared. The c. m. i. differs from the "terminalization coefficient' in that its determination requires scoring the number of chiasmata in two succeeding stages of prophase.

chiasmata position interference (*Carter & Robertson* 1952) — = → chiasma interference.

chiasmate — of a → meiosis with normal → chiasma formation (→ achiasmate; cryptochiasmate).

chiasma terminalization (*Darlington* 1929) — a progressive shift between diplotene and metaphase of the first meiotic division in the distribution of chiasmata along the arms of paired chromosomes from their points of origin to more distal positions (Fig. 10). The terminalization process may be complete, partial, or entirely lacking. In the last case the chiasmata are called localized (*Darlington* 1931).

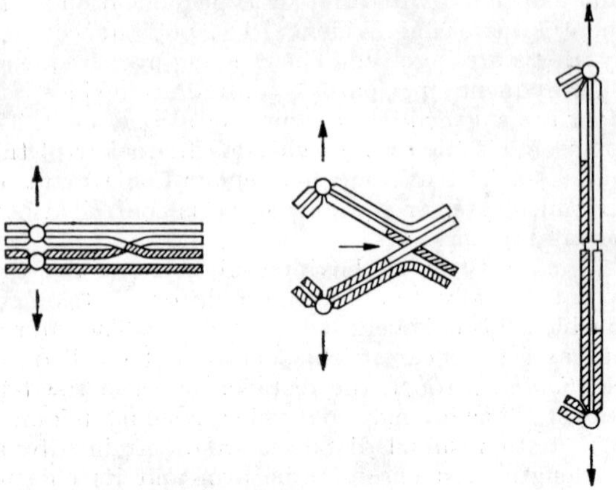

Fig. 10. Diagrammatic representation of three successive stages in the terminalization of a chiasma.

The "terminalization coefficient" is a measure of the degree and progress of c. t. and denotes the fraction of the chiasmata that are terminal at a given stage between diplotene and metaphase I of meiosis (the coefficient is 1 when all chiasmata are terminal). Generally terminalization coefficients seem to be higher in organisms with small chromosomes and in those showing some distal chiasma localization initially.

chiasmatype theory (*Janssens* 1909, 1924; *Darlington* 1929) — a theory postulating that true chiasmata are the direct result of → crossing-over, being formed at precisely the points where the exchange of segments between (non-sister) chromatids took place. In contrast to this "one-plane theory" the "two-plane theory" postulates (*Sharp* 1934) that the cross-figures appear because the chromosomes of the pairing configurations split in contiguous regions along perpendicular cleavage planes. Both the "two-plane" and its variant, the "neo-two-plane theory" (*Matsuura* 1937), are only of historical interest now since all experimental evidence favors the chiasmatype theory.

chimera (*Winkler* 1907) — in most cases a plant, more rarely an animal, composed of tissues of two or more → idiotypes owing to → mutation, somatic → segregation, or grafting (= mosaic).

Chimeras form a heterogeneous group and in plants may be classified in three different ways (*Tilney-Basset* 1963):

1. **According to their origin** into those arising by:

a) spontaneous mutation,

b) induced mutation,

c) sorting out from variegated seedlings (→ variegation), or

d) grafting.

2. **According to their structure** into:

a) sectorial (different tissues grow side by side and occupy distinct sectors of varying size),

b) periclinal (different tissues are disposed one with the other, one occupies the center of the plant structure as a core, the second tissue grows around it as covering layer),

c) mericlinal (actually an interrupted periclinal c. with one kind of tissue of limited extent placed in a peripheral position and covering a central core).

3. **According to their behavior** into:

a) Species c. or "graft c.", produced artificially by grafting together different species or genera. At the place of union a callus tissue is formed from the stock and scion and this gives rise to adventitious buds of chimerical structure. A. c. of this type is recognized by the general differences between the cells of the species concerned.

b) Chromosomal c., the tissue layers of which differ in their chromosome number. They are artificially produced by treatment of germ layers with → spindle poisons, e.g. colchicine, and sometimes arise spontaneously.

c) Gene-differential c., which arises by somatic mutation of a gene to its recessive allele, or back-mutation to its dominant allele in one of the germ layers. This type of c. occurs spontaneously and may be induced by → mutagens.

d) Plastid-differential c., which arises either by spontaneous → plastid mutation or by sorting out of two kinds of plastids from a mixed egg or zygote. This type of c. is recognized by the → variegation of the leaves and distinguished from a similar gene-differential c. by the different, non-Mendelian mode of inheritance.

In animals, the following types of chimerism may be distinguished:

1. Zygotic chimerism: fertilization by one sperm of an ovum and by another sperm of the first or second polar body derived from the same primary oocyte. A single individual results in both cases.

2. Postzygotic chimerism: fertilization by two sperms of two separate ova and fusion of the two zygotes at a very early stage of development. A single individual results.

3. Twin chimerism: fertilization by two sperms of two separate ova. Two individuals result.

chloroplast (*Schimper* 1883) — in plants, a cell organelle (→ plastid) characterized by a limiting double membrane and an internal, chlorophyll-containing lamellar structure embedded in a protein-rich stroma (Fig. 73). The chloroplasts are capable of light-dependent photophosphorylation and of photosynthetic CO_2 fixation.

Chloroplasts are endowed with their own DNA complement and are capable of *in vitro* DNA-directed RNA and protein synthesis. They contain DNA polymerase and a unique class of 70 S → ribosomes (forming → polysomes) that are distinct from the 80 S particles of the surrounding cytoplasm. These data suggest that chloroplasts with highly specialized functions are semi-autonomous systems, capable of self-replication.

Chloroplasts contain about 56% protein, 32% lipids, and 8% chlorophyll; they are of various shapes (diameter 4 to 6 μm), spherical or ovoid. Their number in a cell varies from one up to 500. The biogenesis and maintenance of a functional c. is under the control of both nuclear genes and genes of the c. genome (→ chloroplast DNA). Nuclear DNA is involved in many steps leading to the synthesis of chlorophyll, of carotenoids, and of the photosynthetic electron transport system.

Chloroplasts originate from submicroscopic ameboid proplastids which have double membranes. When a proplastid reaches a diameter of about 1 μm, its inner membrane invaginates to form lamellae which in light continue developing and differentiating to the mature condition found in the c. (→ plastid).

chloroplast DNA (*Chun* et al. 1963; *Sager & Ishida* 1963) — → deoxyribonucleic acid (organelle DNA) localized in → chloroplasts (contour length 40–45 μm); its information content is about 4 to 8×10^8 daltons (10^{-15} to 10^{-16} g or approximately 2×10^5 base pairs) and equals that of a bacterial cell. The c. DNA is circular and one chloroplast may contain as many as 15 to 30 molecules. About 2 to 3% of the c. DNA codes for chloroplast → ribosomal RNA, about 1% codes for c.-specific → transfer RNAs. Other c. cistrons code for c. ribosomal protein, for the proper organization of c. membranes, and one or more cistrons code for the synthesis of a functional photosynthetic apparatus. Calculations suggest that a chloroplast contains more potential genetic information than a → mitochondrion.

chondrioid (*Kellenberger & Huber* 1953) — in bacteria, a cell organelle functionally equivalent to the → mitochondrion of higher organisms. Chondrioids originate by a process of invagination from the cell membrane surrounding the bacterial cell and in addition are assumed to be capable of reproduction by division.

In morphology chondrioids resemble mitochondria of the tubular type and consist of a covering membrane which surrounds numerous tubular and vesicular interior structures. They contain the same enzymes (*inter alia*: dehydrogenases, phosphorylases, cytochromes) as the mitochondria of higher organisms. In addition to the "normal' mitochondrial functions, the chondrioids appear to perform tasks involved in cell wall formation, in the reduplication of the nuclear equivalents (→ nucleoid; endosome) and in the formation of endospores.

chondriokinesis (*Nicoloci-Roncati* 1912) — the reproduction of → mitochondria and their distribution during mitosis and meiosis.

chondriome (*Meves* 1907) — 1. collectively the → mitochondria of the cell; 2. all hereditary determinants localized in the mitochondria (→ cytoplasmon; plasmon; inheritance).

chondriosome (*Benda* 1902) — = → mitochondrion.

choriheterosis (*Dodge* 1945) — → heterosis effects connected with the → heterokaryotic state.

chreod (*Waddington* 1961) — a stabilized pathway of change, i.e., a time-extended trajectory of a system that, if it at some time diverts from its path, tends to change in such a way as to return to a later section of the trajectory. Systems with this capability are called homeorhetic.

chromatid (*McClung* 1900) — one of the two visibly distinct longitudinal subunits of all reduplicated → chromosomes (a half-chromosome) which become visible between early prophase and metaphase of → mitosis and between diplotene and the second metaphase of → meiosis. After these stages the chromatids are known as "daughter-chromosomes". The chromatids are those longitudinal units of the chromosome which become separated during anaphase of mitosis and anaphase II of meiosis. Sister chromatids are the two chromatids derived from one and the same chromosome during its replication in interphase, as opposed to nonsister chromatids derived from homologous chromosomes. Normally, chromatids of homologous chromosomes carry identical → genetic information (identical linkage groups), apart from allelic differences between nonsister chromatids, and are similar morphologically. In all probability, each c. contains a continuous DNA double helix.

chromatid aberration — a category of chromosome structural changes (→ chromosome mutation) produced in one and not the other → chromatid of a chromosome as opposed to chromosome and subchromatid aberrations in which both chromatids or subunits of one chromatid are involved. Chromatid aberrations arise spontaneously or are induced experimentally by → mutagens during and after chromosome reduplication in the interphase nuclei.

chromatid break — a discontinuity in only one of the two chromatids of a chromosome (→ isochromatid break).

chromatid bridge (*Smith* 1935) — a dicentric chromatid with the two centromeres passing to opposite poles during anaphase (→ bridge-breakage-fusion-bridge cycle). Chromatid bridges arise primarily from spontaneous or experimentally induced chromatid aberrations and secondarily (secondary structural change) through crossing-over between the chromosomes of → structural heterozygotes in meiosis.

chromatid interference (*Mather* 1933) — → interference.

chromatid non-disjunction (*Sansome* 1933) — 1. the distribution of both sister chromatids of a chromosome to one cell pole during anaphase of mitosis or in anaphase II of meiosis;

2. the passing of homologous segments of chromatids to the same pole in → translocation heterozygotes (→ adjacent distribution).

chromatid segregation — the term c. s. or "double reduction" as opposed to "chromosome segregation" is used in the case of polysomy (autopolyploidy, various forms of aneuploidy) and heterozygosity for one or more biallelic loci (BBBb in Fig. 11), when two sister-chromatid segments of a chromosome are included in the same meiotic product (gone or gamete), the chromatid segments having identical alleles ("equational exceptions").

The following are prerequisites for c. s. with respect to a given locus (*Haldane* 1930; *Mather* 1936):

1. The paired chromosomes must possess different alleles at the locus concerned (B, b in Fig. 11). During meiosis I, → multivalents must be formed and chromosomes with the "critical" allele configuration (Bb)

must be distributed to the same cell pole ("genetic non-disjunction").

2. There must be at least one crossover (or any odd number of crossovers) between the centromere and the locus concerned, so that the alleles of the locus are not reductionally (BB to one pole, bb to the other) but equationally (alleles Bb to each pole) distributed in meiosis I. In meiosis II the two chromatid segments with the same allele (that had become associated with different centromeres through crossing-over) are distributed to the same pole and are included in one meiotic product, thus completing the c. s. for this locus.

The probability of a specific locus experiencing c. s. is expressed by α, the product of the parameters e and a ($\alpha = e \times a$). e represents the average frequency with which sister-chromatid segments with identical alleles enter the same gamete; this is a function of the distance between the locus concerned and the centromere; a represents the probability of occurrence of genetic non-disjunction (distribution of the two chromosomes taking part in the "critical" crossing-over (with alleles B or, resp., b) to the same pole in meiosis I).

The value of c. s. (α) varies between certain limits from locus to locus and depends *inter alia* upon the crossing-over frequency between the particular locus and the centromere, on the frequency of multivalent formation and the frequency of genetic non-disjunction (which in turn is determined by the mode of orientation and disjunction of the multivalents).

With diploid species, c. s. leads to gametes and zygotes with abnormal chromosome complements; in the case of polysomics with more than two homologous partners for one, a few, or all chromosomes of the basic chromosome set, c. s. leads to an increase in the proportion of gametes homozygous for the dominant and recessive alleles of the relevant locus, and thus changes the genotypic segregation ratios owing to the increased proportion of recessives.

chromatid tetrad — a → bivalent in first meiotic division consisting of two paired chromosomes and thus of four chromatids.

chromatid translocation — → translocation.

chromatin (*Flemming* 1882) — a complex ensemble of DNA, basic chromosomal protein (→ histone), non-histone chromosomal protein (→ hertone), plus a small amount of RNA in interphase nuclei of eukaryotes. It represents the interphase form of the cell's nuclear genetic material. C. usually appears as an extended network of fibrils ranging in width from 10 to 15 nm and consisting of a single strand of double helical DNA plus protein which is further coiled to form a supercoil (→ chromosome).

According to *Kornberg* (1974) c. consists of a repeating unit (the → nucleosome) containing two each of the main histones (excluding F 1 histone) and about 200 base pairs of DNA forming a flexibly jointed chain. This indicates a regular spatial arrangement of proteins in c., the proteins protecting stretches of DNA from nuclease attack.

Between 5 and 20% of the DNA of c. (template active c.) is transcribed into RNA (→ ribosomal RNA, transfer RNA, messenger RNA), but much more of the DNA is available as template for *in vitro* → genetic transcrip-

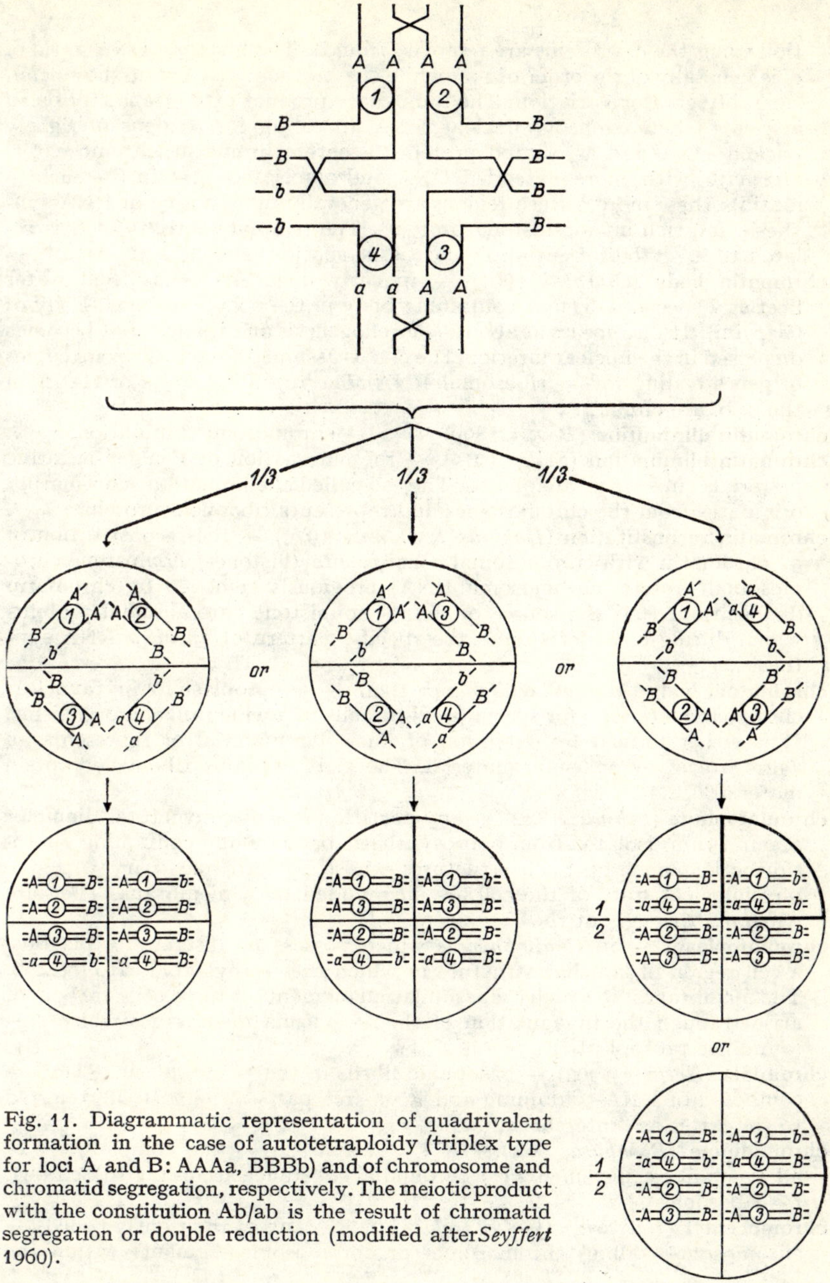

Fig. 11. Diagrammatic representation of quadrivalent formation in the case of autotetraploidy (triplex type for loci A and B: AAAa, BBBb) and of chromosome and chromatid segregation, respectively. The meiotic product with the constitution Ab/ab is the result of chromatid segregation or double reduction (modified after *Seyffert* 1960).

tion when the c. proteins are removed from it. The histone: DNA ratio of c. is generally of the order of 1 : 1, while the hertone: DNA ratio shows considerably greater variation. The immediate product of c. transcription *in vivo* is → heterogeneous nuclear RNA (6 to 100 S). Regions of tightly packed DNA and associated protein (→ heterochromatin; chromocenter) alternate with more extended DNA and associated protein (→ euchromatin); these more diffuse regions are generally more active in RNA synthesis and rich in non-histone proteins. The template activity of c. is related to → cytodifferentiation and tissue-specific → gene activation.

chromatin body (*Giardini* 1901) — in early oogenesis of dytiscid water beetles a special → rDNA-containing body in the oocyte nucleus (Body of Giardini). It disappears at about → vitellogenesis and its material becomes dispersed in the nuclear interior. The c. b. is assumed to contain extra copies of genes coding for → ribosomal RNA. The repeating units of rDNA in the c. b. are circular.

chromatin diminution (*Boveri* 1899) — = → chromosome diminution.

chromatin elimination (*Seiler* 1914) — the elimination of Feulgen-negative materials in early anaphase. The so-called "elimination chromatin" originates from the chromosomes and represents ribonucleoprotein.

chromatin reconstitution (*Dahmus & Bonner* 1970) — the reconstitution of → chromatin with chromosomal constituents (histones, nonhistone chromosomal protein, chromosomal RNA) previously removed by chromatin dissociation. It is the source of the → nonhistone chromosomal proteins which during c. r. determines the specific pattern of *in vitro* RNA synthesis.

chromatoid body (*Morgan & Uzmann* 1966) — a cytoplasmic juxtanuclear clump characteristic for spermiogenetic cells in a wide range of organisms. It is either formed by extrusion of nucleolar material or represents an aggregate of cytoplasmic material. The c. b. contains ribonucleoprotein material.

chromatophore (*Schmitz* 1882) — any particle (vesicular structure; diameter about 15 nm) isolated from photosynthetic bacteria and containing photosynthetic pigments. Such structures whose pattern may vary with the physiological state of the cell are surrounded by a membrane. They are scattered throughout the cytoplasm.

chromatoplasm — in *Cyanophyceae* which possess no discrete → plastids, a cell region of lamellar structure in which the → thylakoids are located. The membranes, to which the assimilation pigments of these cells are bound, arise through the invagination of the → plasma membranes which surround the protoplasts.

chromidia (*Hertwig* 1902) — basophilic fibrils in the → cytoplasm. They are composed of RNA-containing and RNA-free parts. *Monné* (1948) referred to the latter as "interchromidia".

chromidiome (*Bernhard, Gautier & Oberling* 1952) — the complete set of all basophilic cell elements (= chromidial substance according to *Cowdry*). (→ ribosomes).

chromocenter (*Baccarini* 1908; *Tischler* 1920) — any of the tightly coiled (→ chromosome coiling) chromosomes or chromosome segments within the interphase nuclei. Chromocenters consist of → heterochromatin and are positively → heteropycnotic (→ allocycly). Their number per nucleus may vary since heterochromatin has a tendency to fuse (→ chromosome pairing).

chromofibril (*Yasuzumi* 1951) — any of the fibrillar subunits of the → chromosome identified by electron microscopy.

chromogene (*Serra* 1955) — any → gene localized in chromosomes, as opposed to cytoplasmic genes.

chromoid — the bacterial chromosome attached to membranes (→ mesosome).

chromomere (*Fol* 1891; *Wilson* 1896) — any of the linearly arranged, bead-like and compacted segments in mitotic and meiotic prophase → chromosomes and in → polytene chromosomes (→ band). The alternation of chromomeres and → interchromomeres results in chromosome-specific c. patterns (→ chromomere size gradient) which may serve to differentiate between the chromosomes of a → chromosome set.

Evidence is accumulating that the c. represents a unit of DNA replication (a → replication unit), chromosome coiling, RNA synthesis (→ genetic transcription), and RNA processing. The transcription products are molecules of → heterogeneous nuclear RNA. These transcripts are supposed to contain the equivalent of one or a few structural → gene sequences at the 3′ end and regulatory information (utilized for the processes of transcription, maturation and translation of the transcripts of structural genes). The DNA content of a c. is sufficient to specify from ten to a hundred average-sized proteins. Regions within which → repetitious DNA sequences are densely crowded are also suggested to be components of the c. (→ g-region).

The chromomeres localized within the → centromere region are called "centromeric chromomeres" (*Östergren* 1947) or "kinetomeres" (*Lima-de-Faria* 1949).

chromomere pattern — the linear order of → chromomeres in chromosomes.

chromomere size gradient (*Lima-de-Faria* 1952) — the phenomenon wherein the chromomeres on both sides of the → centromere are large and become progressively smaller towards the chromosome ends (the c. s. g. constitutes one component of the socalled → chromosome gradient).

The c. s. g. represents a system of both structural and replicating order in the chromosomes of eukaryotes. Larger chromomeres which contain more DNA not only replicate later but also require more time for replication than smaller ones (*Lima-de-Faria* 1972).

chromonema (*Wilson* 1896) — The smallest light microscopic strand in → chromosomes and chromatids.

chromoplast — a → plastid containing (yellow or orange) pigments other than chlorophyll (→ chloroplast).

chromosite — → lysogenization.

chromosomal — pertaining or referring to the structure, constituents, and function of chromosomes.

chromosomal aberration — = → chromosome aberration.

chromosomal aneuploid (*Dyer* et al. 1970) — → aneuploid.

chromosomal inheritance — → inheritance of → genetic information contained in the chromosomes (= Mendelian inheritance) as opposed to extrachromosomal inheritance.

chromosomal mosaic — an individual showing at least two cell lines with different (either numerical or structural) → karyotypes, derived from a single zygote by post-fertilization events.

chromosomal RNA — 1. RNA transcripts (→ ribonucleic acid; genetic transcription) of chromosomal DNA, i.e., RNA complementary (cRNA) to specific chromosomal DNA sequences. A fraction of cRNA represents precursor molecules of → messenger RNA, → transfer RNA, or → ribosomal RNA (→ heterogeneous nuclear RNA).

1. A special class of RNA molecules assumed to introduce sequence-specificity into chromosomal protein-DNA interaction (*Huang & Bonner* 1965). Unique properties of this class of molecules comprise: 1. Covalent association with chromosomal → histones,

2. small molecule size of 30 to 50 nucleotides (molecular weight less than that of transfer RNA),

3. in some instances a content of up to 27% dihydrouridylic acid,

4. hybridization, mostly to → repetitious DNA s quences, to a high percentage of the genome (approximately 2 to 5%).

The existance of this class of RNA as a separate species is questionable. There are indications that it might be a degradation product of → heterogeneous nuclear RNA.

chromosomal structural change — a change in chromosome structure (→ chromosome mutation) arising spontaneously or induced by → mutagens.

chromosome (*Waldeyer* 1888) — a linkage structure (→ linkage) consisting of a c.-specific linear sequence of → genetic information (→ gene). Chromosomes of eukaryotes and prokaryotes are essentially involved in two main activities:

1. Those concerned with the transmission of genetic information from cell to cell and generation to generation,

2. those concerned with the ordered release of this information to control cellular function (→ gene expression) and development.

The principal synthetic activities related to these functions are the synthesis of DNA and RNA, respectively.

Chromosomes are autoreduplicating structures of differential complexity in eukaryotes and prokaryotes whose number (per cell), morphology and organization are organism-specific characteristics.

1. **The chromosomes of eukaryote organisms:** Eukaryote chromosomes are characterized by their complex structure, their ability to produce a color reaction with basic dyes and their characteristic behavior during → mitosis and → meiosis, in which they undergo regular changes in their degree of compaction (→ chromosome coiling). During interphase of the → cell cycle they are included in the membrane-bounded cell → nucleus. The differentiation of chromosomes into → chromomeres, → heterochromatin, → euchromatin, → constrictions (→ centromere; → nucleolus organizer) are expressions of the differential coiling along the length of →

chromatids and diagnostic tools for the identification of individual chromosomes of the → chromosome set. Replicated chromosomes consist of two such chromatids which are the functional subunits of chromosomes, i.e. the smallest distributable units.

Number and morphology of chromosomes are species-specific. If the centromere is not strictly terminal (telocentric chromosomes), each c. is divided by the centromere region into two equal (metacentric chromosomes) or unequal (acrocentric chromosomes) arms. The individual chromosomes of a species-specific set are mutually nonhomologous (each bears a different → linkage group). In → diploids each c. finds in the other set a structurally similar, homologous partner (provided no heterozygous → chromosome mutations are present in the diploid complement) with which it may pair in meiosis (→ chromosome pairing). Cells produced from diploid cells in the course of meiosis (germ cells, gametes, or gonia) contain only one chromosome set; they are → haploid. Somatic cells of a → polyploid species contain more than two chromosome sets.

The → sex chromosomes are causally related to → sex determination and are called X- and Y-chromosomes, in contrast to → autosomes. The standard chromosomes of a chromosome complement are called → A-chromosomes, as distinguished from → B-chromosomes or supernumeraries, which in certain species are present in addition to the A-chromosomes.

a) **Chemical components of the chromosomes of eukaryotes.** The chromosomes contain almost equal amounts of DNA and → histones, a variable amount of → non-histone chromosomal proteins, and a small quantity of chromosomal RNA. In addition, lipids, polysaccharids and metal ions have been reported to be components of the c.

Chromosomal DNA forms the backbone of the chromosome fibers. Each chromatid probably contains a single DNA double helix associated with chromosomal proteins and condensed or coiled along its length. Nonrepetitive (unique DNA) and → repetitious DNA sequences (serially repeated sequences) are interspersed in a generally ordered manner (→ g-region). This is reflected by the fact that eukaryotic chromosomal DNA after denaturation reassociates in a heterogeneous manner over a wide range of rates: some fractions associate rapidly (→ satellite DNA is represented in this fraction), others at an intermediate rate; a final portion reassociates very slowly and is called unique or single-copy DNA (→ Cot-value). since it associates at a rate expected for non-repetitious DNA. The other portions represent repetitious DNA. Frequently part of the repetitive DNA differs from the bulk of chromosomal DNA in G + C value and hence appears as "satellites" (satellite DNA) when the DNA is banded in a CsCl density gradient. Such satellite sequences are located predominantly in the centromeric → heterochromatin.

Chromosomal DNA (→ deoxyribonucleic acid) is replicated semiconservatively (Fig. 12) and bidirectionally in each → replication unit (Fig. 32). Each c. consists of numerous replicating units arranged in tandem which vary in length between 20 and 70 μm (average about 30 μm). The replication rate of chromosomal DNA (0.5 to 2.0 μm/min.) is less than one tenth of the rate of replication of bacterial DNA (30 μm/min.). There is

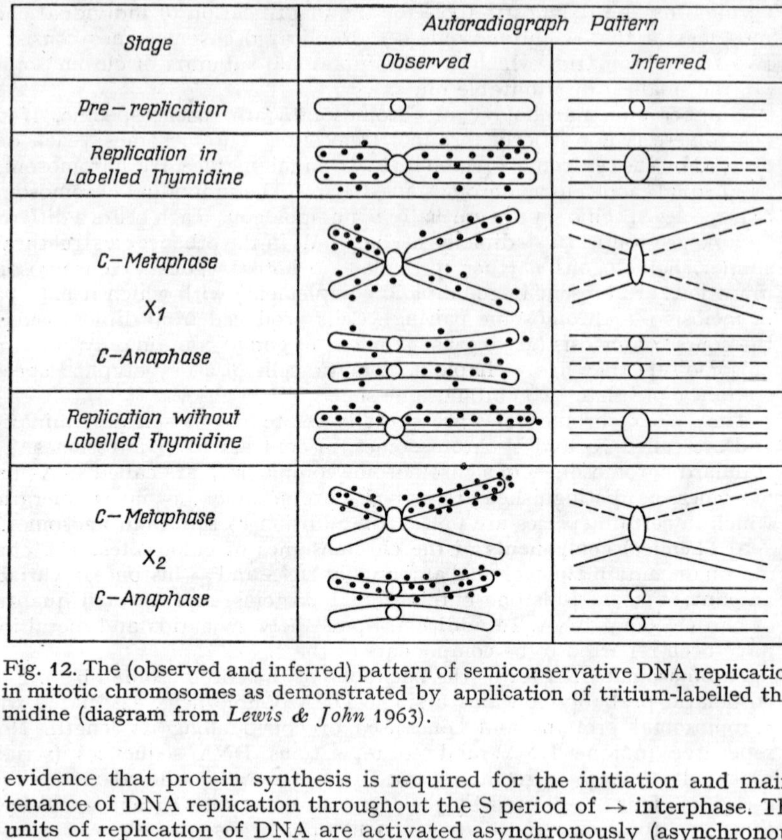

Fig. 12. The (observed and inferred) pattern of semiconservative DNA replication in mitotic chromosomes as demonstrated by application of tritium-labelled thymidine (diagram from *Lewis & John* 1963).

evidence that protein synthesis is required for the initiation and maintenance of DNA replication throughout the S period of → interphase. The units of replication of DNA are activated asynchronously (asynchronous replication). Their number is about 2 to 10×10^4 per cell. Early replicating DNA is relatively rich in G + C and late replicating DNA (usually localized in constitutive → heterochromatin) relatively rich in A—T base pairs. There are indications that the replication of chromosomal DNA might occur in association with the inner membrane of the → nuclear envelope. Possible relationships between the nuclear membrane and DNA replication are (*Comings & Okada* 1973):

1. DNA synthesis may be initiated at the nuclear membrane.

2. The replication fork may remain at the nuclear membrane both during initiation of replication and subsequent chain growth.

3. Initiation and continued replication may be membrane associated only in the late replicating heterochromatin.

Chromosomal proteins include two classes, the basic chromosomal proteins or → histones, and the acidic or → non-histone chromosomal proteins (→ hertones). Both interact with DNA by forces which include ionic ones. The non-histone proteins vary in quantity and there appears to be a limited heterogeneity between different tissues and species. Histones (which are absent in some fungi and dinoflagellates, and replaced by protamines in the sperm chromosomes of some animals) may be separated into five major fractions which do not vary significantly in type or quantity over a large range of organisms.

There is evidence that non-histone chromosomal proteins may interact with histones, negating the latters inhibitory effect on DNA transcription (→ gene expression).

The newly synthesized chromosomal DNA becomes quickly associated with newly synthesized histone. Possibly, histone enters the chromosome very near the to replication fork.

b) **The ultrastructure of the chromosomes.** Chromosomes have before replication two longitudinal subunits of DNA which are conserved during replication (→ polytene chromosome). They are uninemic, i.e., the DNA is one continuous double helix before replication (from anaphase through G_1 of interphase). Evidence for non-DNA linkers is absent. Measurements of chromosome fibers range from 3 to 50 nm in diameter and there is frequently a correlation between diameter and specimen preparation (thin-sectioning and surface spreading).

The diameter of genetically active chromatin fibers is about 3 to 5 nm and the next order in chromatin fiber thickness (probably due to secondary coiling of the basic 3 to 5 nm fibers) is 10 to 15 nm. There are indications of core-like structures in chromosomes resulting from looping fibers (→ lampbrush chromosome). Eu- and heterochromatin do not differ in fiber structure and dimension but in the degree of compactness of fibers.

The phosphate groups of the DNA in chromosome fibers are bound by ionic bonds to the basic groups of histones. The DNA stretches covered by non-basic parts of histones have a much larger proportion of free phosphates. Probably the histones lie in the major groove of DNA. Interaction of histone with DNA as well as interhistone interaction is of importance in maintaining the condensation of DNA in chromosome fibers in the form of a superhelix. Such superhelices may, in turn, pack together into a large coil via torsion of individual superhelices. Processes of this kind may be involved in the differential compaction (coiling) and decompaction of chromosomes which take place cyclically. The expression of that potential is influenced by many factors, extrinsic and intrinsic to the c. *per se*.

Many details of chromosomal ultrastructure in eukaryotes are presently not understood and a field of speculation.

c) **Functional structures of chromosomes.**
In addition to the regularly repeated changes of structures by chromosome coiling (compaction and decompaction), interphase chromosomes undergo changes of structure associated with chromosome reduplication and DNA transcription via differential gene activation. These are reversible structural relations which involve individual replication and transcription units (chromomeres). In → lampbrush and → polytene chromosomes such changes associated with DNA-dependent RNA synthesis are clearly recognizable (→ puff; Balbiani ring).

d) **Pairing structures of chromosomes.**

In → chromosome pairing during meiosis, a special pairing structure is formed between pairing partners. This is called the → synaptonemal complex.

2. **The chromosomes of prokaryotic systems and cell organelles in eukaryotes:** The chromosomes or genophores of prokaryotic systems differ in structure and complexity from those of eukaryotes. Usually, the entire genetic information (→ plasmid) is contained in a single linkage structure which is double- or single-stranded DNA (in several → viruses RNA takes the place of DNA) of various length and molecular weight. Each of these molecules is a single → replicon.

The genomes of bacteria, plasmids, mitochondria (→ mitochondrial DNA), and chloroplasts (→ chloroplast DNA) represent circular double-stranded DNA molecules of system-specific size and structural complexity (→ genetic circularity).

The genomes (chromosomes) of viruses and bacteriophages occur in several geometric forms. Genetic information may be encoded in either DNA or RNA and these linear or circular molecules may be single- or double-stranded dependent upon the type of virus. Single-stranded DNA or RNA is rearranged into a double-stranded → replicative form (→ replicative intermediate) during replication which is semiconservative as in the case of chromosomal DNA of eukaryotes.

Phages with linear duplex chromosomes may be divided into three broad classes on the basis of their chromosome type:

1. One class has chromosomes with unique ends and short, single-stranded → cohesive extensions (coliphages lambda, P2, 186 and phi 80). The cohesive extensions are complementary in nucleotide sequence and may give rise by pairing to circular molecules.

2. A second class has circularly permuted ends (→ circular permutation) and substantial duplex terminal repetition (T-even coliphages, coliphage P1 and 15, Salmonella phage P22).

3. Phage chromosomes exemplified by the T-odd coliphages (T1, 3, 5, and 7) where interruptions occur in the polynucleotide chain at specific points.

chromosome aberration — 1. In the broadest sense all types of changes in chromosome structure and chromosome number. A standard → karyotype is used for comparison and identification of the type of change.

2. All types of chromosome structural changes or → chromosome mutations, collectively (= chromosomal aberrations).

3. A category of chromosome structural changes in which both chromatids of the chromosomes are involved (= presplit aberrations), as opposed to → chromatid and → subchromatid aberrations. Presplit aberrations normally arise (spontaneously or by application of → mutagens) in the not yet reduplicated chromosomes of the presynthetic (G_1) interphase nucleus.

chromosome arm (*Navashin* 1912) — one of the two main segments (= chromosome limbs) of the → chromosome; they are separated by the → centromere and their length depends upon the centromere position. If the two arms are of the same length, i.e., the centromere occupies a median position, the chromosome is said to be "isobrachial" (*Sorokin* 1929) or metacentric, otherwise (arms of unequal length) the chromosome is described as "heterobrachial" (*Sorokin* 1929), "cephalobrachial" (*Levitsky* 1931) or acrocentric (*White* 1945). Chromosomes with terminal centromeres are "monobrachial" (*Sorokin* 1929) or telocentric.

chromosome assortment — → assortment; chromosome movement.

chromosome banding — the experimental production in → metaphase chromosomes of differentially stained or fluorescing regions which serve in chromosome identification. Banding patterns can be divided into two major categories:

1. C-banding with Giemsa staining after specific pretreatment which indicates regions of constitutive → heterochromatin on both sides of the → centromere.

2. Banding of certain sites along the chromosome arms due to quinacrine fluorescence (Q-banding) or Giemsa and other staining after various pretreatments (G-banding). A specific variant of this technique reveals c. b. which is the reciprocal of Q- and C-banding and is called R-banding.

chromosome break — any discontinuity in chromosome structure occurring across the entire cross-section of the chromosome; chromosome breaks are either of spontaneous occurrence or are produced by the action of → mutagens. They are recognizable during nuclear division (mitosis, meiosis) and lead to the formation of centric and acentric fragments. Discontinuities which involve only one of the two chromatids or subunits of the chromatids are referred to as chromatid and subchromatid breaks.

According to the "breakage—reunion model" (which serves to interpret the production of chromosomal structural changes), spontaneous or experimentally induced breaks of the chromosomes are assumed to be the primary lesions which ultimately give rise to → chromosome mutations.

chromosome bridge — a double bridge linking the cell poles, formed by dicentric chromosomes (with two → centromeres) during anaphase of mitosis and meiosis (→ bridge—breakage—fusion—bridge cycle). The dicentric chromosomes are produced as a result of primary or secondary → chromosome mutations. If only one of the two chromatids or chromatid subunits participate in bridge formation, the terms chromatid bridge and subchromatid bridge are used.

chromosome chimerism (*Chu, Thuline & Norby* 1964) — the presence of cell populations of various → karyotypes in the same individual. C. c. is the result of transplantation or of double fertilization, if both fertilized meiotic products share in the development of the embryo (→ chromosome mosaicism).

chromosome coiling — the spiral or helical coiling of the chromonemata of the chromosomes (→ coil) — one of the most striking phenotypic changes in the chromosomes during → mitosis and → meiosis. The coiling cycle transforms the functional form of the chromosome (maximum despiralization) into the transport form (→ allocycly, spiralization coefficient, packing factor). Two types of coils may theoreticall be formed between two or more chromonemata: the "paranemic coil" has freely separable subunits, the "plectonemic coil" has intertwined subunits (Fig. 13).

In certain Protozoa the spiral structure may be recognized without any prefixation treatment, but with other material pretreatment is necessary for the analysis of the spiral structure.

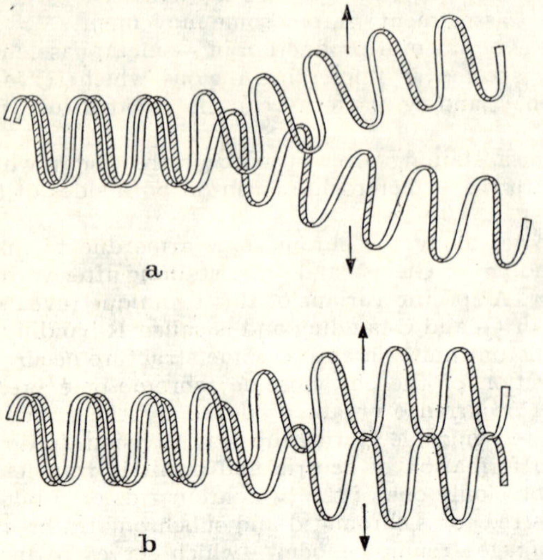

Fig. 13. Diagram of the two types of coils (a) paranemic coiling, b) plectonemic coiling which may be formed between two (or more) chromonemal subunits of a chromosome.

The direction of c. c. (right- or left-handed) is random for particular chromosomes. Reversals of the direction of c. c. may occur at the → centromere and (during → meiosis) at the → chiasmata and, occasionally, elsewhere along the length of the chromosome. During → interphase of the cell cycle, c. c. is not necessarily lost completely, i.e., the coils of one mitosis may carry over to the prophase of the next one as a loose coiling. The two chromatids of a chromosome are independently coiled at prophase.

1. **Relational coiling:** The twisting of one strand (chromatid or individual chromosome) about another strand ("external coiling") as in a rope. The coils interlock in such a way that the strands are not separable without unwinding the coils. Relational coiling is characteristic of sister chromatids during mitosis and paired homologous chromosomes in the meiotic prophase.

2. **Internal coiling:** This type of coiling concerns individual chromosome threads and is considered to be responsible for the shortening and thickening of the chromosomes during mitosis and meiosis. This takes place by an increase in the diameter of gyres of an internal coil, thus reducing the number of turns. There may be more than one order of internal coiling and mitotic chromosomes can be distinguished in this respect from chromosomes in meiosis I.

The latter show at least two orders of internal coiling, major and minor coils. The internal coils develop as a result of internal forces the nature of which is not yet understood.

3. **Relic coiling:** This concerns relaxed spirals recognizable in the chromosomes of mitotic and meiotic prophase. They are regarded as the incompletely resolved internal coils of the preceding nuclear division ("supercoils").

The mitotic (Fig. 14) and meiotic coiling cycles show a series of differences. In mitotic prophase, the chromosomes occur as irregular, undulating threads; as prophase continues, these threads become coiled. The fine coils increase in diameter and the chromosomes become shorter and thicker to give the characteristic maximally spiralized metaphase chromosomes. There is no change in coil structure during anaphase. During telophase coiling is relaxed, the coils pull away from one another and despiralization commences. In interphase, the chromosomes are extensively despiralized and in this state are capable of the DNA-dependent RNA synthesis (functional form of the chromosome).

As compared to mitosis, the relic coils of the last premeiotic mitosis

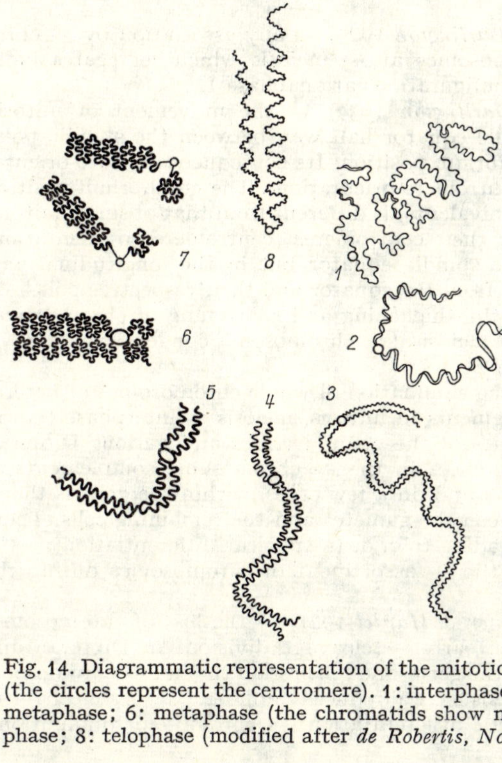

Fig. 14. Diagrammatic representation of the mitotic coiling cycle of a chromosome (the circles represent the centromere). 1: interphase; 2, 3 and 4: prophase; 5: prometaphase; 6: metaphase (the chromatids show major and minor coils); 7: anaphase; 8: telophase (modified after *de Robertis, Nowinski, Saez* 1965)

are much more completely smoothed out in meiotic prophase (leptotene and zygotene). In pachytene, chromosome contraction begins by formation of major and minor coils. The coils of the meiotic metaphase I have a greater diameter than those in mitotic metaphase and occur in smaller number. Maximum contraction is generally in diakinesis.

The causes and mechanisms of chromosome coiling are still but little understood (see *John & Lewis* 1965).

chromosome complement (*Darlington* 1932) — the group of chromosomes derived from a particular gametic or zygotic → nucleus. It may be composed of one (haploid or monoploid nucleus), two (diploid nucleus), or more (polyploid nucleus) → chromosome sets.

chromosome complex — a group of chromosomes which in the extreme case may consist of an entire → chromosome set and which is distributed as a unit in meiosis. Chromosome complexes occur in → complex heterozygous species (which are heterozygous for several reciprocal → translocations) if alternative chromosomes of the configuration are regularly distributed to the same pole after ring or chain formation by meiotic → chromosome pairing between the chromosomes involved in the translocations and their structurally unchanged homologues (→ alternative distribution).

chromosome configuration (*Darlington* 1929) — any association by → chromosome pairing of chromosomes at → meiosis which segregates independently of other such configurations at anaphase I.

chromosome congression (*Darlington* 1937) — the movement of mitotic chromosomes to the spindle equator half way between the spindle poles where they reach an equilibrium position. Its consequence is auto-orientation of centromeres (→ centromere orientation). The equilibrium position for meiotic chromosomes (bivalents) is different from that of single mitotic chromosomes in so far as their centromeres (centromere co-orientation) do not come to lie at the spindle equator but in the longitudinal axis of the spindle, equidistant from the equator and their respective poles.

chromosome contraction — the thickening and shortening of chromosomes during → mitosis and → meiosis (= chromosome condensation) by → chromosome coiling.

chromosome diminution — the elimination of whole chromosomes or heterochromatic chromosome segments in mitosis, meiosis or interphase (*i.a.* in *Ascaridae,* in certain species of the genus *Cyclops* and various Diptera). The process leads to differences in the → chromosome complements of the germ line and the soma and, in a few cases, within the gonads themselves, to differences between the gametes and the remaining cells. There are various mechanisms leading to c. d. (→ nuclear differentiation).

chromosome disjunction — the → assortment of chromosomes during the first meiotic division.

chromosome elimination (*Seiler & Haniel* 1921) — the loss of chromosomes from nuclei during certain early → cleavage divisions in Dipteran and Hemipteran insects. Only the germ-line cells retain the full → chromosome complement; all somatic cells have a 'reduced' number. C. e. is a mechanism of somatic versus germ-line differentiation (→ chromosome diminution).

chromosome erosion (*Levan* 1948) — the occurrence of a number (variable but often very high) of constrictions in the chromosomes due to the action of C-mitotic or mutagenic agents (= chromosome mottling). Characteristic of this condition are sharply defined alternatively stained and unstained segments.

chromosome field (*Lima-de-Faria* 1954) — → chromosome gradient.

chromosome fusion (*Seiler & Haniel* 1921) — the union of two (or more) chromosomes by means of chromosome structural changes to form a single chromosome (→ chromosome mutation). Among chromosomes with localized → centromeres, stable unions are possible only if the fusion takes place in the centromere region (→ centric fusion).

Otherwise di- or polycentric chromosomes are formed; these behave anomalously during nuclear division, resulting in the formation of anaphase bridges (→ chromosome bridge) which are broken. Chromosome fusions of the centric fusion type play a part in the evolution of karyotypes.

chromosome gradient (*Lima-de-Faria* 1954) — the gradual decrease (or increase) in → chromomere size, stainability, and thickness of the chromosomal fibrillae observed in specific regions of the chromosome. The gradients exhibit a variation of defined form which bears definite relations to particular regions of the chromosomes:

1. The origin of the gradients is related to the → centromere. The chromomeres are generally larger and the fibrillae more stainable near the centromere.

2. The rate of decrease of the constituents of the gradients bears a definite relation to the position of the chromosome ends. Size of chromomeres and stainability of fibrillae decrease gradually towards the chromosome ends. The slope of the gradients is a function of the length of the chromosome arms.

3. The shape of the gradients is influenced by intercalary knob formations and their adjacent regions. Particular seriations located at regular intervals in the chromosome body represent regions where the gradients are locally disturbed. This means that the chromosome gradients may vary in a defined way under the influence of specialized chromosome regions.

The term "chromosome field" refers to the fact that the variations observed within the gradients take place in such a way that the properties exhibited by certain entities in one part of the chromosome bear a definite relation to the properties exhibited by other entities in other parts of the chromosome (*Lima-de-Faria* 1954).

chromosome hybridity — → structural heterozygosity.

chromosome interference — → interference.

chromosome map — the graphic representation (in linear form) of a chromosome in which the genes (genetic markers) belonging to a particular → linkage group are plotted according to their relative distances. A distinction is made between → genetic maps and cytological maps.

1. **Genetic maps:** the frequency of intergenic → crossing-over between any two linked markers is used as a measure of their distance (→ genetic recombination) which is expressed in → maps units. The accuracy of

mapping is dependent upon the precision with which the crossing-over frequency can be estimated from the recombination frequency.

2. **Cytological maps:** genes are located on the basis of cytological findings obtained with the aid of → chromosome mutations. Detailed cytological maps can only be prepared if the species concerned possesses large and visibly differentiated chromosomes which can be thoroughly analyzed genetically. To prepare such a map, → deletions, → inversions or → translocations are experimentally induced by → mutagens. They result in specific changes in the linkage relationships which may be defined precisely by genetic analysis. In addition, the location and extent of such structural changes may be determined microscopically (usually during meiosis, sometimes during mitosis) and in this way it is possible to correlate the genetic and structural findings.

Another method of preparing cytological maps is based upon microscopic analysis of the → giant chromosomes of Diptera in which the position and extent of the chromosomal structural changes may be determined exactly. Here again the genes localized by recombination analysis may be correlated with chromosome structure.

The linear order of genes in cytological maps constructed in the above manner agrees completely with that in genetic maps. However, the distances between the genes under consideration may differ appreciably in the two types of c. m. In general, gene distribution is more regular on the cytological map than on the genetic map. The explanation lies in the nonrandom distribution of crossing-over and the effects of double crossing-over and → interference.

chromosome matrix (*Sharp* 1929) — a controversial matrix material (= kalymna) which is supposed by some authors to surround the components of any chromosome and to follow cycles of addition and loss during the process of nuclear division. There is no definite evidence for the existence of a c. m. (said to be RNA and lipoprotein in nature) in the sense of a **"chromosome sheath"** added to the chromosome surface in prophase and covering the coiled chromosome up to telophase when the material is thought to diminish or disappear (→ lampbrush chromosome).

chromosome mobilization — operationally, the production of a donor chromosome in bacterial Hfr-cells which is ready for linear sequential transfer to a recipient cell during → conjugation. C. m. is under → sex factor control. C. m. requires protein synthesis for the production of an F-coded initiator followed by obligate requirement for DNA synthesis. DNA synthesis in the Hfr parent acts as a driving force for the initiation of chromosome transfer into the recipient parent. Effective chromosome transfer (more than several percent of the leading extremity of the donor chromosome) requires tight homologous pairing between the lead region of the donor chromosome and the appropriate segment of the recipient chromosome. This allows pairing of homologous regions of the donor and recipient chromosomes and provides for the orderly and efficient integration of genetic material from the donor into recombinants.

During c. m. the fully replicated circular donor chromosome becomes linear.

chromosome mosaic — → chromosome mosaicism.

chromosome mosaicism — the presence of cell populations of various karyotypes in the same individual. The different karyotypes are produced from an originally "pure" karyotype by → chromosome mutations, mitotic chromosome loss, → non-disjunction, etc. (→ chromosome chimerism). Such individuals (chimeras) are termed "chromosome mosaics". The various karyotypes which they exhibit may differ both in chromosome number and chromosome structure (→ gynandromorph).

chromosome mottling (*Tjio & Levan* 1948) — = → chromosome erosion.

chromosome movement — the movements of chromosomes during mitosis and meiosis which are a prerequisite for the anaphase separation of chromatids (mitosis and meiosis II) or chromosomes (meiosis I). There is generally little movement of chromosomes during interphase. The random behavior of the chromosomes at anaphase which gives rise to chromatid or chromosome distribution to the cell poles depends upon the movements which take place on the spindle prior to mitotic and meiotic metaphase.

During mitosis, the chromosomes undergo orientation (→ centromere orientation) relative to the poles (auto-orientation) leading to the "congression" of single chromosomes on to the metaphase plate. After that the polar movement of the chromatids takes place.

During meiosis the pairing of homologous chromosomes (→ chromosome pairing) involves a certain amount of movement. Furthermore, there is a twisting of homologues about one another during pairing and at diplotene a separation of the pairing partners takes place. The latter process begins at the centromeres and has been regarded as a repulsion. The movements which the paired chromosomes undergo on the spindle prior to metaphase I of meiosis are of three kinds (*John & Lewis* 1965):

1. orientation of the chromosomes relative to their partners and the poles (co-orientation);

2. → pre-metaphase stretching and the subsequent reapproach of co-oriented centromeres;

3. congression movement determining the distribution of the pairing configurations on to the metaphase plate.

Only after the completion of these movements are the meiotic chromosomes able to undergo anaphase separation leading to their → assortment to the poles.

The congression and polar movements during mitosis and meiosis depend upon the joint action and interaction of the centromeres and spindle fibers. The mechanisms of c. m. and the forces underlying them are not yet accurately understood. The numerous hypotheses advanced to explain pre-metaphase and anaphase movements embrace traction fibers, electrostatic forces, tactoid and sol-gel transformations as well as jet-propulsion mechanisms.

All recent hypotheses of c. m. via → spindle mechanics attribute an important role in force generation to the microtubular apparatus. Mechanisms proposed assume as their main factors the assembly and disassembly of → microtubules, sliding of antiparallel microtubules or lateral interaction (zipping) of non-parallel microtubules.

chromosome multiformity (*Tobias* 1953) — the existence of widely different → chromosome complements within the members of a given taxon.

chromosome mutation — any structural change (= intergenic mutation or chromosome aberration) involving the gain, loss or relocation of chromosome segments. Chromosome mutations arise spontaneously or are induced experimentally by chemical or physical → mutagens (→ mutation)

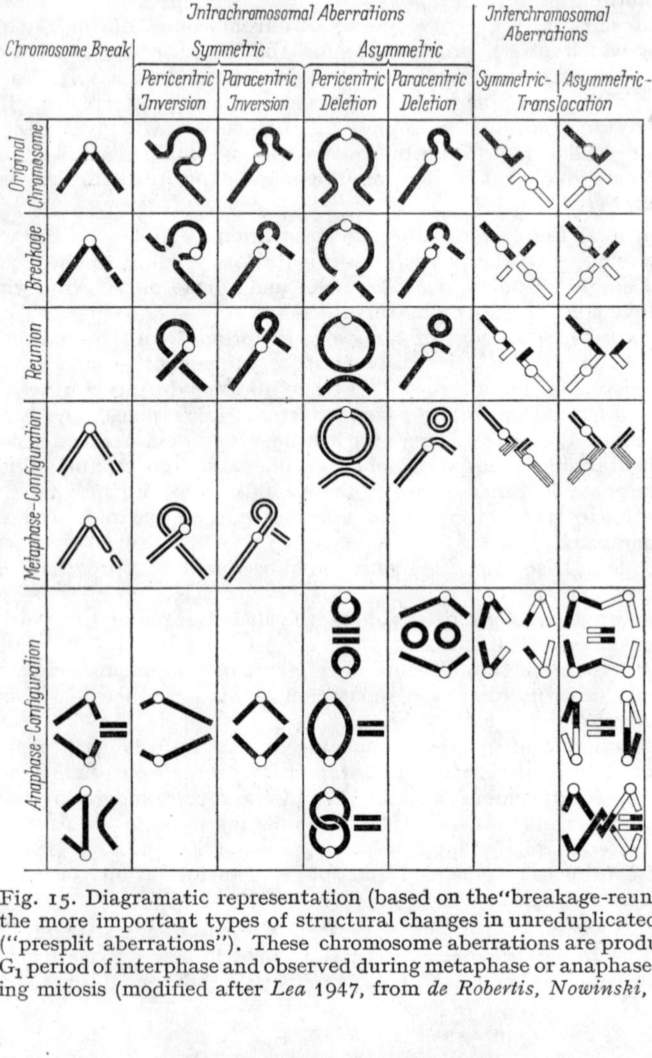

Fig. 15. Diagramatic representation (based on the "breakage-reunion model") of the more important types of structural changes in unreduplicated chromosomes ("presplit aberrations"). These chromosome aberrations are produced during the G_1 period of interphase and observed during metaphase or anaphase of the succeeding mitosis (modified after *Lea* 1947, from *de Robertis, Nowinski, Saez* 1960).

and can be detected cytologically during mitosis, meiosis or in giant chromosomes of Diptera. The ability to detect different kinds of chromosome mutations will depend on the size, structure and number of chromosomes concerned and the ease with which they can be handled.

Chromosome mutations are either intrachromosomal (= homosomal) and then called "intrachanges", or interchromosomal and then called "interchanges" (allelosomal when in homologous, or heterosomal, when in nonhomologous chromosomes). Intrachanges are either in one arm (= paracentric or homobrachial) or involve both chromosome arms (= pericentric or heterobrachial).

Descriptively, all such structural changes include → **deletions** or deficiencies (the actual loss of terminal or intercalary portions of the chromosome and the included genes), → **duplications** (the presence of a group of genes more than once in the genome), → **inversions** (a chromosome segment becomes inverted and reinserted into its original chromosomal position), and → **translocations** (relocations of chromosome segments within one or between chromosomes). Deletions and inversions are intrachanges, duplications and translocations may be intra- or interchanges. All these types of structural changes may be basically of three sorts:

1. **Chromosome type structural changes** (Fig. 15): Both chromatids at identical loci are engaged in the aberration formation. The whole chromosome is the unit of aberration production ("chromosome aberrations").

2. **Chromatid type structural changes** (Fig. 16): The single chromatid is the unit of aberration formation ("chromatid aberrations").

3. **Subchromatid type structural changes** (Fig. 79): A fibrillar subunit (in most cases a half chromatid) is assumed to be the unit of aberration formation ("sub- or half-chromatid aberrations").

In all these kinds of structural changes, rejoining must take place in such a way that the polarity of the two strands of the DNA (the structural backbone of eukaryotic chromosomes) is conserved, i.e., if the two strands of double-stranded DNA are called C and W and, e.g., an inversion occurs, then the segment of strand C between the two inversion points must be inserted into strand W and *vice versa*. Thus neither strand will have the same base composition as it did before rearrangement.

With heterozygosity for the chromosomal structural changes (= structural hybridity), "secondary chromosome mutations" may arise from the primary ones by → crossing-over between the normal chromosome and the structurally changed pairing partner during first meiotic division.

The chromosome mutations are the final result of complicated multiphase processes the course of which is largely unknown as yet. Two model-hypotheses have been developed to interpret the occurrence of chromosome structural changes: the "breakage-reunion" and the "exchange" models.

1. **Breakage-reunion hypothesis** (*Stadler* 1932, *Sax* 1938, *Wolff* 1961, *Evans* 1962): According to this hypothesis, the primary lesions which lead to chromosome structural changes are breaks in the sense of discontinuities which occur either spontaneously or as a result of mutagen action. The vast majority of these breaks in the chromosome (90–99%)

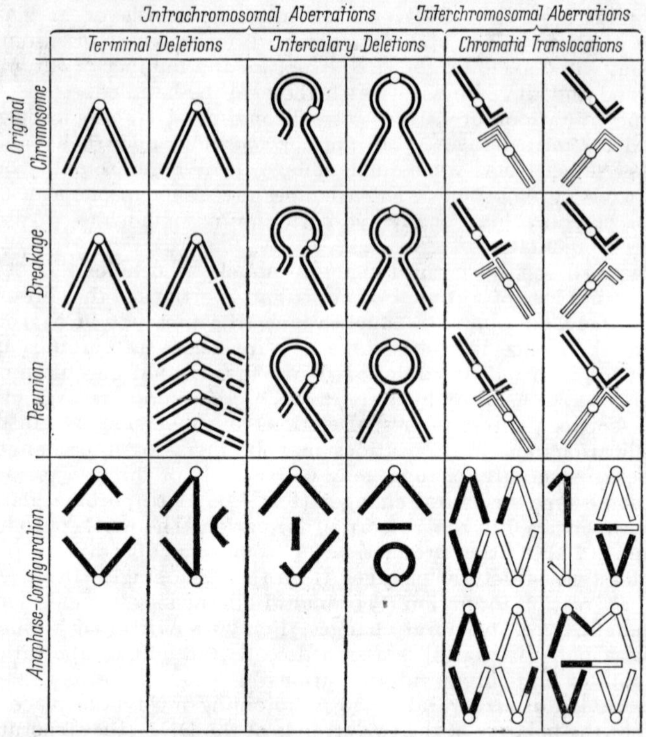

Fig. 16. Diagrammatic representation (based on the "breakage-reunion model") of the more important types of structural changes in reduplicated chromosomes ("postsplit aberrations"). These chromatid aberrations are produced during and after chromosome reduplication and are observed during metaphase or anaphase of the succeeding mitosis (modified after *Lea* 1947, from *de Robertis, Nowinski, Saez* 1960).

are rejoined in the original order by "repair processes". In this case ("restitution"), the chromosomes suffer no visible structural changes, but such changes can occur as intra- or interchanges should restitution to the original structure fail. This may result either from a "stabilization" of the breakage surfaces causing the discontinuity to remain, or through a process termed "reunion". In reunion, two or more breaks interact and unite in a new order, thus producing a chromosome structural change (cf. Fig. 15–16).

Restitution and reunion processes compete with each other; in this competition the spatial arrangement of the breaks with respect to one another and the period during which the breaks remain "open" are decisive.

2. **Exchange hypothesis** (*Revell* 1959, 1963, 1966; *Evans* 1962; *Rieger* 1966): According to this hypothesis the primary lesions leading to the formation of aberrations are not breaks but local regions of instability (labile sites) in the chromosomes. All aberrations of intra- and interchromosomal nature are considered to be the result of exchange processes which involve interaction between two local instabilities in close proximity. This stage is called "exchange initiation" and represents a secondary, more stable state. The "exchange initiation" is followed, either imme-

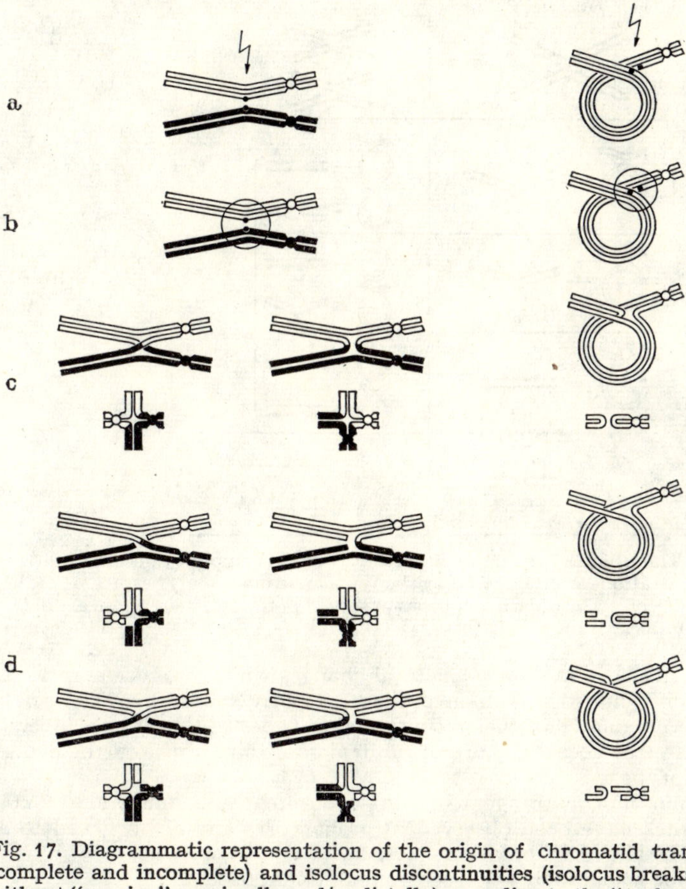

Fig. 17. Diagrammatic representation of the origin of chromatid translocations (complete and incomplete) and isolocus discontinuities (isolocus breaks with and without "reunion" proximally and/or distally) according to the "exchange model" of aberration production. The process of aberration production is assumed to proceed in three phases: Occurrence of local "instabilities" (a), exchange initiation (b), and complete (c) or incomplete (d) exchange owing to interaction between two instabilities (modified after *Rieger* 1966).

diately or after a delay, by a mechanical exchange process. If on spatial or temporal grounds two primary lesions fail to interact to produce "exchange initiation", these lesions may be repaired. According to the exchange model, the occurrence of chromosome structural changes is based on a three-phase system:

a) the occurrence of local instabilities; b) the pairwise co-operation of the instabilities in "exchange initiation"; c) the exchange process which is analogous to crossing-over (Fig. 17, 18).

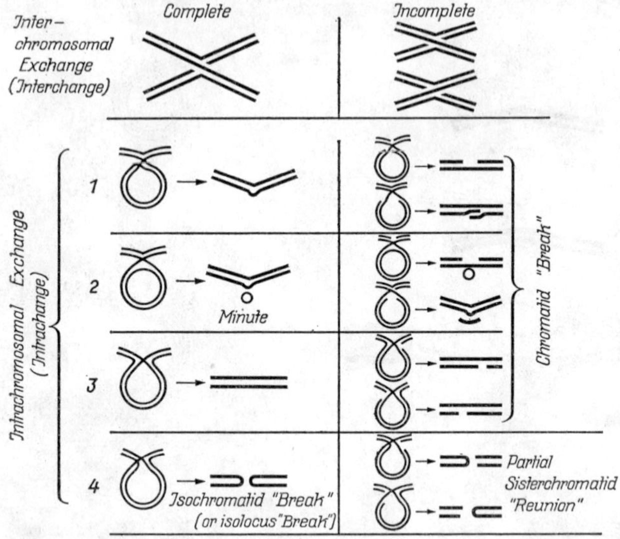

Fig. 18. The origin of the more important types of complete and incomplete chromatid aberrations as interpreted by the "exchange model" of aberration production in somatic cells. The aberration types observed in mitotic metaphase are indicated by heavy lines (from *Revell* 1959).

As a rule, chromosome structural changes which produce acentric fragments and dicentric reunion products rapidly succumb to cell selection and are eliminated. Other reconstruction types (duplications, inversions, eucentric translocations) are of fundamental importance in the evolution of karyotypes.

Chromosome rearrangements analogous to the c. mutations found in eukaryotes have been observed in prokaryotes and are believed to arise from aberrant recombinational events.

chromosome orientation — → centromere orientation.

chromosome pairing — 1. the highly specific side-by-side association of homologous chromosomes. **Homologous c. p.** can be classified into 4 types defined as follows (*Riley & Law* 1965):

a) **Meiotic synapsis or pairing:** The intimate association of homologues easily recognized during first prophase of meiosis and leading to → bivalent (pairing of two homologous in diploids) or → multivalent formation (pairing if more than two homologues as in poly- or aneuploids). The bi- or multivalents resulting from pairing of zygotene chromosomes are maintained to metaphase I if → crossing- over and → chiasma formation took place (→ desynapsis). Meiotic pairing is the prerequisite for the regular anaphase separation of meiotic chromosomes. When the pairing process is viewed under the electron microscope the visibly single zygotene chromosomes form a tripartite ribbon called → "synaptonemal complex" (*Moses* 1958) which is absent in achiasmate meiosis.

Meiotic pairing begins at "contact points" (→ zygomere) from which it runs along the chromosomes in zipper-fashion. The contact points are at the chromosome ends, at the centromeres, or at both concurrently. "Localized pairing" is a form of pairing confined to those chromosome segments near the first contact point. In all cases, the pairing affinities of chromosomes are satisfied by associations in twos. For multivalent production between more than two homologues a regional change of pairing partners takes place (Fig. 19).

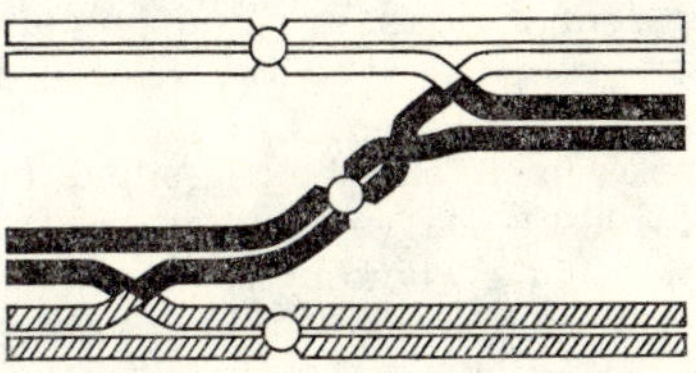

Fig. 19. Change of partners during meiotic chromosome pairing giving rise to a pairing configuration, a trivalent, consisting of three homologous chromosomes,

According to *Grell* (1962) two pairing phases may be distinguished in meiosis: "exchange pairing" and "distributive pairing". Exchange pairing is specific pairing leading to crossing-over between homologues. If exchange has taken place, exchange pairing is directly followed by distributive pairing and the chromosomes with chiasmata remain associated up to the time of anaphase separation. Those chromosomes which have not undergone exchange by crossing-over may take part in associations during the phase of distributive pairing and under these circumstances even non-homologues may pair.

The mechanism of meiotic pairing is still unclear. Two distinct aspects are under genetic control. The extent to which pairing is realized within the chromosome complement is controlled by the activities of major and polygenes. Mutants of major synaptic genes can so disorganize the pairing process that no pairing occurs and quantitative modifications of a wide magnitude can result from other genetic conditions. Furthermore, pairing specificity can be increased or decreased by the action of specific genes permitting or preventing the pairing of chromosomes distantly or closely related genetically and phylogenetically.

Meiotic c. p. possibly occurs in two steps (*Hotta & Stern* 1971): During zygotene relatively small sequences of DNA interspersed along the homologous chromosomes begin to pair ("gross pairing") and chromosome pairs may be stabilized by the synaptonemal complex. A second step follows gross alignment of chromosomes and consists in the formation of single stranded DNA regions by action of an endonuclease and possibly other enzymes. These could pair and recombine ("molecular pairing"). Duplex formation between specialized zygotene regions or single-stranded regions produced by endonuclease might be facilitated by a specific DNA-binding enzyme. There is evidence that meiotic c. p. is temporally associated with a semiconservative replication of DNA (→ DNA binding protein). This means that a distinctive component of DNA evidently does not replicate during premeiotic S period but does so at the zygotene stage (→ colchicine-binding protein; distributive pairing).

b) **Somatic pairing:** The close juxtaposition and coorientation of homologous chromosomes as, for example, during the prophase and metaphase of somatic divisions in *Drosophila* and other Diptera.

c) **Salivary gland pairing:** The intimate association of homologous → polytene chromosomes in the salivary glands and other tissues of species of Diptera This pairing process may be a special expression of the somatic pairing seen in other cells of these organisms.

d) **Secondary association or pairing of bivalents:** The nonrandom distribution of bivalents at first metaphase of meiosis in some polyploid species. The bivalents occur in pairs or groups if they are related by genetic and evolutionary factors.

2. **Nonspecific pairing** of chromosomes which can be classified into 3 types defined as follows (*Riley & Law* 1965):

a) **Heterochromatic fusions:** Associations of chromosomes, leading to the formation of → chromocenters or to the random association of heterochromatic chromosome regions.

b) **Terminal nonspecific associations:** Terminal associations between univalent chromosomes at first metaphase of meiosis; these may also be a consequence of heterochromatic fusions.

c) **Nonspecific pairing at prophase of meiosis:** The pairing of nonhomologous chromosomes or chromosome regions if homologous synapsis is interrupted by → structural heterozygosity or in the absence of homologous partner chromosomes (as in some monoploid individuals: "intrahaploid pairing").

chromosome polymorphism — the occurrence of one or several chromosomes in two or more alternative structural forms within a population. The structurally changed chromosomes are the result of → chromosome mutations which may be fixed, homo- and heterozygous, in a certain percentage in the population ("balanced chromosomal polymorphism"). The chromosomal structural types then fluctuate around a mean, roughly constant from generation to generation and possess → adaptive value. The various chromosomal structural types present in c. p. are, in a sense, homologous chromosomes, as they can mutually replace the alternative structural form in the karyotype. The c. p. may be eliminated if one of the structural variants proves to be superior under the prevailing environmental conditions and its proliferation is strongly furthered by natural selection at the expense of the alternative type (→ genetic polymorphism).

chromosome puffing — → puff; polytene chromosome.

chromosome pulverization (*Levan* & *Tjio* 1948; *Nichols* et al. 1964) — the destruction of chromosome structure, varying from an apparently total fragmentation of the chromatin continuity to various degrees of defective condensation and erosion (→ premature chromosome condensation; prophasing).

chromosome rearrangement — = → chromosome mutation.

chromosome reduplication — the synthesis of all those compounds which give the chromosome form, control its mechanical activity during the cell cycle, and regulate its metabolism. The primary event in c. r. in pro- and eukaryotes is the replication of DNA.

chromosome reinitiation — in bacteria, the beginning of a round of chromosome replication in an already replicating chromosome.

chromosome set — in eukaryotes, a minimum viable complement of indispensable (→ B-chromosome) chromosomes, each represented once. The number of chromosomes in a basic c. s. is called the **basic number.** The gametic complement of a fertile, functionally diploid, polyploid is a secondary b. c. s.; the number of chromosomes in such a set is called the secondary basic number (*Dyer* et al. 1970).

The genes contained in the chromosomes of one c. s. represent one → genome; each chromosome of the set is a carrier of a specific → linkage group.

Diploid organisms with sexual reproduction possess two c. sets in the → diplophase of their life cycle, one in the → haplophase. This → alternation of nuclear phases is the result of → fertilization and → meiosis. n the case of → polyploid organisms, the number of c. sets is increased proportional to the degree of polyploidy.

chromosome sterility → hybrid sterility.

chromosome stickiness (*Beadle* 1932) — → stickiness.

chromosome substitution (*Karpechenko* 1935) — the exchange of single chromosomes or chromosome pairs by other chromosomes of the same chromosome complement ("autosubstitution") or by chromosomes of the complements of other species of genera ("allosubstitution" or "alien substitution").

chromosome theory — the chromosome theory of → Mendelian heredity states that the → chromosomes, as carriers of → genetic information, represent the material basis of "nuclear inheritance". The regularity of the gene distribution from cell to cell (→ mitosis) and from generation to generation (→ meiosis) and the concomitant → genetic recombination depend on the characteristic behavior of the chromosomes in the course of nuclear division.

The first clear formulation of the chromosome theory was made by *Sutton* (1903). He referred to the relationship between the separation of homologous chromosomes in meiosis and the separation of character differences in gamete formation as postulated by *Mendel*.

chromosomin (*Stedman* & *Stedman* 1943) — = → nonhistone chromosomal protein.

chromotype (*Battaglia* 1956) — = → chromosome set.

chronocline — a character gradient in the time dimension (→ cline).
chronogenetics (*Gedda* & *Brenci* 1973) — a branch of genetics devoted to the study of the fundamental, endogenous genetic time (hereditary biological time) with its theoretical and practical, normal and pathological implications.
cilium — in many algal, protozoan and metazoan cells, any of the whiplike organelles (= undulipodium) with a diameter of the order of 200 nm. Their dry weight composition is roughly 70 to 84% protein, 13 to 23% lipid, 1 to 6% carbohydrate, and 0.2 to 0.4% adenine and uracil nucleotides. Cilia contain ATPase and adenylate kinase and are composed of 11 fibrils, two of which are centrally located and somewhat smaller and more fragile than the 9 peripheral ones (diameter 20 to 25 nm, length 35 to 37 nm). The peripheral fibrils have a double tube substructure. Both cilia and → flagella, which have the same chemical constitution, are enclosed by a semipermeable membrane (about 9 nm wide); within this membrane the peripheral fibrils are evenly spaced around a circle some 160 nm in outside diameter (Fig. 20).

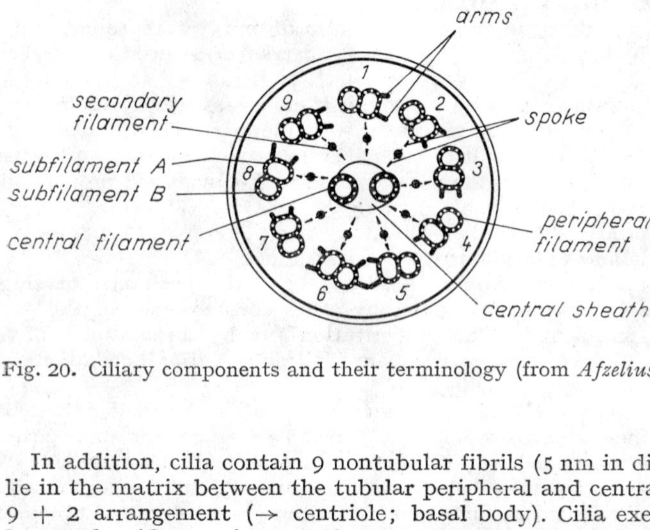

Fig. 20. Ciliary components and their terminology (from *Afzelius* 1969).

In addition, cilia contain 9 nontubular fibrils (5 nm in diameter) which lie in the matrix between the tubular peripheral and central fibrils of the 9 + 2 arrangement (→ centriole; basal body). Cilia exert a unilateral beat and a sideways force on the water to move the cell in a direction perpendicular to the organelle. Frequently, cilia are aggregated into compound structures within which the units act in synchrony.
cin-duction (*Ozeki* & *Howarth* 1961) — = → colicinoduction.
circadian rhythm — any cyclic quantitative variation regularly occurring with 24-hour periodicity (e.g., mitotic indices of proliferating cells).
circular permutation — of physically linear DNA molecules in bacteriophages with the same set of genes arranged in the same order which are related to each other as if each had once been in a circle, and the circles had been

opened by randomly distributed breaks:

A B C D E .... X Y Z A B
D E F G H .... Z A B C D E
F G H I K .... C D E F G

In addition to their fixed length and circular p., these DNA molecules have a terminal redundancy. Such molecules reveal unusual → recombination frequencies suggesting → genetic circularity.

cis-configuration (*Haldane* 1942) — the → alleles of two linked genes or the → heteroalleles of a → cistron are said to be in the cis-configuration (coupling phase) if in doubly → heterozygous or → heterogenotic cells the two (dominant) wild-type alleles are located in one, the two mutant alleles in the other of two homologous chromosomes (AB/ab or ++/ab).

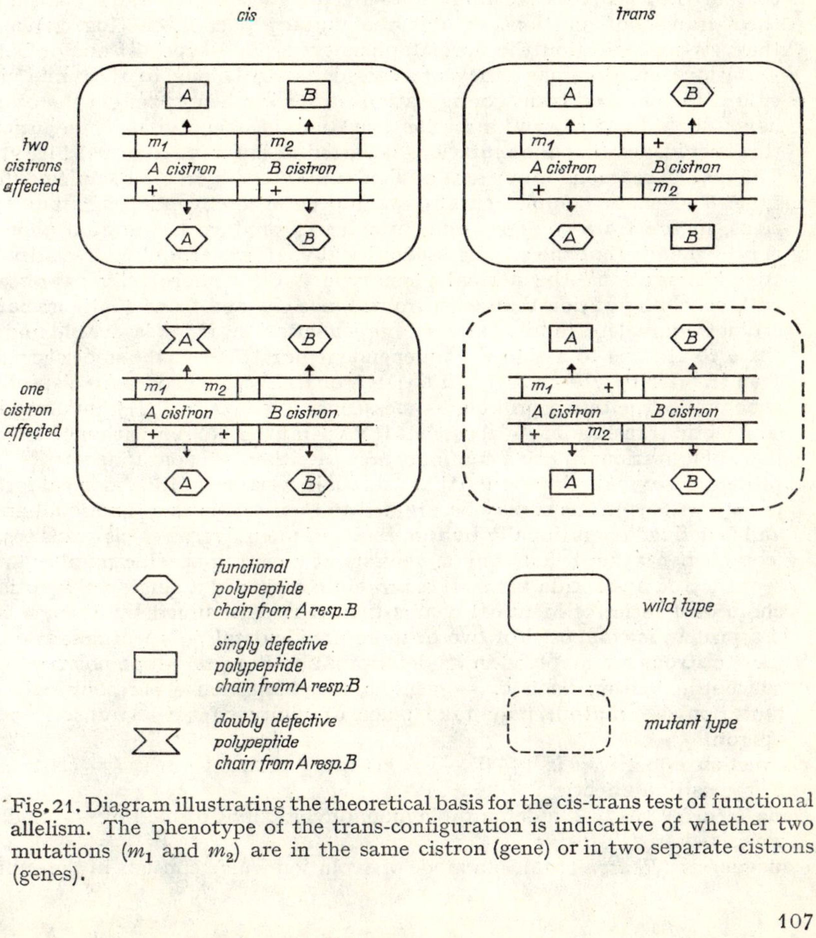

Fig. 21. Diagram illustrating the theoretical basis for the cis-trans test of functional allelism. The phenotype of the trans-configuration is indicative of whether two mutations (m_1 and m_2) are in the same cistron (gene) or in two separate cistrons (genes).

A double heterozygote in the repulsion phase, with the dominant alleles each in one of the homologous chromosomes (Ab/aB or +b/a+), is termed "trans-configuration" or "trans-heterozygote" (→ cis-trans test).

cis dominant effect (*Epstein & Beckwith* 1968) — → operon.

cisterna — any of the spaces (sacs or cavities) between the membranes of the →endoplasmic reticulum and filled with fluid contents (→ nuclear envelope).

cis-trans test (*Lewis* 1951; *Benzer* 1957) — a genetic test (= complementation test) to decide whether two → gene mutations (m_1 and m_2) occur in the same functional gene (→ cistron) and to establish the limits of this genetically active region. To this end, the heterozygotes (or → heterogenotes) in which the concerned mutations are located in the same chromosome (cis-configuration: $++/m_1m_2$) and those in which they are present in separate chromosomes (trans-configuration $+ m_1/m_2 +$) are compared. Two recessive mutations are assigned to the same cistron if their trans-configurations exhibit the mutant phenotype (Fig. 21) and their cis-configuration the normal phenotype (wild-type). If one or both mutations are dominant they are considered to belong to the same cistron if the cis- and trans-configurations differ in phenotype. On the other hand, should the cis- and trans configurations be phenotypically indentical this is evidence that the mutations occurred in different cistrons (Fig. 21). The fact that among mutations of a cistron "intra-cistronic complementation" (→ genetic complementation) can occur whereby independent mutations, in the trans-configuration, interact to produce the normal phenotype demands that the c. t. be used critically. However, in "intra-cistronic complementation" the normal phenotype is often more fully expressed with cis-configuration than with trans. Since cis- and trans-configurations are not equivalent in all cases of complementation, it is reasonable under these conditions to assign the independent mutations to the same cistron.

cistron (*Benzer* 1957) — a segment of DNA or RNA which codes for a specific gene product, either a protein (→ messenger RNA) or an RNA involved in → genetic translation (→ ribosomal RNA; transfer RNA); within cistrons, pairs of mutations in trans-configuration are either deficient for a particular protein (enzyme), or produce that protein in a structurally abnormal form (*Fincham* 1959). The c. may be equated to the → gene as a functional unit and is defined operationally by the results of unequivocal → cis-trans tests (complementation tests). Any c. consists of a cistron-specific number and sequence of nucleotide pairs. If a protein consists of a single polypeptide chain or a series of identical chains then it is determined by a single c.; if a protein is composed of two or more nonidentical polypeptides, two or more cistrons are involved in its determination ('one c. — one polypeptide concept'). Within each c. → genetic recombination (→ recon) and → mutation (→ muton) may take place at many sites (→ transcripton; operon).

cis vection effect (*Lewis* 1955) — = cis-trans position effect (→ cis-trans test; position effect).

clade (*Huxley* 1957) — a delimitable monophyletic unit formed by → cladogenesis.

cladogenesis (*Rensch* 1947) — a mode of evolution which consists in the splitting of an evolutionary (phylogenetic) line, such as a species. The two or more descendent species may then follow independent evolutionary courses. C. not only involves species splitting and species divergence but also the origination of genera, families, and other higher taxonomic categories. C. and → anagenesis (phyletic evolution) are combined in varying proportions in most lines of descent.

clastogenic (*Shaw* 1970) — of agents (clastogens) capable of inducing chromosome structural changes (→ chromosome mutation; mutagenic).

cleavage — 1. the furrowing of the cell cytoplasm after completion of nuclear division. The furrow originates in the → cortex immediately peripheral to the equator of the → spindle. The deepening of the cleavage furrow involves actual growth of the cortex in this region, which continues until c. is complete (→ cytokinesis).

2. The division of the egg cell after fertilization (zygote) into a number of cells during a time interval called the "period of c.". It lasts to the moment when a shifting of regions of the egg relative to one another (blastulation, gastrulation) sets in. Henceforth the individual is usually referred to as an embryo.

At the beginning of gastrulation the cleavage cells are arranged into a "blastula", which in its typical form is a hollow sphere. The main physiological function of c. is the restoration of the balance between the nuclear size and the cytoplasmic volume with which it is associated.

Cleavage types may be classified as 1. **equal c.** in eggs with little yolk, 2. **unequal c.** in eggs with somewhat more yolk, and 3. **superficial c.** in extremely yolky eggs in which c. occurs in only a small area of the egg.

cleavage map — a → chromosome map of viruses based on the fragments produced by → restriction endonucleases of bacteria (→ DNA modification-restriction system).

cleavage nucleolus — a → nucleolus which differs from normal nucleoli in that it lacks the granular outer region. C. nucleoli, observed in the interphase nuclei of cleaving amphibian eggs in the form of nucleolar bodies, are almost devoid of basophilia and do not incorporate appreciable amounts of radioactive uridine. Possibly, they are sites of accumulation of proteins of cytoplasmic origin.

cleavage nucleus — 1. the nucleus of the fertilized egg cell or zygote;

2. the nucleus of egg cells which develop parthenogenetically (→ parthenogenesis).

cleistogamy — → fertilization within closed flowers making selfing obligatory.

cline (*Huxley* 1939) — a gradient (decrease or increase) within a continuous population in the frequencies of different → genotypes in different localities ("genocline") or → phenotypes ("phenocline") (*Darlington & Mather* 1949).

clone (*Webber* 1903) — a population of cells or organisms derived from a single cell or common ancestor by mitoses. The mode of → reproduction giving rise to a c. is asexual. A c. is not necessarily homogeneous and therefore the terms c. or cloned should not be used to indicate homogeneity in a population.

cloning — → clone; molecular cloning.
closed regulatory loop (*Thomas* 1971; *Kourilsky* & *Gros* 1975) — a regulatory system of gene expression controlled by two regulatory proteins, each of which controls the rate of synthesis of the other. → Autoregulation is the simplest case of a c. r. l.
C-meiosis (*Levan* 1939) — a modified form of → meiosis which occurs following the action of colchicine ("C-meiosis" = "colchicine meiosis") and other spindle poisons. The spindle poisons cause partial or complete blockage of the spindle mechanism which normally directs chromosome distribution. At the same time, → chromosome coiling and → chiasma formation are *inter alia* functions which are indirectly affected. The complete suppression of the spindle mechanism, which may involve only the first meiotic division or only the second, or both divisions, gives rise (via the formation of → restitution nuclei) to meiotic products with more than one chromosome set.

During the normal course of meiosis in diploid organisms, four haploid meiotic products are produced from each meiocyte. As a rule, spindle inhibition during the first or second meiotic division yields two diploid meiotic products. If the spindle cannot function in both meiotic divisions, the result is a tetraploid meiotic product (→ C-mitosis).
C-mitosis (*Levan* 1938) — a modified form of → mitosis, caused by partial or complete inactivation of the spindle mechanism and concomitant disturbance of the → chromosome movement normally leading to chromatid distribution in anaphase. The modification is produced by the action of colchicine ("C-mitosis" = "colchicine mitosis") and other spindle poisons. Parallel with the effect on chromosome movement is an increased chromosome contraction. The → centromeres divide late and the chromatids associated at the centromere mutually repel one another. In this way cross-shaped, so-called "C-pairs" are formed.

Complete spindle inactivation leads to the formation of → restitution nuclei with doubled chromosome numbers. Partial spindle inactivation hinders the regular anaphase movement of the chromatids and frequently gives rise to multipolar anaphases.

C-mitoses, just as C-meiosis, exhibit a typical threshold reaction and never occur unless concentrations of the spindle poison exceed a critical value. The appearance of typical C-metaphases after complete spindle inactivation can vary (*Östergren* 1950); the chromosomes may be distributed over the entire cell, may be clumped together in the cell center, may be arranged in a starshape configuration, or may be strewn about the peripheral regions of the cell.

Colchicine inhibits spindle function by binding to the protein subunits of spindle → microtubules, and thereby blocking their assembly into larger, functional structures.
coadaptation (*Dobzhansky* 1951) — the → selection process (= integration) by which harmoniously collaborating genes become accumulated in the → gene pool of a population. The integration of genes is of two kinds (*Mather* 1953): One is a balance among alleles leading to overdominance and through it to a balanced polymorphism. This is called "relational

balance". The other is a balance among different genetic loci and is referred to as "internal balance" or "epistatic balance".

coancestry (*Falconer* 1960) — the degree of identity by descent of the two parents of a diploid individual. Values of c. are obtained from the probabilities of occurrence of genes identical by descent in a pair of gametes taken at random from a pair of parents. C. may be used in place of → coefficient of relationship and → coefficient of parentage.

coated vesicle — any of the → vesicles contained in the → Golgi apparatus which may serve in the transport of specific substances from the → endoplasmic reticulum. C. vesicles appear to arise from the plasma membrane or in the interior of the cell near the Golgi area.

co-conversion — the linked conversion of two or more genes (→ gene conversion).

code — → genetic code.

coding ambiguity — the phenomenon whereby one nucleotide triplet (→ codon) codes for more than one amino acid. Ambiguity in the → genetic code may result from: 1. A codon may be misread owing to environmental conditions existing around the tRNA-mRNA-ribosome complex which interfere with the codon recognition process (→ mistranslation). 2. A codon may be misread because of a specific change in the primary, secondary, or tertiary structure of either the → transfer RNA or the → ribosome such that it affects the codon-anticodon interaction. 3. An → aminoacyl-tRNA synthetase may misrecognize either an amino acid or a tRNA and form a wrong → aminoacyl tRNA.

coding ratio (*Crick* 1963) — the number of bases in nucleic acids divided by the number of amino acids whose sequence they determine in a particular polypeptide (→ genetic code).

coding-recognition site — → transfer RNA.

coding triplet — → codon.

codogenic — of that strand (= sense strand) of double-stranded DNA which is used for → genetic transcription. The c. strand or nucleotide sequence is usually identified by its ability to specifically hybridize (→ nucleic acid hybridization; molecular hybridization) with its complementary → messenger RNA. In prokaryotes, two types of organisms have been recognized by this kind of investigations: 1. those in which only one DNA strand is c. for all genes, and 2. those where some genes are coded for by one, others by the complementary DNA strand and where overlap may occur.

codominant — of → alleles not mutually related in terms of dominance-recessivity; instead, their gene products occur independently, and both are manifest phenotypically. If, for example, the heterozygote of the alleles A^1 and A^2 belonging to a multiple allelic series and → dominant to a, shows no phenotype intermediate to that of A^1A^1 and A^2A^2 then the alleles are termed codominant.

codon (*Crick* 1963) — any triplet of → nucleotides (coding unit) in DNA or RNA (if RNA is the carrier of primary → genetic information as in some viruses) that codes for a particular amino acid or signals the beginning or end of the message (→ anticodon). The term c. is also used for the corresponding (and complementary) sequences of three nucleotides in the →

messenger RNA (→ nodoc) into which the original DNA sequence is transcribed (→ genetic transcription).

Of the 64 possible codons in the → genetic code, the mRNA triplets UAA (ochre), UAG (amber), and UGA (opal) serve as → terminator codons. AUG and GUG are → initiator codons (→ regulatory codon). Synonymous (= degenerate) codons or synonyms are different codons for the same amino acid. It is uncertain whether there are ambiguous codons (→ coding ambiguity), i.e., triplets which represent more than one amino acid. A c. may be changed by → mutation so that it now codes for a different amino acid (→ missense codon) or for no amino acid at all (→ nonsense codon or terminator codon).

codon recognition — the relationship between the anticodons of → transfer RNA and the codons in → messenger RNA. A site on the ribosome-mRNA complex (→ initiation complex) at which codon-dependent binding of → aminoacyl-tRNA occurs during → genetic translation is called the codon recognition site (*Watson* 1964).

coefficient of aberration production (*Lea* & *Catcheside* 1942) — the aberration frequency (→ chromosome mutation) per cell per roentgen in the case of experimental induction of aberrations by means of ionizing radiations.

coefficient of coancestry (*Malecot* 1948) — The probability that two homologous genes, one from individual I and the other from L, are identical by descent, i.e., are descended from the same ancestral gene. The complementary probability, $1\text{-}f_{IL}$, is the probability that these two genes come from unrelated ancestors.

coefficient of coincidence (*Muller* 1916) — → coincidence.

coefficient of hybridity (*Morton* et al. 1971) — a measure of genetic distance between populations, i.e., a measure of the relatedness of two populations as estimated from the allele frequencies in the ensemble of populations (→ coefficient of kinship).

coefficient of inbreeding (*Wright* 1929) — → inbreeding coefficient.

coefficient of integration — the relationship between the frequency of transfer of a donor marker to recipients and the frequency of its transmission from recipients to recombinants following $F^- \times$Hfr conjugal mating in bacteria (→ F^- factor).

coefficient of kinship (*Malecot* 1948) — a measure of the genetic relatedness of two individuals or populations. The c. of k. of individuals is the probability that a gene taken at random from individual A is identical by descent with a gene at the same locus taken at random from individual B (→ coancestry). The c. of k. of two populations is the probability that two genes at a locus carried by an individual randomly chosen from the → F_1 after crossing the populations are indentical by descent (→ coefficient of hybridity).

coefficient of parentage — the c. of p. (for one genetic locus) between two individuals X and Y is the probability that a random gene from X is identical by descent with a random gene from Y (*Malecot* 1948). (→ inbreeding coefficient).

coefficient of relationship — the probability that two individuals have inherited a certain gene from a common ancestor, or the proportion of all their allelic genes that have been inherited from common ancestors.

coefficient of reunion (*Darlington & Upcott* 1941) — the empirical coefficient of reunion is the ratio of the total number of identifiable chromatid and chromosome breaks to the number of breaks which participate in a discernable reunion (→ chromosome mutation).

coefficient of selection — the intensity of → selection as measured by the proportional reduction in the gametic contribution of a particular → genotype compared with a standard genotype (symbol s). The contribution of the favored genotype is taken to be 1, the contribution of the genotype selected against is 1 — s. This expresses the → fitness or adaptive value of one genotype compared with the other. A coefficient s = 0.2 means that for every 100 zygotes produced by the favored genotype, only 80 are produced by the genotype selected against.

coenocyte — a multinucleate cell produced by repeated nuclear division without cytokinesis (= multikaryon).

coenogamodeme (*Gilmour & Heslop-Harrison* 1954) — = → coenospecies.

coenospecies (*Turesson* 1922) — a group of indivudals of common evolutionary origin comprising more than one taxonomic species. The c. is made up of → ecospecies which following hybridization may participate to a limited extent in mutual exchange of genes. Gene exchange between different c. does not occur. Sterile hybrids are produced by hybridization.

coenzyme — usually a nonprotein compound of low molecular weight neccessary for the action of some → enzymes.

cog region (*Angel* et al. 1970) — any group of similar sequences of nucleotides in DNA molecules which may specifically be recognized by endonucleases or other enzymes.

cohesive end — (*Ris & Chandler* 1963; *Hershey* et al. 1963) — in various → temperate bacteriophages and → mitochondrial DNA, a single-stranded polynucleotide extension (10 to 20 nucleotides) at one end of a linear, monomeric and double-stranded DNA molecule which is complementary to a sequence of the same length at the other end of the molecule. C. ends can cohere specifically and provide a mechanism to join the two ends of one molecule (cyclization of the molecule; → genetic circularity), or to join two or more such molecules (dimers, trimers, etc.).

coil — one complete revolution of the internal spiral of a chromosome (→ chromosome coiling). (*Darlington & Mather* 1949).

coiling — → chromosome coiling.

coimmune — of → temperate phages if the → repressor made by each of them can repress lytic growth of the other (→ heteroimmune), i.e., recognizes the operators of the others.

coincidence (*Muller* 1916) — the coefficient of c. denotes the proportion of observed double → crossovers to the number expected from the random combination of single crossovers among three or four linked genes (*Darlington & Mather* 1949).

The coefficient of c. is the most commonly used genetic measure of the → interference intensity. For positive interference it is less than 1, for negative interference greater than 1. In the absence of interference, the coincidence value equals 1. In cases of absolute interference (no double crossing-over), the coincidence value equals 0.

co-inducer — a molecule which interacts with → repressor protein to free the DNA of the → operon from restraints on its transcription into → messenger RNA.

colchicine-binding protein (*Hotta & Stern* 1973) — a protein found in mitotic and meiotic nuclei which might be involved in meiotic and mitotic → chromosome pairing. An increase of c.-b. p. coincides with the leptotene to pachytene stages of → meiosis.

colchiploidy (*Dermen* 1953) — polyploidy induced following the application of the spindle poison colchicine (→ polyploid).

colicin (*Frederiq & Betz-Bareau* 1953) — any of a class of bactericidal protein antibiotics which are specified by extrachromosomal → plasmids (→ colicinogenic factor) in certain bacterial strains, can adsorb to the surface of susceptible bacteria and, in the majority without penetrating the surface, kill these cells. The receptor sites on the bacterial surface can also bind certain bacteriophages; hence colicins may have evolved from bacteriophages by a process of gene elimination. Three types of action are known by which different colicins may kill sensitive cells: 1. Arrest of DNA synthesis and DNA degradation; 2. inhibition of protein synthesis which can be traced to a specific alteration of some component (16 S → ribosomal RNA) of the ribosomes; 3. inhibition of oxidative phosphorylation and overall arrest of macromolecular syntheses.

Cells producing a c. are themselves immune to its effects due to the formation of a specific c. inhibitor which, probably by reversible association with the c., inhibits its action. Bacterial mutants resistant to c. which nevertheless still adsorb c. are called "mutationally immune", "tolerant" and "refractory". Different colicins have different host specificities and different antigenic properties.

colicin factor — = → colicinogenic factor.

colicinoduction (*Ozeki & Howarth* 1961) — transfer of baterial genes from donor (col^+) to recipient (col^-) cells due to the presence of → colicinogenic factors which act as a → conjugon and mediate bacterial → conjugation (→ sex factor).

colicinogenic (*Fredericq* 1953) — of bacterial strains able to produce → colicins.

colicinogenic factor (*Fredericq* 1953) — any of the extrachromosomal genetic elements (plasmids, symbolized by Cf) responsible for the production of → colicins in colicinogenic strains of *Enterobacteriaceae* (= colicin factors; col-factors). Col-factors are transferable from colicinogenic (col^+) to non-colicinogenic (col^-) strains by means of cell contact or by → transduction. Each col-factor represents an independent → replicon and consists of double-stranded (circular) DNA.

Some of these col-factors have been reported to be → conjugons mediating bacterial → conjugation or to interfere with bacterial fertility while others do not affect the fertility of donor strains.

Col-factors are genetic units of varying complexity in which the structural genes for the synthesis of one or several different colicins may be associated with genes governing other properties (as, e.g., repression, immunity, fertility, susceptibility to malespecific phages, inhibition of fertility, restriction of phages, and resistance to UV). As a rule, Col-factors are independent of the bacterial chromosome. Some Col-factors of *E. coli* K 260 have, however, been found associated with the chromosome or with detached chromosomal genes. Recombination with other plasmids is obtained by transducing Col-factors to recipients already carrying the → F factor Fd, F′lac, or some other Col-factors. Recombination with → resistance factors occurs when the recipient cell already carries a certain Col-factor (Col V) in addition to the R factor. The fertility properties determined by Col-factors are due to an F-factor called F_x and closely attached to the colicinogenic determinants. F_x can be physically or functionally separated from these. Like the original Col-factor, F_x is transferred with efficiency to F^- (recipient) cells and imports to them a low level of chromosomal fertility. F_x is itself capable of autonomous multiplication and transfer.

colicinogeny → colicin.

colicin tolerant (*Hill & Holland* 1967; *Nomura & Witten* 1967) — of bacterial mutants that are tolerant to the killing action of a number of → colicins. Such mutants still possess normal colicin receptor sites but the sequence of biochemical changes that normally follow colicin adsorption does not occur. The mutation results in widespread changes in the structure of the cell surface.

colinearity (*Crick* 1963) — the exact point-by-point relationship between the order of amino acids along the polypeptide chain and the order of the corresponding → codons along the polynucleotide chain of the nucleic acid (→ sequence hypothesis; genetic transcription; genetic translation).

collaterals — individuals of a family not related by direct descent. Inheritance is spoken of as "collateral" if particular → characters (traits) appear in collaterals (e.g., uncle and niece, etc.) as is characteristic for some irregularly manifesting recessive traits.

collochore (*Cooper* 1941) — a small nonspecific heterochromatic region causing chromosome association in → achiasmate meioses. Collochores are points of close contact between homologous chromosomes. They fulfill the function of → chiasmata in achiasmate meiosis without being the result of genetic → crossing-over and exchange of chromatid segments.

combinant (*Lenz* 1938) — 1. = → codominant;

2. of nonallelic genes showing functional interaction in the production of a particular phenotypic character (→ gene interaction).

commiscuum (*Danser* 1929) — a group of individuals all of which can actually or potentially exchange genes (a species).

commitment — the fixation of a cell to a unique sequence alternative to its previous state (predominantly in microorganisms). Examples of c. are → sporulation in bacilli, meiosis and sporulation in yeast, or meiosis in explanted microsporophytes.

compact X-chromosome — → dosage compensation.

comparate chiasmata (*Darlington* 1937) — → chiasma.

comparium (*Danser* 1929) — a group comprising all → coenospecies which may participate in mutual hybridization, whether direct or indirect.

compartmentation — the subdivision of cells into parts (→ cell compartments) by double or single → unit membranes.

competence (*Waddington* 1932; *Thomas* 1955) — 1. → induction.

2. In → genetic transformation, the ability of bacterial cells to take up exogeneous DNA and to become transformed by it. C. is a transient property of bacterial cultures and requires genetic information for expression of this ability and special physiological conditions. Three hypotheses have been put forward to explain the temporary activity of competent cells to take up DNA:

a) Competent bacteria attain during growth a partially "naked" surface through which DNA molecules of considerable size can penetrate. b) Enzymatically active receptor sites are present at the bacterial surface. c) The mesosomal vesicles constitute an active absorption site which would be initially exposed to the surrounding medium by the action of a lytic competence factor (a protein). The DNA could then pass into the → mesosome and be integrated at the point where the mesosome is joined to the recipient chromosome (the replicating point).

3. In → transfection, the ability of bacterial cells to incorporate bacteriophage DNA and to produce phage progeny.

competence factor (*Charpak & Dedonder* 1965) — → competence (= competence inducing factor; competence provoking factor).

competition — in ecology, genetics, and evolution, the endeavor (which may take various forms and have various, direct or indirect, deleterious results) of two (or more) organisms of the same or of different species to gain the same particular factor or thing, or to gain the measure each wants from the supply of any factor or thing when that supply is not sufficient for both (or all) (*Milne* 1961).

competitive ability — the success of one species in a mixture which is generally attributed either to the ability of that species to produce a greater proportion of offspring in the next generation at the expense of the other(s) or to its ability to reduce the numbers of the other species without itself changing (the success of one component of the mixture may also depend on its ability to avoid the other(s) without necessarily having any effect on it).

complement (*Darlington* 1932) — → chromosome complement.

complementary — 1. of two structures, each of which defines the other, e.g., the two strands of a → DNA double helix;

2. of four-strand double → crossing-over;

3. of genes which by interaction produce an effect qualitively distinct from the effects of any of them separately (→ gene interaction).

complementation — 1. all cases in which interaction between viral gene products or gene functions in multiply or mixedly infected cells results in an increased yield of infective virus of one (or both) parental types (→ helper virus).

2. The complementary action of homologous sets of genetic material (→ genetic complementation).

complementation analysis — determination of whether cells with the same phenotype have the same genotype (→ genetic complementation).

complementation group — a number of mutants which all have the same qualitative properties of → genetic complementation. A number of c. groups each of which consisting of similar mutually non-complementing mutants represents a 'cluster of c. groups', A cluster of c. groups has the following properties:

1. No member of the cluster complements with any other member of the cluster.

2. The cluster is formed from those c. groups which are most similar to each other with respect to the other groups with which they will complement.

3. The cluster must consist of those groups with those properties which will make the cluster contain the maximum number of groups. A c. g. may belong to more than one cluster but a cluster is characterized by those groups which belong uniquely to it.

complementation map (*Giles* 1958) — → genetic complementation.

complementation test — = → cis—trans test.

complementation unit (*Demerec & Hartman* 1959) — a subunit of the complementation map of a → cistron as determined by → cis-trans tests (complementation tests) which show that → genetic complementation may also occur between two mutants of one cistron ("intracistronic complementation"), neither of which complements a third.

This result is evidence that some mutations involve one c. u., other mutations another c. u., while still others affect both complementation units or "complons" simultaneously.

complex heterozygous (*Renner* 1917) — of a special type of genetic system in plants (called complex heterozygotes) based on heterozygosity for multiple reciprocal translocations (as exemplified by *Oenothera*). C. h. species are heterozygous for numerous allele pairs linked to different chromosomes but segregating as a unit during → meiosis.

The best example of "complex heterozygosity" is the plant genus *Oenothera*. Species of this genus have 6, 8, 10, 12 or all 14 of their chromosomes linked in a chain or ring at meiosis as a consequence of reciprocal translocations and the specific type of chromosome pairing in case of translocation heterozygosity (*Cleland* 1922, 1962). Within this pairing configuration paternal and maternal chromosomes alternate.

During anaphase I of meiosis, a regular distribution of adjacent chromosomes to opposite cell poles (→ alternative distribution) occurs. Thus all paternal chromosomes and genes of the pairing configuration are carried to one pole, all maternal chromosomes and genes to the other pole. With a ring of 14 chromosomes only two kinds of gametes are produced, identical genetically with the two that united to form the plant. The group of chromosomes and genes distributed as a unit during meiosis and identifiable by a particular group of alleles is called a "complex". It acts as one → linlage group (= *Renner* complex). Since in different *Oenothera* species different numbers of chromosomes are included in the ring, the size of the complexes is species specific (Fig. 22).

Selfing of a complex heterozygous *Oenothera* does not produce a Mendelian combination of 1:2:1 (→ segregation). The two expected classes of "complex" homozygotes normally fail to appear; only complex heterozygotes are recovered thus representing a system of "enforced heterozygosis". Their hybrid nature becomes evident when "heterozygotes"

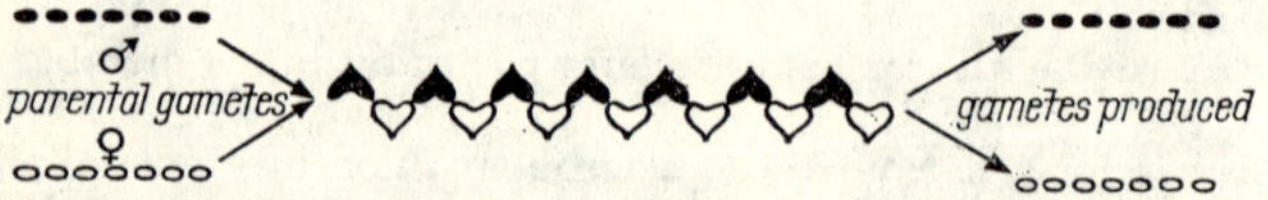

Fig. 22. Diagram showing a chain configuration (14 chromosomes in metaphase of first meiotic division) in a complex heterozygous *Oenothera*. The configuration consists of two "*Renner* complexes", one from the mother (white) and one from the father (black). During anaphase I, the maternal complex of chromosomes is distributed to one pole, the paternal complex to the other (from *Cleland* 1962).

are crossed to another heterozygous or homozygous species resulting in the production of "twin hybrids". Complex heterozygosity is maintained in most cases by one or more → lethal factors per complex which are lethal in the homozygous state. The lethals are either zygotic or gametophytic. Zygotic lethals cause the death of zygotes; gametophytic lethals kill or inhibit either the male or female gametophyte. In most *Oenotheras*, these lethals are "balanced", i.e. both complexes possess nonallelic lethals. A species with a balanced set of gametophytic lethals gives rise to only one kind of functional egg and one kind of sperm. Plants with zygotic lethals produce two types of functional eggs and sperm, but half the zygotes produced by selfing will die since they receive one or the other of these lethals in double dose.

Occasionally, homozygous *Oenothera* forms are produced from complex heterozygotes; these have been classified by *Renner* (1941) into 4 types:

1. **Primary homozygotes:** They are produced from forms in which one complex is free of lethals from the outset.

2. **Hybridogenic homozygotes:** During gamete formation the chromosome carrying the lethal of one complex is replaced by a chromosome lacking a lethal.

3. **Crossing-over homozygotes:** As a result of crossing-over the lethal of a complex is replaced by a nonlethal portion of the partner chromosome.

4. **Translocation rearrangements:** A lethal free complex is formed by translocation; this complex, together with the two original complexes, gives rise to so-called → half mutants from which homozygous "full mutants" are produced by selfing.

The origin of complex heterozygosity in the genus *Oenothera* and the regular alternative distribution of the chromosomes in the translocation rings appears to be favored by the following conditions (*Cleland* 1957):

1. The *Oenothera* chromosomes are all of the same size and possess → centromeres surrounded by heterochromatin; these heterochromatic segments are preferentially involved in structural changes.

2. The translocated segments are normally of equal size, so that the

centromeres in the translocation configuration are almost equidistant from one another.

3. The localization of the centromeres and the restriction of the chiasmata to the ends of the chromosomes favor the → alternative distribution of the chromosomes included in the ring configuration.

4. The maintenance of translocation heterozygosity ultimately leads to hybrids with different translocations; following crossing between open-pollinated populations all the chromosomes (14) of the hybrids may form a ring in meiosis.

5. Following ringformation by all chromosomes, the lethals acquire positive selective value, if their negative effects are compensated by normal alleles.

6. Finally self-pollination in combination with the lethals of complexes provides for the maintenance of maximal heterozygosity and the true breeding of the complex heterozygotes.

complex locus (*Dunn* 1954) — a cluster of two or more closely linked and functionally related → genes or → cistrons constituting a pseudoallelic series (→ pseudoalleles). A c. l. may contain one → operon (→ locus).

complon — any complementation subgroup within a linear complementation map (→ genetic complementation).

compound — a genotype heterozygous for two mutant → alleles (m^1/m^2) of the same locus. Together with the wild-type allele, the mutant alleles form a series of multiple alleles (+, m^1, m^2, etc.).

compound chromosome (*Novitsky* 1954) — in *Drosophila,* any monocentric chromosome in which the material from one chromosome arm is represented twice (→ X chromosome). Gametes of the compound-bearing individuals generally carry either two or no doses of the chromosome arm (*Lindsley & Grell* 1968). The two arms comprising a c. c. may join 1. by attachment of the base of one to the terminus of the other to form an acrocentric chromosome, or 2. by attachment of both proximally to a single → centromere to form a metacentric chromosome; the ends of either an acrocentric or a metacentric chromosome may join to form a compound → ring chromosome.

The compound arms may be in the same sequence or one may be inverted with respect to the other. The elements of a c. c. may, therefore, pair as a spiral (tandem configuration) or as a hairpin (reversed configuration).

Compton effect — in an atom, the interaction of a high energy photon with a single electron resulting in the ejection of the electron and secondary emission of a lower energy (longer wave length) photon.

concatemer (*Thomas* et al. 1968) — in viruses, a tandemly-repeating nucleotide sequence with the entire genome constituting the repeating length. Concatemers are assumed to be the immediate precursor to the mature virus genome.

concordance — the identity of members of matched groups or pairs (such as twins) in any one or in a given → trait. The opposite of c. is → discordance.

concordant orientation — → translocation.

conditional lethals — → lethal factor.

conditional mutation — any of a class of → mutations whose viability is dependent on a set of (permissive) conditions. They perish under non-permissive or restrictive conditions. A special class of c. mutations is the temperature-sensitive mutation (→ lethal factor). At the restrictive temperature, the mutant dies or further development is arrested; at permissive temperatures the mutant survives and reproduces. In micro-organisms, the majority of temperature-sensitive mutations are single → basepair substitutions of the → missense type of mutation. At the permissive temperature, the mutant protein is functional; at the restrictive temperature it is not.

conditioning — a direct alteration in the development of the → phenotype of an individual resulting from the action of external agents during early, critical stages of development. Examples of c. are → phenocopies, → dauer-modifications, and other changes transmitted partially or wholly through successive asexual or sexual generations in the absence of the inducing stimuli (*Hill* 1967).

conduction (*Clark* & *Adelberg* 1962) — in bacterial → conjugation, the transfer of the bacterial genetic material and of the → conjugon (= "promoter") as linked elements (→ donation).

configuration (*Darlington* 1929) — = → chromosome configuration.

conformation — of the different spatial arrangements of the atoms that may arise by rotation about bonds in a molecule. The term conformational change refers to a change in the average positions of the atomic nuclei not including covalent bond changes.

congenital — of a → phenotype (character, trait) recognizable at birth or referable to constitution at birth. C. is a descriptive term neither including all inherited conditions, nor excluding those characters which result from environmental influences alone. Malformations evident at birth are therefore c. and may be caused by genetic or extrinsic factors.

congression (*Darlington* 1937) — → chromosome congression.

conidiospore — an exospore of fungi which, in contrast to the endospores (→ ascospore) formed within a cell or an organ, is constricted singly or in chains. Conidiospores represent "aplanospores" (= conidia).

conjugation — the term c. denotes several events that connect, couple, or unite chromosomes, nuclei, cells or individuals.

I. **Conjugation of chromosomes:** = → chromosome pairing.

II. **Conjugation of nuclei:** a) = → karyogamy, b) = → dikaryotization.

III. **Conjugation of cells:** a) in general, → syngamy, the union of sex cells (gametes) during fertilization;

b) the fusion of a number of cells into a multinucleate plasmodium or syncytium.

V. **Conjugation of individuals:** the union of unicellular organisms or of filaments of equivalent cells;

a) **in certain algae,** the formation of connections between pairs of laterally juxtaposed cells leading to the fusion of the cells and their nuclei;

b) **in ciliated protozoa** a temporary association of two individuals during meiosis and reciprocal cross-fertilization;

c) **in bacteria** (bacterial conjugation or bacterial mating), a parasexual mode of unidirectional (polarized) transfer of genetic information involving direct cellular contact between a donor ("male") and a recipient ("female") cell. The contact is followed by the formation of a cellular bridge physically connecting both cells. Subsequently a segment (sometimes all) of the male's chromosome may be transferred into the female and may undergo → genetic recombination with a homologous chromosomal segment of the recipient cell. The conjugation process invariably occurs between cells of different mating type (→ sex factor) with or without transfer of chromosomal material or, following transfer, with or without subsequent recombination. With chromosome transfer almost always a temporary diploid, partial zygote ("a → merozygote") is produced. Only those portions of the donor segment which following recombination become part of the recipient's → genome will participate in the subsequent replication processes.

The entire processes of bacterial c. may be separated into several distinct and successive stages (*Clark* & *Adelberg* 1962):

1. The formation of specific "mating pairs" after random collisions between cells of opposite mating types. Special genetic elements called → "conjugons" are required for the establishment of the cell-to-cell contact which is prerequisite for transfer and recombination of genetic material.

2. The supply of energy sources necessary for conjugation bridge formation at the site of donor pilus attachment to the recipient cell surface and the transfer of genetic information (→ chromosome mobilization).

3. The polarized transfer of at least a portion of a "male" chromosome into the recipient, the precise order of transfer depending on the donor strain used. DNA is synthesized both in the donor and in the recipient cell. One reformed strand of DNA is transferred (starting from the 5′ end) from the donor, the complementary strand is synthesized in the recipient.

4. The integration of some (or all) of the transferred material into the recipient's chromosome.

5. The segregation of the recombinant chromosome from the residual genetic information within the merozygote with the subsequent production of a recombinant clone.

conjugational DNA synthesis — the synthesis of plasmid DNA during bacterial (*E. coli*) conjugation mediated by F-like or I-like conjugative → plasmids. A specific single strand of DNA is transferred to the recipient bacterium where the complementary strand is synthesized. DNA homologous to the transferred material is synthesized in the donor upon the template of the remaining strand. The bacterial functions necessary for this c. DNA s. differ from the requirements for bacterial DNA synthesis.

conjugative plasmid — any F-like or I-like → plasmid which mediates bacterial → conjugation. A specific single strand of DNA is transferred during conjugation to the recipient bacterial cell where the complementary strand is synthesized. DNA homologous to the transferred material is synthesized in the donor upon the template of the remaining strand.

conjugon (*Luria* 1963) — any of a number of different genetic elements (→ plasmids) in bacteria that promote bacterial → conjugation and gene transfer. These include the → F plasmid, → colicinogenic factors (Col), and → resistance transfer factors (R). Conjugation due to the presence of F and certain R-factors is brought about by the synthesis of a special kind of pilus (→ sex pilus) at the bacterial surface which also acts as receptor for F-specific (male-specific) bacteriophages.

Cells with any one of these factors (donor cells) can conjugate with related cells (usually lacking such a factor) and during conjugation the conjugons are transferred from donor (male) to recipient (female). In some cases, the conjugons may transport from the donor to the recipient male chromosomal material that has become temporarily associated with the conjugons which are DNA elements (→ replicon).

Conjugons have three functions:

1. determine the surface properties and synthetic abilities which lead the formation of effective contacts between conjugation partners;

2. mobilize the "genote" to be transferred from donor to recipient;

3. provide the immediate energy source necessary for the transfer of chromosomal material.

Conjugons are transferable from cell to cell by → transduction via bacteriophage as well as by bacterial conjugation.

conjunctive segment (*Cooper* 1944) — = → collochore.

consanguinity — genetic relationship by descent from at least one common ancestor in the preceding few generations. The effect of consanguineous mating is to increase the probability that the offspring will be homozygous at any locus, i.e., will be on the average less heterozygous than the offspring of a non-consanguineous mating.

constitutive — of → enzymes (proteins) produced by the cell in fixed amounts independent of need, in constrast to → inducible and → repressible enzymes.

constriction (*Agar* 1911) — an unspiralized segment of fixed position in the metaphase chromosome.

1. **Nucleolar constriction** (*Darlington* 1937): A secondary constriction in the region of which the → nucleolus becomes organized (→ satellite).

2. **Primary or centric constriction** (*Darlington* 1937): A constriction associated with, and determined by the → centromere region.

3. **Secondary constriction** (*Darlington* 1929): Any noncentric constriction either nucleolar or non-nucleolar.

consultand (*Murphy* 1970) — the person whose genotype is of primary relevance to the → genetic counselling problem at hand. It may or may not be the same person as the counselee (frequently c. is used synonymously with counselee).

contact guidance — in cell culture, the elongation of cells and movement along fine surface grooves when cultured on glass (→ contact inhibition).

contact inhibition (*Abercromble & Heaysman* 1954) — in cell culture, the cessation of cell locomotion and/or growth when freely growing cells from a multicellular organism come into physical contact (on a plane substrate) with each other (→ contact guidance). C. i. requires the presence of actively metabolizing cells and probably involves inhibitory factors associated with the → cell membrane. As a result of c. i., cells in culture tend to form confluent monolayers with negligible overlap (→ contact retraction).

contact paralysis — the cessation of forward extension of the pseudopods of a cell as a result of its collision with another cell (→ intercellular invasion).

contact retraction (*Weiss* 1958) — in cell culture, intercellular collision in the form of an abrupt retraction of the contacting cells, often causing them to spring apart (→ contact inhibition).

contracomplementation (*Portin & Ruodhonen* 1972) — in *Drosophila melano gaster*, the lethality of certain heterozygous combinations of alleles of sex-linked genes, each allele being homozygous viable.

controlling element (*McClintock* 1956) — in eukaryotes, transposable elements which make themselves visible through abnormal control of the activity of standard genes. A c. e. may, in the simplest case, inhibit the activity of a gene through becoming integrated in, or close to, that gene. Occasionally, either in germinal or somatic tissue, it may be excised from this site and, due to this, the activity of the gene is more or less restored, while the element may become reintegrated elsewhere in the genome where it may affect the activity of another gene (*Fincham & Sastry* 1974). Usually the presence of c. elements (in maize and *Drosophila*) leads to a number of cytogenetic changes with at least three uniquely characteristic features: 1. the c. e. appears to be integrated at the site of the gene affected and manifesting the property of high "mutability"; 2. the kind or kinds of changes at a locus under c. e. regulation are particular and non-random; 3. the c. e. is capable of transposition from one locus to another, the latter located either on the same chromosome or on a nonhomologous chromosome.

Alleles expressing a mutant phenotype due to the presence of a c. e. are called controlled alleles. C. elements in eukaryotes are similar in many details of behavior to → temperate phages involved in → lysogeny.

In "one-element" systems, the c. e. acts autonomously with regard to its inhibition of gene action and its occasional transposition. In nonautonomous, "two-element" systems, the element inhibiting gene action (the socalled operator or receptor) is itself controlled by another element (the regulator), in some way complementary to it, located elsewhere in the genome; the excision of the first element and the release of gene activity then occurs only in the presence of the second element assumed to supply some missing excision function.

controlling gene — any → gene which controls the time and rate of action of → structural genes (→ operon).

conversion (*Winkler* 1930) — → gene conversion; phage conversion.

convertant — the product of → gene conversion.

convertogenic — of agents able to induce → gene conversion.

convivium — a group of genotypes within one → commiscuum, which is geographically isolated from other such groups.

coordinate repression (*Ames & Garry* 1959) — → repression.

co-orientation (*Darlington* 1937) — → centromere orientation.

copolymer — a polymeric molecule containing more than one kind of monomer unit (→ homopolymer).

copulation — the union of sexual units. In a broader sense, the sexual act. According to *Kniep* (1928) the following types of c. are to be distinguished:

1. **Gametic c.:** The fusion of differentiated, uninucleate → gametes.

2. **Gametic-gametangial c.:** The fusion of a differentiated uninucleate gamete (male or female) and a differentiated gametangium which produces no differentiated gametes of the opposite sex.

3. **Gametangial c.:** The fusion of two gametangia which can be distinguished morphologically as male or female. The gametangia do not differentiate gametes. One or many pairs of nuclei may participate in gametangial c.

4. **Somatic c.:** The fusion of undifferentiated vegetative cells.

copy-choice recombination (*Lederberg* 1955) — a hypothesis for the interpretation of intrachromosomal → genetic recombination which is not regarded as a physical exchange of preformed genetic strands (chromosomes, chromatids). Copy-choice implies the pairing of two parental genomes or their subunits (if more than one linkage group is present at the time of their replication). After a part of the genetic information of one strand has been copied, the copy process switches to the other of the paired strands and begins to incorporate the existing genetic information in the growing copy. If the parental strands contained different alleles, a recombinant chromosome is produced possessing some alleles of the one, some of the other. In this way the process leads to a nonreciprocal recombinant. If the copy process alternates between both parental strands, reciprocal recombinants may be produced. (→ partial replica hypothesis.)

copy error — an error in the → DNA replication process giving rise to a → gene mutation.

copy error lag (*Braun* 1965) — in the case of → copy errors giving rise to → gene mutations after incorporation of a fraudulent base into nucleic acid, time will elapse between the base incorporation and the permanent change in base sequence. This is called c. e. l. (→ phenomic lag; segregation lag).

co-repressor (*Jacob & Monod* 1961) — an → effector which activates a → repressor (→ operon).

correlated response (*Wigan & Mather* 1942) — a change in one → character (phenotype) occurring as an incidental consequence of → selection for a seemingly independent character.

correlation coefficient — a coefficient which measures the degree of association between two or more variables.

cortex (*Chambers* 1940) — a peripheral cytoplasmic region of animal cells (the "ectoplasm") including the → cell membrane by which the cells are bounded. It mediates the cell's interactions with its environment. In egg cells and *Paramecium*, the cortex has been proved to play an important role in early development and → differentiation (*Raven* 1961; *Sonneborn* 1963, 1964); it is a carrier of a system of developmental factors — the "cortical field" — which is of importance for the determination of early steps in the differentiation of the embryo from the fertilized egg cell. The cortical field is the seat of a system of morphogenetic factors which determine the polarity and symmetry of eggs. The developmental information in the fertilized egg is presumably stored in the nucleus, the cytoplasm as well as the cell cortex.

cortical granule (*Allen & Hagström* 1955) — any of the round to elliptical, membrane-bound bodies (approximately 0.5 to 0.8 µm in diameter) in the → cortex of animal oocytes, especially just before fertilization. They contain mucopolysaccharides and participate in the formation of the fertilization membrane by the cortical reaction, i.e., the ganules are released into the space which lies immediately outside the oocyte → cell membrane. C. granules are probably derived from the → Golgi apparatus and may be formed by coalescence of tiny vesicles.

cost of natural selection (*Haldane* 1957) — the amount of selective elimination which accompanies the process of substituting one allele for another in evolution. The process of substitution involves lowering of population → fitness and thus creates a →genetic load (substitutional or evolutionary load).

cotransduction — the → transduction of more than one genetic marker in a single event. This suggests that transducing bacteriophages can sometimes carry a segment of the bacterial chromosome long enough to transduce closely linked genetic loci concurrently.

Cot value (*Britten & Kohne* 1968) — an expression for the rate of DNA (or RNA) renaturation (annealing). The value is used for definition of genomic organization.

The Cot values are defined as $C_0 \times t$ (C_0 is the initial concentration of DNA expressed in moles of DNA phosphorus or moles of nucleotides per liter, and t is the time of renaturation under standard conditions in seconds). It is expressed on a logarithmic scale ranging from 10^{-4} to 10^{4} ($Å_{260}$ units/ml $\times$ hr/2 = Cot).

$Cot_{1/2}$ is defined as $C_0 \times t_{1/2}$ where $t_{1/2}$ is expressed as time for 50% renaturation. The unity of the C. v. are moles sec^{-1}, $liter^{-1}$, or alternatively absorbance units (generally read at 260 Å) $hours^{-1} \times 0.5$.

DNA renaturing at low Cot (10^{-4} to 10^{-1}) is composed of highly → repetitive sequences, DNA renaturing at intermediate Cot values is moderately repetitive (10^{0} to 10^{2}), and DNA renaturing at high Cot values is minimally or nonrepetitive.

coupling (*Bateson, Saunders & Punnett* 1905) — → cis-configuration.

covalent bond — a chemical bond in which electrons are shared in a molecular orbital involving both the bonded atoms.

covarion (*Fitch & Markowitz* 1970) — in a given → gene, those codons which are concomitantly variable resulting in favorable mutations, or in mutations leading to amino acid substitutions of little or no effect.

C-pair (*Levan* 1938) — → C-mitosis.

criss-cross inheritance (*Bridges* 1913) — → inheritance.

cristae (*Palade* 1953) — folds (crests or ridges) of the inner mitochondrial membrane (cristae mitochondriales) to which ATPase knobs are attached. → Mitochondria with c. are usually found in animal cells with intensive respiratory metabolism and are uncommon in plants.

cRNA — RNA molecules complementary to specific chromosomal DNA sequences (→ chromosomal RNA).

cron (*Huxley* 1957) — the time unit for evolutionary processes (1 cron = 1000000 years; 1 kilocron = 10^9 years; 1 millicron = 1000 years).

cross — in genetics, the bringing together of genetic material from different individuals (genotypes) in order to achieve → genetic recombination.

Genetic crossing is associated with a number of processes:

1. **In higher organisms** with sexual → reproduction, genetic crossing is achieved by the processes of nuclear fusion (karyogamy) and → meiosis (cross-fertilization).

2. **In bacteria** genetic crossing is achieved by the → parasexual phenomena of → conjugation, → tranduction, → sexduction, and → transformation.

3. **In viruses and bacteriophages,** genetic crossing is achieved by multiple infection of the host cells with particles of differing genotypes. If bacteria are infected simultaneously with 2 or 3 different virus particles these crosses are referred to as "biparental" and "triparental", respectively.

crossbreeding — = → outbreeding.

cross-feeding — → syntrophic.

cross-fertilization — → fertilization.

cross-induction (*Borek* & *Ryan* 1958) — → induction.

crossing barrier — any of the genetically controlled mechanisms which either entirely prevent or at least significantly reduce the ability of individuals of a population to hybridize with individuals of other populations. The formation of c. barriers is a fundamental by-product of natural → selection altering the genetics of reproduction and setting a population on its own evolutionary pathway.

crossing-over (*Morgan* & *Cattell* 1912) — the events leading to → genetic recombination between linked markers in both pro- and eukaryotes. Formally, it is the reciprocal exchange of segments at corresponding positions along pairs of homologous linkage units (chromosomes) by symmetrical breakage and crosswise rejoining.

In eukaryotic organisms, crossing-over may be meiotic or mitotic:

1. **Meiotic crossing-over:** The exchanges are observable microscopically in suitable material by the formation of → chiasmata between homologous chromosomes. In a population of cells going through meiosis, the incidence of exchanges between any two given points in one chromosome pair is highly correlated with the physical distance between these two points (→ genetic map). Genetically, the process of crossing-over is inferred from the proportion of recombinant individuals among the progeny of heterozygotes. From the fraction of recombinants one can calculate the fraction of recombinant gametes from which those individuals originated (→ crossing-over value; recombination value). In haploid organisms, the fraction of recombinant individuals and the fraction of recombinant gametes coincide.

Any model of meiotic crossing-over must satisfy three essential requirements:

a) Recombination is always reciprocal (Fig. 23),
b) each event involves only two of the four chromatids (normally non-

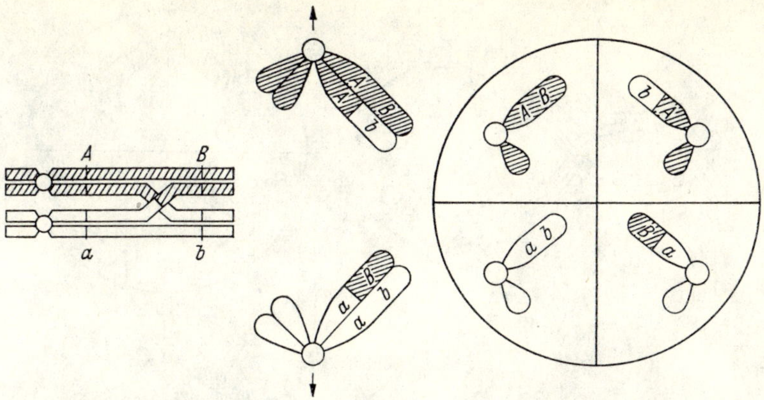

Fig. 23. Meiotic crossing-over (exchange of nonsister chromatid segments between paired homologues), chromosome disjunction in anaphase I and the genetic constitution of the four haploid products which result at the end of the second meiotic division.

sister chromatids) of two homologous chromosomes and therefcre only two of the four products of meiosis,

c) double or multiple crossing-over ("compound crossing-over") may involve all four, three, or only two of the four chromatids. If two crossing-over events within homologous linkage structures involve the same two chromatids (→ coincidence; interference), they are said to be "regressive" ("two-strand exchange", Fig. 24a). If they involve completely different pairs of chromatids, they are said to be "digressive" ("four-strand exchange", Fig. 26b), and if they involve one common chromatid they are referred to as "progressive" ("three-strand exchange", Fig. 24c and d).

C.-o. involves the breakage and reunion of complementary DNA strands (in eukaryotes, probably during the genesis of the → synaptonemal complex). The enzyme mechanism includes single-strand breakage, strand digestion, end-to-end joining, and repair. These activities are prerequisite to hybrid formation (→ polaron hybrid DNA model), and accomplishment of reciprocal recombination.

Special types of crossing-over are called "unequal" and "illegitimate". **Unequal crossing-over** (*Sturtevant* 1925) produces one chromatid containing a gene twice (duplication) and another lacking that gene (deletion). **Illegitimate crossing-over** (*Darlington* 1932) gives rise to secondary structural changes in heteromorphic (structurally heterozygous) chromosomes and takes place in haploid or polyploid individuals between homologous segments or homoeologous (= partially homologous) chromosomes.

2. **Mitotic or somatic crossing-over** (*Stern* 1936): Crossing-over at mitosis in either somatic or gonial tissue which like meiotic crossing-over takes place in the four-strand stage of chromosomes (Fig. 25).

The consequences of somatic crossing-over depend upon:

a) the location of the crossing-over site, and

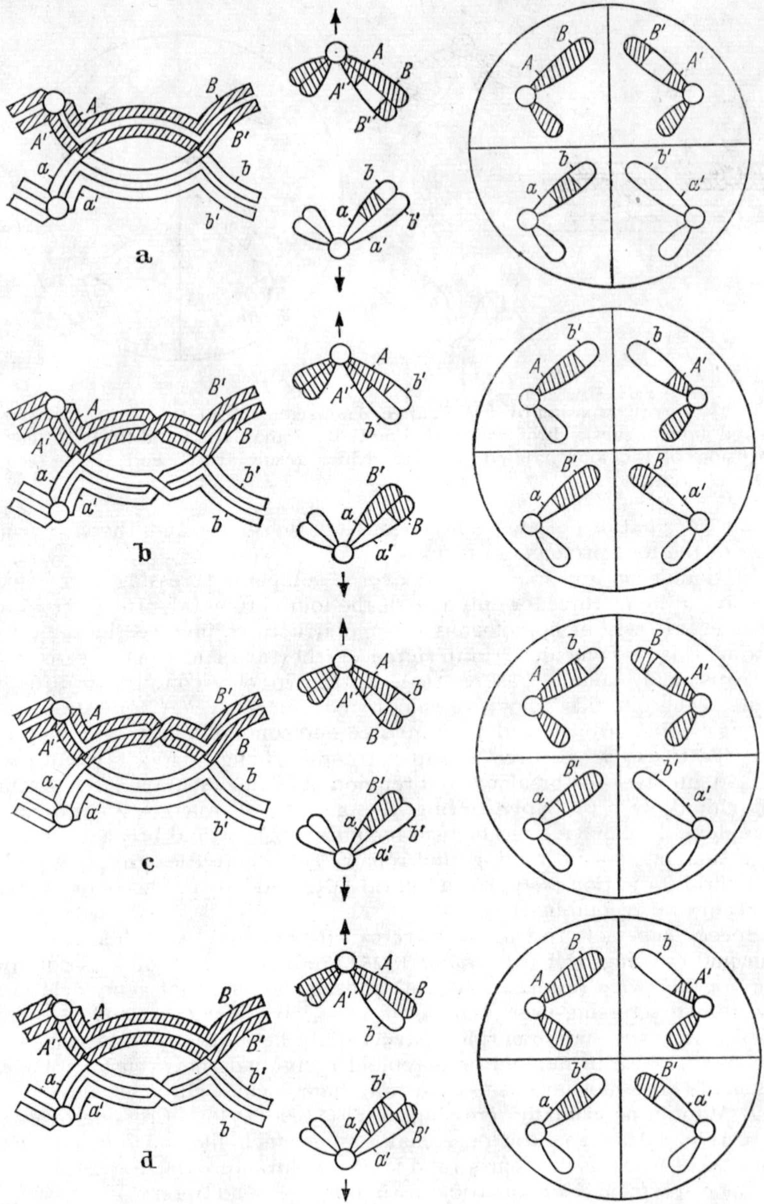

Fig. 24.

Fig. 24. Diagrammatic representation of meiotic double crossing-over and the sequences of chiasmata, chromosome disjunction in anaphase I and the genetic constitution of the four haploid products which result at the end of the second meiotic division. The two crossovers are regressive (reciprocal chiasmata) in a, digressive (complementary chiasmata) in b, and progressive (diagonal or disparate chiasmata) in c and d.

---

b) the mode of orientation and, thus, distribution of the centromeres of the chromosomes engaged in somatic crossing-over. A single crossover between two linked heterozygous loci (A/a and B/b) and distribution of the centromeres 1 + 2 to one and 1′ + 2′ to the other cell pole in anaphase results in homozygosity for genes distal to the point of exchange (B/B and b/b) for which the genotype was originally heterozygous (B/b). This event is detectable phenotypically by the production of a b/b-spot (Fig. 25 a), provided that it is the right part of the body and that the expression of the gene is autonomous.

With a single somatic crossover between the centromeres and the first of the two heterozygous loci a twin patch aa and bb may result if the centromeres 1 + 2 are distributed to one and 1′ + 2′ to the other cell pole (Fig. 25 b).

Patterns of distribution of the type 1 + 2′ (one pole) and 1′ + 2 (other pole) leave the original heterozygosity unchanged.

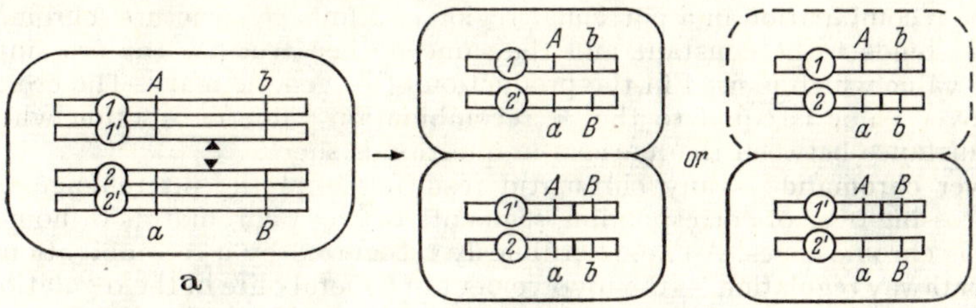

Fig. 25. Diagram showing somatic (mitotic) crossing-over and its genetic consequences in a double heterozygote Ab/aB. a) Crossing-over between the two loci which results in a bb-spot; b) Crossing-over between the centromere and the first locus which results in twin spots (in one of the two spots the locus has become homozygous for a, in the other for b).

crossing-over homozygote (*Renner* 1941) — → complex heterozygous.

crossing-over map — a → genetic map made by utilizing crossing-over frequency as a measure of the relative distances between genes in one linkage group.

crossing-over modifier — any gene or chromosomal structural change increasing or decreasing the crossing-over frequency. Generally, reduction in crossing-over is easier to achieve than its enhancement. With rare exceptions all structural heterozygotes ("heterokaryotypes") have lower crossing-over values for regions included in or immediately adjacent to the structural change. A compensatory increase often takes place in those segments farther removed from the aberration.

crossing-over position interference (*Whitehouse* 1965) — = → chiasma interference.

crossing over potential (*Sybenga* 1965) — the basic number of cross-overs of the chromosome segment under observation, which is only realized in the absence of → interference.

crossing-over suppressor — any gene or heterozygous structural change of chromosomes suppressing or reducing (in most cases locally) the frequency of meiotic crossing over (= C-factor). (→ crossing-over modifier).

crossing-over unit — a frequency of exchange of one percent between two pairs of linked genes; one percent of crossing over is equal to one unit on a linkage map (= → map unit).

crossing-over value — the frequency of crossing-over between two linked genes (→ markers). For a fixed set of genetic and environmental conditions, recombination in a particular region of a linkage structure (chromosome) tends to be constant and the same is then true for the crossing-over value which is used in the production of → genetic maps. The crossing-over value is equal to the → recombination value or fraction when the distance between the markers in question is short.

crossover chromatid — any chromatid resulting from the interchange by → crossing-over of corresponding segments between chromatids of homologous chromosomes. A c. c. is ordinarily recognized by a recombination.

cross-pathway regulation — the involvement of a metabolite in the regulation of enzymes (genes) of an unrelated pathway (= supraoperon control), as opposed to cases in which only those genes are regulated that are concerned with the synthesis of the regulatory metabolite (→ operon; regulon).

cross reacting material — a protein produced by mutant genes that is enzymatically inactive but shows serological properties similar to the protein of the not mutated (wild-type) gene (symbol CRM). Mutants producing CRM are referred to as CRM positive (CRM^+), mutants that do not are called CRM negative (CRM^-).

crossreactivation (*Luria* 1947) — the reappearance in the phage progeny of genetic markers of inactive (irradiated with UV or ionizing radiation) bacteriophages after the mixed infection of a bacterium with inactive and active phages that differ from each other in one or more of their genetic loci (= "marker rescue"). The interpretation of c. (symbol CR) is as follows: If an irradiated phage genome enters a bacterial cell simultaneously with a fully active, unirradiated phage genome which is able to supply the necessary physiological functions required for the production of phage progeny, then gene loci residing in lesion-free regions of the irradiated genome can be rescued from the damaged genome by → genetic recombination with the replicating viable phages (*Doermann* et al. 1955).

The probability of c. of a particular marker decreases with increasing radiation dose. The probability of c. of two linked markers after irradiation with a specific radiation dose increases with decreasing distance between them.

In fact c. is not → reactivation of inactivated phages, but rather the incorporation by recombination of parts of an inactivated phage genome into a viable one.

cross-sterility — the failure of fertilization because of genetic or cytological conditions (→ incompatibility) in crosses between pairs of individuals despite the formation of normal, functional gametes.

cryptic — of 1. → structurally heterozygous individuals not identifiable on the basis of abnormal meiotic → chromosome pairing configurations ("cryptic structural hybrids");

2. a form of → polymorphism controlled by recessive genes ("cryptic polymorphism");

3. any mutation which is exposed by a → sensitizing mutation and otherwise poorly detected (such mutations probably escape detection because of the plasticity of composition of the corresponding polypeptide) (*Koch & Drake* 1970);

4. phenotypically very similar → species (cryptic species) which do not hybridize under normal conditions.

cryptochiasmate (*White* 1974) — of a type of → meiosis characterized by delayed opening-out of → bivalents in spermatogenesis. Under such circumstances, the chiasmata are revealed only for a brief period of time just before anaphase I (→ achiasmate).

cryptochimera — a → chimera which is outwardly recognizable as such only after regeneration.

cryptoendomitosis (*Bauer* 1952) — an → endomitosis which takes place in the interphase nucleus and shows none of the stages comparable to mitosis. The polytene → giant chromosomes of Diptera are produced by c. .

crystal-containing body (*Thornton & Thimann* 1964) — any of the membrane-bounded organelles in plant cells containing crystalloids. They are predominantly localized in storage cells and other metabolically less active cells and may represent a specialized type of → microbody.

C-terminus — that end of the peptide chain which carries the free alpha-carboxyl group of the last amino acid (→ N-terminus; genetic translation).

culture alteration — any persistent change in the properties or behavior of cultured cells (→ cell culture; tissue culture), e.g., altered morphology, chromosome constitution, virus susceptibility, nutritional requirements, proliferative capacity, malignant characters, etc. (*Feodoroff* 1967).

cumulative genes (*Nilsson-Ehle* 1911) — = → polymeric nonallelic genes (→ gene interaction).

cuticle — (*Brongniart* 1834) — a lipophilic layer of material superimposed on to the ordinary epidermal → cell walls in the outer parts of the plant. From the exterior to the interior of the cell wall, a distinction may be made between the epicuticle, the cuticle proper, and the cuticular layer.

C value (*Swift* 1950) — in eukaryotes, the DNA quantity per → genome, i.e., per chromosome set. The DNA content of diploid nuclei is usually referred to as the 2 C value. In cells which are about to divide by → mitosis, the amount of DNA is gradually doubled (during the S-period of → interphase) until it reaches the 4C value. Haploid nuclei, e.g. those of sperm and egg nuclei or the spore nuclei in higher plants, will contain the 1C amount of DNA.

cyclic adenosine monophosphate — a regulatory compound (adenosine-3′, 5′-cyclic monophosphate) involved in the regulation of the activity of enzymes, hormones and genes. It acts as an intermediate (→ second messenger) for the action of hormones on cells in target organs. It is also involved in many genetic control systems and able to reverse → catabolite represssion (→ catabolite activator protein). cAMP is found in cells as diverse as bacteria and animal cells and is produced from adenosine triphosphate (ATP) by action of adenylate cyclase.

cyst (*Davis* 1908) — 1. a group of → spermatogonia, spermatocytes, or early spermatids of common origin and surrounded by a septal or cyst membrane;

2. any enveloping structure, usually a membrane;

3. a resting spore.

cytaster (*Flemming* 1882) — one of the aster-like figures containing the → centrioles which is formed in the cytoplasm outside the nucleus prior to the onset of mitosis and meiosis (→ mitotic center).

cytocatalytic (*Lewis* 1967) — of a mode of → evolution initiated by an abrupt gross mutation, usually doubling of the chromosome complement or, perhaps less commonly, the gain or loss of individual chromosomes.

cytochimera (*Derman* 1947) — a → chimera in which different tissue sectors or tissues differ in chromosome number (= "chromosomal chimera").

cytodifferentiation — the process by which during the development of the individual from the fertilized ovum specialized cells, tissues, and organs are formed. C. consists in the progressive restriction of the developmental potential and in the specialization of function. A biologically equivalent process takes place in some unicellular organisms in which part of the cell may undergo morphological and functional specialization.

Among the important characteristics of c. are (*Grobstein* 1959):

1. Initiation of change; 2. stabilization of change after cessation of stimulus; 3. the efficacy of many substances, exogeneous and endogeneous, as inductive stimuli; 4. a limit of five or six as the number of cell types which may differentiate directly from any cell type; 5. progressive limitation in the number of developmental pathways; 6. restricted periods during which a cell is competent to respond to an inductive stimulus; 7. the discreteness of cell types produced in c.; 8. a requirement for a minimal and preferable heterogeneous cell mass to initiate c. in many instances, and to maintain it in some.

A prerequisite for an orderly c. is an overall regulatory program (→

messenger recognition factor) which insures that the developmental alterations

1. occur in a definite chronological order,
2. are spatially restricted to certain cells of the multicellular assembly, and
3. are of definite extent (→ modulation).

Two aspects of c., called sequential and branching c., may be distinguished:

1. Sequential c.: one cell type gives rise to one single but different cell type although the total number of cells may have increased by mitosis.
2. Branching c.: a single cell type yields a multiplicity of cell types.

In all types of c. both the cell's inner controls and influences from its environment play a role. The environment is assumed to trigger the activation of genes or gene products previously inactive, or to inactivate genes that were previously active (differential → gene activation). The pattern of genetic expression in differentiated cells can be inferred from the cellular distribution of RNA and proteins. Branching c. may be due to environmental influences acting unequally on identical cells or due to non-random segregation of cytoplasmic elements to daughter cells during mitosis, thought to cause unequal positioning of inducing material to the two daughter cells.

According to the model of differential gene activation the basic instructions contained in the genomes of differentiating cells remain unchanged and the differentiated cells are genetically equivalent to each other and to cells in the germ line. In some specific cases, c. has been found to be accompanied by elimination of certain chromosomes or chromosome segments (→ chromosome elimination).

cytogamy (*Wichterman* 1940) — the fusion or → conjugation of cells.

cytogenetics (*Sutton* 1903) — a field of investigation which developed from the (originally) separate sciences of → genetics and → cytology and is concerned with problems based on the correlation of genetic and cytological (especially chromosomal) features characterizing a particular genetic system under investigation. Essentially, the field of study comprises the behavior of the chromosomes during mitosis and meiosis, their origin and their relation to the transmission and recombination of genes. Modern c. ("molecular cytogenetics") has a distinctly chemical accent.

cytogony (*Hartmann* 1904) — the → reproduction by single cells (gametes or agametes) (→ somatogony).

cytohet (*Sager* 1973) — a cell containing two cytoplasmic genomes which differ in one or more genes (= cytoplasmic heterozygote). Replicative dominance is the term proposed for the emergence from a c. of one of the two parental alleles as sole or preferential product.

cytokinesis (*Whitman* 1887) — the process of division or segmentation of the cytoplasmic portion of a cell and the segregation of daughter nuclei to separate cells, as distinct from nuclear division or karyokinesis; the cell becomes divided into two or more (multiple division) cells. During c. the cytoplasmic components are distributed to the daughter cells.

C. represents a basic process of development and, as a rule, proceeds in conjunction with nuclear division or karyokinesis (→ mitosis, meiosis). In certain cases, c. is omitted after nuclear division and multinucleate plasmodia or syncytia are formed.

The period of c. constitutes only a small interval of the → cell cycle in most organisms. C. is regulated by the → mitotic apparatus. The plane of division is correlated with the equatorial plane, the position of the mitotic apparatus; it is usually perpendicular to the center region of the → spindle.

The mechanism of c. differs in animal and plant cells (Fig. 26). The former generally divide by furrowing (→ cleavage) which is initiated by relaxation of the polar → cortex. Owing to the concentration of an actinomycin-like contractile substance in the gelated cortex, the cells become rounded off before the division. Bond cleavage in ATP acts as energy source for the contraction. The cytoplasm constricts; this constriction becomes accentuated and deepened until division is complete. Formation of the new → cell membranes takes place in the stalk at the base of the furrow (*Roberts* 1961).

The following types of furrowing may be distinguished (*Mazia* 1956):

1. Furrowing at the equator (egg cells in animals) which may take place simultaneously around the circumference or may begin to one side.

2. Furrowing displaced towards one pole (occasional animal cells) often associated with asymmetry of asters or centrosomes (smaller aster towards the pole where the small cell will form).

3. Furrowing accompanied by pulling away of daughter cells (many types of animal cells, especially ameboid cells) sometimes with a stalk between daughter cells.

4. Furrowing along an irregular path that seems to be determined by surface differentiation of the cell (some ciliate protozoa).

In plant cells, the formation of the → phragmoplast and the → cell plate leads to c. without furrow formation. The movements of the → endoplasmic reticulum and the → dictyosomes, and the fusion of vesicular material at the equator are important components of normal c. (→ phragmosomes).

The term "equal" c. is used when the mother cell divides into two daughter cells which show a close agreement in their sizes. "Unequal" c. describes the process when qualitative and/or quantitative differences in the daughter cells are produced in the course of c.

cytokinin — a class of substances which, regardless of other activities, promote → cytokinesis in plants. Cytokinins, together with other plant growth substances, play a major role in ensuring orderly growth and development. They are composed of one hydrophilic group of high specificity (adenine) and one lipophilic group without specificity. The structure needed for cell division stimulating activity is that of adenine bearing on the amino group in the 6 position a non-polar side chain.

C. activities include: 1. induction, promotion, and regulation of DNA, RNA, protein and thiamine biosynthesis; 2. regulation of organ formation, apical dominance, and branching; 3. enhancement of flowering and of seed germination; 4. regulation of phloem transport and mobilization of metab-

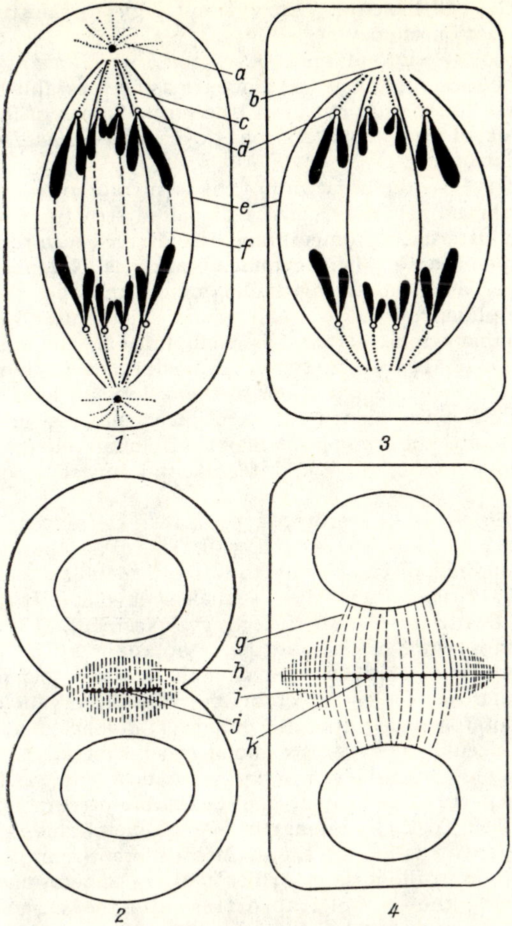

Fig. 26. Diagram illustrating the organization of the mitotic spindle and the different modes of cytokinesis in animal (1 and 2) and plant cells (3 and 4). a) centriole; b) polar cap or polar center; c) chromosomal spindle fiber; d) centromere; e) continuous spindle fiber; f) interzonal connection; g) and h) spindle during cytokinesis; i) phragmoplast; j) intermediate bodies; k) cell plate; 1 and 2: anaphase and telophase, respectively, with cytokinesis of the constriction (furrowing) type; 3 and 4: anaphase and telophase, respectively, with cytokinesis by formation of a cell plate (from *de Robertis*, *Nowinski*, *Saez* 1965).

olites; 5. preservation of flowers, fruits, vegetables, and leaves through prevention of senescence.

cytology — the study of structure, function, development, reproduction, and life history of cells (= cell biology). "Enzyme cytology" is the study of the intracellular localization of enzymes.

cytolysosome (*Novikoff* 1960) — any enlarged → lysosome which also contains mitochondria or other cell components. Since it has been established that these bodies do not function in cytolysis, the terms "areas of focal degradation" (*Hruban* et al. 1963) or "autophagic vacuoles" (*de Duve* 1963) are more appropriate.

cytomembrane (*Sjöstrand* 1955) — the basic unit of the → membrane system of the cell (→ **unit membrane**)). It may consist of paired structures separated by a space called cisterna. Cytomembranes are divided into rough (granular) surface cytomembranes (**alpha cytomembranes,** as exemplified by the → cell membrane), and smooth (agranular) cytomembranes (**beta cytomembranes,** as exemplified by the → endoplasmic reticulum and the → nuclear envelope; **gamma cytomembranes,** as exemplified by the membrane system of the → Golgi apparatus). Beta cytomembranes are membranes which appear to represent invaginations of the cell membrane.

cytomixis (*Gates* 1911) — the phenomenon of extrusion or passage of → chromatin from one cell into the cytoplasm of an adjoining cell (as in pollen mother cells, meristematic, tapetal, integumental, nucellar, and ovary cells).

cytoplasm (*Strasburger* 1882) — all the protoplasm of a → cell excepting the → nucleus ("karyoplasm") and the → cell wall. The c. is separated from the external surroundings of the cell by the → cell membrane and from the nucleus of eukaryotic cells by the → nuclear envelope. It is a carrier of heredity in its own right (→ inheritance; cytoplasmon). It may be subdivided into the cytoplasmic matrix and the cytoplasmic organelles.

1. **Cytoplasmic matrix:** The polyphasic, colloidal "ground plasm" excluding all electron microscopically detectable cytoplasmic organelles and particles. It is a heterogeneous system composed of long macromolecular chains and molecular aggregates; among other substances, it contains numerous enzymes, metabolic reserve substances and, in its aqueous phase, the main part of the cellular pool of soluble precursors.

2. **Cytoplasmic organelles:** To these belong the → cell membrane, the → endoplasmic reticulum which divides the c. into different compartments, the → plastids, → mitochondria, → ribosomes, → spherosomes, → lysosomes, → centrioles, the → Golgi apparatus, also the vacuoles and the → tonoplast.

Most of the metabolic and biosynthetic functions of the cell are completed in the c. It is the substrate for → gene action and exerts a regulatory effect on the expression of genetic information. In the course of → embryonic development it is organized into different, organ-forming regions occupied by equivalent nuclei (→ cytodifferentiation). The nucleus and c. are so interdependent that an isolated nucleus without c. is as incapable of prolonged living as is a c. without nucleus.

cytoplasmic inheritance — → inheritance of → characters or traits whose determinants are not located on the chromosomes (= extrachromosomal or extranuclear inheritance). (→ cytoplasmon.)

cytoplasmic matrix — → cytoplasm.

cytoplasmon (*Renner* 1929) — all cytoplasmic hereditary determinants (plasmids, plasmogenes or nonchromosomal genes) excepting those localized in the → plastids (→ plasmon; plastom) and → mitochondria (→ chondriome).

cytopon (*Cone* 1969) — a cytoplasmic bridge connecting two or more cells which represent a → syncytium. Cytopons may arise *in vitro* and *in vivo* as a result of incomplete → cytokinesis following mitosis.

cytosegresome — an area of cytoplasm which has become screened off by a membrane from the remainder of the cell (= cytolysosome). The segregated portions may either be derived from the cell's own cytoplasm (autophagy), or may stem from other cells (heterophagy). Cytosegresomes are involved in lysosomal digestion (→ lysosome).

cytosis (*Novikoff* 1961) — a general term for cellular ingestion processes by pinocytosis or phagocytosis. "Endocytosis" for intake and "exocytosis" for extrusion were proposed by *de Duve*.

cytosome (*Lundegårdh* 1922) — a nonspecific term for any of a number of dense and polymorphous bodies in electron micrographs. Cytosomes are membrane-bounded granules in the cytoplasm of the cell and probably represent → lysosomes, → microbodies, → multivesicular bodies, and → spherosomes.

cytostatic → of any physical or chemical agent capable of inhibiting cell growth and cell multiplication.

cytosterility — cytoplasmic male-sterility with maternal inheritance. It involves the interaction of genes for male-sterility and cytoplasmic factors.

cytotaxis (*Sonneborn* 1963) — a process which specifies the arrangement of macromolecules into organelles and these into supra-organellar structures. These arrangements serve as sources of information for the assembly and organization of similar structures.

cytotaxonomy — the study of the natural relationships of organisms by a combination of cytology and taxonomy.

cytotubulus (*Ledbetter & Porter* 1963) — a delicate, unbranched tube (diameter 23—27 nm) of unspecified length. Cytotubuli are usually evenly spaced in the cortical plasm (→ cortex) of the cell. Their function is still unclear. The outer layer of each c. consists of an annular bundle of lengthwise oriented elementary fibrils. The fine structure corresponds with that of the → undulipodia. Cytotubuli of somewhat smaller diameter are detectable in the region of the → spindle during unclear division and are identical with the spindle fibers.

cytotype (*Müntzing* 1953) — any variety (race) of a species whose chromosome complement differs quantitatively (in chromosome number) or qualitatively (in chromosome structure) from the standard complement of that species.

D

dam — the female parent (applicable to mammals) (→ sire).

Danforth equilibrium — ref. to a theory which states that the frequency of occurrence of a given mutant allele within a population is equal to the frequency with which the allele arises by → mutation, multiplied by the average number of the generations in which it may manifest itself before it becomes eliminated owing to its unfavorable action.

dark holding recovery (*Patrick* et al. 1964) — in yeast, an increase in viability of UV-irradiated wildtype cultures due to storage in water or saline for periods of up to eight days. D. h. r. is assumed to result from the corrective action of intracellular enzymes upon UV-induced lesions prior to DNA replication and cell division in these cells.

dark reactivation = → dark repair.

dark repair — any repair of DNA lesions which does not require light (= dark recovery). The basic types of d. r. are → excision repair, → post-replication repair, → reinitiation recovery, and replicative repair (→ photo-reactivation).

Darlington rule (*Darlington* 1937) — according to this rule which governs chromosomal sterility, the fertility of allopolyploid forms (→ alloploid) is inversely proportional to the fertility of the hybrid from which they arose. Sterile diploid hybrids with little or no chromosome pairing (→ chromosome pairing) give rise to fertile allopolyploids which form bivalents almost exclusively during meiosis.

Conversely, allopolyploid species produced from diploid forms which form bivalents exhibit an irregular chromosome distributin (→ centromere orientation) due to the formation of → multivalents which often leads to sterility.

Darwinian evolution — → evolution.

Darwinian fitness — operationally, the average contribution which the carriers of a → genotype or of a class of genotypes, make to the → gene pool of the following generation relative to the contribution of other genotypes. D. f. (= adaptive value) is a function of both the genotype and the environment (*Dobzhansky* 1962).

dauermodification (*Jollos* 1921) — a character change usually induced by extreme environmental conditions which survives for a long time in the vegetative or generative offspring in the absence of the inducing stimulus. It gradually weakens, however, and finally disappears. By reciprocal crossing, it has been shown that the persistence of d. is effected by the cytoplasm (= persistent modification).

daughter chromosome — any of the two chromatids (sister chromatids) of which the replicated chromosome consists after mitotic metaphase or meiotic anaphase II.

daughter nucleus — any of the two cell nuclei resulting from a nuclear division (→ mitosis; meiosis).

deactivation (*Muller* 1954) — → photoreactivation.

decondensation stage (*Marquardt* 1937) — a stage of → mitosis between interphase and prophase characterized by a short term decondensation of the heterochromatin.

deconjugation — = → desynapsis.

dedetermination — the loss of the → determination of cells which consequently recover their totipotent state.

dedifferentiation (*Butler* 1933) — the loss of recognizable specializations that define a differentiated cell (→ cytodifferentiation) and equip it for a particular array of functional activities. D. does not necessarily imply that any reversal of the process of differentiation has taken place. It is frequently observed when differentiated cells are explanted and grown in cell culture.

dedikaryotization (*Mittwoch* 1954) — = → dediploidization.

dediploidization (*Buller* 1941) — in Basidio- and Ascomycetes, the formation of haploid cells or hyphae from a dikaryotic diploid mycelium or a dikaryotic diploid cell (→ dikaryon). *Mittwoch* (1954) used the term "dedikaryotisation" in the same sense to avoid the concept "diploid"(→ haploidization).

defective virus (*von Magnus* 1954) — a virus particle characterized by the following features: 1. a normal viral structural protein; 2. loss of part of its genome (deletion mutants); 3. capability of reproduction only in the presence of a → helper virus; 4. interference with the intracellular replication of non-defective homologous virus (defective interfering virus).

deficiency (*Bridges* 1917) — a structural change (→ chromosome mutation) resulting in the loss of a terminal acentric chromosome, chromatid, or subchromatid segment and the → genetic information which it contains (Symbol Df) (→ deletion).

deficiency-exaggeration — → exaggeration.

degeneracy (*Crick* 1963) — → genetic code.

degree of genetic determination — the fraction of the total variance (→ genetic variability) that is genetically determined.

delayed inheritance (*Boycott & Diver* 1923) — → inheritance.

deleterious — of mutations which in wild species lower the → fitness of their carriers to survive or to reproduce and in cultivated species impair the traits considered desirable by the breeder (*Dobzhansky* 1952).

deletion (*Painter & Muller* 1929) — a structural change (→ chromosome mutation) resulting in the loss of a section of the genetic material and the genetic information contained therein from a chromosome (linkage structure) of pro- or eukaryotes. In eukaryotes, the loss of an intercalary chromosomal segment may be distinguished from the loss of a terminal chromosome region by applying of the term deletion to the former and deficiency to the latter. In practice, deletion is frequently used for both types of structural change. In both cases a centric and an acentric fragment is produced in those eukaryotes whose chromosomes possess a localized → centromere (Fig. 27).

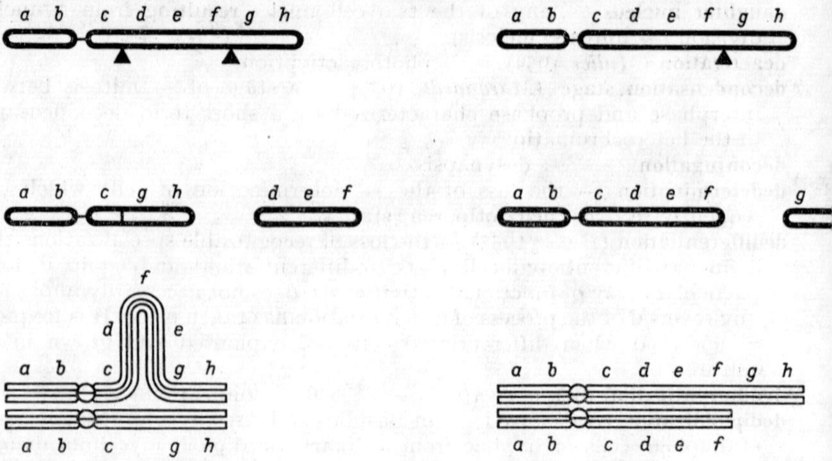

Fig. 27. Intercalary (left) and terminal (right) deletions and the meiotic paring configurations in karyotypes heterozygous for the deletions (bottom).

The size of the deleted material can vary from a single nucleotide to sections containing a number of genes or whole chromosome regions. Evidence for a deletion may be obtained by genetic or cytological (microscopic) methods.

The existence of relatively large deletions in the chromosome complement of eukaryotes may be proved cytologically by the occurrence of centric and acentric fragments in mitosis or by the absence of regional pairing in first meiotic prophase and in (somatically paired) → polytene chromosomes of Diptera in that pair of homologues which is heterozygous for the segment lost.

Genetic proof of a d. rests on the operational definition of deletions as mutants which do not back-mutate to the original form and do not recombine in genetic crosses with two or more point mutations which do back-mutate and recombine (→ deletion mapping).

The genetic consequences of deletions are primarily due to the loss of genetic information and secondarily due to quantitative changes in the genotype as well as changes of the genetic balance. Depending upon the size of the deleted chromosome segment, a d. may act as a recessive or dominant → lethal.

deletion mapping — the genetic localization (mapping) of the positions of deletions in the linkage structures (chromosomes) of eu- and prokaryotes. The genetic positions of deletions in eukaryotes may be estimated without extensive linkage studies, simply according to whether or not they phenotypically "disclose" certain recessive mutations. All segmental losses which permit the manifestation of a recessive mutation in the heterozygous state must at a minimum include the locus of this mutation.

Deletion mapping in prokaryotes exploits the following principles:

1. 3 mutants (a, b and c), each differing from the wildtype by deletion of a part of the linkage structure are tested for recombination.

2. If two mutants (a and c) mutually recombine and yield the wildtype (+), but neither recombines with the mutant b, then 3. these results may be described by a matrix based upon the results of the particular recombinations tests (+ = recombination; o = failure of recombination):

	a	b	c
a	o	o	+
b	o	o	o
c	+	o	o

Recombination results permit a unique ordering of the three mutants in the linkage structure. In this case, the deletions fall in the serial order a, b, c such that a and b overlap on one side and b and c on the other:

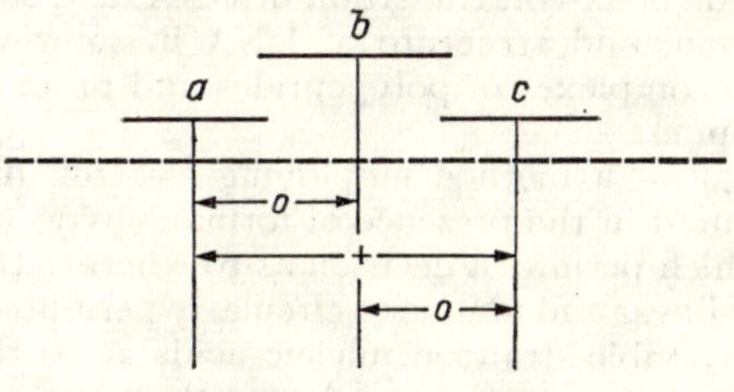

deme (*Gilmour & Gregor* 1939) — a local interbreeding group ("panmictic unit") consisting of a series of individual genotypes. The term is always used with a prefix which characterizes the d. more precisely. Among others the following combinations are possible:

1. with regard to the spatial relationships:

topodeme: a group of individuals existing in a certain geographic region (e.g. all individuals of *Armeria maritima* in the region of the British Isles);

ecodeme: a group of individuals associated with a specific habit (e.g. a particular mountain peak);

2. with regard to character differences:

phenodeme: a d. differing phenotypically from others, assuming that it is not yet experimentally possible to establish whether this is due to genotypic or environmental factors;

genodeme: a d. differing from others in genotype; it needs not also be characterized by divergent phenotypic characters;

plastodeme: a d. differing phenotypically from others owing to the effect of the environment;

3. with regard to propagation relationships:

gamodeme: a d. whose individuals are so distributed spatially and temporally that within the limits of the mating system they can all mate with one another (= "breeding population"). The suffix N% is used for a more precise, quantitative characterization. Thus, N%-gamodeme means that each individual has N% chance of mating within the d.;

autodeme: a d. composed predominantly of autogamous individuals;

endodeme: a "gamodeme" of predominantly endogamous (→ dioecious) plants (or bisexual animals);

agamodeme: a d. of predominantly apomictic plants or asexual animals;

clonodeme: a d. with predominantly vegetative reproduction;

4. with regard to trends in variation:

clinodeme: several d. which together form a gradient (→ cline).

Secondary derivatives may be made from these primary word forms by assembling a series of prefixes in the order corresponding with the required meaning. This allows a more exact characterization of the d.; thus: e.g. "phenotopodeme" (a "topodeme" which differs phenotypically from others).

denaturation — the loss of the native configuration of a macromolecule resulting from specific treatment (e.g., heat treatment in thermal d.). D. is extensively used to study the helix-coil transition in DNA, and as a probe of the stability, composition, and structure of DNA in solution. The same technique is applied to complexes of polypeptides and proteins with DNA (→ denaturation mapping).

denanturatio mapping (*Inman* 1967) — a method employing electron microscopy of DNA partially denatured in the presence of formaldehyde (to keep regions from renaturing), which permits a decision as to whether the gene sequence in a viral genome is linear and unique or circularly permuted, and the characterization of viral double-stranded nucleic acids as to the distribution of specific nucleotide sequences rich in adenine-thymine base pairs (these melt at lower temperatures).

The DNA molecules are photographed in the electron microscope and the d. map is constructed by measuring the location of denatured regions along each molecule. The denaturation map so constructed charts the extent of denaturation along the molecule. The first sequences to denature have the lowest G + C content. Compared with → deletion mapping, d. m. requires no second (mutant) DNA. D. m. identifies genes by using their characteristic denaturation map as a fingerprint.

deoxyribonuclease — any enzyme that degrades → deoxyribonucleic acid. Endonucleases produce breaks within the DNA strands; exonucleases remove nucleotides from the ends of strands.

deoxyribonucleic acid — a polymer of subunits called deoxyribonucleotides which is the primary genetic material of all cells, i.e., the material equivalent to → genetic information (symbol DNA). DNA of prokaryotic and eukaryotic organisms is normally found in a double-stranded state (*Watson-Crick* double helix), yet several important biological processes (replication of DNA, → genetic transcription, → genetic recombination, repair processes, certain stages of the life-cycle of some viruses) transiently involve single-stranded regions.

In eukaryotes, nuclear DNA contained in the chromosomes and organelle DNA (→ mitochondrial DNA, → chloroplast DNA) may be distinguished. Nuclear DNA is composed of three major classes: highly, moderately, and nonrepetitive DNA (→ Cot value).

The content of these classes may vary considerably from organism to organism. In general, highly → repetitive DNA (simple sequence DNA) constitutes 3–30%, non-repetitive (single copy) or → unique DNA 30–80%, and intermediate the remainder. Not all highly repetitive DNA is → satellite DNA; a significant portion is dispersed throughout less repetitive sequences in → main band DNA.

1. **Structure of DNA:** Single- or double-stranded DNA molecules can exist as threads, circles or chains, and may be coiled or supercoiled. Circular DNA has been found in bacteria, bacterial viruses and some animal viruses; it is characteristic for the DNA of bacterial → plasmids, → mitochondria, → kinetoplasts, and → chloroplasts of eukaryotes. This DNA may have various structural forms including supercoiled and open circles (one or both strands have breaks) that are either monomeric or multiple length (catenated oligomeric molecules) and also interlocked cyclic molecules.

Each nucleotide consists of the pentose sugar 2-deoxy-D-ribose, phosphoric acid (joined to the sugar moiety by an ester linkage and conferring on DNA its acid properties) and one or another of several different nitrogenous bases linked to the deoxyribose at carbon atom number one of the sugar. The glycosidic combination of base and sugar is called a **deoxyribonucleoside** and their phosphate esters are the **deoxyribonucleotides.**

All DNA samples isolated from a variety of organisms contain as the major components four nitrogenous bases (Fig. 28): **adenine** and **guanine** (purines with a double ring structure), and **thymine** and **cytosine** (pyrimidines with only a single ring). These four bases predominate but in addition DNA may contain minor components such as 6-methyl adenine (in bacteria) or 5-methyl cytosine (in plants and animals). Exceptions to these generalizations exist: DNA of T-even phages of *Escherichia coli* contains 5-hydroxymethyl cytosine instead of cytosine, and certain phages of *Bacillus subtilis* have other bases apparently replacing thymine in their DNA. In spite of these exceptions there are generally four types of nucleotides, depending on which of the bases is attached to the sugar-phosphate ester. These are known as deoxyadenylic acid, deoxyguanidylic acid, deoxythymidylic acid, and deoxycytidylic acid.

The DNA macromolecule built up of a long chain of nucleotides is joined together by means of diester links in which one phosphoric acid molecule forms bonds between the 3′ and 5′ positions of consecutive molecules to form a chain referred to as a **polydeoxyribonucleotide.** The sequence of nucleotides in the chain makes up the **primary structure** of DNA, which can be thought of as having a sugar-phosphate backbone to which is attached at each level one or another of the nitrogenous bases (Fig. 29).

DNA usually consists of two polynucleotide chains joined by hydrogen bonds between pairs of bases that occupy the same level on the two strands. The two chains are coiled plectonemically (in an interlocked form) about the same axis. They follow right-handed helices, with 10 bases in each for one turn (3,4 nm) of the spiral. The phosphates are on the outside of the double helix; the bases are perpendicular to the long axis of the molecule on the inside and spaced at intervals of 0.034 nm along it. The two chains run in opposite directions in terms of the 3′—5′ phosphatedeoxyribose

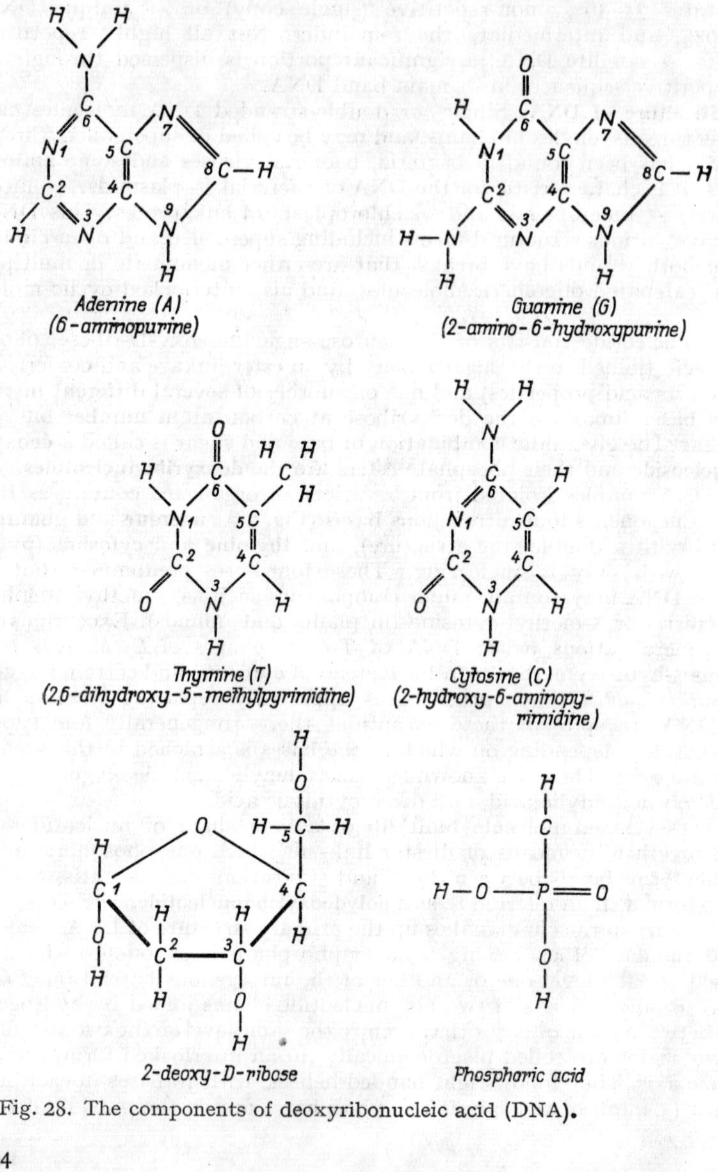

Fig. 28. The components of deoxyribonucleic acid (DNA).

linkages and thus have a definite polarity, which is relevant to the genetic function of the molecule. If the sequence of atoms in the backbone of one chain is ... $-C_3-C_4-C_5-O-P-O-C_3-C_4-C_5-$ ... then it will be ... $-C_5-C_4-C_3-O-P-O-C_5-C_4-C_3-$ ... in the other (Fig. 29).

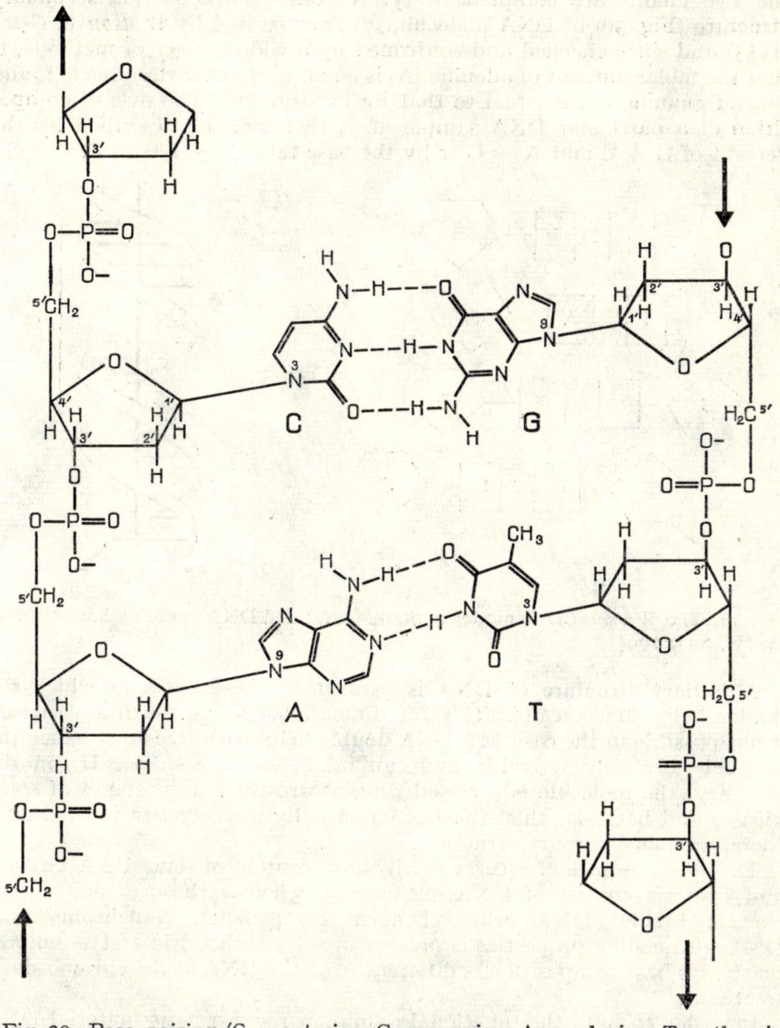

Fig. 29. Base pairing (C = cytosine; G = guanine; A = adenine; T = thymine) in a DNA double helix as proposed by *Watson* and *Crick*. The sugar-phosphate backbones of the two polynucleotide chains run in opposite directions (indicated by arrows) (from *Herskowitz* 1965).

The pairing between bases of the two chains is highly specific. Adenine (a purine) is always hydrogen-bonded to thymine (a pyrimidine) and guanine (another purine) to cytosine (another pyrimidine). The existence of these specific base pairs means that the sequences of nucleotides along the two chains are complementary. A consequence of this **secondary structure** (Fig. 30) of DNA molecules, first suggested by *Watson & Crick* (1953) and since checked and confirmed by a wide variety of methods, is that the molar amount of adenine (A) is equal to that for thymine (T), and that of guanine (G) is equal to that for cytosine (C). The overall composition of a particular DNA sample may, therefore, be described by the percent of G + C and A + C or by the **base ratio** A + T/G + C.

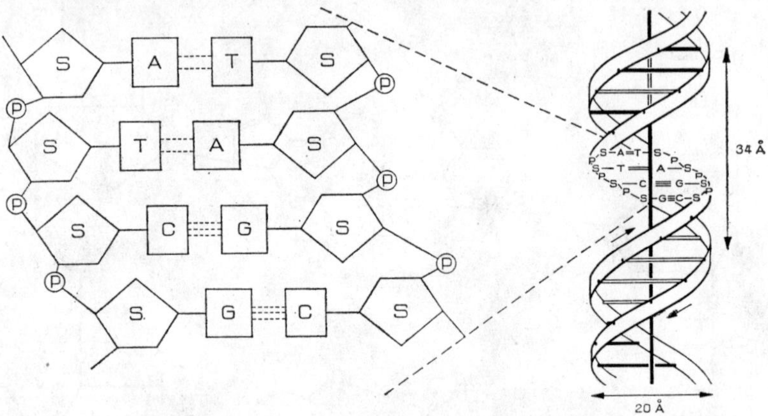

Fig. 30. The *Watson-Crick* model of double-stranded DNA (from *de Robertis, Nowinski, Saez* 1965).

A **tertiary structure** of DNA is present in those cases in which the double helix shows a definite three dimensional shape. Such a structure is not possible in the case of a DNA double helix with free ends, since the molecule is freely movable and can take over any shape. If, on the contrary, the molecule is a closed ring, as are the chromosomes of some viruses and bacteria, then the DNA is usually more or less twisted and, therefore, has a tertiary structure.

The chromosomes of viruses (with the exception of some RNA viruses) and bacteria consist of DNA exclusively, whereas those of eukaryotes contain besides DNA other chemical components. Nonchromosomal DNA with coding properties is present in → mitochondria and → chloroplasts. Its base composition is different from the DNA of the chromosomal fraction.

In order to fulfil the functional requirements of genetic material DNA must carry → genetic information (instruction), must be able to replicate (to reproduce itself) and to undergo → mutation (to alter the instructions occasionally). The genetic information necessary for directing all the specific activities of the cell resides in the sequence of the four primary bases (→ cistron, gene) in form of a → genetic code which is read by the processes of → genetic transcription and → genetic translation.

2. **Replication of DNA:** DNA replication is a complex series of events in which diverse multienzyme systems function whose operations concern three stages of the process: initiation, elongation, and termination. Replication occurs with the two strands of the double helix undergoing partial separation (breakage of hydrogen bonds and regional unwinding of the chains), thus allowing each strand to act as a template for the synthesis of a complementary strand by specific base pairing ("semi-conservative replication" according to *Delbrück & Stent* 1957). Separation of strands goes hand in hand with replication and is not complete before the synthesis of the new polynucleotide chains. Deoxyribonucleoside triphosphates are the monomeric precursors for DNA synthesis. They are enzymatically (→ DNA polymerase) joined together in the presence of a single-stranded DNA sequence which acts as template. The biosynthesis of DNA involves three main stages: 1. the synthesis of deoxyribonucleosides; 2. their phosphorylation to triphosphates by appropriate kinases, and 3. the polymerization of these triphosphates to form polynucleotide chains (Fig. 31).

The replication of DNA in both pro- and eukaryotes (→ replicon; replication unit) proceeds bidirectionally (Fig. 32) in a discontinuous manner and is characterized by the following main features: 1. As the parental strands of a double helix separate, periodically short stretches of RNA are initiated by an → RNA polymerase. 2. After the RNA primer has reached a certain length (50 to 100 nucleotides), which can vary from one system to another, it is extended by a DNA polymerase. The transition from RNA to DNA appears to be base specific in some systems but not in others. 3. The further processing of primed DNA chains probably involves removal of the RNA primer, gap filling by a DNA polymerase and finally joining of the pieces to a neighboring DNA chain by a DNA ligase to achieve continuity of the progeny DNA strand (Fig. 31). 4. Since replication occurs along both parental strands, and since DNA polymerase catalyzes addition of nucleotides only at free 3'-OH groups, elongation along the strand ending in a 5'-phosphate ester is accomplished by synthesis "backwards" in the 5' to 3' direction for lengths of about 1000 nucleotides (in bacteria). These lengths, called *Okazaki* pieces, are then joined together. They are associated with replicating DNA by base pairing to the template strands. Their synthesis is probably initiated by the RNA primer. It is not clear, however, whether both progeny strands at one replication fork are elongated by the synthesis and joining of Okazaki fragments (totally discontinuous synthesis) or whether this type of synthesis is restricted to only one side of the growing fork (semi-discontinuous synthesis). 5. DNA chain growth in prokaryotic and eukaryotic cells proceeds bidirectionally from a single initiation point of replication. Each virus or bacterial genome represents one → replicon. The replication units of eukaryotic chromosomes are tandemly arranged. Replication is initiated at the center of a replication unit and chain growth proceeds in opposite directions towards outlying termini where the newly replicated chains join with those synthesized in adjacent regions of DNA (Fig. 32). The minimal distance between initiation sites is about 40 to 50 nm. 6. In eukaryotes, DNA replication occurs at different times in different regions of individual chromosomes.

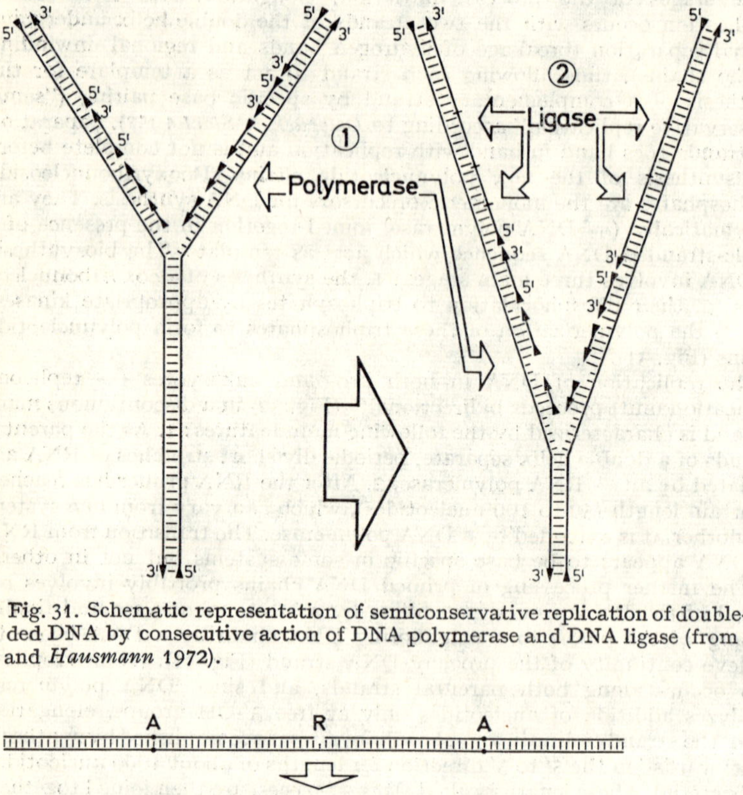

Fig. 31. Schematic representation of semiconservative replication of double-stranded DNA by consecutive action of DNA polymerase and DNA ligase (from *Bresch* and *Hausmann* 1972).

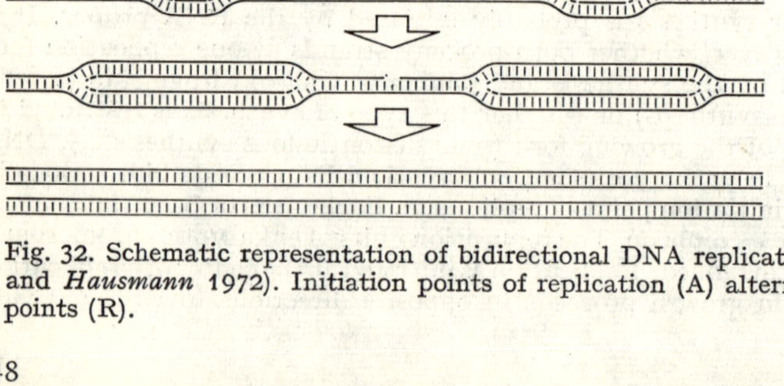

Fig. 32. Schematic representation of bidirectional DNA replication (from *Bresch* and *Hausmann* 1972). Initiation points of replication (A) alternate with swivel points (R).

There are indications that early replicating DNA is relatively GC-rich and late replicating DNA is relatively AT-rich. The one-way rate of bidirectional DNA replication in bacteria is some 20 to 30 nm per minute at 37 °C; the one-way rate of DNA replication in eukaryotes is some 0.5 to 1.2 nm per minute. 8. In prokaryotes, protein synthesis is required only for the initiation of DNA synthesis. In eukaryotes, protein synthesis seems to be required for both the initiation and completion of DNA synthesis. 9. Both asymmetric and symmetric replication have been found at different stages of the same bacteriophage or under different conditions in the same bacterium. In other cases, only symmetric replication or, in the case of single-stranded DNA synthesis, only asymmetric replication is found. 10. Nuclear and organelle DNAs replicate at different times in the eukaryotic cell cycle, but their replication cycles are tightly coupled.

Single-stranded instead of double-stranded DNA is found in some viruses. The DNA strand present in the infectious virus particle is called the "plus" strand. After its entry into a host cell by infection, this strand serves as a template for the formation by DNA polymerase of a complementary "minus" strand, whereby a double helix (the → "replicative form", symbol RF) is formed. This double helix is then replicated in the conventional manner using host cell DNA polymerase to produce many new double-helical copies. Synthesis of single-stranded DNA for progeny viruses needs a special device to insure that only one of the two potential replicas is produced.

As expected from its genetic function, the quantity of chromosomal DNA is dependent upon the number of chromosomes present in the cell. This relative constancy of DNA means that both the cycle of DNA replication and chromosome reduplication must keep pace. There are, however, indications of variations in the incorporation rate of DNA precursors in the DNA content in interphase nuclei, in → polytene chromosome loci, or in particular heterochromatic regions (such as "nucleolus-associated" chromatin). Because of these variations two types of DNA have been inferred: a stable one with genetic function, and a labile one with a metabolic function ("metabolic DNA"); the latter presumably being synthesized when the cell needs an intensification of the information.

The evolution of the higher, multicellular organisms from more primitive ones has been accompanied by a large increase in the total amount of DNA per cell. This probably reflects an increase in the total genetic information content necessary for the development of increasing structural and behavioral complexity. The range of complexity in the → genome, provided that this is represented by the total amount of DNA, increases by about 10^3 from bacteria to mammals. The range becomes even greater if viruses are included. The number of nucleotide pairs ranges from about 5×10^3 (in small viruses), 1.9×10^5 (in *Escherichia coli*), 3.0×10^7 (in *Bacillus megaterium*) to 5.6×10^9 in human cells. (For comparison one → cistron or gene contains about 1 500 nucleotide pairs.) In general, it is assumed that most chromosomes in eukaryotes are composed of one, or at most a few, DNA molecules extending the whole length of the chromatid. All the genetic information encoded in chromosomal DNA is contained in one chromosome (linkage structure of a definite number of linearly arranged genes) in viruses and bacteria, and distributed among a species-specific number of chromosomes (the chromosome set) of eukaryotes.

Seven classes of direct and indirect evidence have established that DNA is the essential genetic material (*Lima-de-Faria & Moses* 1966):

1. DNA is found mainly in the → nuclei of eukaryotes;

2. the DNA content of a cell is directly related to its chromosome number;

3. DNA is a stable macromolecule;

4. DNA structure fits well into the replication properties of the → gene;

5. genetic → transformation in bacteria is mediated by DNA;

6. in viral infection, it is the virus DNA that is responsible for the transmission of the genetic information leading to the production of a new generation of viruses;

7. the → genetic code is a DNA code.

Only in the RNA viruses has the role of DNA as carrier of genetic information been taken over by → ribonucleic acid.

deoxyribonucleoside — → deoxyribonucleic acid.

depressor effect (*Delbrück* 1945) — following the mixed infection of a bacterial cell by two bacteriophage types and successful → mutual exclusion, the numerical reduction of the offspring of the reproducing phage type freed by lysis of the bacterial cell.

derepression — the release of → represesion of enzyme (protein) production after removing a specific metabolite, no → inducer being necessary (→ trans-derepression; induction).

dermatoglyphics — the pattern of the ridged skin of palms, fingers, soles and toes, and the study and the systematic classification of such patterns into a number of different types. The study of d. is important in medical genetics chiefly because of the characteristic combinations of pattern types associated with some chromosomal disorders. Furthermore, since these pattern are heritable, d. may be used as one method of measuring resemblance between twins in the determination of twin zygosity.

dermatoplast (*Küster* 1935) — → gymnoplast.

descendants — all the individuals resulting from the sexual reproduction of one pair of individuals.

desmosome (*Bizzozero* 1871) — any of the specialized areas of the → cell surface that serve to bind the surface to another structure. The latter may be another area of the cell surface (intercellular contact), or some extracellular structure (= cell junction). In general desmosomes appear to be permanent cell structures and probably remain intact during cell division.

According to *Bennet* (1969) desmosomes may be classified in several ways:

1. **According to shape:** a) Macular d. which have a discrete, spotlike shape (maculae adherentes); b) Zonular d. which encircle a cell like a belt or zone and bind it to its neighbors over its entire circumference at the level of the zonular d. (terminal bars or zonulae adherentes).

2. **According to the structure to which the cell surface is bound at the d.:**

a) Autodesmosomes which bind an area of the cell surface to another d. on another area of the surface of the same cell; b) Homodesmosomes which bind an area of the cell surface of one cell to a d. on the surface of another cell (iso- or allodesmosomes); c) Heterodesmosomes which bind an area of the cell surface to some non-cellular structure (hemi-isodesmosomes).

The d. (= node of *Bizzozero*) consists of parallel → unit membranes separated by an interspace which is bisected by a dense midplate and backed by a thick dense plaque. Protein (approximately 76% of the dry weight) is the major constituent of desmosomes. The presence of approximately 17% carbohydrate and 10% lipid suggests that glycoproteins, lipoproteins, glycolipids, and mucopolysaccharides may also be components of the desmosomal junction.

desmotubule (*Robards* 1968) — a tubule (diameter about 20 nm) traversing each → plasmodesma. The wall of the d. probably consists of eleven subunits arranged around a central cavity through which runs a 4 nm diameter rod. This is connected to the inside of the tubule wall by fine filaments. At the ends of each plasmodesma the plasmalemma and cell wall are closely appressed to the tubule, thus precluding direct continuity between the cytoplasm of adjacent cells. Desmotubules are assumed to be nuclear spindle fibers which became trapped in the wall during cell plate formation.

desynapsis (*Li* et al. 1945) — the falling apart during diplotene or diakinesis of → meiosis of chromosomes which paired during zygotene and remained paired at pachytene (= desyndesis). Due to d. most or all chromosomes may be present as → univalents at metaphase of meiosis I. An accompanying decrease in → chiasma number and chromosome contraction, precocious division of univalents, imbalanced segregation and reduced fertility may ensue. D. is often genetically controlled. If so, one or a few recessive genes are involved (→ asynapsis). Mutant genes resulting in d. are called desynaptic genes.

desyndesis (*Sharp* 1934) — = → desynapsis.

detached-X (*Kaufmann* 1933) — a metacentric chromosome produced from → attached X-chromosomes of *Drosophila melanogaster* by → interchange or crossing-over; one arm contains the X-chromosome while the other arm consists either of the long or the short arm of the Y-chromosome (Fig. 33). The process leading to d. is termed "detachment".

detachment (*Belling* 1924) — 1. the loss of bi- or univalents in meta- or anaphase I of meiosis by their exclusion from the → spindle;

2. the production of → detached X-chromosomes.

determinants (*Weismann* 1891) — hypothetical units of the → germ plasm (= hereditary determinants).

determinated disjunction — → disjunction.

determination (*Roux* 1905) — a process which initiates a specific pathway of development by singling it out from among various possibilities for which a cellular system is competent. In multicellular organisms it is useful to distinguish between determinations resulting in immediate → differentiation and those resulting in a reproducible cell state propagated in the nondifferentiating phase by cell heredity (*Hadorn* 1965).

Provided the external conditions are favorable, d. confers an entirely

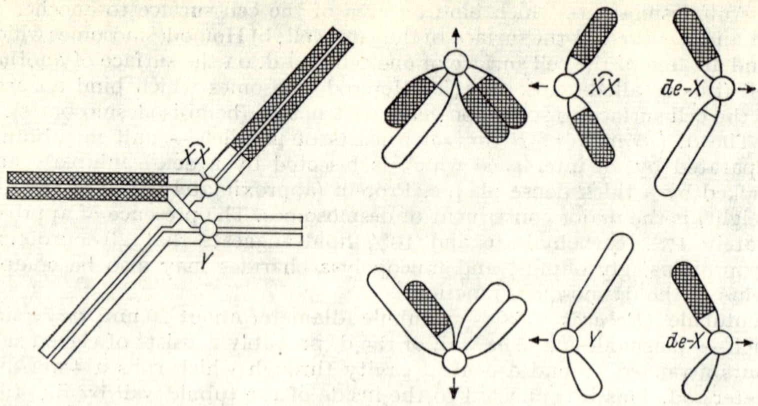

Fig. 33. The formation (by crossing-over between an "attached X-chromosome" and the Y-chromosome) of "detached X-chromosomes".

specific differentiation upon each part of the embryo. The d. may be stable (definitive), in which case there is only one developmental pathway open to the cell or embryo part, i.e. an automatic course of differentiation is prescribed; or it may be labile. Different parts of the developing organisms may acquire d. at different periods (→ transdetermination).

During d., an event takes place which decides which of several types of development will actually be realized. This "event" may be regarded as the activation of one or another set of gene actions, i.e., in the course of development, genetic information is selectively transformed into phenotypic effects (→ gene activation; embryonic development).
Three main types of mechanisms resulting in d. may be distinguished (*Waddington* 1956):

1. **Oöplasmic segregation:** different regions of the cytoplasm of the egg possess specific properties. Thus, certain cytoplasmic regions called "oöplasms" or "organ-forming substances" (the older name) can only develop in one direction. These cytoplasmic regions may be highly localized, showing definite boundaries, or they may overlap into one another.

2. **Evocation:** two adjacent regions of an egg or embryo mutually affect each other so as to change the direction of development of one or both. Such processes are very frequent after the → cleavage period when parts of the embryo which were previously separated come into contact owing to positional changes and folding during gastrulation. Thus, a region determined very early in the development, e.g., by ooplasmic segregation, may come in contact with an as yet undetermined part of the embryo and so affect it that it develops to form a definite type of tissue.

3. **Field action:** in many embryological processes, the development of a particular point in a region of the egg depends upon its relationship to other adjacent points or to its position within the region. The mutual

influence between the parts of an embryonic region is of such a nature that it is possible to remove parts of the embryonic tissue and replace them in an inverted position. Owing to the field action, the development of every one of th points is so influenced by the points of the rest of the tissue that a normal embryo can be formed ("field phenomena"), i.e. any piece of embryonic tissue is so capable of regulation, that the original and complete field condition may be restored fully after experimental interference. The term "field" is chosen by analogy with physical field theories (as, e.g., magnetic field, gravitational field) although no physical forces operate in the embryonic field; nevertheless, the biological events have similar general characteristics to the physical. In both cases certain activities are effective over tho whole field.

detrimental (*Muller* 1934) — any mutation which lowers the viability (= subvitals). The survival rate of the individuals concerned is above 10% but less than 100%. The great majority of spontaneous or induced mutations which produce visible changes are d. (→ lethals).

deuterohermaphroditic (*Correns* 1928) — → hermaphroditic.

deuterotoky — → parthenogenesis.

deutoplasm (*van Beneden & Bessels* 1868) — products elaborated by the cell and stored in the cytoplasm as lipid droplets, yolk bodies, pigment and secretion granules (= inclusion, paraplasm).

development — a process of regulated growth and differentiation (→ embryonic development) that results from the interaction of → genome with cytoplasm, internal cellular environment, and external environment.

D. is a programmed sequence of phenotypic changes, under temporal, spatial, and quantitative control which is irreversible or difficultly reversible at least under ordinary environmental conditions. The sum total of these modifications constitutes the life cycle of an organism (*Sussman* 1965).

Four major processes interacting with one another make up the complex process of d.:

1. Genetic replication, i.e. the means by which (in the course of → mitosis) the sets of genetic information (the genes in the chromosomes) are duplicated;

2. growth, i.e. the increase in mass of the organism which is intimately associated with cellular metabolic activity;

3. cellular differentiation, i.e. the process by which cells of common origin, and thus genetically identical, diverge in their structure and/or function and give rise to morphologically and physiologically distinguishable cell lineages;

4. histo- and organogenesis, i.e. the processes by which differentiated cells aggregate to form tissues of unitary function. These tissues then associate to make up organs.

developmental cycle — the complex progression of phenotypic alterations in a cell or a multicellular organism during → development. This progression occurs in a fixed chronological order at specific intracellular sites and/or within specific cells of a multicellular organism, and to fixed amounts and extents. Initial stages of development of a multicellular organism commence while nearly all of its nuclear → genes are inactive, i.e., the initial stages are sustained by stable gene products (→ messenger RNA) formed during → oogenesis.

developmental flexibility (*Thoday* 1953) — → flexibility.

developmental genetics — a branch of genetics primarily concerned with questions dealing with the manner in which genes control or modulate the process of development (→ phenogenetics).

developmental homeostasis (*Lerner* 1954) — → homeostasis.

deviation (*de Beer* 1951) — a mutation-controlled (→ mutation) shift of a developmental process in an alternative direction. The extent of the phenotypic effects produced by the mutation in such a case is variable. As a rule, the effect is greater the earlier the shift takes place.

diagynic — of genes localized in the → X-chromosome (X-linked genes) which are transferred from mother to sons in the case of female heterogamety (→ diandric).

diakinesis (*Haecker* 1897) — → meiosis.

diallelic (*Atwood* 1944) — → monoallelic.

diandric — of genes localized in the → X-chromosome (X-linked genes) which are transferred from father to daughters in the case of male heterogamety (→ diagynic).

diaphoromixis (*Burnett* 1956) — → heteromixis.

diauxy (*Monod* 1942) — the adaptation of microorganisms to culture media which contain two different carbohydrates. Initially, only one of the carbohydrates is used, later the second as well. The organism possesses constitutive → enzymes for the first and requires induced enzyme synthesis before it can metabolize the second carbohydrate.

dibasic (*Darlington & Janaki Ammal* 1945) — → polybasic.

dicentric (*Darlington* 1937) — of chromosomes or their longitudinal subunits (chromatids, subchromatids) having two localized → centromeres (= dikinetic). Dicentric derivatives (→ isodicentric; heterodicentric) of originally monocentric chromosomes are the result of chromosomal structural changes (→ chromosome mutation).

Dicentric chromosomes can be produced directly as a consequence of asymmetric reciprocal → translocations prior to replication of the chromosomes in the interphase nucleus. Dicentric chromatids may occur *inter alia* as a result of → isochromatid breaks, asymmetric chromatid translocations (Fig. 16), after → crossing-over within paracentric heterozygous → inversions (Fig. 56) and after crossing-over in heteromorphic bivalents heterozygous for a → duplication. They can lead to dicentric chromosomes following replications subsequent to their corresponding distribution during mitosis or meiosis.

If in the anaphase following their formation, the two centromeres of the dicentric chromatids are distributed to opposite cell poles, the "intercentric segment" lying between the centromeres is stretched between the poles as a bridge (→ chromosome bridge). If the dicentric chromatid arrives at one of the two cell poles, it will be reduplicated in the following interphase and may become a dicentric chromosome. In mitosis, the centromeres of a dicentric chromosome orientate themselves independently of one another and the events occurring during the distribution of the two dicentric chromatids (daughter chromosomes) depend on the type of orientation. If the two "intercentric segments" lie parallel to one another during metaphase, the dicentric daughter chromosomes are separated from one another without complication but if the chromatids become twisted in the intercentric region, criss-cross bridges are formed during anaphase.

The bridges between the centromeres of dicentric chromatids or chromosomes are either broken during anaphase or they survive. In the latter case, they complicate → cytokinesis (formation of → restitution nuclei). If the bridge is broken, the break may occur in any position, resulting in → deletions and → duplications, giving rise to daughter cells with unequal amounts of → genetic information. If the breakage is followed by a reunion, the dicentric chromatids and chromosomes may be the starting point for a → bridge-breakage-fusion-bridge cycle.

dichogamous (*Sprengel* 1793) — of flowers or hermaphroditic animals whose male and female sex organs are active at different times. "Dichogamy" is sub-divided into:

1. **Protoandrous or proterandrous dichogamy**: the male sex organs are active before the female ones;

2. **Protogynous or proterogynous dichogamy**: the female sex organs are active before the male ones.

Organisms in which both protoandrous and protogynous dichogamy occur are termed "heterodichogamous".

dichopatric (*Smith* 1965) — of → allopatric populations whose geographic ranges are separated to the extent that individuals of the involved populations never meet (→ parapatric).

dichophase (*Bullough* 1963) — that phase of the mitotic cycle in which a cell is faced with the choice between → differentiation for mitosis and differentiation for ageing and tissue function (→ apophase).

diclinous — having male and female germ cells in separate flowers either on the same plant, which is said to be "monoecious", or on different plants, which are said to be "dioecious" (→ monoclinous).

dictyokinesis (*Perroncito* 1910) — the distribution of → dictyosomes during cell division.

dictyosome (*Perroncito* 1910) — any of the parts of the → Golgi apparatus (= Golgiosomes) isolated by differential centrifugation.

Dictyosomes consist of 2—7 (sometimes up to 20 or more) layers of flattened membranes (cisternae) surrounded by small vesicles. The → cisternae are isodiametric in the plane of flattening (diameter between 0.2 and 1.2 μm; in conjugates up to 5 μm). The vesicles surrounding the stacks of cisternae are formed by localized dilatations which are probably a product of dictyosome activity. The cisternae are joined by some bonding constituent within the intercisternal matrix which contains intercisternal elements (diameter 7—8 nm). The center to center spacing of intercisternal elements is about 15 nm.

Dictyosomes dispersed throughout the cytoplasm make up the Golgi apparatus of plants and invertebrates. At telophase (→ mitosis), the dictyosomes become grouped at the periphery of the → cell plate in plants and form small vesicles which fuse to produce the plate. In vertebrates, the dictyosomes are either aggregated together in a loose or compact juxtanuclear mass or they are scattered through the cytoplasm either as single elements or, more often, as small aggregates. Eventually, they occupy a cortical position (→ cortex).

The component cisternae of a d. are not equivalent. They differ structurally and probably have different functional capabilities.

Dictyosomes are somehow replicated to form new such elements, but the mode of reproduction is not yet known.

dictyotene — a stage that interrupts meiotic prophase in mammlian oocytes. During d., the chromosomes become decompacted; the oocyte grows in size and accumulates cytoplasmic organelles or their precursors to provide structures for the meiotic divisions and a maternal supply of RNA and proteins for the early embryo.

didiploid (*de Litardière* 1925) — = → amphidiploid.

differential affinity (*Darlington* 1928) — the failure of two chromosomes to pair at meiosis in the presence of a third, although they pair in its absence (*Darlington & Mather* 1949). D. a. is the result of different homology relationships between the chromosomes. Pairing of two partially homologous chromosomes does not take place in the presence of a third chromosome if the latter is completely homologous to one of the former. The degree of d. a. is largely determined by the size of the homologous and nonhomologous segments of the partially homologous chromosomes.

differential distance (*Mather* 1936) — the interval between the first → chiasma and a fixed point in a bivalent relative to which this chiasma is believed to originate (→ centromere interference; interstitial distance; residual distance).

differential mitosis (*Bauer* 1952) — a mitosis leading to daughter cells of different chromosome number during the chromosomal differentiation of soma and germ tract of certain Diptera species.

differential reactivity (*Darlington & LaCour* 1938) — the different behavior of hetero- and euchromatic chromosome segments during mitoses taking place under more or less long-term action of low temperature (+2 to −3 °C). As a result so-called → special segments are formed.

differential segments (*Darlington* 1931) — 1. on a cytological basis, chromosome segments in sex chromosomes and interchange (translocation) heterozygotes wherein the two chromosomes (→ chromosome pairing) differ structurally (structural heterozygosity), in contrast to "pairing" segments, within which the chromosomes pair and crossover;

2. on a genetical basis, chromosome segments which are generally transmitted as unrecombined units by heterozygotes. The genes localized there tend to be inherited *en bloc* (as "supergenes").

A segment can be genetically differential because (*Lewis & John* 1963):

a) Meiotic pairing and thus recombination by crossing-over is limited for genetic or structural reasons, although there is a homologous segment with which to pair.

b) The cross-over products (→ crossing-over) are inviable.

c) Lack of homology or absence of a partner prevents meiotic pairing.
All segments differential on a cytological basis are genetically differential also. But the reverse is not true.

differentiation — the origin of differences during → embryonic development between the spatial parts of an originally homogeneous whole. The primary features of embryonic d. are: 1. selective control of cell proliferation; 2. distinctive intracellular and supracellular morphology; 3. metabolic diversification in cells or cell groups; 4. modulation of cellular activities by intercellular regulators. All these processes aresu bject to genetic regulation.

Two phases of d. may be distinguished:

1. Intracellular d., i.e., changes within a cell through interactions among nucleus, cytoplasm, and the environment thereby leading to specialization of cell functions (→ cytodifferentiation);

2. Intercellular d., i.e., a divergence among cells which depends upon intracellular d. to accomplish those changes in each of the cell lines which then diverge from one another (→ histogenesis).

In multicellular organisms, d. is the most spectacular aspect of → development involving the formation of distinctly different types of cells and tissues from a → zygote by the processes of specific regulation of gene activity (→ cytodifferentiation). Differentiated cells are not only morphologically different, but show from a functional point of view chemical, immunological, and behavioral differences as well. These changes are the result of factors operating both within and outside the cells. Although d. in higher organisms may be examined at increasingly complex levels of cellular organization, it ultimately reflects a change in the cell's biochemical repertoire in response to specific regulatory stimuli which either affect the production (by → genetic transcription) or utilization of → messenger RNA or the formation of a functional protein from the polypeptide determined by → genetic translation.

Through progressive changes in cell structure and cell chemistry d. leads to the formation of different types of tissue and is accompanied by the organization and localization of differentiated cells. This process leads to morphogenesis, i.e. to the production of definitive structural types. D. is usually but not always associated with growth (→ modulation). It is a two-phase process. The future of an egg region is more or less permanently determined in the early stages of development, so that it may only be modified experimentally within narrow limits. Later, the region concerned develops to a definitive end product. The process by which the end product achieves its "finality" is called → determination.

Correlative or dependent d. is the term used to signify the d. of an embryo part as a result of the influences of neighboring parts and not solely of internal factors.

Normally, a cell which has achieved a particular functional state in the course of d. proves to be stable in that state and does not revert back to its previous condition (does not "dedifferentiate"). As a result of d., different tissues show different relative enzyme activities. This may also be the case for cells of the same tissue. Such a difference may be quantitative as well as qualitative.

D. is frequently accompanied by changes in the cell nucleus but such changes (→ chromosome diminution; loss of whole chromosomes or genomes; endomitotic polyploidization; polytenization of chromosomes) should not be regarded as causes but, generally, as specific paths of d. (→ nuclear differentiation).

diffuse stage (*Iyengar* 1945) — in some organisms, a stage between pachytene and diplotene in the first meiotic division during which much of the DNA of bivalents becomes reorganized into Feulgen-positive lateral loops (→ lampbrush chromosome). The process of decompaction of chromosomes continues until the bivalents disappear, producing a nucleus without clearly defined chromosomal strands. During the d. s., the DNA that was folded into the chromomeres of pachytene bivalents is almost completely dispersed into the nucleoplasm. The d. s. is probably a period of synthetic activity of the chromosomes which resembles mitotic → interphase.

digametic (*Wilson* 1911) — = → heterogametic.

digenesis — = → alternation of generations.

digenomatic (*Winkler* 1920) — of → euploid organisms with two sets of chromosomes (→ polygenomatic).

digonic — of organisms which produce male and female gametes in different gonads within the same individual.

digressive (*Weinstein* 1936) — → crossing-over.

dihaploid — of an individual produced from a tetraploid form and which possesses half the tetraploid number of chromosomes.

dihapllophase (*Bonnet*) — = dikaryophase (→ dikaryon).

diheterozygous = → dihybrid.

dihybrid — → heterozygous in respect to two pairs of alleles.

diiso-compensating trisomic (*Kimber & Sears* 1968) — a compensating → trisomic in which the missing chromosome is compensated for by two → isochromosomes, one for each arm of the missing chromosome (= double-isotrisomic; *Kimber & Sears* 1968). (→ ditertiary compensating trisomic; telotertiary compensating trisomic; ditelotrisomic; teloisotrisomic).

diisosomic (*Kimber & Sears* 1968) — of a cell or individual which lacks one chromosome pair but has two homologous isochromosomes for the same arm of the missing pair (→ monoisosomic; triisosomic).

diisotrisomic (*Kimber & Sears* 1968) — an individual deficient in one chromosome but which has a pair of homologous → isochromosomes for one arm of the missing chromosome (→ diiso compensating trisomic).

dikaryon (*Maire* 1902) — a binucleate cell, spore, or mycelium consisting of such cells in Asco- and Basidiomycetes, which may be homokaryotic (containing two genetically identical nuclei) or heterokaryotic (containing two genetically different types of nuclei).

"Dikaryotization" is the process by which a dikaryon is produced: Two sexual cells or organs, each containing one or more haploid nuclei of unlike → mating type fuse and the nuclei from the different strains remain associated in one or more pairs, each called a d., while retaining their individuality during → dikaryophase. Plasmogamy (fusion of cytoplasms) is in this way separated from karyogamy in a kind of "interrupted fertilization" because the two members of each d. may repeatedly undergo mitotic divisions (which are generally simultaneous) before performing karyogamy (→ di-mon matings).

dikaryophase — a third nuclear phase (→ alternation of nuclear phases) in addition to the haploid and diploid phases in Euascomycetes and Basidiomycetes which is interposed between plasmogamy and karyogamy (= dihaplophase). The → dikaryon characteristic for the d. originates (following plasmogamy) by the intracellular association of pairs of compatible nuclei, which may be of identical or different origin in self-fertile and cross-fertile species. The pairs of haploid nuclei divide synchronously during vegetative growth, thus establishing and maintaining an equal ratio of the two nuclear types. In Euascomycetes, the d. occurs as the ascogeneous hyphal system, in Basidiomycetes, as the secondary mycelium. The d. leads to and is terminated by karyogamy and meiosis in homologous specialized cells called asci and basidia (*Raper* & *Esser* 1964).

dikaryosis — in higher fungi (Euascomycetes and Basidiomycetes), a substitute for diploidy. It differs from diploidy primarily in the spatial separation of the two genomes into a genetically balanced system (→ dikaryophase; heterokaryosis).

dikinetic — = → dicentric.

dilution gene — any modifying gene which reduces the action of a major gene and shows no other phenotypic manifestation (= diluting factor).

dimegaly (*Henking* 1891) — a condition in which two types of female and/or male gametes are produced in the same organism.

diminution (*Herla* 1895) — = → chromosome diminution.

dimixis — → heteromixis.

di-mon mating — mating between di- and monokaryons in Basidiomycetes, as first described by *Buller* (1931) (*Buller* phenomenon). Three types may be distinguished:

1. **compatible di-mon's:** the monokaryon (e.g. A^1B^1) is compatible with both components of the → dikaryon (e.g. $A^2B^2 + A^3B^3$);
2. **hemicompatible di-mon's:** the monokaryon is compatible only with one component of the dikaryon (e.g. $A^1B^1 \times (A^1B^1 + A^2B^2)$);
3. **noncompatible di-mon's:** the monokaryon is compatible with neither component of the dikaryon (e.g. $A^1B^1 \times (A^1B^2 + A^2B^1)$).

As a rule, the monokaryon changes to a dikaryon in all three cases.

dimorphism — 1. the occurrence of two forms (differing in genotype or chromosome morphology) within a population (→ polymorphism);

2. the occurrence of two different types of flower on one plant.

dinokaryon (*Chatton* 1920) — a type of nuclear organization in dinoflagellates which differs from that of other eukaryotes. A striking feature is the absence of a → chromosome coiling cycle. Chromosomes maintain the same appearance throughout the cell cycle and are distinctly visible as rod-shaped bodies within the interphase nucleus. The DNA is not associated with protein. Dinoflagellate mitosis (dinomitosis) is without a typical spindle. During division a complex process of invagination of the intact nuclear envelope takes place at one side of the d. and results in the formation of parallel cylindrical channels through the d. which contain bundles of → microtubules. Nuclear constriction occurs perpendicular to these channels which may be involved in the separation of daughter chromosomes. Dinoflagellates thus resemble prokaryotes in the manner of chromosome separation, chemistry and ultrastructure and the d. may represent an intermediate (mesokaryotic) type in the evolution of the typical eukaryote nucleus.

dioecious — of plants with either male of female sex organs, in contrast to → monoecious organisms. D. plants require a partner of opposite sex for the completion of their sexual cycle. Plants with hermaphroditic and male flowers or hermaphroditic and female flowers on different individuals are called androdioecious and gynodioecious (*Darwin* 1877), respectively. Species which, in addition to hermaphroditic offspring, also produce monoecious offspring with male and female inflorescences are referred to as androgynodioecious.

In fungi, individuals which function as either donor or recipient of a nucleus are called d. Dioecious fungi include: 1. all species in which mycelia bear either female or male sex organs, but not both; 2. all species with two mating types, in which zygote formation follows the fusion of isogametes or isogametangia. Since sexual behavior is determined purely through the physiological difference between the two mating types, they are called physiologically d. (*Esser* 1971).

diphasic — of chromosomes with one arm euchromatic, the other heterochromatic.

diplobivalent (*Barber* 1940) — a bivalent composed of two → diplochromosomes and hence 8 chromatids.

diplochromosome (*White* 1935) — a → chromosome which has reduplicated twice since the last effective mitosis, the centromere region being undivided (*Darlington & Mather* 1949). It consists of four chromatids, as opposed to normal chromosomes with two chromatids.

diplo-haplontic (*Svedelius* 1915) — of organisms ("diplo-haplonts") in whose life cycle a haploid phase exists between meiosis and fertilization including all higher plants, many algae and fungi (→ diplontic, haplontic). Meiosis in diplo-haplontic organisms is intermediate and its products are not gametes but meiospores. These spores do not fuse to form a zygote but undergo mitotic divisions and develop into haploid individuals. These then eventually contain the gametes and hence they are called gametophytes. The diploid individual arising from the zygote is the site of meiosis, and since it forms the meiospores it is called the sporophyte (Fig. 34). In diplo-haplontic lifecycles two generations alternate (→ alternation of generations), one reproducing sexually and one reproducing asexually.

diploid (*Strasburger* 1905) — of cells, phases of life-cycles (→ diplo-haplontic, diplontic) and organisms ("diploids") with two homologous → chromosome sets (one paternal, the other maternal) in which each type of chromosome (autosome) except the → sex chromosomes of the heterogametic sex is always represented twice (symbol 2x or 2n), as opposed to → haploid (one chromosome set; 1x or 1n) and → polyploid (more than two chromosome sets; 3x, 4x, etc.).

"Diploidy" is initiated with the fusion of two haploid gametes during

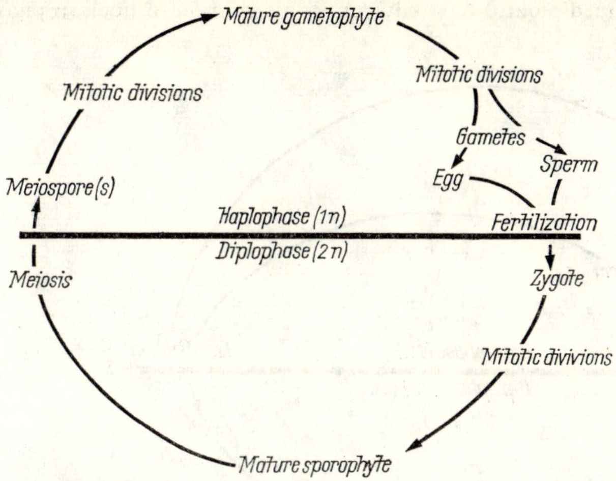

Fig. 34. Diagrammatic representation of a diplo-haplontic life cycle in plants (from *Cook* 1965).

→ fertilization. By this process a "holozygote" (→ zygote) is formed. The adaptive advantage of a dominant phase of diploidy in higher organisms may result from: 1. diploidy masks recessive deleterious mutations and 2. allows interactions between alleles to produce heterozygotes superior to either homozygote.

Diploidy for part of the → genome is only known in viruses and bacteria and may result from several mechanisms (→ heterozygous). "Zygotes" in bacteria which are diploid only for part of the genome and haploid for the remainder are referred to as "merozygotes" and are usually unstable.

The process of partial genetic transfer producing a bacterial merozygote by incorporation of a → "merogenote" into a recipient bacterial cell is known as "meromixis". In case of heterozygosity for the → alleles in question partial diploids show the phenomena of dominance, recessivity and → genetic complementation.

The term **"functional diploid"** (*Darlington* 1928) denotes an → alloploid organism behaving in → segregation like a diploid one.

diplontic — of organisms ("diplonts") having a life cycle in which the products of meiosis behave directly as gametes (→ diplo-haplontic, haplontic) including all multicellular animals. Only the gametes of diplonts are haploid. The organism arising from the zygote and all derivatives produced by vegetative reproduction or parthenogenesis are diploid and contain two (in case of polyploidy more than two) chromosome sets in their somatic cells. As opposed to diplo-haplonts, there is no → alternation of generations in diplonts, but only an → alternation of nuclear phases (Fig. 35).

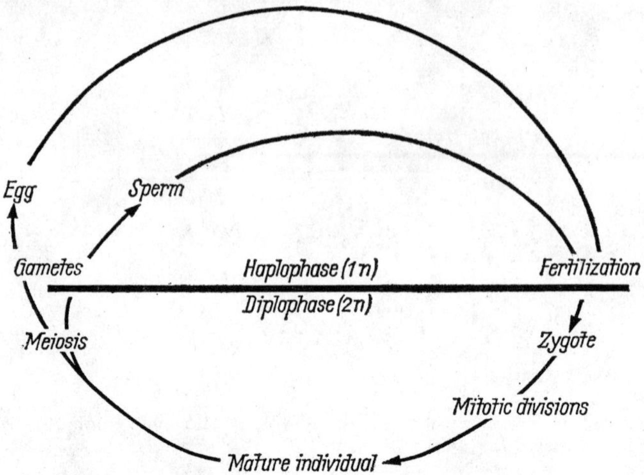

Fig. 35. Diagrammatic representation of a diplontic life cycle as found in higher animals and some algae (from *Cook* 1965).

diplophase (*Buder* 1916) — 1. the diploid phase or generation between fertilization zygote formation) and the beginning of meiosis (→ haplophase);

2. = → diplotene in the prophase of the first meiotic division (*Belling* 1928).

diplosis (*Renner* 1916) — the doubling of the gametic chromosome number.

diplosome — a pair of → centrioles.

diplospory — → apomixis.

diplotene (*v. Winiwarter* 1900) — the fourth stage of meiotic prophase (→ meiosis) where the paired chromatids begin to separate.

diplounivalent — a → diplochromosome in pollen grain mitosis, consisting of four chromatids. Diplounivalents are produced following the failure of the second meiotic division by chromosome reduplication preceding pollen grain mitosis.

directional selection (*Mather* 1948) — → selection.

disassortment — preference for → mating between dissimilar individuals (→ assortment).

discordance — the nonidentity of matched groups or pairs (such as twins) in any one or in a given → character, as opposed to concordance.

By means of "discordance analysis", which may be used in human genetics, it is possible to establish that a pair of twins originated from the same egg.

discordant orientation — → translocation.

disjunction — the separation of daughter chromosomes (chromatids) during anaphase of mitosis and second meiotic division and of paired chromosomes (→ chromosome pairing) during anaphase of first meiotic division (→ assortment).

dislocation (*Navashin* 1926) — a structural change of chromosomes (→ chromosome mutation) produced by loss or displacement of chromosome segments. Homologous pairs of chromosome segments differing in their linear sequence with other segments are referred to as "dislocated" segments (*Darlington* 1937).

disomic (*Blakeslee* 1921) — 1. of cells and individuals (2n) with two → chromosome sets whose members are represented by pairs of homologues, as opposed to → monosomic, → nullisomic, → trisomic, and → tetrasomic ones;

2. of cells containing one chromosome set plus a second representative of one member of the set (n + 1).

disparate chiasmata (*Darlington* 1937) — → chiasma.

dispermy — the entrance of two spermatozoa into an egg cell (= dispermic fertilization) (→ polyspermy).

displacement loop (*Kasamatsu* et al. 1971) — in covalently closed molecules of → mitochondrial DNA, a short three-stranded region. It appears to be formed by displacement synthesis of a short progeny strand with a specific region of the mitochondrial DNA chains (L chain) serving as a template. The corresponding region of the other (heavy) parental chain is displaced. D. loop molecules are interpreted to be at an early stage of replication.

disruptive selection (*Mather* 1948) — → selection.

dissociation — 1. a type of structural change in chromosomes which can increase the chromosome number (*White* 1957). D. is a special type of translocation (→ centric fusion gives the opposite result) whereby a metacentric large chromosome and a small supernumerary (supplying one → centromere and two → telomeres) are translocated so that two acrocentric or submetacentric chromosomes are produced (Fig. 96b);

2. the spontaneous or induced separation of the nuclear components of a heterokaryotic → dikaryon; in fungal colonies this may result in changed sectors (*Pontecorvo* 1953).

dissociation factor (*Kohler* et al. 1968; *Bade* et al. 1969) — a protein factor (in *E. coli*; mol. weight about 9000) which plays a role in the cycling of → ribosomes. By complexing with the 30 S subunit the d. f. (DF) splits the free 70 S runoff ribosome at the end of polypeptide synthesis into its 30 S and 50 S subunits. After that DF is cyclically released at a subsequent stage in the formation of a → polyribosome. DF is identical with one of the → initiation factors (IF_3), i.e., it has, in addition to its role in separating ribosomes into subunits, other functions in the initiation of protein synthesis (→ genetic translation).

distal (*Navashin* 1912) — of a part of a chromosome arm, which is farther from the → centromere than another (proximal) part (*Darlington & Mather* 1949).

distortional segregation — → segregation; B-chromosome.

distributive pairing (*Grell* 1962) — that phase of the meiotic cycle (→ meiosis) during which chromosomes that had previously failed to undergo → exchange pairing may enter into segregational associations with other non-crossover chromosomes, homologous or nonhomologous.

ditelomonotelosomic (*Kimber & Sears* 1968) — a cell or individual in which one chromosome pair is missing but which has a pair of telocentric chromosomes for one arm and an unpaired telocentric chromosome of the other arm of the missing chromosome pair (→ double-monotelosomic).

ditelosomic (*Kimber & Sears* 1968) — of a cell or individual deficient in two homologous chromosome arms (→ tritelosomic; diisosomic).

ditelotrisomic (*Kimber & Sears* 1968) — an individual deficient for one chromosome but which has a pair of homologous telocentric chromosomes for one arm of the missing chromosome (→ diiso-compensating trisomic).

ditertiary compensating trisomic (*Kimber & Sears* 1968) — a compensating → trisomic in which the missing chromosome is compensated for by two tertiary chromosomes, one having one arm of the missing chromosome and the other the second arm (→ diiso-compensating trisomic).

divergence index (*Ginsburg* 1954) — → intergradation index.

dizygotic — of twins produced by two separate ova, fertilized by separate sperm, as opposed to → monozygotic twins arising from a single fertilized ovum which develops into two embryos at some relatively early developmental stage. Mono- and dizygotic twins are also called identical and fraternal twins, respectively. Monozygotic twin pairs normally have identical genotypes, they are like-sexed, and usually closely similar in physical appearance. Dizygotic twins are as similar genetically as ordinary sib pairs, having, on the average, half their genes in common. Zygosity of twins may be investigated in two ways: 1. by examining the fetal membranes, or 2. on the basis of similarities (concordance) and differences (discordance) between the cotwins.

DNA — a shorthand for → deoxyribonucleic acid.

DNA-binding protein (*Hotta & Stern* 1970) — 1. in meiotic cells, a protein with broad phyletic distribution and unique cellular localization which shows a high binding efficiency for single-stranded DNA and the property of catalyzing the → renaturation of thermally denatured DNA. The protein is assumed to play a role in meiotic recombination and to promote → chromosome pairing and → crossing over between homologous chromosomes It is absent from somatic tissues.

2. In *E. coli*, a protein (= DNA unwindung protein) which specifically stimulates the synthetic activity of → DNA polymerase II and reduces the synthetic activities of both *E. coli* DNA polymerases I and II (*Signal* et al. 1962). Analogous stimulations of DNA polymerase activity have been reported for phage T7-induced DNA polymerase and the T7 DNA-b. p.

The DNA b. p. of *E. coli* appears to have two sites, each capable of binding to a macromolecule. One site is specific for single-stranded DNA regardless of sequence, the other is specific for DNA polymerase II. A DNA coated with DNA b. p. makes that DNA a substrate for polymerase II but not of DNA polymerase I. It also prevents → RNA polymerase from transcribing single-stranded DNA. Therefore, the protein may play a key role in the cell in reactions involving repair, replication, recombination and transcription.

DNA body — → gene amplification.

DNA dependent RNA polymerase — → RNA polymerase.

DNA-DNA hybridization — hybridization (annealing) of purified, denatured DNA as used to investigate homologies between DNA molecules from different sources and to examine the relation of particular DNA fractions to the rest of the genome. Reannealing kinetics are used as a measure of genome complexity (→ Cot value).

DNA ligase (*Gellert* 1967) — any of a number of enzymes (= polynucleotide ligases, DNA sealases) which catalyze the closing of single-strand breaks (nicks) in native double-stranded DNA and the circularization of DNA by the formation of phosphodiester bonds between 3′ OH groups and 5′ monophosphate groups of successive nucleotides in the DNA chain. DNA ligases from various sources require either ATP or NAD as a cofactor. They are supposed to play an important part in the replication of DNA (covalent joining of *Okazaki* pieces), → repair of DNA lesions (→ excision repair) and genetic recombination.

DNA-like RNA (*Scherrer* et al. 1963) — = → pre-messenger RNA.

DNA methylation — the specific methylation of DNA bases by DNA methylases. DNA m. occurs in virtually all organisms examined and involves the enzymatic transfer at the nucleotide level of the methyl group of S-adenosyl methionine to specific bases in DNA. It is the basis of bacterial → DNA modification-restriction systems. Other possible functions, such as a punctuation system for genetic trancription or a method of regulating the number of times a → messenger RNA is translated, have been proposed. The methylation pattern is a unique feature of DNA from different species.

DNA modification-restriction system — in bacteria, enzyme systems that determine the acceptance or rejection (restriction) of foreign DNA which has entered the cell. Each system carries its own specificity and bacteria may possess one or more genetically determined modification and restriction specificities. The essential components of DNA m.-r. systems are: 1. a modification enzyme that methylates (or glycosylates) the host DNA at a small number of sites determined by specific base sequences, and 2. a → restriction endonuclease that recognizes as foreign and cleaves any DNA not carrying the host-specific modification pattern. Strains with a given restriction specificity, also have the corresponding modification specificity protecting their DNA from their own restriction activity.

When a strain possesses more than one DNA m.-r. s., each system functions independently. Restriction and modification activities are probably components of a single complex enzyme molecule which may be able to dissociate, splitting off a functional modification methylase without restriction activity.

Bacterial mutants with altered restriction and modification properties have been isolated: $r^- m^+$ mutants fail to restrict but can still modify DNA; $r^- m^-$ mutants neither modify nor restrict.

DNA packing ratio — the length of DNA double helix per unit length of chromosome fiber in eukaryote chromosomes.

DNA plasm (*Kellenberger* 1962) — any of the areas in prokaryote cells in which the DNA is concentrated (= nucleoid). The shape of such areas is highly variable and covers a range from a delicate network of branches to a compact nuclear body.

DNA polymerase — any of a number of enzymes catalyzing the formation of → deoxyribonucleic acid from deoxyribonucleoside triphosphates and using DNA (DNA-dependent DNA polymerases) or RNA (RNA-dependent DNA polymerase) as template. A second type of reaction catalyzed by DNA p. is called → repair synthesis in which a partially single-stranded template is converted to a completely double-stranded form. Thus repair synthesis restores a damaged DNA molecule to its native state (→ excision repair). Mutant alterations of DNA-dependent DNA p. can affect the specificity of DNA replication (→ mutator gene).

All known DNA-dependent DNA polymerases require a primer with a free 3′ hydroxyl group available to make a phosphodiester bond. Since the two chains of a parental DNA molecule have opposite chemical polarities, synthesis of at least one strand is discontinuous as a series of so-called Okazaki pieces. In bacteria, three DNA p. species may be distinguished:

1. **DNA p. I (Kornberg polymerase)** has been implicated in repair of DNA lesions (both in the repair replication step and in the excision step in → excision repair). It consists of one polypeptide chain (about 1000 amino acids and mol. weight of 109,000) and has exonuclease activity. The rate of synthesis is 17 nucleotides/sec. Each bacterial cell contains about 400 molecules of this enzyme.

2. **DNA p. II** is localized in the plasma membrane of the bacterial cell and has a mol. weight of 70,000. The maximal number of molecules per cell is 50. No clear idea of the possible role of this polymerase in replication is at hand.

3. **DNA p. III** is essential for bacterial DNA replication. Its mol. weight is 150,000. The number of molecules per cell is maximally 20. Its *in vitro* rate of synthesis is more than 1300 nucleotides/sec (about the rate found in *in vivo* replication).

Each of the three DNA polymerases of *E. coli* has an associated nuclease that catalyzes the exonucleolytic digestion of single-stranded DNA in the 3′ → 5′ direction.

Distinct molecular species of DNA polymerases have been reported for the nucleus, cytoplasm, mitochondria, and chloroplasts of **eukaryotic cells.** The nucleus contains two DNA polymerases designated D-DNA p. I and II (β and α, resp.). The former is a low mol. weight protein (approximately 3 to 4×10^4), while D-DNA p. II has a much higher mol. weight (in the range of 2 to 3×10^5). D-DNA p. II is also found in the cytoplasm.

Another type of DNA p. (γ) with the unique feature of copying polyribonucleotide strands, such as poly-adenine, is found in both the nucleus and cytoplasm and has been called R-DNA polymerase. Mitochondria and, possibly, chloroplasts contain other types of DNA polymerases with their own unique properties.

RNA-dependent DNA polymerase (→ reverse transcriptase) is a polymerase contained in virions of RNA tumor viruses. It catalyzes the synthesis of a DNA copy (a → provirus) of the resident or endogeneous RNA and also the polymerization of deoxyribonucleotides using exogeneous polynucleotides as templates in the presence of poly- or oligodeoxyribonucleotides. The enzyme has been used successfully to obtain DNA copies (cDNA) of → messenger RNA (→ antimessenger). (→ terminal deoxyribonucleotidyl transferase).

DNA puff (*Breuer & Pavan* 1955) — a unique and exclusive type of → puff in sciarid → polytene chromosomes which gives rise to an excessive and disproportionate synthesis of DNA. DNA puffs constitute an unusual mechanism for amplifying (→ gene amplification) the activity of particular chromosomal sites. Disproportionate DNA synthesis in DNA puffs involves a number of additional rounds of replication. The excessive DNA is not shed from the chromosome but remains integrated. DNA p. formation is intimately linked to metabolic changes in salivary gland metamorphosis.

DNA reiteration — the presence in eukaryotes of DNA sequences which exist in many similar, probably not identical, copies in the → genome (→ gene amplification; repetitious DNA).

DNA repair — the enzymatic reconstruction of a continuous two-stranded DNA molecule without mismatch from a molecule which contained damaged regions (single strand breaks, damaged bases, or other structural defects). The general mechanisms of DNA r. are → excision repair, → photoreactivation, and → post-replication (recombinational) repair. DNA r. promotes survival after genetic damage and is involved in DNA replication and → genetic recombination. Inhibition or interference with repair processes might lead to (*Legator & Flamm* 1973): 1. reduced cell survival with fewer mutations; 2. reduced cell survival with more mutations; 3. normal cell survival with fewer mutations. The first effect may be accounted for simply by non-selective inhibition of both repair synthesis and repair recombination. The second can be attributed to a selective inhibition of repair synthesis which could lead to greater recombination repair of existing lesions and hence to the induction of more mutants. The third may be due to specific inhibition of repair recombination under certain conditions.

DNA replication — → deoxyribonucleic acid; chromosome; replicon.

DNA restriction → DNA modification-restriction system.

DNA-RNA hybridization — a process based on the fact that isolated RNA on being mixed with denatured (single-stranded) DNA under appropriate conditions forms hydrogen bonds with the complementary DNA sequences (from which they were transcribed) leading to DNA-RNA hybrids. The rate of DNA-RNA hybridization is limited by the collision frequency of complementary nucleotide sequences. DNA-RNA h. is a useful tool for localizing genes in chromosomes (→ DNA-DNA hybridization).

DNA sealase — = → DNA ligase.

DNA steresis (*Bloom & Leider* 1962) — the loss of parts of the chromosome material after local irradiation with heterochromatic ultraviolet light. The irradiated segments stain but weakly or not at all when treated according to the *Feulgen* or methyl green-pyronine methods. Only a weak color reaction is obtained with the *Alfert-Geschwind* reagent for basic proteins. It is assumed that the effect is due to the lack of DNA-containing material in the segment concerned.

DNA-unwinding protein (*Alberts* & *Frey* 1970) — in *E. coli*, a protein (mol. weight about 22,000; approximately 800 copies per log-phase cell) which binds tightly and cooperatively to single-stranded DNA and much less tightly, if at all, to RNA or double-stranded DNA. The protein facilitates DNA unwinding within a cell by lowering helix stability by virtue of tight cooperated binding to exposed single strands of DNA. It is required for both → genetic recombination and replication of → deoxyribonucleic acid.

DNP — abbreviation for deoxyribonucleoprotein.

deoxyribonuclease — → endo -and exonucleases degrading DNA.

dominant (*Mendel* 1865) — of genetically controlled → characters and the corresponding → alleles which are manifest in all → heterozygous, heterokaryotic or heterogenotic members of the first filial generation (F_1) from a cross between two homozygous (pure-breeding) strains differing in respect to these characters. The dominant character is evident in three quarters of the individuals of the second filial generation (F_2) in diploid sexually reproducing organisms (→ segregation). Those characters and alleles being masked by the dominant ones are referred to as "recessive".

"Dominance" and "recessiveness" are not properties of the genes *per se*, but the result of the action of the genetic locus in question within the total reaction system of the particular → genotype. Both may be modified in degree by environmental influences as well as by other genes ("dominance modifiers") of the genotype. These modifications are brought about by influences on the gene-controlled processes and not on the genetic material proper. In genetic nomenclature, recessive characters and genes (alleles) are generally symbolized by small letters, the corresponding dominant ones by the same letter or letters, but beginning with a capital.

Complete dominance and complete recessiveness are the extreme cases between which all transitional "degrees of dominance" are possible.

Degree of dominance is the ratio of deviation of a heterozygote to that of the dominant homozygote from the midparental value. It is a most important genetic parameter in studying quantitative inheritance concerned with → heterosis and represents a measurement of the mean effects of genes.

1. **complete dominance**: the heterozygous form (Aa) is phenotypically identical with the homozygous form (AA);

2. **semidominance** (= **partial dominance**; incomplete dominance): the phenotype of the heterozygous form (Aa) lies between that of AA and aa, i.e., it is "intermediate";

3. **conditioned dominance** (*Goldschmidt* 1938) (= irregular dominance): the characters and alleles concerned show variable manifestation in the heterozygous state, i.e., in one genotypic environment and under certain environmental conditions they are dominant, under others they are recessive;

4. **conditional dominance**: the dominance of the characters and alleles

in question can be observed in the heterozygous state but the homozygous phenotype is not known;

5. **alternating dominance (= change of dominance; reversal of dominance):** in the course of ontogenetic development of a heterozygous individual the d. effect of one allele of a heterozygous allele pair may be transferred to the other ($A_1a_2 \rightarrow a_1A_2$). In such a case, the alleles are manifest one after the other at various phases of development of the organism, i.e. first one allele is d., and then the other. Alternating dominance can also occur in the phylogeny of some species so that the dominance relationships change from generation to generation. Numerous external and internal factors may cause this phenomenon, e.g., changes in genotypic environment, → position effects, polyploidy, labile genes, etc.;

6. **delayed dominance:** the delayed manifestation of the dominant allele of a heterozygous allele pair so that the alternative allele only relinquishes its phenotypic action at a later stage of ontogenesis;

7. **shifting dominance:** variable dominance of an allele occasioned by external influences;

8. **sex-influenced dominance:** the dominance of the allele concerned differs in the two sexes.

Dominance modifiers (*Fisher* 1928) or **domini genes** (*Goldschmidt* 1935) are genes influencing the degree of dominance of homozygous and/or heterozygous non-alleles and their corresponding phenotypic characters.

dominance deviation — → variation.

dominance modifier (*Fisher* 1928) — → dominant.

dominant lethal — any → mutation (mainly structural and numerical chromosome aberrations) which kills an individual heterozygous for it. It is not possible in any particular case to confirm its genetic nature or to allocate it to a particular chromosome by tests such as are applicable to other kinds of mutations (→ lethal factor).

dominant lethal assay (*Russell* 1951; *Bateman* 1966) — a test for detecting environmental mutagens which induce mutations in germ cells of animals. The d. l. a. is a component of → mutagenicity testing systems.
The genetic basis for dominant lethality is mainly the induction of structural and numerical chromosomal aberrations which may give rise to the preimplantation loss of inviable zygotes, to early fetal deaths, and to sterility and semisterility in F_1 progeny. Thus, in the conventional d. l. a., male rodents are dosed singly with subtoxic concentrations of the agents under test and then mated during weekly periods with groups of untreated virgin females. Matings in weeks 1—3, 4—5, and 6—8 following treatment of male mice, and in weeks 1—5, 6—8, and 8—12 in the case of treatment of male rats, respectively, represent samplings of cells at the postmeiotic and premeiotic stages of spermatogenesis at the time of treatment. Females are inspected daily for vaginal plugs, dissected on day 12 or 13 pregnancy, and scored for corpora lutea and for total implants (comprising early and late fetal deaths and living fetuses). Mutagenic effects are expressed as the mutagenic index, i.e., early fetal deaths/total implants times 100. (*Epstein & Röhrborn* 1971).

domini gene (*Goldschmidt* 1935) — → dominant.
donation (*Clark & Adelberg* 1972) — in bacterial → conjugation, the transfer of the bacterial genetic material and the → conjugon (= promotor) in an independent manner as with → colicinogenic factors (→ conduction).
dosage — → gene dosage.
dosage compensation (*Muller* et al. 1931) — a regulatory mechanism for → sex-linked genes which differ in dose between females and males in organisms with an XX–XY or XX–XO mechanism of → sex determination. It operates such that the resultant phenotypes of these genes are equivalent in the two sexes. Mechanisms of d. c. may differ in different organisms.

In *Drosophila*, three theories of d. c. have been put forward to explain d. c.:

1. The X-activation theory proposes that the single male X is much more active than either one of the two female X's.

2. Each sex-linked gene is assumed to have a set of dosage modifiers (compensators and anticompensators) which are also sex-linked and these act to equalize male and female expression for a particular gene. Some genes are thought to have more than one set of compensators although the compensator genes themselves are not excluded from having other effects.

3. Sex-determining genes themselves are assumed to act as dosage modifiers. It is suggested that these genes in determining sex also set the developmental pattern for each individual and this in turn affects gene expression.

D. c. in mammals (*Lyon* 1966) results from the random inactivation of major X-linked genes on all but one X-chromosome. This inactivation is effected by facultative → heterochromatization (during early embryonic development) of all X chromosomes in excess of one in somatic cells as evidenced by precocious condensation during prophase, late replication, and → sex chromatin formation during interphase.

dosage effect (*Stern* 1929) — 1. the quantitative differential action of the → alleles of a gene on the phenotypic expression of the corresponding characters. The extent of the genetic dosage effect is correlated with the frequency of the particular allele in the genotype.

2. The deviation of phenotypic expression from → dosage compensation towards an effect proportional to the gene dose.

double crossing-over (*Sturtevant* 1914) — → crossing-over.
double diploid (*Warmke & Blakeslee* 1939) — → alloploid cells and organisms with the diploid chromosome complement of two diploid species (= amphidiploid; allotetraploid) (→ double haploid).
double fertilization (*Navashin & Guignard* 1899) — → fertilization.
double haploid (*Warmke & Blakeslee* 1939) — → alloploid cells and organisms with two chromosome sets, each from one species (→ double diploid).
double monoisosomic (*Kimber & Sears* 1968) — a cell or individual which lacks one chromosome pair but has two isochromosomes, one for each arm of the missing chromosome pair (→ monotelomonoisosomic).
double monotelosomic (*Kimber & Sears* 1968) — a cell or individual in which one chromosome pair is missing but two telocentrics, one for each arm

of the missing pair, are present (→ monotelomonoisosomic; ditelomonotelosomic).

double reduction (*Darlington* 1929) — → chromatid segregation.

doublet — an especially thick → band in → polytene chromosomes, probably formed by a → duplication (one-band tandem repeat). Doublets are presumed to be composed of a number of recombinationally separable subunits with a common or related information content.

double telotrisomic (*Kimber & Sears* 1968) — an individual deficient in one chromosome but which has two telocentric chromosomes, one for each arm of the missing chromosome (= pseudotrisomic).

drift — (*Wright* 1921) — → genetic drift.

drift-region (*Blattner* et al. 1972) — the area between the start site and the entry site(s) where DNA-bound → RNA polymerase can move without producing RNA chains by → genetic transcription. The start site is the nucleotide sequence which triggers RNA synthesis at the start point nucleotide (→ promoter).

D-RNA (*Georgiev* 1961) — in eukaryotes, RNA species with nucleotide sequences complementary to nuclear DNA sequences and not representing → ribosomal RNA or → transfer RNA (D-RNA = DNA-like RNA) (→ pre-messenger RNA; heterogeneous nuclear RNA).

drug resistance factor — = → resistance factor.

drumstick (*Davidson & Smith* 1954) — the → sex chromatin of polymorphonuclear neutrophil leukocytes of human blood which is contained in a specific drumstick-like nuclear appendage that has its head (about 1.5 μm in diameter) attached to the nucleus by a thread-like stalk. Drumsticks differ from the sex chromatin of other cells in being extruded from the nucleus and in being visible in only a relatively small proportion of cells (in about 1 in 40 neutrophiles in normal females and in less than 1 in 500 cells in normal males).

ds RNA — abbreviation for double-stranded RNA (ss RNA = single-stranded RNA).

Dubinin effect (*Stern & Kodani* 1944) — a form of variegated-type → position effect which leads to the reduction or inactivation of the action of a normal allele and is conditioned by translocated → heterochromatin in the immediate vicinity of the allele. This type of position effect was first found for the cubitus interruptus gene in *Drosophila* by *Dubinin & Sidorov* (1935).

duplex (*Blakeslee, Belling & Farnham* 1923) — of a polyploid having two dominant alleles for a given genetic locus (as AAa in triploids, AAaa in tetraploids) (→ nulliplex).

duplicate genes (*Shull* 1914) — two identical allele pairs showing the same phenotypic action but localized in different chromosomes. D. g. are of relatively frequent occurrence and are probably due to secondary polyploidy (→ polyploid). In gene formulae, the d. g. are generally given the same symbol but characterized by different suffixes or subscripts.

For the case of heterozygosity and complete dominance, A_1 is dominant over a_1 and A_2 (identical with A_1) over a_2 (identical with a_1). Since A_1 and A_2 have the same phenotypic action, the combinations A_1a_2 and

A_2a_1 are phenotypically equal and different from a_1a_2. For these reasons, the normal F_2-segregation (9:3:3:1) does not occur; instead the segregation ratio is 15:1 provided that the dominant alleles of the d. g. do not act additively or cumulatively ("cumulative d. g."). In this case, the 15:1 segregation becomes 9:6:1 [$9A_1.A_2.$: 6 ($3a_1a_1A_2.$ and $3A_1.a_2a_2$) : $1a_1a_1a_2a_2$].

By analogy with d. g., the presence of 3 identical allele pairs is referred to as triplicate, of 4 identical allele pairs as quadruplicate genes, and more than 4 allele pairs generally as polyplicate genes (→ gene interaction).

duplication (*Bridges* 1919) — a chromosomal structural change (→ chromosome mutation) resulting in the doubling of a section of the genome of pro- and eukaryotes. The size of the duplicated segment may vary considerably.

Duplications are considered to play a role in origin of new genes through functional diversification of duplicate members.

Cytologically, the existence of relatively small duplications in the chromosome complements of eukaryotes may be proved by the appearance of regional alternations of pairing (→ chromosome pairing) in first meiotic prophase and (somatically paired) → polytene chromosomes of Diptera in that member of a pair of homologues which is heterozygous for the duplicated segment.

The smallest duplications that can be investigated cytologically are those of single bands of polytene chromosomes. However, whole chromoomes arms may be duplicated, thus giving rise to → isochromosomes. Where whole chromosomes (linkage groups) are doubled in the haploid chromosome set, this constitutes a → genome mutation (→ aneuploid). Duplications may occur, *inter alia*, as primary structural changes of chromosomes, as a result of disturbances in the crossing-over process (unequal → crossing-over), or as secondary chromosome mutations due to crossing over in inversion- or translocation heterozygotes.

A distinction is made between "inter- and intrachromosomal duplications" according to whether the duplicated chromosome segment is incorporated in another chromosome of the → chromosome set or whether the duplicate region is present in the same chromosome.

1. **Interchromosomal duplications:** the duplicated segment is incorporated into a nonhomologous chromosome (Fig. 36a) or it occurs as a fragment in the chromosome set (Fig. 36b).

2. **Intrachromosomal duplications:** the duplicated segments are contained within a single chromosome (Fig. 36c—h) and may be non-contiguous, i.e. separated by other segments (Fig. 36c, e), or contiguous, i.e. immediately adjacent (= repeats or tandem-duplications) (Fig. 36d, f). Also, they may occur in the same chromosome arm ("intraarm duplications", Fig. 36c—f), or in different ones ("interarm duplications", Fig. 36g, h), with the same order of loci with respect to the centromere as in the original ("direct duplications", Fig. 36c, d), or with the order inverted ("reversed" or "inverse duplications", Fig. 36e, f). In nature, the most frequent type of d. is the tandem d. or repeat.

The genetic consequences of duplications depend on the genetic infor-

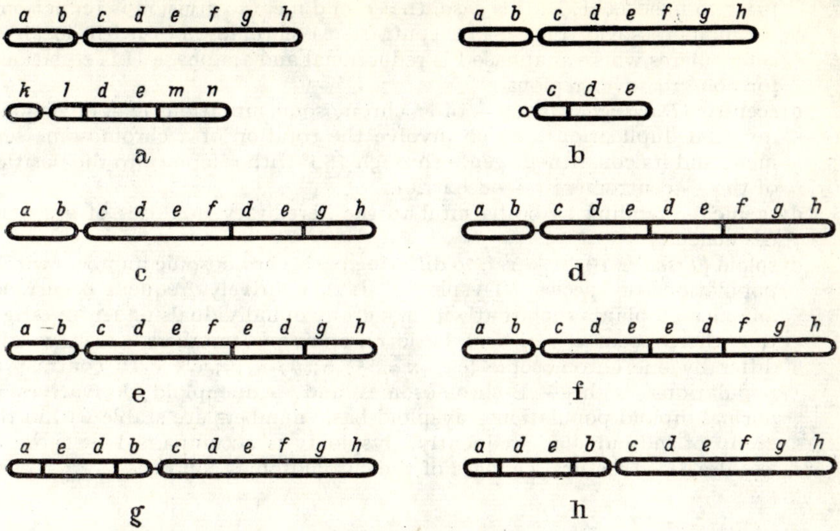

Fig. 36. Diagram showing several types of interchromosomal (a and b) and intrachromosomal (c to h) duplications.

mation which they contain and the change in the gene balance effected by them. When homo- and heterozygous, they may cause an increase or decrease in the viability of their carriers and in extreme cases may act as lethals. Under evolutionary conditions, small duplications may provide a basis for the mutational differentiation of genetic material.

"Complementary Duplication-deficiencies" occur in the case of unbalanced gametes (e.g. AB/CB and AD/CD) produced as a result of translocation heterozygosity (→ translocation) and → adjacent distribution if the chromosome segment duplicated in the one gamete is lacking in the other and *vice versa*. The gametes mutually complement each other and their combination in the zygote $\frac{\text{AB}}{\text{AD}}\frac{\text{CB}}{\text{CD}}$ leads to viable offspring.

duplicon (*Sibatani & Hiai* 1964) — = → replicon.

durability (*Dobzhansky* 1968) — the probability that a → unit of evolution (i.e., a strain, Mendelian population, or a species) will leave descendants after a given long period of time.

dyad (*Nemec* 1910) — 1. a pair of cells resulting from the first meiotic division or from an aberrant course of both meiotic divisions;

2. the two chromatids connected at the → centromere which make up one chromosome in first meiotic division (*Nemec* 1910);

3. in the case of the special type of pure "postreductional" → meiosis (in organisms having chromosomes with non-localized centromeres), the two chromatids go to each pole as independent bodies with no centromeric connections and undergo a secondary pairing during interphase

prior to meiosis II. In this case, the second meiotic division is reductional for non-crossover regions in contrast to chromosomes with localized centromeres where anaphase I is reductional and anaphase II is equational for non-crossover regions.

dyscentric (*Darlington* 1926) — of → chromosome mutations (e.g. inversions, inverted duplications) which involve the rotation of a chromosome segment and its constituent genes through 180° with reference to the position of the → centromere (→ eucentric).

dysgenic — tending to be harmful to the hereditary qualities of a species (→ eugenic).

dysploid (*Tischler* 1937) — ref. to differing basic chromosome numbers within populations or species. "Dysploidy" is of relatively frequent occurrence among seed plants such that within a group of individuals under investigation there occur a series of basic chromosome numbers which usually differ by one chromosome (e.g. $x = 5; 6; 7; 8; 9;$ etc.) In contrast to populations with → B-chromosomes and → aneuploid derivatives of normal diploid populations, dysploid basic numbers are stable within the group of individuals. Frequently, dysploidy is accompanied by more or less drastic structural changes of the chromosomes concerned.

E

early enzyme — any of the bacteriophage-induced enzymes (= early protein; phage-induced enzyme) that are synthesized *de novo* after phage infection under the direction of the invading viral genome. Early enzymes direct the synthesis of phage DNA or its precursors and are of two types: 1. An enzyme activity which was absent in the uninfected bacterial cell; 2. a protein synthesized after infection having an enzyme activity very similar or identical to an enzyme found in uninfected cells, whose rate of reaction is many times greater than that of the host enzyme or with different properties or requirements.

early gene — any of those bacteriophage or virus genes expressed before DNA synthesis in the host cell begins (→ late gene).

echylosis (*La Bella & Krass* 1968) — the release of non-particulate material from the cell through an apparently intact → cell membrane (→ inchondriosis; inchylocytosis).

ecophene (*Turesson* 1922) — the range of → phenotypes produced by one → genotype within the limits of habitat under which it is found in nature (*Darlington & Mather* 1949).

ecophenotype — any nongenetic modification of the → phenotype in response to an environmental condition.

ecospecies (*Turesson* 1922) — a group of → ecotypes capable of interbreeding without loss of fertility or vigor in the offspring.

ecological density — the number or mass of individuals per area or volume of habitable space.

ecological niche — an abstraction that includes all of the physical, chemical, physiological, and biotic factors that an organism needs in order to survive.

eclipse — the decrease or disappearance of infective virus particles which accompanies the release of the viral genome into a host cell.

ecliptic period (*Luria* 1950) — the period (= dark period) which elapses between the infection of a bacterium with bacteriophages and the occurrence of the first infectious phage offspring. It corresponds to the first half of the so-called latent period.

ecoclines — grades of variation within a species. They occur as reactions to different ecological zones occurring within the species distribution (→ cline).

ecosystem — the complex of living and nonliving components in a specified location that comprise a stable system in which the exchange of material follows a circular path (→ biome).

ecotone — → biome.

ecotype (*Turesson* 1922) — a local race (= ecological race) with genotypes adapted to a particular restricted habitat as a result of natural → selection within the local environments. Ecotypes of the same → ecospecies are interfertile and produce fertile offspring.

ectodesma (*Lambertz* 1954) — any of the → plasmodesma-like structures in the outer epidermal cell walls of many plants. They extend from the cell lumen to the undersurface of the cuticle.

ectogony (*Waller* 1917) — → xenia.

ectopic pairing (*Slizinski* 1945) — the tendency of specific regions (→ bands) of *Drosophila* → polytene chromosomes to stick to other regions in different chromosomes.

ectoplasm (*Haeckel* 1873) — the external, peripheral layer of the → cytoplasm of a cell which is more rigid and devoid of granules (= cortex).

eduction (*Sunshine & Kelly* 1971) — the loss of host genetic material concomitant with the exit of a → plasmid (episome) integrated into the host chromosome. The resulting deletion-containing cells are called eductants.

effective breeding population (*Wright* 1931) — the effective number of breeding individuals of a population (= effective population size). (→ breeding population).

effectively dominant (*Muller* 1950) — → dominant.

effector (*Jacob & Monod* 1961) — 1. in genetics, a small molecule (metabolite) that interacts with the → repressor of an → operon, activating or inactivating the repressor molecule with respect to its ability to combine with the → operator. In the case of enzyme induction, "positive effectors" or "inductors" decrease the activity of the repressor. In the case of → repression of enzyme formation, "negative effectors" or "co-repressors" enhance the activity of the repressor and inhibit the synthesis of the repressible protein.

2. In enzymology, a regulatory metabolite of an enzyme (= modifier or modulator) that typically modifies the affinity of an enzyme for its substrate and frequently for other reaction components as well. Effectors need not to have similarity with the substrate of the enzyme whose activity is influenced. In some cases, the substrate has been found to act as an activator. This situation is referred to as substrate co-operation. A positive e. increases the activity or binding of a second ligand, a negative e. decreases it (→ allosteric).

egg — in the case of oögamy, the female → gamete specialized for the storage of food, as opposed to a sperm (male gamete) which is specialized for motility and fertilizes the e. in case of sexual → reproduction. The e. is a cell system which has reached a definite step in organization. Its cytoplasm is ordered in the form of a gradient system. Certain cell particles can be enriched and secreted in specific regions and a pattern may be discernable in the → cortex which is decisive for the further course of development (→ predetermination).

egg mother cell — a megasporocyte (symbol EMC) from which an → egg cell is derived during → megasporogenesis in higher plants.

elaioplast (*Wakker* 1888) — a → plastid in which oil predominates as the storage product. Elaioplasts are specialized leukoplasts.

electron transport particle — a particle (symbolized ETP) derived from → mitochondria, and capable of carrying out electron transport from suitable substrates to oxygen.

elementary membrane (*Sitte* 1961) — = → unit membrane.

elementary particle (*Fernandez-Moran* 1962) — a repeating particulate subunit (symbol EP) of the mitochondrial membranes (→ mitochondrion). 10^4 to 10^5 of these particles, disposed in regular arrays, are associated with the cristae mitochondriales and the inner membrane of the mitochondrial envelope. Each EP may exist in an extended or in a compressed form. The extended form consists of a spherical or polyhedral head piece (diameter 8 to 10 nm), a cylindrical stalk (about 5 nm long and 3 to 4 nm wide) and a base piece (4 to 11 nm), which is an integral part of the outer dense layers of the cristae. In the presence of ATP the extended form is converted to the compressed form which is a single spherical unit of about 15 nm in diameter. The molecular weight of an EP, which is lipoprotein in nature, is about 1.3×10^6. EPs are assumed to represent the functional units of electron transport and to contain a complete supply of those enzymes which are responsible for oxidative phosphorylation. The functional parts form a unit with the structural protein.

The average distance between two adjacent EPs is about 11 to 11,5 nm. Particles similar in dimensions to the EPs of mitochondria have been isolated from chloroplasts and are designated → quantasomes (*Fernandez-Moran* et al. 1964).

elimination chromatin (*Seiler* 1914) — → chromatin elimination.

elimination coefficient — the frequency with which certain genotypes (carrying specific genes or gene combinations possessing deleterious effects) die prematurely or, resp., are hindered during reproduction and are genetically eliminated as a consequence (→ genetic death). If, for example, a certain gene has an average e. c. of 5% then this means that of 20 individuals possessing this gene among their genotypes, one will perish before the gene concerned can be transmitted to the offspring. Under these conditions, the gene will survive in 20 generations before it is eliminated. Its → persistence is 20 generations.

elongation factor (*Lucas-Lenard & Lippmann* 1966) — soluble protein factors (= transfer factors) involved in the polymerization of peptides (elongation of peptide chains) during → genetic translation. Unlike → termination and → initiation factors which catalyze single steps in protein synthesis, e. factors are required for the formation of every peptide bond.

In *E. coli*, two major proteins (called EF—G and EF—T) are involved in the elongation step for peptide chains. EF—T is split into two subfractions, EF—Tu and EF—Ts. Energy released from the hydrolysis of GTP by EF—G or by EF—T is thought to be essential for chain elongation. EF—Tu catalyzes the GTP-dependent binding of aminoacyl tRNA to the A site of the → ribosome. EF—Tu seems to have one binding site for aminoacyl tRNA and perhaps another for GTP, GDP or EF—Ts. EF—Tu and EF—Ts correspond to two of the four subunits of bacteriophage Qβ replicase.

EF—Tu — GTP reacts with aminoacyl tRNA to form an aminoacyl tRNA—EF—Tu—GTP complex, which transports aminoacyl tRNA to the ribosome-messenger RNA complex where release of inorganic phosphate and EF—Tu — GDP occurs. EF—Ts is a catalytic factor involved in the production of EF—Tu — GTP from EF—Tu — GDP. The sequence of reactions is (*Leder* 1973):

1. EF—Tu · GDP (inactive) + EF—Ts →
   EF—Tu ·EF—Ts (active) + GDP;
2. EF—Tu · EF—Ts + GTP → EF—Tu · GTP + EF—Ts;
3. EF—Tu · GTP + aminoacyl tRNA →
   aminoacyl tRNA · EF—Tu · GDP;
4. aminoacyl tRNA · EF—Tu · GTP $\xrightarrow[\text{mRNA}]{\text{ribosome}}$
   aminoacyl tRNA—mRNA— ribosome + EF—Tu · GDP + P_i.

EF—G promotes the translocation of peptidyl tRNA from the acceptor site (A site) to the peptidyl (P) site on the ribosome. Concomitant with this translocation is the expulsion of deacylated tRNA from the peptidyl site, coordinate movement of the mRNA along the ribosome, and hydrolysis of GTP (GTPase activity). EF—G is bound to the larger subunit of ribosomes. The sequence of reactions is:

1. ribosome + EF—G + GTP → ribosome · EF—G · GTP;
2. ribosome · EF—G · GTP → ribosome · GDP + P_i;
3. ribosome · EF—G · GDP → ribosome + EF—G + GDP.

Mitochondrial e-factors are similar and interchangeable with their bacterial counterparts.

In eukaryotes, two e. factors (EF—1 and EF—2) have been found. EF—1 appears to have the properties of bacterial EF—Tu in binding aminoacyl tRNA to the ribosome. EF—2 is directly analogous to the prokaryotic EF—G but specific for eukaryotic cytoplasmic ribosomes.

embryo — in animals, the young organism arising from the fertilized → egg cell. In seed plants, the young → sporophyte arising from the union of the female and male sex cells (→ embryonic development). The origin, development and evolution of the e. is referred to as "embryogenesis" or "embryogeny".

embryonic development — the canalized process of histogenesis of multicellular organisms whereby fertilized egg cells (zygotes) are transformed into a restricted number of alternative states among which there are few if any intermediates. During e. d. the cells arising from the zygote by mitosis become different (→ induction) from one another although they are presumably endowed with identical genomes (→ cytodifferentiation). Programming for e. d. begins as early as oögenesis and the cells of an embryo differ from one another at their earliest delimitation. Each of the canalized pathways of e. d. is called a **"creode"** by *Waddington* (1962), and is the result of essentially all the available gene-reaction systems whose intensities are mutually adjusted by interlocking control systems (→ genetic regulation).

E. d. is a highly complex process. In the first phase the various parts of the germ come to exhibit clearly apparent differences ("regionalization", "embryonic segregation" or "arealization"). By a second process (histogenesis) any one particular region gradually changes its character. A third major process consists in the moulding of a region of the egg into a definite shape and is called "morphogenesis".

E. d. involving all 3 major processes may be called "individuation" since it confers on the developing material the particular and specific individual character of the organ into which it is developing.

Three major types of developmental systems may be distinguished: 1. those depending on segregated oöplasms, 2. those depending on gradient systems, and 3. those depending on inductive relationships.

Early e. d. begins with → fertilization and proceeds by a series of stages (cf. *Waddington* 1956):

1. **Fertilization:** in this stage, two processes are important for the inception of e. d. in sexually reproducing organisms: the union of the haploid egg nucleus with the sperm, and the activation of the egg leading to the start of division. The processes are separate from one another and it may happen that activation occurs without fusion of the nuclei.

2. → **Cleavage:** the egg cell divides into ever smaller descendants by → mitosis. In this process, the eggs of each species follow a quite definite cleavage pattern at least for the first 3 or 4 cleavages (some species considerably longer). The position of the cleavage planes is largely a consequence of the arrangement of the mitotic spindles which in turn is controlled by local differences in the egg cortex, including the organization of the cytoplasm.

3. **Blastulation:** the newly formed cells migrate to the periphery of the "cell cluster" producing in the simplest case a cavity filled with liquid and surrounded by a cell layer – the blastula. This marks the end of cleavage period. Depending on the amount of yolk, the blastula may be formed asymmetrically. In eggs very rich in yolk, e. g. bird's eggs, it may be reduced to a flat plate of cells – the blastoderm – which swims to the upper pole of the yolk mass.

4. **Gastrulation:** This takes place immediately after the formation of the blastula. During this short and extremely critical period of e. d., the different regions of the blastula are so folded and pushed into each other that the actual embryo is produced, generally consisting of three separate layers (in *Coelenterata* and lower forms, only two layers – an outer and an inner). These three fundamental layers are called the **ectoderm, endoderm** and **mesoderm.** The ectoderm is the outermost layer, from which derive the skin and nervous tissue. The inner layer, the endoderm, gives rise to the intestines and their appurtenances, while the middle layer, the mesoderm, builds the muscles, skeleton, etc.

In the different groups of organisms, the folding which brings the individual layers into their correct positional relationship may differ greatly, but apart from this, the formation of an outer, a middle and an inner layer is common to all animal species.

During gastrulation a new and widely distributed cell differentiation begins. From this point onward the e. d. is organized by the controlled activity of the cell-borne genetic apparatus. Although the genome of the zygote and the blastula cells stemming from it show a slowly increasing level of activity throughout cleavage and blastulation, the indispensible nature of embryonic gene action is not observable prior to gastrulation. Gene activation sets in suddenly in a short period preceding the onset of gastrulation and most of the RNA synthesized during this activation appears to be → messenger RNA (*Bacharova* et al., 1966). Likewise, ribosomal RNA occurs shortly after the beginning of gastrulation. Nevertheless, the ribosomal RNA necessary during cleavage and blastulation is synthesized very much earlier (mainly during oögenesis, as has been shown in studies with → lampbrush chromosomes).

5. **Formation of basic organs:** Very soon after gastrulation, the fundamental pattern of the embryo begins to emerge. In most cases those organs are produced which remain during subsequent stages of development and form the most important organs of the adult animal. However, in some animals the embryo first develops into a larva in which organs are produced which will undergo further radical changes during the transition to the adult animal, so that it is very difficult to establish a general scheme. The e. d. of the vertebrates proceeds more or less according to the following scheme. First, the ectoderm forms the epidermis covering the entire body. A thickened plate, differentiated from the epidermis, folds itself into a groove and finally becomes enclosed into a tube forming the central nervous system. Cells from the ectodermal sheet inside of the embryo migrate to the juncture between neural and epidermal parts of the ectoderm and there form nerve ganglia and other organs. The mesoderm sheet is divided lengthwise in a series of individual segments. Along the middle of the embryo there forms a long rod-shaped structure, the notochord, the first skeletal element to form. The mesoderm thickens on either side of the neural tube and becomes transversely segmented so that a series of block-shaped tissues (the somites) are formed. These form the main musculature and the innermost layers of the skin. At each side of the somites, nephri or kidneys are produced from a mesoderm segment. Within the embryo the endoderm is folded into a tubular structure from which the gut and the digestive system arise during the further course of development. The formation of this organ always begins earlier at the anterior than at the posterior end of the embryo.

embryonic differentiation — → differentiation resulting in embryonic development.

embryonic field — a chemo-differentiated region of embryonic tissue during early → embryonic development. The development of any given point of an e. f. depends on its relations with the other points in its neighborhood or on its position within the region as a whole. The term "field" is used to refer to the coordinated and integrated character of a whole complex of processes proceeding within such a region. An e. f. giving rise to the formation of a definite organ with characteristic individual shape may be referred to as as "individuation field" (*Waddington & Schmidt* 1933).

embryonic induction — → induction.

embryonic lethal — → lethal factor.

embryo sac (*Hofmeister* 1849) — the mature female gametophyte in higher plants (angiosperms) (→ megasporogenesis).

embryo sac competition — → *Renner* effect.

embryo sac mothercell — → megasporogenesis.

emiocytosis (*Lacy* 1961) — the process by which cellular constituents are released from cells (= reverse → pinocytosis or exocytosis).

empiric risk — in human genetics, the risk that a certain child will have the condition or trait under study in those cases where genetics is obviously involved but no definitive pattern of inheritance can be demonstrated.

endochromocenter (*Geitler* 1953) — an enlarged → chromocenter produced by → endomitosis and the concomitant increase in chromosome number.

endocytosis (*de Duve* 1963) — the bulk uptake of material through the cell membrane by formation of a vacuole (e. = pinocytosis and phagocytosis, collectively). For e., a cell puts forth groups of undulating membranes at the surface which infold globules of the surrounding fluid (pinocytosis) or solid particles adherent to the cell (phagocytosis). These folds are sealed off to form membrane-bound intracellular vesicles (pinosomes and phagosomes).

endoduplication (*Jorgensen* 1928) — the regular doubling of the haploid chromosome number in andro- or gynogenetic development owing to the failure of cell-wall formation in the first division of the zygote and the union of the spindles in the subsequent mitosis.

endogamy — all systems of sexual → reproduction (= inbreeding) in which the mating partners are more closely related than in random mating (panmixia) within the population (→ mating system).

endogenote (*Morse, Lederberg & Lederberg* 1956) — that part of the bacterial chromosome which is homologous to the genome fragment (the "exogenote") transferred from the donor into the recipient cell during the formation of a → merozygote. A bacterial cell which contains an "exogenote" in addition to its genetic material is referred to as a "syngenote" and is hyperploid for the fragment concerned. The syngenote may be → heterogenotic or homogenotic.

endomembrane system (*Morré* et al. 1971) — the functional continuum of membraneous cell components consisting of the → nuclear envelope, → endoplasmic reticulum and → Golgi apparatus as well as vesicles and other structures such as → annulate lamellae derived from the major components. All these are local specializations of a single interassociated membrane system peculiar to eukaryotic cells.

endomitosis (*Geitler* 1939) — a form of somatic polyploidization (→ polyploid) which is fairly common in differentiated or differentiating tissues. It takes place within an intact → nuclear envelope and gives rise to endopolyploidy, i.e., duplication of the chromosomes without cell or nuclear division in the usual sense. E. is characterized by an increase in DNA content.

In some organisms, e. has been found to resemble a normal mitotic cycle. Endoprophase is followed by endometaphase. In endoanaphase the chromatids separate within the nucleus, and in endotelophase an interphase state with double the original chromosome number is attained. The formation of the → mitotic apparatus and of a metaphase plate is omitted. The autonomously separated chromatids normally remain in a typical parallel arrangement.

An e. is termed "masked e." (*Resch* 1952) or **"endoreduplication"** (*Levan & Hauschka* 1953), where no stages formally comparable with the normal mitotic cycle are observed. An endoreduplication may be established with certainty if it is followed by a normal mitosis in which the products of endomitotic chromosome reduplication reveal themselves as → diplo- or quadruplochromosomes which separate before anaphase producing a doubled or quadrupled number of individual chromosomes, as the case may be.

Endopolyploidization associated with e. may lead to very high degrees of ploidy. In general, fixed degrees of ploidy are characteristic of certain tissues. In most cases, endopolyploidy represents an ultimate state of → cytodifferentiation and is not a factor of cytodifferentiation.

If separation of the reduplicated chromosomes products formed by e. fails to occur, so-called polytenic or polynemic chromosomes result (→ polytene chromosomes).

endonuclease — any enzyme capable of making internal cuts in DNA chains (→ exonuclease).

endophenotype (*Lewis & John* 1963) — → phenotype.

endoplasm (*Pringsheim* 1854) — the granular central portion of the → cytoplasm as opposed to → ectoplasm. Both e. and ectoplasm are defined by light microscopic studies (→ endoplasmic reticulum).

endoplasmic reticulum (*Porter* et al. 1945) — a complex membraneous (tubular and vesicular) system (symbol ER) in the cytoplasm of eukaryotic cells which is a component of the cell's → endomembrane system, delimits → cell compartments, and conveys metabolites, macromolecules, and perhaps electrolytes into, out of, and within the cytoplasm. The ER consists largely of lipid (30 to 50%) and protein (40 to 60%) with some RNA. The lipids are mostly phospholipids (50 to 66%) among which lecithin predominates. The protein is partly structural and partly enzymatic.

The lipoprotein membranes (→ unit membrane) differ in configuration and constitution from the → cell membrane and bring the → nuclear envelope into close association, if not in continuity, with the → Golgi apparatus and the cell membrane. Many of the ER membranes are studded with ribosomes **(rough ER)**; others are devoid of ribosomes **(smooth ER).** Membranes of the rough ER measure approximately 5 to 6 nm in

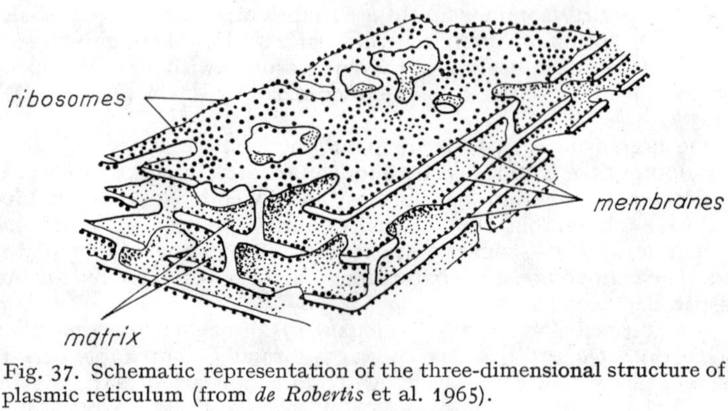

Fig. 37. Schematic representation of the three-dimensional structure of the endoplasmic reticulum (from *de Robertis* et al. 1965).

width, whereas the cell membrane is 7.5 to 10 nm wide. The membranes of the ER are symmetrical, but show dense cross bridges probably representing protein bands separating lipid micelles (Fig. 37).

The ER is the most variable cell organelle and forms a dynamic, rapidly changing structure. Its formation is cell-type specific. It develops moderately in very young meristematic cells, but can develop strongly in growing cells and during mitosis. ER elements are sparse in mature, metabolically relatively inactive cells, but highly developed in cells actively engaged in protein synthesis.

A variety of biochemical activities encompassing the entire gamut of metabolic events take place within the ER, many of which may be related to its canicular structure. Such activities include (*Breswick & Schwartz* 1968):

1. Carbohydrate metabolism.

1.1 Biosynthetic systems responsible for glycogen synthesis are intimately associated with the → microsomes.

1.2 Degradative enzymes (e.g., glucose-6-phosphatase), are located exclusively in the membraneous portion.

2. Protein metabolism.

2.1 The synthetic mechanisms by which the proteins are elaborated are located in the rough ER.

2.2 Degradative enzymes (e.g., peptidases and amino acid oxidases) are contained within the smooth ER.

3. Lipid metabolism.

3.1 Fatty acid, steroid, and phospholipid biosynthesis are functions of the smooth ER.

3.2 The metabolism of steroids (including oxidation and reduction reactions) also occurs in the smooth ER.

3.3 Esterases (e.g., cholinesterase) have been found in the smooth ER.

4. Electron transport and related processes.

4.1 Certain elements of the electron transport system (e.g., NADH and NADPH-cytochrome b_5 reductases) are present in the smooth ER.

4.2 The microsomal membranes possess an active ATPase complex which may be related to ion transport in this system.

endopolyploid — of cells whose chromosome number has been increased by → endomitosis and for which the degree of ploidy is proportional to the number of endomitoses which have taken place.

endoreduplication (*Levan & Hauschka* 1953) — a type of endomitotic → chromosome reduplication in the interphase stage of the → cell cycle. E. takes place without any mitosis-like manifestation and consists of a double, triple or multiple reduplication of interphase chromosomes. It is widespread in eukaryotes, represents the most important process of somatic polyploidization, and may originate either spontaneously or be induced by different treatments (e.g., low temperature, irradiation, chemical treatment). When more than one extraduplication occurs, this may be called repetitive e.

In cells and tissues where → differentiation takes place in the absence of mitosis, nuclear growth occurs quite commonly by endoreduplication (→ endomitosis; polyploid).

endosome (*Novikoff* 1963) — a "young" → vacuole containing material that is to be digested, but lacking, as yet, the necessary digestive enzymes (= phagosome).

endosperm — a tissue specialized for nourishing the developing embryo of flowering plants. The e. arises from double → fertilization by the second pollen (male) nucleus of two of the eight nuclei of the embryo sac (primary e. nucleus) and is typically triploid. Since it does also originate from the fusion of one female nucleus with one or more others, or with a male nucleus, or with both, it may be diploid, triploid, tetraploid, pentaploid or even higher ploid depending on the species in question.

end point mutation (*Demerec* 1946) — any experimentally induced → gene mutation in bacteria which due to delayed manifestation (→ cell division lag; mutational lag) becomes recognizable only after a series of cell divisions (→ zero point mutation).

end product inhibition — a biological control mechanism in sequential → enzyme systems in which the accumulation of the final product of a sequence of metabolic reactions causes the inhibition of its own formation. End product inhibition is an example of "negative feedback control-mechanisms", which may be grouped into competitive and non-competitive ones.

1. **Competitive mechanisms:** The inhibition of an early enzyme in a sequence by the end product of the enzyme sequence is competitive with the substrate of an early enzyme. There is structural similarity between end product and substrate.

2. **Non-competitive mechanisms:** The end product of the enzyme sequence inhibits an earlier enzyme in the sequence in some manner other than by competition with the substrate for the catalytic site. The end product which is inhibitory becomes bound to an independent site of the enzyme. This causes a conformational change in the enzyme, altering its catalytic activity. The second site is specific for binding the end product of the multienzyme sequence. Enzymes of this types are called → allosteric.

enforced heterozygosity (*Muller* 1917, 1918) — → heterozygous.

enhancement — increased yield of viruses or increased cytopathic effects, or both, after mixed infection of cells by unrelated animal viruses, at least one of which is noncytocidal.

enhancer — → modifier gene.

enneaploid — of a → polyploid with nine chromosome sets in the somatic cells.

entry exclusion — → surface exclusion.

entry site — the ribosome site available for the initial binding of → transfer RNA during → genetic translation. In the process of initiation (→ initiation complex), → initiator tRNA binds first at this socalled A site and is then translocated to a second (peptidyl or P) site, in which state it would be reactive with a second aminoacyl tRNA molecule which could now bind to the e. s.

enucleate — of a → cell lacking a → nucleus. The removal of the nucleus with very little cytoplasm is called enucleation and represents a classical method for investigating interactions between nucleus and cytoplasm (→ nuclear transplantation).

environment — those conditions, external or antecedent to an organism, a genotype, or a particular locus of the genotype which are related to its development. The interaction of the e. with the → genotype determines the → phenotype. The components of the e. may be subdivided into those which are genetic (→ residual genotype) and those which are nongenetic (→ environmental variance).

The heterogenicity of the e. can have a profound effect on evolutionary processes and can influence directly the genetic structure of → populations.

environmental mutagen — any → mutagenic agent which is part of the human environment (→ mutagenicity testing).

environmental resistance — the sum of the physical and biological factors which prevent a → species from reproducing at its maximum rate (→ biotic potential).

environmental variance — that portion of → phenotypic variance caused by differences in the → environment to which the individuals in a population have been exposed.

enzyme — a → protein which even in low concentration speeds up, enables or controls chemical reactions in living organisms without being used up in the reactions. Enzymes are typically composed of a protein part (the apoenzyme) which confers specificity and a nonprotein part (the coenzyme) necessary for activity. Enzymes bind to substrates thereby affecting their breakdown or union with other substances. They are molecules of complex chemical constitution produced in living cells under genetic control (→ genetic transcription; genetic translation).

Most enzymes have molecular weights substantially higher than 17000 and contain more than one peptide chain. Normally, these polypeptide subunits show no catalytic activity by themselves. It is essential that the subunits be fitted together exactly in a specific, three-dimensional conformation to preserve enzyme activity.

Two fundamental processes are characteristic for e. action:

1. a **"recognition reaction"** during which an e. selectively binds a substrate and exhibits discrimination between closely related molecular structures (e.g. d- and l-forms);

2. a **"catalytic reaction"** in which covalent bonds are made or broken to form products.

A particular region of the e. possesses or can assume a conformation complementary to its substrate. This provides for binding which presents the substrate in an optimal configuration to the catalytic site of the e. or coenzyme. The catalytic site is that portion of the e. where the substrate is bound and where it is transformed into its reaction products. The catalytic sites contain specific charged groups, as well as hydrophobic groups, contributed by specific amino acid residues in the peptide chain of the enzyme. Different specific binding points of the active site may arise from widely separated points on the peptide chain, since the latter is ordinarily in a folded conformation in the intact enzyme structure.

Changes in its secondary or tertiary conformation may greatly alter the enzyme's catalytic activity (→ allosteric).

During or after binding of specific substrate molecules, enzymes undergo a conformational change called the "induced fit" of the e. to its substrate. **Catalytic power** designates the ability of an e. to accelerate a chemical reaction. **Catalytic factor** is the term used to define the quantitative ratio of the enzymatic and nonenzymatic rates under some conditions. **Catalytic residues** are the amino acid residues directly involved in the covalent bond changes during enzyme action. **Specificity residues** designate those residues involved both in the binding of substrate and in subsequent processes of the chemical reaction which are not directly involved in covalent changes. **Contact residues** designate those residues which contain at least one atom within a bond distance (about 0.2 nm) of the ligand under observation and may include both catalytic and specificity residues. **Kinetic specificity** refers to the specificities in steps following the binding step. **Binding specificity** is the specificity of the binding process and **active site** designates that general region in the neighborhood of the catalytic residues identified with the binding and reaction of substrates.

Two or more enzymes acting in a linear sequence so that the product of the first is the substrate of the second, and so on, make up the simplest enzyme systems in a single homogeneous aqueous phase. In the cell such systems may consist of as many as 12 or more separate enzymes acting linearly. Sometimes there is "branching" of enzyme systems allowing metabolic channelling. Some systems have been shown to be cyclic or closed.

Enzymes that are synthesized in fixed amounts, irrespective of the growth conditions and in absence of an → inducer are referred to as **"constitutive enzymes".** Those enzymes which are synthesized only in the presence of their substrates are called **"inducible enzymes".** In this case, the substrate acts as an inducer of enzyme synthesis ("enzyme induction"). In the absence of substrate the gene determining the enzyme in question is inactive (→ genetic transcription; genetic translation). According to the → operon model, the inducible enzyme synthesis is started when the inducer (→ effector) interacts with definite sites of the → repressor molecule inactivating it and thus activating the gene responsible for enzyme synthesis. Mutations of the → regulator gene or of the → operator gene of an operon may transform inducible enzyme synthesis into constitutive enzyme synthesis. Enzyme induction (as well as enzyme → repression) involves not a single e., but a sequence of enzymes acting in successive metabolic steps. Enzymes whose rates of production are observed to vary together show "coordinated enzyme synthesis". Enzyme synthesis is called "repressible" if it is inhibited by the product of the reaction which the enzyme catalyzes (→ end product inhibition; repression).

Five categories of reactions encompass virtually all the chemical reactions catalyzed by enzymes:

1. hydrolysis, i.e. breaking a bond with the addition of the elements of water, and separation into two molecules;

2. group transfer, i.e. transfer of an integrated group of atoms from one molecule (donor) to another molecule (acceptor);

3. oxidation and reduction, i.e. transfer of one or more electrons, or hydrogen atoms (electron + proton), from one molecule (the one undergoing oxidation) to another molecule (the one undergoing reduction);

4. isomerization, i.e. rearrangement of the position of some atom, or group of atoms, within a molecule;

5. condensation, i.e. linking, by covalent bonds, of two molecules (like or unlike) to form a new molecule.

enzyme cytology — → cytology.

enzyme induction (*Monod* et al. 1951) — an increase by → derepression of the rate of synthesis of an enzyme as determined by the synthesis of a specific → messenger RNA. E. i. is a rather rapid monophasic response which is dependent upon the presence of an → effector, usually a metabolite. The opposite process is called **enzyme repression,** i.e., the disappearance or nonsynthesis of an enzyme as determined by the absence of a specific mRNA synthesis.

An increase in enzyme activity in response to a change in a specific environmental factor is termed **apparent e. i.**

enzyme repression — → enzyme induction.

eobiogenesis (*Pirie* 1937) — term used to designate the first instance of life (→ biopoesis; neobiogenesis).

eobiont (*Pirie* 1937) — → biopoesis.

eosome (*Mc Carthy, Britten & Roberts* 1962) — → ribosome.

epigenesis — in embryology, the theory that new structures and organisms develop from an originally undifferentiated mass of living material in the course of → embryonic development (→ preformation).

epigenetic — of all processes relating to the expression (→ genetic transcription; genetic translation) and the interaction of the genetic material. E. mechanisms may act at three levels of cell organization: 1. They may directly regulate gene function (turn on and turn off) or may modulate the synthesis of specific kinds of proteins. 2. They may regulate cell differentiation by modifying the translation of RNA into proteins. 3.They may regulate the topographic distribution and function of proteins (→ epigenetics; epigenotype).

epigenetics (*Waddington* 1940) — that branch of biology which deals with the causal analysis of development.

epigenotype (*Waddington* 1939) — the total developmental system consisting of a series of interrelated developmental pathways through which the adult form of an organism is realized. It comprises the totality of interactions among genes and between genes and the nongenetic environment resulting in the phenotype (→ epiphenotype). The e. of a cell is a stable, heritable (during at least many cell generations) character whose mode of impression is over and above or in addition to the classical genotype, i.e., the base sequence in DNA.

epinucleic (*Lederberg* 1966) — of a type of → heredity based upon the restrictive information transmitted in tissue lines of multicellular organisms. It cannot sensibly be attributed to DNA-sequence codes.

epiphenotype (*Cahn* 1969) — The aggregate of synthetic capabilities and other properties resulting from the expression of recognizable → cytodifferentiation (→ epigenotype).

episite — → lysogenization.

episome (*Thompson* 1931; *Jacob* & *Wollmann* 1958) — any bacterial → plasmid which can exist in two states within the cell, either independently in the cytoplasm or, following insertion, as an integral part of the host's chromosome.

epistasis (*Bateson* 1907) — a form of → gene interaction whereby one gene interferes with the phenotypic expression of another nonallelic gene (or genes), so that the phenotype is determined effectively by the former and not by the latter when both genes occur together in the genotype. Genes whose expression is altered by nonallelic genes are said to be "hypostatic", or to exhibit "hypostasis". Epistatic action may depend upon the presence of a dominant allele A, the recessive allele a, or the heterozygous genotype Aa. The hypostatic reaction of nonalleles may depend upon the presence of B, b or Bb.

1. **E. of dominant genes:** if gene A is dominant over a and gene B is dominant over b, then A is recognized as the epistatic gene if the expression of B and b is suppressed and the genotypes A.B. and A.bb appear the same. In this case, the normal F_2 dihybrid → segregation ratio (9:3:3:1) is modified to 12 (9A.B. + 3A.bb):3aaB.:1aabb.

2. **E. of recessive genes:** in this case recessive gene a suppresses the manifestation of B and b. Under this condition the F_2 segregation ratio of 9A.B.:3A.bb 4(3aaB. + 1aabb) results.

In population genetics and quantitative genetics, the term epistasis is sometimes used to refer to **all** nonallelic gene interactions (→ interaction deviation).

epistatic deviation — = → interaction deviation.

epistatic disequilibrium (*Moran* 1967) — = → linkage disequilibrium.

epithelial — in tissue culture, of cells whose morphology is compact and resembles that of squamous epithelium as opposed to spindleshaped, fibroblast-like cells.

epivirus (*Brinton* 1972) — any → plasmid whose infectious form is obligatory cell attached. The infectious plasmid system is virus-like, i.e., it has a double-stranded → replicative form, a single-stranded infectious form, and an extracellular (but not cell-free) infectious particle; in contrast to virion viruses, epiviruses can introduce DNA into a cell that is not limited by the size of the virion "package".

equational division (*Weismann* 1887) — 1. = → mitosis;
2. → reductional division.

equational exception (*Bridges* 1916) — in the case of autopolyploid forms, of XXY flies of *Drosophila*, of crossing-over in → attached—X females of *Drosophila*, or of crossing-over and primary → nondisjunction, a gamete with chromosomes which contain sister chromatid segments (→ chromatid segregation).

equational separation — the separation of sister chromatids or sister chromatid segments and their distribution to different cell poles (= equation) (→ reductional separation).

equational split — the "split" (perpendicular to the reductional split) in homologous chromosomes, paired during → meiosis I (→ chromosome pairing), which separates the two chromatids of a reduplicated chromosome.

equator — the periphery of the → spindle on a plane perpendicular to the spindle axis and equidistant from the poles in a dividing cell; the periphery of the → metaphase plate.

equatorial plane — the plane midway between the two daughter nuclei of a dividing cell.

equatorial plate (*van Beneden* 1875) — = → metaphase plate.

equilibrium population (*Crow* 1948) — a population in which the → gene frequencies have reached an equilibrium between mutation pressure and selection pressure. The → genotype frequencies correspond to those expected for random mating and free combination between the loci.

E. R. — an abbreviation sometimes used for → endoplasmic reticulum, or ergastoplasm.

ergastic (*Meyer* 1906) — of cell materials such as accumulations of nutritive materials and products of metabolic activity.

ergastoplasm (*Garnier* 1897) — the ribosome carrying components of the → cytoplasm which represent basophilic areas (storing basic dyestuffs) of the groundplasm of the cell. In the electron microscope the e. shows up as an accumulation of → ribosomes situated either on the parallel lamellae of the → endoplasmic reticulum or accumulated free in the groundplasm. The basophilic behavior of the e. is due to the → ribosomal RNA.

ergosome (*Wettstein, Staehelin* & *Noll* 1963) — = → polysome or polyribosome.

escape synthesis — the derepression (constitutive expression) of a bacterial → operon as mediated by at least two separate mechanisms: 1. Enzyme synthesis is dependent upon phage DNA replication (classic e. s.) and may be due to depletion of free repressor by phage-induced replication of the operon's → operator. 2. Extension of → genetic transcription into the bacterial operon from a promotor site of a → prophage during prophage induction. This type of e. s. is independent of phage replication, but is affected by mutations involving phage DNA transcription.

ethnic group — a group of individuals or a population racially or historically related and having common cultural characteristics.

etiology — the demonstrated cause of a → trait or of a disease.

etioplast — a → plastid of dark-grown plants (angiosperms) containing crystalline prolamellar bodies, i.e., an arrangement of the membrane material which has been observed only under conditions that lead to protochlorophyll(ide) accumulation.

euapogamy (*Farmer* & *Digby* 1907) — a form of → apomixis in which the sporophyte develops from a gametophyte without fertilization and zygote formation (= diploid apogamy or unreduced apogamy).

eucell — the → eukaryotic cell.

eucentric (*Darlington* 1936) — of structural changes of chromosomes of the interchange type (→ chromosome mutation) in the course of which the transposed segments, as regards their linear gene order, do not change their positions relative to the centromere.

euchromatin (*Heitz* 1928) — those chromosomes or regions of chromosomes are said to consist of e. and are called "euchromatic" which show the normal cycle of → chromosome coiling, normal staining properties and do not become → heteropycnotic, as opposed to → heterochromatin.

The euchromatic portions of the → chromosome appear to be almost free of rapidly-reassociating, → repetitious DNA. They show the same DNA-histone ratio as → heterochromatin but three to four times more acidic chromosome proteins (→ hertone).

E. is the genetically active, dispersed part of → chromatin in eukaryotic → interphase nuclei.

euchromatization — the replacement of → heterochromatin by → euchromatin.

euchromocenter (*Gregoire* 1932) — in interphase nuclei, a → chromocenter which consists of the heterochromatic segments (→ heterochromatin) located on both sides of the → centromere of one or more (in case of fusion of chromocenters) chromosomes.

euchromosome (*McClung* 1902) — = → autosome.

eugenic (*Galton* 1883) — of measures or trends which improve the genetic endowment of a human population, as opposed to → dysgenic.

eugenics (*Galton* 1883) — the study of agencies under social control that may improve or impair the hereditary qualities of future generations of man, either physically or mentally (*Stern* 1960).

E. may be subdivided then into negative (preventive) and positive (progressive) eugenics. The aim of negative e. is to prevent the increase or the presence of alleles which produce undesirable phenotypes. Positive e. is concerned with furthering the increase of alleles which cause desirable phenotypes or, at least, guarding against the decrease of such alleles (→ euphenics).

euhaploid (*Katayama* 1935) — → haploid.

eukaryon (*Dougherty* 1957) — the → nucleus of → eukaryotic organisms (→ prokaryon).

eukaryotic (*Chatton* 1925) — of those plants and animals ("eukaryota" or "eukaryotes") whose cells have in contrast to prokaryotes typical → nuclei with nuclear envelopes, chromosomes, as well as nuclear divisions in form of → mitosis and → meiosis during which the chromosomes are visible by light microscopy in species-specific number and form.

eumeiosis (*Battaglia* 1945) — → meiosis.

eumitosis — (*Sagan* 1967) — → premitosis, mitosis.

euphenics (*Lederberg* 1963) — that part of → euthenics concerned particularly with the engineering of human development. E. means the improvement of phenotypes based on defective genetic constitutions and can compensate for, or redeem certain genetic defects.

euploid (*Täckholm* 1922) — of cells, tissues and individuals with one complete → chromosome set ("monoploidy") or with whole multiples ("diploidy", "polyploidy") of the basic, monoploid number of chromosomes characteristic of the species. Each chromosome is represented once in each chromosome set (→ aneuploid).

eupsychics (*Lerner* 1968) — educational and psychological engineering as a means of managing human biological resources.

eupycnotic — of chromosomes or chromosome segments which are normally coiled (→ chromosome coiling) and stained as opposed to → heteropycnotic.

euselectivity (*Sedlmayer* 1956) — true selective → fertilization as opposed to → paraselectivity.

eusexual — of organisms that show regular alternation of → karyogamy.

eutelegenesis — a means of positive → eugenics which proposes the use of preserved sperm from selected donors for voluntary artificial insemination, thereby increasing the frequency of desirable genes in the human → gene pool.

euthenics — improvement in the human environment as a means of managing human biological resources and an indispensable partner of → eugenics (→ euphenics; eupsychics).

eversporting (*Bateson* 1906) — of organisms that exhibit variations of a particular character from generation to generation, resulting (usually) from unstable, mutable genes (→ gene mutation). The organisms never breed true, but produce in every succeeding generation the same segregations.

eversporting displacements (*Muller* 1930) — chromosomal structural changes which give rise as a consequence of position effect to → variegation or mosaicism in somatic tissues.

evocation (*Needham* et al. 1934) — the → induction of → cytodifferentiation in embryonic cells (→ individuation).

evocator — a chemical substance which acts as an internal stimulus inducing developmental processes (→ differentiation; embryonic development). The agent responsible for e. action originates from an → organizer of the embryo and initiates the → determination of neighboring embryonic regions that are in the competent phase.

evolution — the transformation of the form and mode of existence of an organism in such a way that the descendants differ from their predecessors (*Zimmermann* 1953).

E. is a change in the genetic composition of a → population the starting point of which is the formation of individuals with different genotypes.

The unit process in e. is → gene substitution and the elementary quantity is the → gene frequency which is the measure of genetic change in a population.

Evolutionary changes are brought about by the primary evolutionary forces which produce and sort out genetic variations and operate in a field of space and time.

The principal causes of biological e. are:

1. → **mutations** supplying the raw materials,

2. → **selection** shaping these raw materials into the biologically fit genotypes of races and species,

3. random → **genetic drift** which can produce rapid changes in gene frequency in small populations,

4. differential → **migration** and → **gene flow** which can shift gene frequencies via exchange of individuals and genetic information between populations,

5. → **isolation** and → **annidation** which may act in a way similar to selection as a directive force in evolution and prevent the mixing of differentiations arisen in the population.

The direction of evolutionary change can be viewed either as a progressive movement of one population from one state to another, or as a divergence between two or more populations.

The indispensable basis for biological e. is the correlated change of genotype and phenotype within a population. Changes of this type are brought about by → gene mutations, → chromosome mutations and → genome mutations. If these genetic changes are not accompanied by changes in phenotype, they do not represent immediate evolutionary material but initially contribute merely to "cryptic genetic variability". This refers to the reservoir of variability of the population (= "potential genetic variability") which has not yet become manifest phenotypically. This part of genetic variability must first be converted to "free variability" by → segregation before selection can be effective as an evolutionary force.

In a relatively large population, the elementary evolutionary process of change is the change of the gene frequency. Three different types of change may be distinguished according to the degree of determinability of the changes for which they are responsible (*Wright* 1949):

1. **Systematic changes** in the population, which in principle may be formulated mathematically, thus allowing the evolutionary trend to be predicted with some certainty. To these belong a) the pressure of recurrent mutations, b) the pressure of immigrations and hybridizations and c) the pressure connected with the selection occuring within the evolving group of individuals.

2. **Random fluctuations** in the gene frequency. In this case only the resultant change can be established. To these belong a) fluctuations which are a consequence of sampling accidents and b) fluctuations in the systematic evolutionary pressures mentioned under 1.

3. **Non-recurrent changes** in the population as a) non-recurrent mutations, b) non-recurrent hybridizations, c) non-recurrent selection processes and d) non-recurrent extreme reductions in the number of individuals of the populations in question (→ population waves).

Evolutionary processes within a species or those leading to the formation of a new species (→ speciation) are termed "infraspecific evolution" (*Rensch* 1947) or "microevolution"; those processes leading to new genera, families, orders, etc. are called "transpecific evolution" or "macroevolution".

Microevolutionary events are usually viewed over a relatively short period of time, macroevolutionary events through the perspective of geological time.

The rate of e. is the time span needed for particular evolutionary changes in an organism or a population. Factors affecting the rate of e. may be the amount of → genetic variability in populations, population structure, the mode of → adaptation, intrinsic mutation rates, and others. Except for mutation rate, all these factors are functions of the → breeding system. Rates of e. are very diverse, not only between different groups of organisms living in different environments or even in the same environment but also between the same line at different periods in its evolutionary history and between different parts of the same organism (*Stebbins* 1950).

Rates of e. falling within the distribution most commonly found when a frequency distribution of such rates is plotted are called **"horotelic"** (*Simpson* 1944). E. at a much slower or faster rate than horotelic e. is referred to as **"bradytelic"** and **"tachytelic"**, respectively.

A contemporaneous group of individuals producing common offspring in the future is a "unit of evolution" (*Thoday* 1953). Depending on the → mating system it may be smaller than, correspond to, or comprise more than an individual species. Changes of the external and internal environmental conditions can lead to considerable changes in the size and complexity of the units of e.

The following **modes of evolution** may be distinguished:

1. **Phyletic e.** (*Simpson* 1944): Any evolutionary change occurring sequentially in a single line of descent without branching of lines. From the point of view of paleontology, this mode of e. is characteristic for the more or less linear e. of phyla when measured against time (Fig. 38).

2. **Quantum e.** (*Simpson* 1944): According to this evolutionary mode, a population which is not in adaptive equilibrium is displaced rapidly towards a new equilibrium by strong selection pressure. Quantum e. leads from one adapative zone to another and no transitional forms occur between them since there is no adaptive equilibrium between the zones (Fig. 39).

3. **Speciation** or **splitting** (*Simpson* 1944): A mode of e., in the course of which species evolve after splitting off and spatial isolation of parts of an initial population (Fig. 40).

4. **Explosive e.:** Relatively rapid production of different lines of descent from a common ancestral group.

Darwinian e. is evolution by ,means of "spontaneous variation" together with the "struggle of life", i.e., by natural selection.

Neo-Darwinian e. is the result of the re-examination of Darwin's theory of e. in the light of discoveries about the physical basis of inheritance. The counterpart of "spontaneous variation" is mutation. A difficulty is that the observed rates of molecular e. and the amount of variability in protein structure in natural populations are apparently too great to be accounted for on the theory of natural selection. This led to the theory of non-Darwinian e.

According to the theory of **non-Darwinian e.**, the central factor in e. is not selection bringing populations into ever greater adaptedness towards their environment, but chance, i.e., random changes (selectively neutral mutations) constantly producing new diversity.

In all probability both neo- and non-Darwinian e. act in nature but their relative importances remain to be determined.

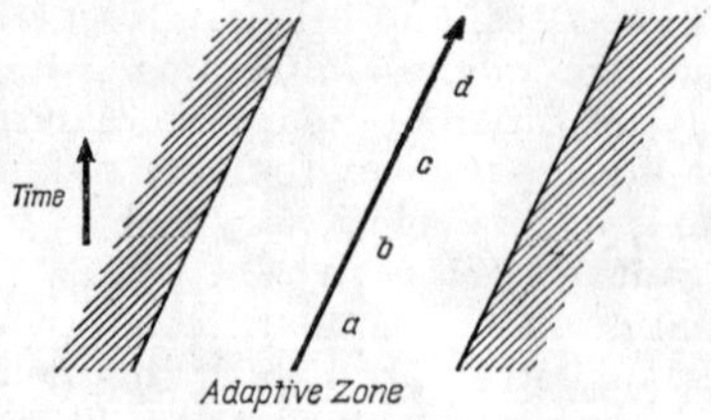

Fig. 38. Phyletic evolution, i.e. transformation of a species in time giving rise to new genera and species (after *Simpson* 1944, from *Merrell* 1962).

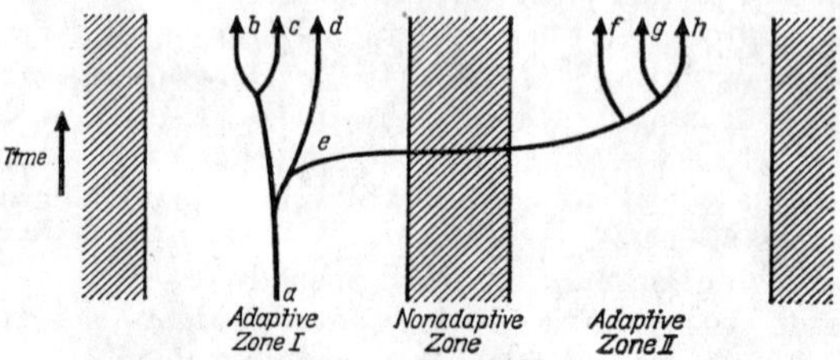

Fig. 39. Quantum evolution, i.e. transformation of a species in time leading to the origin of higher systematic categories such as orders (after *Simpson* 1944, from *Merrell* 1962).

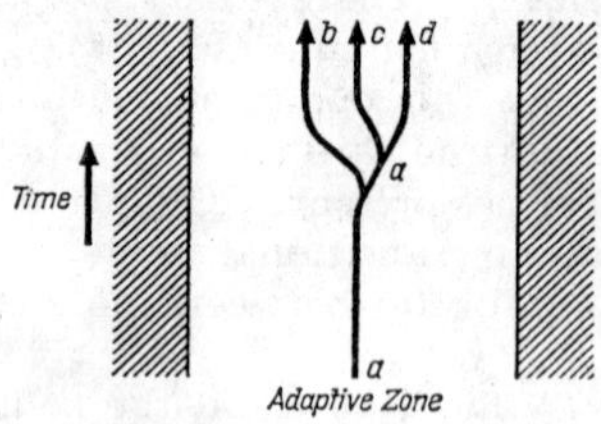

Fig. 40. Speciation or splitting, i.e. an increase in the number of species owing to reproductive isolation of populations (after *Simpson* 1944, from *Merrell* 1962).

evolution pressure — the result of the combined action of → mutation pressure, → immigration and hybridization pressure and of → selection pressure. It gives rise to systematic changes in the → gene frequency of a population. These changes, which are the basis for all evolutionary processes, may in principle be expressed mathematically.

evolutional load (*Kimura* 1960) — = substitution load (→ genetic load; cost of natural selection).

evolutionary divergence — the degree of divergence, at the intra- and interspecific levels, of two (or more) populations, which presumably have evolved from a common ancestor. A direct test of the degree of evolutionary divergence of two compatible populations is the relative variability of their F_2 generation.

evolutionary plasticity — the genetic adaptibility of populations or lines of descent. It depends on the presence in the population of → genetic variance in adaptive traits that are available for natural and artificial → selection to work on.

evolvon (*Edström* 1968) — the operational unit in → evolution. It is assumed to consist of a DNA master sequence with a series of redundant sequences which constitute a repository of genetic information. The morphological correlate of the e. is assumed to be the → chromomere.

exaggeration (*Mohr* 1923) — the expression of a hypomorphic → allele placed opposite a deficiency or → deletion. In case of e., the recessive allele gives rise in single dose (hemizygous) to a more extreme mutant phenotype than in double dose (homozygous).

exchange — 1. transposition of subchromatid, chromatid or chromosome segments within one ("intrachange") or between two or more chromosomes ("interchange") giving rise to chromosome structural changes (→ chromosome mutation).

2. Reciprocal exchange of chromatid segments between chromosomes due to meiotic or mitotic → crossing-over giving rise to → genetic recombination.

exchange pairing (*Grell* 1962) — → chromosome pairing; distributive pairing.

excision repair (*Setlow & Carrier* 1964; *Boyce & Howard Flanders* 1964) — the repair of DNA lesions (single strand breaks, damaged bases, and other structural defects) by a → dark repair mechanism which has been found in many biological systems, including human cells (Fig. 83). E. r. consists in the excision of defective regions (oligonucleotides) in one of the two DNA strands and the subsequent resynthesis of the excised nucleotide sequence (repair replication), utilizing the complementary base pairing information in the intact strand. In most cell types, the amount of DNA replaced by e. r. involves less than 1% of the genome. → Unscheduled DNA synthesis in eukaryotes is an expression of e. r.

E. r. probably occurs in the following steps:

1. Recognition of the damaged DNA region.

2. Single-strand endonucleolytic incision adjacent to the DNA lesion if the lesion is not itself a break.

3. Exonucleolytic removal (excision) of the lesion and several adjacent nucleotides (broadening of the gap).

4. Repair replication by → DNA polymerase of the degraded DNA segment with the information on the complementary strand as the template (gap filling).

5. Covalent joining (rejoining) by → DNA ligase of the last 3′ to 5′-phosphodiester bond to restore the integrity of the two-stranded molecule.

The incision may precede the excision step but it has not been ruled out that the two might normally occur as a single enzymatic process. Usually, efficient e. r. occurs only before the first DNA replication; → post-replication or recombinational repair may take place after and/or during replication of damaged DNA. In this sense, these two types of DNA repair are complementary mechanisms.

In wild type cells of *E. coli* at least two pathways of e. r. may be present: one is DNA polymerase I dependent, the other DNA polymerase III dependent.

Repair-defective mutants may be defective in one or more of the (early) steps of e. r.

exclusion — → surface exclusion.

exocytosis (*de Duve* 1963) — extrusion (by reverse → pinocytosis and → emiocytosis) of material from a cell (→ endocytosis). E. occurs by fusion of the membrane delimiting a secretary granule with the → cell membrane and concerns particles which do not diffuse through the cell surface.

exogamy (*McLennan* 1865) — includes all those systems of sexual → reproduction (→ mating system) in which there is a greater frequency of mating between unrelated or only distantly related partners than would be the case for purely random mating (panmixia) within the population (cross-fertilization or outbreeding). The genetic consequence of e. is an increased incidence of heterozygosity in the population (→ endogamy).

exogenote (*Morse, Lederberg & Lederberg* 1956) — → endogenote.

exomutation (*Imai* 1936) — → plastom mutation.

exonuclease — any enzyme that degrades DNA starting only at free ends of the DNA molecule.

exophenotype (*Lewis & John* 1963) — → phenotype.

explant — an excised fragment of a tissue or an organ used to initiate an *in vitro* culture (→ cell culture).

exoplasmosis (*Zucker-Franklin & Hirsch* 1964) — the release of enzymes from cells or into phagocytic vacuoles.

exosome (*Fox* et al. 1971) — a sequence of exogeneous DNA which after uptake by eukaryotic cells results in → genetic correction without integration of the exogeneous information into the linear structure of the chromosome (→ genetic transformation; transgenosis). The e. is assumed to replicate in step with the chromosome and can be lost or transmitted with it at the time of cell division. Either the exosomal gene or the corresponding chromosomal gene is transcribed, leading to phenotypic mosaicism.

expressivity (*Vogt* 1926) — the phenotypic expression (kind or degree) of a penetrant gene or genotype which may be slight, intermediate or severe and may be described in either qualitative or quantitative terms (→ specificity). E. like → penetrance depends both on the genotype and on the external environment. It may be constant or variable; it may be similar or different in both sexes.

expressor (*Metzenberg* 1972) — a chemically undefined positive regulator of → gene expression in eukaryotes.

extension gene — → modifier gene.

external (*Darlington* 1937) — of → chromosome mutations involving two nonhomologous chromosomes (→ fraternal; internal).

extrachromosomal — of structures or processes that are not part of the chromosomes or proceed outside the chromosomes, respectively (→ inheritance; plasmid).

extrachromosomal inheritance — → inheritance.

extranuclear — of structures or processes which are located, proceed or are active outside the → nucleus in contrast to intranuclear structures or processes.

F

F_1, F_2, etc. (*Bateson & Saunders* 1902) — → filial generation.
factor (*Mendel* 1865) — in genetics, a determinant of a → character (nowadays a → gene) with respect to which an organism may show a Mendelian difference.
F-agent (*Lederberg, Cavalli & Lederberg* 1952) — = → F-plasmid.
family — 1. a group of individuals bearing a certain genetic relationship to each other by virtue of common descent (*Lerner* 1958);
2. a group of closely related → species of plants or animals.
F duction (*Jacob* et al. 1960) — transfer of bacterial genetic markers (fragments of the bacterial chromosome) by the F-prime (F′) plasmid (→ F plasmid), derived from Hfr strains (= sex duction). The markers became part of the F element by → genetic recombination and stem from those chromosome regions with which the F element was previously associated. The F′-strains, like F^+-strains, transfer their F element to F^--cells with high efficiency.

The bacterial marker or group of closely linked markers transferred with the F′-element into F^--cells persists and replicates in the recipient cell without necessarily recombining with its chromosomal structure. The nucleotide sequences involved are under these cricumstances present in duplicate, one set as part of the recipient cell's chromosome, and the other in the recipient cell's F′-element. Cells of this type are → heterogenotes and can serve in → complementation tests as efficiently as a transductional heterogenote.
fecundity — potential fertility, i.e., the ability to produce functional gametes as measured by the quantity of gametes, particularly eggs, produced.
feedback (*Warburton* 1955) — a system in which a deviation of any size whatsoever from a certain value calls forth control factors which counteract (negative feedback) or increase (positive feedback) the deviation.
feedback control — a control system in which the accumulation of the product of a reaction leads to a decrease in its rate of production, or a deficiency of the product leads to an increase in its rate of production (→ feedback inhibition).
feedback inhibition (*Umbarger* 1956; *Yates & Pardee* 1956) — the inhibition of an early enzyme in a biosynthetic pathway by the end product of that pathway. In f. i., substrate and inhibitor may bind at two different sites on the enzyme (→ allosteric) and the inhibitor binding appears to produce a change in the tertiary or quaternary structure of the → protein. The inhibitor specific site and the substrate binding site may be on different polypeptide chains of a multichain enzyme. The obligatory requirement for excesses of two or more end metabolites to achieve inhibition is called "concerned" (*Datta & Gest* 1964) or "multivalent feedback inhibition" (*Paulus & Gray* 1964).

F. i. is a means by which cells adapt to their environment. In contrast to → induction and → repression of enzyme synthesis, f. i. acts at the enzyme level and thus is a rather direct and rapid means of regulating metabolic activity in response to an environmental change. By induction-repression, on the other hand, the expression of a gene is apparently altered, with the result that specific enzyme production may or may not occur, depending upon the particular environmental condition (*Helsinski & Yanofsky* 1966).

F episome — = → F plasmid.

fertility — the reproductive potential of an individual or population as measured by the ability to produce viable offspring. "Normal" f. results when viable → gametes are produced in normal numbers, when union of gametes is normal, and when normal viability of → zygotes, embryos and fetuses is followed by the production of viable offspring. When this chain of events is interrupted, or when deviations from average performance are great enough to be evident, the result is infertility or reduced f. Degrees of reduced f. vary all the way from that which is barely detectable down to complete → sterility (*Hutt* 1969).

"Effective f." (or probability of reproduction) is the term used in human genetics for the average number of children from patients with a particular hereditary disease in relation to the average number of children from comparable individuals not subject to the disease. The effective f. is a measure of the selective disadvantage associated with a hereditary illness. Effective f. may be reduced on biological, psychological and sociological grounds in the case of a hereditary illness.

fertility inhibition (*Egawa & Hirota* 1962) — inhibition of fertility by other → plasmids (→ fi^+ plasmids) in bacterial (*E. coli* K 12) strains carrying the → F plasmid. The plasmids inhibiting fertility are called f. i.$^+$ (fi^+).

fertilization — the fusion of two → gametes of opposite sex (♂ and ♀, + and −) to form a zygote (= syngamy, zygosis, or cytogamy). Karyogamy or the fusion of nuclei of gametes is the culmination of fertilization. By this process the homologous chromosomes that were separated in the course of → meiosis, which is in the sexual cycle necessarily correlated with the process of f., are brought together again. F. is carried on in different organisms by different mechanisms.

1. **Self-fertilization** (selfing): fusion, of male and female gametes from the same haploid, diploid, or polyploid organism.

2. **Cross-fertilization** (crossing): fusion of male and female gametes from different haploid, diploid, or polyploid individuals.

3. **Selective or preferential fertilization:** fusion of germ cells of different genotypes from one or both sexes in combinations having nonrandom frequencies. S. f. is a consequence of a direct reaction between egg cells and sperms or between embryo sacs and pollen tubes. This is called **"euselectivity"** by *Owen* (1945) in contrast to **"paraselectivity"**. In the case of paraselectivity of fertilization, a selective fertilization process is imitated by decreased functioning or complete sterility of particular gamete types.

4. **Double fertilization:** in angiosperms, the fusion of one male gamete

with the egg nucleus at the same time as a second male gamete fuses with a second female gamete to produce nutritive tissue (endosperm) (*Darlington & Mather* 1949).

fertilization-stasis (*Boyes & Walker* 1954) — an all inclusive concept for different abnormalities which occur following (species-) hybridization at the time of → fertilization and immediately thereafter. They are the result of disharmonies between the → genomes of the parent forms.

F factor — = → F plasmid.

fibroblast — a type of cell found in connective tissue of mammals and characterized by an irregular branching morphology. Fibroblasts are usually elongated, spindle-shaped cells which synthesize collagen.

filament — a universal component of the cytoplasmic matrix which is 4—5 nm in diameter and of indefinite length. Filaments vary greatly in abundance in different cells.

filial generation — any → generation (offspring) following the parental generation. The generation resulting from a cross of the parental individuals is called the first filial generation or F_1. From selfing or crossing of F_1 individuals the second filial generation or F_2 arises. The progeny of F_2 individuals makes up the third filial generation (F_3), etc.

fimbrium (*Duguid, Smith, Dempster & Edwards* 1955) — = → pilus.

F incompatibility — incompatibility which operates after an F-prime factor (→ F plasmid) has been successfully transferred to a recipient bacterial cell already carrying an F plasmid. If the resident F plasmid is in the autonomous state, either this or the incoming F-prime element is inherited unilaterally at cell division, ultimately giving a population of cells carrying one F plasmid or the other. If the resident F plasmid is integrated into the bacterial chromosome, it is always the incoming F-prime element that is inherited unilaterally. Both F. i. and → F surface exclusion are determined, at least in part, by the F plasmid itself.

fingerprint — 1. the pattern of the ridged skin of the distal phalanx of a finger;

2. a method for combining electrophoresis and chromatography to separate the components of a protein.

fi$^+$ plasmid (*Watanabe* et al. 1964) — any of a class of → resistance factors (R plasmids) able to repress the bacterial → F plasmid or sex factor. Those plasmids which do not repress the fertility of F cells are called fi$^-$ (→ fertility inhibition). fi$^+$ plasmids produce a cytoplasmic → repressor which switches off synthesis of → sex pili.

first division — the first of the two meiotic divisions (→ meiosis), formerly called the heterotypic or reductional division (→ second division).

first division segregation — the separation (→ segregation) of heterozygous → allele pairs during first division of meiosis (→ prereduction for the locus).

fission — 1. **binary f.:** the division of a one-celled organism by cell cleavage into two parts (daughter cells) in asexual reproduction immediately after → mitosis (or → endomitosis). The two daughter cells arising by binary f. are genetically like each other and like the original.

2. **Multiple f.:** division of the → nucleus several times within the cell

body. The cytoplasm fragments terminally into as many parts as there are nuclei present. Multiple f. may take the form of "**schizogony**" in which a number of new active individuals are produced, or of "**sporogony**" which results in a host of usually dormant spores.

fitness — the survival value and the reproductive capability (the proportionate contribution of offspring to the next generation) of a given genotype as compared with the average of the population or other genotypes in the population (= "adaptive value" or "selective value").

If differences in f. are in any way associated with the presence or absence of a particular gene in the individual's genotype, then → selection operates on that gene causing a change of → gene frequency, and consequently also of → genotype frequency. Differences of f. that give rise to selection are an aspect of phenotype.

The f. of a genotype with respect to any particular locus is not necessarily the same in all individuals. It depends on the environmental circumstances in which the individual lives, as well as on the residual genotype, i.e. the genes at other loci.

flagellin — aggregates of protein molecules of which bacterial flagella consist.

flagellum — a cell organelle capable of autonomous mobility. The propagation along its axis of a symmetrical undulation produces a force on the water and results in movement along the long axis of the f. Flagella have with rare exceptions a characteristic 9 + 2 arrangement of fibrils (a central pair of singlet tubules surrounded by nine parallel doublet tubules). (→ basal body; cilium).

flexibility (*Thoday* 1953) — 1. **Genetic f.** (*Thoday* 1953): the ability of a genotype to vary and adapt itself to changing conditions and thereby survive. This faculty stems from the cryptic or potential → genetic variability (→ flow of variability).

2. **Phenotypic f.** (*Thoday* 1953): the ability of an individual to remain functionable in a whole series of different environmental conditions. Phenotypic f. may be subdivided into "developmental f." and "behavioral f.". **Developmental f.** (*Thoday* 1953) is the capacity of an individual to adapt reversibly to local conditions. The genotype of a developmentally flexible organism may either permit different phenotypes to be formed under different environmental conditions, such that each phenotype is best adapated to the surroundings in which it forms, or to be so balanced that development is extensively buffered against changes of environment. In the latter case, the same adaptive phenotype is produced in a series of different environments.

Waddington (1957) describes the first form of developmental f. as "adaptive flexibility", the second as "developmental canalization".

Behavioral f. (*Thoday* 1953) is the faculty of an individual to adapt itself reversibly to local or time-associated factors, or for a mobile individual, the ability to seek out that environment to which it is best adapted.

Salisbury (1940) used the term "plasticity" in the same sense as phenotypic f. In a relatively uniform and stable environment, selection towards genetic and phenotypic f. is slight; instead selection is directed towards adaptation and stability. These relationships are reversed under unstable environmental conditions which result in a strong selection towards genetic f.; the weaker this selection is, the stronger is the selection towards phenotypic f.

floating (*Darlington & Gairdner* 1937) — of a chromosomal structural or gene change for which a mating group is not uniform (→ genetic polymorphism).

flow of variability (*Fisher* 1930) — the movement of → genetic variability within a population. It is the result of hybridization and → segregation. In the course of the movement, cryptic (potential) variability is transformed into free variability. The flow of variability can be influenced by selection if the allele frequencies, and with them the phenotypes, undergo a quantitative change.

F-mediated transduction — = → F-duction or sex duction.

foldback DNA (*Britten & Smith* 1969) — in annealing experiments (→ Cot-value), single-stranded DNA which folds upon itself and renatures with some similar but reversed repeats in the same molecule.

formenkreis — an aggregate of geographically representative (allopatric) species or subspecies.

formylmethionine tRNA — a → transfer RNA molecule ($tRNA_f^{Met}$) that participates in the initiation of bacterial protein synthesis (→ genetic translation; initiator tRNA).

forward mutation — a → gene mutation from wild-type to the mutant condition, as opposed to → back mutation. The terms f. m. and back mutation are purely descriptive.

F. mutations may occur at any one of the large number of sites (→ muton) within a → gene or a → gene cluster. Secondary mutations (→ reversions or back mutations) may restore whatever activity was lost due to the f. m. Such mutations may occur at the originally mutated site (true reversion of the f. m.) or at some other site (false reversion, pseudoreversion, or partial reversion). When the reversion definitely occurs at a site separable by recombination from the site of the f. m., it is said to occur by → suppression, or by a → suppressor mutation.

F. m. rates (→ mutation rate) are the sum of a variety of types of molecular events; these include → base pair substitutions, base pair additions and deletions (→ frameshift mutation), and at a higher level chromosome rearrangements of various types.

founder effect (*Mayr* 1942) — the establishment of a new → population by a single or a few individuals. The limited number of founders of a new group can give rise to a large drift effect (→ genetic drift). The f. e. is an important event in determining the genetic make-up or → gene pool of natural populations. The f. e. and subsequent isolation of the new colony may lead to drastic reorganization of the gene pool. It may precede → speciation, especially in insular situations.

FP factor — any → sex factor promoting bacterial host chromosome transfer.

F-pilus (*Brinton, Gemski & Carnaham* 1964) — any of a group of filamentous appendages (→ pilus) of "male" strains of *Escherichia coli* which are consistently associated with the presence of the → F-plasmid (F-factor) or a similar genetic element acting as a → sex factor. F-pili (diameter about 8 nm, length variable up to 20 μm) are defined by their specific attachment of the spherical RNA phages (that adsorb laterally along the whole length of each pilus) and assumed to act as conductors of nucleic acids in various situations (male-specific phage infection, bacterial conjugation).

The synthesis of F-pili has been shown to be directed by the F-plasmid or F-factor. F-pili can be distinguished electron microscopically from other types of pili and from flagella by several criteria, especially by the fact that male-specific phages attach to them during infection. The F-pilus is probably the structure of f^+ antigen which is specific for *E. coli* strains carrying the fertility factor, F, whether in the autonomous (F^+) or integrated stages (Hfr).

F plasmid — a genetically distinct unit (= F episome; F factor; sex factor) of circular, duplex DNA in *E. coli* with a molecular length of approximately 94,500 base pairs (→ plasmid). Its presence in the cell, either extrachromosomal or integrated into the bacterial chromosome, determines the "sex" of the bacterium, the host → chromosome mobilization and transfer via → conjugation (→ conjugon) of bacterial marker genes, and the formation of → sex pili.

Based on their constitution with respect to the F p. (Fig. 41), bacterial cells may be subdivided into four types (*Demerec* et al. 1966):

1. **F^- cells:** Bacteria lacking the sex factor F. This confers on the cells activity as a genetic recipient ("female") in conjugation, lack of activity as a genetic donor ("male"), failure to transmit F, ability to be infected with F and resistance to male-specific phage.

2. **F^+ cells:** Bacteria harboring an autonomous sex factor F which does not carry any genetically recognizable chromosome fragments. Such cells transfer F with high frequency to F^- cells which become converted to the F^+ state.

3. **F′ (F prime) cells:** Bacteria harboring an autonomous sex factor F which carries a genetically recognizable segment of the bacterial chromosome. F′ factors are also called "substituted sex factors" (*Hayes* 1964), "F-merogenotes" (*Clark & Adelberg* 1962), or "F-genotes" (*Ramarkrishnan & Adelberg* 1965).

F′ plasmids are formed in two steps from a F^+ bacterium:

1. Insertion of F into the bacterial chromosome to give Hfr;
2. a subsequent aberrant excision of a circular DNA molecule containing the essential genes of F and some bacterial genes.

Since F′ carries a fragment of the bacterial chromosome, it interacts at a specific site on the bacterial chromosome of the recipient cell, which is the region bearing the same genes as those found on the chromosome fragment carried by the F′ element.

4. **Hfr cells:** Bacteria harboring a sex factor F which is integrated with the bacterial chromosome. This state may or may not (defective sex factors) confer on the cell the phenotype of "high frequency genetic donor" since this depends on the functioning of many loci on the F plasmid and possibly on the chromosome as well. Hfr derivatives of F^+

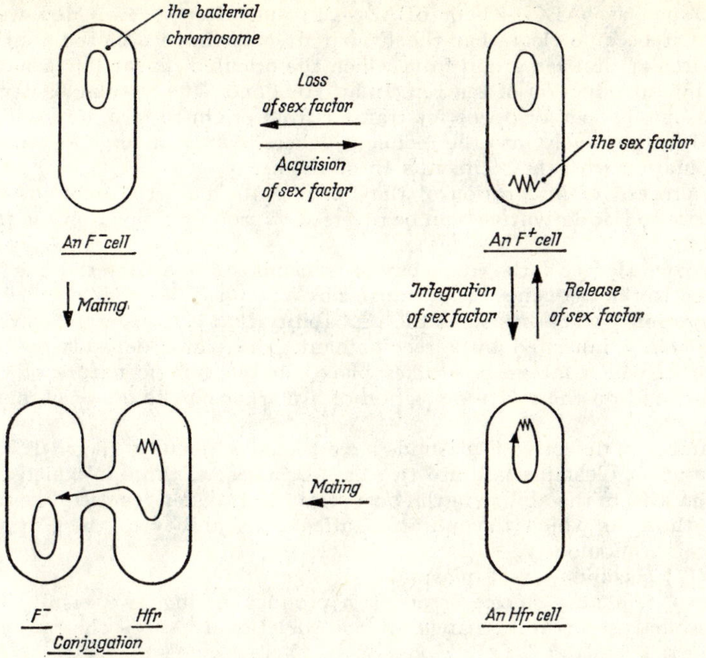

Fig. 41. The main sexual types of *Escherichia coli* and their interrelations during bacterial conjugation (from *Scaife* 1963).

cells acting as high frequency genetic donors are capable of transferring (in conjugation) the entire bacterial linkage group (chromosome) into a recipient, usually an F^- cell. Transfer of the bacterial chromosome by Hfr is rarely complete; the majority of the F^- cells normally receive only a segment of the bacterial chromosome.

The properties of Hfr cells may be accounted for by postulating that they arise by integrating the Fp. (pairing between F and the chromosome, followed by a reciprocal genetic exchange) into the continuity of the (circular) bacterial chromosome at any one of a number of possible points. The genetic transfer mediated by Hfr is polarized .An Hfr genome has an origin (o), the point on the linkage structure at which transfer is initiated, and transfers successive loci ABC···Z sequentially until transfer is interrupted spontaneously or artificially or the terminus of the linkage group is reached. The F.p. is linked at or near the terminus and behaves as a chromosomal constituent. As a consequence of spontaneous interruptions during the transfer process, most transfers result in partial zygotes (merozygotes).

Different Hfr derivatives inject the linkage group in different arrangements (such as oABC···ZF or oBAZ···CF) and after several Hfrs were checked it became clear that the F^+ parent must have a closed continuous (circular) linkage group from which the oriented, linear Hfr genome arises by introduction of the F.p. In an Hfr donor, the circular bacterial chromosome is open or opens for transfer from origin to terminus, and is injected progressively into the recipient, where it may undergo → genetic recombination with the recipient's chromosome.

The process of integration of the Fp. into the bacterial chromosome giving rise to Hfr derivatives can be reversed to produce F^+ or F-prime (F′) strains.

In crosses Hfr × F^-, the frequency of transmission of a bacterial marker is related to the frequency of its transfer by a factor called the **"coefficient of integration"** which measures the probability that a transferred marker will be stably inherited by a recombinant. The factor depends on the medium in which integration takes place, on the genetic nature of the recipient, and on the relative position of a marker on the donor chromosome.

A number of defective F plasmids have been described (in *E. coli* K 12). These mutant plasmids fall into two broad classes: 1. those in which the mutation affects the ability of the host strain to transfer genetic material, and 2. those in which the mutation affects the ability of the F. p. to replicate autonomously.

F-prime (F′) plasmid — → F plasmid.

fragment — 1. a new centric or acentric product arising as a result of a chromosomal structural change of the deletion type (→ chromosome mutation);

2. a small → B-chromosome.

frameshift mutagen — any mutagenic agent (→ mutagen) with the property of causing a shift in the reading frame of the → messenger RNA synthesized from the altered DNA template (→ frameshift mutation). F. mutagens show strong preference for particular sites in DNA which are probably either long monotonous sequences or tandemly repeated base sequences. This means that particular genes might be strongly mutagenized by a particular f. m. while others would be insensitive.

Common to any f. m. is a heterocyclic ring structure and the ability to bind to DNA by intercalating into the base pair stack of DNA. There are indications that f. mutagens may exert mutagenic effects (in bacteria) in indirect ways, i.e., by producing a situation for error-prone → repair synthesis. This action may be superimposed on more direct effects.

frameshift mutation (*Brenner* et al. 1961; *Crick* et al. 1961) — a class of mutations (= reading frame mutations, sign mutations) that arise from the insertion (+) or deletion (−) of a nucleotide, or any number of nucleotides other than three or multiples of three into or from DNA (→ frameshift mutagen). F. mutations displace the starting point of → genetic transcription of the → genetic code and the resulting mRNA is "misread" by the translation process (out-of-phase translation) from the point of nucleotide addition or deletion. Thus once a f. m. is introduced into a gene the reading frame is shifted so that all codons distal to the mutation are read out of phase and reading frame, or unless the particular genotype also carries a → f. suppressor mutation (frameshift suppression).

A closely positioned f. m. of opposite sign results in a double f. m. (+ −) and restores the normal reading frame in mRNA except for the segment between the two frameshifts (Fig. 42).

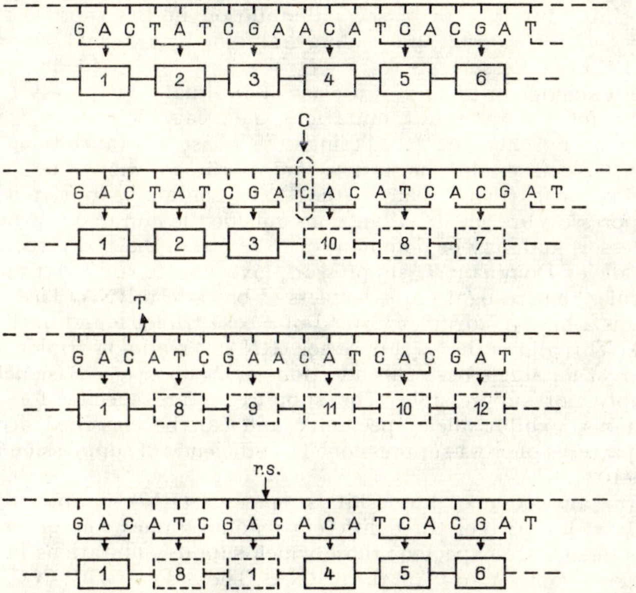

Fig. 42. Diagram illustrating the effect of "reading frame mutations" that shift the reading of the genetic code. a) The normal sequence of nucleotides in DNA (only one of the two complementary strands shown) that codes for an amino acid sequence in a functional polypeptide (the amino acids are symbolized by 1, 2, 3, etc., and each is coded for by three nucleotides); b) Insertion of a single nucleotide (cytosine = C) resulting in a mutant nucleotide sequence and a polypeptide without biological activity (incorrect amino acids indicated by broken lines); c) deletion of another nucleotide (thymine = T) resulting in another mutant nucleotide sequence; d) partial restoration of the correct reading frame by genetic recombination of sequences b and c (r. s. = recombination site) in the interval between the deletion and insertion site.

The codons which result after f. m. fall into three categories:

1. sense codons, which are read or are translated the same as before f. m.;

2. missence codons, which code for a different amino acid;

3. nonsense codons, which code for no amino acid.

Three models have been proposed to explain the origin of f. mutations: 1. errors of DNA replication; 2. errors of → genetic recombination due to unequal crossing over; 3. errors of → DNA repair.

There is evidence that intercalating agents (f. mutagens) induce f. mutations at repeating DNA sequences **(hot spots for f. m.)** which are presumably subject to mispairing during recombination or repair replication. Mispairing produces f. mutations either by adding (+) or deleting (−) base pairs.

frameshift suppression (*Yourno* et al. 1969) — the reversion of the effects of a frameshift mutation by a second mutation in the same gene (internal suppression) restoring the reading frame or by external suppressors (→ frameshift suppressor).

frameshift suppressor — any of a class of external → suppressor mutations specific for → frameshift mutations, i.e., capable of suppressing such mutations by changing (correcting) the phase of translation. Almost all suppressible frameshift mutations are of the addition type, most likely including an extra GC pair in the DNA, containing repeated GC pairs. F. suppressors are genetic alterations outside the mutated cistron (external suppressor) and may be dominant or recessive to their corresponding wild-type allele. Dominant f. suppressors give rise to altered transfer RNA molecules that recognize quadruplets of bases in mRNA. Their anticodon contains a base quadruplet instead of a base triplet found in the wildtype tRNA. This allows the → suppressor tRNA to read non-triplet codons and to correct for single base pair additions in the message. Base deletions are probably not suppressible. The suppressors are specific for frameshift mutations, exhibit allele specificity and can be classified according to their patterns of cross-suppression. The efficiency of suppression is between 1 and 10%.

If the anticodon of frameshift suppressor tRNA is able to base pair equally well in two or three different reading frames, four general types of f. suppressors are expected: those which suppress mutations in sequences of repeated G, C, A, or U (in the mRNA). The codons GGG, CCC, AAA, and UUU specify glycine, proline, lysine, and phenylalanine, respectively (→ genetic code).

Mutations of ribosomal proteins may also lead to the suppression of f. mutations. In this case, the structure of the → ribosome may promote or restrict the frequency of translation of quadruplets of bases by wildtype tRNAs.

frameshift suppressor tRNA — → frameshift suppressor; suppressor tRNA.

freemartinism — a modification of bovine gonadogenesis resulting in sterility of the genetic female co-twin of a male and accompanied by blood admixture between the twins *in utero*.

F surface exclusion — F specific → exclusion (→ F plasmid) of the formation by bacterial → conjugation of an F-prime derivative of a cell that already carries an F plasmid and is, therefore, a poor recipient in conjugational crosses.

full mutant — → half-mutant.

full sibs — individuals that have both parents in common (→ half sibs). Full sib mating is the most extreme form of → inbreeding that can occur in bisexual diploid organisms.

fundamental number (*Matthey* 1945) — the number of chromosome arms (symbol FN) of a chromosome set (karyotype). Change of f. n. may occur by → chromosome mutations, e.g. → centric fusions or → dissociations.

furrowing — → cleavage and → cytokinesis.

fusion nucleus — a nucleus produced in the → embryo sac by the union of a nucleus from the chalazal end with a nucleus from the micropylar end. Usually the f. n. later unites with one sperm nucleus to give a triploid nucleus which the → endosperm is derived.

fusion translocation — = → centric fusion.

fusome (*Hirschler* 1948) — intercellular bridges in various tissues which undergo a common, synchronous development. Unlike the other intercellular complexes such as → desmosomes and → plasmodesmata, the fusomes merely consist of single, large, annular openings in the cell wall. Electron microscope observations support the theory that these bridges allow the free passage of cytoplasmic materials, including the mitochondria. The fusomes are probably the result of incomplete cell division where the annular furrow of the spindle residual-body does not divide completely during → cytokinesis. The fusomes show a strongly osmophilic annular thickening around the opening.

G

gametangiogamy (*Kniep* 1928) — the fusion (copulation) of whole → gametangia in certain zygo- and ascomycetes. The pairwise union of multinuclear cells is followed after a more or less long time (→ dikaryophase) by a pairwise fusion (karyogamy) of sexually different nuclei. In this case, karyogamy takes place simultaneously between the nuclei of many pairs of nuclei, not as in → gametogamy between two gametic nuclei ("polyfertilization" according to *Hartmann*).

gametangium — → germ cell.

gamete (*Strasburger* 1877) — a mature reproductive cell (in some cases only a nucleus), capable of fusing with a cell of similar origin but of opposite sex to give a → zygote. The gametes are produced from gametocytes during → gametogenesis. Normally, the chromosome number of the gametes is half that of the somatic chromosome number of the organism from which they derive; the reduction of the chromosome number is brought about by → meiosis. The genetic composition of gametes is determined by the genetic composition of the parent organism in which the gametes are formed and by the behavior of the chromosomes in meiosis. Gametes possessing the somatic chromosome number are known as "unreduced gametes". They normally occur in individuals which undergo diploid → parthenogenesis, and occasionally in sexually reproducing organisms, owing to irregularities in meiosis. The fusion of two unreduced gametes yields a polyploid zygote (individual). In many one-celled organisms the gametes scarcely differ from the vegetative adult organism in size and form. This situation is termed "hologamy". In "merogamy", the gametes are smaller than the parent cells and are frequently motile and different in shape. They are usually produced from the vegetative parent cell by multiple division. The following distinctions are made on the basis of a difference in size and form between the male and female (+ and —) gametes:

1. **Isogametes:** male and female (+ and —) gametes are strikingly similar to each other. The mutual fertilization process is called "isogamy".

2. **Anisogametes:** male and female (+ and —) gametes are slightly different in size and structure. The fertilization process is referred to as "anisogamy".

3. **Heterogametes:** male and female gametes are strikingly different in size and form; the male gamete (sperm, spermatozoon) is usually very small and motile, the female (egg, ovum) relatively large and immobile. The fertilization process is referred to as "heterogamy".

Gametes are spoken of as balanced and unbalanced, respectively, according to whether they contain a complete chromosome complement or a complement in which there are → deletions or → duplications of genetic information.

gametic excess — the non-random distribution of alleles as measured mathematically by a determinant, a cross-product ratio, or by transformation of either one (= linkage disequilibrium or gametic phase unbalance). If the gametes ab, aB, Ab, and AB have frequencies q_0, q_1, q_2, and q_3 then the g. e. is defined as $D = q_0q_3 - q_1q_2$ or the difference between the coupling and repulsion phases. D is zero when the alleles are distributed at random, positive with an excess of coupling, and negative with an excess of repulsion type gametes.

gametic mortality — → zygotic mortality.

gametoblast — archesporium tissue which is not yet differentiated (→ archesporium).

gametocyte — → germ cell.

gametogamy (*Kniep* 1928) — the sexual union (→ copulation; fertilization) of two single-celled gametes of different sex and the fusion of the gamete nuclei to give the zygotic nucleus.

gametogenesis — the formation of female and male → gametes, that is sex cells (→ oogenesis; spermatogenesis).

gametogonium — → germ cell.

gametophyte (*Hofmeister* 1851) — the haploid, sexual generation which produces the → gametes in → diplo-haplontic plants characterized by having an → alternation of generations. In diplo-haplonts, the products of meiosis are → meiospores (→ megasporogenesis; microsporogenesis) which undergo mitotic divisions and develop into haploid individuals, → the gametophytes (→ sporophyte). Micro- (bearing sperms) and macro-gametophytes (bearing an egg cell)may be distinguished.

gametotoky — = amphitoky (→ parthenogenesis).

gamobium — the sexual generation in those organisms which exhibit an → alternation of generations.

gamogenesis — = sexual → reproduction or gamogony.

gamogony (*Hartmann* 1904) — = sexual → reproduction.

gamone — any of a group of biological agents which initiate the processes leading to → fertilization. They are subdivided into:

a) **Gynogamones:** the female fertilization substances liberated from the egg cells. The gynogamone-I-complex effects 1. the chemotactic attraction of the sperm, 2. the activation of sperm movement, 3. the antagonism towards androgamone I. Gynogamone II (probably identical with "fertilicine") causes sperm agglutination.

b) **Androgamones:** male fertilization agents. Three complexes are distinguished. Little is yet known of the activity of androgamone I. Androgamone II acts 1. to relesae the egg jelly, 2. to precipitate the substance of the egg membrane, and 3. to neutralize the agglutinating action of gynogamone II. Androgamone III probably exerts a liquefying action on the egg surfaces.

gamont (*Hartmann* 1904) — → germ cell.

gamontogamy — = → hologamy.

gamophase (*Winkler* 1920) — = → haplophase.

gap — 1. a discontinuity in variation;

2. a narrow, unstained (Feulgen-negative) region in the chromosome

which may occur in one chromatid or more rarely isolocally in both; it is induced by the action of mutagens which cause chromosomal structural changes. The nature of the gaps is still unclear; they do not represent discontinuities in the chromosome.

gap junction (*Revell & Karnovsky* 1967; *Revell* 1968) — any of the specialized areas (= nexuses) in the cell membranes of two apposed vertebrate or invertebrate cells in which fusion occurs, in contrast to occluding → tight junctions which act to seal intercellular spaces. G. junctions participate in electrotonic and metabolic coupling and are considered as low resistance pathways for intercellular communication.

In cross section the g. c. represents a seven-layered structure which results from the parallel apposition of two 7.5 nm thick → unit membranes (trilaminar in appearance) separated by a 2—4 nm electron-lucent space or gap. The overall thickness of the g. c. comprising the cell membranes is about 15—19 nm.

gap repair — the filling of gaps in one chain of a DNA duplex. The gaps arise during → excision repair.

gastrulation — → embryonic development.

G + C ratio — the proportion of guanine (G) + cytosine (C) in DNA, expressed as a molar percentage of total bases (→ base pair ratio).

geitonogamy — → idiogamy.

gene (*Johannsen* 1909) — a particular sequence (→ cistron) of nucleotides along a molecule of DNA (or on occasion RNA: certain viruses) which represents a functional unit of → inheritance (defined operationally by the → cis-trans test). Genes can be divided into those which code for polypeptides (structural genes), those which are transcribed into RNA (→ transfer RNA, ribosomal RNA) but not translated (→ genetic transcription, genetic translation) into proteins, and possibly those whose functional significance does not demand that they are transcribed at all (DNA sequences which may govern the punctuation or regulation of genetic transcription).

A g. consists of a linear array of potentially mutable units (mutational sites or → mutons) between which intragenic → genetic recombination (→ recon) can occur and permits the → genetic mapping of mutational sites.

As a result of → gene mutation, alternative forms of a particular g. are produced which are referred to as → alleles. The existence of a gene is extrapolated from these alleles which generally influence the same phenotypic → character or trait (being the product of → gene action). In the absence of information concerning the primary gene product (the polypeptide coded for by a cistron) and its mutational alterations, the operational definition of a g. is usually derived from → cis-trans- or complementation tests. Cells containing two functioning → genomes with different → recessive mutations in trans-configuration are expected to show a mutant phenotype if the mutations are in the same g., a wild type phenotype if the mutations are in different genes (the interpretation of such tests is complicated by the phenomenon of intragenic → genetic complementation).

Genes in → diploid organisms occur as pairs of alleles. The members of a pair segregate (→ segregation) during → meiosis so that any offspring receives only one member of a pair from each parent. The phenotypic expression of genes depends upon allelic and nonallelic interactions (→ gene interaction).

Genes are the basis of both continuous (qualitative) and discontinuous (quantitative) characters and usually produce effects on a wide variety of biochemical and morphological characters. So-called **"pleiotropic" effects** are the result of a variety of effects originating from a single primary action, i.e. coding for a definite polypeptide ("one cistron-one polypeptide model").

The collection of genes contained within one chromosome in linear sequence (→ genetic map) and in definite positions constitutes a → linkage group. There is evidence in some microorganisms at least, that genes which control a series of related biochemical reactions are often adjacent to each other in the linkage structure (→ gene cluster). The same seems to be true for the case of the mutational sites within one gene. Sites with similar properties show a tendency to cluster and are not distributed randomly within the nucleotide sequence defined functionally as a gene.

With respect to its specific location in higher organisms, a gene may be contained in a → sex chromosome ("sex-linkage") or in one of the autosomes. According to their relative location two nonallelic genes may be members of the same or different linkage groups.

In relation to its alleles the manifestation of a gene may be → dominant, → recessive, intermediate, or → combinant; in relation to nonalleles → epistatic, hypostatic, → complementary, or indifferent to interaction processes. The → penetrance of a gene may be complete or incomplete, equal or different in both sexes, or confined to one sex only. Likewise, the → expressivity of a gene may be stable or variable, equal or different in both sexes.

Nucleated cells of eukaryotes may contain besides the chromosomal **extrachromosomal or extranuclear "genes"** showing non-Mendelian inheritance. In some cases these extranuclear hereditary determinants seem to be foreign organisms (e.g., viruses, kappa particles in → killer strains of *Paramecium*); in others, they appear to be associated with normal constituents of the cell (plastids, mitochondria, kinetosomes, and kinetoplasts). Chromosomal and extrachromosomal genes are interrelated in two ways; the former are able to mutate the latter, and both may interact to produce particular phenotypes.

gene action — the expression of → genes by → genetic transcription into complementary RNA sequences and, in the case of structural genes, the subsequent translation of messenger RNA into polypeptide chains which form the ultimate protein products of g. a. (primary g. a.). The analysis beyond the primary action of a gene is greatly complicated by the integrated state of cellular and developmental metabolism, by the remoteness of the → phenotype from primary gene action, and the number of intervening steps which can be influenced by other genes (→ gene interaction) and by environmental factors (→ gene activation). In contrast to prokaryotes, where the trancription and translation of genetic information take place in one compartment (→ cell compartment), in eukaryotes the two processes are separated topologically by the → nuclear envelope. In addition, in these cells genetic information is also present and utilized within certain cytoplasmic organelles (→ organelle DNA), notably → mitochondria, and → chloroplasts (in plants). The operation of the nuclear and extranuclear → genomes is coordinated by some yet unknown mechanisms.

The regulation (→ genetic regulation) of primary g. a. can, *sensu strictu*, occur only at the level of transcription (transcriptional control). The onset of transcription, the rate at which it proceeds and its cessation are regulatory functions common to all living organisms.

The whole series of biochemical processes which lead from a gene to the → phenotype by which it is recognized is referred to as the → **"gene-action system"** (*Waddington* 1962).

Gene action is called **"cell-specific"** if the expression of a particular gene is limited to a definite cell system, and **"phase-specific"** if the manifestation is limited to a definite developmental period. Both phenomena are consequences of differential → gene activation.

The action of a particular gene is **"sex-limited"** if the corresponding phenotypic character is realized in only one of the sexes. **"Sex-controlled"** g. a. is a type of action characterized by the different expression of a gene in male and female organisms. Genes showing either sex-limited or sex-controlled manifestation may be localized either in sex-chromosomes or autosomes.

gene activation — the differential expression of → gene action at the chromosome level brought about via mechanisms of → repression and derepression. Specific products of the cell determine which genes of the → genome will function. The cell, in turn, is modified by the action of these genes with the result that particular genes are then derepressed in succession (→ operon). In higher organisms, differential g. a. is a fundamental component of the processes of → differentiation and → development and follows a definite pattern in time and space.

gene amplification (*Brown & David* 1968) — the selective production of multiple gene copies without proportional increase in others. G. a. is known to occur in the → DNA puffs of *Rhynchosciara* and a few other species of flies, in the ovaries of several insects (crickets and water beetles), and during oogenesis of many organisms such as fish, mollusks, worms, insects, and amphibians.

The best known example of g. a. is selective replication of the genes coding for → rRNA (rDNA amplification) in amphibian oocytes. It occurs by a chromosome copy mechanism with at least the first copy of amplified rDNA derived from the corresponding chromosomal DNA sequence. The 1000—5000 fold amplification of this sequence could occur by 1. excision of a copy of the repeating sequence from the chromosome and its extrachromosomal replication; 2. utilization of the preexisting transcription machinery to produce an RNA transcript of the entire repeating sequence of rDNA followed by the → reverse trancription of this molecule and its replication by a DNA-dependent mechanism.

gene cluster (*Demerec & Hartman* 1959) — any group of two or more closely linked (contiguous) genes on a chromosome that are related functionally (= gene complex). Genes of a g. c. are usually structural cistrons coding for various steps (enzymes) in a particular metabolic pathway. A g. c. may behave as a unit in the transfer of information under the control of an → operator and then represents an → operon. However, not all g. clusters are operons.

gene complex (*Brink* 1932) — = → gene cluster (→ Renner complex).

gene conversion (*Winkler* 1930; *Lindegren* 1949) — unequal recovery of genetic markers (alleles) in the region of the exchange during → genetic recombination in those cases in which all products of meiosis can be recovered (as in fungi). In principle, unequal recoveries of alleles (departures from the 2A:2a or, if the haploid products of meiosis undergo a post-meiotic division, from the 4A:4a pattern) could result from asymmetry in the mechanism of genetic exchange or from revision of a → heteroduplex joint.

At the molecular level, g. c. inolves the accurate transfer of DNA regions of 100 to 200 nucleotides from one chromatid to another. This results in homozygosis of small chromosome regions which originally were → heteroallelic. Meiotic g. c. is assumed to be a two-step process: 1. the formation of short hybrid DNA segments in corresponding parts of two homologous chromatids in meiotic prophase; 2. correction of mismatched base pairs in these segments by excision and repair (mismatch repair) of regions of → heteroduplex DNA either during or immediately after intergenic recombination by crossing over during which the damage requiring such repair may arise.

Intragenic polarity patterns of g. c. (→ polaron) reflect influences of the position of a mutation in the region of conversion on the freuency of g. c. and are of two types: 1. Unidirectional polarity where pairwise interallelic crosses of heteroalleles show that the right hand allele consistantly converts more frequently than the left hand allele; 2. bidirectional polarity where the alleles giving lowest conversion frequency in two-factor crosses are not at one end but lie somewhat between the two ends of the gene involved in conversion.

Conversion polarity may be understood as a variation in the efficiency of formation or correction of heteroduplex regions as a function of the distance away from special fixed points at which the formation of heteroduplex regions begins. As a consequence, mutants near fixed points would be included more frequently in heteroduplexes compared to mutants that are further away from these points.

G. c. is associated in roughly 50% of the conversional events with a high frequency of outside marker recombination by crossing over (Fig. 47). Processes analogous to meiotic g. c. in eukaryotes, i.e., changes from heterozygosity to homozygosity by mechanisms other than segregation, occur in mitotic cells as well as in prokaryotes (→ phage conversion).

gene diversity (*Nei & Roychoudhuri* 1974) — = → heterogeneity index.

gene dosage — the number of → alleles in a particular genotype; this depends *inter alia* on the degree of → ploidy. In haploids, each allele is repesented once, in diploids (with two chromosome sets) twice, in polyploids or polysomics more than twice. The completely → recessive genotype for a particular locus in the case of autopolyploidy or polysomy is referred to as **"nulliplex"** (aaa, aaaa, etc.). **"Duplex"**, **"triplex"**, **"quadriplex"**, etc. refer to genotypes in which the corresponding → dominant alleles are present once (e.g. Aaa, Aaaa), twice (AAa, AAaa), three times (AAA, AAAa), and four times (AAAA in the case of "tetraploidy").

The alleles of sex-linked genes located in the → X-chromosomes of organisms with an XX—XY- or XX—XO-mechanism of → sex-determination are present twice in the homogametic (XX) sex, and only once in the heterogametic (XY or XO) sex, that is, show a dosage difference connected with the sex of the organism (→ dosage compensation).

gene duplication — the duplication of DNA sequences representing genes which seems to have played a dicicive role in the → evolution of higher organisms. G. d. may occur by: 1. unequal → crossing over between two homologous chromosomes during → meiosis; 2. unequal exchange between two sister chromatids of the same chromosome during mitosis; 3. chromosome structural changes of the intrachange type; 4. redundant duplication of DNA in certain regions; 5. duplication of whole genomes resulting in → polyploid cells and individuals; 6. duplication of whole individual chromosomes resulting in → aneuploid cells and individuals.

If g. d. has been accomplished according to mechanisms 1 to 4 the species is expected to maintain the overall karyotype structure despite a noticeable increase in DNA content.

G. d. is of importance in → evolution since the existance of a surplus copy of the same gene enables one of the copies to accumulate mutations and to eventually emerge as a new gene, while another copy retains the old function required for survival through the transitional period.

gene expression — the phenotypic manifestation of genes by the processes of → gene action.

gene fixation — the establishment of a mutant allele in a population. The probability of g. f. is the probability by which a mutant allele becomes eventually established in a population (→ substitutional load; cost of natural selection; genetic load).

gene flow (*Birdsell* 1950) — the spread of genes from one breeding population to others owing to the dispersal of gametes or zygotes. G. f. may give rise to changes of allele frequency and is thus a factor of → evolution. Its extent in any given species is dependent on the size and structure of the breeding populations and on the range of dispersion of the gametes and individuals, factors which vary widely among different kinds of organisms.

gene frequency — the proportion of one particular type of → allele to the total of all alleles at this genetic locus in a breeding population, or the probability of finding the specific gene under consideration when a gene is randomly chosen from the population. The genetic composition (the → gene pool) of a population is described by the array of gene frequencies, i.e. by specifying the kinds and numbers or proportions of different alleles at each locus. Among a group of individuals, gene frequencies at a particular locus may be estimated from the → genotype frequencies. Both gene and genotype frequencies are constant from generation to generation, in the absence of migration (gene flow), mutation and selection (→ *Hardy-Weinberg* law). Analyses of gene frequencies are an important tool of population genetics and to some extent permit a genetic reconstruction of the genetic history of populations. Changes in the g. f. depend mainly on three types of processes:

1. **Systematic processes,** e.g. recurrent mutation, recurrent selection, and recurrent immigration.

2. **Random processes** such as various fluctuations in g. f., in particular those due to accidents of sampling in selection as well as migration.

3. **Unique processes** such as non-recurrent mutation, selection, migration, and accidental fixation of an allele.

Yet another category consists of changes in the system of coefficients determining the magnitude of the processes mentioned under 1–3. These changes occur because the gene pool of the population has been undergoing transformation or the environment has changed.

Systematic processes tend to bring the g. f. to stable equilibria at particular values. Dispersive processes tend to distribute the g. f. away from equilibrium values. If not held in check by the systematic processes this would in the end lead to all genes being either fixed or lost in populations not infinite in size. In fact the opposing tendencies of both types of process reach a point of balance at which the dispersion of g. f. is held in check by the systematic processes.

The gene frequencies of a certain locus may be inferred from the genotypic frequencies in the following manner: suppose that two alleles (A_1 and A_2) are present and that 100 genotypes have been classified, then an example situation is:

		A_1A_1	A_1A_2	A_2A_2	total	
number of genotypes		30	60	10	100	
number of genes	A_1	60	60	0	120	200
	A_2	0	60	20	80	

Thus the frequency of A_1 is 60 or 0.6, that of A_2 40 or 0.4. In general:

genes		genotypes		
A_1	A_2	A_1A_1	A_1A_2	A_2A_2
frequency p	q	P	H	Q

$p + q = 1$ and $P + H + Q = 1$. Since each diploid individual carries 2 genes of each locus, the frequency of A_1 alleles is $1/2(2P + H)$, and the relation between g. f. and genotype frequency in the individuals, $p = P + 1/2H$, $q = Q + 1/2H$.

gene interaction — interaction between allelic or nonallelic genes of the same genotype in the production of particular phenotypic → characters. Dominance is the major form of interactions between → alleles. On the basis of interaction with their wild-type alleles, mutant alleles are designated amorphs, hypomorphs, hypermorphs, antimorphs or neomorphs.

Nonallelic interactions involving two loci or allele pairs result in variations of the normal dihybrid → segregation scheme (Fig. 43). Conventional modifications include the following segregation ratios together with the groupings from 9A.B.: 3A.bb: 3aaB.: 1aabb with which they can be equated:

1. 9A.B.: 3A.bb: 4(3aaB. + 1aabb) in the case of recessive epistasis (Fig. 43, I).

2. 12(9A.B. + 3A.bb): 3aaB.: 1aabb with modifier gene interaction (dominant suppressor or dominant → epistasis) (Fig. 43, II).

3. 9A.B.: 7(3A.bb + 3aaB. + 1aabb) with complementary gene interaction (Fig. 43, III).

4. 13(9A.B. + 3aaB. + 1aabb): 3A.bb with modifier gene interaction (recessive → suppressor) (Fig. 43, IV).

5. 9A.B.: 6(3A.bb + 3aaB.): 1aabb in the case of additive effects of the dominant alleles (Fig. 43, V).

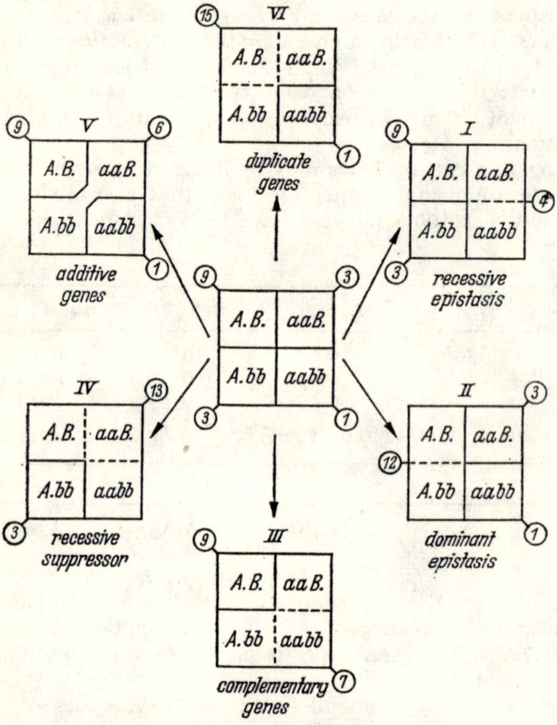

Fig. 43. Modifications (due to gene interaction) of the 9:3:3:1 phenotypic F_2 segregation ratio (center of the diagram) in the offspring of a double heterozygote (AaBb). The outer squares (I to VI) show the consequences of different types of gene interaction (modified after *Darlington* and *Mather* 1949).

6. 15(9A.B. + 3A.bb + 3aaB.) : 1aabb in the case of duplicate loci with dominance of both (Fig. 43, VI).

gene location — the determination (by various methods) of the distances between genes of the same chromosome (→ chromosome map, genetic map), or their degree of → linkage.

gene locus (*Morgan, Sturtevant, Muller & Bridges* 1915) — the position occupied by a → gene in the chromosome or → chromosome map. A g. l. may be established by using cytological (cytologic mapping) and genetic methods (→ genetic mapping). By means of intragenic recombination (→ genetic recombination) it may be subdivided into linearly arranged units called "sites" (→ gene map). Allelic gene loci are termed homologous if they occur in homologous chromosomes or chromosome segments (→ allele).

Actions and processes are "locus-specific" (*Hadorn* 1955) if they can be associated exclusively with a certain place on the chromosome.

gene magnification (*Ritossa* 1968; *Ritossa & Scala* 1969) — the heritable reparative process within the genome (e.g. of *Drosophila melanogaster* males homozygous for the bobbed mutation) whereby organisms deficient in ribosomal DNA (→ rDNA) are capable of regenerating new nucleolus organizer material. G. m. is a phenomenon related to → gene amplification and probably involves a mechanism for the interstitial lengthening of chromosomal regions.

gene map — by analogy with the → chromosome map, the graphical linear arrangement of small chromosome segments; the → mutational sites belonging to the gene concerned ("cistron") are plotted to correspond with their relative distance from one another. As in the preparation of chromosome maps, the frequency of → genetic recombination (representing an intragenic recombination) between these sites, serves a as measure of the distance which separates them (→ genetic mapping).

gene mutation — any heritable change within the limits of a single → gene ("intragenic mutation" or "point mutation"), as opposed to changes of chromosome structure (→ "chromosome mutation") or chromosome number (→ "genome mutation"). Discrimination between gene mutations and chromosomal structural changes (especially minute ones) is, in practice, frequently impossible.

As a result of g. m. alternative states of a gene — called → alleles — arise. Gene mutations may arise spontaneously (in the absence of any definable cause) or may be induced experimentally by a variety of physical and chemical → mutagens.

Spontaneous g. mutations undoubtedly arise from a multiplicity of causes and contain a multiplicity of different mutational lesions (base pair substitutions being the most frequent). The types of spontaneous mutations and their characteristic frequencies depend upon the organism, upon the genes surveyed, and to a certain extent upon environmental and other modifying factors.

A cell in which a primary mutational event has taken place may give rise either to mutant cells, or to both mutant and wildtype cells. In the former case, the mutation is said to be complete; in the latter case it is referred to as a mosaic mutation.

Since the → genetic information carried by a gene is encoded in a specific nucleotide (base) sequence of DNA (RNA takes the role of DNA in some viruses) and is replicated by forming complementary nucleotide chains (→ deoxyribonucleic acid), intragenic mutation may result from any change in the normal sequence of entire nucleotides or their component bases: e.g., substitutions of nucleotides or bases called transitions and transversions, insertions, deletions, inversions or transpositions to new positions (with or without inversion) of nucleotides (Fig. 44).

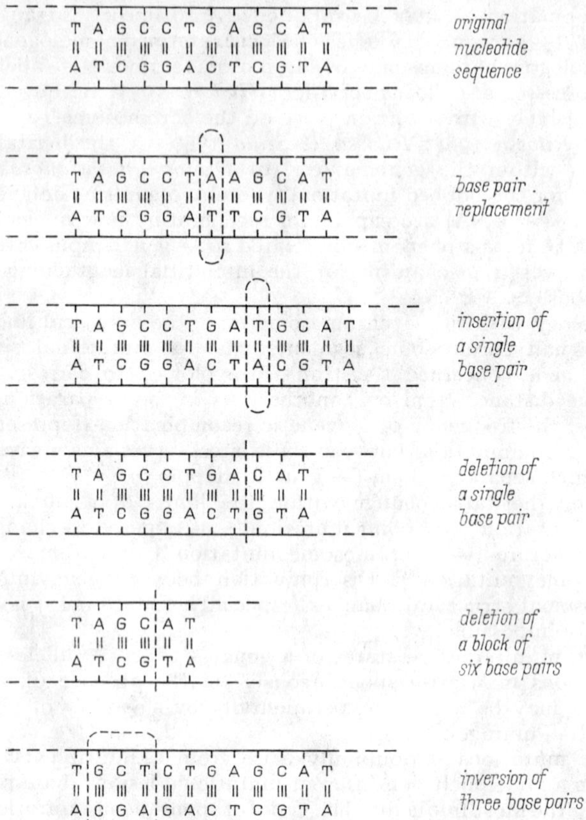

Fig. 44. Different types of changes in the nucleotide sequence of a gene giving rise to gene mutations (modified after *Watson* 1965).

"Transition" (*Freese* 1959) means the replacement of one pyrimidine base by another pyrimidine, or of one purine by another purine at some site of the DNA polynucleotide:

Adenine — Thymine   Thymine — Adenine
    ↓↑   or   ↑↓
Guanine — Cytosine   Cytosine — Guanine

"Transversion" (*Freese* 1959) is the replacement of a pyrimidine by a purine, or a purine by a pyrimidine, in the DNA polynucleotide:

Adenine — Thymine ⇄ Thymine — Adenine
    ↑↓   ↑↓
Cytosine — Guanine ⇄ Guanine — Cytosine

Both kinds of substitutions should be possible at both the nucleotide and subnucleotide (base) level.

The consequences of base pair substitution mutations are → same sense, → missense and → nonsense codons in the messenger RNA transcribed from a structural → gene.

1. **Whole nucleotide changes** may be brought forth by breaking the DNA sugar-phosphate backbone at two or more positions, followed by rearrangement of the molecular fragments, thus giving rise to deletions and other rearrangements within the limits of the preformed material of the gene. It is possible that single nucleotide changes may also arise without breakage by mechanisms which spread the polynucleotide strand lengthwise. When functioning as a template during replication, an entire nucleotide may then be added to the complementary chain. This process involves changes in both the old *and* new strands of DNA.

2. **Changes of the base portion of nucleotides** may arise in a number of ways in both "old" (template gene) and "new" genes.

a) **Base changes in "old" genes** may lead to faulty base incorporation into the new polynucleotide; they are produced by tautomeric shifts, deamination, depurination, rotational substitution and dimerization.

Faulty base incorporation may result from a perturbation of the normal attraction between A and T or G and C. A change in one of the bases of the template chain may lead to the attraction of the wrong bonding partner thereby altering all following events at this particular site. Such a new type of attraction may be caused by a rare, spontaneous shift in the tautomeric state of the old base or by the incorporation of a base analogue (→ heteroduplex). Change of tautomeric state denotes a rearrangement in the distribution of the electrons and protons in the base, giving rise to structures which are unable to make hydrogen bonds with their usual partners, but able to do so with certain other bases: a tautomer of T will pair with G instead of A, a tautomer of C with A instead of G, a rare tautomer of A with C instead of T, and a tautomer of G with T. In each case the tautomeric shift has made a new purine-pyrimidine base pair possible (Fig. 45).

Changes in the old bases can also occur after treatment with chemical mutagens (e.g. deamination of purines and pyrimidines whereby A is changed to hypoxanthine, pairing with C; C becomes uracil, pairing with A; G becomes xanthine which still is able to pair with C, but forms only two H bonds) or physical mutagens (e.g. dimerization of thymine and cytosine by UV-light). Rotational substitution leading to intranucleotide base change means that the members of a base pair rotate through 180° after breaking their bonds and rejoining (C—G becomes G—C for instance and gives rise to a double transversion).

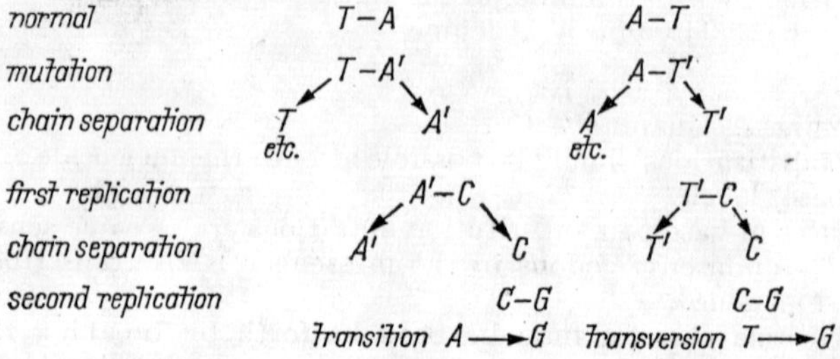

Fig. 45. The sequence of processes leading to transition and transversion as a consequence of tautomeric shifts in adenine (A) and thymine (T), respectively.

b) **Base changes in "new" genes** are brought forth by mistakes in incorporation ("incorporation errors") and replication ("replication errors"). Such changes may be the consequence of the incorporation of base analogues (as, e.g., bromouracil (BU) into DNA). The usual tautomer of BU is in the keto state like T and thus usually incorporated as the complement of A. The rare enol tautomer of BU can pair with G, as does the enol form of T and forms BU—G. The result is two kinds of incorporation error: formation of A—BU and G—BU. The BU of the A—BU pair can then continue to specify A and no error of replication takes place. If, however, BU in the enol state pairs with G, then the grand parental A is replaced by G in a transition which is the consequence of a replication error. Furthermore, since BU in the G—BU base pair is normally in the keto form at the next replication, it pairs with A: the result is a G to A transition. BU may thus give rise to purine transitions in both directions (Fig. 46).

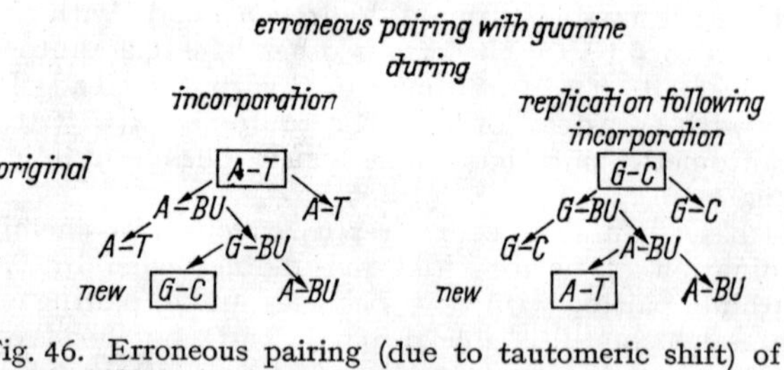

Fig. 46. Erroneous pairing (due to tautomeric shift) of bromouracil (BU) with guanine (G) during initial incorporation (left) or during replication following correct incorporation (right).

Induced mutations are produced via one or more intermediate steps that are initially reversible: Initially reparable DNA lesions ("mutation initiation") pass through a "premutational phase" and then reach a final phase ("mutation fixation") which begins with a new round of DNA replication. DNA replication and → DNA repair can lead from premutational changes to permanent changes (fixed mutations) in the nucleotide sequence of DNA (→ reversion analysis).

The smallest alterable fragment of genetic material, and hence the basic unit of mutations is called the **"muton"** (*Benzer* 1957); it constitutes a mutational or mutable site within a gene or cistron. The number of **mutational sites** (*Pontecorvo* 1957) at which → heteroallelic mutations can occur is 500—1 500 per average gene. Each site can exist in several alternative forms. These sites are the deoxynucleotides from which DNA regions making up the genes are constructed. Just as with the order of genes in a chromosome, the rearrangement of mutational sites within a gene is strictly linear.

Not all intragenic mutations are deletions or substitutions of a single nucleotide or base. A commonly detected form is the **"multi-site mutation"** involving several or many adjacent sites in the genetic region in question. Multisite mutations fail to recombine with two or more mutations that recombine with each other and their extent is revealed by recombination analysis. They do not revert to the original unmutated (wild-type) condition (→ reversion, backmutation) either spontaneously or after treatment with mutagens and are probably always due to the → deletion of the sequence of mutational sites with which they fail to recombine.

The average frequency (chance) with which a particular mutational event takes place per cell (or individual) per generation is referred to as the **"mutation rate"** and may be estimated by various methods from the number of **"mutants"**. A mutant is a phenotypically changed cell or organism resulting from a mutation (the standard of reference is the wild-type condition). **"Mutation frequency"** is the frequency of certain mutants in a population; it does not necessarily reflect the rate at which such mutants arose.

A number of factors are responsible for difficulties (especially in bacteria) in obtaining accurate mutation rates: 1. inhibitory effects (produced by interference of the parent cells with the multiplication of mutant cells) and delayed increase in the number of descendents of a mutated cell **("cell division lag")**; 2. delay in the phenotypic expression of mutational changes (→ end point mutations) due to the time required for changes to be manifest in the affected biochemical reactions **("phenomic lag")**; 3. delay of phenotypic expression of a recessive mutation occurring in one nucleus of a multinucleate cell or in only part of the chromosomal subunits; this is due to the time required for the mutant nucleus or mutant strands to segregate from the nonmutated ones **("segregation lag")**; 4. selective effects causing differential growth of mutant and parental cells.

Application of different mutagens usually results in different **"mutation spectra"** within one gene. The term mutation spectrum denotes the distribution of mutant sites within the limits of a gene after application of a number of mutagens, as determined by mapping a large number of mutations induced by each agent. Generally, the spectrum of spontaneous intragenic mutations is significantly different from that of any of an induced series both in respect to distribution of so-called **"hotspots"**, where many mutations are produced at a particular site, and in the virtual absence of coincidence of spontaneous and induced sites. This is shown by the fact that crosses between them always yield recombinants. The mutation spectra of various mutagens are in fact characterized by overlaps (many of the sites of each series coincide) but usually each differs from the other, especially in the distribution of "hot spots". "Hot spots" and mutagen specific mutation spectra clearly show that mutational sites within one gene react differently towards different mutagens **("intragenic electivity")**. Furthermore, different genes may react to the same mutagenic treatment with strikingly different mutation rates **("intergenic electivity")**.

The **primary effect of a g. m.** is to cause an alteration in the amino acid sequence of the polypeptide molecule, the primary product of the nonmutated structural gene. Since the polypeptide is synthesized via → messenger RNA as the specific template, the sequence of events is: one mutated gene, one mutated messenger RNA, one mutated polypeptide chain. A single-site mutation causes the substitution of one amino acid for another in the polypeptide. The biological effect of such amino acid substitutions is dependent both on their type and their particular location within the polypeptide. Mutational changes in the polynucleotide sequence of DNA (→ missence mutation) and the amino acid sequence of the polypeptide are colinear (→ co-linearity).

The immediate effect of altering a protein with enzymatic properties is a change in the efficiency of a particular step in a metabolic sequence (→ gene action) causing a complete or incomplete → genetic block. Gene mutations may affect enzymes both quantitatively and qualitatively. Quantitative changes of constitutive as well as inducible → enzymes may result in a complete inhibition of synthesis or in the production of a decreased amount of the particular enzyme. Qualitative changes give rise to structurally altered enzymes often accompanied by changes in enzym function. In extreme cases, the catalytic activity of an enzyme may be completely lost by gene mutation (→ cross-reacting material; nonsene mutation). The seemingly simple primary effect of a g. m. is followed by a multitude of secondary events of which the mutant (mostly → recessive) phenotype is the ultimate result. In the large majority of mutant phenotypes that have been studied, it is only the secondary effects which have been analyzed.

Gene mutations are the ultimate source of → genetic variability and thus one of the primary evolutionary forces. They supply the raw material for → evolution which is then sorted out by → selection, → genetic drift and by recombination. In connection with the → adaptive value which is conveyed by a g. m. to its carrier, three types of gene mutations may be distinguished:

1. The adaptive value of the mutant is less than that of the normal

form under all environmental conditions; the g. m. is ultimately eliminated.

2. Under the environmental conditions operative at the time of occurrence of the g. m., the adaptive value of the mutant is higher; the mutant type gradually supplants the original type.

3. Under certain environmental conditions the adaptive value of the mutant is higher than that of the original type, under other conditions lower; neither the mutant nor the original type are eliminated; natural selection establishes a state of equilibrium between the two types.

The majority of gene mutations are of negative selective value and of more or less unfavorable action. Their effect may be lethal or lethal only under certain conditions (→ lethal factor); their presence may be detrimental to the adaptive value of the bearer to a variable degree. In a particular genotypic milieu, they may have a neutral action and in rare cases they may even benefit the bearer organism.

gene pool (*Dobzhansky* 1951) — the total → genetic information encoded in the sum total of the genes in a breeding → population existing at a given time. The gametes of all breeding individuals in a population furnish a pool of genes from which the genes of the next generation are chosen.

A g. p. which preserves a constant composition throughout a series of generations is denoted as existing in equilibrium. Such a g. p. possesses a balanced structure and its components are "coadapted", i.e. changes of the allele frequencies of one genetic locus may involve changes of the allele frequencies of other loci until either the original balance is restored or a new one established.

The g. p. and the → genotype frequencies may remain unchanged only in successive generations if 1. the population size is large enough to exclude → genetic drift, 2. no mutations are produced preferentially in any given direction, 3. there is no differential selection with regard to certain genotypes, and 4. the genotypes of immigrating individuals correspond to the native genotypes.

If one of these conditions is not fulfilled under the given environment, changes of the gene and genotype frequencies result and these continue until a new equilibrium is established. Biological → evolution is based on changes in the gene pools of populations.

gene redundancy — the presence of genes in multiple forms in the DNA of eukaryotes. G. r. can be brought about by 1. a high degree of ploidy of the entire genome; 2. the presence of → polytene chromosomes with lateral multiplicity of genes; 3. extra copies of parts of the genome (→ gene amplification); 4. linear multiplicity of genes in the chromosomes (→ gene reiteration).

gene reiteration — linear (not lateral as in → polytene chromosomes) multiplicity of genes in the → genome (→ gene redundancy). Among the genes that are known to code for proteins (structural genes) in eukaryotes, only the histone cistrons are clearly reiterated. Multiple copies (100 to 100000) of genes coding for → ribosomal RNA (18 S and 28 S) are present in all eukaryotes and the same is true for the 5 S ribosomal RNA cistrons and the cistrons coding for → transfer RNA.

Bacteria exhibit from 10 to 50 cistrons for each of the two major ribosomal RNA components. G. r. is most probably an adaptation for the production of a large amount of specific transcription products.

gene substitution — the replacement in a population of one allele by a mutationally arisen new allele (→ gene mutation). G. s. plays a key role in the evolution of species. It confers a load on the population which is called substitutional load or "cost of natural selection" (→ genetic load). The total variance loss for selection in all the generations involed in a g. s. is equal to the relative increase in → fitness brought about by that substitution (*Crow* 1970).

genealogy — a family → pedigree or lineage.

genecology (*Turesson* 1923) — the hereditary structures that adjust natural races to the ecology of their environment.

generation — 1. in the case of life cycles with an → alternation of generations, that phase which extends from one to the immediately successive reproduction.

2. In genetical research, a g. corresponds to a whole reproductive cycle and comprises those individuals of a population who are equally removed from a common ancestor.

generation time — the average time between two successive → generations; the time required by a cell population to double (→ cell cycle).

generative — → somatic.

generative nucleus — a haploid nucleus in the → pollen grain which produces two sperm nuclei by mitosis (→ microsporogenesis).

gene regulatory system — → genetic regulation.

gene repression — → genetic regulation.

genes in common — genes inherited by two individuals from a common ancestral source.

gene symbols — → genetic nomenclature.

gene tagged — of chromosomes or → chromosome mutations marked genetically by marker genes (→ genetic marker).

genetic — ref. to the structure, mutation, replication and transmission of the genetic material (→ epigenetic; paragenetic).

genetically heterogeneous — of traits or characters where mutations at different loci can independently produce the same trait.

genetic antipolarity (*Ito & Crawford* 1965) — production of reduced amounts of an enzyme specified by the nearest unmutated gene on the operator side of a → polar and deletion mutant in an → operon (→ genetic polarity). The extent of g. a. is a function of the map location of the → nonsense codon. Nonsense mutations on the operator-proximal side of the gene have the most extreme polar and antipolar effects, i.e., a gradient of degree of antipolarity exists (its orientation runs in the same direction as the gradient of polarity). Antipolar effects may be partially relieved by introduction of a → nonsense suppressor gene.

genetic assimilation (*Waddington* 1942) — the "genetic fixation" (via selective processes) of "acquired" characters (phenotypes). Prior to fixation they seem to appear solely as a response to specific environmental stimuli, i.e. they are → modifications whose range of variability is genetically controlled; after fixation (assimilation) they are also produced in the absence of the particular environmental condition ("pseudoexogeneous adaptations"). Originally, the phenomenon of g. a. was supposed to result from three steps:

1. The organism reacts to the new environmental stimuli (Y) with modifications which are of adaptive value;

2. selection of genotypes which promote the constancy and expressivity of these particular phenotypes in the new environmental (stabilization);

3. automatic transformation of step 2. in such a way that the critical phenotypes are also produced in environments other than Y (genetic fixation).

The phenotypes assumed to be somatically "acquired", stabilized and genetically assimilated may in fact be a consequence of "threshold effects" (*Stern* 1958). In the case of "threshold selection", the special environmental conditions may reveal which individuals (from among a number) already carry polygenes and modifiers of the phenotype in question. The character differences may be subthreshold and thus not discernable in the original and above threshold in the new environment where they are endowed with positive selective value. Under these circumstances genotypes whose action is above threshold in both the original and the new environment may arise by crossing.

genetic background — = residual → genotype.

genetic balance (*Bridges* 1922) — the coordination and balance between coadapted genes and gene systems of a particular genotype. The internal integration of the genotype is prerequisite if the organism is to develop and function as an integrated whole.

genetic block — the reduction in → enzyme activity effected by a → gene mutation. A **complete g. b.** results in the complete absence of a particular enzyme, an **incomplete or leaky** g. b. is the formation of an altered enzyme that has a limited activity. In the case of genetic blocks, a particular product is not produced or is produced in insufficient quantity for normal cellular metabolism (→ gene action). As a consequence of such a block, precursor products of chemical reactions in a biosynthetic sequence preceding the blocked one may be accumulated.

Cells (or individuals) which are inviable in consequence of one or more genetic blocks are called "auxotrophic", in contrast to "prototrophic" ones. Their viability may be restorted by identifying and supplying the lacking growth factors. Such cells are "monoauxotrophic" if the defect may be compensated by one factor and "polyauxotrophic" if more than one growth factor is necessary. Polyauxotrophy may be the result of one or several genetic blocks. A single block may cause polyauxotrophy 1. if it is located immediate before a branching point of a biosynthetic sequence (inhibiting all reaction chains starting at that point in the absence of the block), or 2. if separate reaction chains are blocked simultaneously by the inhibition of the biosynthesis of one particular enzyme.

genetic carrier — an individual heterozygous for a recessive gene. Such a person may be clinically normal, but may participate in matings that produce offspring with the homozygous disease condition (→ genetic counseling).

genetic circularity — the presence of linkage structures (maps) without ends in bacteriophages, some animal viruses, bacteria, plasmids, mitochondria and chloroplasts. G. c. may be of various types:

1. The chromosomes are physically circular, i.e., closed molecules of double-stranded DNA. This DNA may have various structural forms including closed (supercoiled) and open circles that are either monomeric or multiple length, and also interlocked cyclic molecules. Both strands of the DNA duplex are covalently closed and free of single breaks in closed circular DNA; the open circular form of DNA is also circular but one or both strands have breaks (physically circular DNA is characteristic of some phages and viruses, bacteria, plasmids, mitochondria and chloroplasts).

2. The chromosomes are linear but have gene sequences which are circular permutations of each other (e.g., bacteriophage T4 and possibly *Streptomyces*).

3. The chromosomes are linear with identical gene sequences, but they always engage in an even number of → crossovers (not reported up to now).

genetic code — all those regularities (coordination principles) according to which the → genetic information is encoded in DNA (or RNA in some viruses), transcribed to → messenger RNA in the course of → genetic transcription and translated into → proteins with specific sequences of 20 amino acids via → genetic translation. The g. c. places the amino acid adaptor molecules (the → transfer RNAs) in the correct positions during protein synthesis.

1. Three successive nucleotide base pairs (joined together by 3′, 5′-phosphate ester linkages) on the DNA polynucleotide chains code for one amino acid. Since there are four common types of nitrogenous bases in DNA (adenine, thymine, cytosine and guanine; in RNA, adenine, uracil, cytosine, guanine) there are 4^3 or 64 possible triplets (→ codons, coding units or code words) for the standard set of 20 amino acids. These various code words have a three-dimensional structural relation with the amino acids for which they code; in all probability, this plays an important role in the mechanism of protein synthesis. It appears that most of the 64 codons are used for coding amino acids that is "make sense" (→ chain terminating codon; initiator codon).

The transcription of the coded instructions takes place when one of the polynucleotide chains of DNA is used as a template in the synthesis of a single strand of RNA, and the sequence of bases in DNA is transcribed into complementary sequences of RNA by specific base pairing. In that process T, C, G and A in DNA direct the respective incorporation of A, G, C and U (uracil) into single strands of RNA. Messenger RNA bears the information for the translation of the genetic message into sequences of amino acids. The code words (in messenger RNA) for the 20 different amino acids are given in Tables 5 and 6.

The middle base of each codon is the one most responsible for the general character of the corresponding amino acid. The third base (→ wobble) most frequently has little discriminatory effect.

If letters N and M represent any of the four bases A, U, G, C in messenger

RNA, the following rules are observed in the codon-amino acid catalogue (*Ninio* 1973):

- NMU and NMC always code for the same amino acid.
- NMA and NMG generally code for the same amino acid; UAA and UAG are terminator triplets; AUA codes for Ileu, while AUG codes for Met; UGA is a terminator triplet, while UGG codes for Trp.
- NMU, NMC, NMA and NMG code for the same amino acid when N is G or C and M is G or C.
- NMU and NMG code for different amino acids when U is N or A and M is U or A.
- The terminator codons are actively recognized by the → release factors; recognition is partially degenerate.

Table 5. *The amino acid code in abbreviated form (from Jukes 1965)*

UUb phenylalanine	CUd leucine	AUb isoleucine	GUd valine
		AUA isoleucine	
UUe leucine		AUG methionine	
UCd serine	CCd proline	ACd threonine	GCd alanine
UAb tyrosine	CAb histidine	AAb asparagine	GAb aspartic acid
UAe gaps	CAe glutamine	AAe lysine	GAe glutamic acid
UGb cysteine	CGd arginine	AGb serine	GGd glycine
UGe tryptophan		AGe arginine	

U = uracil; C = cytosine; A = adenine; G = guanine; b = U or C; d = U, C, A or G; e = A or G.

2. As is evident from the rules cited and from the tables **the code is degenerate,** i.e., most amino acids can be specified by more than one codon. Almost all of the degeneracy is confined to the III (third) codon position. With very few exceptions, the degeneracy rules are $U_{III} = C_{III}$ and $A_{III} = G_{III}$, with all four bases being equivalent in the third codon position in a number of cases. Thus most amino acids possess either two or four codons.

Amino acids with hydrocarbon residues have U or C as the second base. Those with branched methyl groups have U as the second base. Basic and acidic amino acids have A or G as the second base. Aromatic amino acids and those derived from a common organic acid are grouped together.

The degeneracy of the code is not evenly distributed. There is a correlation between the frequency of occurrence of an amino acid and the number of codons coding for it.

3. The code is **nonoverlapping:** adjacent codons do not overlap.

4. The code is **commaless:** there are no special signals ("commas") that give advice how to select the right codons for reading.

5. Correct reading of a long nucleotide sequence is provided by starting the genetic transcription in one direction from fixed starting points (→ initiator codon). Displacement of the starting point by mutation displaces the reading of all triplets and produces a totally incorrect transcription (→ frameshift mutation).

Table 6. *The nucleotide sequences of RNA codons (in Escherichia coli) (from Singer & Leder 1966, modified)*

First letter (5′-OH Terminal Base)	Second letter (Middle Base)				Third letter (3′-OH Terminal Base)
	U	C	A	G	
	Phe	Ser	Tyr	Cys	U
U	Phe	Ser	Tyr	Cys	C
	Leu, Phe?	Ser	terminator codon (ochre)	terminator codon	A
	Leu*, F-Met	Ser	terminator codon (amber)	Try	G
	Leu	Pro	His	Arg	U
C	Leu	Pro	His	Arg	C
	Leu	Pro	Gln	Arg	A
	Leu	Pro	Gln	Arg	G
	Ileu	Thr	Asn	Ser	U
A	Ileu	Thr	Asn	Ser	C
	Ileu	Thr	Lys	Arg	A
	Met*, F-Met	Thr	Lys	Arg	G
	Val	Ala	Asp	Gly	U
G	Val	Ala	Asp	Gly	C
	Val	Ala	Glu	Gly	A
	Val*, F-Met	Ala	Glu	Gly	G

* codons for the initiation of polypeptide chains by f-met-transfer RNA when present at the 5′-hydroxyl terminus of a polynucleotide.

6. **The code is essentially universal,** i.e., one and the same codon does not code for different amino acids in different organisms. The sole differences between the dictionaries of code words for amino acids in different organisms seem to be confined to the punctuation signals: The AUG codon which initiates protein synthesis responds to a formylated, methionyl tRNA (→ initiator tRNA) in bacteria but to a methionyl tRNA with a free amino group in the cells of higher organisms. The termination signals are probably the same in all organisms, although there is some evidence suggesting that the UGA codon may be used for the amino acid cysteine in eukaryotes.

The type of binding patterns observed with → isoaccepting tRNAs can vary widely with the organism and with the particular amino acid considered.

Ambiguity of the g. c. refers to the fact that one codon codes for more than one amino acid and suggests that a given codon can be recognized by more than one species of tRNA. Three types of ambiguity have been identified: 1. A codon may be misread due to environmental conditions around the tRNA-mRNA-ribosome complex which interfere with the codon recognition process (→ misreading in translation). 2. A codon may be misread because of a specific change in the primary, secondary, or tertiary structure of either the tRNA or the ribosome. 3. An → aminoacyl-tRNA synthetase may misrecognize either an amino acid or a tRNA molecule, leading to the formation of the wrong aminoacyl-tRNA.

genetic compensation (*Lewis & John* 1970) — = intercistronic → genetic complementation.

genetic complementation (*Fincham* 1966) — the complementary action (cooperation) of homologous sets of genetic material (genomes) involving the interaction of mutant genes, or their products, in double mutants. Those combinations that resulet in marked improvement in the function under study or in the development of a character which cannot be realized by the individual action of single mutants are said to complement each other.

A prerequisite for complementation studies is the collation within the same cell of complete or incomplete homologous genomes in doubly → heterozygous, → heterokaryotic, or → heterogenotic condition. This is achieved in different genetic systems by different means:

1. **In eukaryotes** (higher organisms) with membrane-bounded → nuclei, different homologous genomes may be combined in the heterozygous condition in one nucleus after meiosis and fertilization, or in fungi characterized by a → dikaryophase they may be combined in haploid heterokaryons (in the heterokaryotic condition), that is, in separate nuclei which are present in a common cytoplasm. In both cases complete genomes coexist.

2. **In prokaryotic bacteria,** a fragment of one genome may be transferred from a donor cell to a recipient cell by the processes of → conjugation, → transduction, or → F-duction (sex-duction).

3. **In bacteriophages,** the combination of two complete virus genomes is achieved by mixed infection of bacterial host cells with two genetically different types of phage.

In the case of diploids, heterokaryons, and bacterial heterogenotes, the two genetic homologues can replicate and remain together for an indefinite number of cell divisions; in the case of bacterial merozygotes and abortive transduction, the complementation state is confined to the cell in which it arose.

G. c. is of fundamental importance in the definition of the → gene and in the analysis of gene function (gene-protein relationship). Complementation studies are based on the intracellular activities of two mutants in → cis-configuration (i.e. introduced by the same genome) or trans-configuration (introduced with different genomes). The cell with a double mutant in cis will contain one complete functional set of wild-type genes and thus be functionally active. In trans, either the two mutants will complement one another and the cell will be functionally active or they will not, and the cell will be deficient in a critical function. This → **cis-trans** or **complementation test** is one criterion for distingushing allelic from nonallelic genes and for identifying physiologically distinct genes or → cistrons. Although it was originally supposed that allelic genes would not complement, while nonalleles would, there are numerous exceptions. A proportion of alleles show some degree of g. c., called intracistronic or interallelic complementation, in contrast to intercistronic complementation.

1. **Intercistronic complementation** (*Schlesinger* & *Levinthal* 1963): Complementation between pairs of heterozygous nonallelic mutations belonging to different functional regions of the → genetic map (different cistrons).

2. **Intracistronic complementation** (*Schlesinger* & *Levinthal* 1963),: Complementation between heterozygous pairs of mutations which are on the basis of other criteria mutations within the same cistron thereby affecting the same polypeptide. Mutants, or classes of mutants, are said to overlap if they do not complement with each other; they do not overlap in those cases where they do complement each other.

G. c. is one phenomenon by which the gene-protein relationship was first studied. Complementation between mutants may occur *in vitro* in the absence of protein synthesis which means interaction at the protein level.

Intercistronic complementation is explained by assuming that each cistron determines the structure of a separate polypeptide chain. The cell or organism functions normally if at least one genome (chromosome) produces each polypeptide in the normal configuration. Complete complementation is to be expected, if the different polypeptides are part of one enzyme (protein) or if different enzymes determine the same biosynthetic sequence. Intracistronic complementation may then be the result of interpolypeptide complementation between aggregates of identical polypeptide chains, coded for by one cistron and which in aggregate (as dimers, trimers, etc.) make up one (hybrid) enzyme molecule; interaction between two or more individually defective polypeptide chains may lead to one functional enzyme molecule, if mutual correction of faulty conformation occurs in the double mutant.

Generally, the amount of active enzyme produced by g. c. is dependent on the pair of mutations under consideration. In most cases, it is much higher with intercistronic than with intracistronic complementation.

The complementation pattern of a series of mutants belonging to a short genetic segment can be represented diagrammatically in form of a **"complementation map"** in which mutually complementing mutants are represented by nonoverlapping, and noncomplementing mutant pairs by overlapping, continuous lines. Such maps are often, but not invariably linear (in some cases circular). In many cases, there is a high degree of correspondence between the positions of mutants on genetic maps and on complementation maps; this is believed to result from → co-linearity of the cistron and the polypeptide chain for which it codes. Some mutations are not strictly correlated on both types of maps. The reason may be, that the final, active protein is a complex, three-dimensional and regularly folded chain. In all probability, complementation maps are maps of polypeptide subunits and functional representatives of the enzyme proteins. The mutationally caused, presumptive damage to the polypeptide chain is then reflected by the extent of the solid line across the complementation map; overlapping lines of the map suggest overlapping, identical defects across small regions of the interacting polypeptide chains (*Fincham* 1965) (→ complementation group).

genetic correction — in higher organisms, correction by exogeneous DNA of genetic defects caused by mutation (→ transgenosis). G. c. may be due to integration of exogeneous DNA into the chromosome or association with the homologous chromosome locus (→ exosome).

genetic counseling — a service which provides people with information concerning the risk of having genetically abnormal progeny (→ recurrence risk). Prediction of recurrence may be based on Mendelian expectations for simple recessive and dominant conditions, or on empirically determined risks for diseases whose mode of inheritance is less well understood (→ amniocentesis). G. c. requires careful evaluation of the total clinical picture of the → genetic disease or malformation and assessment of the family history or kindred.

genetic death (*Muller* 1950) — preferential elimination (by → selection) of genotypes which are carriers of mutationally arisen → alleles (→ genetic load) which reduce their → adaptive value or fitness (= genetic extinction). G. d. removes mutant genes from the gene pool of the population and may be the consequence of a reduction in physical vigor, sexual drive and/or fertility, or a high mortality rate in embryonic or juvenile stages of individuals possessing the critical genotype. G. d. is described as **"overlapping"**, if

1. an individual is eliminated as a consequence of the presence of one particular gene, but would also have suffered g. d. by one or more other independent genes in its genotype **("independent overlapping")**, or if 2. g. d. results from the synergistic effect of two or more mutant genes **("dependent overlapping")**.

The number of g. d. which may be involved in the complete substitution of one allele by another varies and may be many times the number of individuals present in any one generation. If a mutant gene (together with all other genes of the genotype) is eliminated from the population in the same generation in which it arose by mutation it is said to have a **"persistence"** of only one generation. On the average a dominant detrimental mutant with a selection coefficient of 0.2 and an adaptive value of 0.8 (as compared to normal), will persist for 5 generations before suffering genetic death. Thus, in each generation there is a 20% chance that the mutant individual will not transmit the gene to its progeny. The rate of g. d. of a particular mutant depends on the effect of the mutation. Its persistence is the shorter the more detrimental its effect. The average persistence of recessive mutations in man is approximately 40 generations or 1200 years.

genetic disease — any genetically conditioned disorder ultimately caused by a particular gene—determined biochemical defect ("inborn error of metabolism" according to *Garrod* (1908) or "molecular disease" according to *Pauling* 1964). In fact, all diseases have genetic and nongenetic (environmental) components whose importance may vary widely. Genetic disorders in the strict sense are only slightly influenced by environmental factors, as opposed to "common" disorders which are largely the result of environmental factors. The relationship between heredity and disease is the subject of "medical genetics".

A genetic d. is not necessarily incurable. Therapy may proceed along various lines depending upon the nature of disease under study.

genetic disoperation (*Bhalla & Sokal* 1964) — → genetic facilitation.

genetic distance — a measure of gene differences between two individuals or populations. The distance between two individuals or populations (i and j), characterized by a number of characters, is the probability that a character chosen at random will be present in i but not in j, or *vice versa* (*Jaquard* 1974). (→ genotypic distance).

genetic drift (*Wight* 1921) — 1. any change, either directed **("steady drift")** or undirected **("random drift")** in → gene frequency in a population (*Wright* 1955).

2. Irregular (random) fluctuations in gene frequency in a population from generation to generation due to finite population size (in "effectively" small populations whose → effective breeding size either remains small or periodically becomes small) or randomly fluctuating selection intensities (g. d. in the sense of random drift only). These fluctuations in gene frequency may lead to the fixation (→ random fixation) of one allele and the extinction of another without regard to their adaptive value. Random drift as a potential evolutionary factor (→ evolution) is referred to as the "*Sewall Wright* effect".

According to *Waddington* (1957), persistent and intermittent drift should be distinguished. In **"persistent drift"**, a population is small enough for statistical fluctuations to exert an important influence on the frequencies of the genes for a number of generations. **"Intermittent drift"** characterizes an occasional, sudden reduction in the number of effective breeding individuals, lasting for only one or two generations. The potential occurrence of random drift in a population depends on the interrelationship between the following factors (*Grant* 1963):

1. The number of breeding individuals (N),
2. the selective value of the allele(s),
3. mutation pressure (u), and
4. gene flow (m).

Before a change in gene frequency is attributable to random drift, it must be established that:

1. The effective population size is small enough,
2. the sub-populations are well isolated, and
3. the genes concerned are subject to very little selection.

genetic engineering — genetic manipulations (by-passing the sexual cycle) by which an individual having a new combination of inherited properties is established. G. e. presently follows two major approaches: 1. The cellular approach involves the *in vitro* culturing of (haploid) cells and the hybridization of somatic cells (→ cell hybridization); 2. the molecular approach involves the direct manipulation of the DNA (*Heyn* et al. 1974). (→ molecular cloning i plasmid e.).

genetic equilibrium (*Hardy* 1908) — in a large randomly mating population, the situation in which both the → gene frequencies and the → genotype frequencies (in the absence of mutation, selection and migration) are in accordance with the predictions of the → *Hardy-Weinberg* law ("*Hardy-Weinberg* equilibrium") and remain constant through successive generations.

Since any chance fluctuations in gene frequency are perpetuated and there is no tendency for the above-specified system to revert to its original state, such an equilibrium is referred to as neutral rather than stable. Owing to systematic processes (mutation, selection and migration) and dispersive processes (random sampling errors leading to differentiation between sub-populations, reduction of genetic variation in small populations, and increase in the frequency of homozygotes at the expense of heterozygotes) which tend to change the gene frequency, the genetic equilibrium of neutral populatons is "labile". It is the tendency of the systematic processes to bring the gene frequencies to stable equilibria at particular values, whereas the dispersive processes tend to scatter gene frequencies away from these equilibrium values. These opposing trends reach a balance which is stable only for the interval during which the conditions remain unchanged.

genetic extinction — = → genetic death.

genetic facilitation (*Lewontin* 1955) — increased fitness for one or more populations (as measured by survival, length of developmental period, body weight, and other characteristics) in the non-overlapping regions of different species or strains competing for limited resources which may diminish the utilization of certain regions of their environmental hypervolume. The contrary response, i.e., reduced fitness of one or both competitors in response to increased competition, is called **genetic disoperation.**

genetic heterogeneity — of apparently similar characters for which the genetic evidence indicates that different genes or different genetic mechanisms are involved in different kindreds.

genetic homeostasis (*Lerner* 1954) — → homeostasis.

genetic homology — identity (complete g. h.) or near identity (partial g. h.) of nucleotide sequences in DNA (→ DNA—DNA hybridization).

genetic inertia (*Darlington & Mather* 1949) — → homeostasis.

genetic information — the information contained in a particular sequence of nucleotides in a DNA or RNA molecule. **Storage of primary g. i.** in permanent form is normally in → deoxyribonucleic acid (in RNA viruses, → ribonucleic acid takes the place of DNA and thus acts as the carrier of primary g. i.). All the g. i. of a genetic system may be contained in one linkage structure (→ chromosome) in an organism, as in viruses and bacteria, or in several such structures, as in higher organisms.

Four main processes are engaged in the **transfer of genetic information** (*Spiegelman & Hayashi* 1963):

1. The replication (duplication, reproduction) of chromosomal DNA, i.e. the genetic material of all cells and some viruses.

2. The replication of RNA in those viruses whose genetic material is RNA and not DNA.

3. The → genetic transcription of g. i. from DNA to RNA (transforma-

tion of g. i. from a permanent (DNA) to a transient form (RNA) which is extant for a limited time interval only).

4. The → genetic translation of g. i. from the transient RNA code or nucleotide sequence into the protein "text", or amino acid sequence.

genetic instability — various types of genetic mechanisms which give rise to phenotypic → mosaicism or → variegation.

genetic interaction — the influence of more than one gene upon the presence or specificity of a particular → character (→ gene action; gene interaction).

genetic isolate — a breeding population which does not exchange genes with any other such group.

genetic load (*Muller* 1950) — for any genetic locus of a genotype (individual) or gene pool (population), the proportion (symbol L) by which the → fitness of the average or optimum genotype at the locus under observation is decreased owing to the presence of deleterious (lethal, semilethal, and other subvital) genes.

G. l. thus denotes the genetic disability sustained by an individual or population and is expressed as the average number of potential → genetic deaths per individual. The total of all contributions to the g. l. of human populations is about four → lethal equivalents on the average and, for example, may be represented by four genes which are lethal as homozygotes or forty genes with a → coefficient of selection of 0.1. Mixing of populations, intermarriage, change of environmental factors which favor a heterozygote advantage, and change of mutation rate are among those factors which may influence the magnitude of genetic load.

The principal components of g. l. are the following:

1. **Mutational loads** due to deleterious mutations at loci which are homozygous for nondeleterious alleles (load maintained by recurrent mutation).

2. **Segregational loads** due to genes segregating from favored heterozygotes and giving rise to unfit homozygotes (load maintained by segregation from advantageous heterozygotes). Minor components of g. l. include **"input loads"**, i.e. the load of inferior alleles in a gene pool caused by mutation and immigration (gene flow); **"substitutional loads"**, i.e. the cost to a population of replacing a particular allele by another in the course of evolutionary change; **"selection loads"** due to selection for an intermediate optimum of a quantitative character, and **"interaction loads"** due to the interaction of deleterious genes. A g. l. is termed **"balanced load"** where decerases in the overall fitness of a population due to the segregation of inferior genotypes occur and whose component genes are maintained in the population because they improve fitness in different combinations (e.g., as heterozygotes).

genetic map — the representation of the genetic (relative) distance separating nonallelic gene loci in a linkage structure (the → genes of one → linkage group or chromosome), or the arrangement of the mutational sites (alleles) of a particular gene by utilizing the frequency of → genetic recombination via → crossing-over between genes (chromosome map) and within genes (→ gene map; fine structure map).

Genetic maps are linear, unbranched structures reflecting the sensitivity of genetic recombination tests in most genetic systems. In some bacteriophages and bacteria, genetic maps take the form of a circle (→ genetic circularity). The units of distance on the map are expressed as probabilities of recombination, or percentages. In many cases, genetic maps and "complementation maps" are parallel (→ genetic complementation). This is assumed to be the result of → co-linearity of the gene and the polypeptide chain whose amino acid sequence it determines.

The classical g. m. is based on the determination of meiotic recombination. Besides this, there is genetic mapping based on mitotic recombination and recombination in (parasexual) bacteria and viruses. To map the genes on the basis of intergenic crossing-over or recombination data, it is necessary to determine all crossover or recombination frequencies for a minimum of three linked loci, since two points are insufficient to determine the specific linear arrangement. Since crossover and recombination frequencies fluctuate, because of variations in sample size and factors that affect either the crossing-over process itself (temperature, nutrition, genotype, age), or exert their influence after crossing-over occurs (e.g. differential viability), standard genetic maps are made under standard conditions optimal for crossing over and hence recombination.

The existence of linear linkage maps suggests that the frequency of recombinations for a given pair of gene loci is positively correlated with their physical distance on the chromosome. Each genetic marker (gene) is represented on the g. m. by a point in space. Distance between many pairs of markers is a function of their respective frequencies of recombination. Mapping of genetic markers is then based on the following facts (*Esser* & *Kuenen* 1965):

1. Frequency of recombination between any two markers is constant under standard experimental conditions.

2. With three linked markers (a, b ,and c) the largest of the three recombination values (r_{a-b}, r_{a-c}, and r_{b-c}) which result from intergenic crossing-over is always either equal to ("additive theorem" of recombination frequencies) or smaller than the sum of both the smaller values (e.g., $r_{a-c} + {}_{b-c} \geq r_{a-c}$ if exchange between a–c is more frequent than between the other two marker pairs). Generally, $r_{a-b} + r_{a-c}$ is larger than r_{a-c}, and only in cases of close linkage or complete positive chromosome → interference may $r_{a-b} + r_{b-c}$ equal r_{a-c}.

On the basis of this relationship and with three markers, it is possible to order each of them in a unique sequence; the third marker is localized between those two others showing the largest recombination (exchange) frequency. In case of $r_{a-b} + r_{b-c} = r_{a-c}$, the marker sequence would then be either a–b–c or c–b–a.

3. Multiple crossing-over is the most frequent cause of non-additivity. If at least one crossover takes place simultaneously in each of two adjacent regions of the chromosome (e.g., a–b and b–c), multiple crossing-over in region a–c may give rise to one recombination between adjacent markers and concurrently to a parental combination for the two external markers (a and c).

4. Crossing-over frequencies are the values needed for exact mapping. In contrast to recombination values, they are always strictly additive; the sum of crossing-over frequencies for two adjacent regions (e.g., a—b and b—c) is equal to the frequency of crossing-over in the whole region (a—c). The crossing-over frequency can be obtained from the observed recombination frequency by correcting for the effects of multiple (double) crossing-over and → interference. Because of interference the correction is mathematically rather complex. The unit measure of chromosome length and distance between any two markers is the → **"map unit"** (= "crossing-over unit" or Morgan unit) which is equal to the corrected recombination frequency, expressed as a percentage. One map unit is defined as the corrected recombination frequency of 1 per cent or one centimorgan. The relative distance between two markers calculated from uncorrected recombination frequencies (the recombination value) is referred to as their **"apparent distance"**, as opposed to their **"actual distance"**, i.e. the distance calculated from the recombination frequency corrected for multiple crossing-over (the "exchange value"). An estimate of actual distances between genes follows from the construction of cytological maps (→ chromosome map) or their equivalents. These are made independently of recombination data.

Mapping according to these principles represents the genes as seemingly dimensionless points on the g. m. The number of offspring tested for recombination is of the order of 10^2-10^3. Under these conditions, it is unlikely that rare recombinations will be discovered. In systems allowing the study of 10^6-10^7 progeny, unique types of (rare) recombinants are recovered as a result of crossing-oer within the limits of single genes (intragenic crossing-over). In such systems, the genetic resolving power **("the genetic resolution power")** is greatly increased. The locus of a gene ceases to appear as a dimensionless point; by "genetic fine structure analysis" via recombination tests of high resolving power (→ deletion mapping), the gene is shown to be composed of a linear sequence of recombinable mutational sites (→ gene mutation) whose positions can be mapped. (→ allele; genetic code).

In systems where for one reason or another recombination-type mapping cannot be effectively employed, some alternative approaches may be used. DNA replication polymerase and transcription polymerase offer an alternative basis for g. mapping. (→ replicational mapping; transcriptional mapping; heteroduplex mapping; temporal mapping).

genetic mapping — any method used in the measurement of positions and relative distances between genes of a → linkage group or sites within a gene (fine scale mapping). (→ genetic map).

genetic marker — any genetically controlled phenotypic difference used in genetic analysis, or, more specifically, any gene difference used in the detection of recombination events (→ genetic recombination) in order to facilitate recognition of a novel (recombinant) genotype by its particular phenotypic expression. In microbial genetics, a more or less arbitrary distinction is made between "selected" and "unselected" markers (*Lederberg* 1954). In recombination analysis, **selected markers** are those which prevent the growth of parental genotypes, manifesting a particular phenotype (e.g., growth factor auxotrophy), in particular environments (media), whereas recombinants for these markers are able to grow freely in the same environment, i.e., have a selective advantage over the parental types, thus facilitating their quick identification.

Unselected markers are without selective advantage for the recombinants as compared to the parental genotypes. Here the identification of recombinants is determined primarily by the internal recombination mechanism and less by the technique applied. Under these conditions segregation is determined as a second step procedure.

In principle, the same markers may be used as selected or unselected ones dependent on the experimental conditions used.

genetic material — the carrier of primary → genetic information: single or double-stranded → deoxyribonucleic acid (single in some, double in most bacteriophages, bacteria and higher organisms) or → ribonucleic acid (in RNA-viruses). G. m. must fulfil at least two fundamental functions: 1. serve as a template for its own → replication ("autocatalytic function"), 2. provide a template for the synthesis of other classes of macromolecules (specifically proteins), i.e., supply the structural and regulatory information it contains to the protein-synthesizing machinery of the cell ("heterocatalytic function").

genetic message — a → messenger RNA molecule coding for a single, functional polypeptide. A → polycistronic mRNA carries several messages.

genetic mimic — = → genocopy.

genetic mobility (*Darlington* 1958) — the ability of biological species to change their habitat or (in plants) the range of distribution of pollen and seeds. It is expressed differently in different individuals. The g. m. largely determines the manner whereby the mechanism of geographic → isolation becomes effective.

genetic mosaic — any individual (called → chimera or mosaic) which is composed of genetically different tissues and displays distinct characters in different sectors of the individual. Mosaic areas may correspond to the expression of different alleles contained in the particular genotype, or may arise by → gene mutation, → chromosome mutation, somatic crossing-over and somatic segregation, aneuploidy, or double → fertilization. Genetic mosaics are called → gynandromorphs or gynanders if part of the individual is male, part is female

genetic nomenclature — the designation of genes by symbols *Mendel* (1865) was the first to use capitals (e. g. A) to represent dominant and small letters (e.g. a) to represent recessive characters without using letters specifically derived from the particular type of expression of the character pair under observation.

Today, genes (and characters) are designated by the letters of the Roman alphabet or by abbreviations of the designations (in Latin or English) given by the person who discovered the phenotypic character pair controlled by the gene pair in question. Generally, the recessive allele of a gene pair is marked by a small letter, the corresponding dominant allele by the same letter or letters, but beginning with a capital. Another notation for alternative alleles employs the same letter or letters with a (+) for the alleles of the standard or wild-type strain or variety. (+) for wild-type alleles is then used irrespective of their being dominant or recessive. Those genes not derived from a standard or wild-type should be designated by letter-symbols which are in accordance with the particular phenotype controlled by them. Letter-symbols should contain no more than two to four letters; important genes may be symbolized by one letter only.

In bacteria and viruses, the prototroph is designated by letters and the auxotroph by the same letters with a superscript (—), e.g., lac, lac^-, try, try^-.

To designate allelic pairs, different systems are used. Most often two alleles are placed in sequence (e.g., Aa or + a, etc.), sometimes they are indicated by "fractions" (e.g., $\frac{A}{a}$ or A/a). Linkage relationships may also be indicated in the form of fractions (e.g., AB/ab or $\frac{AB}{ab}$ and Ab/aB or $\frac{Ab}{aB}$ for two pairs of alleles linked in → cis- and trans-configurations, respectively).

genetic polarity (*Jacob* & *Monod* 1961; *Franklin* & *Luria* 1961) — the phenomenon wherein certain → gene mutations (→ polar mutation) in bacterial or viral → operons, in addition to inactivating the product (protein) of the gene in which they are located, also reduce the expression of all the operator-distal genes of the operon. G. p. seems to be a general characteristic of all operons transcribed into → polycistronic mRNA. Mutational changes which produce polar effects (in most cases → nonsense mutations, in some cases → frameshift, insertion, and deletion mutations) appear to have as a common alteration the introduction, either directly or indirectly, of a polypeptide chain terminating codon (→ terminator codon) within the gene.

This results in a decreased amount of mRNA for the operator distal genes. Much of the operon mRNA is shorter and corresponds to genes operator-proximal to the mutation site. The loss of mRNA from beyond the mutation site might result either from premature termination of transcription at or near the mutation or from an accelerated degradation of the messenger corresponding to the operator-distal genes. Missense mutations do not exert g. p.

The polar effect of a nonsense (or frameshift) mutation depends on its location in the gene, i.e., is a function of the distance of its site from the nearest operator-distal cistron boundary. The gradient of g. p. is such that the farther the mutation site from the end of the gene in which it is located, the greater is the reduction of proteins coded for by the subsequent genes. There are indications that the degree of g. p. depends, in part, on the distance from the site of the critical mutation to the next translational initiation site.

The mechanisms of g. p. may be as follows: 1. Nonsense codons by stopping the movement of ribosomes translating nascent RNAs prevent further transcription because transcription and translation are coupled.

2. Nascent mRNA chains continue to be elongated after a nonsense codon at some early position has prevented ribosomes from translating the messenger. Since the newly made stretches of mRNA are devoid of ribosomes they are rendered susceptible to attack by nucleases and are degraded (→ genetic antipolarity).

genetic polymorphism (*Ford* 1940) — the regular and simultaneous occurrence in the same population of two or more discontinuous variants or genotypes ("morphs" according to *Huxley* 1955) in frequencies (even for the rarest type) which cannot be accounted for by recurrent mutation ("morphism" according to *Huxley* 1955). (→ chromosome polymorphism).

1. **Balanced polymorphism** (*Ford* 1940): a g. p. usually maintained by a selective superiority of the heterozygote over either type of homozygote and characterized by a stable and optimal relationship between genetically different morphs produced by an equilibrium of selection effects on the different allelic combinations.

2. **Transient polymorphism** (*Ford* 1940): a g. p. limited to that period of time in which one allele undergoes replacement by a superior one.

3. **Neutral polymorphism** (*Ford* 1940): a g. p. dependent on a gene action which is roughly neutral in its effect on the survival rate of the genotype in which it is contained.

4. **Regional or geographic polymorphism**: a polymorphism characterized by two or more genetically different morphs inhabiting different regions of the habitat of the population.

5. **Unisexual polymorphism**: a g. p. confined to one sex only owing to sex-limited manifestation of genes which can recombine in both sexes.

6. **Cryptic polymorphism**: a g. p. in which the genetically different morphs are not identifiable by their phenotype.

According to *Huxley* (1942), there are three possibilities which may help to install a balance of the selective advantages between different morphs:

a) heterozygotes may be more efficient than homozygotes;

b) one or more of the morphs may show a selective advantage only if the number of individuals in the population falls below a certain threshold value;

c) each of the different morphs is most efficient under specific environmental conditions which are bound to different places at different times.

Cases a) and b) are associated with changes in the respective selective values.

G. p. in any form has been considered adaptive by providing an increased diversity of genotypes thus leading to reduced intraspecific competition and increased buffering against environmental changes.

genetic recombination — a set of processes that result in new → linkage relationships of → genes or recombinational units (→ recon) of genes. G. r. gives rise to normal genophores or chromosomes carrying genes derived from different parents (recombinant individuals, cells, or molecules). The products of g. r. (the recombinants) may be detected after meiosis (meiotic recombination), after mitosis (in the case of mitotic recombination) or after equivalent processes occurring in bacteria and during multiplication of viruses.

G. r. requires the intimate association of all or part of the genetic material from two (or more) different genotypes. This is achieved by different mechanisms in different genetic systems:

1. **In higher (eukaryotic) organisms,** the intimate association is achieved by nuclear fusion and meiosis. Recombination may occur by three different processes:

a) → chromosome assortment via the spindle mechanism during meiosis which leads to "interchromosomal recombination" of unlinked markers,

b) "intrachromosomal recombination" of linked markers by → crossing-over, and

c) → gene conversion of one allele in the presence of the other resulting in nonreciprocal recombinants (Fig. 47) as opposed to the other two mechanisms which give rise to reciprocal recombination products (→ recombination system).

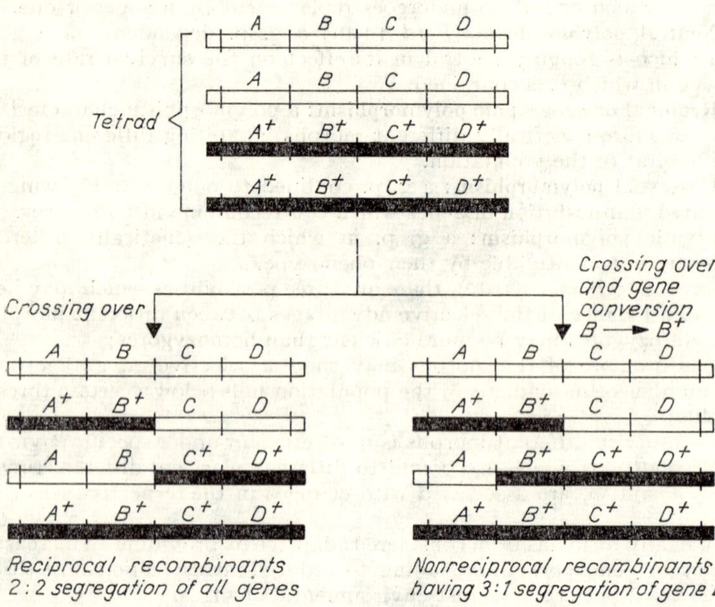

Fig. 47. Schematic representation of reciprocal and non-reciprocal recombination following crossing over (from *Watson* 1970).

2. **In bacteria,** the association of different genetic material may occur

a) by → genetic transformation,

b) by → conjugation,

c) by → transduction, and

d) by → F-duction, and may be followed by intrachromosomal recombination or gene conversion.

3. **In viruses** (bacteriophage), the same result may be achieved by mixed infection with two (or more) viruses differing in genotype.

"Interchromosomal recombination": This type of recombination is limited to eukaryotic organisms whose genetic information is contained in more than one linkage structure (chromosome) and represents the assortment of whole chromosomes in meiosis, leading to the independent → segregation of unlinked genes (allele pairs localized in nonhomologous chromosomes). Reduction of the chromosome number to one-half generally occurs during the first meiotic division and chromosome assortment is due to the random orientation of pairing configurations at metaphase in diploid organisms. The polarized distribution of particular chromosomes to one pole is of rare occurrence and leads to nonrandom assortment and nonrandom, segregation of markers at loci on different chromosomes (→ affinity, meiotic drive, sex-ratio).

Apart from meiotic recombination, interchromosomal recombination may also occur in somatic cells during mitosis via the production of aneuploid nuclei which may ultimately become regulated down stepwise to yield balanced (euploid) descendants (→ haploidization).

"Intrachromosomal recombination of the intergenic type": Recombination of allelic markers at different loci on the same linkage structure in both eu-(higher organisms) and prokaryotic (bacteria, viruses) systems. Recombination of this type is by → crossing-over, i.e., a process involving breakage and reunion of chromosome parts or segments, and gives rise to reciprocal recombination products. The frequency of recombination between any two markers ("the degree of → linkage") is constant under constant experimental conditions. The intergenic type of g. r. may take place during meiosis as well as mitosis (somatic or mitotic recombination of linked markers). Mitotic recombination occurs by a mechanism, which like the meiotic one, gives rise to reciprocal recombinants and is thus called "mitotic crossing-over". It is possible to map genes belonging to one linkage group (→ genetic map) by both meiotic and mitotic recombination.

"Intrachromosomal recombination of the intragenic type": Recombination between the subunits (mutational sites) of a gene; this may take place at many sites (→ recon). On the basis of the recombination frequencies, the mutational sites can be arranged in a linear order as a → gene map which is one-dimensional, like a linkage map. Intragenic recombinations are rare events with a tendency to show negative → interference. They are either reciprocal or nonreciprocal. Nonreciprocal recombination leads to numerically unequal segregation (→ gene conversion) products. It may be associated with numerically equal or unequal postmeiotic segregation in eukaryotes. Negative interference in intragenic r. is highly localized and polarity may be observed in respect of outside marker recombination. Intragenic recombination often appears to be impeded by heterozygosity and is frequently correlated with reciprocal r. (crossing over) in its immediate neighborhood (Fig. 47). Recent hypotheses on the molecular mechanism of g. r. have postulated the formation of segments of hybrid DNA (→ polaron hybrid DNA model), followed both by the correction of mismatched base pairs (gene conversion) and by crossing over of chromatids with a certain probability within or at the end of each hybrid segment (→ heteroduplex repair). G. r. may then be considered as a two-step process: 1. formation of a structure capable of producing recombinants (close physical contact between homologous DNA segments). 2. The resolution of recombinants from their structure.

The molecular events in 2. probably include breakage of DNA strands, formation of regions of hybrid DNA, localized degradation and synthesis of DNA (correction of mismatches in DNA), and rejoining of ends. These events are expected to require the participation of recombinases, i.e., nucleases, polymerases, and ligases. Cells lacking any of these activities might be expected to be → recombination-deficient.

Unequal g. r., as opposed to equal g. r., is intermolecular r. of homologous segments (chromosomes) in nonaligned molecules and generates tandem → duplications, triplications, etc., and corresponding → deletions.

genetic rectification — a mechanism supposed to make redundant or reiterated genes (→ gene reiteration) evolve in parallel and behave as a single Mendelian factor.

genetic regulation — the overall regulation of the type (specificity) and rate of cellular processes by regulation of the activity of specific genes controlling individual biochemical reactions, or of a block of functionally related genes which control a particular reaction pathway (→ gene action).

Since the major steps in the production by "structural genes" of → enzymes (proteins) that control the biosynthetic, metabolic and catabolic processes are → genetic transcription (DNA → RNA), → genetic translation (RNA → polypeptide) and assembly (polypeptide → protein), gene activity as well as the resultant enzyme activity may be controlled at any of these stages. G. r. as described from experimental work with bacteria may involve:

1. Control of enzyme synthesis by stimulation (positive control) or → repression (negative control) at the transcription of → operons, i.e. → messenger RNA formation from DNA. In microorganisms, this is interpreted according to the "operon concept". Any of the four steps of RNA synthesis (initiation, continuation, termination, and release) may be controlled.

2. Control of enzyme synthesis at the translation of → cistrons, i.e. the polypeptide formation from messenger RNA.

3. Control of enzyme activity after synthesis.

Without doubt, the main mechanism of g. r. in prokaryotes is → transcriptional control (→ sigma factor).

In eukaryotes, g. r. probably involves selective transcription of different regions of the genome in response to stimuli, varying temporally (induction and repression) and more or less permanently (cytodifferentiation and cell transformation). In the general sense, g. r. at the transcriptional level may occur in three ways: 1. regulatory factors can determine which portions of the genome are available for transcription by nonspecific → RNA polymerase (→ nonhistone chromosomal protein); 2. alternatively, or in addition, regulatory factors may act on RNA polymerase determining which portions of the DNA it transcribes (→ gene action; cascade regulation); 3. multiple RNA species may discriminate between various chromosomal sites at which transcription begins. Processes regulated posttranscriptionally may involve (*Tomkins* et al. 1969): 1. specific protection or degradation of nascent molecular messenger (→ pre-messenger RNA; heterogeneous nuclear RNA); 2. transport of messengers from nucleus to cytoplasm; 3. association of functional messengers bound to polysomes with specific regulatory proteins (→ informofer); 4. activation or inactivation of ribosomes; 5. alterations in the folding of nascent polypeptide chains either by association with other proteins or with some small molecules; 6. alterations in the function of specific factors involved in the initiation or termination of protein synthesis. In addition to "gene-specific" g. r., another type of regulation is manifest as a differential degree of interphase coiling (→ allocycly, chromosome coiling) of various chromosomal segments or whole chromosomes. Tightly coiled ("compacted") chromosome regions and chromosomes, referred to as consisting of → heterochromatin, seem to be genetically inactivated and transcription-inactive in this state (→ dosage compensation).

genetic relationship — the extent to which the parents of a forthcoming generation are genetically related to each other. G. r. is a variable quantity measured by the → coefficient of relationship.

genetic replication — the negative-positve image formation, by the master method through hydrogen bonding between pairs of purine and pyrimidine bases (→ deoxyribonucleic acid).

genetics (*Bateson* 1905) — the science of heredity and variation. It uses viruses ("virus genetics"), microorganisms ("microbial genetics"), plants ("plant genetics"), animals ("animal genetics"), and man ("human genetics") as objects of study. The subject matter is the phenomenology and physiology of heredity **("classical genetics")** as well as the nature of the genetic material and the storage of → genetic information, its replication, mutation, transmission, recombination, and translation into systems by which the genetic material mediates its control over metabolism and development and determines the reappearance of parental characters among progeny **("molecular genetics"). "Population genetics",** as distinguished from studies of inheritance at the familial level, describes in mathematical terms the consequences of inheritance on the population level and attempts to predict the behavior of future generations. It deals with the frequencies and interactions of genes in interbreeding populations and studies the agencies (e.g., mutation, natural and artificial selection, gene flow, migration, and chance factors) which tend to alter gene frequencies and thus to cause evolutionary changes.

genetic screening — the testing of individuals for gene products or resulting metabolites with the aim of identifying human disorders due to mutant genes. G. s. may be of any order of magnitude, ranging from testing selected individuals to testing all individuals regardless of age or clinical state (*Levi* 1973).

genetic segregation (*Bateson & Saunders* 1902) — the separation of → allele pairs from one another and their distribution to different cells, usually at → meiosis ("meiotic s.") and sometimes (as a consequence of mitotic → crossing-over) at → mitosis ("mitotic s."). G. s. can be observed only in genotypes with → heterozygous pairs of alleles.

Meiotic s. is the substance of Mendel's **second** law (→ inheritance): The paired (diploid) condition of chromosomes and genes becomes unpaired (haploid) as a result of random chromosome → assortment during meiosis. Meiosis specifies that one member of each pair of alleles is transmitted to each offspring whose diploidy is restored because a haploid → genotype is contributed by each of the two meiotically produced gametes taking part in fertilization. G. s. of a particular heterozygous pair of alleles may occur at the first ("first division s.") or second (→ "second division s.") meiotic division owing to either "pre-" or "postreductional" separation of the alleles under observation (→ meiosis).

A genotype heterozygous for one pair of alleles (Aa) gives rise by meiotic segregation to two types of gametes which are genotypically either A or a. By random union of these gametes during fertilization three kinds of "segregants" are produced by crosses between two such heterozygotes. Their genotypes will be either AA, Aa, or aa in a ratio of 1AA:2Aa:1aa. This ratio is referred to as the **"genotypic segregation ratio"**.

If allele A is dominant over a, the expected **"phenotypic segregation ratio"** (the proportion of phenotypic characters controlled by the pair

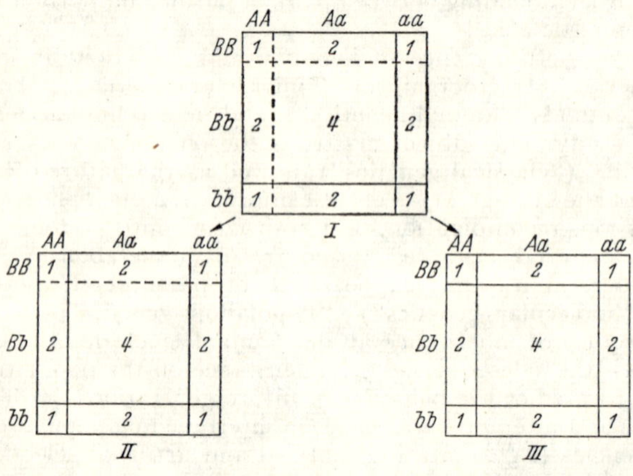

Fig. 48. Modifications of the 9:3:3:1 F_2 phenotypic segregation ratio in the offspring of a double heterozygote (selfing or intercrossing of the genotype Aa, Bb) owing to the absence of dominance in one or both heterozygous allele pairs. With dominance of one allele at each locus and independent action of nonalleles, the nine genotypes give only four phenotypes in the ratio of 9:3:3:1 (I). More genotypes are phenotypically distinguishable, if dominance (e.g., dominance of A over a, or B over b) is absent in one (II) or both pairs of alleles (III).

of alleles in question present in the progeny of the heterozygote) approximates 3A (the homozygous and heterozygous dominant genotypes AA and Aa): 1a (the homozygous recessive genotype aa). Three phenotypes are produced in those cases where no dominance between the alleles occurs (e.g. a^1 and a^2 giving rise to the genotypes a^1a^1, a^1a^2 and a^2a^2 in the approximate ratio 1:2:1 and all manifest distinct phenotypes).

Heterozygosity for more than one independent pair of alleles forms the substance of Mendel's **third** law of → inheritance. The segregation ratios of such polyhybrids result from the combinations of the independent monohybrid segregations for each of the heterozygous allele pairs (→ linkage). A hybrid with two pairs of alleles (e.g., Aa, Bb) with dominance of A over a and B over b forms combinations as follows:
(3/4A + 1/4a) × (3/4B + 1/4b) = 9/16A.B. + 3/16A.bb + 3/16aaB. + 1/16aabb.

Modifications of this dihybrid phenotypic segregation ratio result when dominance is absent in one or both allele pairs (Fig. 48), when phenotypes result from → gene interaction between nonalleles (Fig. 43), when different combinations of alleles or nonalleles have different viabilities, or when the → gene loci under observation show → linkage, i.e., are located on the same chromosome (in the same linkage group), and, therefore, need not show independent segregation.

If two pairs of alleles (Aa and Bb) are linked in the cis-configuration (AB/ab) the frequency of the gamete genotypes is 1/2(1 − p) AB: 1/2pAb: 1/2paB: 1/2(1 − p) ab, and with linkage in trans-configuration (Ab/aB) their frequency is 1/2pAB: 1/2(1 − p) Ab: 1/2(1 − p) aB: 1/2p ab in which p symbolizes the recombination percentage.

Effective s. (*Darlington* 1931) is that which gives rise to viable gametic and zygotic combinations.

Secondary s. (*Darlington* 1928) is the s. (in an → allopolyploid) of differences between the ultimate diploid parents due to → chromosome pairing between chromosomes introduced by the same gamete (autosyndesis), as opposed to primary s. which is due to "allosyndesis" between chromosomes introduced by different gametes.

Preferential s. (*Rhoades* 1942) is due to nonrandom → assortment of a particular chromosome or chromosome segment (and the genes located therein) with respect to the four cells produced at meiosis (→ affinity; meiotic drive; segregation distortion). Preferential s. may be inferred from the cytological behavior of chromosomes or from the distortion of genetic segregation ratios, or from both. True preferential s. occurs, e.g. when during → oögenesis a certain chromosome or chromosome segment is included preferentially in the egg cell, while its homologue goes to the other nuclei (polar nuclei, or the equivalent nuclei in plant → megasporogenesis) which do not participate in zygote formation. Cases of preferential s. occur during male gametogenesis when meiosis does not produce four spermatids, or when the four cells arising are not equally functional.

Somatic (or mitotic) s. is the mitotic formation of cells differing from one another in respect to their genetic make-up, by mutation, mitotic crossing-over, or unequal assortment of extrachromosomal → hereditary determinants.

genetic step — in human genetics, a term used to define the relationship between one parent and one offspring (= one genetic step). By this single g. s. the genetic resemblance is reduced to 0.5 through the intervention of meiosis: a child receives only 1/2 its chromosomes and autosomal genes from one parent. Thus, two steps separate grandfather to grandson and consequently they share an average of 0.25 genotypic identity. In the case of full sibs, a special g. s. exists. Their similarity through the father involves two genetic steps and their total resemblance for paternal genes is thus 0.25. Two first cousins are separated by three genetic steps.

genetic suppression — → suppression.

genetic system (*Darlington* 1939) — any of the species-specific ways of organization and transmission of the genetic material in pro- and eukaryotic organisms, which determine the balance between coherence and recombination of genes and control the amount and type of gene combinations. Evolution of genetic systems means the evolution of those mechanisms effecting and affecting → genetic variability. Factors which characterize a g. s. include the mode of → reproduction, the type of population dynamics (breeding size, sex ratio, degree of panmixia), the mode of chromosome organization (genetic information all in one linkage group or distributed to several such groups), the chromosome cycle (normal meiosis in both sexes, or abnormal in one in the case of eukaryotes), the → recombination index, and the presence or absence of genetic and → chromosome polymorphism.

The genetic system and its components determine the capability of a population to undergo evolutionary changes. Any g. s. is under genetic control.

genetic transcription — the synthesis of RNA (messenger RNA, transfer RNA, ribosomal RNA, 5 S RNA) from a DNA template (→ transcription error). G. t. is mediated by DNA-dependent → RNA polymerase (transcriptase); its essential features include the selection and recognition of initiation sites, correct copying of the codogenic DNA strand, the recognition of signals indicating the end of the transcription unit (→ transcripton, operon), and the subsequent release of RNA. Several protein factors are involved in g. t. (→ sigma factor; psi factor; rho factor).

In prokaryotes, the immediate product of g. t. of DNA sequences coding for polypeptides is → polycistronic messenger RNA. The transcription and translation systems (→ genetic translation) are closely coupled in prokaryotes, i.e., g. t. can proceed only if the already transcribed part of the message is being translated. In eukaryotes, there is no close coupling of the transcription-translation system and the immediate product of g. t. is a large → pre-messenger RNA molecule which is subsequently processed to yield monocistronic mRNA (→ heterogeneous nuclear RNA).

The t. of double-stranded DNA is asymmetric in the sense that only one of the two DNA strands of a given gene serves as a template for RNA. Two modes of asymmetric g. t. are known: 1. t. occurs from only one (the codogenic) DNA strand; 2. there is a switch from one strand to its complement. In either case the RNA product (transcript) should not be self-complementary. In → mitochondria both strands are transcribed (symmetric g. t.) but one of the two transcripts rapidly disappears.

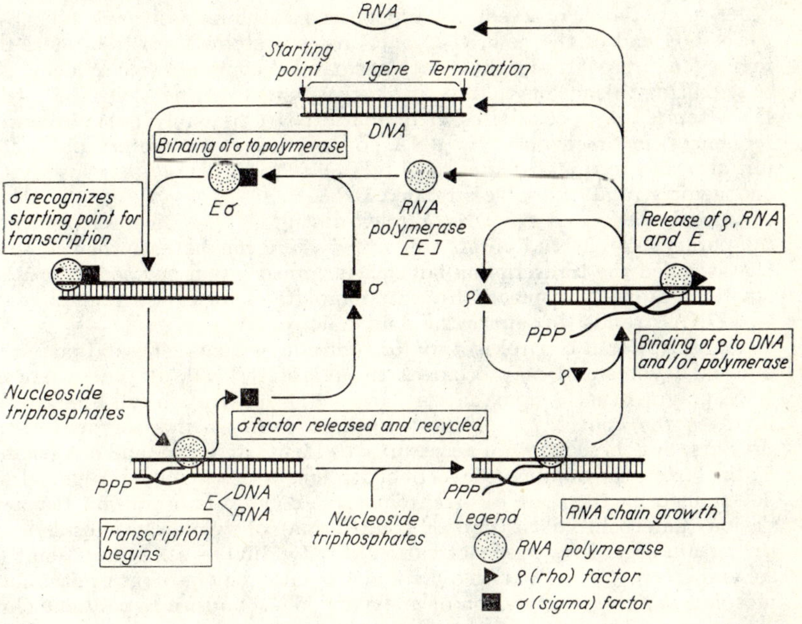

Fig. 49. Schematic outline of the process of genetic transcription in *Escherichia coli* (from *Watson* 1970).

G. t. is a multi-step process which involves the interaction between RNA polymerase, intact template and four different substrates (purine- and pyrimidine triphosphates), and, in addition, the cooperation of special → initiation, → elongation, and termination factors (Fig. 49):

1. Recognition of the DNA template by RNA polymerase and formation of a preinitiation complex

$DNA + E \rightleftarrows E - DNA$

(E = RNA polymerase).

2. Initiation of polymerization

$E - DNA + XTP \rightleftarrows DNA - E - XTP_1$ (initiation complex)

(XTP = substrate, i.e., purine- or pyrimidine triphosphate).

3. Propagation of the RNA chain

$DNA - E - XTP_1 + XTP_2 \rightleftarrows DNA - E - XTP_1XTP_2$

$DNA - E - XTP_1 - XTP_2 \rightleftarrows DNA - E - XTP_1XMP_2 + PPi$

(PPi = inorganic phosphate).

4. Termination of elongation
$DNA - E - XTP_1 - XTP_2 + nXTP_n \rightleftarrows DNA - E - XTP_1 - XTP_2 \cdots XMP_n + PP_i$.
5. Release of product
$DNA - E - XTP_1 - XTP_2 \cdots XMP_n \rightarrow DNA + E + RNA$.

RNA synthesis proceeds in a polarized fashion. Synthesis begins at the 5′-OH end of the new RNA and proceeds stepwise to the 3′-OH terminus (i.e., from left to right). With antiparallel copying by complementary base pairing this means that chain synthesis starts at the 3′-OH end of the DNA template. Special sites in template DNA (possibly pyrimidine-rich sequences) are recognized by RNA polymerase (→ promoter) and act as initiator and terminator sites in g. t. Such sequences may also specify the codogenic strand of double-stranded DNA.

A local opening of the DNA duplex during g. t. in those places where temporary DNA-RNA hybrids are formed as intermediates in the transcription is called the **transcription bubble.** As transcription proceeds, the RNA chain is gradually displaced from its template chain by the reunion of the two DNA strands (to reform the double helix).

Transcriptional control is a major route of → genetic regulation in pro- and eukaryotes and may occur at the template level, at the level of the RNA polymerase (→ sigma factor), and at the substrate level. Fine control involves the control of specific transcriptons by → inducers and → repressors, but these may not necessarily be identical in pro- and eukaryotes. In the latter, the dominant level of control (coarse control) is structural and determined by the state of → chromosome coiling. It occurs by 1. inactivation (heterochromatization or compaction) of whole chromosome sets (in certain organisms), 2. inactivation of individual → sex chromosome (in some species), 3. inactivation of heterochromatic regions of interphase chromosomes, and 4. inactivation of unexpanded → bands of → polytene chromosomes. In bacteria, about 80 to 100% of the genome is transcribed; in eukaryotes, it is between 2 and about 20% depending on the organism and the cell type. The time at which a particular gene in eukaryotic cells is transcribed into RNA has no immediate connection with the time at which this RNA is translated into protein. The templates for protein synthesis (messenger RNA) pass from the nucleus to the cytoplasm in a form that is, on the whole, resistant to intracellular degradation; they may persist there for relatively long periods (→ translational control).

A special mode of g. t. is → **reverse transcription** by viral RNA dependent → DNA polymerase (reverse transcriptase). It takes place in two steps: 1. Synthesis of a single DNA strand complementary to the viral DNA genome (→ antimessenger DNA) to yield a RNA-DNA hybrid. 2. Synthesis of a double-stranded DNA using this hybrid as a template.

genetic transformation (*Griffith* 1928; *Avery* et al. 1944) — the intraspecific and interspecific (between members of different but related species) transfer of → genetic information by means of "naked" extracellular DNA in bacteria. With g.t., in contrast to → transduction, the DNA not integrated into the recipient cell's genome cannot function and manifest its genetic information phenotypically. Uptake of transforming DNA requires that its minimum molecular weight be of the order of about 10^5. Single-stranded (denatured) DNA is much less readily incorporated into the recipient cell than undenatured, double-stranded DNA.

The general features of t. as an experimental system are as follows:

1. Extraction of DNA from bacterial donor cells, resulting in the breakdown of the bacterial chromosome into a relatively large number of fragments each large enough (10000–20000 nucleotide pairs) to include several genes.

2. Reversible binding of exogenous DNA to bacterial cells.

3. Irreversible uptake (penetration, incorporation, fixation) of one or a few DNA fragments by competent cells.

4. Association (pairing, synapsis) of an absorbed fragment of transforming DNA with the homologous region of the recipient's genome.

5. Integration (by → genetic recombination) of part of the transforming DNA molecule into the recipient's chromosomal DNA.

6. Replication of the integrated new information and segregation of the transformants.

During bacterial t. a donor-recipient complex is formed. In this structure a single-stranded segment of recipient DNA is replaced by a segment of donor DNA which becomes covalently continuous with the recipient material. This recombinant structure may be preceeded in time by another form of the complex in which the donor and recipient moieties are held in association by noncovalent bonds. The state of the bacterial cell during which it is able to take up DNA irreversibly and subsequently undergoes g. t. is called → competence (→ transfection; RNA transformation).

There are indications that DNA-mediated bacterial transformation occurs in nature by the release of genetically active DNA from untreated bacterial cells.

Recently, DNA-mediated corrections of genetic defects (→ genetic correction) have been shown to occur in eukaryotes. Occasionally bacteriophages have been used as an enriched source of specific genes and there is evidence for the synthesis of enzymes coded for by the phage DNA (→ transgenosis).

genetic translation — the second step in the "read out" of → genetic information, whereby the → genetic code contained in specific nucleotide sequences of → messenger RNA molecules (synthesized during → genetic transcription) is translated into specific amino acid sequences, i.e., directs the order of the amino acids incorporated into polypeptides during protein synthesis. During g. t. of the genetic code, the mRNA is read sequentially in sets of three nucleotides. Each nucleotide triplet, or → codon, specifies the amino acid which will be added to the growing polypeptide chain. The reading frame (→ frameshift mutation) of the codons is established at the initiation of polypeptide synthesis, and is maintained by a mechanism which moves the → ribosome along the message three nucleotides a time (→ mistranslation).

A variety of enzymes and → translation factors (→ initiation, transfer, and termination factor) is required for g. t. and for the unique events which serve in its accurate initiation and termination.

G. t. starts from the amino-terminal end of the polypeptide. The sequence into which the constituent amino acids are built is determined by the mRNA which is translated from its 5′ end to its 3′ end. The two reacting species in peptide bond formation are → aminoacyl tRNA and the growing polypeptide itself esterified to the terminal adenosine residue of a tRNA molecule. The usual recognition between the → anticodon of aminoacyl tRNA and mRNA is between the first two bases of the codon (from the 5′ end) and the last two bases of the anticodon. Between the last base in the codon and the remaining base in the anticodon the pairing appears to be more flexible (→ wobble).

The unit of g. t. is the → cistron; the translation product a specific polypeptide. There is linear correspondence (→ colinearity) between the cistron (gene) and the polypeptide it codes for. The main steps in g. t. are the following (Fig. 50):

1. **Initiation of g. t.:** The initiation of protein synthesis (i. e. the synthesis of the first peptide bond) involves the formation of an → initiation complex containing the smaller subunit of a ribosome, → initiator tRNA (f. Met-$tRNA_F$ in prokaryotes, chloroplasts and mitochondria; Met-$tRNA_F$ in eukaryotes) and mRNA. Specific → initiation factors bind the initiator tRNA to the ribosome. The initiator tRNA anticodon is bound to an initiator codon (AUG or GUG) at or near the 5′ end of the mRNA. After addition of the larger ribosomal subunit the ribosome — mRNA — initiator tRNA complex is ready to synthesize the first peptide bound (→ translation reinitiation).

2. **Peptide chain elongation:** This process can be divided into three distinct and consecutive steps: codon-directed binding of aminoacyl tRNA, peptide bond formation, and translocation. Upon binding of the initiator aminoacyl tRNA to the initiation complex (→ transfer factor), at the A site of the ribosome, peptide bond formation occurs. For protein synthesis to continue, translocation has to occur and involves three processes: 1. removal of the deacylated tRNA from the P site of the ribosome; 2. movement of the peptidyl tRNA from the A to the P site so that a new aminoacyl tRNA can bind at the A site; 3. movement of message by one codon. → Elongation factor G and GTP hydrolysis are required for all three processes. (→ pre-translocation complex; post-translocation complex).

In principle, initiation of g. t. and elongation are similar in prokaryotes and eukaryotes, but there are subtle differences.

3. **Polypeptide chain termation:** Both prokaryotic and eukaryotic cells utilize remarkably similar processes to terminate protein synthesis. Termination, i.e., the release of the completed polypeptide from the → polyribosome and disengagement of ribosomes from mRNA, is directed by specific → terminator codons (UAA, UAG, and UGA) in mRNA which are recognized by special → release factors. When translocation places one of the terminator codons in the ribosomal A site, the ribosome does not bind an aminoacyl tRNA-elongation factor-GTP complex. Instead it binds a release factor which activates peptidyl transferase, which then hydrolyzes the bond joining the polypeptide to tRNA in the ribosomal P site.

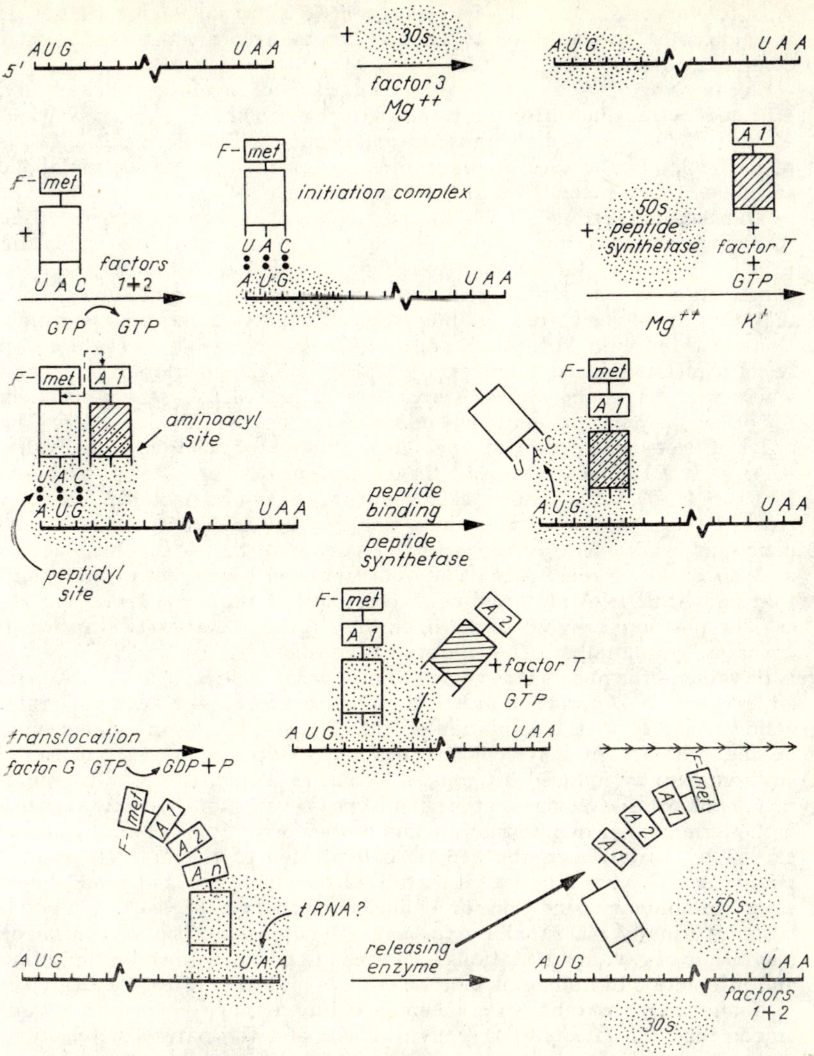

Fig. 50. Outline of genetic translation in *Escherichia coli*; AUG or GUG: initiator codons; UAA, UGA, or UAG: terminator codons (from *de Man* and *Noorduyn* 1969).

The amino acid sequences of the polypeptides produced during g. t. determine how the amino acid chains will fold to generate biologically active three-dimensional protein structures.

In prokaryotes, g. t. starts before → genetic transcription is terminated; in eukaryotes, transcription and translation take place in different cellular compartments.

Mechanisms of → genetic regulation that act by regulating the rate and efficiency with which mRNA is translated are mechanisms of → translational control, as opposed to transcriptional control. Translational control is of particular importance in eukaryotes because of the relative metabolic stability of messenger RNA in these organisms.

Translational errors (→ mistranslation) may occur during g. t. due to codon-anticodon mispairing and represent transient errors which are principally of two kinds: 1. misreadings of a codon in mRNA, i.e., the substitution of one amino acid for another; 2. errors that involve the maintenance of the correct reading frame. Translational errors are a function of the individual bases and not of the overall composition of the codon being translated. Their frequency depends upon the position in the codon where the base occurs, while the error pattern, i.e., what bases are mistaken for others, is more or less independent of the position in the codon. The order of decreasing error-proneness as a function of position in the codon is III > I > II. In each position the error pattern is UmC (i.e., U mistaken for C), > UmA, while UmG is least frequent or even undetectable. A similar pattern exists for C in that CmA ≫ CmG, while CmU is a rare mistake.

genetic unit — an operationally definable portion of the → genome, such as a → codon or → gene. Any genetic u. is defined by the set of mutations that fall within it. A mutation used to define and map some g. u. is a marker. The position assigned to a g. u. on the → genetic map is its locus which covers a finite number of → mutational sites.

genetic value — the phenotypic value an individual would have if there were no environmental → variation and no dominance effects (→ genetic variance).

genetic variability — the formation of individuals differing in → genotype, or the presence of genotypically different individuals, in contrast to environmentally induced differences which, as a rule, cause only temporary, nonheritable changes of the phenotype (→ variation). **Genetic variance** is the term used to denote that portion of the phenotypic variance which is caused by variation in the genetic constitution of the individuals in a population. G. v. is a universal feature of breeding populations and a necessary preliminary condition for evolutionary change. The g. v. of a population is usually measured by the average heterozygosity per locus. The magnitude of g. v. in populations is assumed to be determined by the action and interaction of numerous factors, such as mutation rates, the degree of → dominance of mutant genes, the mode of interaction (→ gene interaction) among loci, the effects of the environment and the nature of selection, stochastic processes (in the case of small population size), breeding systems, and population structure.

The chief sources of g. v. are the following processes:

1. → **mutation** as the ultimate source of all genetic variability; it gives rise to changes in the molecular structure of the genes (→ gene mutation), in the linear sequence of chromosome segments (→ chromosome mutation) or in the number of chromosomes (→ genome mutation);

2. → **gene flow,** i.e. migration of individuals or gametes carrying new alleles from one population to another;

3. → **hybridization** between individuals carrying mutations in different genes followed by the production of offspring with new gene combinations via → genetic recombination.

Processes 1. and 2. may directly alter the → gene frequency in a population and are thus the basis of evolutionary change (→ evolution). Genetic recombination assembles an existing array of alleles of different genes into a variety of combinations without altering their frequencies. Although genetic recombination is, therefore, not an evolutionary force, it is by far the most effective source of individual g. v. in sexually reproducing organisms. Process 3. generates most of the intrapopulation genotypic differences which then become available for → selection and → genetic drift.

Regulation of g. v. in a population is primarily determined by its chromosome behavior and system of reproduction. Equilibrium conditions providing for the retention of g. v. can be established by the interaction of mutation and selection or through selection for heterozygotes.

According to *Fisher* (1930) and *Mather* (1943) two types of g. v. are to be distinguished.:

1. **Free genetic variability:** that part of the total g. v. which is manifested by the phenotypes in a population and which is, therefore, exposed to the action of → selection.

2. **Potential (= "concealed" or "cryptic") genetic variability:** That part of the g. v. not yet manifested in the phenotypes, and hence not exposed to selection in any given generation. Potential g. v. can become phenotypically manifest in later generations on passing from the potential to the free state. Both the store and the rate at which this type of g. v. becomes transformed into free g. v. are controlled by natural selection and are dependent on the breeding system of the population under observation. Asexual reproduction limits, or even prohibits, the realization of potential g. v.; in endogamic populations the store of potential g. v. is always small.

Potential g. v. is present in heterozygous genotypes; the potential g. v. of homozygotes is nil and thus their flexibility slight. The type and quantity of potential g. v. may become modified by particular cytogenetic mechanisms.

The underlying factors determining the store of concealed g. v. are:

a) unrealized mutations and gene combinations;

b) the separate presence of genes with complementary action;

c) genes whose activity is reduced or prevented by modifier genes or epistatic interaction;

d) recessive alleles not manifest in the heterozygous state;

e) limitations by linkage, and in polygenic systems limitation in the production of recombinants, and

f) genes and gene systems giving rise to similar phenotypes under the prevailing environmental and developmental conditions.

Release of concealed g. v. is brought about by different mechanisms:

in a) and b) by genetic recombination, in c) by inactivation or elimination of the modifiers and epistatic genes, increased penetrance and recombination, in d) by segregation or increased penetrance, in e) by recombination via crossing-over, and in f) by changes of environmental conditions or phenotypic differentiation by placing the genes in question into different residual genotypes.

A store of concealed g. v. will contain variants which, when set free, may prove unsuccessful under all conditions, useful under a definite set of circumstances, neutral or even harmful at the time of realization but useful later on. The significance of concealed g. v. lies in the provision of new material for selection when transformed into free g. v.

Decay of g. v. means the reduction of heterozygosity as a consequence of → genetic drift, leading to loss and fixation of alleles at various loci. Its rate is intimately tied to the population size.

genetic variance — that portion of the phenotypic variance which is caused by differences in the genetic constitution of the individuals in a population (→ variation). G. v. can be divided into two components: 1. g. v. resulting from differences between homozygotes **(additive g. v.)**; 2. g. v. resulting from specific effects of various alleles in heterozygotes **(dominance variance).** Dominance here refers to deviations of the heterozygote from the two homozygotes.

genetic variation — → variation.

genocline → cline.

genocopy (*Nachtsheim* 1957) — the production of the same → phenotype by different genes ("mimetic genes").

genome (*Winkler* 1920) — in eukaryotes, the basic (monoploid) → chromosome set of an organism, consisting of a species-specific number of → linkage groups; hence the sum total of its genes. In prokaryotes (bacteria, viruses) all → genetic information is contained in one linkage group and g., therefore, denotes the complement of all hereditary factors included therein.

The smallest possible unit of the g. in mutation (→ gene mutation) and recombination (→ genetic recombination) is the individual nucleotide pair of → deoxyribonucleic acid (in some viruses of → ribonucleic acid), and is referred to as a muton and a recon, respectively. These nucleotides are arranged in a definite linear order into larger functional units of the g.:

1. → **cistrons** (usually consisting of several hundred nucleotide pairs) are units (genes) which specify by → genetic transcription the nucleotide sequence of various species of → ribonucleic acids, and by → genetic translation, the amino acid sequence in specific polypeptides;

2. → **operons** (or transcriptons) are the units of transcription of DNA into → messenger RNA and include several functionally related cistrons;

3. → **polarons** are units in terms of the points where crossing-over can be initiated and probably coincide with operons;

4. → **replicons** are units of replication and possibly consist of a series of operons;

5. whole linkage structures (chromosomes) may behave as units of → genetic regulation in eukaryotes under particular circumstances.

Two genomes are said to be exactly homologous if both contain the same genetic loci in identical sequence in their linkage structures, and homeologous, if only a part of the segments is identical in localization, due to structural rearrangements.

The diplophase of the life-cycle of diploid eukaryotes is characterized by the presence of two, the haplophase by the presence of only one genome. This → alternation of nuclear phases is brought about in the case of sexual reproduction by → fertilization and → meiosis. In genetic systems with parasexual modes of reproduction, the bringing together of two genomes or one genome and one genome fragment into the same cell is achieved by a variety of other means (→ genetic recombination).

Organisms having more than two genomes in their cells are called → polyploid.

In eukaryotes, nuclear and organelle genomes may be distinguished (→ organelle DNA). Nuclear DNA is represented by two groups of nucleotide sequences: → unique DNA and → repetitious DNA (→ gene reiteration). The ratio of unique to repetitive DNA varies from species to species.

genome allopolyploid (*Stebbins* 1947) — → alloploid.

genome mutation — any spontaneous or experimentally induced change in the number of complete chromosomes giving rise to heteroploid cells or individuals (which are either → polyploid or → aneuploid) as compared with the number of → chromosome sets or individual chromosomes in the species under observation.

genome segregation (*Bauer* 1943) — the separation of complete genomes (sets of chromosomes) during mitosis in eukaryotes; this is supposed to bring about "somatic → reduction" of the chromosome number in polyploids.

genomic exclusion (*Allen* 1963, *Nanney* 1963) — in crosses of certain inbred strains (A) of *Tetrahymena pyriformis* with clones of other strains (B), the recovery in the progeny of genes from one parent (A) only, although both exconjugants are present. The genes from the other parent strain (B) are excluded.

G. e. is characterized by two consecutive conjugations. Cells of the strains (B) whose genes are excluded have a defective micronucleus and in crosses with A cells, all B meiotic products disintegrate. During the first (abnormal) conjugation, the A conjugant undergoes meiosis normally generating sister haploid nuclei (the male and female pronuclei); the male pronucleus migrates to the B conjugant and then each pronucleus becomes diploid (probably by endoreduplication). In each conjugant the diploid synkaryon divides twice by mitosis and gives rise to two new macronuclei and two new micronuclei, and the old macronucleus starts to resorb. Once the exconjugants separate, the old macronucleus is retained and the new ones are resorbed. The progeny of the first conjugation are, therefore, heterokaryons with a micronucleus homozygous in genotype, express the macronuclear genes of the old macronucleus (distinct genotypically from the micronucleus), and are sexually mature. The second round of conjugation is normal and gives rise to a functional macronucleus similar in genotype to the micronucleus. The synkarya are derived from two meiotic products, one from each conjugant. If the two exconjugants from the same round 1 pair are remated, the (immature) progeny are, therefore, genetically homozygous at all known loci for the genes of the normal parent (B).

By g. e. the population becomes cleaned of defective cells and the development of new amicronucleate lines is prevented. Thus, g. e. may be understood as a means for the reacquisition of a normal micronucleus (*Allen* 1967).

genoneme (*Lewis* & *John* 1970) — = → genophore.

genophore (*Ris* 1961) — the physical equivalent of a linkage group (the linkage structure or chromosome) in prokaryotes, bacterial → plasmids, → mitochondria and → chloroplasts (= genoneme) which is represented by a "naked" nucleic acid molecule. The term g. was proposed to emphasize the differences in structure of → chromosomes in prokaryotes and eukaryotes.

genospecies (*Ravin* 1963) — a genetically defined → species, i.e., a group of individuals potentially able to contribute to, or share in, a common gene pool.

genotroph (*Durrant 1962*) — plants transformed into new distinct heritable types by nutritional → conditioning

genotropic (*Waddington* 1962) — embryonic → induction.

genotype (*Johannsen* 1909) — 1. the sum total of the → genetic information (genes) contained in the linkage structures (chromosomes) of the pro- and eukaryotes, as distinguished from their → phenotype (→ idiotype). The g. determines not a unique phenotype, but a range of phenotypic capacities referred to as an individual's "norm of reaction" to the environment. The share of one particular g. in the → gene pool of the next generation is determined by the viability and reproductive success of its carriers.

2. The genetic constitution in respect to the → alleles at one or a few → genetic loci under observation. If specific loci primarily responsible for the appearance of a particular → character are studied, the rest of the genotype is referred to as the "residual genotype or background genotype" (→ epigenotype).

genotype-environment interaction — a third class of interaction parallel to allelic and nonallelic → gene interaction which takes place between genes and their environment. If the environmental factor can be controlled and applied as a treatment to different genotypes, the variance contributed by genotype-environment i. may be calculated. This is not usually possible, and thus this interaction is generally regarded as part of the environmental variance (→ variation).

genotype frequency — the proportion or frequency of any particular genotype among the individuals of a population. Genotype frequencies are a function of → gene frequencies. Changes in frequencies of both are brought about by systematic and/or dispersive processes.

genotypic — ref. to any phenomenon or process connected with or controlled by the → genotype.

genotypic distance — the g. d. between two individuals A and B, for a given locus, is the probability that the genotype A is not the same as that of B (→ genetic distance).

genotypic expression — the manifestation by → gene action of specific genotypic activity.

genotypic mixing (*McBride* 1962) — in animal viruses, after mixed infection with genetically different types of a particular virus, the enclosure within the virus envelope of two or more different genomes. Such genotypically mixed virions are different from → heterozygous bacteriophages in which the heterologous DNA is part of a single viral DNA molecule.

genus — a systematic unit consisting of a group of → species which are of common phylogenetic origin and appear to be clearly differentiated from other such groups.

geographic isolate — → isolate.

germ cell (*Engler & Prantl* 1897) — any of the "reproductive" (generative) cells in multicellular organisms, as distinct from somatic cells. Germ cells are ultimately either → gametes or → agametes and arise from primordial germ cells or gonocytes. The term g. c. covers both the ancestral forms and their ultimate differentiation products. Germ cells may be classified as follows (*Luckhaus* 1965):

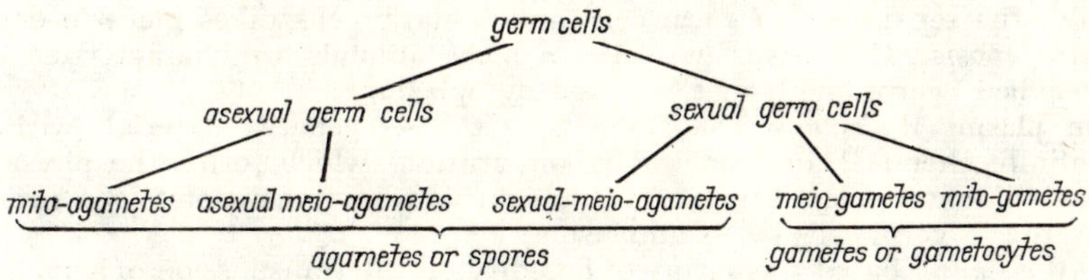

1. **Agametes (= spores)** are noncopulating germ cells (in sporozoa and plants), usually asexual (in some cases sexually differentiated), which may arise by meiosis (**"meio-agametes"**, "gonospores", "meiospores", "tetra-spores" or "gonia") or by mitosis (**"mito-agametes"** or "gonidia"). Organs producing agametes are called agametangia (= sporangia, gonidangia); in cases of heterospory, microsporangia (= androsporangia) and megasporangia (= macrosporangia, gynosporangia) are distinguished (→ megasporogenesis, microsporogenesis).

2. **Gametes** are sexually differentiated copulating germ cells which may arise by meiosis (**"meio-gametes"**, "gonogametes" or "gones") or by mitosis (**"mito-gametes")**. The cells giving rise to → gametes are referred to as "gametocytes" (= "gamonts", "gamentangia", "gametogonia"). In metazoa, gametocytes are undifferentiated precursors ("primordial germ cells") of egg cells (ova) and sperm which are produced in the gonads (ovaries in females, testes in males). The process of egg cell production during gametogenesis proceeds via oögonia, primary oöcytes, secondary oöcytes and oötids (→ oögenesis), the process of sperm production via spermatogonia, primary and secondary spermatocytes and spermatids (→ spermatogenesis).

Gametocytes in lower plants are in most cases sex organs and are called gametangia (isogametangia in thallophyta producing isogametes; anisogametangia in thallophyta producing anisogametes; oögonia and antheridia in thallophyta producing egg cells and spermatozoids; archegonia and antheridia in mosses and ferns; karpogonia and spermatangia in red algae; ascogonia and antheridia in fungi).

germinal spot (*Purkinje* 1825) — the nucleolus of the egg cell.

germinal vesicle (*Purkinje* 1825) — the → nucleus of the → oöcyte at the end of the growth period and before the initiation of meiosis.

germ layer — one of the (two or three) fundamental cell layers referred to as ecto-, endo-, and mesoderm in an early embryo of a multicellular animal from which tissues and organs of the adult are formed (→ embryonic development).

germ line (*Weismann* 1865) — a lineage of "generative" cells (= germ track) ancestral to the gametes (sperm and egg cells proper) which, during the development of an organism (particularly in animals), are set aside as potential gamete-forming tissues. These ancestral cells, together with the gametes, are called → germ cells, as opposed to somatic cells. Location, nature, and time of formation of potential gamete-forming tissues are species-specific and may vary greatly from one species to another. Generally, the separation of germ cells and somatic cells takes place in early ontogenesis. The germ line cells of some animals are characterized by so-called "germ line bodies" in their cytoplasm.

germ plasm (*Weismann* 1883, 1885) — the → genetic material (with a definite chemical and molecular constitution) which forms the physical basis of inherited qualities and is transmitted from generation to generation by the → germ cells (= Keimplasma).

Weismann based the nature of heredity on the transmission of a nuclear substance which in his theory of the continuity of the g. p. is equated with the "Keimplasma". He thought of the g. p. as that substance of specific molecular constitution which "is reserved unchanged for the formation of the germ cells of the following generation" during the development of the body or soma of an individual from the fertilized egg cell.

germ track — = → germ line.

G factor (*Watson* 1964; *Nishizuka* & *Lipman* 1966) — a protein (tRNA translocase) involved in the → translocation during bacterial → genetic translation of the new peptidyl tRNA from the A to the P site of the → ribosome (→ tRNA releasing factor). It is involved in the hydrolysis of GTP during polypeptide elongation which is thought to assist the rearrangement of tRNA, mRNA, and ribosomal subunits. These rearrangements must occur between successive rounds of peptidyl transfer. Probably G enters and leaves the ribosome during each cyle of amino acid addition.

giant chromosome — → polytene chromosome.

glycocalyx (*Bennet* 1963) — a component of the → cell surface located outside the → cell membrane and, in many instances, attached wholly or in part to its outer surface (= cell coat). The g. always contains sugars of some sort frequently in company of, and often covalently bound to, other species of substances (such as amino acids, lipids, lignins, proteins, or other polymers). In contrast to the cell membrane, the g. components are readily pervaded by water, ions and small molecules and show low electrical resistance. The structural and chemical characteristics of the g. vary widely.

One of the variants of the g. is the → cell wall of plant cells.

Golgi apparatus (*Golgi* 1898; *Ramon & Cajal* 1908) — in animal and plant cells, a series of concentrically bent double membranes enclosing submicroscopic spaces (flattened, membrane-bound sacs also referred to as "cisternae", "saculles", or "lamellar units") which can be individualized and may be called → dictyosomes or golgiosomes. Size and number of these "cisternae" are subject to considerable variation; their relative length may differ from species to species, with stage of ontogeny, and in relation to activity. The extent of the G. a. is determined by the extent to which its component dictyosomes are interassociated. The G. a. may be localized in the vicinity of the cell nucleus or the cell periphery (cortex), or the dictyosomes may be distributed throughout the cell.

The G. a. appears to be concerned with the elaboration of secretory materials which include a variety of compounds:

a) in animals, the zymogen of pancreatic cells, mucous secretions, lactoprotein in mammary gland cells, compounds of thyroxin, melanin granules and perhaps other pigments, tropocollagen and even collagen.

b) in plants, at least part of the substance of the primary → cell wall, some of the components of secondary walls and the slimes.

Materials concentrated in parts of the G. a. may move through cavities of its structure in continuity with the outside and thus leave the cell, or they may become budded off in form of Golgi vesicles which move through the cytoplasm to the plasma membrane of the cell where they release their contents to the cell exterior.

The G. a. is differentiated into what is called its "forming face" and "maturation face". In the cavities of the latter, processes such as accumulation, condensation, transformation, and synthesis result in the formation of Golgi products. It buds off vesicles that carry the Golgi products away from its structure.

The G. a. or some functional equivalent should be regarded as an organelle with the same degree of integrity as, e.g., the → mitochondria and → plastids. Hypotheses concerning the origin of the G. a. fall into 3 general categories:

1. *de novo* origin;
2. development from a precursor structure, and
3. replication from an existing G. a.

The distribution of Golgi bodies during mitosis is called "Golgio-kinesis".

"Golgio-lysis" and "Golgio-rhexis" (*Sosa* 1930) are terms denoting, respectively, the dissolution and fragmentation of the G. a.

gonad — in animals, a reproductive organ (ovary in females, testis in males, ovotestis in hermaphrodites) in which gametes (ova and sperm) are produced).

gone (*Lotsy* 1904) — any → germ cell arising by meiosis.

gonia — primordial sex cells such as oögonia and spermatogonia (→ oögenesis, spermatogenesis).

gonidia (*Renner* 1916) — → germ cell.
gonochoristic — of individuals having functional gonads of one sex only, either male or female (= gonochoric or dioecious).
gonocyte (*Renner* 1916) — any unripe → germ cell.
gonogenesis (*Levitzki* 1925) — the formation of → gones during meiosis.
gonomery (*Haecker* 1895) — the separate grouping of maternal and paternal chromosomes during → cleavage stages of some organisms.
gonosome (*Plate* 1913) — = → sex chromosome.
gonospore (*Renner* 1916) — → germ cell.
gonotokont (*Lotsy* 1904) — any cell (= meiocyte) producing → gones by → meiosis.
gonozoospore (*Oehlkers* 1956) — a → zoospore produced in the course of meiosis.
G period (*Howard & Pelc* 1953) — periods of the interphase nucleus in eukaryotes.

G_1 is the period during which the chromosomes are released from their condensed mitotic state and protein and RNA synthesis resume at a rapid rate. Particular enzymes have been shown to be specifically synthesized in G_1, which ends with the beginning of the → S period of interphase.

G_2 is the interphase period which bridges the internal between the end of DNA synthesis (S period) and the initiation of mitosis. Condensation of chromosomes presumably begins during G_2 by supercoiling involving complexing → histones with DNA and subsequent folding up of the fibers.

grade (*Huxley* 1958) — a unit of biological improvement from an evolutionary point of view comprising a group of individuals similar in their level of organization.
gradient — levels or degrees in polarized systems (developmental potency, character formation, metabolism, etc.).
graft-hybrid — → grafting.
grafting — the production of a → chimera by vegetative fusion of parts from different individuals. **Graft-hybrids** are plants in which tissues or cells of different → genotypes exist side by side, as some cell layers come from the scion and some from the stock.
granum (*Meyer* 1883) — a structure within the chloroplast (→ plastid) which is recognized as a green granule by the light microcsope and as a column of flattened discs piled one on the top of the other by the electron microscope. Each disc consists of a pair of membranes appressed back to back (thickness 6 nm or more). Often grana are interconnected by double-membraned structures (which also extend into the stroma: stroma lamellae). The light absorbing pigments are contained in the grana.
g-region (*Lee & Thomas* 1973) — relatively short DNA regions along the chromatids of eukaryotes which are characterized by clustering of → repetitive DNA sequences. The average length of such regions is about 5 nm to 10 nm (15,000 to 30,000 nucleotide pairs), or about one-half the average length of DNA per (*Drosophila*) chromomere. The g.-regions are assumed to be composed of approximately regularly spaced, intermittently repetitious sequences of tandemly-repeating sequences in which some portion (less than half) is non-repetitious.

groundplasm — the "submicroscopic" ground substance or cytoplasmic matrix (revealed by electron microscope) which remains after exclusion of the cell organelles and cell particles. The g. (the "hyaloplasm" of light microscopy) is a highly complex, polyphasic system in which are suspended all the resolvable elements of the cytoplasm, including the larger organelles such as the → ribosomes, the → mitochondria, the → plastids (in plant cells), → lipid droplets, and → vacuoles. It contains (in unknown degrees of organization) most of the enzymes of intermediary metabolism of the cell (including those involved in preliminary or alternative pathways of ATP synthesis), accomodates the metabolic reserves of the cell, and its aqueous phase contains the main cellular pool of soluble precursors. The g. also contains a population of contractile protein molecules responsible for most cell movements.

gymnocyte — a cell without a → cell wall, as distinct from a lepocyte surrounded by a cell wall.

gymnoplast (*Küster* 1935) — in higher plants, a cell without → cell wall (= gymnocyte), in contrast to dermatoplasts, i.e. plant cells with a cell wall. Yeast cells and bacteria devoid of their cell wall are frequently called protoplasts. Gymnoplasts or protoplasts, the cell wall of which is incompletely removed, are termed spheroplasts or semi-gymnoplasts.

gynaeceum — the female reproductive organ of higher plants; it is composed of pistil, style, and stigma.

gynander — = → gynandromorph.

gynandrism — a type of → genetic mosaicism; the individuals showing g. (= gynanders or gynandromorphs) are mosaics of XX and X cells (→ X chromosome).

gynandroid (*Whiting, Greb & Speicher* 1934) — → gynandromorph.

gynandromorph (*Goldschmidt* 1915) — an individual of a dioecious (unisexual) species which is a sexual mosaic (a chimera), typically male in certain portions (sectors) of the body and typically female in others. Gyandromorphs result from the presence of chromosomes of both sexes in different parts of the body.

Gynandromorphs have been observed in animals (not yet in plants) where sex differentiation is apparently intracellular and not hormonal and may arise in several ways:

1. by loss of an → X-chromosome during early cleavage in cases of an XX—XY-mechanism of sex-determination, provided XO-cells are male in character (as in *Drosophila*);

2. by suppression of polar body formation giving rise to a binucleate egg cell. In cases of female heterogamety (as in silkworms), one of these nuclei contains an X-chromosome, the other a Y-chromosome. Fertilization of each of the nuclei by a sperm containing one X-chromosome may result in a g.;

3. by fertilization of only one of the two nuclei present in a binucleate egg cell in cases where the diploid nucleus gives rise to female and the haploid nucleus to male tissue (as in honey bees);

4. by loss of a whole chromosome set during the first cleavage division.

Both the size and position of the male and female sectors are dependent on the time and position of the onset of these processes. In animals with sex hormones, the production of clear-cut gynandromorphs is prevented since these hormones give rise to an intersexual development irrespective of the different constitution of the tissues.

An exceptional kind of g. in *Habrobracon* is called **"gynandroid"** (*Whiting, Greb & Speicher* 1954). In *Habrobracon*, sex is determined by a system of multiple sex alleles (xa, xb, xc, etc.). Females arise in association with heterozygosity for the sex alleles, males in association with hemi- or homozygosity. Gynandroids are haploid and arise from unfertilized, binucleate eggs in which the two nuclei contain different sex alleles. In the region of the genitalia at the juncture between tissues of different genotypes for x, small female reproductive appendages arise, possibly by diffusion of gene products peculiar to one of the x tissues and interaction with locally produced substances of the other tissue; this may give rise to female differentiation.

gynephoric (*Waardenburg* 1932) — ref. to → sex-linked, recessive modes of inheritance whereby phenotypically normal females act as heterozygous → carriers of the genes in question.

gynoautosome (*Yamamoto* 1938) — an → autosome carrying → sex realizers for femaleness (→ androautosome).

gynodioecious (*Darwin* 1877) — → dioecious.

gynoecious (*v. Uexküll-Syllenband* 1901) — of plants having only female flowers (→ androecious).

gynogenesis (*Wilson* 1925) — female → parthenogenesis. After fertilization of the ovum, the male nucleus is eliminated and the haploid individual (described as "gynogenetic") so produced possesses the maternal chromosome set only (= pseudogamy; merospermy) (→ androgenesis).

gynomerogony — the development of an egg fragment containing only the female nucleus (→ andromerogony), i.e. with exclusively maternal chromosomes.

gynomonoecious (*Darwin* 1877) — → monoecious.

gynospore (*Battaglia* 1955) — = → megaspore.

gynosporogenesis (*Battaglia* 1955) — = → megasporogenesis.

gyre — = → coil (→ chromosome coiling).

H

habitat — that place within a → biotope inhabited by a particular population, species, or species group.

"Habitat selection" is the selection by a dispersed individual of an appropriate species-specific h. and the avoidance of unsuitable ones. It is one of the most important factors leading to the localization of populations and to the restriction of species to their specific ecological niches (*Mayr* 1963).

"Habitat exclusion" is the mutual exclusion of the habitats of two related species.

habituation (*Binns & Meins* 1973) — a type of heritable cellular change in tissue cultures of various plant species characterized by a newly acquired capacity of cells to produce growth regulatory substances. Induced reversal of h. has been reported indicating that h. is not due to a permanent genetic change. H. may result from heritable alterations in the pattern of gene expression.

hairpin loop — any double-strandet region in single-strandet RNA and DNA caused by base pairing between complementary base sequences. Such regions might play a role in the stability of the molecule and their formation is due to the ability of the nucleotide chain to turn back on itself and form a particular secondary structure. (→ palindrome).

Haldane's rule — when among the offspring of crosses between species and varieties, one sex is absent, rare, or sterile, it is the → heterogametic sex.

half-chromatid — one of the two hypothetical fibrillar subunits which constitute a → chromatid. A single → chromosome of eukaryotes thus consists of four half-chromatids after reduplication during interphase.

Chromosomal structural changes of the → intrachange and → interchange type (→ chromosome mutation), in which half-chromatids are the units of change are referred to as "half-chromatid aberrations".

half-heterogamy (*Renner* 1918) — → heterogamy.

half-mutant (*de Vries* 1917) — a "mutant" in → complex-heterozygous species of *Oenothera* (*Oe. Lamarckiana*) which carries a modified nonlethal genome plus an unmodified (*Lamarckiana*) genome in both sperm and egg. In each generation such mutants give rise to so-called "full-mutants" containing the modified nonlethal genome in double dose (*Cleland* 1962).

According to *Darlington* (1931), half-mutants are the product of rare crossing-over between homologous segments in otherwise nonhomologous chromosomes, thus giving rise to effective exchange between two "complexes" (retranslocation).

half sibs — individuals that have only one parent in common.

half-tetrad — two of the four strands (chromatids of a → bivalent) from a single meiotic event (→ tetrad). Recombination studies in *Drosophila* involving → attached X-chromosomes permit the analysis of half-tetrads. The same is true for mitotic recombination where the recovery of two of the four chromatids that may be engaged in mitotic → crossing-over is possible.

half-translocation (*Herskowitz & Muller* 1953) — half of a reciprocal → translocation. It may result from:

1. the absence of one of the reciprocal fragment fusions giving rise to an incomplete translocation (loss or behavior of the unjoined fragments frequently causes descendent cells to die or to be abnormal);
2. segregation in heterozygotes for a symmetrical translocation whereby only one of the two reciprocal exchanges is included in a gamete.

haplochromosome (*Darlington* 1965) — any → chromosome arising in those cases where chromosome reduplication and chromosome splitting are inhibited. Each such chromosome (→ hemichromosome) consists of one chromatid of normal size. Haplochromosomes may be induced experimentally by irradiation and are characteristic of discordant or precocious → polymitosis.

haplo-diplontic — → diplo-haplontic.

haploid (*Strasburger* 1905) — of cells or individuals with a single genome or → chromosome set provided more than one linkage structure (as in eukaryotes), distinct in homology and function, is present. Viruses and bacteria whose genome consists of only one linkage structure are haploid, and so are parts of the life cycle (referred to as "haplophase") of eukaryotes. In diploid species of eukaryotes, individuals may occasionally occur whose **somatic** cells are haploid. In most cases they result from spontaneous or induced parthenogenetic development of unfertilized eggs (→ parthenogenesis). Such individuals are referred to as **"monohaploids"**. They are to be distinguished from so-called "polyhaploids", i.e. haploid individuals arising from → polyploid species.

Exceptional haploids may be classified as follows, based on the special cytological features (*Kimber & Riley* 1963):

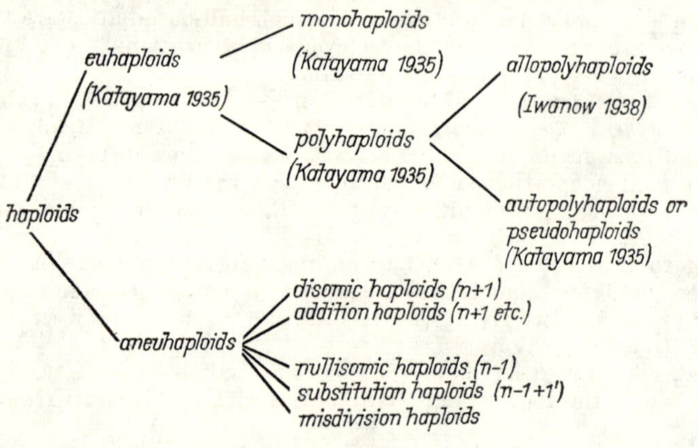

The cytogenetic constitution of polyhaploids varies and depends on the type of polyploid from which they arose. The chromosome numbers of **aneuhaploids** are not exact multiples of one of the basic numbers of the group, as compared with euhaploids. Aneuhaploids may have extra chromosomes, or may be deficient in chromosomes, relative to the euhaploids. Furthermore, the extra chromosomes may be members of the chromosome set of the species or they may be alien (as, e.g., in addition h.). Haploids deficient in a chromosome of the normal set (subhaploids or nullihaploids) have been rarely found. Substitution haploids have one or more alien chromosomes substituted for the species-specific ones. Misdivision haploids are those with telocentric chromosomes and misdivision (→ centromere misdivision) derivatives of some chromosome(s). In accordance with their chromosomal constitution, the meiotic behavior (→ chromosome pairing) of these various types of haploids may vary considerably.

haploidization — the transformation of → diploid into → haploid cells (or strains) in certain fungi. H. is a process of the → parasexual cycle and apparently proceeds by the progressive loss of chromosomes, one chromosome at a time, via → nondisjunction until a haploid (euploid) nucleus is formed from the → aneuploid intermediates. During h., the chromosomes assort independently of one another and thus a variety of different haploids are obtained.

haplontic — of organisms ("haplonts") with a life cycle in which meiosis occurs in the zygote giving rise to four haploid cells, e.g., most primitive unicellular or filamentous algae and protozoa (→ diplo-haplontic; diplontic). Only the zygote of haplonts is diploid. In some species, this cell becomes a resistant spore tiding the organism over hard times (Fig. 51).

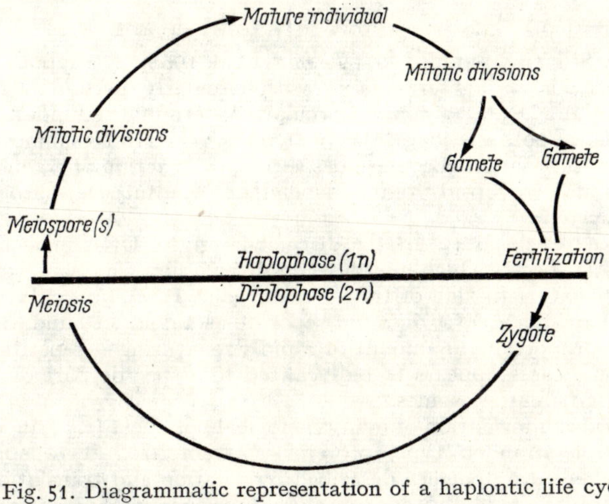

Fig. 51. Diagrammatic representation of a haplontic life cycle as found in many algae (from *Cook* 1965).

haplophase — the haploid phase or generation (= gamophase according to *Winkler* 1920) of the life cycle, lasting from meiosis to fertilization (zygote formation), as distinct from → diplophase.

haplosis — the meiotic reduction (→ meiosis) of the somatic chromosome number.

haplosomic — = → monosomic.

haplospory (*Battaglia* 1947) — apomictic development (→ apomixis) of a female gametophyte from a mother cell produced in the course of a normal → meiosis.

haplo-triplo-disomic — of cells in which the number of chromosomes is unmodified (species-specific) but one normal chromosome is replaced by one of its → isochromosomes. Such variants are → monosomic for one, and → trisomic for the other arm of the same chromosome.

hapten — a substance (partial or incomplete antigen) that can react specifically with an → antibody, but cannot by itself evoke its appearance.

Hardy-Weinberg law (*Hardy* 1908; *Weinberg* 1908) — a statement (= "H.-W. formula" or "H.-W. equilibrium principle") in mathematical terms concerning the theoretical relationship between → gene frequencies and → genotype frequencies within populations. It describes the genotypic equilibrium produced by a static gene pool. In a large random mating → population, both the gene frequencies and the genotype frequencies are constant from generation to generation in the absence of mutation, migration, and selection (the population is in H.-W. equilibrium). If the frequencies of allele A and a (of equal selective value and occurring together in a population) are q and $1 - q$ and mating is at random, then male gametes $qA + (1 - q)\,a$ fertilize eggs of $qA + (1 - q)\,a$ producing offspring (genotypes) according to the following binomial law:

$$[qA + (1 - q)\,a] \times [qA + (1 - q)\,a]$$
$$= q^2AA + 2q(1 - q)\,Aa + (1 - q)^2\,aa.$$

headful hypothesis (*Streisinger* et al. 1967; *Thomas* et al. 1968) — a hypothesis explaining replication of bacteriophages with circularly permuted and terminally redundant DNA (→ genetic circularity; terminal redundancy). According to this hypothesis long → concatemers of a viral genome are converted into mature virion chromosomes through the action of a genetic site-independent and → capsid volume-dependent (headful size) chromosome breakage system.

heavy shoulder DNA (*Comings* 1972) — a distinct family of DNA molecules in some eukaryotic species which manifests itself by a non-Gaussian skewing to the heavy (GC-rich) side of the → main band DNA (in analytical ultracentrifugation). H. s. DNA does not separate as a distinct band, does not contain a disproportionate amount of rapidly renaturing → repetitious DNA, and in some cases appears to be localized to a specific part of the nucleus or to specific chromosomes.

helix — the natural conformation of many biological polymers (e.g., nucleic acid of the double-stranded type, proteins) characterized by a spiral structure with a repeating pattern described by rotation and translation.

helper virus — a virus supplying one or more functions that a so-called defective virus is unable to perform. By infection of a cell with both a defective and helper virus the former is enabled to multiply.

hemi-alloploid (*Löve & Löve* 1949) — = segmental allopolyploid (→ alloploid).

hemi-autoploid (*Löve & Löve* 1949) — of → polyploids which either started as autopolyploids and later became differentiated in the direction of allopolyploidy, or arose from more or less sterile intervarietal or intersubspecies hybrids.

hemichromosome (*Darlington* 1965) — any → chromosome arising when reduplication at interphase is inhibited but splitting into chromatids occurs. Such chromosomes and chromatids are half-size as regards the number of their fibrillar longitudinal elements when compared with regularly reduplicated chromosomes of the same organism (→ haplochromosome; diplochromosome).

hemicompatible — → di-mon mating.

hemihaploid — of cells, cell—complexes, or individuals with only half the normal gametic ("haploid") chromosome number. Hemihaploids may arise from polyploids.

hemiploid (*F. v. Wettstein* 1932) — of cells or individuals with half the normal somatic chromosome number (→ haploid).

hemizygous — of genes present only once in the genotype and not in the form of pairs of → alleles (as in haploids, in → differential segments of → sex chromosomes, or in diploids as a result of aneuploidy or loss of chromosome segments).

hereditary determinant — a genetically functional unit which is replicated with conservation of specificity. A h. d. may be chromosomal (nuclear) or nonchromosomal (extranuclear, extrachromosomal) and its presence in an organism's → idiotype is essential for the appearance of a particular → character (→ inheritance).

hereditary disease — = → genetic disease.

hereditary factor (*Mendel* 1865) — = → hereditary determinant.

heredity (*Spencer* 1863) — that process which brings about the biological similarity between parents and progeny. It consists of the conservation of specificity during replication of the → genetic material (the storage of → genetic information) and is carried out by means of transcription and translation of the genetic information (→ genetic transcription; genetic translation; genetic code). Genetics is the science of heredity.

heritability (*Lush* 1949) — The proportion of the total phenotypic variance for which genetic differences are responsible (**broad-sense h.**), or the proportion of the total phenotypic variance that is due to additive genetic variability (**narrow-sense h.**). H. is a measure of the amount of → genetic variability, excluding that expressed by heterozygotes, and decreases with an increasing environmental component of → variance for the character under observation.

hermaphroditic — of individuals ("hermaphrodites") bearing recognizable male and female tissues and producing mature male and female gametes either at the same ("simultaneous" or "synchroneous hermaphroditism") or at different periods ("consecutive hermaphroditism") of the life cycle.

Protogynous hermaphrodites function first as females then transform into males, protandrous hermaphrodites transform from males into females. The gametes of h. animals may be produced in separate male and female gonads (testes and ovaries) or in the same gonad (ovotestis). Plant gametophytes are called h. when the two sexes are represented in the same flower (or prothallus in the case of cryptogams). They are generally referred to as monoecious when male and female gametes mature in different flowers on the same plant.

A population which is composed exclusively of h. individuals may be called "balanced hermaphroditic". A population composed of a majority of hermaphrodites with varying degrees of maleness or femaleness and a minority of pure male and female individuals may be designated "unbalanced hermaphroditic" (*Bacci* 1950).

hertone (*Cole* 1972) — = → non-histone chromosomal protein.

heteroallelic (*Roman* 1956) — of genes ("heteroalleles") having mutations at different mutational sites, as distinct from homoallelic ones with mutations at the same site (= nonidentical and identical → alleles, respectively). Pairs of heteroallelic genes yield recombinants by intragenic recombination via → crossing-over, pairs of homoallelic genes do not.

heterobrachial (*Sorokin* 1929) — → chromosome arm.

heterocentric (*Sears & Camara* 1952) — of → dicentric chromosomes or chromatids (= "heterodicentric") whose → centromeres are of unequal strength. Heterodicentrics frequently behave like monocentrics, i.e. they do not give rise to a → chromosome or chromatid bridge during anaphase.

heterochromatic (*Heitz* 1928, 1929) — originally of chromosomal segments or whole chromosomes which possess a dense, compact structure in telophase, interphase and early prophase, as distinct from euchromatic segments. Today, this phenomenon is referred to as "positive heteropycnosis" and h. chromosomal regions or chromosomes are defined as those which become → heteropycnotic at some stage of the cell cycle, i.e. show "allocyclic" (→ allocycly) behavior.

Heterochromatin is generally recognized by the following features:

1. A condensed state throughout → interphase (positive heteropycnosis giving rise to → chromocenters); 2. out of phase replication of DNA (late replication); 2. transcription inactivity in the compacted state; 4. underreplication in → polytene chromosomes; 5. susceptibility to chromosome structural changes. Heterochromatin is subclassified into two types (*Brown* 1966):

1. **Constitutive heterochromatin** is present in all cells at identical positions on both the homologous chromosomes and forms a permanent structural characteristic of a given chromosome pair. Generally, the DNA of constitutive heterochromatin consists of → repetitious DNA, non-repetitive, AT-rich → mainband DNA, and non-repetitive → heavy shoulder DNA (in some species). → Satellite DNA tends to be localized to centromeric heterochromatin.

2. **Facultative heterochromatin** varies in its state in different cell types, at different developmental stages, or even from one homologous chromosome to another, i.e., is heteropycnotic in special cell types or at special

stages. It represents inactivated and condensed segments of euchromatin (→ dosage compensation).

A multitude of functions have been ascribed to heterochromatin which may act (*Cooper* 1959):

1. On genes, to control → mutation, to modify specific → gene action, → penetrance, and → specificity, and to affect quantitatively the action of some or all of the genes.

2. Within chromosomes, to stabilize → centromeres and → telomeres, to affect the processes connected with the production of → chromosome mutations, to bring about or prevent meiotic → chromosome pairing, to regulate → crossing-over and → chiasma localization, to produce specific, nonspecific, and reversed pairing and conjunctive properties, and to cause → variegation and chromosomal → sticky effects.

3. Trans-chromosomally, by controlling the dimensions of all the chromosomes of a nucleus, by regulating crossing-over, pairing, and disjunction of other chromosomes, and by controlling variegation brought about by genes located on other chromosomes.

4. Metabolically, by performing or mediating special syntheses of nucleic acids, proteins, nucleolar material, and energy-rich substances, and by controlling the transfer of substances across the nuclear envelope.

5. On the cell, by playing a role in the control of cell size and by governing mitosis.

6. On development, by regulating rates of growth and differentiation, by regulating the very nature of differentiation itself, and perhaps also by playing a role in → sex determination.

7. In speciation, by providing neutral anchorage and supplementary chromosomal parts for chromosome mutation, increases in the number of chromosomal arms and, in addition, duplicate genes that may acquire new functions.

8. In theory, as the special seat of the unorthodox in genetic systems.

Some of the functions suggested above are well documented others not.

heterochromatic fusion — → chromosome pairing.

heterochromatinized — of euchromatic chromosome regions which, when juxtaposed to → heterochromatic ones, become cytologically transformed ("heterochromatinization") in the direction of or to a heterochromatic state (→ position effect) (*Cooper* 1959).

heterochromosome (*Montgomery* 1904) — originally, any chromosome differing from the → autosomes (or "euchromosomes") in size, shape or behavior. Today the term is generally used as a synonym for → sex-chromosome (= allosome).

heterocytonic (*Catcheside* 1958) — of a cell or mycelium (in fungi) consisting of genetically different cytoplasms, characterized by different extranuclear hereditary determinants (→ plasmon), as distinct from "homocytonic". The terms hetero- and homocyton are parallel to hetero- and homokaryon which designate equivalent nuclear states (→ heterokaryotic).

heterodichogamous — → dichogamous.

heteroduplex — a double-stranded DNA molecule with a region of non-complementarity due to chemical or physical influences which alter bases of one of the two strands of the molecules or to occasional faults in DNA replication, or to → genetic recombination between molecules within a region of genetic difference (→ heteroduplex mapping).

H. molecules (as opposed to homoduplex molecules with completely complementary base sequences) do not persist as such. During replication, each strand serves as a template for the synthesis of two genetically distinct homoduplex molecules, provided h. regions are not eliminated prior to DNA replication by a correction mechanism (→ polaron hybrid DNA hypothesis). Molecules corrected in this way would only yield one type of homoduplex molecule (Fig. 52).

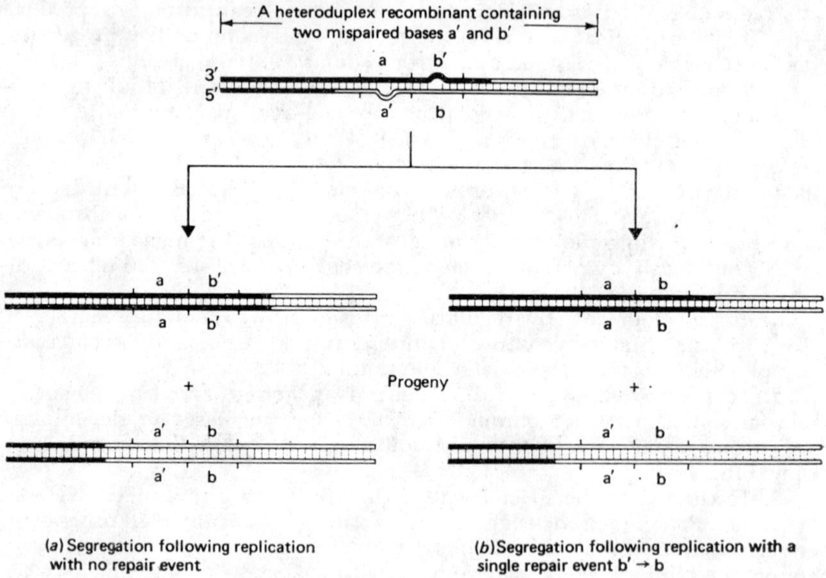

Fig. 52. Heteroduplex segregation with and without repair event (from *Watson* 1970).

1. **Mutational h.:** a double-stranded DNA molecule in one of whose polynucleotide chains a mutational event occurred. Such a molecule is incapable of generating a uniform progeny by replication. Instead, a mixed progeny of mutated and nonmutated molecules is produced in the following way:

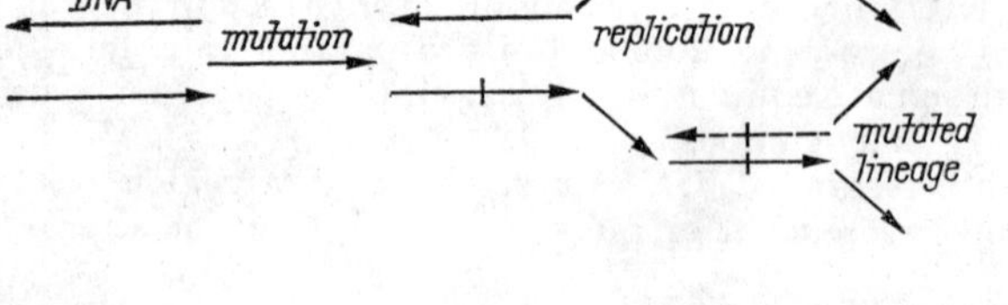

2. **Recombinational h.**: a double-stranded DNA molecule which gives rise to a parental and a recombinant molecule in the replication cycle immediately following its production. According to the breakage-reunion model of → genetic recombination, breaks occur at identical sites along the length of confronting pairs of DNA molecules and recombination results from the reunion of pieces of one broken molecule with pieces of the other. This process may take place between two native DNA helices or between two strands (one from each of the confronting molecule pair). The newly formed recombinant strands could then serve as templates for the synthesis of their complements, or they could form helices with strands of the parental molecules which were not engaged in breakage-reunion. In the latter case recombination heteroduplexes are produced. Each double-stranded molecule consists of one parental and one recombinant polynucleotide chain.

heteroduplex mapping (*Davis* & *Hyman* 1971) — the physical mapping of base sequence homology of two DNA molecules by constructing *in vitro* a double-stranded molecule in which one of the DNA strands is from one DNA molecule and the complementary DNA strand from another. Regions of insufficient homology exist as unpaired single-stranded regions (heteroduplexes) and can be recognized, measured and located with respect to one end of the molecules by electron microscopy (→ sequence homology map).

heteroduplex repair — correction of heteroduplex regions in double-stranded DNA assumed to be involved in genetic recombination and various phenomena connected with it (e.g., → gene conversion, → high negative interference, → map expansion). The sequence of events involved in h. r. may be the following: 1. single strands of recombining double-strands of DNA are cut; 2. base pairing occurs between different parental complementary strands; if these carry different genetic information, a heteroduplex DNA region may form in the course of the first step of genetic recombination; 3. conversion of the heteroduplex region by a repair-like mechanism (→ DNA repair) to a homoduplex structure; during this procedure genetic information from one strand could be transferred to its complementary strand.

H. r. accounts for gene conversion, map expansion, and high negative interference..

heterodynamic (*Waddington* 1953) — → homodynamic.

heteroecious (*Correns* 1928) — 1. producing male and female gametes on different individuals (= dioecious): a) diploheteroecy, i.e. dioecy in the diplophase, and b) haploheteroecy, i.e. dioecy in the haplophase (*Hartmann* 1929);

2. of fungi spending different stages of their life cycle in different hosts.

heterofertilization — the → fertilization of the endosperm- and embryo-forming nuclei by gametes of different genetic constitution (*Darlington* & *Mather* 1949).

heterogameon (*Camp* & *Gilly* 1942) — a → species comprised of varieties (races) which give rise to morphologically stable populations by selfing and several types of viable and fertile progeny after crossing.

heterogametic (*Wilson* 1910) — that one of the two sexes which during → meiosis gives rise to two types of gametes (male- and female-determining), in contrast to the "homogametic" sex. In the case of the XX—XY or XX—XO-types of → sex determination, the h. sex is that with both an X- and a Y-chromosome or with only one X-chromosome (= digametic). In most cases, the male sex is h., the female homogametic with respect to the → sex-chromosome constitution.

heterogamous (*de Vries* 1911) — of → complex heterozygous forms whose male gametes transmit genes or gene-complexes different from those of the female gametes, as distinct from iso- or homogamous hybrids.

heterogamy — 1. = → anisogamy;

2. alternation of two sexual generations, one being truly sexual, the other parthenogenetic (→ alternation of generations; parthenogenesis);

3. the mating of unlike individuals, i.e., the preference of an individual for mating with another of unlike phenotype or genotype, as distinct from "homogamy" (→ mating system).

heterogeneity index (*Nei & Roychoudhuri* 1974) — a measure of genic variance of a population (= gene diversity) equal to the probability of nonidentity of randomly chosen genes.

heterogeneous nuclear RNA (*Scherrer & Darnell* 1962) — a heterogeneous high molecular weight (10^5 to 2×10^7 daltons), metabolically unstable (lifetime 5 to 15 min.) RNA fraction (symbol Hn RNA) synthesized in the nuclei of eukaryotes and representing about 3% of the total cellular RNA.

The main characteristics of Hn RNA are the following:

1. Sedimentation values from 30 S to 100 S with no evidence of uniformity of size; 2. base composition similar to DNA (40 to 45% GC), in contrast to the high GC content (~70%) of → ribosomal precursor RNA; 3. location mostly, if not entirely, outside the nucleolus; 4. derivation of messenger RNA from Hn RNA (→ premessenger RNA).

20 to 40% of Hn RNA as well as eukaryotic mRNA contain regions of polyadenylic acid about 200 nucleotides long. Poly(A) is posttranscriptionally added at the 3′-OH end of the Hn RNA molecules (→ terminal riboadenylate transferase). After poly(A) is added, processing, i.e., enzymatic steps leading to a breakage of the Hn RNA, occurs. After processing, part of the Hn RNA molecules are transferred from the nucleus to the cytoplasm where they presumably perform the messenger function (mRNA).

Hn RNA is complexed with protein in the form of ribonucleoprotein particles. The proteins, themselves heterogeneous, are cell specific. The proteins may be concerned with the processing of Hn RNA; others may serve as site-specific signals directing the action of endonucleases, or may promote the selection of certain Hn RNAs (about 10%) for transport to the cytoplasm.

The poly(A)-terminated Hn RNA contains transcripts of intermediate → repetitious DNA covalently linked to nonrepetitive transcripts. A generalized structure for Hn RNA locates mRNA (3,000 to 4,000 from 5,000 to 50,000 nucleotides) adjacent to the poly(A) stretch at the 3′-OH end. At or near the junction between the mRNA and the turnover region of the Hn RNA (lost during processing) there are sequences that derive from DNA sites repeated many times. They are distributed more or less uniformly over the 5,000 to 25,000 bases of the turnover region and may exist as intramolecular loops (hairpins). These and oligonucleotides rich in uridylate may serve as signals for posttranscriptional modification of the Hn RNA molecules.

heterogeneric tRNA (*White* et al. 1973) — → transfer RNA.

heterogenesis — = → alternation of generations.

heterogenetic (*Waddington* 1939) — ref. to meiotic → chromosome pairing in hybrids (alloploids) when the pairing partners are derived from different original ancestors, as distinct from "homogenetic" pairing where the association is between chromosomes derived from one of the original ancestors. The frequency of h. pairing as compared with homogenetic, depends on the relative degree of affinity between the completely homologous, as compared to the partly homologous ("homoeologous"), chromosomes, i.e. on their → differential affinity.

If the parental genomes of a segmental allopolyploid are symbolized as $A_1A_1A_2A_2$, and the individual chromosomes as $1A_11A_11A_21A_2$, $2A_12A_12A_22A_2 \cdots nA_1nA_1nA_2nA_2$ (n being the basic haploid number), then pairing in meiosis between $1A_1$ and $1A_2$ chromosomes is h., and pairing between two $1A_1$ is homogenetic. The parallel (but not synonymous) terms → allo- and autosyndesis refer to the pairing between chromosomes derived from different parental gametes (allosyndesis) or from the same gamete (autosyndesis) (*Stebbins* 1950).

heterogenic (*Lewis* 1947) — 1. of a population, or

2. a gamete containing more than one → allele of a particular gene or genes, as distinct from a gamete which is "homogenic" in this respect.

heterogenotic (*Morse, Lederberg & Lederberg* 1956) — of bacterial "syngenotes", i.e., partially diploid bacteria **("heterogenotes")**, arising by → tranduction, → F-duction, or → conjugation which contain a → genome fragment (called an **exogenote**) from a donor cell in addition to their own (recipient) genome (called **endogenote**) and which are → heterozygous for pairs of marker genes localized in the segments concerned, as distinct from "homogenotic" cells ("homogenotes") which are homozygous in this respect. The heterozygous condition set up by the introduction of an exogenote into the recipient may be referred to as **cis-** ($++/m^1m^2$) or **trans-heterogenotic** ($+m^2/m^1+$) depending on whether two different mutational sites (either belonging to different → cistrons or to the same cistron) were introduced together (exogenote m^1m^2, endogenote $++$) or not (exogenote either $+m^2$ or m^1+, endogenote m^1+ and $+m^2$, respectively) by the genome fragment (exogenote) from the donor cell.

heterogony — 1. → parhenogenesis,

2. differential growth rates for two organs or parts of an individual, the rates showing a constant relationship to one another (*Huxley* 1932).

heterograft — a tissue graft from a donor of one species to a host of a different species (= xerograft). As a rule, heterografts are rejected faster than are → allografts.

heteroimmune — of phages that have a → repressor which recognizes its own operator but not the operator of another phage, and an operator sensitive to its own repressor but not the repressor of another phage. Phages that share the same repressor-operator specificity are homoimmune or coimmune.

heterokaryon — a bi- or multinucleate cell, or a tissue composed of such cells, whose nuclei are of more than one genetic type, as distinct from a homokaryon in which the nuclei are genetically identical. The property of being hetero- or homokaryotic is referred to as heterokaryosis and homokaryosis, respectively. In certain fungi (Asco- or Basidiomycetes), heterokaryosis constitutes the principal vegetative phase of the life cycle.

In contrast to → heterozygous diploids, the different alleles of a particular genetic locus in binucleate heterokaryons are separated by → nuclear envelopes.

heterokaryosis (*Hansen & Smith* 1932) — the association of genetically different nuclei in a common cytoplasm. H. is widespread in fungi and may initiate → plasmogamy leading to sexual as well as to → parasexual recombination.

In fungi, h. represents an alternative to diploidy. Two types of heterokaryons may be distinguished and may occur in the same species (*Raper & San Antonio* 1954):

1. **Indefinite heterokaryons** in which the ratio of component nuclear types is indefinite and subject to change. Heterokaryons of this type are basically vegetative.

2. **Dikaryons** represent genetically balanced heterokaryons, each containing two haploid genomes in a ratio of 1:1. Heterokaryons of this type constitute an integral phase of the sexual cycle of all higher fungi, and in the Basidiomycetes are capable of indefinite vegetative propagation.

heterokaryotype — a → karyotype which is heterozygous for a → chromosome mutation (symbolized HTK), as opposed to a structurally homozygous "homokaryotype" (symbolized by HOK).

heterolabeling — → sister chromatid exchange.

heteromeric — of genes which in combination control the manifestation of a particular character, each gene having a definite, but individually different share in the production of that character, in contrast to homomeric genes whose individual share is quantitatively similar (→ gene interaction).

heteromixis (*Burnett* 1956) — mating systems in (heteromictic) fungi; sexual reproduction results from the fusion of genetically different nuclei derived from different thalli, in contrast to → "homomixis". H. includes:

1. **Dimixis:** two types of complementary nuclei control mating. The nuclear types are determined by two alleles at a single locus (= heterothallic; haplodioecious; two allele morphological and physiological → incompatibility; bithallic).

2. **Diaphoromixis:** several types of complementary nuclei control mating. The nuclear types are determined by multiple alleles at one or two loci, the bipolar and tetrapolar conditions, respectively (= multipolar sexuality; bipolar sexuality; multiple allele physiological heterothallism; bithallic; incompatibility factors at one or two loci).

3. **Homo-heteromixis:** sexual reproduction results only from the fusion

of genetically different nuclei normally derived from the same thallus; reproductive mode derived from 1. or 2. and hence "homo-dimictic" or "homo-diaphoromictic" (= secondarily homothallic; amphithallic; pseudomonothallic).

heteromorphic (*Carothers* 1917) — of "homologous" chromosomes differing in size or form (→ bivalent).

heteromorphosis — = → homoeosis.

heteronuclear (*Krooth* 1965) — of → cell cultures showing a marked variation (numerical and structural) from cell to cell in their chromosome complement. If h. cultures are cloned, each colony will usually regenerate the same spectrum of chromosome complements as was present in the parental mass culture. Homonuclear cultures show almost no variation from cell to cell in the chromosome complement, except when such cultures are developed from chromosomal mosaics.

heterophenogamy (*Strandskov* 1953) — → isophenogamy.

heteroplasmic — = → heteroplasmonic.

heteroplasmonic — of cells containing two or more types of cytoplasmic hereditary determinants (= heteroplasmic), in contrast to homoplasmonic. The term h. is the equivalent to → heterozygous, → heterogenotic, and → heterokaryotic in the case of extrachromosomal hereditary determinants. A "heteroplasmon" (→ plasmon) may arise by mutation of some but not all the homologues of an extrachromosomal determinant which are present in a cell or by fusion of cells containing alternative forms of such a determinant. A heteroplasmon may give rise to → somatic segregation.

heteroplastidic (*Michaelis* 1957) — of cells containing → plastids of different shape, in contrast to homoplastidic ones.

heteroploid (*Winkler* 1916) — of all chromosome numbers deviating from the normal chromosome number of the diplophase (in organisms with a predominating diplophase), or of the haplophase (in organisms with a predominating haplophase) of typical representatives of a given species (*Levan* & *Müntzing* 1963). H. chromosome numbers may be either → euploid or → aneuploid.

heteropycnotic (*Gutherz* 1907) — of chromosomes or chromosome regions which, in comparison with "isopycnotic" ones (*Östergren* 1950), are out of phase (→ allocycly) in respect to their coiling cycle (→ chromosome coiling) and staining properties. (→ heterochromatic).

1. **Positively h.:** chromosomes or chromosome segments are tightly coiled by prophase and remain so as others (isopycnotic) are undergoing decondensation at anaphase. During interphase they show up as the → chromocenters. Positive heteropycnosis is characteristic of heterochromatin.

2. **Negatively h.:** chromosomes or chromosome segments condense more slowly or to a lesser extent during prophase than do the positively h. and undergo decondensation more rapidly at anaphase.

Chromosomes and chromosome regions which are positively h. at one stage of the cell cycle, or in one kind of cell or tissue, or in one of the sexes, may be negatively h. at another stage, in another cell, or in the other sex.

hetero-R state (*Hashimoto* & *Hirota* 1966) — → resistance transfer factor.

heterosequential (*Carson* et al. 1967) — → band.

heterosis (*Shull* 1911) — the superiority of heterozygous genotypes with respect to one or more characters in comparison with the corresponding homozygotes (= hybrid vigor). H. is the phenotypic result of → gene interaction in heterozygotes and is thus confined (at least in its maximal amount) to that state. It can be disrupted by → inbreeding and restored by interbreeding of the inbred lines.

Dobzhansky (1952) proposed to distinguish between **euheterosis** (true heterosis) and **luxuriance** according to whether the heterozygote is adaptively superior to the homozygotes or not.

1. **Mutational euheterosis:** H. resulting from the sheltering of → deleterious recessive mutants by their adaptively superior dominant alleles in populations of sexually reproducing and cross-fertilizing organisms.

2. **Balanced euheterosis:** Heterosis due to the occurrence of mutations and gene combinations which confer a higher adaptive value or a higher agricultural usefulness on heterozygotes than is to be found in the corresponding homozygotes (→ overdominance). While 1. is a protective device of a sexual species with a particular population structure against mutation pressure, balanced euheterosis permits the maintenance of a multiplicity of genotypes in a population and these may be adaptive in the different ecological niches occupied by the population.

3. **Luxuriance:** Both 1. and 2. are adaptive states in outbreeding sexual species; luxuriance, however, is the term proposed to designate those cases in which hybrids between species, varieties, or strains are larger, faster growing, or otherwise exceed the parental forms in some character(s) when this is evidently neither the result of the sheltering of deleterious genes nor of balanced gene combinations. Luxuriance is an evolutionary accidental condition produced by the complementary action of genes present in both parental forms and combined in the hybrid after crossing. Luxuriance is more frequent in domesticated than in wild species and is not connected with any adaptive superiority of the hybrids as compared with their parents.

heterosomal — of chromosomal structural changes (→ chromosome mutation) involving two or more nonhomologous chromosomes (→ allelosomal; homosomal).

heterosome — = → sex-chromosome, as opposed to → autosome.

heterospory (*Ernst-Schwarzenbach* 1939) — the production of micro- and megaspores (→ spore), as opposed to homospory.

heterostyly (*Darwin* 1877) — a combination of morphological, physiological, and genetic mechanisms which promote outbreeding in certain angiosperms. H. means the presence in a species of two (distyly) or three (tristyly) different types of individuals, distinguished by the relative positions of stigma and anthers in the flowers, as distinct from homostyly. In the case of distyly, one type of individual has short styles and long stamens the other long styles and short stamens. With tristyly, the alternatives are "short-styled", "medium-styled" and "long-styled" individuals. Usually, pollination and fertilization is between flowers differing in the lengths of their styles ("heteromorphic → incompatibility").

heterosynkaryon (*Harris* 1970) — → synkaryon.
heterothallic (*Blakeslee* 1904) — → heteromixis.
heterotropic (*Wilson* 1906) — of a → sex chromosome which does not find an exactly homologous partner in the heterogametic sex (as, e.g., in XX–XY- and XX–XO-systems of sex-determination).
heterotypic (*Flemming* 1887) — ref. to the first meiotic division (→ meiosis).
heterozygous (*Bateson & Saunders* 1902) — 1. **in higher organisms** (eukaryotes), of → diploid or → polyploid individuals ("heterozygotes") having different → alleles at one or more genetic loci (genes) in homologous chromosome segments, in contrast to homozygotes with identical alleles at these loci.

Heterozygosity provides a store of genetic reserves and potential plasticity and permits a large proportion of the individuals to exhibit combinations of phenotypic properties near the optimum. Underlying both is the greater → homeostasis of heterozygotes.

Structural heterozygotes (structural hybrids) are individuals heterozygous for → chromosome mutations, i.e. → "heterokaryotypes". Their parental gametes differed in the arrangements of genes in corresponding linkage structures, in contrast to "homokaryotypes".

Permanent (enforced) heterozygotes are individuals in which the heterozygous state of specific parts of the genome is fixed or made obligatory by factors contributing to the maintenance of heterozygosity. Fixation of heterozygosity may be a function of balanced or partially balanced → lethal factors, heterozygosity for → translocations (→ complex heterozygous) or → inversions, or → chiasma localization. → Parthenogenesis is an important ingredient in many cases of permanent heterozygosity.

2. **In bacteriophages and bacteria,** of partially diploid particles and cells, respectively, which carry two copies (containing different alleles of one or more genes) of at least some of their genetic information.

Heterozygous bacteriophages are called HETs. Single factor crosses (mixed infections) produce two distinct types of h. progeny. One of these probably corresponds to a "terminal redundancy" of the phage linkage structure (*Streisinger* et al. 1964) resulting from the incorporation of two double-stranded segments of a homologous sequence of genetic information into a single genome. The segments in question are localized at opposite ends of the linear genome ("overlap HETs"). When this redundant information stems from genetically different parents the result is a region of heterozygosity corresponding to the length the of redundancy:

DNA molecule 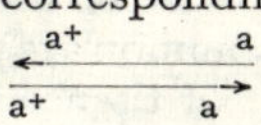

(duplicated information terminally represented by different alleles).

The other type is suggested to be a → heteroduplex structure (*Levinthal* 1959) formed by a breakage-reunion recombination (also called "internal heterozygote" or "recombinational heterozygote"). The information on each strand of the double-stranded DNA molecule is then genetically distinct within the h. region and should be lost by semiconservative replication:

DNA molecule 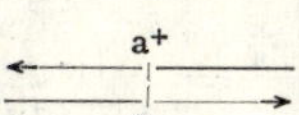

(different information in the two polynucleotide chains).

Heterozygosity in phages concerns only a very limited region of the whole genome.

The heteroduplex HETs cannot persist throughout a cycle of DNA replication, but the overlap HETs may persist for several replication cycles.

Heterozygous bacterial cells may arise by → transduction, → F-duction, and → conjugation. They generally contain an additional genome fragment (an → "exogenote") and are h. (→ "heterogenotic") for one or more marker genes where the donor and recipient cells carried different alleles. In contrast to the HETs they are quite comparable to the heterozygotes in higher organisms which arise by the union of gametes differing in genotype.

HETs (*Hershey & Chase* 1952) — a shorthand for partially → heterozygous phages.

hexaploid — of a → polyploid (allo- or autoploid) with six chromosome sets in the somatic cells.

hexasomic — of an otherwise diploid cell or individual having one chromosome represented six times instead of twice ($2n + 4$) (→ polysomic).

Hfr cell — a cell harboring the → F plasmid integrated into the bacterial chromosome which may or may not (in case of a defective F plasmid) confer on the cell the phenotype of **h**igh **f**requency of **r**ecombination. Each Hfr strain transfers the bacterial chromosome with a definite polarity to F^- recipient cells during bacterial → conjugation. Hfr strains originate by opening up the circular chromosome of the bacterial (F^+) cell at one of many possible sites and insertion of the F plasmid at this place.

high negative interference (*Chase & Doermann* 1948) — positive correlation of recombination events (→ genetic recombination) restricted to very small regions (adjacent intervals) of paired linkage structures (chromosomes), within which there is an excess of double → crossing-over as compared to expectation based on a random distribution of single crossover events (→ heteroduplex repair).

histocompatibility gene — any gene regulating the immunological properties of histocompatibility antigens. Such cellular antigens, which represent the transcription products of h. genes, differentiate individuals and cause the rejection of tissue transplants. Histocompatibility and histo-incompatibility (acceptance and rejection of transplants, respectively) depend upon the genetic relation of donor and host. Transplants exchanged between members of the same inbred strains (isografts) which have h. alleles in common survive permanently, while grafts exchanged between members of different strains (allografts) which have no h. alleles in common are promptly rejected.

histogenesis — the formation and development of tissues (= histogeny).

histone (*Kossel* 1884) — any of the basic proteins complexed with DNA (= nucleohistone) in the chromosome of eukaryotes (→ nonhistone chromosomal protein). The close correlation between h. synthesis and DNA replication is presumably related to the fact that within cells there exists a strict requirement for newly synthesized DNA to form DNA-h. complexes (→ chromatin; nucleosome).

The association by electrostatic bonds of h. with DNA in eukaryotic chromosomes has implicated histones in two possible functions:

1. a structural role in maintaining and controlling the conformation of the chromosome through the cell cycle (coiling and condensation of chromosomes);

2. involvement in mechanisms of → genetic regulation (coarse control resulting in tissue-specific gene repression patterns during development) by forming the supercoiled structure of DNA which in turn sterically inhibits transcription by → RNA polymerase.

There are five main types of h. in the → chromatin of eukaryotes. These are F 1 (= H 1) and F 2b (= H 2 B) (lysine-rich h.), F 3 (= H3) and F 2 a 1 (= H 4) (arginine-rich h.), and F2 a2 (= H 2 A) (an intermediate type). Usually, there are nearly equimolar amounts of all the histones except for F 1, of which there is about half as much. The five main types are all fully characterized by amino acid analyses and are not — in a given organism — tissue-specific. The sequence studies of histones have delineated well defined regions of the chain rich either in basic and helix-stabilizing residues or rich in apolar acidic and aromatic residues. The latter regions are the sites of secondary structure and h.-h. interaction, while the former are the sites of DNA interaction. The amino acid sequences of these proteins have, during the course of evolution, remained extraordinarily stable (F 1 shows greater variability than the other h. fractions).

Three types of *in vivo* chemical modifications (post-transcriptional) have been found to occur in histones, namely methylation, acetylation and phosphorylation. These modifications, especially the acetylation of amino groups, have been postulated to function in modulating the interactions of h. with DNA. Most of these modifications occur in cells actively synthesizing DNA and may be involved in: 1. the attachment of newly synthesized h. to DNA; 2. changes in chromosome coiling; 3. changes in chromosomal activity during the cell cycle or differentiation; or 4. maintenance of the correct conformation of h. to DNA.

Chromatin of fungi does not contain histones such as are characteristic of the chromosomes of other eukaryotic organisms. In very specialized cells which do not synthesize RNA or protein, other basic proteins are found. Some spermatozoa contain protamine instead of h. and the nucleated erythrocytes of birds contain a unique h. designated F 2C. Thus h. is thought to be responsible for the final repression of the DNA in the avian erythrocyte and may be an evolutionary alternative to the loss of the nucleus which occurs in the mammalian erythrocyte.

The genes coding for histones are reiterated, i.e., present in multiple copies per genome. Spacer DNA seems to be quite abundant. One DNA segment may contain coding sequences for all major classes of histones.

hitch-hiking effect (*Smith & Haigh* 1974) — the increase in frequency of genes closely linked to a gene which is selectively favored. This effect in a large population may be more important than random → genetic drift in determining the level of heterozygosity for selectively neutral alleles.

HNI — a shorthand for → high negative interference.

Hn RNA — abbreviation for → heterogeneous nuclear RNA.

holandric (*Enriques* 1922) — of a type of → inheritance controlled by genes completely linked to the → Y-chromosome. The genes in question are transferred exclusively from father to sons in case of XX(♀)—XY(♂)-systems of → sex-determination, and thus their manifestation is limited to the male sex (→ hologynic).

holocentric — of → chromosomes with diffuse kinetic activity due to the presence of nonlocalized (diffuse) instead of localized → centromeres (= holokinetic). A chromosome with a diffuse centromere has no localized organelle which elaborates chromosomal fibers, but the entire body of the chromosome forms a sheet of such fibers on the poleward surface and these converge to the pole. Sister chromatids are not associated at any point by centric connections and a complete spatial separation is possible which may make the chromatid, or in some species even the half-chromatid, an autonomous entity (→ reductional division; meiosis). Such chromosomes move broadside to the poles during anaphase and do not have the V- or J-shaped configuration of chromosomes with a localized centromere.

holoenzyme — an → enzyme composed of apo- and coenzyme.

hologamy — a mode of → reproduction characterized by the fusion (plasmogamy and karyogamy) of whole (unicellular) individuals (= gamontogamy, macrogamy).

hologynic (*Enriques* 1922) — of → sex-limited characters which are manifest only in the female sex (→ holandric).

holokinetic (*Bauer* 1952) — = → holocentric.

holozygote — → zygote.

homeorhesis (*Waddington* 1957) — an equilibrium property characterizing the regulatory pathway of development which is the result of the activity of many genes controlling different synthetic processes. These processes interact in a way which assures that their combined activity delimits a pathway exhibiting homeostatic properties and is referred to as h. or canalization (→ homeostasis).

homeostasis (*Cannon* 1929) — the tendency of a system to be specified by context (individual, population, process, etc.) to maintain a dynamic equilibrium and in case of disturbance to restore the equilibrium by its own regulatory mechanisms.

1. **Genetic or collective h.** (*Lerner* 1950; *Lewontin* 1956): The tendency of a population to equilibrate its → gene pool and to maintain a genetic composition leading to an optimum balance (= "genetic inertia" of *Darlington & Mather* 1949) in the environment in which the population lives. The concept comprises all mechanisms of autoregulation in populations, including processes of → coadaptation, and is intended to convey an analogy with the phenomenon of h. at the level of the individual organism. G. h. determines the extend to which a gene pool can respond to → selection. It is manifest by the tendency of phenotypes to return to the original condition when selection is relaxed after a population has been exposed to severe selection pressure for a particular phenotypic character. Since phenotypes are the product of genotypes, selection for a new phenotype is coupled with the trend to abandon the previously integrated genotype, thus leading to lowered fitness. Decrease of → selection pressure for the new phenotype then allows at least partial return

(by natural selection) to the original gene combination with maximal fitness accompanied by a (partial) restoration of the original phenotype.

2. **Epigenetic h.** (*Nanney* 1958): The persistence of differences at the cellular level in the absence of genetic or environmental "differentials", i.e. their relative stability to disturbances. The term may also cover what has been called **"developmental h."** (*Lerner* 1954), "canalization" or "homeorhesis" (*Waddington* 1957), namely the capacity of the developmental pathways to produce a normal phenotype in spite of developmental or environmental disturbances.

3. **Physiological h.** (*Cannon* 1932): A type of phenotypic reaction buffering the organism against fluctuating external conditions by internal self-regulatory mechanisms which enable the individual to adjust to variable environments not by conforming to changes in the external environment but by resisting those changes (either by varying phenotypically or by remaining constant in the face of environmental changes). Physiological h. is frequently associated with heterozygosity and then confers a higher average fitness in various environments to the carriers of such genotypes, as compared to their less homeostatic relatives (homozygotes) (*Grant* 1963).

homeostat (*Danielli* 1956) — any of the self-reproducing cytoplasmatic organelles (e.g., mitochondria, centrosomes, plastids, kinetosomes, etc.) or processes determining or controlling one or more characters of a cell lineage.

homeotypic (*Flemming* 1887) — ref. to the second meiotic division (→ meiosis).

homoallelic (*Roman* 1956) — → heteroallelic.

homodynamic (*Waddington* 1953) — of genes simultaneously influencing the same developmental process, in contrast to heterodynamic genes.

homoeoallelic (*Washington* 1971) — in → polyploids, of duplicated, triplicated, etc. genes similar to each other in function but located on → homoeologous chromosomes with different evolutionary origin.

homoeologous (*Huskins* 1932) — of partially homologous chromosomes. "Homoeology" designates the residual homology of originally completely → homologous chromosomes.

homoeosis (*Bateson* 1894) — the alteration of one organ of a segmental or homologous series from its own characteristic form to that of some other member of the series (*Darlington & Mather* 1949). Mutant genes giving rise to such alterations, by interference with primary processes in embryonic development, are called "homoeotic" mutants (*Goldschmidt* 1945) (homoeosis = heteromorphosis).

homoeotic — of mutants that probably represent lesions in regulatory genes and cause one body part to develop into another.

homogametic (*Wilson* 1911) — of that sex which produces only male or female determining gametes in sex digamic organisms, in contrast to the heterogametic sex.

homogamous (*Sprengel* 1793) — of hermaphroditic flowers and animals whose male and female sex organs function at the same time, in contrast to → dichogamous ones.

homogamy — the preference of a mating individual for another with similar pheno- or genotype, as opposed to heterogamy.
homogeneric tRNA (*White* et al. 1973) — → transfer RNA.
homogenetic (*Waddington* 1939) — → heterogenetic.
homogenic (*Fisher* 1928) — → heterogenic.
homogenotic (*Morse, Lederberg & Lederberg* 1956) — → heterogenotic.
homograft — a tissue graft between two members of the same species. Normally homografts are rejected ("homograft reaction"), but accepted if the donor and host are isogenic or nearly so (identical twins, highly inbred animal strains).
homo-heteromixis (*Burnett* 1956) — → heteromixis.
homoimmune — → heteroimmune.
homokaryon — → heterokaryon.
homokaryotype — → heterokaryotype.
homologous — 1. of chromosomes or chromosome segments ("homologues") which are identical with respect to their constituent → genetic loci (the same loci in the same sequence) and their visible structure, as distinct from nonhomologous and → homoeologous chromosomes or parts of chromosomes;

2. of structures similar in the same or different organisms due to similar hereditary and developmental origin (*Owen* 1840);

3. of similar variations ("law of homologous genes" (*Vavilov* 1922)).

homomeric — → heteromeric.
homomixis (*Burnett* 1956) — sexual reproduction in homomictic fungi resulting from the fusion of genetically similar nuclei normally derived from the same thallus (= homothallism, monothallism). (→ heteromixis; amixis).
homonuclear (*Krooth* 1965) — → heteronuclear.
homoplasmic — = homoplasmonic (→ heteroplasmonic).
homoplasmonic — → heteroplasmonic.
homoplastidic (*Michaelis* 1957) — → heteroplastidic.
homopolymer — a long chain macromolecule (a polymer) composed of linked identical monomeric subunits as distinguished from a "copolymer" made up of more than one type of subunit.
homosequential (*Carson* et al. 1967) — → band.
homosteric (*McElroy, Deluca & Travis* 1967) — of enzymes which combine at the catalytic site with their normal substrate or a structurally similar compound, resulting in a modified reaction of the bound intermediate. Homosteric enzymes are to be distinguished from → allosteric ones where the modifier molecule combines at a site other than the catalytic site.
homothallic (*Blakeslee* 1904) — → heteromixis.
homozygote disadvantage load (*Dobzhansky* 1965) — = segregational load (→ genetic load).
homozygous (*Bateson & Saunders* 1902) — → heterozygous.
horizontal evolution (*Brown* et al. 1972) — → ribosomal DNA.
hormone — any type of chemical messenger responsible for the timing and regulation of growth and development and the homeostatic maintenance

of the fully developed organism. Many hormones achieve their effects by (indirectly) altering gene activity.

horotelic (*Simpson* 1944) — ref. to standard rates of → evolution (→ bradytelic, tachytelic).

host cell — a cell whose metabolism is used for the growth and reproduction of a virus.

host-cell reactivation (*Garen & Zinder* 1955) — enzymatic → dark repair (Symbol HCR) of lethal and nonlethal lesions induced by UV and some chemical agents (damage by ionizing radiation is only slightly repaired) in the DNA of bacteriophages after infection of host cells (→ reactivation). The same processes also repair lesions in bacterial DNA. Bacteria performing h.-c. r. are designated HCR(+), mutants which do not do so (they are highly sensitive to UV) HCR(−). The repair process is ineffective in the case of 5-bromouracil substituted DNA, and is inhibited by caffeine and acriflavine. It overlaps with photoenzymatic repair (→ photoreactivation). HCR is, in all probability, the result of excision of damaged portions from DNA and their replacement by material newly synthesized in conformation with the information contained in the undamaged complementary DNA strand. With single-stranded DNA and RNA, HCR fails to occur (*Rupert & Harm* 1966).

host-controlled DNA modification — → DNA modification-restriction system.

host-controlled DNA restriction — → DNA modification-restriction system.

host-controlled variation (*Luria & Human* 1952, *Bertani & Weigle* 1953) — a general phenomenon in which DNA may be modified when it is synthesized in one cytoplasm and then undergoes → restriction upon entering another cytoplasm (= host-induced modification, host-controlled modification). H.-c. v. has been observed in viruses (extension or restriction of the host-range), in restriction of both → F plasmids and col-I in bacterial strains carrying a certain → prophage, in decreased → zygotic induction and in the formation of recombinants when crosses between specific strains of bacteria were performed. H.-c. v. is not inherited.

host-killing efficiency (*Luria & Delbrück* 1942) — the capability of "inactivated" bacteriophages to kill their host cells.

host-mediated assay (*Gabridge & Legator* 1969) — an indirect method for evaluating potential mutagenic agents (→ mutagenicity testing) introduced to bridge the gap between *in vitro* microbial studies and tests in mammals. The mammal (usually the mouse) is injected intraperitoneally with a microorganism (bacterium, yeast) as an indicator for induced → gene mutations and then treated with a potential mutagen. The comparison between the action of the compound on the microorganism directly and in the h.-m. a. indicates whether the host can detoxify the compound or mutagenic products can be formed as a result of host metabolism. To get an even more precise evaluation of persistance, organ distribution, retention and metabolism of a potential mutagen some modifications of the initial procedure are being used, e.g., incubation of microorganisms in extracts of organs and diffusion bags, in urine and in blood and liver.

host range mutant (*Luria* 1945) — any mutant of bacteriophages able to overcome the resistance of phage-resistant mutant bacteria. Host range mutants differ from wild-type phages in the structure of their adsorption organs. Such a structural change permits the stereo-specific fit with the phage receptors in the envelope of the phage-resistent bacteria in spite of the modification of the cell surface which prevents attachment of the wild-type phage. Most of these mutations extend the host range beyond that of wild-type (*Stent* 1963).

hot spot (*Benzer* 1955) — any region in a gene that mutates at very much higher frequencies than neighboring regions of the same gene. Possibly h. spots are DNA regions containing a sequence of identical bases.

HT-transducer (*Schmieger* 1972) — phage mutants with increased ability to transduce bacterial markers (→ transduction).

human genetics — the study of the genetically controlled similarities and differences (physical and mental, normal and abnormal) in human beings, in respect to their causes and modes of transmission (inheritance) from generation to generation (→ medical genetics, pharmacogenetics).

hyaloplasm (*Pfeffer* 1877) — the ground substance (= cell sap; cytoplasmic matrix; ground plasm; soluble phase of the cell) of the cell which lies between the canals of the → endoplasmic reticulum and contains the cell organelles (mitochondria, plastids in plants, Golgi apparatus, etc.). The h. is structureless under the electron microscope except for the presence of scattered → ribosomes and very fine → microfilaments and → microtubules. It contains some soluble RNA and the enzymes concerned with anaerobic metabolism.

hybrid — any offspring of a → hybridization or cross between two genetically unlike individuals.

1. **Genetic h.** (*Darlington* 1937): The result of the fusion of two gametes differing in their genetic constitution (in the → alleles they carry) or produced by mutation in a homozygote.

2. **Structural h.** (*Darlington* 1929): The result of fusion of two gametes which differ in the arrangement of their genes owing to chromosomal structural changes (heterozygosity for → translocations, inversions, deletions, etc.). **Cryptic structural hybrids** (*Stebbins* 1945) are those whose structural differences are so small that they do not generally lead to disturbances in meiotic → chromosome pairing.

3. **Numerical h.** (*Darlington* 1931): The result of fusion of two gametes which differ in respect to the number of their chromosomes.

4. **Permanent h.:** A stable hybrid maintained as a heterozygote of balanced lethals by the loss of homozygous lethal genotypes (→ complex heterozygous).

hybrid breakdown — → isolation.

hybrid complex — any group of plant species in which the morphological differences of the basic diploids have been masked by hybridization. They may be classified as follows (*Grant* 1953):

1. **homogamic c.:** Usually, the hybrids are diploid and exhibit normal meiosis. Structural heterozygosity (if present in the F_1) is eliminated by natural selection;

2. **clone complex:** The hybrids reproduce mainly or exclusively by asexual means;

3. **heterogamic c.:** The hybrids are permanent structural heterozygotes, as, e.g., in *Oenothera* (→ complex heterozygous);

4. **polyploid c.:** The hybrids are polyploids and undergo sexual reproduction;

5. **agamic c.:** The hybrids or their derivatives reproduce mainly or exclusively via unfertilized seeds or bulbils.

hybrid DNA — a double-stranded DNA molecule with → heteroduplex regions.

hybrid incapacitation — a collective term comprising → hybrid sterility and → hybrid inviability.

hybrid inviability — the lowering of somatic vigor due to aberrant developmental processes in hybrids (→ isolation).

hybridity — the state of being → heterozygous or hybrid. **"H. equilibrium"** (*Darlington & Mather* 1949) is the term proposed to designate the equilibrium state of hybridity in a stable breeding population which is dependent on:

1. the rate of recovery of homozygosity (itself depending on the type and amount of → inbreeding in the population),

2. the frequency of lapses from inbreeding, and

3. the amount of heterozygosity produced by lapses (this in turn depends on the amount of genetic variation in the breeding population). The **"h. optimum"** (*Darlington & Mather* 1949), i.e., the optimal amount of h. in a particular population, is thus normally related to the mode of breeding. It is low in inbreeders, and high in outbreeders.

hybridization — 1. in the broadest sense, any cross-mating (crossing) of two genetically different individuals which leads to hybrid progeny (→ cell hybridization).

2. The crossing of individuals belonging to two unlike natural populations that have come into contact secondarily. Five kinds of h. may be distinguished when h. is considered as a population phenomenon (*Mayr* 1963):

a) occasional h. of sympatric species, usually resulting in hybrid individuals that are either inviable or sterile (no back-crossing possible with the parental species);

b) occasional or recurrent h. resulting in more or less fertile hybrids between sympatric species (some hybrids may backcross to one or both of the parental species);

c) partial interbreeding between two formerly isolated populations in a secondary zone of contact (→ intergradation) if the populations failed to acquire complete reproductive isolation during the period of geographic isolation;

d) complete local breakdown of reproductive isolation between two sympatric species, resulting in the production of → hybrid swarms that may include the whole range of variability of the parental species;

e) production of a new specific entity in plants due to h. and subsequent doubling of chromosomes, giving rise to allopolyploidy (→ alloploid).

hybrid sterility — the impairment (partial sterility) or suppression (complete sterility) of the reproductive capacity of F_1 or of later generation hybrids between genetically different populations (usually belonging to different species). H. s. is an isolating mechanism (→ isolation) which may partially or completely prevent the → gene flow between populations. The basis of h. s. may be genic, chromosomal, or cytoplasmic (*Dobzhansky* 1937; *Ehrman* 1962):

1. **Genic h. s.:** The result of abnormal developmental processes caused by specific gene complexes in the hybrid. A special type of this category is h. s. due to intersexuality where the interference occurs with those developmental processes leading to the normal differentiation of male and female sexes.

2. **Chromosomal h. s.:** The result of structural differences between parental chromosomes which interfere with their pairing (→ chromosome pairing) and disjunction at meiosis.

3. **Cytoplasmic h. s.:** The result of discordance between the chromosomal genes and the cytoplasm. The cytoplasmic component of the system may be due either to autonomous hereditary determinants in the cytoplasm or to formative influences exerted by chromosomal genes which were present in the egg cell before meiosis (→ predetermination).

Another classification distinguishes between h. s. due to disturbances of the haploid or the diploid phase of the life cycles, i.e. between **haplontic** and **diplontic sterility.**

hybrid swarm — a collection of hybrids produced by complete, or local breakdown of isolating barriers between two → sympatric species (= hybrid flock). The h. s. replaces the parental species and serves as a continuous bridge between the two parental extremes (*Mayr* 1963).

hybrid vigor — → heterosis.

hybrid zone — a geographic zone in which two populations, formerly separated by geographic → isolation, hybridize after the breakdown of the geographic barrier provided there is no reproductive isolation between them. A hybrid population or zone of secondary → intergradation is produced by hybridization. The hybrid populations are highly variable; some individuals resemble one or the other subspecies but most are hybrids more or less intermediate between the two original populations.

hyparchic — of genes in → mosaics, prevented from phenotypic manifestation by genes from genetically different neighboring tissues, as opposed to "autarchic" genes.

hyperchimera — a plant → chimera with both sectorial and periclinal distribution of genetically different tissues.

hypermorph (*Muller* 1932) — → allele.

hyperplastoid (*Martin & Sprague* 1973) — of cell lines from mammalian tissues which have finite replicative life-spans and (in the majority of cases) display the → karyotype of the donor (→ neoplastoid). Cessation of replication of h. cell lines and clones may be due to the accumulation of errors in macromolecule synthesis or due to differentiation to death (terminal differentiation).

hyperploid (*Bělăr* 1928) — of → cells or individuals with one or more added chromosomes or chromosome segments in their complements, as opposed to **hypoploid** cells and individuals which are deficient for one or more chromosomes or chromosome segments. Depending on the degree of ploidy in the original cell or individual, hyper- and hypohaploids, hyper- and hypodiploids as well as hyper- and hypopolyploids may be distinguished (*Winkler* 1916).

hypomorph (*Muller* 1932) — → allele.

hypoploid (*Bělăr* 1928) — → hyperploid.

hypostasis (*Bateson* 1907) — → epistasis.

hyteresis (*Darlington* 1935) — a lag in the movement (action) at one level in response to stress at another level, e.g., in the adjustment of the external form of a chromosome to its internal stress during the cycle of → chromosome coiling (*Darlington* 1937).

I

identical by descent (*Malecot* 1948) — of two genes if they are both derived from the same gene in some ancestor.

idiochromosome (*Henking* 1891) — = → sex chromosome.

idiogamy — = → self-fertilization.

idiogram (*Navashin* 1922) — a diagrammatic representation of chromosome morphology used diagnostically for comparison of the → karyotypes of different species and varieties. The construction of idiograms is based on measurements (in several or many cells) of total chromosome length and arm length ratio as well as on → centromere position, → nucleolus organizer position, position of → heteropycnotic segments, etc. for any one chromosome during metaphase of → mitosis. Centromere position is usually calculated from the total length, c, and the lengths of the long and short arms, l and s, respectively, of the chromosome in question and expressed either as a difference $d = l - s$, as a ratio $r = l/s$ or as a centromere index $i = 100s/c$.

Chromosomes in an i. may be classified according to their centromere position as follows (*Levan, Fredga & Sandberg* 1964):

Centromere position	Arm ratio	Chromosome designation		
Median *sensu strictu*	1.0	M	atelocentric	(metacentric)
Median region	1.7	m		
Submedian	3.0	sm		(submetacentric)
Subterminal	7.0	st		(subtelocentric)
Terminal region		t		(acrocentric)
Terminal *sensu strictu*	∞	T		telocentric

idioplasm (*Naegeli* 1884) — the sum total of → hereditary determinants (→ idiotype).

idiotype (*Siemens* 1921) — the sum total of → hereditary determinants of an organism, consisting of its → genotype (all determinants localized in the chromosomes; chromosomal genes) and → plasmotype (all determinants localized extrachromosomally).

I-line — = → inbred line.

illegitimate crossing-over — → crossing-over which is unequal in the sense that the reciprocal products of → genetic recombination are not strictly homologous with respect to each other or their parents (recombinational error). I. c.-o. generates → duplications and → deletions, respectively, in the crossover products.

immigration — → gene flow from one population into another ("genetic input") as mediated by immigrating individuals capable of interbreeding with members of that population. The percentage of individuals in a population who become members by immigration is referred to as the **"immigration coefficient". "Immigration pressure"** is the rate of change of → gene frequency in a population subject to i. It depends on the i. rate and on the difference in gene frequency between immigrants and natives. **"Immigration load"** is that part of the → genetic load which is due to the incorporation of those alien genes into the → gene pool that reduce fitness in their new environment, irrespective of whether these genes had been advantageous or → deleterious in their former environment. The load of inferior alleles in a gene pool caused by → gene mutation and i. is referred to as its "input load" (*Mayr* 1963).

immune response — a complex series of events (initiated by the interaction of immuno-competent cells and → antigen. The i. r. in multicellular organisms provides the capacity to distinguish and eliminate or suppress both foreign (extrinsic) material and, in addition, intrinsic alien patterns arising by somatic mutation or an equivalent process. The immune system therefore plays a key role in resistance to pathogenic microorganisms, "surveillance" against malignant cells, and control of autoaggressive processes. The cells taking part in i. responses are 1. lymphocytes and 2. various accessory cells (e.g. macrophages, cells of the reticuloendothelial system, various types of blood leucocytes).

immunity substance (*Jacob & Wollman* 1956) — in → lysogenic bacteria, a temperature-sensitive substance (repressor) which prevents the transcription of → prophage genes whose transcripts are needed for vegetative replication of the → prophage. The same substance is responsible for the immunity of lysogenic cells to infection by phage of the same type as their prophage.

immunogenetics — a branch of genetics encompassing all the areas of overlap between immunology and genetics. I. includes the application of methods of immunology to the recognition of individual differences for genetic studies (e.g., blood groups, and protein allotypes), and the application of genetic methods to the study of immunological phenomena (e.g., the mechanisms determining antibody induction, structure and specificity).

inactivating DNA alteration (*Freese* et al. 1969) — DNA lesions which block its replication. If not eliminated by mechanisms of → DNA repair, such lesions are mostly lethal. Occasionally they give rise to → gene mutations and frequently (misrepair) to large DNA and chromosome alterations (→ chromosome mutation).

inborn error of metabolism (*Garrod* 1902) — a genetically determined biochemical disorder in which a specific enzyme defect produces a metabolic block. The block may have pathological consequences (= molecular disease). Nearly all the i. e. o. m. studied in human genetics are inherited as autosomal recessives, a few as sex-linked traits.

inbred — the progeny of matings between relatives. An inbred individual is connected through both the female and male parent to the same recent ancestor and can thus receive two copies of a gene that was carried by this ancestor (these copies are called "identical by descent").

inbred line — any more or less → homozygous line derived from an outbreeding population by repeated → inbreeding (self-fertilization, sib-mating, etc.) of an individual and its progeny. A population in which inbreeding is inevitable, because the population size (→ breeding size) is limited thereby forcing relatives to mate to each other, is equivalent to an i. l. The rate of loss of → genetic variability is then a function of the size of the population.

inbreeding — a → mating system (= endogamy) which, in contrast to → out- or crossbreeding (= exogamy), involves the breeding together of individuals more closely related than mates chosen at random from a population. The degree of i. in any particular population or mating is a function of the relationship between the mating partners (→ inbreeding coefficient). The closest form of i. is self-fertilization; other forms are sib matings and half-sib matings.

All forms of i. result in an increase of homozygosity, i.e. lead to genetic fixation. When i. is imposed on a normally outbreeding population, the population is broken up into smaller groups accompanied by a decrease in total heterozygosity as fixation occurs in the subgroups. If no selection for particular homozygotes occurs, → genetic variance as a whole is increased in the population (transformation of potential into free → genetic variation). I. in normally outbreeding populations generally leads to a loss of fitness referred to as → inbreeding depression.

In plants, morphological and physiological mechanism favoring i. are common, e.g., failure of flowers to open (cleistogamy) or opening of flowers only after the pollen has been shed. The relative restriction of outbreeding mediated by these and other mechanisms ranges from facultative self-pollination to obligate self-fertilization.

inbreeding coefficient (*Wright* 1929) — 1. in diploid individuals, the probability that the two genes (alleles) at a locus are → identical by descent **(i. c. of an individual).**

2. In populations, the probability that the two genes of a random member of the population are identical by descent **(i. c. of a population)**.

The rate of reduction in heterozygosity per generation due to inbreeding is $^1/_2$ with self-fertilization, $^1/_4$ with sib matings, $^1/_8$ with half-sib matings and $^1/_{16}$ with cousin matings. If the frequencies of two alleles (A and a) in a population are pA and qa and the i. c. is F, then the chance that an individual will be genotypically AA is pF and that it will be genotypically aa is qF. The chance that the two alleles at that locus will not be identical by descent is 1 — F. Thus, for a population with an i. c. of F the following → genotype frequencies are expected:

Genotype	Frequency
AA	$p^2(1 - F) + pF$
Aa	$2pq(1 - F)$
aa	$q^2(1 - F) + qF$

inbreeding depression — a decrease in → fitness and vigor, resulting from → inbreeding imposed on individuals that are normally outbreeding (= inbreeding degeneration). The extent of i. d. is not the same in all lines produced by inbreeding. That point after several generations of inbreeding at which no further decrease of fitness and vigor occurs is referred to as the "inbreeding minimum". I. d. is the consequence of increased homozygosity for → deleterious recessive genes (→ genetic load) and the break up of balanced polygenic systems (*Mather* 1941). The crossing of inbred lines which have reached their inbreeding minimum frequently results in → heterosis.

inbreeding load (*Morton, Crow & Muller* 1956) — the extent to which an equilibrium population under random mating is impaired by inbreeding which ordinarily increases → genetic load.

incest taboo — in human societies, the restrictions against → consanguineous matings, i.e., the mating of certain categories of relatives (→ inbreeding).

inchondriosis (*La Bella & Krass* 1968) — ingestion by cells of particulate material with or without extracellular medium (= phagocytosis; endocytosis; heterophagy).

inchylocytosis (*La Bella & Krass* 1968) — ingestion by cells of fluid material by convolution of the cell membrane and vesiculation of engulfed bulk fluid.

inchylosis (*La Bella & Krass* 1968) — penetration of water and solutions across an apparently intact cell membrane (→ inchylocytosis).

incidence — in human genetics, the frequency of new occurrence of a genetically caused medical condition among individuals of a specific population, within a certain period of time (e.g., incidence of → genetic disease). From the i. (symbolized by x) of a character and its mode of inheritance, the frequency, q, of the "abnormal" allele may be calculated. If each of the affected individuals is → heterozygous, then the frequency of the allele in question is equal to one half the i. of the character: $q = {}^1/_2 x$. (→ prevalence).

incipient species — → superspecies.

incompatibility (*Stout* 1918) — the selective restriction of mating competence by extrasexual "incompatibility genes" which limit or prevent zygote formation (fertilization) by certain classes of gametes and thus in an otherwise freely interbreeding group limits either self-fertilization ("self-incompatibility") or cross-fertilization ("cross-incompatibility") between certain self-incompatible or self-compatible individuals if these carry genes for self-incompatibility. Members of the same i. class are typically strictly self-sterile and cross-fertile and the various i. systems which occur in the plant kingdom usually promote outbreeding. Mating between individuals of compatible classes is usually reciprocal.

In general, homo- and heterogenic i. systems may be distinguished (*Raper & Esser* 1964):

1. **Homogenic i.:** Zygote formation is prevented if the mating partners carry identical i. genes. It may be controlled by one or two gene loci. In the first case, there are at least two mating types (bipolar i.), in the second, at least four (tetrapolar i.). The number of mating types equals the number of alternative alleles in bipolar systems. Inbreeding is reduced to 50% and 25%, respectively, while outbreeding is practically unrestrict-

ed. Homogenic i. may be based either upon two alleles of one series, multiple alleles in one series, or multiple alleles in each of two series.

2. **Heterogenic i.:** This system consists of more than two gene loci. Mating partners homozygous at all loci interact freely, but those heterozygous for the loci in different combinations either interact unilaterally or do not interact at all. It is a unique system which, in contrast to homogenic i., limits the sexual compatibility of heterogenic partners and thus promotes inbreeding, while homogenic i. restricts the mating between genetically similar organisms and thus favors outbreeding. Heterogenic i. is known only in fungi.

A more detailed classification of i. systems in angiosperms is as follows (*Crowe* 1964):

1. **Heteromorphic i.** characterized as follows:

a) each mating type is morphologically distinct and recognizable without breeding tests (e.g. heterostyly);

b) the mating type of the pollen is determined sporophytically, i.e. the mating behavior of the pollen is controlled by the sporophyte of the pollen producer irrespective of the pollen's incompatibility genes;

c) dominance between the alleles of the incompatibility gene(s) is expressed both in the pollen and the style;

d) the style inhibits the growth of incompatible pollen tubes. This type of system may be controlled by one gene with two alleles (dimorphy) or by two genes each with two alleles.

2. **Homomorphic i.** characterized as follows:

a) there are no morphological differences between mating types, thus they can be identified only by breeding tests;

b) there are numerous mating types;

c) there are numerous alleles of the incompatibility gene(s).

The control of the mating type of the pollen in homomorphic systems of i. may be either gametophytic or sporophytic. In the case of gametophytic i., alleles of the i. gene(s) act individually in the style and incompatible pollen tubes are inhibited there. The control may be by alleles at one or two loci. In the case of **sporophytic i.,** alleles of the i. gene can exhibit either dominance or individual action in both male and female organs. Inhibition of incompatible pollen tubes may be on the style or i. may be expressed between gametes after fertilization (post-fertilization i.).

A special system of i. is **cytoplasmic i.** (*Jones* 1951) or **protoplasmic i.** (*Garnjobst & Wilson* 1956) which consists in deleterious interactions of the genome or cytoplasm of one species, strain, hybrid, or sex with the cytoplasm of another.

Unilateral i. is the phenomenon whereby the pollen of a population is inhibited from functioning on the plants of another population while in the reciprocal cross no inhibition occurs. It occurs between populations of self-incompatible plants as well as populations of self-compatible plants, both in the cross of self-incompatible x self-compatible and self-compatible x self-incompatible. Unilateral i. is based on an incompatibility gene system. (→ F incompatibility; plasmid incompatibility).

incorporation error — → gene mutation.

incubation medium recovery — → liquid-holding recovery.

independent assortment — → assortment; genetic segregation.

index case — = → proband.

individual — from a genetic point of view, a unit of life comprising all the cellular derivatives of one fertilization up to the succeeding meiosis, or all the derivatives of one of the four spores produced by meiosis up to the succeeding fertilization (*Darlington & Mather* 1949).

individuation (*Waddington & Schmidt* 1933) — in embryonic → development, the process whereby a spatially-organized tissue, organ, or whole embryo comes into being as a result of → induction.

individuation field (*Waddington & Schmidt* 1933) — → embryonic field.

inducer (*Jacob & Monod* 1961) — any positive → effector that binds to the → repressor and causes a conformational change which relaxes repressor-operator binding, i.e., induces the enzymes of an → operon. Anti-inducers stabilize the repressor-operator complex.

inducible — of enzymes (proteins) not synthesized in the absence of their substrates. The substrates act as → inducers whose presence in the cell increases the rate of production of i. enzymes ("enzyme → induction"). Such enzymes fluctuate widely in amount as a function of changes in the cellular environment. Most i. enzymes are catabolic, i.e., involve either the degradation of some substance derived from the environment and utilized as a source of energy, or the production of molecular fragments for synthetic processes (→ repressible, operon).

induction — 1. **Enzyme induction** (*Monod, Cohen-Bazire & Cohn* 1951): The induced synthesis of so-called → inducible enzymes (proteins) following the addition of a metabolite referred to as → inducer. In contrast to→ constitutive enzymes, inducible as well as → repressible enzymes fluctuate widely in their amount as a function of changes in the cellular environment. Enzyme i. is probably a common phenomenon found in all types of organisms.

Enzyme induction is said to be "sequential i." if cells that can respond with induced synthesis of a particular enzyme, in the presence of its specific substrate, produce intracellular intermediates that in turn serve as substrates for the i. of certain other inducible enzymes. In this case, the availability of only one substrate may initiate a sort of chain reaction resulting in the successive induction of a series of specific enzymes each of which reacts with only one of a number of related substrates.

2. **Prophage induction** (*Lwoff* 1953): The breakdown of the functional integration of the → prophage into the genetic system of a → lysogenic bacterium followed by the vegetative replication and maturation of the phage progeny and the ultimate lysis of the bacterial cell. Prophage i. occurs spontaneously with a low probability, $10^{-3}-10^{-6}$, per cell division, and can be experimentally increased in practically all cells of a lysogenic population by a variety of physical and chemical agents, or by transfer of prophage DNA to a nonlysogenic cell ("zygotic" or "transfer i."). Prophage i. requires two events: 1. release of repression (accomplished by the inducing effect) and 2. phage-directed excision of the prophage DNA from the bacterial chromosome.

In more general terms, prophage i. means the initiation of any event which depends on the presence of the prophage in a lysogenic cell (*Lieb* 1966).

Cross-induction (*Borek* & *Ryan* 1958) is the i. of phage development in lysogenic bacteria by the transfer of radiation sensitive metabolic products from UV irradiated to nonirradiated cells during → conjugation. This transfer is unidirectional from F^+ to F^- cells.

3. **Zygotic induction** (*Jacob* & *Wollmann* 1956): The conversion of a prophage to the vegetative state and the eventual lysis of the "zygote" after transfer of a prophage to a nonlysogenic recipient bacterium. Zygotic i. takes place after → conjugation between a lysogenic (ly^+) → Hfr cell and a nonlysogenic (ly^-) F^- cell (→ F plasmid). Owing to the absence in the F^- cell of the → immunity substance which permits the maturation of lytic phage following the transfer of the chromosome-associated prophage into the F^- cell, this recipient cell is lysed.

Z. i. is recognizable by the appearance of → plaques or lytic centers on plates containing the "zygote" together with a multitude of nonlysogenic cells. Z. i. is limited to crosses of the type Hfr (ly^+) $\times F^-$ (ly^-) and does not occur in Hfr (Ly^+) $\times F^-$(Ly^+) or Hfr (ly^-) $\times F^-$(ly^+) crosses (*Braun* 1965).

4. **Embryonic induction:** The determination of the developmental fate of one cell or tissue by another, the latter designated as the "inductor". Embryonic i. is the result of an interaction, in which one group of cells (inductor or evocator) evokes and controls the differentiation of other groups of cells. The kind and extent of embryonic i. depend on the embryonic region from which the inductor stems as well as on the region to which it is transferred in the case of transplantation. The inductive stimulus (probably a chemical one) exerted by the inductor seems to result in the activation ("derepression") of the → genetic material of the reacting system and may either impinge directly on the → structural genes ("genotropic action") or on the → regulator gene products ("plasmotropic action").

The propensity of a group of cells or of a tissue to form a particular organ in reaction to embryonic i. is referred to as **"competence"**. During → embryonic development there is a sequential appearance and loss of competence, i.e. a time-regulated changing pattern of competence occurs. The appearance of competence may be understood as a priming of some genes of the reacting system. Embryonic i. may proceed over a long time span with cumulative effects in the competent tissue, or it may start from a high level of → "predetermination", i.e. very early determinative events. In the latter, the time course is shorter than in the former.

inert — of chromosome segments (usually → heterochromatic) which are supposed to be genetically inactive as judged by the absence of → gene mutations or effects on genetic balance.

infertility — → fertility.

informatin (*Krichevskaya* & *Georgiev* 1969) — a protein which is assumed to associate with nascent nuclear RNA and to be involved in its processing (→ heterogeneous nuclear RNA; informofere).

information — → genetic information.

informational macromolecule (*Vogel* et al. 1963) — any macromolecule (DNA, RNA, protein) with a unique role in determining the properties of living matter (= semantide) by affecting either short-term biochemical reactions, medium-term ontogenetic events, or long-term evolutionary events.

informational suppressor — a → suppressor that shows allele and not gene specificity.

informaton (*Samarina* et al. 1968) — → informofer.

informofere (*Samarina* et al. 1967) — in eukaryotes, the association product (30 S) of → pre-messenger RNA and specific globular proteins (informatin). Each i. (mol. weight ~800,000) is composed of up to 20 subunits (informatons with a mol. weight of ~40,000). Pre-mRNA is localized on the surface of informofers and remains bound during processing, i.e., attack by specific endo- and exonucleases. The function of i. proteins is probably related to the liberation of nascent pre-mRNA from → chromatin, in stabilization and eventually processing of pre-mRNA (→ informosome).

informosome (*Spirin* et al. 1964) — any of the cytoplasmic ribonucleoprotein (RNP) particles in eukaryotes which contain pre-messenger RNA or messenger RNA. Possibly, mRNA on the outer membrane of the → nuclear envelope interacts with a protein (differing from informatin in → informoferes), to form informosomes, which possibly are involved in → translational control, in the transport of mRNA from the nucleus into the cytoplasm, or in the stabilization of the labile mRNA molecule.

Informosomes exhibit a great heterogeneity in size (from 20 S to 110 S), evidently due to the heterogeneity in the sizes of the mRNA which they contain.

inheritance — the transmission of → genetic information from parents and ancestors to offspring.

1. **Chromosomal or Mendelian inheritance (heredity):** I. as controlled by the replication, mutation, recombination, transmission, and read-out of the genetic information stored in form of a → genetic code in particulate, chromosomal → genes whose sum total makes up the genotype (*Johannsen* 1909). Chromosomal i. is based on principles first discovered experimentally by *Gregor Mendel* (1865) and rediscovered by *Correns, Tschermak* and *de Vries* (1900). **Mendel's first law:** When purebred (→ homozygous) strains differing in a particular → character and the corresponding → allele pair (AA and aa, respectively) are crossed, the F_1 individuals (Aa) are uniform (independent of the direction of the cross: either AA♀ × aa♂ or aa♀ × AA♂) and usually express one of the two characters (that controlled by the → dominant allele A) to the exclusion of the other (the → recessive one) (**"principle of uniformity in F_1"**).

Mendel's second law: Recessive characters which are masked in the → heterozygous F_1 arising from a cross between two purebred strains, reappear in a specific proportion of the F_2, that is, the members of an allele pair (Aa) separate from each other when an individual forms → germ cells (**"principle of → segregation"**).

Mendel's third law: Members of different allele pairs (e.g. Aa and Bb) assort independently of each other when germ cells are randomly formed provided the genes in question are unlinked (→ linkage) **("principle of independent assortment").**

The basic deductions of "Mendelism" which have been correlated with the behavior of → chromosomes ("chromosome theory of i.") may be summarized as follows:

a) Hereditarily determined characters are controlled by hereditary determinants, or genes, which occupy definite → genetic loci on the chromosomes and are transmitted from parent to progeny (→ genetic information).

b) In → diploids, these genes exist as pairs of alleles in somatic cells. Each germ cell produced by → meiosis receives only one of the two alleles.

c) Fusion of the germ cells at → fertilization is usually random, i.e., germ cells carrying a particular allele do not preferentially unite with others containing particular alleles.

d) Allelic gene pairs controlling different characters are transmitted independently. During germ cell formation, particular genes do not remain preferentially associated, provided the genes concerned are not linked, i.e., are not localized in the same → linkage group or chromosome (→ genetic recombination).

e) When an organism receives during fertilization different alleles for a particular character, only one of these is usually (→ codominant) expressed (the → "dominant") and the other (the "recessive"), though maintained and later transmitted, is not manifest phenotypically.

"Monogenic" or "oligogenic" i. is manifest by characters which are mainly controlled by the alleles at one genetic locus (gene). → Variation is discontinuous.

"Multigenic", "polygenic", or **"quantitative" i.** is manifest by characters or components of the → phenotype which are controlled and affected by several interacting genes (→ gene interaction). Variation is continuous.

Unilateral i. (*Winge* 1927) is associated with → linkage in the Y-chromosome of those genes in question and implies the resemblance of offspring and parent of the same sex (→ holandric). **Criss-cross inheritance** (*Bridges* 1913) is associated with linkage in the X-chromosome (sex-linkage) of the genes concerned and implies the resemblance in particular characters of sons to their maternal and of daughters to their paternal parent.

Delayed i. (*Boycott & Diver* 1923) is associated with → predetermination, i.e., the determination of particular characters prior to fertilization by the genotype of the female parent which are manifested in the following generation. The phenotype of a hybrid is thus "matroclinous" in respect to these characters and corresponds to the genotype of the premeiotic oögonium (= "maternal effect").

2. **Extrachromosomal, non-Mendelian, or cytoplasmic inheritance (heredity):** I. is controlled by extrachromosomal, cytoplasmic hereditary determinants whose sum total makes up the "plasmotype" (*Imai* 1936, *Jollos* 1939), in contrast to the genotype or all hereditary determinants contained in the chromosomes.

Extrachromosomal i. may be effected by means of nonparticulate parts of the cytoplasm, or by means of inclusion particles (mitochondria, plastids in plants, plasmids). Cytoplasmic hereditary determinants in their totality represent the plasmon or cytoplasmon (those contained in the plastids represent the plastom).

The following constitute the main criteria of extrachromosomal i.:

a) Nonsegregation or non-Mendelian segregation of character differences;

b) indifference to nuclear substitution, i.e., constancy of the characters in a series of backcrosses to the paternal strain;

c) maternal inheritance of the characters;

d) differences in the results of reciprocal crosses in the case of oögamy;

e) mapping of the hereditary determinants not normally possible (→ genetic map);

f) infection-like transmission of the hereditary determinants;

g) exclusion of spurious extrachromosomal i. by, e.g., → dauermodification or → predetermination.

All systems of extrachromosomal i. detected thus far interact with chromosomal genes or their products. In many cases, such systems are concerned with the expression and integration of → gene action or the determination of → gene activation. They are then referred to as "epigenetic" (*Nanney* 1958).

inhomologous — = → nonhomologous.

initial spindle — → spindle.

initiation complex (*Nomura & Cowry* 1967) — an obligatory intermediate of protein synthesis (→ genetic translation) in pro- and eukaryotes which consists of → messenger RNA programmed with → initiator codons at the 5'-OH end of the smaller ribosomal subunit (→ ribosome), → initiation factors and → initiator tRNA.

1. **In prokaryotes,** mitochondria, and chloroplasts the initiation of protein synthesis requires three initiation factors and the formation of a 30 S ribosomal subunit — mRNA — fMet—tRNA complex that is converted to a 70 S i. c. by the addition of a 50 S ribosomal subunit. Subsequent to this the i. c. is competent to bind the → aminoacyl tRNA specified by the second codon.

2. **In eukaryotes,** the major components of the i. c. include the smaller ribosomal subunit (40 S), mRNA with an appropriate initiator codon, initiator tRNA (Met-tRNA), and initiation factors. The main differences from prokaryotic i. complexes are the lack in eukaryotes of formylation of initiator tRNA and a slight alteration in the specificity of the initiation factors in the 80 S ribosomal system.

initiation factor (*Stanley* et al. 1966) — any of a class of protein factors which are requisite for the formation of the → initiation complex and the initiation of mRNA translation (→ elongation factor; termination factor; interference factor). In the presence of these factors (which belong to the translation factors), ribosomes bind to specific ribosome binding sites on mRNA which correspond to the initiation sites for polypeptide synthesis. In their absence, the mRNA cannot be translated (→ genetic translation).

1. **Prokaryotic i. factors:** At least three i. factors (IF-1, IF-2, and IF-3) are required for the initiation of polypeptide synthesis. They participate both in the formation of the mRNA-ribosome complex and in the binding of N-formylmethionyl-tRNA to the ribosome. All three have to be present to form the complete → initiation complex.

IF-1 (mol. weight 9,400) promotes catalytic use of IF-2; it binds to the 30 S subunits of ribosomes only in the presence of IF-2, GTP, fMet-tRNA, and mRNA. Like IF-2 it is released from the initiation complex following union with a 50 S ribosomal subunit. Unlike IF-2, release of IF-1 does not require GTP hydrolysis. IF-1 consists of fewer than 90 amino acids and at least must have binding sites for guanine nucleotides and for IF-2, the 30 S subunit, or both.

IF-2 directs binding of initiator fMet-tRNA and has been isolated in two forms (mol. weights 80,000 and 100,000, respectively). The complex containing the 30 S ribosomal subunit, IF-3, and mRNA is joined by IF-2 and GTP. fMet-tRNA is bound to the 30 S initiation complex by IF-2. The IF-2s have three, and perhaps four, specialized sites: one for guanine nucleotides, one for fMet-tRNA, one for the 30 S subunit, and perhaps one for IF-1.

IF-3 binds mRNA to the 30 S ribosomal subunit and consists of two forms (IF-3α with a mol weight of 21,500 and IF-3β with a mol. weight of 23,500). Both have → dissociation factor activity and prevent the sequestering association of a 30 S subunit with a 50 S subunit, while promoting the proper binding of mRNA to the 30 S subunit. IR-3 binds to a segment of the 16 S → ribosomal RNA in the 30 S subunit and shows the unique property of releasing aminoacyl-tRNA (with the exception of the initiator fMet-tRNA) from their complexes with the 30 S subunit. The IF-3s have at least two binding sites: one for the 30 S subunit and one for regions, perhaps sequences (initiation signals) of mRNA. Different species of *E. coli* IF-3 factors can discriminate between different mRNAs and therefore regulate which of these are translated (→ interference factor).

2. **Eukaryotic i. factors:** Their function is similar to the prokaryotic i. factors. At least five protein i. factors are involved in mammalian peptide chain initiation. Initiation requires the selection of the initiator Met-tRNA$_f$ by IF-E$_2$ for the formation of a [Met-tRNA$_f$ · IF-E$_2$ · GTP] complex. This complex is bound to the 40 S ribosomal subunit (independent of mRNA) by IF-E$_3$. Subsequently, mRNA binding and 60 S subunit joining is promoted by the cooperative action of IF-E$_4$, ATP, IF-E$_1$ and IF-E$_5$. Binding of Met-tRNA$_f$ by natural mRNA requires IF-E$_4$ and ATP. IF-E$_5$ promotes the joining of the 40 S complex with the 60 S subunit, IF-E$_1$ inhibits the [80 S · Met · tRNA$_f$] complex formation in the absence of mRNA binding.

initiation signal (*Steitz* 1969) — → initiation factor.

initiator (*Jacob* et al. 1964) — 1. according to the → replicon model of bacterial DNA replication, a gene product that interacts specifically with a particular site of the replicon, the replicator. The result of this interaction is the initiation at the replicator site of a new round of DNA replication.

2. = a → promoter, i.e., the starting point for the transcription and/or translation process (*Margolin* & *Bauerle* 1966).

initiator codon — any → codon (AUG, GUG, and GUA) in → messenger RNA that directs initiation of polypeptide synthesis (→ genetic translation; initiation complex) by stimulating the binding of → initiator tRNA and by phasing the translation of messenger RNA. The isolation and sequence analysis of ribosomal binding sites of natural mRNA molecules demonstrated the exclusive use of AUG as the i. c. in all currently known initiation sequences. In *in vitro* systems, all three codons may initiate translation when their position is at or near the 5′ terminus of the message. When the codons are in internal positions of the mRNA and preceded by an i. c., they code (→ genetic code) for methionine (AUG) and valine (GUG and GUA). (→ translational reinitiation).

initiator RNA (*Reichard* et al. 1974) — short RNA sequences (iRNA) which initiate the discontinuous synthesis via → Okazaki pieces of → deoxy ribonucleic acid. iRNA is covalently linked to the 5′ ends of DNA progeny strands.

initiator tRNA — any → transfer RNA species which ensures the initiation of → genetic translation. In **bacteria, viruses, mitochondria,** and **chloroplasts,** N-formylmethionyl-tRNA ($tRNA_f^{Met}$) is the initiator tRNA. Methionine is formylated at the NH_2 position after its attachment to the tRNA molecule by an enzyme called transformylase. $tRNA_f^{Met}$ is distinguished from other tRNAs by 1. formylation of methionine attached to the RNA; 2. ability to initiate protein synthesis *in vitro*; 3. recognition by two codons (AUG and GUG) which differ in the first letter of the codon (→ wobble in the first base); 4. inability of this tRNA to insert methionine into internal positions of a polypeptide (this is due to a protein factor T which forms a ternary complex with GTP and all other aminoacyl-tRNAs).

In eukaryotes, a methionine-specific tRNA serves the purpose of fMet-tRNA in prokaryotes. Although *in vivo* it is not formylated, this eukaryote tRNA species can be formylated by *E. coli* transformylase (hence $tRNA_f^{Met}$ also here). The other Met-tRNA present in eukaryotes donates its amino acid into internal positions of proteins and is designated $tRNA_m^{Met}$. Probably, the absence or replacement of the sequence (G—T—ψ—C—GCA) in eukaryotic cytoplasmic i. tRNA enables this tRNA species to initiate protein synthesis without the prerequisite of formylation.

input load — → genetic load, immigration.

insertion — → translocation.

insertion model — a model explaining → genetic correction by integration of exogeneous homologous or heterologous DNA into the DNA of the recipient cell (→ exosome).

insertion mutation (*Shapiro* 1969; *Michaelis* et al. 1969) — in *E. coli* a class of extreme → polar mutations which are the consequence of the insertion of pieces of foreign DNA (→ insertosome) into a structural gene. A ribosome travelling along a mRNA molecule transcribed from such an altered gene will almost certainly encounter a → nonsense codon soon after passing the site of insertion. If the inserted fragment is sufficiently large and contains no initiation signals in the proper reading fame, a strong polar effect

(→ genetic polarity) will result (polar insertion). I. mutations will not be suppressible by extra- or intragenic suppressors (→ motator phage, transposon).

insertional duplication — → translocation.

insertosome (*Malamy* et al. 1972) — a sequence of DNA nucleotide pairs inserted into a bacterial or viral operon. The insertion may lead to complete inactivation not only of the gene into which the DNA is inserted but in most cases also of genes within an operon distal to the site of insertion (→ polar insertion; insertion mutation).

instability (*Mather* 1953) — a type of variation which appears to be random in direction. It is not genetic in origin and has no identifiable environmental cause. I. is sometimes explained as being due to developmental errors ("developmental noise") arising from sudden changes during development, unconnected with particular environmental influences. "Stability" refers to the opposite condition, i.e. non-occurrence of variations of the type described above, or more generally, to any condition characterized by the absence of → plasticity (*Bradshaw* 1965).

integrase (*Zissler* 1967) — a site-specific recombinase system responsible for the integration of a → prophage into a bacterial chromosome. I. is a phage gene product and mediates reciprocal recombination at a site on the phage genome (→ terminase). Integrase-less phage mutants are defective in integration, i.e., in both attachment of the phage DNA to the bacterial chromosome (insertion) and its detachment therefrom (excision). Such mutants can establish abortive → lysogeny.

integration system — in the case of the lysogenic response upon infection of bacteria with → temperate phages, an → integrase mediated mechanism for → prophage insertion at a specific site on the host DNA. The i. s. includes a region required for recognition of the proper prophage site in the DNA and one or more cistrons controlling proteins which catalyze the integration process (→ terminase system).

intensifier — → modifier gene.

interaction deviation — → variation.

interaction of genes — → gene interaction.

interarm fiber — any of the DNA-containing fibers as seen in electron micrographs of whole mount mitotic chromosomes which cross at the → centromere and at the ends of arms in chromosomes and in some cases all along the length of the sister chromatids.

interarm pairing — → chromosome pairing between the two identical, completely homologous arms of one → isochromosome (= internal pairing).

interband (*Painter* 1939) — → band.

interbreeding — of individuals capable of actual or potential gene exchange by hybridization. I. is the process that holds individuals together in → populations, and populations together in subspecies (races) and species.

intercalary — ref. to chromosome segments not terminally localized and distinguished from other segments terminal in position.

intercalation — → translocation.

intercellular invasion — the active migration of cells of one type into the interiors of tissues consisting of dissimilar cell types. A necessary condition for a cell type to be invasive to a given host tissue is that it lacks → contact paralysis of locomotion during collision with cells of that host tissue.

intercellular junction — any junctional structure involved in intercellular bonding (→ tight junctions, septate junctions, desmosomes and zonulae adherens); → gap junctions and → synaptic junctions are thought to be responsible for intercellular communication.

intercentric (*Sears & Camara* 1950) — ref. to the intercentromeric segment localized between the two → centromeres of a → dicentric chromosome or chromatid.

interchange (*Belling* 1925) — = reciprocal → translocation (→ intrachange).

interchange trisomic (*Belling* 1927) — → trisomic.

interchromomere — any of the regions (more or less uncoiled fibrils) connecting adjacent → chromomeres in the chromosomes. In → polytene chromosomes of Diptera, interchromomeres are called "interbands".

interchromosomal — ref. to effects, phenomena, or processes in which two or more chromosomes participate as, e.g., a chromosomal structural change in one chromosome influencing the gene action (→ position effect) or crossing-over frequency in another chromosome, as opposed to intrachromosomal.

intercistronic divide (*Rechler & Martin* 1969) — a base sequence (= intercistronic stretch, intercistronic space, or intercistronic region) between termination of one and initiation of translation of the next cistron in polycistronic mRNA.

intercistronic stretch (*Steitz* 1969) — = → intercistronic divide.

interdeme selection (*Wright* 1956) — → selection.

interference (*Muller* 1916) — 1. **Chromosome interference** (*Muller* 1916): the effect of one → crossing-over event in either decreasing **(positive chromosome i.)** or increasing **(negative chromosome i.)** the probability that a second crossover will take place in its vicinity. In both positive and negative chromosome i., the frequency of multiple (double) crossovers is not in accordance with a random distribution of subsequent crossovers. Chromosome i. is also called "chiasma i." (*Mather* 1931), "chiasma position i." (*Carter & Robertson* 1952), "crossover i.", "crossover position i.", or "position i." (*Whitehouse* 1965).

Strengths of chromosome i. are commonly measured as coefficients of → coincidence (*Muller* 1916), i.e. as the quotient of the number of observed double crossovers and the number expected from the Poisson distribution:

$$K = \frac{\text{actual frequencies of doubles}}{\text{expected frequencies of doubles}}$$

Coincidence varies inversely with interference, and coincidence values usually range from 0 (complete positive i.) to 1 (no. i. whatsoever). In those cases where the occurrence of one crossover increases the probability of a second nearby (localized negative i.), the number of double crossovers is in excess of the random expectation and the coincidence values are greater than 1. It is improbable that this kind of i. has anything to do with chiasma formation (as is assumed in positive i., hence "chiasma i."). One explanation for localized negative i. is the assumption that pairing of linkage structures, which actually gives rise to recombination (so-called "effective pairing"), is usually restricted to short regions which pair discontinuously and within which the probability of recombination is high. According to this hypothesis, recombination in two intervals of the linkage structures is greater than expected on the basis of random probabilities, provided these intervals are so close together that they frequently belong to the same effectively paired region.

2. **Chromatid interference** (*Mather* 1933): The nonrandom participation of the four chromatids of two homologous chromosomes in successive crossovers. This type of i. results in a deviation in the frequencies of so-called two-strand, three-strand, and four-strand double crossovers (→ crossing-over) from the 1:2:1 ratio expected if crossing-over is random with respect to the strands (chromatids) involved in two successive crossovers. Positive chromatid i. increases four-strand double crossovers, whereas negative chromatid i. increases two-strand doubles. Positive i. of this type has not yet been found; negative chromatid i. has been reported in some fungi. Chromosome i. influences the crossover frequency, chromatid i. the mode of crossing-over. Both types of i. are usually connected with changes in the recombination frequency as compared with random expectation.

3. **Interchromosomal interference:** The compensation of a decreased crossover frequency (positive chromosome i.) in particular paired chromosomes (which may result from, e.g., heterozygosity for chromosomal structral changes) by an increased crossover frequency (negative chromosome i.) in others.

interference distance (*Mather* 1936) — the average distance within which successive crossovers and chiasmata are formed after the first has been produced in the → differential distance from the centromere. The i. d. is genome-specific and controlled by chromosome → interference.

interference factor (*Groner* et al. 1972; *Lee-Huang* et al. 1972) — any of a class of protein factors (translational repressors) which change the cistron specificity of → initiation factor IF-3 during → genetic translation in prokaryotes. I. factors (mol. weight 30,000 to 100,000) affect the recognition of mRNA initiation sites by the ribosome and stimulate the translation of certain cistrons while inhibiting that of others (cistron-specificity of action).

interference range — the minimal genetic distance sufficiently large for two crossing-over events not to interfere with each other, i.e. where their coincidence value becomes 1. The i. r. seems to be different in different linkage groups and in different regions of the same linkage group.

interferon (*Isaacs & Lindemann* 1957) — a class of cell-coded proteins that is formed in response to infection with most DNA and RNA animal viruses. I. inhibits viral multiplication and acts by inducing the synthesis of a second cell protein which prevents the association of viral mRNA with ribosomes to form viral polysomes.

intergenic (*Muller* 1941) — of → mutations (chromosomal structural changes or → chromosome mutations) involving more than one gene, in contrast to intragenic mutations (→ gene mutation).

intergradation (*Mayr* 1942) — a decsriptive term for the coming into contact and hybridization of phenotypically different populations or groups of populations along a zone of primary or secondary contact, thus giving rise to character gradients (clines).

Primary intergradation: takes place while all the populations involved are in continuous contact. The populations are interconnected by gradients of phenotypic characters and → gene frequencies. The gradients developed gradually.

Secondary intergradation (→ introgression): the populations now connected by a character gradient were originally completely separated and came into contact a new. Since their reproductive isolation is incomplete, on becoming sympatric gene exchange occurs. In this case, the character gradient usually has a steep slope. Zones of secondary i. are hybrid belts or hybrid zones between populations which had become differentiated during a preceding period of isolation. They are usually characterized by a high degree of individual variation (*Mayr* 1963).

intergradation index (*Ginsburg* 1954) — a measure of the range of variability in all populations of a species with respect to a particular character. The total variability of the character under observation is referred to as the "divergency index". The indices may be calculated mathematically and are used in species diagnosis.

interkinesis (*Gregoire* 1905) — a short "resting stage" which may occur between the first and second meiotic division (→ meiosis). In contrast to the mitotic interphase, there is no chromosome reduplication during i. Each chromosome consists of two chromatids, which do not lie parallel but diverge due to a lapse of attraction. The regions where the sister-chromatids are in association are localized to both sides of the → centromere.

interlocking (*Mather* 1932) — 1. **True i. of meiotic chromosomes:** The intertwisting of different (nonhomologous) chromosomes which ensues during meiotic pairing (→ chromosome pairing) if a foreign chromosome passes through a loop between associations of two homologous chromosomes (Fig. 53). True i. takes place during zygotene-pachytene of the first meiotic division and chiasma formation is a prerequisite for its maintenance up to metaphase I.

According to *Darlington* (1937), simple i. in a species with → chiasma terminalization may be classified at metaphase into three types according to whether the chiasmata formed by the interlocked chromosomes are moving away from the point of i. in both bivalents, in one, or neither.

a) **Proximal i.:** The chiasmata move away from the point of i. in both pairs of chromosomes, since the i. is within the centromere loop.

b) **Distal i.:** The chiasmata on either side of the i. move in the same direction, i.e., towards one end, in both of the interlocked pairs, owing to the i. being distal to the most proximal chiasma.

c) **Proximal-distal i.:** The interlocking is proximal with respect to one pair of chromosomes and distal with respect to the other.

These different types of i. (in practice types b) and c) are not usually distinguishable and both may be described as distal) may occur between pairs of chromosomes (bivalents) and between separate chromosomes of multiple rings, in cases of heterozygosity for reciprocal → translocations (see Fig. 53) or in polyploids.

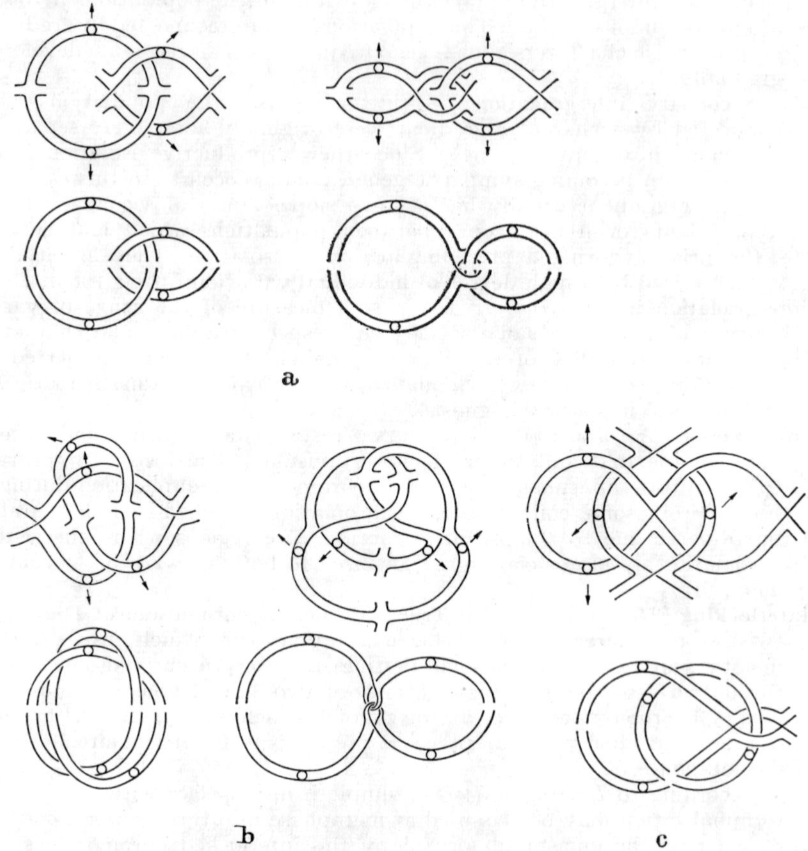

Fig. 53. Diagrammatic representation of different types of interlocking chromosomes as observed during diplotene (upper row of a, b and c) and metaphase I of meiosis (lower row of a, b and c), a) proximal (left) and proximal-distal (right) interlocking within a ring-of-four pairing configuration; c) double interlocking of two bivalents (modified after *Darlington* 1965).

Another type of i. is known as "double" i., i.e., two successive loops of a bivalent have chromosomes interlocked with them. Under these circumstances each loop separates the original parental chromosomes (Fig. 53).

2. **Spurious i. of meiotic chromosomes:** I. characterized by a whole bivalent passing through a loop between the pairing partners of another bivalent.

3. **I. in dicentric chromosomes:** I. during mitosis in → dicentric chromosomes with long intercentromeric segments resulting in single or double anaphase bridges (→ chromosome bridge). This type of i. is due to single ("criss-cross separation") or multiple torsion ("interlocked separation") in the intercentromeric region.

intermediary — of characters controlled by a → heterozygous pair of alleles (aa') which are intermediate in phenotype in comparison with the phenotypes controlled by the corresponding homozygous pairs of alleles (aa and a'a'). Intermediacy is a consequence of incomplete dominance (→ codominant). The mode of → inheritance in such cases is referred to as "intermediary inheritance" or "intermediary heredity" as distinguished from → "dominant" and → "recessive inheritance".

intermitotic — ref. to the interphase of the mitotic → cell cycle.

internal (*Darlington* 1937) — 1. ref. to an intrachromosomal structural change (e.g. → inversion) in contrast to → external structural changes, in which two or more nonhomologous chromosomes are involved;

2. of meiotic pairing (→ chromosome pairing) between the two homologous arms of an → isochromosome;

3. of a coil (→ chromosome coiling) within the single chromatid between mitotic pro- and anaphase, or at meiosis jointly for two sister chromatids, as distinguished from the "external coil" (including relic and relational coils) which is recognizable when the internal coiling is hidden (*Darlington* 1935).

internal balance — → coadaptation.

internal coiling (*Darlington* 1935) — → chromosome coiling.

interphase (*Lundegardh* 1912) — that part of the → **cell cycle** (= "resting stage") during which metabolism and synthesis occur without visible evidence of division. The → nucleus of a cell not visibly engaged in division (→ mitosis, meiosis) is referred to as an **"interphase nucleus"**, "resting nucleus", or "metabolic nucleus".

interphase nucleus — → interphase and → nucleus.

interreduplication (*Hsu & Moorhead* 1956) — polyploidization during interphase (= endoreduplication) as distinguished from reduplication during prophase (proreduplication by → endomitosis), during metaphase (metareduplication via → C-mitosis or C-meiosis), during anaphase (anareduplication by formation of a → restitution nucleus) and during telophase (teloreduplication by fusion of two nuclei after karyokinesis).

intersex (*Goldschmidt* 1915) — any individual of a unisexual (dioecious) species whose reproductive organs and/or secondary sex characters are partly of one and partly of the other sex without showing genetically different parts (→ gynandromorph), i.e., sex-intergrades who are neither clearly male nor clearly female but represent a mixture of male, female and intermediate characters. Intersexes are generally infertile or at most produce mature gametes of one sex only. Hence it is always possible to distinguish them from → hermaphroditic individuals which produce mature gametes of both sexes. Intersexes may start their development in conformity with the genetically determined sex and then switch over to the other sex. The degree of intersexuality is determined by the timing of the switch-over in differentiation.

Intersexes may arise (in mammals) by:

1. aberrations of genetic and chromosomal origin;
2. aberrations of gonadogenesis;
3. reversal of the sex of the accessory genital organs;
4. gene mutations.

intersterility — cross-sterility confined to specific genetically determined groups of individuals ("intersterility groups"). Under these circumstances, a group of individuals (A) may be incapable of mating with another group (B), but may successfully mate with group C which in turn may be unable to mate with group D. A and B on the one hand, and C and D on the other are then said to represent intersterility groups.

interstitial distance (*Mather* 1936) — the distance between successive chiasmata in those cases where more than one is present in a bivalent (→ differential distance).

interstitial segment — in the case of heterozygosity for a reciprocal → translocation between two nonhomologous chromosomes, any of the four chromosome regions between the → centromeres and the loci of exchange within the characteristic cross-configuration, consisting of four partially homologous chromosomes and produced by strictly homologous meiotic pairing (→ chromosome pairing), as distinguished from the so-called "pairing segments", i.e. the chromosome arms not involved in the translocation, together with the regions distal to the points of exchange in those arms which are involved (→ differential segments).

interzonal connection (*Schrader* 1932) — any connection between the chromatids during their separation in mitotic anaphase (= connecting fiber or interzonal fiber).

intrabreeding — mating only between individuals of the same population (= endogamy).

intrachange — any → exchange of segments within one chromosome (either within the same or between chromosome arms) resulting in chromosomal structural changes (→ chromosome mutation), as distinguished from the interchange of segments between two (or more) chromosomes. The units of structural change in case of i. and interchange may be both chromatids of the chromosome, only one chromatid, or a fibrillar subunit (usually a half-chromatid) of the chromatid.

intrachromosomal — → interchromosomal.

intracodon recombination — → genetic (intragenic) recombination between adjacent nucleotides within a → coding unit (a nucleotide triplet).

intragenic (*Muller* 1941) — → intergenic.

intrahaploid pairing (*Darlington* 1937) — meiotic → chromosome pairing between chromosomes of the basic (haploid) chromosome set. I. p. may be homologous pairing between duplicated chromosome segments or (less often) nonhomologous torsion pairing of genetically and structurally nonhomologous chromosomes or chromosome segments. I. p. of the homologous type may give rise to chromosomal structural changes by → crossing-over between duplications.

intranuclear — → extranuclear.

introgression (*Anderson & Hubricht* 1938) — the incorporation of genes of one species into the → gene pool of another species by hybridization ("introgressive hybridization") and backcrossing. According to *Anderson* (1953), sympatric i. (in which old and new genotypes produced by i. coexist in the same populations) and allopatric i. (in which new genotypes produced by i. exist separately from the old genotypes) may be distinguished.

inversduplication — → duplication.

inversion (*Sturtevant* 1926) — a chromosomal structural change of the "intrachange" type (→ chromosome mutation), characterized by the reversal of a chromosome (or chromatid) segment, and the gene sequence contained therein, relative to the standard arrangement of the linkage group in question. There is no experimental evidence for the existence of terminal inversions. According to the "breakage-reunion-model", intercalary inversions are the result of two breaks (referred to as inversion points) in one chromosome, 180° rotation of the segment localized between them and reunion at the sites of breakage.

The following i.-types may be distinguished, according to the number of inverted segments within a chromosome and the location of the i. points with respect to each other.

1. **Single inversions:** only one segment of the chromosome is inverted.

a) **paracentric inversions:** both i. points are located in the same chromosome arm (Fig. 54b) (Synonyms: acentric, dyscentric, parakinetic, asymmetrical inversions).

b) **pericentric inversions:** the two i. points are located in different chromosome arms and the inverted segment includes the centromere (Fig. 54a, c) (Synonyms: transcentric, eucentric, transkinetic, symmetrical inversions).

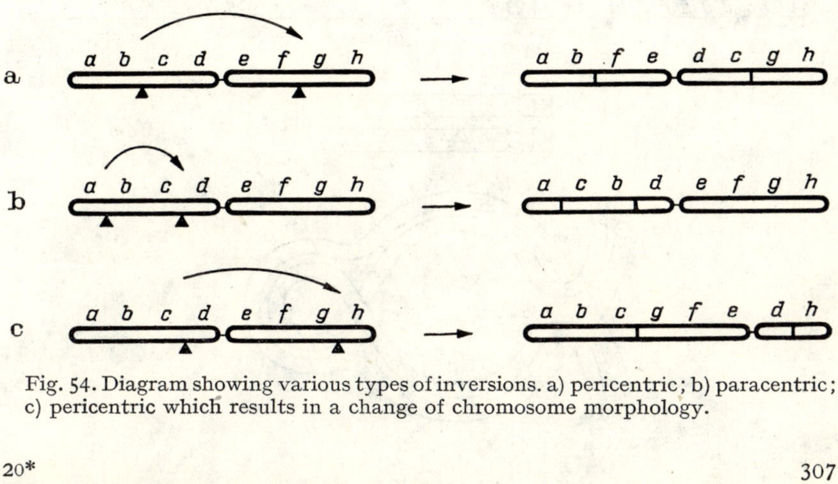

Fig. 54. Diagram showing various types of inversions. a) pericentric; b) paracentric; c) pericentric which results in a change of chromosome morphology.

2. **Complex inversions:** several segments of the chromosome are inverted.

a) **independent inversions:** the inverted segments are separated from one another by an uninverted segment (a b c d e f g h → a/cb/de/gf/h) (Fig. 55a).

b) **direct tandem inversions:** two inverted segments are mutually adjacent (a b c d e f g h → a/cb/ed/fgh).

c) **reversed tandem inversions:** two inverted segments are adjacent but mutually interchanged (a b c d e f g h → a/ed/cb/fgh).

d) **included inversions:** a segment within an inverted segment is inverted again (Fig. 55b) (a b c d e f g h → a/gfedcb/h → a/gf/de/cb/h).

e) **overlapping inversions:** part of an i. is inverted a second time to-

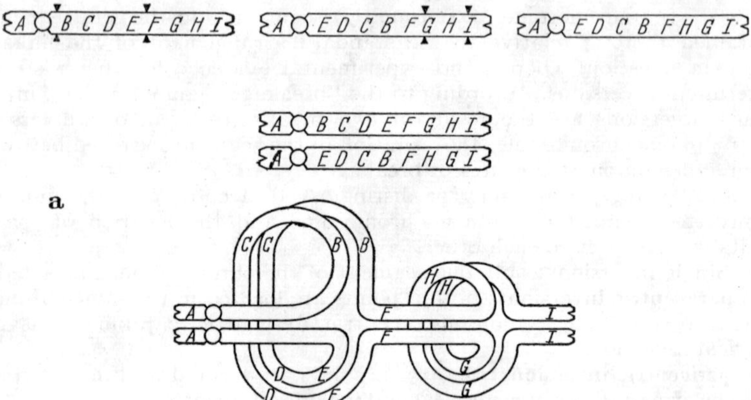

a

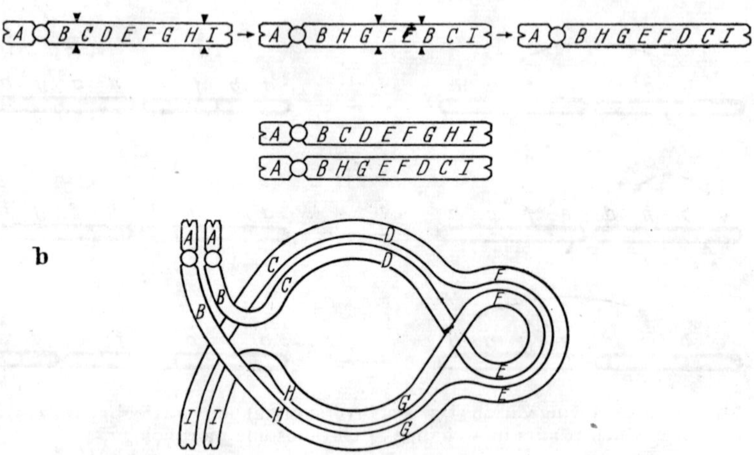

b

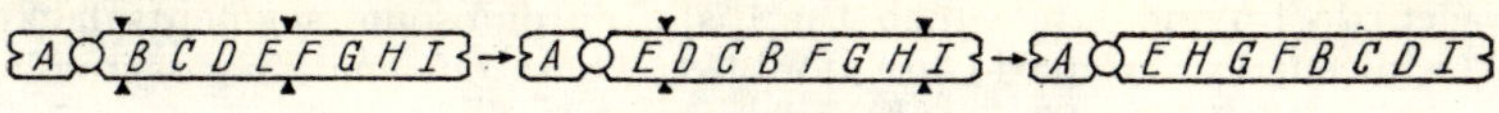

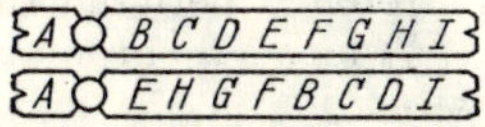

c

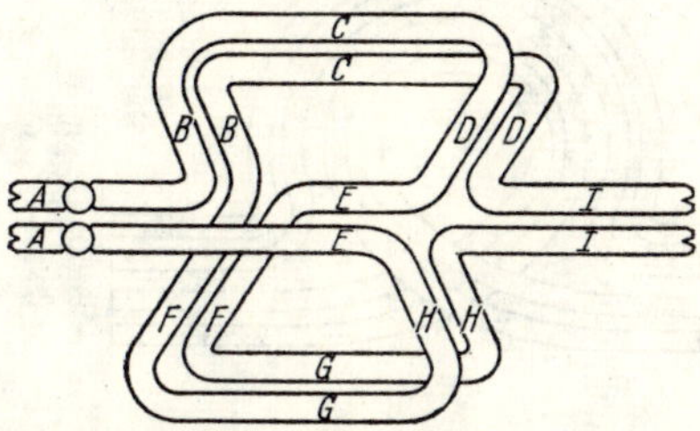

Fig. 55. Pairing configurations in karyotypes heterozygous for complex inversions: a) two independent inversions: b) two included inversions; c) two overlapping inversions (modified after *Dobzhansky* 1951).

gether with a segment which was not previously inverted (Fig. 55c) (abcdefgh → a/edcb/fgh → a/ed/gf/bc/h).

Complications in meiosis arise in conjunction with heterozygous inversions, i.e., if a structurally normal homologous chromosome is present in addition to the i. chromosome. The meiotic pairing behavior of the i. chromosome and its structurally normal partner depends on the length of the i. and the longitudinal relationship of the inverted and uninverted chromosome segments. If the i. is long, chromosome pairing involves the formation of a retrograde loop in the homologous, structurally unchanged chromosome; the i. chromosome affixes itself to its partner in such a way that homologous loci pair with one another. (Similarly, loop formation occurs in i. heterozygosity in the somatic pairing of → polytene chromosomes of Diptera).

If the inverted segment is so short that loop formation is not possible, either the i. segment remains unpaired or pairing of nonhomologous segments may take place. If the i. segment is very long, it may pair without loop formation and the uninverted terminal segments then remain unpaired.

Crossing-over and chiasma formation within (i. chiasma) and outside the inverted segment give rise to secondary structural changes (duplications and deletions), depending on 1. the i. -type, 2. the number of chiasmata, and 3. the localization of the chiasmata. Such structural changes lead in turn to meiotic products (gones) with unbalanced sets of chromosomes.

In the case of a **heterozygous, paracentric i.**, a chiasma within the loop (1 or 2 in Fig. 56) results in the formation of a dicentric chromatid and an acentric fragment in which the distal chromosome segments lacking in the dicentric chromatid are doubled. Those chromatids not involved in crossing-over remain unchanged, i.e. one possesses the gene order of the i. chromosome, the other that of the normal partner. During anaphase I, the dicentric chromatid is stretched between the centromeres which have migrated to the poles and thus forms a chromatid bridge (Fig. 56a).

Fig, 56, Consequences of crossing-over and chiasma formation in a heterozygous paracentric inversion, Top: the meiotic pairing configuration (pachytene). The numbers indicate several sites of crossing-over and chiasma formation, Middle: anaphase-I-configurations resulting from crossing-over at 1 or 2 (a), 1 + 2 (b), 2 + 3 (c), or 1 + 2 + 3 (d) Bottom: anaphase-II-configurations resulting from the anaphase-I-configurations indicated in the second row of the figure.

If chiasmata are formed within the i. loop at 1 and 2, a double bridge and two fragments are produced in anaphase I (Fig. 56b). Chiasma formation at 2 and 3 results in one fragment and a monocentric loop chromatid ("loop univalent") in anaphase I. Both ends of the loop univalent are joined to the centromere. This behavior may occasion the formation of a bridge in anaphase II (Fig. 56c).

Chiasma formation at 1, 2 and 3 leads to the production of two monocentric loop univalents and two fragments and to two bridges in anaphase II (Fig. 56d).

In the case of **heterozygous, pericentric inversions,** chiasma formation does not result in the production of dicentric chromatids, fragments or loop-univalents. However, the crossover chromatids differ owing to the formation of single (1 to 5 in Fig. 57a) or diagonal (1 + 3 and 1 + 4 in Fig. 57c) chiasmata in the i. loop through duplication and deletion of the terminal segments.

Reciprocal chiasmata (1 + 2 in Fig. 57b) in the i. loop do not give rise to aberrant chromatids. Complementary chiasmata (1 + 5 in Fig. 57d) result in duplications and deletions in all four chromatids.

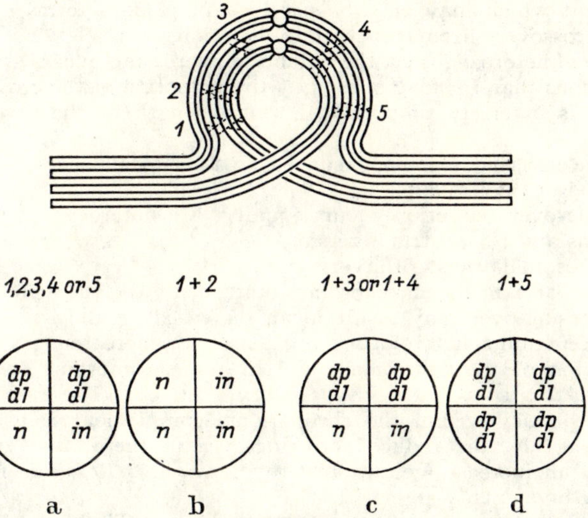

Fig. 57. Consequences of crossing-over and chiasma formation in a heterozygous pericentric inversion. Top: The meiotic pairing configuration (pachytene). The numbers indicate various sites of crossing-over and chiasma formation. Bottom: The consequences of crossing-over at different sites as manifested in the genetic constitution of the meiotic products (n = normal chromosome; in = chromosome with inversion; dp = chromosome with duplication; dl = chromosome with terminal deletion).

Inversions may be detected in the case of i. homo- and heterozygosity in various ways, using genetic and/or cytological methods:

1. by detecting changed → linkage relations with the aid of genetic linkage studies;

2. by the detection of eventual changes in chromosome morphology during the metaphase of mitosis;

3. by observing changes in the band pattern of → polytene chromosomes.

The following are diagnostic of inversions in the case of i. heterozygosity:

1. i. loops during the prophase of meiosis;

2. formation of bridges, fragments, and loop-univalents in the anaphase of meiosis;

3. aberrant meiotic products which may be uncovered by means of → tetrad analysis;

4. decreased fertility resulting from the production of genetically unbalanced meiotic products via → crossing over.

Inversions often affect the crossing-over frequency. In this context a distinction is made between:

1. effects within the i. bivalents (according to experiments of *Sturtevant* and *Beadle* 1936 on *Drosophila*).

a) Heterozygous inversions lead to a localized crossing-over reduction in the i. bivalent which may only be partly explained in terms of the elimination of crossover chromatids with duplications and deletions.

b) In the case of heterozygosity, the reduction in crossing-over is greater for short inversions than for long ones, i.e., the reduction of the crossing-over frequency is inversely proportional to the length of the inverted segments.

c) Outside the inverted segment, reduced crossing-over occurs only in the neighborhood of the i. points.

d) The crossing-over reduction occurs equally with heterozygous paracentric inversions and pericentric inversions.

2. Interchromosomal effects of inversions:

The presence of an i. in the chromosome complement and the associated reduction of crossing-over may result in an increased crossing-over frequency in the remaining heterologous bivalents. An increased crossing-over in the region of the centromere and at the chromosome ends is characteristic of this so-called "*Schultz-Redfield* effect". The resultant crossing-over frequency within the chromosome complement is usually unchanged. As a rule, the reduced crossing-over in the i. bivalent is compensated by an increase (conditioned by the *Schultz-Redfield* effect) in the remaining bivalents.

Inversions represent a factor in → karyotype evolution. Their occurrence may result in a → chromosome polymorphism within a population especially among insect species (e.g. *Drosophila*), less often in plants. The result is that in addition to a heterozygous i. type (heterokaryotype) two homozygous forms (homokaryotypes) occur: one containing the standard pair of chromosomes, the other the inversion pair.

The general evolutionary significance of inversions is as follows: Because of the extensive crossing-over reduction within heterozygous i., balanced gene complexes are built up ("super-genes"). These frequently confer a greater biological → fitness upon the heterokaryotypes. In comparison with the homokaryotypes, the heterokaryotypes often prove to be heterotic (→ heterosis) and homeostatic (→ homeostasis) with respect to characters which favor their → adaptation. If the karyotypes with inversions prove to be adaptable to certain environmental conditions they can be the starting point for the production of new local populations.

invisibles (*Goldschmidt* 1938) — → gene mutations or chromosomal structural changes (→ chromosome mutation) without visible effects.

in vitro — of stucies under artificially controlled conditions outside the organism as opposed to *in vivo* studies carried out in the intact organism.

***in vitro* genetic assay** — any method of → mutagenicity testing which is based on qualitative plate tests. Procedures of screening of genetic activity of chemical agents have been described for the detection of lambda phage → induction, reverse mutation in bacteria, and mitotic → gene conversion and reverse mutation induction in fungi. Some technical modifications of plate tests involve the utilization of hepatic drug-metabolizing enzymes for the potentiation of certain chemicals, construction of highly sensitive indicator strains through alterations in their capacity to repair genetic damage or limit permeability of specific molecules, and the development of nonrestrictive assays such as mitotic gene conversion or forward mutation (→ host-mediated assay).

in vivo — → *in vitro*.

I pilus — → sex pilus.

isoacceptor tRNA (*Doctor* et al. 1961) — → transfer RNA species which accept (are acetylated by) the same amino acid. The number of isoaccepting tRNAs found is greater than would be predicted from the → genetic code whose degeneracy requires their presence. The multiplicity of i. tRNA species and their widely varying levels have led to the postulate that such tRNAs may be involved in regulatory functions (translational control). In addition, isoacceptor tRNAs are involved in the production of → suppressor tRNA species.

isoallele (*Stern* & *Schaeffer* 1943) — any → allele so similar in its phenotypic effect to that of other independently occurring alleles as to require special techniques to distinguish among them.

iso-anisosyndetic (*Lilienfeld* 1951) — → isosyndetic.

isobrachial (*Sorokin* 1929) — → chromosome arm.

isochromatid break (*Sax* 1941) — a discontinuity in both chromatids of a chromosome at apparently identical sites as identified by genetic methods or by cytological observation in mitotic and meiotic meta- or anaphase (= isochromatid deletion, isolocus break, or isolocus deletion).

isochromocentric (*Dangeard* 1946) — of nuclei showing as many → chromocenters as they contain chromosomes (→ oligosomal).

isochromosome (*Darlington* 1940) — a monocentric or dicentric (→ isodicentric) chromosome with homologous (equal and genetically identical) arms which are mirror images of one another. Stable monocentric isochromosomes may arise by → centromere misdivision (transverse breakage and union in the inner zone of the centromere region) and reduplication of the telocentric fragments in the following interphase, provided the chromatids of these fragments do not become separated during mitotic anaphase but are distributed together (united at the terminally localized half of the original centromere) to one of the two cell poles.

Meiotic → chromosome pairing of isochromosomes may be either internal, fraternal, or internal-fraternal (*Sen* 1952):

a) **Internal pairing:** The two arms of an i. pair with each other to produce a ring → univalent.

b) **Fraternal pairing:** One of the arms of an i. pairs with a homologous arm of another chromosome.

c) **Internal-fraternal pairing:** Pairing of the i. is partly internal, partly fraternal.

isocoding — of genes coding for one and the same polypeptide, as, e.g., those coding for → histones (→ gene reiteration; repetitious DNA).

isodicentric (*Darlington* & *Wylie* 1953) — of a → dicentric chromosome or chromatid in which the free arms on both sides of the two centromeres are of equal length and homologous (→ isochromosome).

isogamete — → gamete.

isogamous (*de Vries* 1911) — 1. of → complex heterozygous species whose male and female gametes transfer the same genes or gene complexes, in contrast to → heterogamous hybrids;

2. of a population derived from parents chosen either at random or because of their coancestry and in which all pairs have the same fecundity (*Malecot* 1948).

isogamy — → anisogamy, gamete.

isogenes — lines on a map connecting points where the same gene frequency is found (→ isophene).

isogenic (*Johannsen* 1926) — of any group of individuals which possess the same genotype, irrespective of their being homo- or heterozygous.

isograft — a tissue graft between two individuals of identical genotypes (= isogeneic graft, syngeneic graft) as, e.g., monozygotic twins, members of inbred strains and the F_1 of crosses between two inbred strains (→ allograft).

isolabeling — → sister chromatid exchange.

isolate (*Wahlund* 1928) — a group of individuals (a population or group of populations) who, because of their isolation, choose their mating partners from menbers of the same group. One i. is completely separated from other such isolates. In human genetics, geographic and social isolates are distinguished. Social isolates may be the consequence of economic, religious, or traditional preferences in marriage.

isolation — the restriction, or limitation, of the extent of → gene flow (exchange of → genetic information) between discrete populations owing to barriers to interbreeding. I. is a factor in → evolution and the various mechanisms of i. may be classified as follows (*Grant* 1963):

1. **spatial i.:** Gene flow between populations is reduced or prevented by the distance between them;

a) **geographical i.:** The populations concerned occupy different territories, i.e. are allopatric;

2. **environmental i.:** Gene flow between populations is limited by the availability of habitats suitable for the growth and subsistence of their hybrid progeny (selection against the hybrids and their derivatives);

a) **ecological or habitat i.:** The populations occupy different habitats in the same territory, i.e. are sympatric;

3. **reproductive i.:** Gene flow is prevented or restricted by genotypically

controlled differences in the reproductive behavior and fertility relationships of the individuals which are members of different populations;

a) **external reproductive i.:** The barrier to gene flow lies outside the individuals and is effective prior to fertilization;

a_1) **mechanical i.:** Male and female reproductive organs are insufficiently coadapted structurally;

a_2) **ethnological i.:** Psychological or behavioral factors prevent the mating of male and female individuals which are members of different populations;

a_3) **seasonal i.:** The period of mating activity is different in the different populations;

a_4) **gametic i.:** In cases of external fertilization, the free-living male and female gametes produced by different populations fail to attract one another;

b) **internal reproductive i.:** The barrier to gene flow operates inside the individuals and results from the unsuccessful interaction between gametophytes, gametes, chromosomes, or genes of the different populations;

b_1) **incompatibility:** In the case of internal fertilization, cross-insemination or pollination takes place between individuals of different populations but does not result in hybrid progeny;

b_2) **hybrid inviability:** Hybrid progeny are produced but are of reduced viability;

b_3) **hybrid sterility:** F_1 hybrids grow to maturity but either their sex organs or their gametes are abortive and nonfunctional;

b_4) **hybrid breakdown:** Hybrids produce F_2 or backcross progeny consisting partly or entirely of inviable or sterile individuals.

I. between species does not generally operate by a single isolating mechanism but by a combination of different mechanisms. Isolating mechanisms function to prevent the breakdown of highly integrated genetic systems (e.g. species) which are adapted to a particular environmental niche. The evolutionary significance of i. for the processes of speciation is its effect on limiting random mating and on population size. I. is a directive force in evolution and prevents or restricts the mixing of genetically controlled differences which have arisen in a population (→ annidation). Breakdown of i. barriers between different breeding populations allows hybridization to occur between members of the two groups. The results of such hybridizations depend among others on the genetic similarity of the populations concerned. If they are genetically rather similar, hybridization facilitates gene flow between contiguous populations and has a tendency to reduce or even to eliminate the genetic differences between them (prevention of local differentiation of populations within a species). Hybridization between species or subspecies may lead to what is called → introgressive hybridization, i.e. the introduction of genes from one population into the other (increase of genetic variation).

Mechanisms of reproductive isolation may originate in two ways (*Grant* 1966):

1. as by-products of evolutionary divergence;
2. as products of selection for reproductive i. *per se*, i.e., by an in-

crease within each species in the frequency of genetic factors which determine a block or aversion to hybridization between two sympatric species over the course of generations. Such selective processes (which may be called the "*Wallace* effect" as proposed by *Grant* 1966) may reinforce reproductive i. which developed as a by-product of divergence but did not necessarily result in a complete barrier to hybridization.

isolation estimate (*Merrell* 1950) — a measure of genetically controlled sexual → isolation between populations (A and B) based upon the relationship between homo- and heterogamic matings that take place. There are four possible kinds of mating between individuals of two populations composed of A♀, A♂ and B♀, B♂ respectively: two homogamic mating types A♀×A♂ and B♀×B♂, and two heterogamic types A♀×B♂ or B♀×A♂. The i.e. is calculated as the quotient

$$i = \frac{(A \times B) + (B \times A)}{(A \times A) + (B \times B)}.$$

It may be used if equal numbers of the types to be compared are at hand. If there is complete sexual isolation (no mating between populations A and B) i is zero; with no sexual isolation i is 1.

isolation gene (*Brieger* 1956) — any pair of alleles causing a reduction in viability or fertility when present in the heterozygous form. Under these circumstances the homozygotes are kept partially or completely isolated by the elimination of the heterozygotes.

isolation index — a measure of genetically controlled sexual isolation between two populations, based on matings of equal numbers of females of each population with males of one population. After an interval large enough to allow approximately 50% insemination, the distribution of all females between the populations is recorded. The percentage of intrapopulational copulation minus the percentage of interpopulational copulations divided by the sum of both, is referred to as i. i. The index is +1 if all copulations were intrapopulational, −1 if all were interpopulational, and 0 if equal numbers of females of both populations were inseminated.

isolecithal — of egg cells in which the yolk is scattered evenly throughout the cytoplasm (→ centrolecithal, telolecithal).

isolocus break (*Thoday* 1953) — = → isochromatid break.

isomeric (*Sirks* 1933) — ref. to equivalent genes which can each produce the same phenotype (→ anisomeric). If several i. genes are present in one genotype they may either be cumulative ("cumulative isomery" or true "polymery") or noncumulative ("noncumulative isomery") in their action.

isomorph (*Serra* 1949) — = hypomorph (→ allele).

isonymous substitution — any DNA base pair substitution (→ gene mutation) that does not change the amino acid coded for.

isophene — a line on a map connecting points of equal expression of a clinally (→ cline) varying character (*Mayr* 1963).

isophenic — of genes with similar phenotypic effects (= → isomeric).

isophenogamy (*Strandskov* 1953) — a → mating system in which mating between individuals of identical or very similar phenotype is more frequent than expected on the basis of random mating, in contrast to "heterophenogamy", in which mating occurs more frequently between individuals of dissimilar phenotype as compared with random mating. If the phenotypes preferentially mating in both i. and heterophenogamy are genetically controlled, then the first system may lead to an increase in homozygotes, the second to an increase in heterozygotes for the alleles determining the phenotypes in question.

isophenous — being of the same → phenotype.

isopycnotic (*Östergren* 1950) — of chromosomes or chromosome regions which do not differ from the majority of chromosomes and chromosome regions of the cell in their appearance, in contrast to → heteropycnotic chromosomes and chromosome regions.

isosteric — of a regulatory mechanism by which enzyme activity is inhibited when the concentration of the end product approaches a certain value. The inhibitor structurally resembles one of the substrates and binds to the same site on the enzyme surface as the substrate. If the binding of the inhibitor occurs at a site other than the substrate site, the mechanism is called → allosteric inhibition.

isostich (*Shapiro & Chargaff* 1964) — any cluster of → oligonucleotides whose lengths are equivalent. Thus dinucleotides pTpT and pCpC are isostichs but not isomers whereas pTpC and pCpT are both isostiches and isomeres.

isosyndetic (*Lilienfeld* 1951) — of allopolyploids (→ alloploid) in which meiotic → chromosome pairing is restricted to the chromosomes of those chromosome sets derived from one species, as opposed to iso-anisosyndetic allopolyploids, in which chromosomes derived from different species are partially homologous and, therefore, able to pair to a limited extent. I. allopolyploids are called "genome allopolyploids" by *Stebbins* (1945); iso-anisosyndetic allopolyploids are referred to as "segmental allopolyploids".

isotelocompensating trisomic (*Kimber & Sears* 1968) — a compensating → trisomic (= monotelo-monoisotrisomic) in which the missing chromosome is compensated for by one telocentric chromosome and one tertiary chromosome (→ diiso-compensating trisomic).

isotertiary compensating trisomic (*Kimber & Sears* 1968) — a compensating → trisomic in which the missing chromosome is compensated for by one isochromosome and one tertiary chromosome (→ diiso-compensating trisomic).

isotrisomic (*Renner* 1941) — → trisomic.

isozygotic — of a genotype homozygous in respect to all genetic loci.

isozyme (*Markert & Moller* 1959) — any of a class of multiple, separable forms of enzymes (= isoenzymes) occurring within the same organism and having similar or identical catalytic activities. Isozymes may have different kinetic properties. The relative concentrations of i. often differ from one tissue ot another and may change during the course of development. The molecular multiplicity plays a role in the differentiation of cells during embryonic development and also increases the evolutionary adaptibility of organisms. Sets of i. can be generated by both genetic and epigenetic mechanisms.

From the genetic point of view isozymes may be classified as follows (*Ogita* 1968):

1. **Unigenic isozymes,** i.e., a set of i. all of which are under the control of one and the same gene. Initially, a single polypeptide chain is produced which gives rise to two or more different molecular species by polymer formation, conformational changes or by modification in various other respects. Unigenic i. are usually immunochemically distinguishable from each other.

2. **Multigenic isozymes,** i.e., a set of i. encoded by two or more genes and subclassifiable into two types:

a) allelic i. each of which is under the control of a codominant allele. If a hybrid molecule is not produced, two isozymes are formed in a heterozygote.

b) Nonallelic i., i.e., two or more i. controlled by two or more genes at different loci. As opposed to allelic i. these i. are immunochemically distinguishable. If they form polymers, hybrid molecules will be generated by random association. Nonallelic i. are different in their enzymological properties.

J

juggernaut polymerase (*Adya* et al. 1974) — a transcriptase (DNA-dependent RNA polymerase) that ignores all signals for termination of transcription due to the → rho factor. J. p. is asumed to originate by attachment of other proteins conferring anti-termination ability.

junctional complex (*Farquhar* & *Palade* 1963) — any specialized area of adhesive contact between cells. Junctional complexes consist of thickened plasma membranes and variable amounts of associated intra- and intercellular materials (→ desmosome). In some cases, the junctional complexes are in the form of rings encircling each cell apex.

Besides linking cells together, two more functions are attributes to the j. c.: 1. they establish a barrier between the apical and basal sides of cells reducing diffusion between the two compartments and allowing active transport. 2. They are responsible for the electrical couplings found in many tissues.

junctional membrane — any membrane region where the cells are in contact. Such regions are much more permeable to small ions than the membrane regions facing the exterior (nonjunctional membrane). In the areas of membrane contact organized regions are localized known as → tight junctions, → gap junctions and → septate junctions.

K

kalymma (*Heitz* 1935) — = → chromosome matrix.
kappa particle (*Sonneborn* 1943) — → killer.
karyoclastic (*Dustin* 1934) — of agents (referred to as "amitotic", → "mitotic poisons", or "mitotic inhibitors") which typically are able to reversibly suppress mitosis, without being lethal to the cell.
karyodesma (*Scott* 1950) — = → nucleodesma.
karyogamy — the fusion of nuclei. K. is one process of the → parasexual cycle in fungi and the ultimate event of → fertilization. By means of k. homologous chromosomes that were separated in the course of → meiosis are brought together again. The result of k. during sexual reproduction is a fertilized egg cell or → zygote (→ plasmogamy).
karyogram (*Chiarugi* 1933) — the sum total of all chromosomes of a cell described in terms of their morphology (→ idiogram).
karyoid — = → nucleoid.
karyokinesis (*Schleicher* 1878) — the division of the cell nucleus (= → mitosis), as distinguished from cytoplasmic division or → cytokinesis. K. represents an orderly system by which the → genetic information contained in the chromosomes of eukaryotes is distributed to dauhghter nuclei which are genetically identical with the mother nucleus provided no disturbances occur in the chain of events which make up the karyokinetic process. The basic mechanisms of k. are the same in most plant and animal cells. Differences involve the formation of the → spindle and the mode of cytokinesis which normally follows k.
karyology (*Trow* 1895) — nuclear cytology.
karyolymph (*Haeckel* 1891) — → nucleoplasm.
karyolysis (*Auerbach* 1874) — 1. the disappearance of the interphase nucleus at the beginning of → karyokinesis (*Auerbach*);
2. in a pathological sense, the dissolution of the nucleus either with or without previous nuclear fragmentation (karyorrhexis).
karyomere (*Fol* 1896) — any of a series of small nuclei ("micronuclei") formed in cells in which the chromosomes diverge at anaphase, i.e. are widely spaced on the → spindle. This phenomenon is common in cleavage divisions in some animals and can be induced experimentally by various chemical means.
karyon (*Haeckel* 1891) — = cell → nucleus.
karyoplasm (*Flemming* 1882) — = → nucleoplasm.
karyoplast (*Strasburger* 1905) — = cell → nucleus.
karyorrhexis — a form of nuclear degeneration (→ karyolysis) consisting of the fragmentation of the nucleus.
karyosphere — that small fraction of the total nuclear volume to which the chromosomes are confined in nuclei extremely rich in → karyolymph (e.g., egg cell nuclei of amphibia).

karyotheca — = → nuclear envelope.

karyotype (*Levitsky* 1924) — the particular chromosome complement of an individual or of a related group of individuals, as defined both by the number and morphology of the chromosomes usually in mitotic metaphase. A diagrammatic representation of a k. based on the morphological features of the chromosomes (the "endophenotype") is called an → "idiogram". The absence, duplication or positional rearrangement of certain morphological features of the chromosomes, as compared with their typical topography (with the standard k.), is visible evidence of → chromosome mutation.

Diploid individuals possessing a k. in which two chromosomes of a pair have different gene arrangements as a result of chromosomal structural changes, are referred to as **"heterokaryotypic"** (or are "heterokaryotypes"), whereas those in which the two chromosomes of a pair are similar in gene arrangement are "homokaryotypic" ("homokaryotypes").

Karyotypes may differ in respect to their basic chromosome number, the shape and relative size of the chromosomes, the number and size of secondary → constrictions, the absolute size of the chromosomes and the distribution and size of → hetero- and euchromatic chromosome segments.

Karyotypes which consist of chromosomes essentially similar in size (all with median or submedian centromeres), and those with chromosomes differing greatly in size and/or (partly) with subterminal or terminal centromeres, may be called **"symmetric"** and **"asymmetric" karyotypes,** respectively (*Stebbins* 1950). In comparison with plants, the karyotypes of animals are usually far more asymmetric. **Karyotype evolution** as a consequence of chromosomal structural changes (inversions, translocations, centric fusions, etc.) is usually associated with a progressive reduction of the basic chromosome number and an increased asymmetry.

Fundamental k. (*Battaglia* 1952) is the term which may be used to refer to the original k. of a species, genus, etc. (symbol k^f). A **derived k.** is any k. evolved from the fundamental k. by the processes of karyotype evolution (symbol k_1, k_2, etc.). **Semikaryotype** (symbol k/2) and **haplokaryotype** are terms proposed to denote the karyotype of the gametes and the haplophase, respectively.

karyotype orthoselection (*White* 1954) — the tendency for the same kind of chromosome structural change (→ chromosome mutation) to occur repeatedly (in one member of the karyotype after another) in a particular evolutionary lineage.

killer (*Sonneborn* 1938) — any of a number of strains of *Paramecium* containing symbionts (kappa) in their cytoplasm. These are relatively large, complex particles composed of DNA, RNA, protein and other substances. This genetically controlled symbiosis is the basis for the production of a toxin by killer strains. When released into the culture medium this substance is able to kill other strains referred to as "sensitives".

The killer character is dependent upon two genetic components, as demonstrated, e.g., for the symbiont kappa. One component is the genotype of the *Paramecium* cell, the other the kappa particles. Kappa particles are reproduced only in cells containing the genes K, s_1 and s_2. In cells of genotype Kk, only half the number of kappa particles is present, in comparison with the homozygous dominant genotype, KK. Kappa is particulate and usually several hundred particles per killer individual are present. Two types of kappa are known: so-called "brights", containing a bright refractile body known as R or R body, and "nonbrights" referred to as N. Nonbrights are self-reproducing and give rise to brights. The killing or P particle is identical with the bright or B particle. Nonbrights are capable of infecting sensitive strains and transforming them into killers.

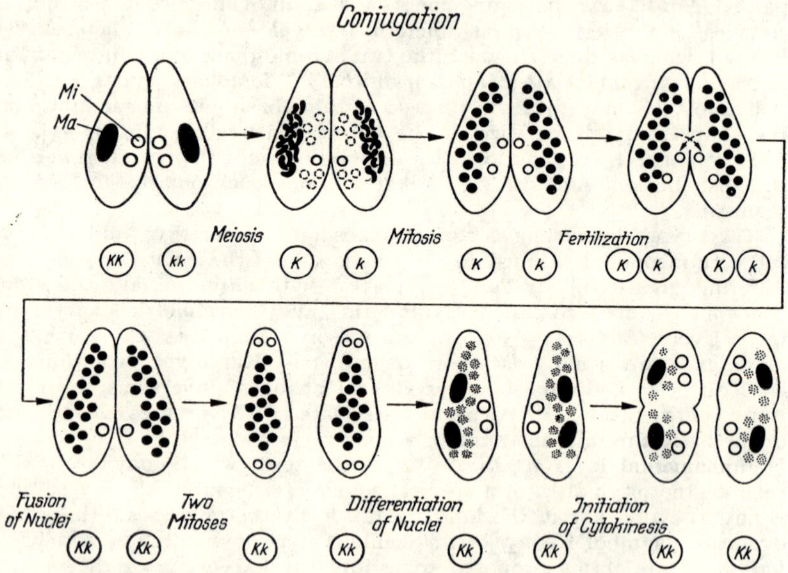

Fig. 58. Conjugation in *Paramecium aurelia* (Mi = micronucleus; Ma = macronucleus; K and k are symbols for the two alleles present in killer individuals and in sensitives) (from *Hagemann* 1964).

→ **Conjugation** (Fig. 58) between a killer (genotype KK and kappa particles in its cytoplasm) and a sensitive individual (genotype kk without kappa) may lead to two different results:

1. If the conjugants separate normally after nuclear exchange, i.e., without exchange of cytoplasm, then the exconjugants possess the same genotype Kk. The progeny of the original killer has the killer character, the progeny of the sensitive mating partner is sensitive (Fig. 59). Kk as well as KK individuals must have kappa particles in order to act as killers.

2. In exceptional cases, the protoplasmic bridge is maintained between

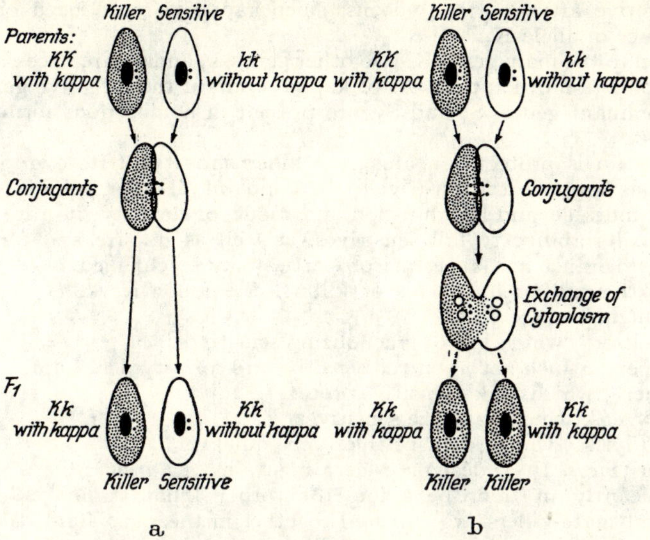

Fig. 59. Conjugation between a killer and a sensitive individual of *Paramecium aurelia*. a) without exchange of cytoplasm; b) with exchange of cytoplasm (from *Hagemann* 1964).

the two conjugants for a considerable time after nuclear exchange allowing an exchange of cytoplasm. Concurrently a transfer of kappa particles may occur. Under these circumstances the sensitive conjugation partner (genotype Kk) becomes a killer.

Autogamy (Fig. 60) in killer individuals of genotype Kk gives rise to 50% homozygous KK animals (killer) and 50% kk individuals which

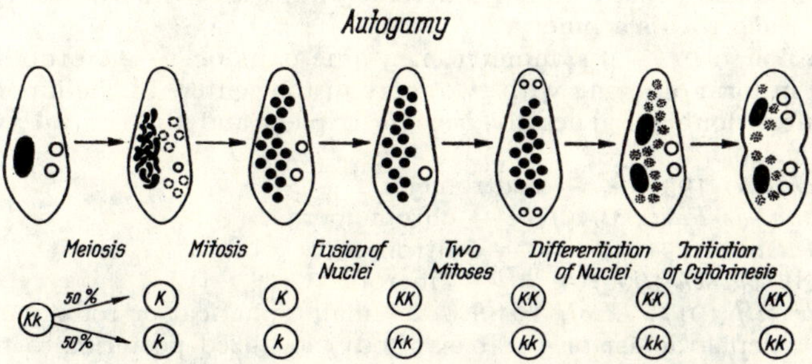

Fig. 60. Autogamy in *Paramecium aurelia* (the individual is heterozygous (Kk) for the alleles K (killer) and k (sensitive). As a result of autogamy in heterozygous Kk individuals, 50% KK- and 50% kk-individuals are produced (from *Hagemann* 1964).

become sensitive after 9—15 divisions, since kappa is reproduced only in the presence of allele K.

In addition to the main gene, K, two other genes, s_1 and s_2, are necessary for the reproduction of kappa. Loss of kappa occurs if the recessive gene, k, or the dominant genes, S_1 and S_2, are present in homozygous form in the genotype..

A killer strain is protected against the killer activity of its own and all homologous kappa but is sensitive to the kappa of all other killer types.

Kappa is mutable and by mutation, its mode of activity, its rate of reproduction, its ability to kill sensitives, as well as its size and form, temperature tolerance and antigenic properties may be changed.

Based on their different modes of killing, the following killer types may be distinguished (*Sonneborn* 1959):

a) "Vacuolizer", which kill by vacuolizing sensitives.
b) "Humper", which act upon the sensitives to form typical humps.
c) "Spinner", which make sensitives rotate.
d) "Paralyzer", which paralyze sensitives.
e) "Rapid lysis", which kill sensitives very quickly.

Mate-killer (*Siegel* 1953) designates certain strains of *Paramecium* which differ significantly in their behavior from other killer strains. When sensitives and mate-killers are cultured together in the same fluid, there is no killing of sensitives by the mate-killers in contrast to the behavior of true killers. In this case, killing of sensitives is confined to direct contact during conjugation. Mate-killers contain so-called mu particles in their cytoplasm. The reproduction of mu (→ metagon) is controlled by a dominant gene (symbol M). Sensitives lack mu particles and usually contain the corresponding recessive alleles of M.

kilo bases — a unit of length (symbol Kb) equal to one thousand bases or base pairs in single- or double-stranded nucleic acids.

kindred — → pedigree.

kinetid — any of the locomotory structures in eucells called → cilia and → flagella. The → basal bodies of kinetids are termed kinetosomes. All of the kinetids in a cell constitute the kinetome and if they are arranged in orderly rows, each row is a kinety.

A k. has a nonad (9 + 2) symmetry, i.e., nine pairs of → microtubules are arranged in form of a ring with two more in the center. In the kinetosomes the outer doublets generally become triplets and the central two disappear (**Fig.** 20).

kinetochore (*Schrader* 1936) — → centromere.

kinetomere (*Lima -de-Faria* 1949) — → chromomere.

kinetonema (*Matsuura* 1941) — = → centromere.

kinetonucleus (*Woodcock* 1906) — = → kinetoplast.

kinetoplast (*Alexeiff* 1917; *Lwoff* 1949) — a small, spherical or rod-shaped structure (= blepharoplast or parabasal body) localized posterior to the basal body of the flagellum of all parasitic flagellates. The k. is a DNA containing, self-replicating or self-dependent cell organelle which ordinarily divides before the nucleus. According to electron microscopy, the k. is a sac, bounded by a double membrane, containing an array of dense

DNA containing fibrils and typical mitochondrial cristae formed from the inner membrane. The k. is a differentiated part of the large, single → mitochondrion in *Trypanosomatidae*. Kinetoplast DNA can be separated from nuclear DNA by CsCl gradient centrifugation and appears to be quite distinct from the nuclear DNA of the same species.

The details of structure, as well as the relative proportions of DNA containing fibrils and of mitochondrial material vary from species to species and with the developmental stage of the life cycle. Flagellates with a k. also have a → basal body from which the flagellum arises.

kinetosome (*Allen* 1912) — = → basal body of a flagellum.

kinetome — → kinetid.

kinety — an orderly arranged row of → kinetids.

knob — a relatively large, heavily stainable → chromomere localized terminally, or (less frequently) interstitially, in the chromosomes of certain organisms (e.g. maize). It may be used as a "landmark" for the identification of particular chromosomes (→ idiogram).

L

lagging — of chromosomes or pairing configurations (referred to as "laggards") which show retarded movement, or no movement at all, in comparison with the bulk of chromosomes during mitotic or meiotic meta- and anaphase. A frequent consequence of lagging is the failure of the chromosomes concerned to be included in one of the two daughter nuclei. Thus lagging may give rise to "aneuploidy".

Lamarckism — the erroneous concept, advocated by *Lamarck*, that organic → evolution is brought about by environmental induction of adaptive changes in living organisms. Such changes were thought to be inherited by the progeny ("inheritance of acquired characters") (→ Darwinism; Neo-Darwinism).

lamella — a double membrane structure (= vesicle), formed by two → unit membranes lying more or less parallel to each other and in close apposition.

lampbrush chromosome (*Rückert* 1892) — a special type of → chromosomes in primary oöcyte nuclei, both vertebrate and invertebrate, during the prolonged diplotene stage of the first meiotic division, and in spermatocyte

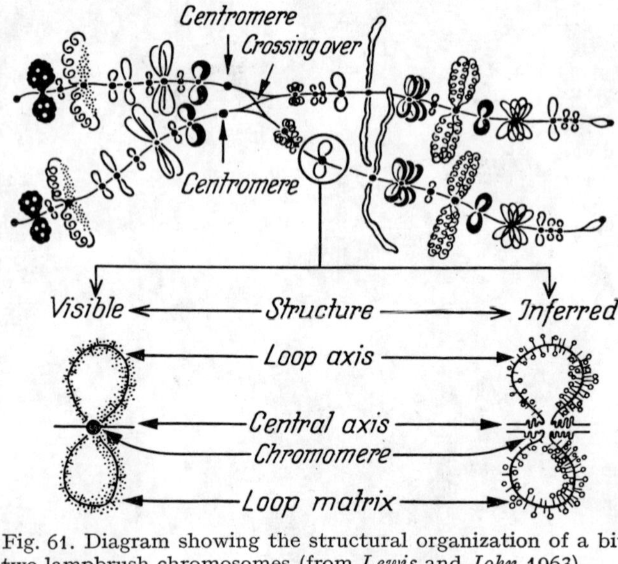

Fig. 61. Diagram showing the structural organization of a bivalent consisting of two lampbrush chromosomes (from *Lewis* and *John* 1963).

nuclei of *Drosophila*. The lampbrush state of chromosomes is phase-specific and reversible. Lampbrush chromosomes (Fig. 61) may even be larger than the (multistranded) → polytene chromosomes of Diptera; the largest with a length of about 1 mm have been found in urodele amphibia and are characterized by loops projecting in pairs from the majority of the → chromomeres which are connected by interchromomeric fibers. Frequently these loops exhibit a thin axis (probably consisting of one DNA double helix) from which fibers project which are covered with a loop matrix consisting of RNA and protein. These loops may vary in size and shape from one chromomere to another and may be used to construct tentative cytological maps of the lampbrush chromosomes. Generally, chromomere size and loop size are inversely correlated.

Since l. chromosomes are in the mid-prophase of meiosis, they are present in the form of → bivalents. The unit of trancription (the → transcripton) in these chromosomes is generally the lateral loop.

Frequently the individual loop is asymmetric, i.e., one of its insertions in the chromomere is thin and devoid of fibers. Following along the loop from the "naked" end, the projecting fibers are first short, becoming gradually longer toward the other insertion. This asymmetry may be a consequence of a polarized system of loop formation, or may result from a restricted zone of RNA synthesis at the inner insertions of the loops and a movement around the loop either of the loop products alone, or of the loop axis plus the products.

The loops are understood to be reversible modifications of the chromosome structure at the loci of active genes (→ puffing); they collapse after inhibition of DNA-dependent RNA synthesis (→ genetic transcription).

In some newts, DNA-containing rings are continously detached from the loops at specific loci. They give rise to numerous small nucleoli. This specific replication process of a small number of particular loci may be the cause for the much higher DNA content of oöcyte nuclei as compared to nuclei of any somatic tissue and is evidence for the independence of single chromomeres during replication. (→ gene amplification).

leaky — of 1. a protein coded for by a mutant gene which shows some residual activity **("leaky protein")**,

2. a → gene mutation which fails to completely prevent the action of a gene and permits some residual expression of the function in question **("leaky mutation")**, and

3. a → genetic block which is incomplete **("leaky block")**.

lepocyte — a cell surrounded by a → cell wall, as opposed to a → gymnocyte.

leptonema (*Gregoire* 1907) — = → leptotene.

leptotene (*v. Winiwarter* 1900) — → meiosis.

lethal — of a gene or genotype which, when expressed, is fatal to its carrier (→ lethal factor).

lethal equivalent (*Morton, Crow & Muller* 1956) — a group of genes (detrimentals) of such number that, if dispersed in different randomly chosen individuals, they would cause on the average one death; e.g., a single gene that is fully lethal in the homozygous condition, two different genes each with 50% probability of causing death if separately homozygous in different genotypes, or three such genes with a killing probability of 33%, etc.

lethal factor — a gene mutation or chromosome structural change (deletion or duplication, as well as inversion and translocation connected with a so-called → position effect) whose presence in the genotype blocks the normal development of the carrier (most often by failure to produce one or more essential gene products) and is fatal to the individual before sexual maturity.

Lethals may be classified in various ways (*Hadorn* 1949):

1. **According to their degree of → penetrance:**

a) lethal factors *sensu strictu* causing the death of all carriers which contain the particular lethal factor in active dose in their genotype (lethality at least 90% as compared to standard).

b) Semilethals causing the death of more than 50% of the carriers (about 50 to 90% lethality of the carriers).

c) Subvitals (*Hadorn* 1948) ("detrimental mutations" according to *Muller* 1948) causing the death of less than 50% of the carriers (about 10 to 50% lethality).

d) Conditional lethals resulting in the loss of some vital function, but only under environmental conditions over which the experimenter may exert some control.

e) Quasi-normals, causing less than 10% lethality.

2. **According to their phase of activity** ("phase specificity of lethals"):

a) gametic (*Mohr* 1926), gonic (*Renner* 1924), or haplophasic lethals (*Hadorn* 1949) which cause the death of gones or gametes and prevent zygote formation, and

b) zygotic lethals (*Mohr* 1926) which cause the death of zygotes at some stage of zygote development. Such lethals may be further subdivided into embryonic, larval, pupal, and early imaginal lethals in insects and into embryonic, postembryonic, and juvenile lethals in birds and mammals.

3. **According to the influences of the external and internal environment** on the action of the lethals:

a) unconditional lethals; their penetrance and expressivity may not be influenced nor compensated by experimental circumstances, and

b) conditional lethals; their action is dependent on the growth or environmental conditions (e.g., temperature-sensitive lethals, or sex-limited lethals, the action of which is expressed in only one of the sexes).

4. **According to their localization:**

a) autosomal lethals localized in the → autosomes, and

b) sex-linked lethals localized in the → sex-chromosomes.

5. **According to their being → dominant or recessive:**

a) dominant lethals, which are lethal in single dose,

b) dominant Mendelian factors with recessive lethal effect, which are lethal only when present in the homozygous form, and

c) recessive lethals without dominance effect, which are ineffective in heterozygous form due to the complete dominance of the wild-type allele.

Synthetic lethals (*Dobzhansky* 1946) are genes that in some combinations cause little or no apparent disadvantage to their carriers, but in other combinations are lethal. Synthetic l. are a class of genes having a direct effect on viability, but where the average effect of a particular gene substitution is less than unity.

Lethals are often maintained at high frequencies in natural populations by various genetic mechanisms. **"Balanced lethality"** (*Muller* 1917) is such a mechanism. It represents a special case of linkage which assures the maintenance of recessive lethals in a system of permanent heterozygosity. The simplest of such systems consists of two linked, recessive lethal factors (l_1 and l_2) in the heterozygous condition (L_1l_2/l_1L_2). The genotype is phenotypically normal, but heterozygotes may segregate lethal homozygotes when crossed. In the case of loose linkage of the two loci or in the absence of factors (heterozygous inversions or translocations) reducing or preventing crossing-over between them, the lethal segregants from a cross of two such heterozygotes approach rather closely the following ratrio:

0.25 l_1L_2/l_1L_2 (lethal) : 0.50 l_1L_2/L_1l_2 (viable) : 0.25 L_1l_2/L_1l_2 (lethal). Systems in which recombination between the two heterozygous loci concerned is not completely prevented are referred to as "partially balanced" and give rise to recombination gametes of the types l_1l_2 and L_1L_2 in additon to the nonrecombinant gametes of type L_1l_2 and l_1L_2. The lethals in systems of balanced lethality may act zygotically or gametically, or partly zygotically and partly gametically in one or the other sex. In all such cases only the heterozygous genotypes l_1L_2/L_1l_2 are viable. If both lethals are zygotic lethals, the fertility is reduced to 50%; if one is zygotic, the other gametic in action, the fecundity (number of zygotes) is 50% but the fertility (number of the individuals which can reproduce) is 25%. **Polygenic lethality** (*Hadorn* 1955) is lethality realized by the combined action of at least two mutated loci.

Effective lethal phase (*Hadorn* 1955) is the term proposed for that phase in ontogeny beyond which a lethal mutant is normally unable to survive. The phenocritic phase (*Haecker* 1918), i.e., that phase during ontogeny at which the developmental pathways of two different genotypes (e.g. wild-type and a mutant genotype) begin to differ, and the effective lethal phase of a genotype containing a lethal, may be widely separated in time, since lethality is frequently the result of particular stresses with which the defective developmental system is confronted. On the basis of their effective lethal phases, lethals may be classified as follows (*Hadorn* 1955):

a) Monophasic lethals, those with only one effective lethal phase.

b) Pluriphasic lethals, those characterized by two or more effective lethal phases separated by developmental periods during which no impairment of development takes place.

c) Aphasic lethals, those which may cause a breakdown of the developmental system of the carrier at any time during the whole of ontogeny.

The causes underlying phase-specificity of lethals may be different:

1. A general biochemical disturbance may block or change indispensable

physiological functions resulting in death during early embryogeny or later, when the system is confronted with specific stresses.

2. Loss or disturbances of a particular function in definite cell systems resulting in a lethal crisis when the specific function is needed during development.

3. Gross disturbances of a morphological nature (e.g. absence of an organ) which are initiated as embryonic disturbances but lead to death much later.

lethal sectoring (*Haefner* 1965; *James & Werner* 1966) — in yeast, bacteria, algae and mammalian cells in tissue culture, the sporadic appearance of clusters (sectors) of dead cells in the post-irradiation vegetative generations of surviving cells. L. s. tends to diminish with advancing generations of an irradiated population.

lethal zygosis (*Alfoldi* et al. 1957) — a lethal event after mixing bacterial recipient (F^-) cells with an excess of Hfr cells (→ F plasmid) that results in a decrease in the number of F^- survivors. L. z. requires a specific contact between F^- and Hfr cells mediated by → F pili.

leucoplast (*Schimper* 1885) — a → plastid containing no visible pigments (→ elaioplast, aleuroplast, amyloplast).

Lewis effect (*Pontecorvo* 1955) — → position effect.

L form — in bacteria, a cell differing from the normal situation in that the cell wall is missing or reduced to a template moiety. L forms arise as more or less stable clones under the action of penicillin with so-called large bodies ("multinucleated" bodies of irregular shape) as an intermediate state. The term stable protoplast (spheroplast) is often used synonymously with L f.

life cycle — all the species-specific, significant stages in those processes by which an organism gives rise to others of its kind. Each life cycle is based upon and controlled by a particular → genetic system. In many plants and some animals, the l. c. involves a regular → alternation of generations. Four basic types of l. cycles are immediately distinguishable:

1. **the haploid l. c.:** the entire life history is passed in the haploid state except for the diploid zygote nucleus.

2. **The diploid l. c.:** the entire life history is passed in the diploid state except for the immediate products of meiosis.

3. **The haplo-diploid l. c.:** alternation of haploid and diploid vegetative phases.

4. **The haplo-dikaryotic l. c.:** the stable association (in higher fungi) of paired haploid nuclei that is propagated vegetatively (the dikaryophase is the substitute for a vegetative diplophase). This type of l. c. proceeds through haploid and dikaryotic phases of varying proportions prior to a single nuclear generation in the diploid state and thence to meiosis.

ligand — a small molecule (including activators, substrates, and inhibitors of enzyme activity) bound to a protein by noncovalent forces. Interactions between like and unlike ligands may be called homotropic and heterotropic, respectively.

ligase — any of the polynucleotide joining enzymes (= DNA ligase; sealase; joining enzyme) which close single-strand breaks (nicks) in DNA.

light repair — = → photoreactivation (→ dark repair).

limb — = → chromosome arm.

limited chromosome — any chromosome that may be either eliminated or diminished during cleavage producing clones of different functional karyotypes (→ nuclear differentiation).

lineage — in human beings, a mating group forming a unit of common descent (→ cell lineage).

lineom (*Kühn* 1961) — the genetic linkage structure of viruses and bacteria (= genophore, virus chromosome, bacterial chromosome).

linkage (*Morgan* 1910) — the association in inheritance of certain genes (and, in turn, of the phenotypic characters they control) due to their being localized in the same chromosome or nucleic acid molecule (linkage structure). Genes borne on → homologous chromosomes belong to the same → linkage group. The number of linkage groups is identical with the number of chromosomes of one → chromosome set.

Linkage of genes is only rarely complete. It is broken (→ genetic recombination) when homologous linkage structures exchange corresponding parts by a process referred to as → crossing-over ("partial l."). Crossing-over in eukaryotes may occur during meiosis (meiotic crossing-over) or mitosis (mitotic crossing-over). In prokaryotes (such as bacteria and viruses) it can occur without meiosis or mitosis by functionally analogous processes at the level of nucleic acid molecules.

Linkage (rather than independent → genetic segregation) of two → genetic markers (e.g., allelic pairs A/a and B/b) is evident if the parental gene combinations (e.g., AB and ab) used in a crossing experiment show a tendency to remain together, i.e., the proportion of so-called recombinants (Ab and aB) is less than the proportion of parental combinations (AB and ab). If two marker genes are borne on nonhomologous linkage structures, independent segregation occurs and the parental and recombination types occur in equal frequency.

In the case of linkage between two markers (A and B), a third marker C may be either linked or unlinked with these two. Lack of l. between two markers is indicated in crossing experiments by their random distribution and with three markers by the impossibilitiy of arranging them in a linear order when the recombination frequencies are used as the basis for their ordering. These relationships allow the unequivocal localization of genes in particular linkage groups. The orderly arrangement of genes in a linkage structure is represented by a → chromosome map which may be either genetic (→ genetic map) or cytological. Strength of l. between two markers is dependent on the distance separating them in the linkage group. Linked genes at loci relatively close together tend to be preferentially transmitted together from generation to generation, while those relatively far apart may frequently become separated by crossing-over and may then segregate nearly as independently as do unlinked genes.

Sex-linkage (*Morgan* 1914) is the special type of l. of those genes localized in the → sex-chromosomes of eukaryotes. According to their position in either nonhomologous (differential) or homologous segments of the X and Y chromosomes, such genes may either be completely (localization in "differential segments" in which no crossing-over takes place) or partially sex-linked. The latter genes, localized in homologous segments of X- and Y-chromosomes, may be exchanged between the sex-chromosomes by crossing-over.

Interchromosomal linkage (*Stern* 1933) or "false linkage" (*Longley* 1945) is a type of spurious linkage between genes belonging to different linkage groups and results from nonrandom (polarized) → assortment of certain chromosomes (and the linkage groups they contain) during meiosis (→ affinity; meiotic drive) or from random assortment, if only particular chromosome distribution patterns give rise to functional gametes or, subsequently, functional zygotes (→ complex heterozygous; translocation).

linkage disequilibrium (*Kimura* 1956) — nonrandom association of genes, i.e., the extent to which genetic frequencies of linked genes deviate from those expected from their allelic frequencies in the population (excess of certain gametic types, deficiency of others). L. d. may affect estimates of dominance effects or of heterozygote advantage. As a result of l. d. (= epistatic equilibrium; gametic excess; gametic phase unbalance) the → genotypic frequencies with respect to the participating loci will, like the gametic frequencies, differ from those expected on the basis of allelic frequencies (→ pseudo-overdominant). L. d. may result from non-additive (epistatic) interactions in → fitness at different closely linked loci (selection) or from some chance event (sampling of gametes in reproduction).

linkage group (*Morgan* 1911) — a group of gene loci which can be placed in a linear order representing the different degrees of → linkage among the genes concerned. Specific linkage groups correspond to specific chromosomes (linkage structures) as shown by special methods. The genes of one l. g. show various intensities of partial linkage, while genes belonging to different groups segregate independently, giving rise to parental and nonparental (recombinant) combinations in statistically equal frequency. The number of linkage groups is limited, the maximum number equalling the number of chromosomes per → genome (the haploid number of chromosomes in diploid organisms).

linkage map — a → chromosome map containing in linear order the genes belonging to the → linkage group associated with a particular chromosome. Crossover and recombination values are utilized to demonstrate a definite serial order of the genes in a chromosome and provide an index of the relative distance between any two genes (→ genetic map).

linked — of two genes (two pairs of alleles) showing less than 50 per cent recombination which is the normal amount for genes of different → linkage groups showing independent → genetic segregation. Linked genes tend to be transmitted together.

lipid — a class of diverse chemical compounds that includes fats, waxes, phospholipids, carotenoids, and sterols. Lipids serve as food reserves and as parts of structures of the cell. All lipids are soluble in fat solvents.

lipochondrium (*Baker* 1951) — a spherical or subspherical lipid body of variable size present in the cytoplasm. In many types of cells, the distribution of lipochondria is restricted to a zone in the vicinity of the nucleus; this zone represents the → Golgi apparatus.

liquid-holding recovery (*Roberts & Aldous* 1949) — one type of bacterial → dark repair process after UV irradiation, which is characterized by an increase in the number of colony forming units when the irradiated cells are held in buffer (in the dark) and plated at intervals on a complex medium (such as nutrient agar). L.-h. r. appears to be due to changes in the proportion of UV-induced lesions repaired by either the → excision repair or the → postreplication repair pathway.

locus (*Morgan, Sturtevant, Muller & Bridges* 1915) — the position of a gene on the → genetic map. Allelic genes are situated at identical loci in homologous linkage structures (chromosomes). The l. of a particular gene is identifiable as a site of an allelic difference arising by → gene mutation. Frequently the term l. is used more or less synonymously with gene, i.e. for a closely linked array of → mutational sites concerned with a single function.

A **compound l.** is one composed of two or more genes (in the sense of → cistrons).

locus-specific (*Hadorn* 1955) — ref. to phenomena or activities correlated exclusively with one particular chromosomal → locus; e.g., l.-s. control of the formation of a particular character, l.-s. → puffing in → polytene chromosomes, l.-s. loop formation in → lampbrush chromosomes, etc.

lomasome (*Moore & McAlear* 1961) — any of the cell's boundary formations (border bodies) which protrude into the interior of the cell (evaginations delimited by the cell membrane) and enclose cytoplasmic and multivesicular structures somewhat variable in form. Lomasomes are located between the → cell membranc and the → cell wall and have been found in fungi, algae and higher plants. They may play a role in secretion, in increasing the surface area for the diffusion of substances involved in cell wall formation or wall breakdown, in membrane proliferation and in endocytosis.

loop chromatid — → inversion.

low molecular weight nuclear RNA — any of a number of RNA species with sedimentation coefficients from 4 S to 8 S in many types of eukaryotic cells. The size of LMWN RNA ranges fom 100 to 300 nucleotides; the total number of LMWN RNA molecules per cell is about $1-2\times10^6$. These RNA species show a specific localization: 1. nucleolus-associated RNA; 2. nucleoplasmic RNA. Some of them exist in ribonucleoprotein complexes. LWMN RNA molecules exhibit specificity of hybridization with nuclear and nucleolar DNA (rDNA) and some contain modified nucleosides.

The only known function of this RNA is the amino acid acceptor and transfer activity of → transfer RNA. Hypothetically, LMWN RNA may be involed in gene regulatory functions (such as, 1. gene activation, 2. derivation from intercistronic or untranslated portions of structural gene products) or may represent structural components of → chromatin.

Ludwig theorem — a theory according to which new genotypes can be added to a population if they are able to utilize new components of the environment (i.e., occupy a new subniche), even if these are inferior to the ancestral niche (*Mayr* 1963). This phenomenon is also called "annidation".

luxuriance (*Renner* 1929) — → heterosis.

lysis — the bursting of a cell by the destruction of the cell membrane.

lysogen — a → lysogenic bacterial cell.

lysogenic (*Lwoff & Gutman* 1950) — of a bacterium possessing the heritable property of bacterial lysis in the absence of free bacteriophage. As the result of lysis, a progeny of infective → bacteriophages is then set free.

Each bacterium of a lysogenic strain harbors and maintains a non-infective structure attached to or integrated into its → genome and perpetuated as part of it. This is called a → prophage and endows the cell with the ability to give rise to infective phage without the intervention of exogenous phage particles. A prophage may enter into a vegetative state which ultimately leads to the production of infectious phage progeny. In a l. cell only a small portion of the integrated prophage (→ lysogenization) genome (the immunity region) is active. Its product (the immunity substance) is a → repressor which prevents the expression of all other prophage genes necessary for phage replication and the production of infectious particles. The repressor is also responsible for the immunity of l. cells to superinfection by (homo-immune) phages similar to the prophage. If deprived of immunity, i.e. of repression, these operons are derepressed. This triggers the whole sequence of events leading to phage maturation and the release after cell lysis of phages. The repressor produced by the prophage may be inactivated spontaneously (rarely) or by external events (induction of prophage). Under such circumstances, phage mRNA synthesis begins and there follows a sequence of events functionally indistinguishable from the lytic cycle. Under special conditions the prophage may be detached from the bacterial chromosome without breakdown of repression, so that a nonlysogenic cell is ultimately segregated (curing of l. cells).

Defective lysogenic strains have been isolated from various bacterial species. They arise by mutation of the prophage and are characterized by their inability to produce structurally intact, infective virus particles. Such strains show the specific immunity pattern of the original strain from which the defective prophage mutant was derived. The nature of the defect in various strains isolated from the same wild-type prophage may differ and the mutational sites responsible for particular defects may be localized by genetic mapping in different regions of the phage chromosome. Defects can be brought about by specific blocks at different positions during the vegetative development of the phage. It normally allows the definite maintainance, by the host, of viral genetic material which can never produce infective virus particles. (→ helper phage).

lysogenization (*Ball* 1925; *Bordet* 1925) — the establishment of → lysogeny by infection of a nonlysogenic bacterial cell with a → temperate bacteriophage and the attachment of the phage to the bacterial chromosome at a specific region, called the attachment region or attachment site. The infecting phage DNA cyclizes and is then linearly inserted into the host chromsome by reciprocal recombination between what is called the bacterial chromosite and the homologous segment of the phage genome, the episite. For the site-specific recombination involved in integration a function specified by a phage gene is needed. Phage mutants unable to perform this function are called integration-defective. Integration of the phage genome into the bacterial chromosome and its passive replication with it is called **stable lysogenization.** If the temperate phage fails to attach to the bacterial chromosome and is inherited unilaterally among the progeny of the infected cell (and eventually diluted out of the population), this is termed **abortive lysogenization.**

lysogeny — the potentiality of bacterial cells for production and release of bacteriophages as a stable, heritable character. It is widespread among strains of different bacterial species. Such strains are referred to as → lysogenic or lysogens.

lysosome (*de Duve* et al. 1955) — any of a class of morphologically heterogeneous cytoplasmic particles (ranging in size from 0.25 to 0.8 μm) in animal and plant tissues characterized by their content of hydrolytic enzymes (e.g. acid phosphorylase, acid ribo- and deoxyribonuclease, cathepsin, β-glucoronase), and the structure-linked latency of these enzymes. The intracellular functions of lysosomes depend on their lytic potential. The single → unit membrane of the l. acts as a barrier between the enzymes enclosed in the l. and the external substrate. The activity of the enzymes contained in lysosomes is limited or nil unless the vesicle in which they are enclosed is ruptured. Such rupture is supposed to be under metabolic (hormonal) control.

Lysosomes start functioning when the cell takes up substances either by phagocytosis or pinocytosis. Insoluble remnants of digestion within the lysosomes usually remain and the particles become packed with different inlucsions. The results of the digestive process are normally amino acids, sugars, and simple organic compounds which may be used as synthetic building blocks by the cell. Under certain circumstances lysosomes start to digest the cell in which they occur by opening their membrane and setting free their digestive enzymes.

The size (about 0.4 μm) and density of lysosomes lies between that of → mitochondria and → microsomes (fragments of the → endoplasmic reticulum). Their fine structure varies from cell to cell and within the same cell type during the life of the lysosomes.

Two main groups of lyosomes may be distinguished (*Novikoff* et al. 1964; *de Duve & Wattiaux* 1966):

1. **Primary lysosomes,** i.e., those which may be formed from small Golgi vesicles or larger vacuoles cut off from the tips of Golgi saccules, and have not yet been involved in digestive processes.

2. **Secondary lysosomes,** i.e., those digestive vacuoles which are or have been sites of digestive activity. Lysosomes belonging to this group may be classified into two separate types, heterophagic and autophagic, based on the exo- or endogenous origin of the material undergoing digestion. Each of these types may be further subdivided into prelysosomal, lysosomal and postlysosomal forms.

lytic — ref. to bacteriophages (= virulent phages) whose multiplication leads to lysis of the host bacterium, in contrast to the behavior of → temperate bacteriophages.

M

M I, M II — symbols for the metaphase of the first (M I) and second meiotic division (M II). Similarly, A I, A II, T I, and T II for anaphase and telophase (→ meiosis).

macroevolution (*Goldschmidt* 1940) — transspecific → evolution which within the span of geological time leads to the origin of higher taxonomic categories such as new genera, families, and orders with new characters and a new adaptive mode. M., also called "quantum evolution" (*Simpson* 1944), implies a more or less rapid shift of a small population to a new equilibrium state (a new adaptive mode) distinctly unlike the ancestral one.

As compared to evolution at the species level, referred to as "microevolution" or → "speciation", m. is neither a fundamentally distinct phenomenon, nor does it employ a different mechanism, although this has frequently been stated by some authors.

macrogametophyte — = → embryo sac.

macromolecule — a large molecule with colloidal properties; the molecular weight is $>1\,000$.

macronucleus (*Bütschli* 1876) — one of two kinds of nuclei (→ micronucleus) possessed by most ciliates and some foraminiferans showing → nuclear dimorphism. During the life cycle of ciliates, periods of vegetative reproduction are interrupted by periods of sexual reproduction. During vegetative reproduction the (diploid, or more often endopolyploid) m. is functionally dominant. This nucleus is discarded during sexual reproduction (see Fig. 58, 60) and is replaced by products of the micronucleus following meiosis, reciprocal fertilization and reestablishment of diploidy. When sexual reproduction is completed the new m. assumes control of most synthetic activities of the cell (with the possible exception of syntheses connected with mitosis, which seem to be related to the micronucleus), i.e., most of the cell's genetic activities seem to be centered in its macronucleus.

macrophylogenesis (*Zimmermann* 1948) — = → macroevolution.

macrospore — = → megaspore.

macula adherens (*Farquhar* & *Palade* 1963) — → junctional complex; → cell junction.

main band DNA — DNA in eukaryotic chromosomes which is not → heavy shoulder DNA or → satellite DNA.

major gene (*Mather* 1941) — any gene (= oligogene) individually associated with pronounced phenotypic effects. Major genes control the production of discontinuous or qualitative → characters, in contrast to "minor genes" or → polygenes with individually small effects. Major genes segregate clearly and are easily subject to Mendelian analysis. The classification of genes into major and minor determinants is more or less arbitrary, since both types are in all probability only end points in a more or less continuous series of gene actions and gene integrations (→ gene interaction).

major spiral (*Sax* 1935) — → chromosome coiling.

male multiple choice — → pair mating.

male-specific bacteriophage — any → bacteriophage specific for bacteria with F-like → sex pili. Such phages are of two types: isomeric RNA phages attaching along the length of the sex pilus and filamentous DNA phages attaching to the tip of the pilus.

malignancy — the ability of tumor cells to grow progressively and to kill their host.

map — → genetic mapping; chromosome map.

map contraction (*Esposito* 1968) — → map expansion.

map distance (*Bridges* 1932) — the distance, in terms of percentages of → crossing-over obtained from genetic experiments, between particular linked genes (→ genetic map). Since m. d. is based on the total frequency of crossing-over in a chromosome region it is directly translatable into units of m. d. or → map units. One unit of m. d. between linked genes is that interval within which 1 per cent crossing-over takes place.

The relationship between "recombination frequency" and m. d. is not one-to-one, since recombination frequency measures only those crossovers that lead to reciprocal exchanges of → genetic markers. The farther apart the genes, the more striking the discrepancies become because double crossovers are scored as nonrecombinants. Expressions that relate m. d. to recombination frequencies and correct for the probable double crossovers which have occurred in the interval under consideration are referred to as "mapping functions".

map expansion (*Holliday* 1964) — the tendency, observed in many linear fine-structure maps (intragenic → genetic recombination), for recombination frequencies between more widely spaced mutants within a gene to exceed estimates of the same intervals obtained by adding the recombination frequencies given by more closely spaced mutants. M. e. is a general → marker effect and the opposite of what is expected from negative → interference.

M. e. has been explained on the basis of an independent correction of base-pair mismatches (→ heteroduplex repair) included in the same hybrid DNA segment. A possible additional source of m. e. may be due to the contribution of reciprocal crossing-over to the total frequency of recombination. This contribution will increase with the square of the distance between the mutant sites of a two-point cross if the pairs of single-strand breaks serving for hybrid DNA formation between the mutant sites of a two-point cross are distributed at random within a given chromosome region.

The oppposite effect is called **map contraction,** i.e., the recombination value observed is less than the value expected by summing adjacent intervals.

map length — → Morgan unit.

mapping — → genetic mapping; deletion mapping; denaturation mapping; heteroduplex mapping.

mappingmark (*Renwick* 1971) — the beginning and the end of a chromosomal lesion that has led to a difference between homologous chromosomes and is used (in human genetics) for estimating the → map distance of a certain locus from one or more mapping marks.

map unit — the unit (= crossing-over unit, crossover unit or Morgan unit) of → map distance between linked genes. It is used to measure chromosome length (→ genetic map) and equals the total crossover frequency between the genes concerned, corrected for the effects of double crossing-over and chromosome → interference and is expressed as a percentage. A corrected crossover frequency of 1 per cent equals one map unit and represents the distance between two gene loci (A and B) when the recombinants ab and AB appear among the progeny of a standard cross (Ab × aB) at a corrected frequency of 1 per cent. The establishment of units of map distance between two loci from recombination frequencies without correction for double crossing-over between them is considered accurate only in those cases where recombination between closely spaced genes is measured, i.e. where interference is complete.

marker gene — a gene of known function and known location on the chromosome (→ genetic marker).

marker effect — any case in which two mutants (→ genetic markers) that are at or near the same site give very different frequencies of → genetic recombination when both are crossed to the same third mutant elsewhere in the gene. An unambiguous test for the occurrence of m. effects can only be made with a system in which the physical positions of the genetic markers are determined by a method other than recombination mapping.

The m. e. may be of two kinds (*Stadler* & *Kariya* 1973). 1. General marker e.: heterozygosity *per se* has a consistent effect on recombination (→ map expansion). 2. Specific m. e.: a certain segregating mutant affects recombination differently from others.

marker rescue — a process of integration (cross reactivation) of genetic markers from irradiated phage into unirradiated helper phage in mixedly infected bacteria. The efficiency of rescue of a marker depends on the gene in which it is located and is independent of the marker itself. The rescue frequency of a marker decreases with increasing radiation (UV) dose.

Marker **co-rescue** is the simultaneous rescue of two markers from a donor strain; it may be due to rescue of two DNA fragments each carrying one marker or rescue of one DNA fragment carrying both markers.

massule (*Bessis* et al. 1958) — a → centriole satellite.

master-slave hypothesis (*Callan* & *Lloyd* 1960) — a hypothesis proposed to explain the large differences in DNA content among various eukaryotes on the basis of a uninemic chromosome model. According to the m.-s. h. each gene may be represented 100 to 10,000 times in a cluster: one copy is the master and the tandem duplications are slave genes. Although each slave gene could mutate independently, it is postulated that the slaves fold back upon the master, match their base sequences, and are corrected wherever incongruities happen to be present.

mastigoneme (*Deflandre* 1934) — any of the fine hairlike appendages (= flimmer) which extend from the shaft of the → flagellum in a variety of motile cells and serve to move water towards the flagellum base.

mate-killer (*Siegel* 1953) — → killer.

maternal effect — any nonlasting effect of the maternal genotype or phenotype on the immediate offspring (→ predetermination).

maternal inheritance — → inheritance controlled by extrachromosomal hereditary determinants (= extrachromosomal, non-Mendelian or cytoplasmic inheritance).

mating — in eukaryotes, the pairwise union of unisexual individuals for the purpose of sexual reproduction (and by extension the "selfmating" of bisexual organisms), ultimately resulting in the formation of → zygotes. Parasexual processes yielding results genetically equivalent to the outcome of sexual mating in eukaryotes occur in prokaryotes; e.g., mating or → conjugation in bacteria as mediated by → conjugons, m. in the sense of pairwise interaction between virus particles within the host cells giving rise to → genetic recombination.

Among sexually reproducing species mating systems do not normally fall into discrete classes, since natural populations may show several of the arbitrarily delimited types of m. in an infinite variety of variations and combinations. Furthermore, the type of mating system may be affected by the selective advantage of particular characters as well as by population size. Nonetheless, in sexually reproducing organisms three main mating systems are distinguished:

1. **Random mating or panmixia:** any individual of one sex has an equal probability of m. with any individual of the opposite sex.

2. **Genotypic assortative mating:** mating involves pairing of males and females on the basis of relationship. M. of individuals more closely related than the average is **"positive genetic assortative m."** or → inbreeding which increases the chances of mating by individuals with like genotypes. M. of individuals less closely related than the average is **"negative genetic assortative m."** or → outbreeding.

3. **Phenotypic assortative mating:** mating involves pairing of males and females on the basis of their phenotypes and may be either **"positive phenotypic assortative m."**, when like is preferentially mated to like, or **"negative phenotypic assortative m."**, when phenotypically unlike individuals are more likely to mate than would be expected on the basis of chance alone.

mating continuum (*Darlington & Mather* 1949) — a group of interbreeding individuals, systematically exchanging chromosomes and genes among one another. Any member of a m. c. may have received genetic material from individuals located anywhere in the area occupied by the group. A m. c. may be divided spatially into various sub-groups.

mating group — a group of individuals, haploid or diploid, characterized by genetic or environmental conditions which favor → mating within the group at the expense of mating externally (*Darlington & Mather* 1949).

mating system — → mating.

mating type (*Sonneborn* 1937) — the mating property of an organism usually genotypically controlled by an allele or by alleles determining mating ability (→ incompatibility). Individuals of the same m. t. do not usually mate among themselves but with individuals of a complementary type.

matrilinear — ref. to a mode of → inheritance mediated by extrachromosomal hereditary determinants which are transmitted only in the female line.

matrix (*Sharp* 1929) — → ground plasm, chromosome matrix.

matrix bridge — a → chromosome bridge which is the result of so-called "chromatic agglutination", i.e., → sticky effects that may impair the separation of chromatids in anaphase of mitosis or of chromosomes in anaphase of the first meiotic division. Matrix bridges are also called "pseudobridges", as opposed to "true bridges" which arise as a consequence of chromosomal structural changes or of crossing-over in structural hybrids (→ inversion).

matroclinal (*Kerner* 1881) — ref. to modes of → inheritance where the offspring resemble the mother more closely than the father ("matrocliny"), as opposed to patroclinal.

maturation — in virology, the overall process that leads to the incorporation of viral genomes into → capsids and complete → virions.

maturation divisions — = → meiosis.

medical genetics — a field of → human genetics concerned with the relationship between heredity and disease.

megaevolution (*Simpson* 1944) — the → evolution of higher systematic categories, such as families, orders, classes, and phyla (→ macroevolution).

megaheterochromatic (*White* 1943) — of populations with a chromosome complement containing a large number of → heterochromatic segments, as distinct from "microheterochromatic" which refers to populations with a chromosome complement containing a small number of heterochromatic segments.

megameric (*White* 1951) — ref. to → autosomes with large → heterochromatic segments.

megasporangium — a sporangium (i.e., a spore sac, capsule, or cell in which spores develop) producing → mega- or macrospores (= macrosporangium).

megaspore — the larger of the two kinds of haploid spores produced by heterosporous vascular plants (= macrospore). On division, it develops into the embryo sac, the gametophytic generation (→ gametophyte; megasporogenesis) of seed plants, as opposed to a → microspore.

megaspore competition (*Renner* 1921) — in the genus *Oenothera*, competition among the four, meiotically produced and genetically different megaspores of a linear megaspore quartet which determines which spore develops and forms the embryo sac. M. c. is also called "*Renner* effect" (*Darlington* 1932) and rcpresents a case of early postmeiotic selection of the functional megaspore.

megasporocyte — the embryo sac mother cell (→ megasporogenesis).

megasporogenesis — in angiosperms, the formation of → megaspores and the production of the embryo sac, i.e. the female gametophyte (megagametophyte), from one or occasionally several cells (megasporocytes, gonotoconts, or embryo sac mother cells) of the subepidermal cell layer within the ovule of a closed ovary (megasporangium). The megasporocyte is normally diploid.

Generally but with numerous species-specific variations, four haploid cells (megaspores) are produced from the megaporocyte in the course of → meiosis. The megaspores are usually arranged in a linear order (e.g. in maize). Three megaspores ultimately degenerate; the fourth enlarges and forms the embryo sac (the megagametophyte) which initially contains a single haploid nucleus. Before fertilization this nucleus undergoes three mitoses resulting in eight haploid nuclei lying free in the embryo sac cytoplasm. After completion of the first mitosis, one of the two nuclei migrates to the chalazal pole, the other to the micropylar pole (the micropyle being to the opening between the integuments through which the pollen tube will enter). Subsequently, two mitoses result in four nuclei at each pole.

Of these, three are enclosed by cell walls. Of the group of cells at the micropylar end, one cell (in most cases the central one) becomes the egg cell, the other two are the synergids without apparent function.

The three cells at the chalazal end of the embryo sac are called antipodals. In most cases they degenerate, at the latest, after fertilization. The two remaining free nuclei either stay adjacent to each other in the middle of the embryo sac or eventually fuse to produce the diploid "secondary endosperm nucleus". This is the condition of the embryo sac at the time of fertilization (→ microsporogenesis).

meiocyte — any cell in which the nucleus divides by → meiosis. In most animals, the primary oöcytes and primary spermatocytes (→ oögenesis; spermatogenesis) are the meiocytes; in most higher plants, the meiocytes are sporocytes giving rise to the embryo sac (female sex) and to pollen grains (male sex) (→ megasporogenesis; microsporogenesis; germ cell).

meiosis (*Farmer* & *Moor* 1905) — in → meiocytes, two successive divisions of the → nucleus preceding the formation of gametes (→ gametogenesis) or meiospores (→ sporogenesis) in eukaryotes. During m., homologous chromosomes pair (→ chromosome pairing), replicate once only (as compared to two replications in the case of two succesisve → mitoses), and undergo assortment so that each of the four meiotic products resulting from one m. receives one representative of each chromosome set. Thus, the zygotic (diploid) chromosome number is meiotically reduced to the gametic (haploid) number characteristic for the gametes or haploid phase of the life cycle. The point of insertion of m. into the life cycle of → haplontic (zygotic meiosis), → diplontic (gametic meiosis) or → diplohaplontic (intermediary meiosis) species differs. The two nuclear divisions of a normal meiosis are called first (meiosis I) and second meiotic division (meiosis II). A special type of m. in some protozoa is referred to as "onestep meiosis" (*Cleveland* 1947).

M. is a genetically controlled process, definitely ordered in both space and time, with a number of variations. These variations, all under some form of genetic control, concern the amount of chromosome pairing, the production, frequency and localization of → chiasmata, the orientation and distribution (→ chromosome orientation) of the pairing partners in the → spindle, the → chromosome movements and the modes of → cytokinesis following the nuclear divisions. Such variations may be either species-specific properties which do not impair the basic functions of m. in any way, or may represent more or less pronounced abnormalities, limiting or preventing the cytological and genetical functions of meiosis (*John & Lewis* 1965).

A normal m., the basis of sexuality of all higher plants and animals, usually serves the following main functions (*Lewis & John* 1964):

1. It provides for the random → assortment of paired chromosomes (→ meiotic drive), the halving (reduction) of the somatic chromosome number and marks the transition from the diplophase to the haplophase of the life cycle (→ alternation of nuclear phases). In the absence of meiosis, fertilization as an integral part of the sexual cycle would lead to a geometric increase in chromosome number. With the intervention of m. the species-specific chromosome number established at zygote formation is maintained.

2. It determines the accurate → segregation of allelic differences (→ allele) as well as the random recombination of unlinked genes (interchromosomal → genetic recombination) and the nonrandom recombination of linked genes (intrachromosomal genetic recombination).

The stages into which a normal m. is typically divided are as follows (Fig. 62):

1. **First meiotic division:**

a) **Leptotene:** The chromosomes appear as long, widely despiralized (→ chromosome coiling) and optically single threads (no differentiation into → chromatids yet recognizable).

b) **Zygotene:** Homologous chromosomes begin to pair (→ chromosome pairing) in a highly specific manner and become twisted around one another ("relational coiling"). Pairing is initiated at one or more so-called "contact points" (→ zygomere) and proceeds from these along the chromosomes.

c) **Pachytene:** The homologues are completely paired, i.e., are in intimate contact along their entire length. In the case of diploid organisms, the number of pairing configurations (→ "bivalents") is equal to half their somatic chromosome number. In → autopolyploids, the pairing configurations may consist of more than two chromosomes (→ "multivalents"). At any one point within these configurations, pairing affinity is evidently saturated by the intimate apposition of two pairing segments. In comparison to leptotene and zygotene, the chromosomes in pachytene are shorter as a result of internal coiling. During pachytene it may be apparent in favorable material that each of the paired chromosomes consists of two chromatids and that there are reciprocal exchanges of segments between homologous nonsister chromatids. Such exchanges are recognizable as cross configurations called chiasmata. Each chiasma is understood to represent the physical result of → crossing-over which, in turn, is the process giving rise to intrachromosomal genetic recombination.

d) **Diplotene:** Chromosome contraction by coiling continues. The homologous chromosomes repel one another and tend to move apart giving a picture of an opening out between homologues. Chiasmata hold the homologues together until their separation in anaphase. At this stage chiasmata are usually clearly observable and scoreable.

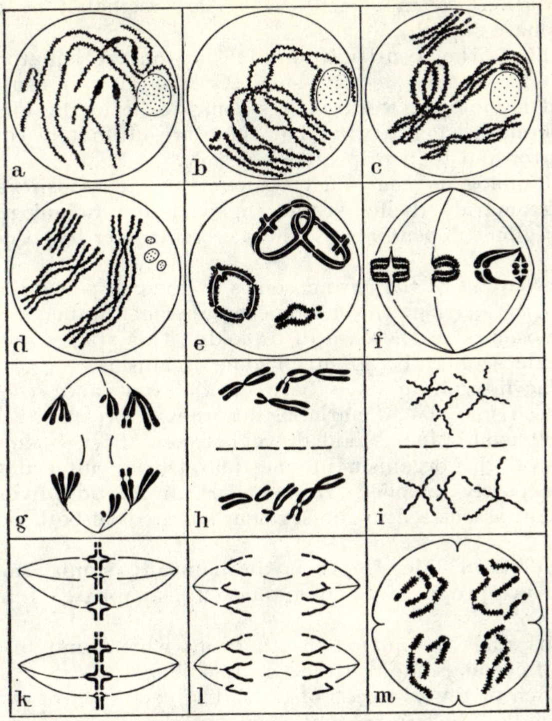

Fig. 62. Diagrammatic representation of the first and second meiotic division in a diploid cell. a) leptotene, b) zygotene, c) pachytene, d) diplotene, e) diakinesis, f) metaphase I, g) anaphase I, h) telophase I, i) interkinesis, k) metaphase II, l) anaphase II, m) telophase II.

e) **Diakinesis:** Chromosome contraction is near maximum and the pairing configurations are well spread throughout the cell with a tendency to lie at the periphery of the nucleus near the nuclear envelope. Successive loops produced by repulsion between pairing partners on both sides of the chiasmata come to lie in planes which are at right angles to each other. If there is only one chiasma in a bivalent this appears cross-like in one plane. Chromosomes at this stage show two types of internal coiling which are referred to as major and minor coils or spirals. During diakinesis the nucleolus normally disappears and the nuclear envelope is disrupted.

f) **Prometaphase I:** Simultaneous with the disruption of the nuclear envelope the → spindle is organized. The pairing configurations become attached to the spindle and their → centromeres begin to coorientate

(→ centromere orientation) by movements which are mediated by the spindle (→ pro-metaphase stretch).

g) **Metaphase I:** The pairing configurations line up on the equatorial plate by movements referred to as "congression". Coorientation of centromeres is finished and homologous centromeres come to lie in the longitudinal axis of the spindle in such a way that these are ultimately equidistant from the equator and their respective poles.

h) **Anaphase I:** Homologous centromeres move toward opposite cell poles pulling two chromatids (being sisters in some and homologous chromatids in other regions, depending on where crossing-over has occurred) behind them.

i) **Telophase I:** Regrouping of the chromosomes at the cell poles. Each pole receives one-half the original chromosome number of the meiocyte, i.e., one set of chromosomes in the case of diploidy. This stage as well as the following interkinesis may be absent in some organisms.

k) **Interkinesis:** The interphase stage between the first and second meiotic divisions, which may be of variable duration if present at all. Whether or not a cell membrane is laid down between the two nuclei at this stage depends on the organism in question. During interkinesis the chromosomes are partly uncoiled. Homologous chromatids diverge considerably but remain associated in the segment situated on both sides of the centromere.

2. **Second meiotic division:** Meiosis II is mechanically similar to a mitosis and proceeds via prophase II, metaphase II, anaphase II and telophase II.

a) **Prophase II:** This stage is omitted in organisms which omit interkinesis. It is a stage of chromosome contraction by coiling.

b) **Metaphase II:** Each of the two cells of a dyad or each area of cytoplasm organizes a spindle which may (but need not) have a consistent pattern of arrangement with respect to the other. The centromeres line up on the equator of the second division spindles.

c) **Anaphase II:** Sister centromeres, together with the component chromatids of each chromosome, separate to the poles.

d) **Telophase II:** Interphase nuclei are reconstituted and cell membranes (or cell walls) are laid down between the four nuclei (each with half the somatic chromosome number of the organism) giving rise to four cells, whose ultimate fate differs in the two sexes of the organism. In males, all four usually develop into → gametes or spores (→ spermatogenesis microsporogenesis). In female animals, three of the four usually abort as "polar nuclei" and only one develops into a female gamete, whereas in flowering plants two or all four take part in the production of the → embryo sac (→ oögenesis; megasporogenesis).

A special variant of this "chiasmate" course of m. is the so-called **"achiasmate meiosis"** which proceeds regularly without crossing-over and chiasma formation. This type of meiosis is generally restricted to only one of the two sexes (e.g., the male of *Drosophila*). Presumably, small heterochromatic regions (→ collochore) fulfil the role of chiasmata by causing chromosome association and mediating the orderly assortment of homologous chromosomes during anaphase I. In cases of achiasmate meiosis the unit of recombination is the whole chromosome. Such meioses provide a means of reducing the extent of genetic recombination and hence variation in the species.

A normal chiasmate m. in organisms with localized centromeres is always "reductional" in its first division as far as the halving of chromosome number is concerned. But it may be "reductional" or "equational" in respect to the disjunction of the alleles at particular heterozygous loci. With random assortment of chromosomes, whether or not for two loci the first division of meiosis is "reductional" or "equational" (prereduction vs. postreduction) depends upon the location of crossing-over (Fig. 63). Prereductional separation for a locus represented by the alleles A and a

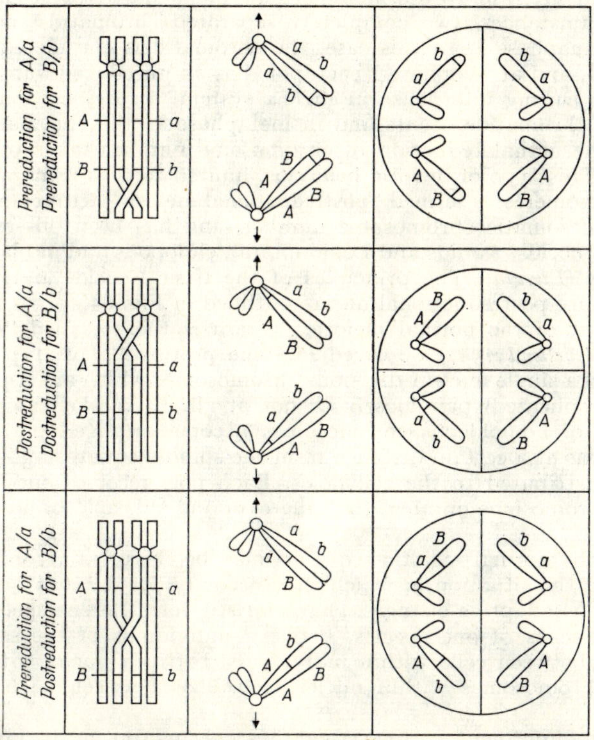

Fig. 63. Nonrandom meiotic assortment of heterozygous loci borne on the same pair of homologous chromosomes. Top: prereduction for loci A/a and B/b; middle: postreduction for loci A/a and B/b; bottom: prereduction for locus A/a, postreduction for locus B/b. First vertical row: points of crossing-over and chiasma formation; second vertical row: anaphase-I-separation of the homologues; third vertical row; the genetic constitution of the four meiotic products (recombination occurs in the case of crossing-over between the two loci).

means that the two chromatids with allele A move to one, the two with the allele a to the opposite pole. However in the case of postreductional separation or disjunction for the same locus, the two chromatids which move together to each of the two poles have one A and one a allele, i.e., are identical in respect to their genetic constitution for that particular locus.

In the case of prereduction, anaphase II will be equational for the locus A, in postreduction it will be reductional for that locus. If there is no crossing-over between the centromere and the locus A/a, anaphase I is always reductional, anaphase II always equational.

The situation is different in organisms with nonlocalized centromeres. In this case, the chromatids are active in the spindle throughout their entire length and are wholly autonomous and independent structures. Under such circumstances, two completely separated chromatids move to each pole at anaphase I. In this case, the chromatid is considered to be a chromosome, and at telophase I each pole has an unreduced number of chromosomes. During interphase in such a system, a reassociation of two homologous chromatids occurs and in metaphase II the chromatids are "paired". The actual reduction of chromosome number takes place at anaphase II. This type of meiotic behavior, limited to organisms with nonlocalized centromeres, is termed **"postreductional meiosis"** with respect to the halving of somatic chromosome number, and has been observed in some insects (Coccids, Aphids and Lepidoptera), chilopods and in plants in the *Juncaceae* (*Luzula*). The principles of the basic Coccid- and the Luzula-type of pure postreductional m. are outlined in Fig. 64.

Another variant of the normal meiotic pattern is the so-called **"one step meiosis"** (*Cleveland* 1947) observed in some protozoa. This type is characterized by a single nuclear division. Chromosomes which are apparently single (unreplicated) pair loosely or not at all. In the latter case, the centromeres of homologous chromosomes become arranged parallel to the division axis at a certain distance from the spindle equator and are then regularly distributed to the cell poles. Each pole receives one-half of the somatic chromosome number. Crossing-over and chiasma formation are absent.

Variations of the normal meiotic sequence may be classified according to their kind and the situation in which they occur (*John & Lewis* 1965). They may occur as adaptive changes, characteristic for different species, between the two sexes of one species, between individuals of the same sex and species, between cells within individuals, between nuclei within cells, between chromosomes within nuclei as well as between parts of chromosomes.

A major period of chromosomal DNA synthesis occurs during the interphase preceding chiasmate m.; unlike the premitotic interphase, premeiotic interphase does not include complete chromosomal DNA replication. About 0.3 to 2% of the total chromosomal DNA fails to replicate by the termination of the premeiotic S phase. Its replication is delayed to the leptotene, zygotene and pachytene. Delayed semiconservative replication of DNA occurs when → chromosome pairing takes place; delayed, nonsemiconser-

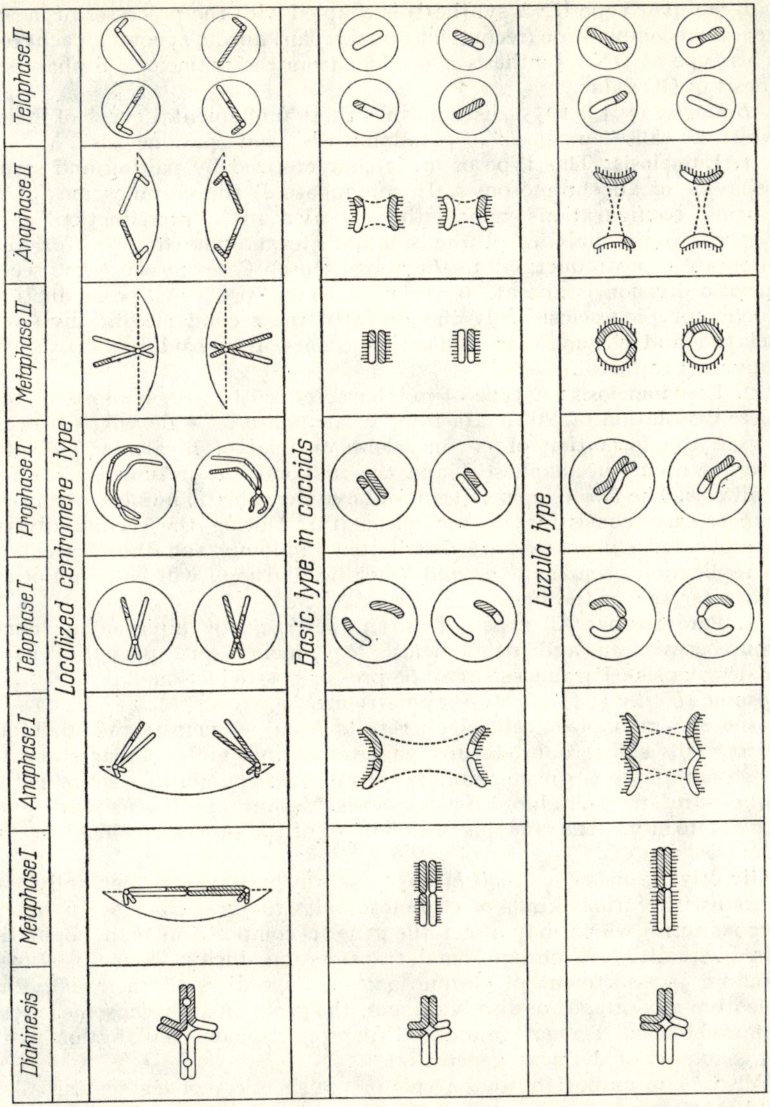

Fig. 64. "Postreductional meiosis" in coccids and in *Luzula* (organisms whose chromosomes have nonlocalized centromeres) as compared with the behavior of a bivalent in organisms with localized centromeres (top) (after *Tristao de Mello-Sampayo* from *Rhoades* 1961).

vative repair type DNA synthesis is coupled with the processes of meiotic genetic recombination (recombinational repair) during zygotene/pachytene. This type of DNA synthesis is distinct from the premeiotic S phase synthesis in GC value.

Battaglia (1945, 1955) distinguishes three fundamental types of meiosis which he calls "eumeiosis", "pseudomeiosis" and "parameiosis".

1. **Eumeiosis:** This type of m. is characterized by pairing and similar behavior of all chromosomes. In metaphase I, the chromosomes of the pairing configurations may either coorientate ("prereduction" with respect to the halving of the somatic chromosome number) or autoorientate ("postreduction" in the above sense). Cases in which the second meiotic division is absent, are referred to as "pre"- or "postreductional apohomotypic meiosis". If the onset of the second meiotic division is delayed and chromosome replication occurs, it is said to be a "diplounivalent mitosis".

2. **Pseudomeiosis:** A type of m. characterized by → asynapsis or complete dissolution of chiasmata prior to anaphase I (→ desynapsis) resulting in the formation of → univalents instead of bivalents. Univalent division occurs in anaphase I and the first meiotic division is, in effect, a mitosis. The two telophase I nuclei have the somatic number of chromosomes, each consisting of one chromatid. During the second division unreplicated chromatids are distributed randomly and thus irregularly. If replication occurs, the second "meiotic division" will be as mitotic in character as was the first.

3. **Parameiosis:** All types of m. characterized by intermediate conditions between eu- and pseudomeiosis (e.g., meiosis with partial asynapsis or desynapsis). Parameiosis may be pre- or postreductional.

meiosome (*Höfler* 1957) — = → spherosome.

meiospore — a haploid cell which results from → meiosis and undergoes mitotic divisions to produce a gamete-bearing, multicellular state, the → gametophyte (= meio-agamete; tetraspore; gonium). (→ germ cell.)

meiotic — ref. to or produced by → meiosis. M. mutants (*Sandler* et al. 1968) are due to mutations affecting one or more of the processes which together comprise meiosis.

meiotic drive (*Sandler & Novitski* 1957) — any mechanism (generally associated with various kinds of chromosomal structural changes in heterozygous form) which may affect the genetic composition of a population via inequalities in chromosomal transmission during → meiosis (nonrandom → assortment of chromosomes). Regardless of their individual selective advantages or disadvantages, the genes on a chromosome favored by preferential segregation make a disproportionate contribution to the → gene pool of the next generation.

M. d. is an exception to the rule of 1:1 recovery of segregating alleles in heterozygous organisms and presupposes two separable phenomena:

1. A mechanism for the directed orientation of a meiotic bivalent on the metaphase I spindle, and

2. the loss of a portion of the meiotic products.

Phenomena of this kind are called m. d. since, apparently, no specific genetic factors are involved in the nonrandom assortment of chromosomes or the unequal frequencies of recovery of the two kinds of gametes produced by heterozygotes. M. d. can result in increased frequencies of alleles despite their strong selective disadvantage, or in decreased frequencies of alleles that would otherwise be selectively favored. The mechanisms leading to m. d. are themselves subject to selection.

Among the best known cases of m. d. is that involving the segregation-distorter (SD) locus in *Drosophila melanogaster* and another system in the same organism called recovery disrupter (RD). Males heterozygous for the SD locus may transmit the SD-carrying chromosome to more than 95% of their offspring (SD is without any effect in the female). RD distorts the sex ratio in favor of females; males from RD-lines may produce offspring consisting of up to 65% females, due to a reduction in the recovery of Y chromosomes (RD is without effect in the female) (→ affinity; automatic frequency response).

meiotic effect (*Magni* 1964) — in fungi, the phenomenon that the frequency of spontaneous mutants recovered is much higher after → meiosis than after mitosis. It is assumed that the m.e. is caused by some inaccuracy of → genetic recombination.

meiotic inversion (*Brown & Cleveland* 1968) — in organisms with → holocentric (or holokinetic) chromosomes, an equational division of the chromatids at the first meiotic division and reductional separation of two chromatids (one from each homologue) at the second division. Without m. i. pairs of chromatids separate from each other reductionally in the first meiotic division and the two members of each pair separate equationally at the second meiotic division. Both normal and inverted meiosis may occur in species with holocentric chromosomes. Among the Hemiptera, m. i. is restricted to the advanced Homoptera (all coccids and aphids so far analyzed).

meiotic synapsis — → chromosome pairing.

meiotrophic (*Engelsberg & Ingraham* 1957) — of mutants which arise from naturally deficient strains of bacteria and can dispense with one of the requirements of these strains, i.e. require fewer growth factors than these. The term permits a distinction between auxotrophs (requiring additional growth factors) and meiotrophs (requiring fewer growth factors).

melanosome (*Fitzpatrick & Quevedo* 1964) — any of the melanogenic granules of pigment producing cells from their earliest recognizable (unpigmented) stage to their completely or partially electron-dense definitive pattern of melanization. M. are specialyzed organelles of differentiated cells.

melting temperature — the temperature at which 50% of the total hyperchromicity of double-stranded DNA is achieved. Tm is the symbol used to denote the midpoint of the 260 nm transition profile of DNA melted out by gradual increase in temperature (→ denaturation).

membrane — → unit membrane.

membrane carrier — any of those proteins which function to facilitate the movement of small molecules across the cell membrane. Without such m. carriers many extracellular substances are inaccessible to intracellular enzymes, and as such they remain functionally cryptic.

membrane system — → endomembrane system.
Mendelian character — a differentiating → character, which is formed under the control of chromosomal genes according to the rules of "Mendelian" → inheritance.
Mendelian inheritance (*Castle* 1906) — the mode of → inheritance of chromosomal genes, in contrast to "non-Mendelian" or "extrachromosomal" inheritance which is controlled by cytoplasmic hereditary determinants.
Mendelian mutation (*Stubbe* 1940) — a → mutation fulfilling "monohybrid" Mendelian segregation. This group of mutations comprises true → gene mutations as well as chromosomal structural changes (→ chromosome mutation) which may be either microscopically visible or submicroscopic.
Mendelian population (*Dobzhansky* 1935) — an interbreeding group of individuals sharing a common → gene pool (→ population). The m. p. is the basic unit of study in population genetics. Usually, there is no clear-cut boundary between two adjacent Mendelian populations owing to → gene flow between them. If gene flow is limited, such populations may show considerable differences with respect of their gene pools. → Gene frequency is the most important single index characterizing a m. p.
Mendelian ratio — a → segregation ratio which is in accordance with the expectation on the basis of Mendel's laws of → inheritance.
Mendelism (*Punnet* 1905) — particulate → inheritance of chromosomal genes according to the chromosome theory of heredity.
mendelize — of genes undergoing Mendelian → segregation (→ inheritance).
Mendel's laws of inheritance — → inheritance.
mericlinal (*Jørgensen & Crane* 1927) — of a → chimera with inner tissue of a particular genetic constitution only partly surrounded by outer tissue of a different genetic constitution.
meristem — an area of rapidly dividing plant cells. The m. may be a single cell (as in ferns), or it may include many cells.
meristic — of a variation in characters which can be counted, i.e., is quantitative in nature.
merodiploid — → merozygote.
merogamy (*Meisenheimer* 1921) — a mode of fertilization in Protozoa brought about by the fusion of so-called "merogametes", i.e., gametes produced by fission or multiple division of vegetative cells and differing from these in size and structure, in contrast to → "hologamy" or syngamy between whole individuals. M. may take the form of iso-, aniso-, or oögamy.
merogenote (*Clark & Adelberg* 1962) — a fragment of a bacterial → chromosome, irrespective of its length or of the mechanism by which it has arisen in the bacterial donor cell.
merogony (*Delage* 1899) — experimentally induced development of an egg fragment containing the sperm nucleus but lacking the egg nucleus.
meromixis (*Wollman, Jacob & Hayes* 1956) — a process in bacteria during which only a fraction of the → genome of a donor cell is transferred to a recipient cell which contributes both its cytoplasm and its entire genome. The result of m. is a → merozygote.
merospermy — fusion of an egg cell with a sperm which has lost its nucleus and does not take part in → karyogamy with the egg nucleus (= → gyno-

genesis). M. is a normal event in some nematodes and may be induced experimentally.

merostathmokinesis (*Gavaudan* 1943) — a → mitosis (= partial → C-mitosis) which gives rise to a → multipolar anaphase due to a partial inactivation of the → spindle. M. results from the action of C mitotic agents applied in doses too low for complete spindle inhibition.

merozygote (*Wollman, Jacob & Hayes* 1956) — in bacteria, an incomplete → zygote which is diploid only for part of its genetic material and haploid for the remainder. The process of m. formation is called → meromixis. A m. contains the whole → genome of the recipient cell and a → merogenote derived from a donor cell. Merozygotes are usually very unstable and may be → hetero- or homogenotic with respect to their genetic constitution. The potential for → genetic recombination in a m. is limited to the diploid part of its genome (→genetic transformation; tranduction; bacterial → conjugation).

mesokaryotic (*Dodge* 1964) — of the nuclear status of dinoflagellates characterized by the absence of deoxyribonucleohistone (which is otherwise characteristic of the → prokaryotic bacteria and Cyanophyceae) and a peculiar type of → dinomitosis.

mesoplasm (*Plowe* 1931) — that part of the → cytoplasm which is surrounded by the cell membrane on the one hand, and by the → tonoplast of vacuoles on the other.

mesosome (*Fitz-James* 1960) — any of the intracytoplasmic membraneous structures of bacteria (= plasmalemmasome), which appear to originate from the cell membrane by an invagination (infolding) and "pinching-off" process. In sections, mesosomes appear as clusters of vesicles and tubules, as membraneous whorls, or as a combination of these types of ultrastructure. The list of proposed m. functions includes their participation in energy production, DNA replication and → nucleoid separation (Fig. 65),

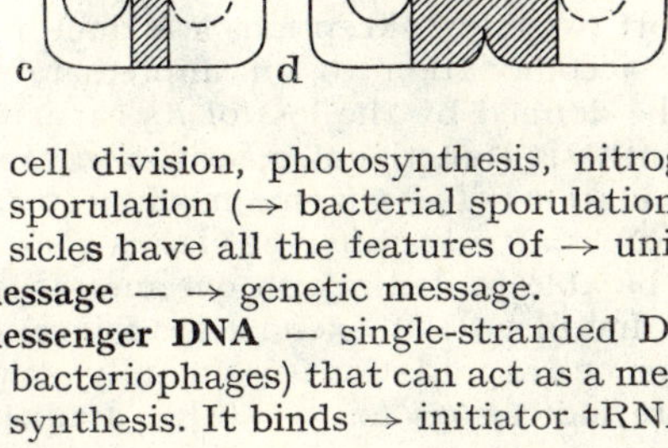

Fig. 65. Replication and segregation of a (circular) bacterial chromosome attached to a mesosome. a) Attachment of the chromosome at a specific site on the membrane; b) replication of the chromosome completed; c) separation of the chromosome sites of attachment owing to membrane growth; d) end of membrane growth at the time the cell is ready to divide (modified after *Jacob* 1966).

cell division, photosynthesis, nitrogen fixation, nitrification, and endosporulation (→ bacterial sporulation). The mesosomes or their derived vesicles have all the features of → unit membrane profiles.

message — → genetic message.

messenger DNA — single-stranded DNA (e.g., DNA isolated from certain bacteriophages) that can act as a messenger in an *in vitro* system of protein synthesis. It binds → initiator tRNA (fMet-tRNA) to → ribosomes in the presence of → initiation factors and GTP. The ability to translate (→ genetic translation) a DNA strand depends critically on the Mg^{2+} concentration.

messenger-like RNA (*Scherrer & Marcaud* 1968) — = → pre-messenger RNA.

messenger recognition factor (*Heywood* 1970) — any specific factor involved in binding → messenger RNA to ribosomes during the formation of the → initiation complex. M. r. factors represent binding factors in both pro- and eukaryotes; they may exert a positive control on protein synthesis and may play an important role in → cytodifferentiation and in the stability of the differentiated state in eukaryotes. Such control could limit the spectrum of proteins a cell at any given time could potentially synthesize on the basis of messenger RNA species contained in the cell.

messenger RNA (*Brenner, Jacob & Meselson* 1961; *Jacob & Monod* 1961) — a population of large → ribonucleic acid molecules of heterogeneous size (= informational RNA; template RNA; unstable RNA, or complementary RNA) synthesized during the → genetic transcription of → deoxyribonucleic acid and serving as a template for protein synthesis (stimulating amino acid incorporation reaction in a cell free polypeptide synthesizing system). The mRNA from a given organism may vary in size over a range of ca. 8 S to 45 S. mRNA is produced by DNA-dependent RNA synthesis (one strand of the DNA duplex is transcribed into a strand of mRNA by the action of the DNA-dependent RNA polymerase, the substrates being 5′-nucleoside triphosphates. The polarity of mRNA synthesis is 5′ → 3′). It has a nucleotide sequence complementary to that of the transcribed DNA strand (specific hybridization with an appreciable fraction of homologous DNA: → DNA—RNA hybrid), and the information content contained in its particular sequence of nucleotides (→ genetic code; genetic information) is translated (→ genetic translation) sequentially to produce specific polypeptide chains.

The chain growth-rate of mRNA in *E. coli* is 55 nucleotides per second (at 37 °C). Both mRNA and → ribosomal RNA grow at the same rate. In bacteria, contiguous, and metabolically related genes form one regulatory unit (an → operon) which, following transcription, gives rise to a polycistronic mRNA coding for several proteins. Shortly after the beginning of bacterial mRNA synthesis, ribosomes attach to it and move along the RNA very closely behind the RNA polymerase. Bacterial mRNAs are not translated uniformly by ribosomes. Discrimination between cistrons in translation may be due to the action of → initiation factors. Generally, the lifetime of bacterial mRNA is short (with few exceptions less than five minutes) thereby permitting rapid accomodation to an unpredictable environment; decay of mRNA can be defined by the loss of its capacity to serve as template for peptide synthesis (functional inactivation) and by the loss of mass (chemical degradation). The enzyme machinery responsible for degrading bacterial mRNAs attacks different classes of messengers at different rates and must be able to distinguish one messenger from another. Overall degradation is due to a 5′—3′ exonuclease following immediately behind the last ribosome, i.e., degradation is intimately coupled to genetic translation. There are indications for a half-life gradient along the operon with the operator-proximal region being degraded most rapidly.

Most **eukaryotic mRNAs** are transcribed from DNA sequences which occur only once per genome, i.e., most of the structural genes belong to the class of nonrepetitious (unique) DNA sequences. The products of DNA transcription in eukaryotes consist not of messenger-sized but giant → pre-messenger RNA molecules (→ heterogeneous nuclear RNA). Pre-messengers (5×10^5 daltons to more than 10^7 daltons) as well as mRNA (about 10^5 to 2×10^6 daltons) produced by processing the pre-messengers (elimination of nucleotides at the 5′-P(PP) end) contain, in many cases at least, tracts of polyadenylic acid (50 to 200 nucleotides) added post-transcriptionally at the 3′-OH end of the molecules. The function of poly(A) tracts in mRNA may be to program its lifetime in the cytoplasm (the tracts in cytoplasmic messengers shorten as a result of metabolic ageing). It has been shown recently that non-histone mRNAs containing and lacking poly(A) may exist. These are transcribed from different genes. Their existence raises the possibility that 1. the mode of transcription of these genes may differ substantially, 2. the processing of nuclear RNA precursors to these two classes of mRNA may proceed by different mechanisms, and 3. the regulation of genetic information may be different for the two classes (*Nemer* et al. 1974). The lifetime of eukaryotic mRNA (which is not polycistronic: → genetic transcription) is much longer than that of bacterial mRNA (typical lifetime a few hours). Highly differentiated cell types, persistently committed to specific synthetic pathways, seem to have unusually long-lived messengers. Messengers of different stability can coexist in the same cytoplasm. Every eukaryotic mRNA molecule seems to carry a repetitive stretch at the 5′ end (linked to the rest of the molecule which is transcribed from unique DNA sequences).

In eukaryotes, newly synthesized mRNA molecules are transferred to the cytoplasm (after processing of pre-mRNA) not as naked strands of RNA but in association with protein, i.e., as nucleoprotein particles (→ informofere; informosome) which may exist for some time in the cytoplasm as free particles (not associated with → polyribosomes or endoplasmic reticulum) until the mRNA is joined with ribosomes to form polyribosomes.

Messenger RNAs for certain proteins needed during early embryonic development of eukaryotes (animals) are laid down in the egg cytoplasm during oogenesis, and are not translated until after fertilization. These mRNAs are called "maternal" or "masked" RNAs and may represent an essential feature of the → determination process.

Degradation of eukaryotic mRNA is far more complicated than degradation of prokaryotic mRNA and very poorly understood. Differential degradation of individual mRNAs is probably an important factor in the regulation of gene expression (→ genetic regulation) and may occur at a rate determined by the rate of genetic translation.

Eukaryotic mRNAs contain about 2.2 methyl groups per 1000 nucleotides, i.e., about 6 or 7 per average cellular messenger of 3000 nucleotides. Methylation is probably a post-transcriptional modification of mRNA precursors; its functional significance is unknown but may be related to the processing of heterogeneous nuclear RNA.

The RNA of RNA viruses has a dual function: it represents the genome of the viruses (formally analogous to DNA) and may, in part, serve itself as messenger (→ antimessenger DNA).

messenger RNA binding factor — = → messenger recognition factor.

messenger RNP particle — in eukaryotes, any of the complexes of cytoplasmic → messenger RNA or its nuclear precursors with particular proteins (→ informofer; informosome). The function of this association might be to process → pre-messenger RNA, to transport mRNA from the nucleus to the cytoplasm or to control its translation.

metabolic DNA (*Pelc* 1968; *Anker* et al. 1971) — non-mitochondrial DNA in the cytoplasm of differentiated animal cells with a metabolic turnover related to cell ageing (= cytoplasmic DNA).

metabolic cooperation (*Subak-Sharpe* et al. 1969) — the phenomenon in which the phenotype of mutant cells in tissue culture is corrected by intimate contact with normal cells. Prerequisites for the study of m. c. are → genetic markers capable of being detected at the cellular level in tissue culture (→ crossfeeding).

→ Gap junctions have been implicated as the structures necessary for m. c.

metabolic nucleus — = → interphase nucleus.

metabolite — a product of metabolism.

metacentric — of → chromosomes whose centromere is localized in a roughly median position (lengthwise). This gives them a V or J shaped appearance at metaphase (→ idiogram).

metafemale (*Stern* 1959) — an individual which has according to the balance theory of → sex-determination a greater dose of female determiners than normal females (e.g., *Drosophila* individuals with 3X-chromosomes and two sets of autosomes (3X 2A)). The term "superfemale", used synonymously with m., should be replaced by m. since the expression of femaleness reaches its upper limit in normal 2X 2A individuals. Metafemale *Drosophilae* show hypoplastic ovaries and both malformed wings and eyes; the majority die as larvae or pupae. Human beings with two sets of autosomes and three X-chromosomes are underdeveloped with respect to their primary and secondary sex characters.

metagamic — 1. → sex determination;

2. of nuclear divisions immediately following → fertilization.

metagenesis (*Haeckel* 1866) — = → alternation of generations.

metagon (*Gibson & Beale* 1961) — a stable messenger RNA with viruslike properties postulated to exist in mate-killing *Paramecium aurelia* (→ killer) and responsible for the maintenance of Mu particles in a stock that contains at least one of two macronuclear genes denoted M 1 and M 2. The m. is able to replicate in another organism (*Didinium*).

metagynic (*Loew* 1895) — → monoecious.

metahermaphroditic (*Bacci* 1965) — of a population which is composed of half hermaphroditic and half gonochoric individuals (→ hermaphroditic).

metakinesis (*Flemming* 1879) — 1. The separation of the two chromatids of each chromosome and their movement to opposite spindle poles during anaphase of → mitosis (*Flemming* 1879);

2. = → chromosome congression to the spindle equator (*Wassermann* 1926).

metandric (*Loew* 1895) — → monoecious.

metaphase (*Strasburger* 1884) — → mitosis, meiosis.

metaphase arrest — the arresting of cell division at metaphase of → mitosis or meiosis by the application of "antimitotic" agents (→ mitotic poisons; mitotic inhibition). Such a block is usually associated with structural abnormalities or the apparent dissolution of th cell's → mitotic apparatus (→ C-meiosis; C-mitosis). In some organisms, the egg nucleus is normally blocked at first or second metaphase (or anaphase) of meiosis until fertilization occurs.

metaphase inhibitor (*Dustin* 1947) — → mitotic poison.

metaphase pairing index (*Tobgy* 1943) — that fraction of the sum total of all cells observed in which two particular chromosomes are found in the paired condition (→ chromosome pairing) at metaphase I of → meiosis.

metaphase plate — the equilibrium position in the equatorial region of the → spindle (= equatorial plate or nuclear plate) reached by the chromosomes via their congressional movement (→ chromosome congression; centromere orientation) at metaphase of → mitosis and meiosis.

metaphase spindle — → spindle.

metaplasm (*v. Hanstein* 1868) — a collective term for the ergastic materials of the → cell (→ paraplasm).

metareduplication (*Hsu* & *Moorhead* 1956) — → interreduplication.

metaxenia — → xenia.

metric — of → characters that are measurable on a continuous scale (→ meristic). Most m. characters are polygenic, i.e., controlled by many genes.

metroclinal — = → matroclinal.

M factor (*Davison* et al. 1969) — a protein factor (in *E. coli*) which stimulates *in vitro* transcription by purified → RNA polymerase (20 to 30 fold) from several templates. The M. f. is distinct from the → rho factor and the → sigma factor but has several properties in common with sigma. Both are thermolabile proteins which sediment at approximately 4 to 5 S and both stimulate RNA synthesis from a variety of templates. M might be a continuation factor preventing abortive termination of initiated RNA chains or might confer additional specificity on sigma-directed intiation (→ psi factor). Stimulation is probably due to interaction of the M. f. with polymerase.

microbody (*Rhodin* 1954) — = → peroxisome.

microevolution (*Goldschmidt* 1940) — infraspecific → evolution or speciation, i.e., all processes of species formation and species differentiation brought about by the combined action of the various evolutionary factors.

microfibril — → cell wall.

microfilament — any of a class of filamentous structures which occur (in addition to → microtubules) in all cells capable of some contractility and in most others as well. They appear to be involved in protoplasmic streaming, cytokinesis, formation of a ruffled membrane, nerve outgrowth and various morphogenetic movements. Microfilaments are of two main types: 1. Filaments with a diameter of approximately 10 nm and generally associated with → desmosomes or distributed in bundles throughout the cell (neurofilaments; tonofilaments). 2. Filaments with a diameter of approximately 5 to 7 nm.

microheterochromatic (*White* 1943) — → megaheterochromatic.

micronucleolus — a small → nucleolus which may arise by nucleolar budding in the course of nucleolar degeneration.

micronucleus — 1. the generative nucleus of some foraminifera and most ciliata (→ nuclear dimorphism);

2. a nucleus, separate from and additional to the main nucleus of a cell, produced during telophase of → mitosis or → meiosis by → lagging chromosomes or chromosome fragments derived from spontaneous or experimentally induced chromosomal structural changes.

micronucleus test (*Boller & Schmid* 1970) — a test proposed for use in → mutagenicity testing and based upon the formation of micronuclei due to chromosome structural changes or spindle dysfunction (e.g., in bone marrow erythroblasts of animals). The m. t. allows rapid screening of potential → clastogenic agents.

microphylogenesis (*Zimmermann* 1943) — = → microevolution.

micropinocytosis (*Yamada* 1955) — → pinocytosis.

micropyrenic — of nuclei smaller than the average nuclei of a particular cell type of the individual or species in question.

microsome fraction (*Hanstein* 1880; *Palade & Siekevitz* 1955) — a subcellular fraction composed of small particles (size 50 to 150 nm) isolated by differential centrifugation and composed of three characteristic structural features: rough vesicles, smooth vesicles, and ribosomes. The microsomal fraction may also contain elements of the → Golgi complex, fragments of the endomembrane system of the cell, mitochondrial components, endocytic vacuoles, and other subcellular entities in addition to vesicles from the → endoplasmic reticulum. Numerous enzymatic activities are associated with the microsomal fraction.

microspecies — a small, morphologically uniform population reproducing mainly if not exclusively by uniparental methods and occupying a definite geographical or microgeographical area. M. are differentiated morphologically from related species and frequently possess a hybrid constitution (*Grant* 1971). Different kinds of m. can be recognized according to the mode of uniparental reproduction (→ jordanon).

microspike (*Taylor* 1966) — any of the microscopic projections of cytoplasm protruding from the cell surface.

microsporangium — a sporangium producing → microspores (a pollen sac or anther lobe of phanerogams) (→ microsporogenesis).

microspore — the smaller of the two kinds of → meiospores produced by heterosporous plants in the course of → microsporogenesis. In seed plants, the m. give rise to the pollen grain, the male → gametophyte.

microsporocyte — the pollen mother cell (→ microsporogenesis).

microsporogenesis — in angiosperms, the formation of → microspores in the microsporangia (anthers) and the production of the male gametophyte (microgametophyte or pollen), normally from the diploid microsporocytes (pollen mother cells). In the archespore of an anther, a large number of microsporocytes are produced. These undergo meiosis which results in the formation of four haploid microspores per microsporocyte. In most cases, the microspores form a tetrad and after they separate from one another, they form pollen. After some time a further division (pollen grain mitosis) takes place in the pollen grain. This results in the production of a smaller "generative" cell (or nucleus) and a larger "vegetative" cell (or nucleus). The generative nucleus is generally very dense, somewhat elongated, and stains darkly, while the vegetative or "tube" nucleus is rounded and takes a lighter stain. The generative cell divides producing two male → gametes, in some cases before the anther has opened, in others at a later stage during the growth of the pollen tube.

microsubspecies (*Huxley* 1940) — → subspecies.

microtubule (*Slautterback* 1963) — any of a class of subcellular filamentous components found in a wide variety of eukaryotic cells. Microtubules are straight cylinders (24 ± 2 nm in diameter) with a hollow 15 nm core. Their wall is 5 nm thick and consists of distinct longitudinal profilaments (12 or 13) composed of globular protein subunits. Any of these subunits represents a dimer (called tubulin) having a molecular weight of 120,000 (6 S). It assumes a globular shape and the subunit peptide chain is about 600 amino acid residues long. The universal characteristics of this protein (containing 150 to 180 glutamic and aspartic acids), regardless of its origin, are: 1. it binds specifically with → colchicine, and 2. it binds with guanosine triphosphat (GTP).

Microtubules are self-assembling polymers. Their assembly is initiated at specific intracellular sites usually regarded as representing local concentrations of subunit material and called m. organizing centers or nucleating sites. Periodic and nonperiodic thin bridges between the component microtubules have frequently been observed and are assumed to be important for maintaining the order of m. arrays.

Microtubules are commonly encountered in six sites: 1. cilia and flagella, 2. nerve processes, 3. the mitotic apparatus, 4. the cortex of meristematic plant cells, 5. elongating cells, and 6. selected structures in protozoa (such as the axostyle of parasitic flagellates, the axoneme, etc.). They are associated with diverse functions, including 1. → chromosome movement, 2. intracellular transport of material, 3. development and maintenance of cell form, 4. cellular motility and 5. sensory transduction. They are without doubt the main agents that establish asymmetries, polarities and changes of shape in higher cells.

Microtubules are highly labile structures assumed to exist in dynamic equilibrium with a pool of readily available subunits; they can be polymerized or depolymerized rapidly at different intracellular loci as conditions change during growth and differentiation.

microvillus — any of the tubular projections from the cell surface (from 0.2 to several µm long and about 0.1 µm in external diameter) which represent labile structures that appear on the cells in particular phases of the cell cycle when the cell is rounding up or respreading. Microvilli are invested by a membrane continuous with the cell membrane and comprising part of it. They are assumed to play a role in cell aggregation, cell fusion and storage organs for excess membrane during the process of rounding up which accompanies cell division.

mictic (*Storch* 1924) — of females producing female and male offspring by → apomixis or by → amphigony (= amphogenous).

mictohaplontic (*Kniep* 1928) — of a haploid organism with genotypically different nuclei.

micton (*Camp* & *Gilly* 1942) — a widely distributed population (produced by → hybridization between two or more species) which is fully fertile in crosses with the parent species.

mid body — a basophilic structure (= spindle bridge or Flemming's body) produced by telophase → spindle fibers constricted in the → cleavage plane of the cell.

middle lamella — the outermost layer of the plant → cell wall connecting two adjacent cells. It is composed primarily of pectic substances.

mid-spindle elongation — the elongation of the → spindle region which lies between the two separating groups of daughter chromosomes in mitotic anaphase (→ mitosis). M.-s. e. takes place in most animal and some plant cells and aids in the movement of chromatids to the poles. This movement ceases when the centromeres have moved about two-thirds the distance between the spindle equator and the poles. Further movement takes place by m.-s. e.

migration — in genetics, the transfer of → genetic information among populations by the movement of individuals or groups of individuals from one population ("emigration") into another ("immigration"). M. may result in changes of → gene frequencies in the populations concerned (→ gene flow) and is thus a factor in → evolution.

The effect of m. will be determined in large measure by the relative sizes of the migrant and non-migrant populations. Its genetic consequences will be functionally related to the migrant's sex, age, and biological relationship to other individuals in the population. The level of m. may itself be under genetic control **(migrational selection** of *Fisher* 1930).

If the probability of m. depends on individual phenotype, this is called **selective m.** (*Fisher* 1930; *Thoday* 1972).

mimetic — = → mimic.

mimic — of nonallelic genes (= mimics) with similar phenotypic effects (= mimetic) (→ genocopy; isoallele).

miniature spindle — → spindle.

minicell (*Adler* et al. 1967; *Freifelder* & *Freifelder* 1968) — any of the small, DNA-deficient bodies formed by misplaced septation near the polar ends of a mutant *E. coli* strain. Minicells do not contain detectable amounts of DNA, cannot synthesize macromolecules of any type, do not grow or divide, but are enclosed by an apparently normal cell surface. Such cells derived from F^- cultures (→ Fplasmid) conjugate with F^+ donors and act as recipients for conjugally transferred DNA.

minimal recognition length — the shortest length of complementary nucleotide sequences forming a stable double-stranded DNA complex in → genetic recombination.

minor gene — = → polygene and → modifier gene.
minor spiral (*Sax* 1932) — = minor coil (→ chromosome coiling).
Minute (*Brehme* 1939) — in *Drosophila,* any of a group of dominant mutants (probably → deletions of sites of → transfer RNA genes) that exhibit (in heterozygous condition) a fairly uniform phenotype. The mutants show a slow rate of development, decreased body size, smaller bristles, and residual fertility. Minutes are scattered throughout the genome and all are lethal in homozygous condition as expected from the absence of a tRNA species.
minute fragment (*Muller* 1940) — a chromosome fragment smaller than the diameter of one chromatid.
miscoding — = → mistranslation.
miscopying (*Beadle* 1957) — → gene mutation.
misdivision (*Darlington* 1939) — → centromere misdivision.
misdivision haploid (*Kimber & Riley* 1963) — → haploid.
mismatch repair — = → heteroduplex repair (→ gene conversion).
mispairing — the presence in one polynucleotide chain of a DNA double helix of a nucleotide which is not the complement of that at the corresponding position in the other polynucleotide chain (→ gene mutation; heteroduplex).
misreading — = → mistranslation.
misrepair — → repair of spontaneous or induced DNA lesions with low fidelity; m. contributes to the production of → mutations. Fidelity of → excision repair is generally high, fidelity of → postreplication or recombination repair considerably lower.
missense codon — → missense mutation.
missense mutation (*Brenner* et al. 1961) — a → base substitution (→ gene mutation) that results, during → genetic translation of the messenger RNA, in amino acid substitution at a specific position in the polypeptide chain (→ samesense mutation; nonsense mutation). Whether a mutant phenotype is readily detectable depends on the particular amino acid substitution. Certain types of amino acid substitutions affect the assigned functions of a → cistron less drastically than others (conservative substitution). Some change the net electric charge of the polypeptide chain, so that a mutant chain becomes distinguishable from the original wildtype chain by electrophoresis. A m. m. may escape detection in those cases where the insertion of the incorrect amino acid does not lead to an appreciable change in the conformation and subsequent function of the protein. A m. m. can be suppressed (→ missense suppressor) by secondary mutations affecting transfer RNA structure and function (→ suppressor tRNA).
missense suppression (*Yanofsky* et al. 1961) — a type of → suppression due to the presence of a → missense suppressor.
missense suppressor (*Yanofsky* et al. 1961) — a → suppressor that acts on a → missense mutation and inserts the correct amino acid at the site of the missense codon. Missense suppressors represent mutationally altered tRNAs (→ suppressor tRNA) which accept the relevant amino acid and insert it into a polypeptide chain in response to an altered (missense) codon. Since a codon is missense at some mutated position and sense everywhere else, suppressor tRNA species cause generalized → mistranslation. They are,

therefore, generally very ineffecient and cells which carry them grow more slowly than normal (→ nonsense suppressor).

mistranslation — the incorporation (mis-insertion) into proteins of amino acids other than the one dictated by the standard codon-anticodon recognition (= miscoding; misreading). M., i.e., infidelity of → genetic translation, occurs in cell-free, protein-synthesizing systems and, less extensively, *in vivo*. A small amount of ambiguity (recognition of a given codon by more than one species of tRNA) is probably an intrinsic property of the translation process. Ambiguity in the → genetic code potentially resulting in m. may be due to: 1. A codon is misread owing to environmental conditions existing around the tRNA-mRNA-ribosome complex which interfere with the codon recognition process. 2. A codon is misread because of a specific change in the primary, secondary, or tertiary structure of either the tRNA or the ribosome such that it affects the codon-anticodon recognition. 3. A tRNA synthetase may misrecognize either an amino acid or a tRNA leading to the formation of the wrong → aminoacyl tRNA.

→ Suppressor tRNAs, besides their major decoding specificity (including → wobble), possess some minor decoding properties which result in m. (→ missense suppressor). A mutation which affects the major decoding specificity of a tRNA species also entails a specific change in its m. specificity.

Streptomycin has been found to cause the m. of certain synthetic messengers.

mitochondrial complementation — increased biochemical activity (as measured by the rate of oxidation and efficiency of phosphorylation) of a 1:1 *in vitro* mixture of → mitochondria isolated from two potential parents over the arithmetic mean of the two preparations when assayed separately. It has been claimed that there is a correlation between m. c. and → m. heterosis in F_1 hybrid plants.

mitochondrial DNA (*Nass* & *Nass* 1962; *Luck* & *Reich* 1964) — organelle DNA contained in → mitochondria and replicated and expressed by a mitochondrion-specific genetic apparatus. M. DNA (abbreviation mt-DNA; mit-DNA) is at least ten times smaller than the genome of the smallest free-living cell (like mycoplasm) and contains information for mitochrondrial rRNA, mitochondrial tRNA, for at least a few mitochondrial ribosomal proteins and for at least a few proteins necessary for the structural and functionl integrity of the ribosomal unit membranes.

The buoyant densities of mtDNA from mammals is within the range of 1.699 to 1.704, those from plants are all found to be 1.706; mtDNA is double-stranded, has a very rapid rate of renaturation and is circular (circular dimers and chains of interlocked circles have been found). The mean contour lengths of the circular molecules from different sources range from 4.74 to 5.45 μm (i.e., correspond to about 15,000 base pairs). mt-DNA from protozoa and fungi is considerably (up to five times) longer than the mt-DNA of vertebrate cells suggesting that the mitochondria of protozoa and fungi may represent an early stage in the reductive evolution of bacterial endosymbionts which culminated in the mitochondria of vertebrates with their small m. genomes. Replicative intermediates of circular mt-DNA (→ displacement hypothesis) have been found. The replication of mt-DNA proceeds with nicking and rapid closure of a covalently closed circular template. A few identical DNAs are present in each mitochondrion. mt-DNA represents the only known mature circular DNA which contains covalently inserted ribonucleotides.

Methylated bases are present in mt-DNA and it may consist of 20 to 50% spacer sequences, probably not transcribed. Transcription of mt-DNA is symmetric over a considerable portion, if not the entire length, of the molecule (→ genetic transcription). The mitochondrial transcription process in vivo can be distinguished from transcription occurring in nuclei by its sensitivity to a specific inhibitor, ethidium bromide, and because it continues during metaphase, unlike nuclear RNA synthesis. Mitochondrial mRNA contains a relatively small poly(A) segment of about 56 nucleotides.

mitochondrial heterosis (*McDaniel* & *Sarkissian* 1966) — the superior activity (due to → mitochondrial complementation) of mitochondria isolated from heterotic (→ heterosis) F_1 hybrids as compared with mitochondria from the two parents of the hybrid. M. h. is claimed to indicate the yield potential in a number of plants.

mitochondrial RNA — RNA complementary to → mitochondrial DNA.

mitochondrial sheath — → spermatozoon.

mitochondrion (*Benda* 1897) — a semiautonomous cell organelle (about 0.5 to 1.0 μm in diameter, and 5 to 10 μm in length), largely made up of protein (60 to 65%) and lipids (35 to 40%), and present in all cells of eukaryotes

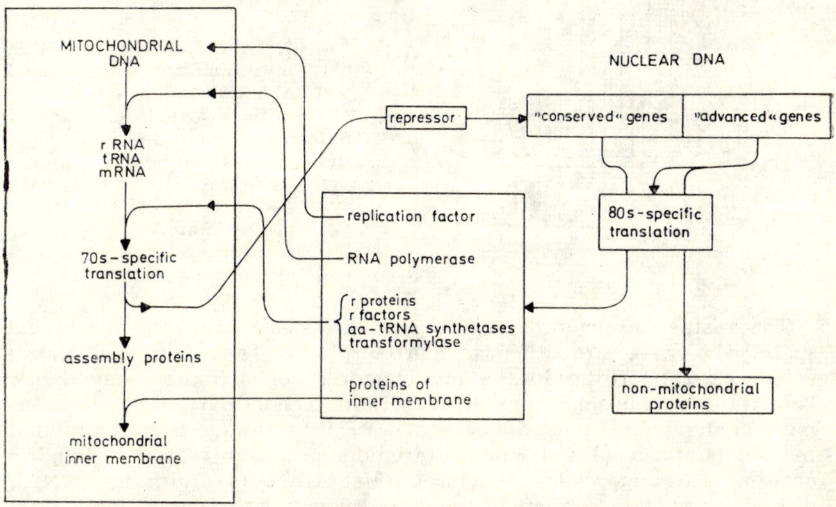

Fig. 66. Cooperation of mitochondrial and nuclear DNA specifying the mitochondrial genetic apparatus of *Neurospora crassa* (from *Küntzel* et al. 1973).

(→ chondrioid). The genesis of the m. requires the function of two genetic systems: the nuclear system and a system localized within the m. (Fig. 66). The m. contains distinctive → ribosomes, → ribosomal RNA (21 S and 12 S), tRNA species, → aminoacyl-tRNA synthetases, → initiation, → elongation, and → termination factors, i.e. the components of a minor protein synthesizing system coded for, in part at least, by → mitochondrial DNA (the mitochondrial genome).

The number of mitochondria per cell is kept rather constant, i.e., there is a tight coordination of cell division and rate of formation of mitochondria which is most likely governed by a strict regulation of mitochondrial DNA replication and transcription.

The primary function of mitochondria in the cell is to synthesize ATP from ADP and P_i and all mitochondria contain the enzymes, coenzymes and other cofactors required by the individual enzyme complexes which catalyze respiration and phosphorylation.

Morphologically, the m. consists of a limiting or outer membrane and an inner membrane separated by an intermembrane space. Outer membrane and intermembrane space make up the mitochondrial outer compartment. The inner membrane encloses an inner space (called the matrix) and is comprised of highly invaginated and structurally distinct cristae. The inner membrane and the matrix space constitute the mitochondrial inner compartment (Fig. 67).

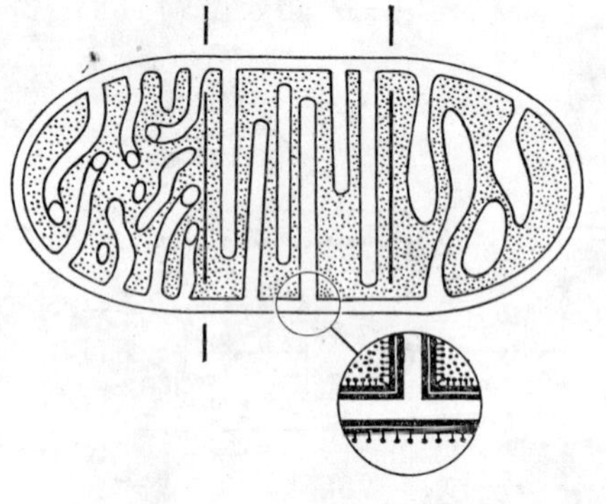

Fig. 67. Schematic representation of a mitochondrion (longitudinal section) showing the main features of the three structural types of mitochondria in combination (left: the tubular type; middle: the cristae type; right: the intermediary type frequently found in plants). Bottom: A detail section showing the arrangement of the "oxysomes" (from *Sitte* 1965).

The cristae are sometimes seen to contain a space, called the intracristal space. The cristae are either lamellar or tubular. There is a good deal of evidence to support the idea that mitochondria evolved from some symbiotic bacteria-like organism. Some features of the mitochondrial protein synthesizing system resemble those of bacterial systems: 1. the synthesis is inhibited by chloramphenicol but not glutarimide antibiotics; 2. mitochondria contain tRNA molecules, including fMet-tRNA (→ initiator tRNA), distinct from those present in the cytoplasm; 3. ribosomes isolated from mitochondria have a lower sedimentation velocity and contain smaller rRNA species than cytoplasmic ribosomes from the same organisms.

mitodepression (*Östergren, Morris & Wakonig* 1958) — a decrease in mitotic frequency which may be experimentally induced by a large number of → mitotic poisons or → antimitotic agents.

mitogenic — of agents (mitogens) which stimulate cells to undergo → mitosis.

mitosis (*Flemming* 1882) — a mode of nuclear division (= indirect nuclear division or karyokinesis) providing for the production of daughter nuclei which contain identical chromosome numbers and are genetically identical to one another and to the parent nucleus from which they arose (→ amitosis; endomitosis; meiosis). M. is a genetically controlled process which is usually followed by → cytokinesis and gives rise to new, genetically equivalent cells in the growing somatic regions of eukaryotic organisms. It maintains the continuity of → chromosome number and type and is a continuous sequence of events which may be divided into five main stages as follows (Fig. 68):

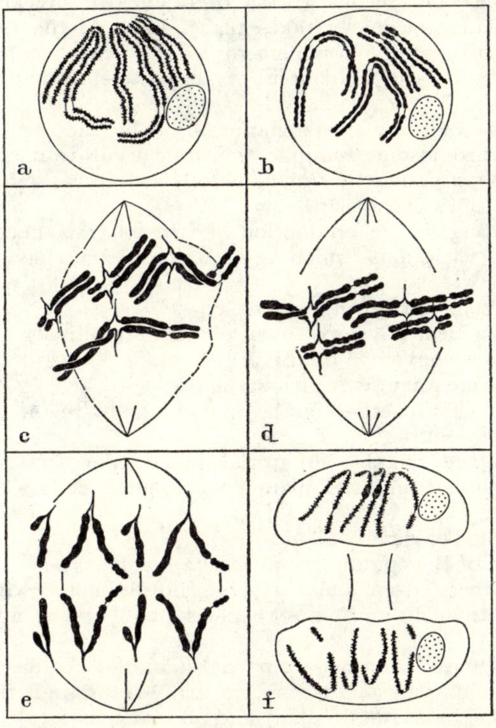

Fig. 68. Diagram showing the main features of the mitotic cycle in a diploid cell with two pairs of homologous chromosomes (2n = 4). a) mid prophase; b) late prophase; c) prometaphase; d) metaphase; e) anaphase; f) telophase.

1. **Prophase:** The chromosomes which reduplicated during → interphase of the → cell cycle make their appearance in species-specific number. At the beginning of this stage they are long threads consisting of two → chromatids each. During prophase which ends with the disruption of

the → nuclear envelope, the chromosomes become progressively shorter and more compact owing to → chromosome coiling which transforms them from the functional interphase form into the condensed "transport form".

2. **Prometaphase:** The → spindle is formed to which the → centromeres of each chromosome become attached. Subsequently the chromosomes undergo congression movement (→ chromosome movement) and finally come to lie at the spindle equator.

3. **Metaphase:** The chromosomes reach their metaphase arrangement on the metaphase plate, an equilibrium position with all centromeres aligned at the spindle equator (→ centromere orientation).

4. **Anaphase:** The two sister chromatids of each chromosome separate by their movement toward opposite poles of the spindle.

5. **Telophase:** The chromatids ("daughter chromosomes") of each chromosome arrive at the poles and reconstitution of a nuclear envelope around each of the two daughter nuclei takes place. Each of the two nuclei is identical in its chromosome complement to the other, to the original complement, and similarly in the content of → genetic information coded in the chromosomal DNA.

The mitotic division cycle of the nucleus characterized by the regular alternation of interphase chromosome reduplication and accurate distribution of the reduplication products is typically followed by → cytokinesis, i.e., division of the cell's cytoplasm.

The mitotic pattern leading to the production of two separate nuclei may show species-specific variations which are classifiable as follows (*Mazia* 1956):

1. According to nuclear behavior during prophase:

a) disappearance of the nuclear envelope (most plants and animals);

b) persistence of the nuclear envelope (many protozoa).

2. According to chromosome alignment at metaphase:

a) all chromosomes collected together in the metaphase plate (many large cells with small chromosomes);

b) centromeres only aligned on the equatorial plane; the arms of chromosomes dangle from the equatorial plane (many small cells with large chromosomes);

c) equatorial alignment not evident (pollen tubes).

3. According to the form of the spindle (→ mitotic apparatus):

a) continuous spindle fibers from pole to pole intermingled with chromosomal fibers connecting chromosomes to poles (many plant and animal cells);

b) continuous "central spindle" distinct from "half spindles" connecting chromosomes to poles; chromosomes may be arrayed around the circumference of an central spindle (many plant and animal cells);

c) spindle functions entirely within the nuclear membrane (many protozoa; some plant and animal cells);

d) each chromosome moves within a distinct spindle element of its own (certain insects).

4. According to → mitotic centers:

a) distinct centers with asters ("astral" or "amphiastral mitosis"). The centers may be represented by → centrioles lying in a larger → centrosome, or large centrospheres without centrioles (most animals, many lower plants);

b) no distinct centers at the poles ("anastral mitosis") (most higher plants and some invertebrates).

5. According to chromosome-to-pole connections and chromosome movement:

a) chromosomes move toward the poles as the chromosomal fibers shorten (most plant and animal cells);

b) chromosomes behave as though they are being dragged by fibers attached along their entire length (diffuse centromere activity) (some animals and plants);

c) chromosomes move with their arms ahead of the centromere ("neocentric activity") (some insects and plants).

6. According to the relative rates of movement of individual chromosomes:

a) all chromosomes move at the same rate, independent of their size (most cases);

b) certain chromosomes move at a different rate in comparison with others (e.g., sex chromosomes).

7. According to elongation of the spindle and shortening of chromosomal fibers:

a) chromosome sets separate by simultaneous shortening of chromosomal fibers and elongation of the spindle;

b) shortening of chromosomal fibers precedes elongation of the spindle.

Further variants concern the process of → cytokinesis, i.e. the segregation of daughter nuclei in separate cells.

Chromosome reduplication and cell growth without formation of daughter nuclei or new cells may take place and is referred to as → "endomitosis" or "endoreduplication". It results in either endopolyploid cells or polynemic chromosomes if the products of endoreduplication do not become individualized (→ polytene chromosome).

Since every sexual cycle consists of two phases, a haploid between → meiosis and → fertilization and a diploid one between fertilization and meiosis, m. can occur during either or both of these phases. In → haplontic organism m. is confined to the haplophase, in → diplontic organisms it is confined to the diplophase, and in → diplo-haplon ones m. occurs during both haplo- and diplophase.

According to the chromosome distribution at anaphase the following types of m. may be distinguished (*Battaglia* 1945):

1. eumitosis, i.e. normal mitosis;

2. mitosis with an irregular distribution of chromosomes giving rise to → aneuploid daughter nuclei;

3. restitutional mitosis resulting in the formation of a → restitution nucleus in meta- or anaphase and thus in a → polyploid cell;

4. endorestitutional mitosis characterized by a restoration of interphase before the end of prophase;

5. mitosis with somatic → chromosome pairing;
6. endomitosis.

mitospore — any haploid or diploid cell resulting from mitotic division, that undergoes mitotic growth to reproduce the kind of organism from which it arose (→ germ cell).

mitostatic — → antimitotic.

mitotic — ref. to or produced by → mitosis.

mitotic apparatus (*Mazia & Dan* 1952) — the ensemble of structures constituting the so-called "achromatic figure" in the classical description of → mitosis. The m. a. includes the asters, which surround the → centriole (if present), and the mitotic spindle plus a matrix in which the spindle fibers are embedded and can be isolated as an intact body from the cell. It is a transient structure present only during the division period of the → cell cycle.

According to electron micrographs, the fibrillar elements of the m. a. represent → microtubules. They are found attached to the → centrioles and → centromeres. Almost all blocks of cell division are associated with either structural abnormalities or dissolution of the m. a. (→ C-mitosis; mitotic poison).

The isolated m. a. consists of about 90% protein (dry weight) and contains 5 to 6% RNA together with polysaccharides and lipids; its water content is approximately the same as that of the surrounding cytoplasm. The isolated m. a. is rich in protein-bound sulfhydryl groups which may be alternatively reduced or oxidized during mitosis. The stability of the m. a. is ascribed to the oxidation of these SH-groups to intermolecular disulfide bonds. ATPase activity is also found in the isolated m. a. and the mitotic microtubule protein (→ tubulin) is capable of an electron transfer reaction.

Morphologically, the isolated m. a. appears fibrous in the phase-contrast microscope and birefringent in the polarizing microscope. The bulk of its structure consists of microtubules, vesicles, ribosome-like particles and "empty spaces" separating the various structures.

mitotic center — a body or agency, visible or not, polarizing the → mitotic apparatus, and defining a mitotic cell pole towards which the chromosomes move, irrespective of the mechanism of movement. In some organisms, the functions of the mitotic centers are associated with the → centrioles (*Mazia* 1961).

mitotic cycle — the sequence of steps by which the genetic material is equally partitioned prior to division of eukaryotic cells into daughter cells (→ mitosis).

mitotic delay — delayed progress of the mitotic cycle characteristically observed after treatment with high energy radiations and chemical mutagens.

mitotic haploidization (*Pontecorvo* et al. 1954) — a process in the → parasexual cycle of fungi, during which whole chromosomes reassort at random during mitosis, eventually producing nuclei with half the diploid chromosome complement by repeated → nondisjunction.

mitotic index (*Minot* 1908) — that fraction of the total number of cells scored in a meristematic tissue which are actively engaged in → mitosis, irrespective of their stage (prophase to telophase).

The m. i. appears to be a function of the generation time for a given species under the conditions provided and the actual length of time a cell of this species remains in mitosis. The m. i. for a cell population in which the period of mitosis is brief will be lower than for a population with a comparable → cell cycle span in which mitosis is prolonged.

mitotic inhibition (*Gavaudan* 1943) — the inhibition of → mitosis (and by extension also → meiosis) accomplished naturally or artificially by (*Mazia* 1961):

1. diverting cells from the preparations for division;
2. inhibiting any one of the preparations for division;
3. arresting cells at the transition stage between interphase and division;
4. disabling the mitotic apparatus (→ metaphase arrest). Theoretically all such blocks are reversible and not fatal to the cell.

mitotic nonconformity (*Bainbridge* & *Roper* 1966) — mitotic instability resulting in parent and daughter cells that do not conform in genotype. In *Aspergillus* strains, m. n. is due to the presence of a chromosome segment in duplicate (one in normal position, one translocated to another linkage group), i.e., a consequence of chromosomal unbalance.

mitotic poison (*Dustin* 1934) — operationally defined as any of many different kinds of noxious agents which act to impair → mitosis (and by extension → meiosis too), irrespective of whether the impairment takes place during or following the treatment of actively dividing cells. The mitotic disturbances may result in morphological changes of the chromosomes and/or inhibition of mitosis as well as cell death.

Mitotic poisons may be classified according to various principles.

Bauch (1947) proposed the distinctions → spindle poisons, cell division poisons, and chromosome poisons. *Ahlström* (1951) recommended the terms spindle poisons, interphase poisons, karyoclastic poisons; *Dustin* (1952) spindle and metaphase poisons, perfect and partial → radiomimetic poisons; *d'Amato* (1954) inhibitors of cytokinesis, inhibitors of the spindle, preprophase and prophase poisons; *Heilmeyer* (1950) interphase poisons, mitotic poisons *sensu strictu*, spindle poisons, and poisons of cytokinesis; and *Loveless* & *Revell* (1949) distinguished mitotic poisons with and without radiomimetic effects and mitotic poisons of the spindle-poison type.

mixoploid (*Němeč* 1910) — of cell populations whose component cells differ in their chromosome numbers, irrespective of whether these numbers are → euploid or → aneuploid. Thus, the term mixoploidy covers all types of mosaics or → chimeras in which the heterogeneity between different elements involves differences in chromosome number. Mixoploidy may stem from a variety of mitotic irregularities, from cellular and nuclear fusions, or even from amitotic processes, if these give rise to viable products. A special case of mixoploidy is endopolyploidy arising by → endomitosis or endoreduplication of chromosomes (*Levan* & *Müntzing* 1963).

mock dominance — → dominance.

modal number (*White* 1945) — the most common chromosome number of a particular taxonomic group (genus, family, class, order), there being no implication that it is the ancestral number of the group in question (→ type number).

modificability — the property of a → genotype to react in response to varying developmental conditions by particular alterations in its → resultant phenotype (→ reaction norm) called → modifications. M. is genotypically controlled with respect to its extent and form (→ flexibility).

modification (*Naegeli* 1884) — any environmentally induced, nonheritable change in the → phenotype of an individual, in contrast to a → mutation (→ dauermodification). That period of time during which the mode of manifestation of a particular → character may be modified is referred to as the "period of modification" (→ modificability) of the character concerned. Many characters are modifiable during a large part of the individual's life cycle, others may be influenced only during one or a few short periods ("sensitive periods") of development preceding their definite determination.

modification methylase — a bacterial enzyme that marks, by attachment of methyl groups to bases at specific sites, the DNA replicated in a cell as DNA belonging to that cell and protects it from the action of → restriction enzymes (→ DNA modification-restriction system).

modification-restriction system — → DNA modification-restriction system.

modifier gene (*Bridges* 1919) — any gene that by interaction (→ gene interaction) affects the phenotypic expression of genes at other loci. Many modifiers can be detected only by their effects on the expression of other nonallelic genes, but others may have phenotypic effects of their own. According to the type of modifiying action such genes may be classified into:

1. **Enhacers** (= intensifiers, extension genes) which intensify the phenotypic effect or increase the → mutability ("mutator genes") of other genes.

2. **Reducers** (= restriction genes) which decrease the effect of other genes or ("inhibitors") completely prevent their manifestation. They are called → suppressor genes when they shift the manifestation of a mutant allele toward the → phenotype of the corresponding wild-type or normal allele (→ reversion).

modulating codon (*Ames & Hartman* 1963) — any → codon responsible (directly or indirectly) for controlling the frequency with which cistrons are transcribed and, perhaps, the frequency with which messengers are translated (→ modulation).

modulation — 1. any of those reversible alterations which cells may undergo in reaction to different environmental conditions without changing their essential nature (e.g., any of the histological types which a given tissue may assume dependent on the nature of its environment) (*Weiss* 1939);

2. the more frequent reading of certain sequences of polycistronic → messenger RNA during → genetic translation. The polycistronic messenger produced by → genetic transcription of an → operon shows a certain constant probability of being read in its entirety; a higher probability

of its first half being read, and an even higher probability of only its initial quarter being read. This influence on the read-out process results in the disengagement of the ribosomes from the messenger or in destruction of the terminal section of the messenger RNA beyond the point where genetic translation was temporarily halted. The result of this phenomenon, which was termed modulation by *Ames & Hartman* (1963), is a greater number of polypeptide chains formed for those genes in the operon which are proximal rather than distal to the → operator end.

Modulation of this type is understood as a genetic adjustment in the relative rates of synthesis through coding for → transfer RNA species of different abundance. According to *Stent* (1964), the regulation of protein synthesis may be achieved by the control of the availability of so-called modulating transfer RNA species which correspond to specific "modulating codons" or "modulating triplets" in the messenger RNA. The "modulating codons" are presumed to concern degenerate representations of particular amino acids, to affect the reading of the messenger by the ribosomes and thus to determine the relative molar quantities of each polypeptide coded for by the operon in question. Translation of any messenger containing one or more of these codons is assumed to proceed only so long as the corresponding modulating transfer RNA species capable of base pairing with the modulating codons are at hand.

molecular biology (*Astbury* 1950) — a branch of modern biology in which biological phenomena and processes are studied not only from a phenomenological point of view but by physical-chemical and biochemical investigations at the molecular level. M. b. is based on developments in genetics, biochemistry, physical chemistry of macromolecules, and chemical physics and represents an effort to account for biological events in terms of the established principles of physics and chemistry.

molecular cloning (*Cohen* et al. 1973) — the cloning of DNA molecules from prokaryotic and eukaryotic sources as part of a bacterial → plasmid replicon. The procedure employed for m. c. requires a plasmid vehicle suitable for replication and selection of foreign DNA (in *E. coli*), a technique for covalently joining DNA fragments from different biological sources, and a method for introducing the resultant hybrid molecules into recipient bacteria without substantially altering bacterial viability (*Cohen & Chang* 1974). (→ nuclear cloning).

molecular disease (*Pauling, Itano, Singer & Wells* 1949) — a genetically determined biochemical disorder in which a specific enzyme defect produces a metabolic block with pathological consequences (= inborn error of metabolism).

molecular genetics — a branch of → genetics which strives to account for genetical events in terms of the established principles of physics and chemistry. In the broadest sense, m. g. is the study of those aspects of genetic systems, or components thereof, which can be described at the molecular level. In a more restrictive sense, the study of the molecular aspects of genetic mechanisms and the control of metabolic processes by genetic material (*Steiner* 1965).

molecular hybridization — = annealing (→ DNA/DNA hybridization).

molecular spiral (*Darlington* 1935) — → chromosome coiling.

monad — an individual cell produced by a meiocyte in place of a → tetrad as a consequence of a meiotic abnormality.

monaster (*Wilson* 1901) — a unipolar → spindle which may arise, e.g., by the suppression of → centriole division, and which results in an irregular (unipolar) nuclear division giving rise to a → restitution nucleus.

monoallelic (*Atwood* 1944) — of → polyploids with multiple allelic loci all of which have the same allele: $A_1A_1A_1A_1$ or $A_2A_2A_2A_2$ in a tetraploid. The other possible genotypes involving a multiple allelic locus in a tetraploid are:

diallelic	$A_1A_1A_1A_2$,	$A_1A_1A_2A_2$ etc.
triallelic	$A_1A_2A_3A_3$,	$A_1A_1A_2A_3$ etc.
tetraallelic	$A_1A_2A_3A_4$,	$A_2A_3A_4A_5$ etc.

monobrachial (*Sorokin* 1929) — → chromosome arm.

monocentric — 1. of chromosomes or chromatids with one localized → centromere (= monokinetic), in contrast to → di- and polycentric chromosomes;

2. of nuclear divisions with a unipolar → spindle (→ monaster).

monochlamydeous — = haplochlamydeous (→ periclinal chimera).

monocistronic — of → messenger RNA molecules that code for one polypeptide chain (→ polycistronic; transcripton).

monoclinous — having male and female germ cells in one and the same flower (= synoecious or hermaphroditic).

monoecious (*Darwin* 1877) — of plants with male and female sex organs on the same individual and producing male and female → gametes (→ dioecious). Plants with male and hermaphroditic flowers or female and hermaphroditic flowers on the same individual are called andromonoecious and gynomonoecious, respectively. M. fungi include (*Esser* 1971): 1. all species which produce male and female sex organs on the same mycelium; 2. all species without sex organs in which each mycelium can serve both as nuclear donor and nuclear recipient.

monogenesis — = → monogony.

monogenic — of character differences controlled by the → alleles of one particular gene, as opposed to di-, tri-, or polygenic control exerted by two, three, or many nonallelic genes.

monogenomatic (*Winkler* 1920) — of cells or individuals possessing one → genome (one chromosome set) in their nuclei, as opposed to di-, tri-, or polygenomatic ones with two, three, or more genomes, irrespective of whether the genomes are genetically identical or different or have equal or unequal chromosome numbers.

monogeny — the production of only male ("arrhenogeny") or female ("thelygeny") offspring.

monogony (*Haeckel*) — asexual reproduction.

monohaploid (*Katayama* 1935) — → haploid.

monohybrid (*de Vries* 1900) — of individuals → heterozygous (e.g., Aa) with respect to the → alleles of one → gene locus and by extension also of crosses (e.g., AA × aa) involving parents that differ with respect to the alleles of one locus, as opposed to di-, tri-, and polyhybrid heterozygotes and crosses involving differences in two, three, or more pairs of alleles.

monoisodisomic (*Kimber & Sears* 1968) — of a cell or individual deficient in one chromosome but which has an → isochromosome for one of the arms of the missing chromosome (= haplo-triplodisomic: *Khush & Rick* 1971). (→ monotelodisomic; monotelosomic).

monoisosomic (*Kimber & Sears* 1968) — of a cell or individual which lacks one chromosome pair but has an → isochromosome for one arm of the missing pair (→ diisosomic).

monokaryon (*Boveri* 1907) — 1. a uninucleate cell, a spore, or a tissue consisting of such cells, as opposed to a → dikaryon (or multikaryon);

2. = → pronucleus (*Boveri*).

monolayer — a single layer of cells growing on a surface.

monolayering (*Abercrombie & Heaysman* 1952) — the tendency of cultured (animal) cells to occupy space evenly and to spread out in thin layers.

monomeric — = → monogenic.

mononeme — of chromosomes which contain but a single DNA double-helix prior to replication, as opposed to → multineme or → polytene chromosomes.

monophyletic — of individuals derived in the course of evolution from a single interbreeding population or phyletic stock, as opposed to polyphyletic.

monoploid (*Langlet* 1927) — of cells or individuals (monoploids) having one chromosome set with the → basic number of chromosomes, i.e., the lowest → haploid number of chromosomes in a polyploid series.

monoplontic — of a → monoploid individual (a monoplont) or phase of the life cycle (→ haplontic).

monosome (*Haselkorn & Fried* 1964) — 1. a single → ribosome bound to messenger RNA (→ polysome);

2. a chromosome in the chromosome complement lacking an homologous partner (→ monosomic).

monosomic (*Blakeslee* 1921) — of an → aneuploid (2n — 1) cell, tissue (in case of → chimeras), or individual with one (or more than one: doubly monosomic, etc.) chromosome missing from an otherwise diploid or allopolyploid (and thus functionally diploid) chromosome complement. Monosomics characterized by the absence of one structurally normal chromosome are called **"primary monosomic"**.

Monosomics may be obtained from several sources (*Khush* 1973): 1. normal disomics, by spontaneous occurrence or by treatment with physical and chemical agents (→ non-disjunction); 2. asynaptic disomics and aneuploids; 3. polyploids, e.g. haploids and triploids; 4. intervarietal and interspecific crosses; 5. translocation heterozygotes; 6. trisomics ; and 7. monosomics.

As a result of their unbalanced chromosome complements, monosomics which are diploid in respect to their other chromosomes are characterized by the presence of one unpaired → univalent in meiosis. The expected meiotic products are theoretically half monoploid and half (n — 1). Frequently however, the univalents lag in chromosome movement and thus more (n — 1) gametes are produced than expected. Normally (n — 1) male gametophytes of diploid plants fail to function, but such female gametophytes of plants, as well as male and female gametes of animals, generally function.

A case of regular "monosomy" is represented by XX—XO-systems of → sex determination and the X-chromosome of the heterogametic sex is said to be a "monosome" (*Montgomery* 1904).

"Tertiary monosomic" (*Rick* 1943) is a term applicable to cells, tissues, or individuals that possess one product of a reciprocal → translocation but lack the reciprocal product and thus are composed of a segment of each of two different chromosomes of the chromosome complement.

"Pseudomonosomic" (*Hiorth* 1948) are cells or individuals which lack one chromosome (one centromere) but contain (due to → translocation) the essential segments of two nonhomologous chromosomes in one of their other chromosomes.

monotelic (*Bauer* et al. 1961) — → centromere orientation.

monotelodisomic (*Kimber* & *Sears* 1968) — of diploid → karyotypes with a chromosome pair represented by one normal and one → telocentric chromosome, the latter consisting of one of the two arms of the normal chromosome (→ trisomic; monoisodisomic; diisosomic; teloisosomic).

monotelomonoisosomic (*Kimber* & *Sears* 1968) — a cell or individual in which one chromosome pair is missing but a telocentric for one arm of the missing pair and an isochromosome for the other arm are present (→ teloisodisomic).

monotelotrisomic — → trisomic.

monozygotic — → dizygotic.

morbid mitosis (*Darlington* & *Thomas* 1941) — → polymitosis.

Morgan unit (*Haldane* 1919) — a unit of genetic map length (= 100 centimorgans), i.e., that length of a chromatid that has experienced, on the average, one crossover event per → meiosis. The map distance between two gene loci, when it is expressed in Morgans, is therefore the number of crossover events between the two loci. For short intervals, the map length in M. units is equal to the → recombination fraction expressed as a decimal, since double-crossovers will have a neglegible frequency.

morph (*Huxley* 1955) — any of the genetic forms (individual variants) that account for → genetic polymorphism.

morphism (*Huxley* 1955) — = → genetic polymorphism.

morphogenesis — the ensemble of processes through which embryos and parts of embryos change their form, and through which groups of cells change their positions with respect to one another in space. M. eventually results in the final form of adult organisms by the establishment of a specific pattern of organs and tissues which bear definite relations to one another in terms of size and cellular content (*Sussman* 1964).

morphosis (*Sachs* 1894) — any nonadaptive and usually unstable variation (→ modification) in an individual's morphogenesis associated with certain changes in its external environment. Morphoses which mimic known mutations in their effect are referred to as → phenocopies.

morph-ratio cline (*Huxley* 1955) — a gradual change in the proportion of two or more distinct varieties from one place to another.

mosaicism — the presence within one individual (called a "mosaic") of cells differing with respect to their → idiotype, chromosome structure, or chromosome number (→ mixoploid). The different sectors of a mosaic may result from the mutation of chromosomal and extrachromosomal → hereditary determinants, somatic → crossing-over (→ twin spot), changes in chromosome number (→ genome mutation), chromosomal structural changes (→ chromosome mutation), or "V-type" → position effects.

A "phenotypic sex mosaic" is an individual whose "mosaic constitution" is due to a development initially directed towards the male and then towards the female phenotype or *vice versa* (→ gynandromorph).

mottling (*Muller* 1930) — 1. = → mosaicism or variegation;
2. → chromsome mottling.

movement index (*Slizynsky* 1955) — → chiasma movement index.

M protein — a protein which stimulates DNA transcription by complete → RNA polymerase under a variety of conditions. Possibly, M associates with RNA polymerase and provides for an increased frequency of initiation of RNA chains. It might function at the binding or initiation stages of the RNA polymerase reaction.

m RNP — messenger ribonucleoprotein (about 40% RNA and 60% protein) complexes released with EDTA under isotonic or low salt conditions from cytoplasmic → polyribosomes (→ messenger RNA).

mt-DNA — abbreviation (= mit-DNA) for → mitochondrial DNA.

multienzyme system — an aggregate of different, functionally related → enzymes bound together by noncovalent forces into a highly organized structure.

multigenic — controlled by several genes, as opposed to → monogenic.

multihybrid — → heterozygous for more than one gene (dihybrid, trihybrid, etc.).

multikaryon — a multinucleate cell (→ dikaryon) or a → coenocyte, as opposed to a → monokaryon.

multimer (*Crick* & *Orgel* 1964) — an active aggregate resulting from the interaction of polypeptides (monomers). It is probable that complementation of the intragenic type generally results from multimer production by differently defective monomers (→ genetic complementation).

multinucleate — of cells with more than one → nucleus. Heterophasic m. cells are the result of fusion between similar cells in different phases of the → cell cycle (→ premature chromosome condensation); homophasic m. cells would be formed by fusion of cells in the same phase of the cell cycle (→ cell hybridization).

multiple alleles (*Morgan* 1914) — → allele.

multiple codon recognition — the recognition of more than one → codon in → messenger RNA by a single → transfer RNA species due to codon-anticodon pairing involving non-standard base pairs (→ wobble) or the use of multiple tRNA adaptors.

multiple genes — two or more nonallelic genes with similar or complementary cumulative effects on a single character.

multiplicity reactivation (*Luria* 1947) — the interaction in a multiply infected cell of inactivated (by, e.g., UV light, X-rays, gamma rays, ^{32}P decay, nitrous acid) bacteriophages or animal viruses, inviable in single infection, so as to produce viable progeny. M. r. results from both functional complementation and genetic recombination between the infecting viruses. Functional complementation works at the level of the early functions and is preliminary to virus replication, recombination or virus DNA repair. Recombination permits the reconstruction of intact virus genomes from undamaged pieces derived from otherwise nonviable virus genomes.

multipolar — of a → spindle with more than two spindle poles which may result from the presence of several nuclei in the cell (each with its centers), from polyspermy in eggs, or from supernumerary divisions of the cell centers (→ centriole). The type of mitosis resulting from the presence of a m. spindle is called a multipolar mitosis and is generally connected with an aberrant mode of chromosome distribution.

M. mitosis appears to be an important, perhaps the most important, mechanism whereby polyploid cells can reduce their degree of ploidy.

multisite mutation — a mutation of one → genetic locus involving changes of two or more adjacent → mutational sites. Multisite mutations (→ gene mutation) occur spontaneously and are inducible experimentally by the application of → mutagens. They do not revert to wild-type spontaneously or after mutagen treatment (→ reversion) and some in all probability are true → deletions (→ deletion mapping).

multivalent — any association of more than two completely or partially homologous chromosomes whose homologous regions are held together by meiotic → chromosome pairing and chiasma formation from prophase to metaphase of the first division of → meiosis (as distinguished from → univalents and → bivalents). Multivalent formation is characteristic of → autopolyploid and some types of → allopolyploid individuals as well as some forms of structural hybrids (e.g., those heterozygous for reciprocal → translocations). Depending on whether three, four, five, six, seven, or eight chromosomes are meiotically paired, the multivalents are called trivalents, quadrivalents, pentavalents, hexavalents, heptavalents, and octovalents, respectively.

The configuration of multivalents in meiosis I depends upon the number of chromosomes participating in pairing, the number and location of chiasmata, and the degree of → chiasma terminalization. Representative types of tri- and quadrivalents (complete chiasma terminalization) are shown in Fig. 69. The arrangement of multivalents within the spindle at metaphase I of meiosis is determined by the particular type of → centromere coorientation, which in turn is dependent on the position of the chiasmata, the relationship in space between the centromeres of the m., and the space available in the spindle.

Oksala (1952) distinguishes between **"primary"** and **"definitive" multivalents.** The former are those held together by the pairing forces between homologous chromosomes or chromosome segments from zygotene up to diplotene, the latter those held together mechanically by chiasmata up to anaphase I of meiosis.

Pseudomultivalents are multivalents formed by nonhomologous chromosomes (e.g., by → sticky effects).

multivesicular body (*Yamada, Muta, Montomura* & *Koga* 1957) — a round to oval structure (diameter from 0.1 to 0.6 μm) which functions as a digestive vacuole for exo- and endogenous materials. Multivesicular bodies (= vesicular conglomerates) are found in a variety of cells and are thought to be self-duplicating organelles (*Sotelo* & *Porter* 1959). They are surrounded by a smooth membrane and their lumina contain a rather large number of round vesicles (diameter from 10—50 nm).

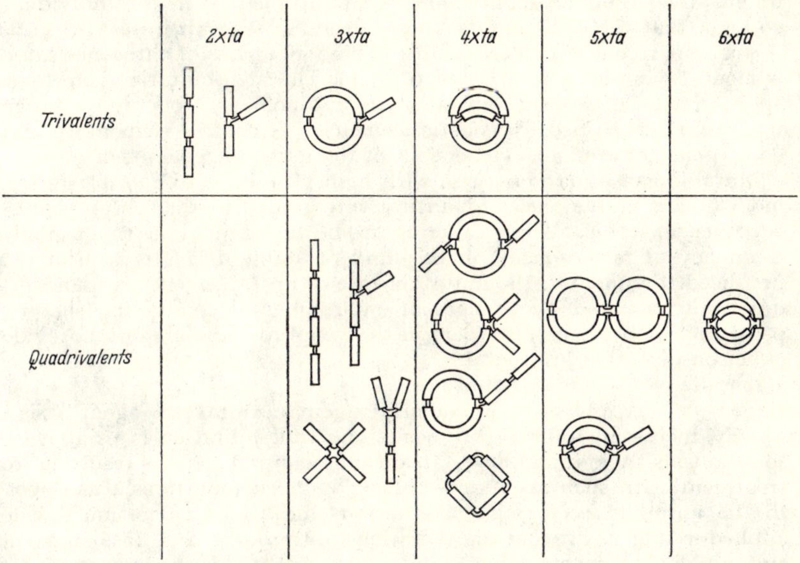

Fig. 69. Diagrammatic representation of different types of trivalents and quadrivalents as observed in diakinesis and metaphase I of meiosis (xta = chiasmata).

mutability — the property of any → gene and → genotype to undergo → mutation. This is an hereditary change not due to → genetic segregation or → genetic recombination. M. provides the ultimate basis for the adjustment (→ adaptation) of → gene pools to changes in the environment. Generally it is advantageous for a species to maintain a constant low level of m., since most mutations are deleterious. This relative constancy of m. is achieved by manifold mechanisms which may consist of hereditary changes that primarily affect the nongenic parts of the cell (e.g., alterations of the pattern of resistance to the admission of → mutagens to the chromosomes, neutralization of the action of mutagens, etc.) or hereditary changes that primarily affect the genetic system *per se* (e.g., changes in the structure of the → genetic material resulting in changes in its susceptibility to mutagens, changes in mechanisms that repair lesions induced by mutagens in the genetic material; → reactivation). Specific genes giving rise to significant increases in the m. of other genes are called → mutator genes.

mutable gene — any of a class of labile or unstable genes which in the course of development mutate spontaneously rather frequently among two or more possible states, in comparison with the majority of genes which as a rule are elements of considerable stability.

Although there is no sharp boundary between "stable" and "unstable" genes, a gene in multicellular organisms is usually referred to as mutable or unstable when its spontaneous → mutation rate in the individual is so high, that the gene and its mutant derivatives result in → variegation. A particular gene may be unstable during all stages of ontogenetic development or in only one or a few of them. The high mutation rate (→ hot spot) of a m. g. is either an inherent, autonomous property of the gene itself or is controlled by genetic elements (→ control element; mutator gene; gene conversion) independent of the mutating gene itself.

Mutable genes may give rise with high frequency to a whole range of more or less stable alleles (quantitatively graded in their expression), or to other unstable alleles, differing from the original in the timing or frequency of mutation, or in the kind of stable derivatives most often produced. Frequently, the mutation rate of unstable genes is found to be drastically affected by changes of environmental conditions (e.g., temperature), or by genes elsewhere in the genotype which exert no evident effect on mutation in general.

mutable site — = → mutational site.

mutagen — any physical or chemical agent ("mutagenic agent") significantly increasing mutational events and thus mutation rates above the spontaneous background level (→ antimutagen). Mutations resulting from treatment with such agents are called "induced mutations", as opposed to the normally occurring "spontaneous mutations". The most widely applied mutagens are ionizing radiations, ultraviolet rays, base analogues, and alkylating agents which can induce both → gene mutations and chromosomal structural changes (→ chromosome mutation).

Most mutagens exert several chemical effects on DNA, some of which induce → mutagenic DNA alterations and others → inactivating DNA alterations.

mutagen specificity — allele- or locus-specific differential induction of → gene mutations or → reversions by different mutagens. M. s. is usually attributed to specific reactions between the mutagens used and the nucleotide sequences of the tested chromosome regions. It may also arise through modification, by the mutagen treatment itself, of some cellular processes that affect different premutational lesions in DNA to different degrees and in different ways.

mutagenic — capable of inducing → mutations (→ mutagen).

mutagenic DNA alteration (*Freese* et al. 1969) — any of those primary DNA alterations which do not prevent DNA replication and may give rise to → gene mutations. Such lesions are not responsible for the production of large chromosome structural changes, i.e., → chromosome mutations (→ inactivating DNA alteration). They may be due to mistakes in base pairing (induced either by base analogues or by minor chemical alterations of DNA bases) or may result from mistakes by the → DNA polymerase.

mutagenic effectiveness — the relation of the mutation frequency to the molar concentration of the mutagen applied.

mutagenic index — → dominant lethal assay.

mutagenicity assay — any test procedure used for the evaluation of potential environmental mutagens and extrapolation of the data to man. Such procedures include the microbial mutagenicity assay (→ spot test; rec assay), the → host-mediated assay, mammalian cells *in vitro, in vivo* cytogenetic studies, and the → dominant lethal assay.

Any m. a. should allow the measurement of: 1. all genetic events of concern to man and 2. the consequences of mammalian metabolism acting upon non-mutagens to yield mutagenic compounds or upon mutagens to inhibit their mutagenic activity.

The tests should be 1. highly sensitive, 2. should invariably be positive for substances that might pose a genetic risk to man, 3. should be simple and inexpensive to conduct, and 4. should serve to provide an assessment of risk or safety to the exposed human population (*Flamm* 1974).

These demands can only be fulfilled by a battery of tests subdivided into different levels or tiers of increasing power of extrapolation to man.

mutagenicity screening — = → mutagenicity testing.

mutagenicity testing — the assessment (= m. screening) of chemicals for mutagenicity by different → mutagenicity assays.

mutant — an individual resulting from → mutation (either → gene mutation, → chromosome mutation, → genome mutation or mutation of cytoplasmic hereditary determinants). The standard of reference is the so-called → wild-type condition, i.e., either the state of organisms as they are found in nature or arbitrarily chosen. Any heritable variation from it which is the result of mutation is a mutant.

mutant frequency — the ultimate proportion of mutants detectable in a population, regardless of how and when they arose (*Zamenhof* 1967).

mutant substitution — a change at the population level due to the spreading of a → mutation to the entire species. Each m. s. takes a long time and only a small fraction of mutations manage to lead to m. s. (→ genetic load).

mutate — to undergo or exhibit → mutation.

mutation (*de Vries* 1901) — any detectable and heritable change in the → genetic material not caused by → genetic segregation or → genetic recombination, which is transmitted to daughter cells and even to succeeding generations giving rise to mutant cells or mutant individuals provided it does not act as a dominant → lethal factor. If the descendants of a mutant cell give rise only to somatic cells in multicellular organisms, a mutant spot or area arises. Mutations in the germ line of sexually reproducing organisms may be transmitted by the gametes to the next generation resulting in an individual with the new mutant condition in both its somatic and germ cells. Since the primary carrier of → genetic information is usually → DNA, a mutation may be any (or a combination of) detectable, unnatural change affecting the chemical or physical constitution, mutability, replication, phenotypic function, or recombination of one or more deoxyribonucleotides; nucleotides may be added, deleted, substituted for, inverted, or transposed to new positions with and without inversion (*Herskowitz* 1965).

The mutational units of a → genotype are, in order of decreasing size, the genome, the chromosome (→ genome mutation), the chromosomal segment involving more than one gene (→ chromosome mutation), and the gene (→ gene mutation). Since a functional gene (cistron) consists of many → mutational sites, a gene mutation may involve the entire functional unit or only one or more of the mutational sites of which it consists.

Mutations may occur spontaneously (→ mutability) and can be induced experimentally by application of → mutagens. There appears to be no qualitative difference between spontaneous and induced mutations, the former being the result of several unknown and ubiquitous mutagenic agents or effects. A considerable fraction of the spontaneous mutations seem to depend upon the intracellular production of mutagens and → antimutagens.

Since the products of mutation are in general deleterious, the process of m. is harmful for a proportion of the individuals in a population (→ genetic load). It is, therefore, advantageous for a population to maintain a relatively constant and low level of → mutability; this may be achieved by various means. When an individual dies or fails to reproduce as a consequence of the reduced → fitness of its mutated genotype, it is a victim of → "genetic death".

A special category of mutations concerns the hereditary determinants which are extrachromosomal and show non-Mendelian → inheritance. These are mutational changes not of the genotypic but of the plasmotypic component of the → idiotype (→ plasmon, plastom).

mutation fixation (*Witkin* 1956) — the development of a premutational DNA lesion into a permanent change in the nucleotide sequence of DNA after treatment with a mutagen. Two cellular processes can lead to m. f.: DNA replication and → DNA repair processes.

mutation frequency — the frequency of certain mutants in a population. This does not necessarily reflect the rate (→ mutation rate) at which the mutations arose.

mutation frequency decline — the significant and irreversible reduction in the frequency of some classes of UV-induced bacterial mutations upon temporary post-irradiation inhibition of protein synthesis. M. f. d. (MFD) appears to be associated only with UV-induced → suppressor mutations, most of which probably originate in genes coding for → transfer RNAs (→ suppressor tRNA), or possibly other elements of genetic translation. MFD is probably a special case of → excision repair (enhancement of the excision of potentially mutagenic damage in certain tRNA coding genes) in which the excisability of premutational UV photoproducts in suppressor genes (but not in the DNA as a whole) is specifically increased under conditions of inhibited protein synthesis. (MFD is absent or drastically reduced in excision-defective bacterial strains.) Possibly, MFD-promoting conditions increase the reparability of suppressor genes by triggering their repression. The repressed state of any gene may favor more efficient repair than the active state. The confinement of MFD to suppressor genes may then reflect the specificity of the conditions required to produce the repressed state (*Witkin* 1969).

mutation initiation — → gene mutation.

mutation isoallele (*Lefevre* 1955) — any one of a number of → alleles with similar genetic activity but different rates of mutation (→ isoallele).

mutation map (*Muller* 1927; *Muller & Painter* 1932) — the frequency of mutation plotted as a measure of chromosome length (→ chromosome map).

mutation pressure (*Wright* 1921) — the continued recurrent production of an → allele by mutation tending to increase its frequency in the → gene pool of a population (→ gene frequency). If no other forces intervene, the mutation of allele A to a leads to changed proportions of these alleles which might ultimately result in the replacement of A by a (the entire population would then be aa in respect to the A locus). However, since → back mutations from a to A also occur (generally at different rates), the population never becomes homozygous for one type of allele by mutation alone. Instead of "fixation" of one allele, an equilibrium will be established at that point where the number of mutations from A to a equals the back mutation rate from a to A.

Although theoretically m. p. may be expected to bring about evolutionary changes in a → gene pool, in fact, the course of → evolution is not controlled to any great extent by mutation. Mutation is a limiting rather than a controlling factor in evolution.

mutation rate — the probability with which a particular mutational event takes place per biological entity (virus, cell, individual) per generation, i.e., the period required for the reproductive cycle. M. r. may be estimated by various methods from the → mutation frequency (→ gene mutation, mutant frequency).

The observed m. r. depends on several factors: 1. the intrinsic mutability of the gene in question (primary rate of damage or error insertion); 2. the probability of → repair; 3. the probability that the mutation will produce a recognizably altered phenotype.

mutation spectrum — the frequency distribution of induced or spontaneous mutations in different genes of a → genome or different sites within a gene. M. spectra are usually mutagen-specific (→ mutagen specificity). Selectivity of mutation induction may be due to the following mechanisms (*Kaplan* 1974): the mutagen may react selectively with certain bases or base sequences of DNA because of special affinities to them (→ mutational hot spot); 2. the attack of the mutagen on DNA may depend on the physiological state of a genome region or its environment, e.g., the state of replication, transcription, repression, host-controlled modification, or certain neighboring cell components; 3. repair systems may act selectively at certain genome regions, or on certain types of premutational lesions distributed non-randomly within these regions; 4. mutational changes at different sites of a gene or in different genes may be phenotypically expressed to differently observable degrees (→ mutagen specificity).

mutational delay (*Newcombe* 1953) — = → mutational lag.

mutational equilibrium — the frequency of mutated genes per individual. It is equal to the frequency of mutations per individual per generation multiplied by the average number of generations during which a mutated gene is maintained before it is eliminated (→ genetic load).

mutational hot spot (*Benzer* 1955) — any site at the nucleotide level within the → gene which is highly mutable in one or the other direction (forward or → back mutation). The intragenic location of mutational hot spots may differ for mutations occurring spontaneously and those induced by various → mutagens, i.e., loci may respond specifically to one of several mutagenic agents. The existence of hot spots is evidence for → mutability at the gene level and presumably reflects the probability that a nucleotide replacement (giving rise to a mutant phenotype) occurs. It implies further that mutation is determined both by the nature of the individual nucleotides of DNA and the interactions among adjacent nucleotides which facilitate or hinder nucleotide replacements or even cause mutation of sites spatially removed from those of the primary chemical change.

mutational lag (*Demerec & Cahn* 1953) — in addition to → copy error lag, → cell division lag, → phenomic lag, and → segregation lag, a delay (= "mutational delay", *Newcombe* 1953) in the appearance of bacterial mutant clones by delayed → gene mutation.

mutational load (*Muller* 1950) — → genetic load.

mutational site (*Pontecorvo* 1958) — any position along the → gene (or chromosome) at which → mutations can occur. The mutational sites are the deoxyribonucleotide building blocks which make up the genes (representing regions of definite length in DNA molecules). Like the order of genes on a chromosome, the order of mutational sites (which are separable by → crossing-over) is strictly linear. The average gene contains about 500 to 1 000 such sites whose linear order and relative spacing in the recombination map of a gene (→ genetic map) corresponds to the order and spacing of amino acid replacements in the polypeptide specified by the gene in question. Since mutations of the same m. s. do not always contain the same amino acid replacement in the corresponding polypeptide, mutable sites can exist in several alternative forms. With four different types of nucleotides in DNA, a m. s. is not expected to show more than four alternative configurations.

Based on the presence or absence of intragenic recombinations (→ genetic recombination) between mutational sites, three types of mutationally produced → alleles of one gene are distinguished:

1. those giving rise to the wild-type by recombination. They are interpreted to be the result of mutation in different sites of the same genetic locus and are referred to as "nonidentical" alleles (*Demerec* 1956) or "heteroalleles" (*Roman* 1956);

2. those giving rise to the wild-type by recombination with certain alleles of type 1., but which do not recombine with all of them. They are interpreted to be the result of changes in two or more adjacent sites, i.e., represent "multisite mutations";

3. those not recombining with each other but with all mutant alleles

of type 1. They are interpreted to be the result of recurrent but not necessarily identical mutation of the same site and are referred to as "identical alleles" (*Demerec* 1956) or "homoalleles" (*Roman* 1956).

mutational synergism — synergism between different mutagens in inducing → mutation. M. s. between chemical agents and UV irradiation has been reported in a number of organisms and in all cases the synergistic effect is explainable in terms of the interaction between the chemical agents and UV repair mechanisms.

mutator gene (*Demerec* 1937) — any gene that produces a generalized increase in spontaneous → mutation rates.The wildtype allele of what is recognized as a m.g.may function in DNA replication or repair, or both (→ mutator phage).The failure of some m. genes to exhibit mutator activity in forward mutation tests, while clearly exhibiting it in reversion tests, suggests that their action is specific to certain alleles or to certain mutational pathways which are not often represented among spontaneous forward mutations.

The following modes of m. g. action may be envisaged (*Yanofsky* et al. 1966):

1. an altered polymerase activity resulting in errors during the replication of normal nucleotide sequences;

2. the production of mutagenic base analogues which lead to replication or incorporation errors;

3. a modification of one or more of the different bases in DNA leading to replication errors.

mutator phage (*Taylor* 1963) — a → temperate virus (e.g., phage Mu of *E. coli*) with the unusual property of not having one → attachment site on the bacterial chromosome, but able to insert itself with high frequency at any site. If the m. p. inserts itself inside a structural gene, the coherence of the genetic message is broken and a mutation results. Phage Mu shows redundancy for certain DNA sequences, making internal crossovers inside the phage genome possible. Insertion of Mu can also induce → genetic recombination between two unrelated DNA molecules, if these both carry an inserted Mu DNA sequence. By exchange between the Mu regions, the two unrelated molecules can be coupled or can exchange parts at either side of the inserted phage DNA (→ insertion motation).

mutein (*Pollock, Fleming & Petrie* 1965) — a mutationally (→ gene mutation) altered protein analogous to the normal or wild-type version. Muteins may or may not possess any of the features (enzymological, immunological, or physico-chemical) of the corresponding wild-type protein. They normally arise by one or more amino acid substitutions in one or more of any of the polypeptide chains which constitute the molecule concerned (→ missense mutation; nonsense mutation).

muton (*Benzer* 1957) — the basic unit of → gene mutation, represented by a → mutational site (a nucleotide pair) within a gene or → cistron. The muton is the smallest alterable element within the one-dimensional array of mutational sites in the → genetic material which when altered may give rise to a → mutant organism (→ recon).

mutual complementation — the production of intact biparental phage chromosomes by cooperation (genetic recombination) between two defective phage chromosomes. M. c. and → marker rescue occur with perfect or nearly perfect efficiency in some bacterial strains but not in others.

N

n, 2n, 3n, etc. (*Blakeslee* 1921) — shorthand for the different degrees of ploidy. The numerical prefix denotes the number of → chromosome sets. In sexual diploids with a normal → alternation of nuclear phases, n is the gametic (haploid or monoploid) as opposed to 2n, the zygotic (diploid) number of chromosomes. 3n, 4n, 5n, etc. refer to triploid (3 chromosome sets), tetraploid (4 chromosome sets), pentaploid (5 chromosome sets) chromosome complements, etc.

nascent — of protein chains in the process of being formed on a → messenger RNA template in polyribosomes.

nebenkern (*Bütschli* 1871) — = → paranucleus.

negative control — prevention of biological activity by the presence of a specific molecule (e.g., inhibition of mRNA initiation by binding of a specific → repressor to a specific operator of an → operon). (→ positive control).

negative liquid holding (*Parry* 1972) — the reduction in cell viability produced by post UV holding treatment in saline. N. l. h. is probably the result of the reduced activity of an enzyme system involved in the → repair of **non-photoreactivatable** UV-induced lesions in some UV sensitive yeast mutants (→ liquid holding recovery).

neighborhood (*Wright* 1946) — a semi-isolated subpopulation of relatively inbred character (= panmictic unit or deme). Neighborhoods arise when a large populations is broken up by → inbreeding.

neobiogenesis (*Keosian* 1965) — a term proposed to describe the presumed repeated origin of life in nature since life began (→ eobiogenesis).

neobiont (*Keosian* 1965) — an organism originating by → neobiogenesis. Neobonts may be commensals, symbionts, or parasites, as well as free-living (*Keosian* 1965).

neocentric (*Rhoades & Kerr* 1949) — → centromere.

neocentric activity — a diffused or terminalized action of the → centromere (usually in first and second meta-anaphase of meiosis, not in mitosis) in individuals with an abnormal genotype. The chromosome arms showing n. a. are pointed like telocentric ends and coorientate with their own centromeres and with other neocentric ends of the same pairing configuration (meiosis I) or daughter element (meiosis II) (*Darlington* 1965).

neocentromere (*Rhoades* 1962) — → centromere.

neo-Darwinian evolution — → evolution.

neomorph (*Muller* 1932) — → allele.

neoplastic transformation — → cell transformation.

neoplastoid (*Martin & Sprague* 1973) — of → cell lines from mammalian tissues (e.g. HeLa) which are immortal and sometimes behave like neoplasms *in vivo*. They are typically → heteroploid, but may be euploid (→ hyperplastoid).

neosome (*McCarthy, Britten & Roberts* 1962) — → ribosome.

neoteny (*Kollmann* 1885) — 1. the retention of juvenile characters in the adult state;

2. the occurrence of adult characteristics in the juvenile state.

neo-XY system — → sex chromosome.

neo-Y-chromosome (*White* 1935) — → sex chromosome.

new reunion (*Newcombe* 1942) — → reunion.

nexus (*Dewey & Barr* 1962) — any of the specialized regions of contact (= gap junction) between the apposed cell membranes of adjacent cells.

N-formyl-methionyl transfer RNA (*Marcker & Sanger* 1964) — the → initiator tRNA in bacteria, → chloroplasts, and → mitochondria. It arises by formylation of the α amino group of methionine ($NH_2 \cdot CH(CH_2 \cdot CH_2 \cdot SCH_3)CO \cdot OH$) which takes place after the methionine has become attached to its → transfer RNA. Formylation is catalyzed by a transformylase. The specificity of N-formyl-methionyl tRNA as a polypeptide initiator during → genetic translation resides in the structure of this tRNA and not in the formyl group.

niche (*Grinnell* 1904) — all the relationships of an organism to its physical and biotic environment, i.e., the genetically determined capacity (range of tolerance) and pattern of biological response of an individual, a species population, or the whole species to environmental conditions (→ ecological niche). The variety of resources a population exploits is called the niche width and may be narrow or wide.

nick — any single strand break in a nucleic acid molecule. An enzyme able to produce such breaks may be called a "nickase", while an enzyme able to heal these breaks may be called a "sealase" (*Hurwitz* et al. 1967).

nodoc (*Cantoni, Ishikura, Richards & Tanaka* 1963) — → transfer RNA.

nombre fondamental (*Matthey* 1949) — the sum total of chromosome arms in the chromosome complement of a species, not including the small arms of acrocentric, rod-like chromosomes.

nonchiasmate — = → achiasmate.

non-congression (*Darlington* 1937) — the failure of pairing configurations (in → meiosis I) or chromosomes (in → mitosis and meiosis II) to come into orderly arrangement on the spindle equator (metaphase or equatorial plate). The configurations or chromosomes concerned are localized outside the equatorial plate and may give rise to an "accessory" plate. Non-congressing chromosomes are usually not included in the nuclei produced by meiosis or mitosis (→ chromosome congression).

non-conjunction (*Belling* 1925) — absence of meiotic → chromosome pairing (→ non-disjunction).

nonconservative — of a mode of DNA replication in which there is no net increase in the amount of DNA but only a replacement of parental DNA with newly synthesized DNA (→ repair replication).

non-Darwinian evolution (*Harris* et al. 1968; *King & Jukes* 1969) — → evolution.

non-disjunction (*Bridges* 1913) — 1. **Cytological (or numerical) non-disjunction:** The irregular distribution of sister-chromatids ("mitotic n.-d.") or homologous chromosomes ("meiotic n.-d.") to the cell poles, as inferred from cytological observation.

a) **Mitotic n.-d.:** The failure of sister-chromatids to be distributed to opposite cell poles during mitotic anaphase resulting in → aneuploid (complementary hyper- and hypoploid) products of mitosis.

b) **Meiotic n.-d.:** The failure of chromosomes contained in pairing configurations (bi- or multivalents) to separate regularly at first anaphase, or the distribution of unpaired (univalent, → asynaptic) homologues to the same cell pole at first anaphase, both cases resulting in aneuploid (hyper- and hypoploid) products of meiosis (gametes or meiospores).

2. **Genetic non-disjunction:** All types of abnormal meiotic behavior resulting in n.-d. as inferred by genetic methods.

a) **Chromatid n.-d.:** (*Sansome* 1933): The distribution of segments of sister-chromatids, distal to a → chiasma, to the same gamete or meiospore at meiosis. In → autopolyploids, this type of n.-d. gives rise to → "chromatid segregation".

b) **Primary and secondary n.-d.** (*Bridges* 1916): In XX−XY-systems of → sex determination, n.-d. leading to the production (by an XX-individual) of egg cells with two X-chromosomes or none" ("primary n.-d.") or to the production of egg cells with two X-chromosomes, X + Y, one X, or one Y-chromosome, by an XXY-individual ("secondary n.-d.").

Primary n.-d. may also occur in male individuals (XY) during first (resulting in XY- and O-sperm) or second meiotic division (resulting in XX- and O-sperm or YY- and O-sperm).

3. **Directed non-disjunction:** The preferential distribution of particular chromosomes (e.g., → B-chromosomes) to one of the two cell poles in mitosis or meiosis.

nonhistone chromosomal protein — in eukaryotes, a class of proteins (NHC proteins or hertones) characterized by pI < 10 which are isolated with DNA in chromatin or chromosomes. The NHC protein fraction includes enzymes of chromosomal metabolism, activator/repressor molecules related to specific gene expression, and protein contributing to the structure of different chromatin forms. NHC proteins (like histones) are synthesized outside the nucleus in the cytoplasm and exhibit a faster rate of turnover than histones with a lifetime of individual fractions varying considerably. The NHC proteins exhibit a large amount of heterogeneity, structurally and functionally. They range in molecular weight from under 10000 Daltons to over 150,000.

There are indications that the NHC proteins, while including many homologous molecules, also include tissue-specific and species-specific proteins. These may be involved in regulating tissue-specific transcription during development. NHC proteins are presumably bound to the chromatin or chromosome by a variety of means (interaction with DNA, → histones, themselves and RNA). Transcribed regions of chromatin are enriched in certain NHC proteins and impoverished in histones. The majority of chromatin, however, is covered with closely spaced histones which exceed the NHC proteins in amount by at least five-fold.

Phosphorylation of n. c. p. is probably involved in the mechanism by which these molecules regulate gene transcription. Phosphorylation possibly functions in histone-replacement from DNA.

nonhomologous — of chromosomes or chromosome segments which contain dissimilar genes and do not pair during meiosis (→ chromosome pairing), as opposed to → homologous or → homoeologous ones. Rare meiotic pairing of such segments in haploids, polysomics or structural hybrids is referred to as "nonhomologous pairing" and does not lead to → crossing-over and chiasma formation.

non-isolabeling — → sister chromatid exchange.

non-Mendelian — the mode of → inheritance shown by extrachromosomal or nonchromosomal hereditary determinants.

nonpermissive — = → restrictive.

nonrandom assortment — → assortment.

nonrecurrent parent — → recurrent parent.

nonreduction (*Belling* 1925) — the failure of meiotic → reduction (halving) of the somatic chromosome number due to disturbed → chromosome pairing (→ asynaptic; desynaptic) resulting in the production of a → restitution nucleus with the somatic number of chromosomes. This abnormal course of the first meiotic division is called by *Rosenberg* (1917, 1926) a "semiheterotypic division" giving rise to a "regression nucleus".

nonrepetitive DNA = → unique DNA or single-copy DNA.

nonsense codon (*Brenner* et al. 1971) — → nonsense mutation.

nonsense mutation (*Brenner* et al. 1961) — any mutation (base substitution or frameshift mutation) that converts a sense (amino acid specifying) → codon into a → chain terminating codon (UAG, UAA, or UGA in mRNA). N. mutations result in premature termination of polypeptide chain synthesis at the site where the nonsense codon occurs in the → messenger RNA.

N. mutations can lead to a pleiotropic polar effect (→ genetic polarity; polar mutation), reducing the level of expression of genes operator-distal to the n. m. in the bacterial → operon. To select for n. mutations in viral functions it is necessary to isolate the mutations in a genotype carrying a → nonsense suppressor (→ nonsense suppression).

nonsense suppression — suppression of the polypeptide chain termination effect of a nonsense codon (→ nonsense mutation) and partial or complete restoration of the normal phenotype. In the case of n. s., nonsense codons can act as sense (amino acid specifying) codons. This may be due to the presence of a → nonsense suppressor gene or a chemical suppressor (such as 5-fluoro-uracil and streptomycin). If, for instance, the nonsense triplet is UAA, then the substitution of 5-fluoro-uracil (5-FU) for U in the mRNA may result in the reading of FUAA as CAA, a sense codon. This will allow translation to continue since the tRNA will mistake FU for C with a higher probability than if U were present.

The efficiency of n. s. depends on the balance between polypeptide chain termination and amino acid insertion and may be drastically affected by the composition of the nucleotides adjacent to the nonsense codon.

Mutations counteracting n. s. are called antisuppressors.

nonsense suppressor (*Capecci* & *Gussin* 1965) — a → suppressor gene that allows the insertion of an amino acid into the growing polypeptide chain

(→ genetic translation) in response to a nonsense codon. In the absence of such a n. s. premature chain termination occurs. Most of the n. suppressors so far tested are dominant over their wild-type alleles. They represent mutations in a gene for a minor → transfer RNA (→ suppressor tRNA), or perhaps in a duplicate of a gene for a major tRNA species, so that a new tRNA is made which can translate a nonsense codon. Such a mutational alteration may affect either the → anticodon or some other part of a tRNA species. A tRNA altered to act as a n. s. normally will be unable to perform at least some of the usual functions of the wild-type tRNA species. Thus only non-essential (usually minor) tRNAs will be detectable as n. suppressor (→ missense suppressor).

In bacteria, three classes of n. suppressors are known to occur:

1. those that phenotypically correct only UAG (amber or N 1 codon) mutants; they represent tRNA having its anticodon changed to CUA;

2. those that correct UAA (ochre or N 2 codons) and UAG mutants (due to → wobble) and have their anticodon changed to UUA;

3. those that correct only UGA (opal or N 3 codon) mutants. In each case the amino acid inserted by the n. s. has at least one codon assignment which differs by a single base residue from the nonsense codon.

The polar effect of a → nonsense mutation is also corrected by n. suppressors. The degree of this polarity alleviation correlates in most cases with the level of phenotypic suppression.

nonsister chromatids — → sister chromatids.

nonsister label exchange (*Craig & Cameron & Jones* 1970) — → sister-label exchange.

nonspecific pairing — → chromosome pairing.

non-unit membrane — any membrane less than 6 nm wide, and appearing in section as a single, electron-dense layer (→ unit membrane).

norm of reaction (*Woltereck* 1909) — the entire range of the variant paths of development (phenotypes) that may occur in the carriers of a given → genotype or → idiotype (the sum total of chromosomal and extrachromosomal genetic information) in all environments, favorable and unfavorable, natural and artificial (*Dobzhansky* 1970).

N-terminus — → genetic translation.

nucellar embryony — a form of → apomixis in which the embryo arises directly from the nucellus, i.e., a parenchymatous tissue (localized between ovule or megaspore and its inner integument) which extends from the chalaza at the base to the micropyle at the apex of the embryo sac.

nuclear — ref. to phenomena, processes and constituents associated with the → nucleus of the cell.

nuclear acidic protein — → nonhistone chromosomal protein.

nuclear association (*Buller* 1941) — a transitory stage during fertilization. It follows plasmogamy and precedes karyogamy. It is characterized by the presence of one or more pairs of nuclei within the same cell (→ dikaryophase).

nuclear basic protein — → histone.

nuclear body — 1. the region in the bacterial cell to which the DNA is confined (= nucleoid);

2. a structure in → meiosis of plant species which seems to originate from the synthetic activities of the chromosomes at the beginning of meiosis. It consists of RNA, proteins and phospholipids (*Walters* 1963).

nuclear cap — in certain organisms, an unusual membrane-enclosed organelle surrounding the anterior part of the → nucleus.

nuclear cloning — surgical placement of nuclei from accessible tissues (such as skin or intestine) into an unfertilized, enucleated egg. By this procedure whole organisms may develop with precisely those features of the organism from which the nuclei where obtained.

nuclear differentiation — any change during → cytodifferentiation with respect to the number, structure and/or function of the → chromosomes which constitute the → nucleus of a particular organism.

1. Changes which, by elimination or diminution of chromosomes, give rise to clones of different functional karyotypes within the individual, as examplified by "chromosome elimination" in *Sciaridae* and "chromosome diminution" in *Ascaris*.

The so-called **"limited chromosomes"** of *Sciara* are large, heterochromatic elements found in the germ line cells. These chromosomes are eliminated from some nuclei during early cleavage of the zygote. Such nuclei develop into the soma whereas those nuclei retaining these chromosomes give rise to the germinal tissues.

Chromosome diminution is typical of the first cleavage division of *Ascaris* in the course of which two daughter cells with different chromosome behavior are produced. In one, the large polycentric chromosomes characteristic of the germ line karyotype break up into several fragments. The centric fragments constitute the karyotype of the descendants of this cell, the acentrics being lost. In the other cell, the chromosomes remain intact until the next division, when one cell undergoes diminution and one does not. The "unitaleral" inheritance of an intact chromosome complement is continued for several divisions after which the cell with the intact chromosomes undergoes equal mitosis resulting in a clone of such cells which will become the germinal tissues. The other cells produce the soma.

2. Changes in the pattern of chromosome duplication leading either to endopolyploid nuclei or polynemic (polytenic) chromosomes. The underlying process is → endoreduplication or endomitosis which gives rise to endopolyploidy by chromosomal reduplication and separation of the reduplication products without nuclear division. Polynemy (or polyteny: → polytene chromosome) results from the same type of endoreduplicational process but the old and newly formed elements remain attached to each other rather than becoming individualized and form multistranded chromosomes which behave as morphological and functional units. Endopolyploidy is frequent both in plants and animals and may even give rise to new races or species if it affects the cells of the germ line. Polyteny is a normal phenomenon in the development of many tissues of the larval and adult Diptera.

3. A third phenomenon of n. d. in multicellular organisms is exemplified by changes in manifestation either at the intrachromosomal (differential

gene activation, → puffing) or interchromosomal level (as, e.g., the genetic inactivation of one of the two → X-chromosomes in mammals associated with → dosage compensation.)

nuclear dimorphism — the presence (in ciliates and some foraminiferans) of two types of → nuclei per cell, designated the → micro- and → macronucleus, respectively. The micronuclei (→ generative or reproductive nuclei) are diploid and capable of unlimited division, whereas macronuclei (→ somatic or metabolic nuclei) are diploid or polyploid and show either limited or no capacity for division. Macronuclei, therefore, will eventually degenerate and have to be periodically reconstructed from the descendants of micronuclei.

Raikov (1958, 1959) distinguishes two types of n. d.:

1. **Primary type:** The macronuclei are diploid and incapable of division. Cell division is associated with a multiplication of micronuclei. Before the separation of daughter cells, some micronuclei develop to produce a new set of diploid macronuclei.

2. **Secondary type:** The macronuclei (most ciliates) are polyploid and incapable of division.

nuclear disruption — a virus-induced rearrangement of the contents of the bacterial host cell which occurs (in *E. coli*) within two to three minutes after infection with T-even bacteriophages. During n. d., the DNA of the host → nucleoid moves from a largely central location in the cell into close association with the cell membrane. N. d. is under the control of a phage gene.

nuclear division — in eukaryotes, the division of the cell → nucleus by → mitosis, → meiosis, or → amitosis. Mitotic n. d. may be open or closed, depending on whether or not the → nuclear envelope persists during division, and centriolar or acentriolar on the basis of the presence of absence of → centrioles.

nuclear DNA — in eukaryotes, the → deoxyribonucleic acid of the cell → nucleus as opposed to → organelle DNA. Nuclear → genes follow the pattern of Mendelian → inheritance, while organelle genes and DNA are inherited in a non-Mendelian way (→ genetic segregation).

nuclear envelope (*Anderson* 1953) — in eukaryotes, the ensemble of constituents making up the double-membraned structure surrounding the interphase and prophase → nucleus of the cell. The n. e. is the universal representative of the cell's → endomembrane system. It is a highly specialized partition between the cytoplasmic and nuclear compartment of the eucell and consists of an inner and outer unit membrane, the perinuclear space, nuclear pores, annular material and internal dense lamellae (Fig. 70).

1. **Nuclear membranes:** the paired membranes of the n. e. run relatively parallel to one another at the nuclear periphery and enclose the nucleoplasm, nucleolus, and chromatin. The membrane in direct contact with the cytoplasm is the outer membrane; the membrane bordering on the nuclear site is the inner membrane. Both are 7 to 8 nm in diameter.

2. **Perinuclear space:** the intermembraneous (electron transparent) zone, about 15 to 30 nm wide, without discernible substructure. Its diameter

varies from cell to cell as well as around the nuclear periphery of the same cell.

3. **Nuclear pores:** Interruptions in the n. e. at sites where both membranes unite to form a circular discontinuity. When seen in a face-on-view the pore is called an annulus. The pore and its associated material is called the pore complex. The annular material, mostly proteinaceous, is probably of importance in the control of pore permeability and endowed with enzymatic activities controlling the state (open or closed) of pores. Pore diameter is in the range of 30 to 100 nm. Their shape can vary between circular and polygonal.

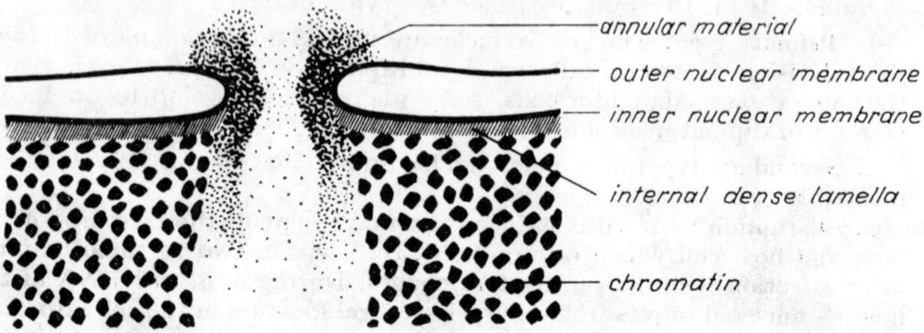

Fig. 70. Diagrammatic representation of a region of the nuclear envelope containing a nuclear pore (from *Stevens* and *André* 1969).

4. **Internal dense lamella:** A prominent thick lamella (thickness ca. 280 nm) of homogeneous and compact material closely apposed to the inner surface of the inner nuclear membrane. It may have a supporting function, may act as a shift substratum of the inner membrane and may also have a physiological role in nuclear permeability.

The n. e. controls the numerous, rapid and constant exchanges between the cell nucleus and the cytoplasm, e.g. entrance of small molecules and ions, exit of macromolecules synthesized inside the nucleus. It is formed by fusion of pieces of the → endoplasmic reticulum around the chromosomes at the conclusion of nuclear division (excepting in fungi and some protozoa). There is some evidence that n. e. fragments remain associated with metaphase and anaphase chromosomes.

The n. e. and the rough endoplasmic reticulum are in direct structural continuity in most cells. → Chromatin is regularly associated with the n. e. The attachment may be temporary and involves specialized parts of the chromatin (attachment points). Association between the n. e. (inner membrane) and initiation and replication sites of DNA synthesis have been postulated for many higher organisms.

nuclear fragmentation — a degenerative breakdown of the → nucleus into two or more parts of equal or unequal size.

nuclear membrane — → nuclear envelope.

nuclear phenotype (*Darlington* 1932) — the genetically controlled → phenotype of the chromosomes, which is species-specific, varies considerably with respect to chromosome appearance and chromosome behavior in the course of the → cell cycle (→ mitosis; meiosis; endoreduplication; nuclear differentiation) and frequently reacts to changes of genotype (by segregation, hybridization, or mutation) in much the same way as do characters belonging to the "exophenotype". Variation of the n. p. depends on genotypic (the chromosomes carrying the genes are subject to the action of these genes) and structural properties (represented by the order rather than the action of the genes in the chromosomes) of the → karyotype, the evolution of which results from the action of natural → selection on the heritable variability of chromosome shape and chromosome function and gives rise to the several different variants of the chromosome mechanism found in nature (→ endophenotype).

nuclear plate — = → metaphase plate or equatorial plate.

nuclear pore — → nuclear envelope.

nuclear pore complex (*Watson* 1959) — → nuclear envelope.

nuclear ribonucleoprotein — RNA-protein complexes in the → nucleus which contain the primary RNA transcripts (→ heterogeneous nuclear RNA; informofer). Electronmicroscopically, n. r. particles appear as a loose arrangement of fine fibrous material which itself is particulate. The protein which constitutes the main mass of the particles is arranged as 10—20 nm subparticles in bead-like fashion on the string or strings which make up the total particle.

nuclear RNA (*Harris* 1963) — = → heterogeneous nuclear RNA; low molecular weight nuclear RNA; ribonucleid acid.

nuclear sap — → nucleoplasm.

nuclear segregation — the segregation of genetically different nuclei contained in a → heterokaryotic multikaryon.

nuclear tetrad — the four haploid nuclei arising from the two meiotic divisions (→ meiosis; tetrad).

nuclear transplantation — the transfer and insertion of the nucleus of a specialized cell into an unfertilized egg whose nucleus has been removed. If the donor-cell nuclei are not taken from the cells of an embryo or larva reared from a fertilized egg but from a young embryo that is itself a result of a nuclear-transplant experiment, this is called serial nuclear t. The effect of n. t. is the production of a clone, i.e., a population of individuals all having identical genomes in their nuclei.

nucleate — having a → nucleus.

nucleating site (*Tucker* 1972) — → microtubule.

nucleic acid (*Altmann* 1889) — → deoxyribonucleic acid; ribonucleic acid.

nucleic acid hybridization — the formation of DNA-DNA and DNA-RNA hybrid molecules by annealing. N. a. h. can be used to determine whether appreciable homology exists between a test sequence and a given fraction of the genome (→ Cot value *situ* nucleic acid hybridization).

nucleo-cytoplasmic interaction — interaction between nuclear genes and the heterogeneous cytoplasm which may be divided into 1. effects of the nucleus on the activity of the cytoplasm, and 2. effects of the cytoplasm on the activity of the nucleus. N.-c. i. involves the passage of molecules from one compartment to another of the eucell; it may result in a wide range of changes in chromosomal and gene activity (→ gene activation) and, thereby, in → cytodifferentiation.

nucleodesma (*Scott* 1950) — a fibrillar connection between cell nucleus and cytoplasm (= karyodesma).

nucleohistone — a complex of → deoxyribonucleic acid and → histone with the histone bound to the large groove of DNA. Two models of n. structure have been proposed: 1. Non-basic residues of the histone are looped away from the DNA, enabling the basic residues to neutralize successive phosphate groups of DNA. 2. One histone is stretched along the length of about 12 turns of the DNA helix, a segment in which about 75% of the DNA phosphate group would not be neutralized by histone basic residues and are, therefore, free for interactions with other cations.

Extraction of n. is always accompanied by extraction of DNA-bound → nonhistone chromosomal protein, suggesting that multimolecular structural complexes exist in native nucleoprotein. Different types of histones vary in their tendency to aggregate and in the closeness of their binding to DNA.

nucleoid (*Piekarsky* 1937) — the region in the bacterial cell to which the DNA (the bacterial → genome) is confined (= karyoid, bacterial nucleus, nuclear body, nucleoplasm, DNA plasm). Folding and supercoiling of DNA makes the n. a more or less compact structure (without membrane) and does not interfere with DNA replication and → genetic transcription. Often there are two or more nucleoids per cell (depending on nutrition). Isolated nucleoids may be membrane-associated (→ mesosome) or membrane-free.

nucleolar chromosome — any chromosome with a → nucleolar organizer.

nucleolar constriction — a secondary → constriction associated with the formation of the → nucleolus.

nucleolar DNA — 1. = ribosomal DNA.

2. A type of single-stranded DNA in eukaryotes. The kinetics of labeling, the base composition, and the lack of complementarity with → ribosomal RNA indicate that this DNA comprises a discrete portion of the genome whose function remains unknown (*Almaric* et al. 1973).

nucleolar interstices (*Bernhard & Granboulan* 1968) — light areas (nucleolar vacuoles) surrounding the nucleolonemas of the → nuclcolus; n. i. vary in size and content and are characterized by a low concentration of structural elements.

nucleolar lacuna (*Chouinard* 1970) — any of the small electron-clear areas within the → nucleolus which contain fibrils of 6 to 10 nm in diameter.

nucleolar organizer (*McClintock* 1934) — = → nucleolus organizer.

nucleolar ribonucleoprotein particle (*Warner & Soeiro* 1967) — → ribosomal precursor RNA.

nucleolar segregation (*Bernhard* et al. 1965) — the redistribution of morphologically discrete nucleolar components induced by various drugs (actinomycin D and agents acting by similar biochemical mechanisms on DNA-dependent RNA synthesis) and characterized by the sorting out into distinct zones of the fibrillar and granular components of the → nucleolus.

nucleolar vacuole (*Chouinard* 1970) — any of the distinct electron transparent areas within the granular zone of the → nucleolus. Such vacuoles contain scattered granules and fibrils and their presence, number and size are correlated with various physiological conditions of the cell. They are assumed to be involved in the release of nucleolar products (RNP precursors of → ribosomes).

nucleolar zone (*Serra* 1942) — any chromosome region, irrespective of whether or not it is a secondary → constriction, that is associated with the formation of the nucleolus during telophase.

nucleolinus — a nucleolar fibrillar center. Nucleolini usually appear as small rounded bodies of different size which are surrounded by dense fibrillar RNP components. In contrast to → nucleolar interstices or light areas, they contain a large amount of fine filaments and dense fibrils. Their function is not clear.

nucleolonema (*Estable* & *Sotelo* 1950; *Bernhard* et al. 1952). — any of the threadlike ribonucleoprotein structures (about 0.1 nm in diameter or larger) contained in the → nucleolus of most cells. They are surrounded by light areas (nucleolar interstices). Nucleolonemas are composed of fibrillar and/or granular elements. The n. may contain chromatin structures in the form of fine filaments or strands.

nucleolus (*Bowman* 1840) — an intranuclear organelle of eukaryotes which represents the chromosomal site for the genes coding for → ribosomal RNA (→ ribosomal DNA). The primary gene products (→ ribosomal precursor RNA), derivatives of these products (ribosomal RNA), their associated proteins, and some, perhaps all, of the enzymatic equipment (RNA polymerase, RNA methylase, RNA cleavage enzymes) required for synthesis, conversion, and assembly of → ribosomes involve the nucleolus. Subsequently ribosomes are transported to the cytoplasm.

Although the chemical components of nucleoli vary in amount depending on cell type and physiological state, generally, three main components are present: DNA, RNA, and protein. The RNA content ranges between 3 and 13% of the dry weight (average 8%); the protein content ranges between 82 and 97% (average 92%). Protein to RNA ratios average at about 11.5:1. Lipids may be present in trace amounts.

Nucleolar shape and size vary widely depending on the organism, cell type, and physiological state of the cell. Two principal regions have been established in the n.: the fibrous central core (pars amorpha) and the granular cortex (pars granulosa). In most cases both are more or less intermingled. Two other nucleolar components are the → nucleolar lacunae and the → nucleolar vacuoles. The central core is composed of DNA fibrils associated with growing strands of RNA, appropriate in size for → preribosomal RNA. This RNA is associated with protein synthesis, and probably forms the RNP particles (15—20 nm) seen in the granular cortex.

N. specific functions in the assembly of cytoplasmic → ribosomes include the transcription of the large (45 S) precursor of 7 S, 18 S and 28 S ribosomal RNA, the methylation of this precursor, and the association of many of the ribosomal proteins with rRNA. 5 S RNA, another component of cytoplasmic ribosomes, is synthesized autonomously, and in several instances independently of nucleolar function. The 45 S ribosomal precursor RNA is associated with the fibrous nucleolar region. The intermediate precursor of rRNA (32 S in mammals, 30 S in lower animals) is formed in the granular cortex. The 15 nm granules found in the cortex are composed of ribonucleoprotein containing 32 S and mature 28 S rRNA. Ribosomal DNA involved in the synthesis of the large precursor to rRNA is finely dispersed in the fibrous component of the n.

Some components of the n., namely those of the fibrous core which contains the 45 S RNA, persist during metaphase; it is associated with the chromosomes during → mitosis.

The disappearance of the n. during mitosis probably arises from disaggregation of its RNP constituents. Nucleoli are reformed at the end of mitotic anaphase from diffused prenucleolar substance which appears as a coating on the surface of the chromosomes or as small spherical droplets between the chromosomes. The → nucleolus organizer is involved in the condensation of the prenucleolar substance to form the n. RNA synthesis begins in the newly formed n., which is mainly composed of fibrillar components; as synthesis continues, granular components gradually accumulate.

The number of discrete nucleoli found in the nucleus of cells varies according to species, strain, tissue type and developmental stage. Interspecies differences in nucleolar number can be accounted for by variation in the number of nucleolar organizers present in the → chromosome set (→ genome).

The n. is frequently surrounded by a layer of condensed → chromatin called nucleolus-associated chromatin. Together with the n. it forms the nucleolar apparatus.

nucleolus associated chromatin (*Thorell* 1944) — → chromatin closely associated with the → nucleolus, and usually isolated with the nucleolar fraction.

nucleolus organizer (*McClintock* 1934) — the chromosome region which is active in → nucleolus formation and contains → ribosomal DNA, the genes from which → ribosomal RNA is transcribed (= nucleolus organizing region; nucleolar organizer). These genes are highly redundant (→ gene reiteration), being present in multiples of 100 to 500 and more per haploid → chromosome set. During metaphase of mitosis, the n. o. of the "nucleolus organizing chromosome(s)" is usually a secondary → construction. The number of n. organizers in a haploid chromosome set is characteristic of the organism. As a rule, at least one n. o. is present per chromosome set and if this is removed, abnormal nuclear behavior (frequently leading to the death of the cell) results.

In the oöcytes of certain animals, the n. o. undergoes differential multiplication (→ gene amplification; ribosomal DNA) to produce a large number of extra-chromosomal n. organizers, nucleoli and rRNA during oögenesis.

The fine structure of the n. o. is similar to that of the → centromere in the following ways: 1. it contains → lampbrush organization (dense axial cores and 5 to 8 nm lateral loops at the time RNA synthesis is occurring); 2. at mid-mitosis, when metabolic activity is shut off, the n. o. becomes more condensed but retains the same filaments and axial core, although less extended; 3. when synthetic activity resumes in telophase, the n. o. again "opens up" displaying the lampbrush-type organization.

nucleolytic — of enzymes (nucleases) that digest nucleic acids. Their specificity ranges from very broad to very narrow; some attack both RNA and DNA, others either RNA or DNA. N. enzymes specific for single- or double-stranded DNA exist. Nucleases may attack a polynucleotide either from an end (exonucleases), or at random (endonucleases).

nucleomixis (*Diamelidis* 1951) — the establishment of connections between the cell → nuclei during → meiosis (→ cytomixis).

nucleoplasm (*Strasburger* 1882) — the nonstaining or slightly chromophilic (liquid or semiliquid) ground substance (the "achromatin") of the interphase nucleus which fills the nuclear space around the → chromosomes and the nucleoli. Little is known of the chemical composition of this ground substance which is not easily defined. It may be called n. or "karyoplasm" when it is gel-like, and "karyolymph" when it is a colloidal fluid, but generally the terms are synonymous. According to electron micrographs, the ground substance consists of a pattern of irregular-shaped particles. Its chemical components are predominantly proteins, some RNA, and a number of enzymes.

nucleoplasmic index (*Hertwig* 1903) — the ratio of nuclear to cytoplasmic volume expressed numerically as

$$NP = \frac{V_n}{V_c - V_n}$$

(NP = nucleoplasmic index; V_n = nuclear volume; V_c = volume of the cell). With increasing volume of the cytoplasm, the volume of the nucleus also increases and *vice versa*. The nucleoplasmic ratio (= nucleo-cytoplasmic ratio) seems to trigger cell division.

nucleoplasmic ratio (*Hertwig* 1903) — → nucleoplasmic index.

nucleoprotamine — in eukaryotes, the saltlike complex between basic → protamine and DNA found in certain cell types (e.g. developing and mature sperms of some organisms), where it is usually the only nucleoprotein present (→ histone).

nucleoprotein — chemical complexes of nucleic acid (DNA or RNA) and specific types of proteins. Marked differences exist between the deoxyribonucleoprotein (RNP) complexes, both with respect to protein composition, three-dimensional configuration and cell function (→ chromosome; nucleolus; informofere).

nucleoside — → nucleotide.

nucleosome (*Navashin* 1912) — a → satellite situated on the surface of the → nucleolus.

nucleotide — any of the monomeric units which are the building blocks of the polynucleotides referred to as → nucleic acids. A n. is a phosphate ester of the N-glycoside of a nitrogenous base and consists of a purine or pyrimidine base, a pentose (D-ribose in → ribonucleic acid, 2′-deoxy-D-ribose in → deoxyribonucleic acid) and a phosphate (PO_4) group. The combination of nitrogenous base and sugar is called a ribonucleoside (when the pentose is D-ribose) and a deoxyribonucleoside (when the pentose is 2′-deoxy-D-ribose). The nucleotides are formed by the addition of a phosphate group to the nucleotides and are either ribonucleotides (in RNA) or deoxyribonucleotides (in DNA):

Nucleic acid	Main purine (PU) and pyrimidine bases (PY)	Pentose	Nucleoside	Nucleotide
		2′-deoxy-D-ribose	deoxyriboside	deoxyribotide
	Cytosine (PY)		Deoxycytidine	Deoxycytidylic acid
	Thymine (PY)		Thymidine	Thymidylic acid
DNA	Adenine (PU)		Deoxyadenosine	Deoxyadenylic acid
	Guanine (PU)		Deoxyguanosine	Deoxyguanylic acid
		D-ribose	riboside	ribotide
	Cytosine (PY)		Cytidine	5′Cytidylic acid
	Uracil (PY)		Uridine	5′Uridylic acid
RNA	Adenine (PU)		Adenosine	5′Adenylic acid
	Guanine (PU)		Guanosine	5′Guanylic acid

nucleotide sequence — a specifically ordered row of → nucleotides in the DNA or RNA of organisms. A n. s. which occurs many times in the DNA of one cell is called a repeated n. s. (→ repetitious DNA). A particular n. s. which occurs only once in a genome is called a unique n. s. (→ unique DNA).

nucleus (*Brown* 1831) — a membrane-enclosed cell organelle (→ nuclear envelope) which represents one of the two main compartments of the eucell and contains the bulk of its → genetic information (nuclear DNA) in the form of → chromatin. Replication of chromosomal DNA, → genetic transcription of DNA and processing of DNA transcripts (→ heterogeneous nuclear RNA; messenger RNA; ribosomal RNA; transfer RNA) occurs in the n.

Three functional classes of proteins are associated with nuclear DNA in the n.: 1. → histones attached to DNA (→ nucleohistone); 2. acidic proteins associated with DNA (→ nonhistone chromosomal protein), present in the → nucleolus and present in the nuclear sap; 3. enzyme proteins which include the enzymes involved in: DNA replication (→ DNA polymerase) → genetic transcription; chemical and physical modification of nucleic acids; energy-yielding specific protein synthesis.

The n. divides by → mitosis (in somatic tissues), by → meiosis (in germinal tissues resulting in the production of → gametes or meiospores) or (rarely) by → amitosis. It may become endopolyploid by the process of → endomitosis (→ nuclear differentiation).

Typical membrane-bounded nuclei are present only in the cell during → interphase **("interphase nuclei")**, prophase, and telophase; **"prophase nuclei"** are dissolved (breakdown of the nuclear envelope) at the end of mitotic prophase, and **"telophase nuclei"** are reformed at the end of mitosis and meiosis.

Interphase nuclei in mitotically active tissues are usually of spherical shape but may vary widely in their appearance in differentiated tissues. They generally show little or no defineable structures except for three components which are contained in a nonstaining or slightly chromophilic mass called the → karyolymph, nucleoplasm or nuclear sap which is essentially protein in character and contains → RNA and a number of enzymes:

1. one or more (usually) spherical bodies referred to as → nucleoli;
2. one or more regions of → heteropycnotic, deeply staining material referred to as → chromocenters;
3. a multitude of faintly stainable fine threads referred to as → chromonemata (or together with the chromocenters as "chromatin"), and representing the despiralized chromosome regions.

With the exception of the chromocenters, i.e., tightly coiled chromosomes or chromosome segments said to be → heterochromatic or positively heteropycnotic, the chromosomes of the interphase nucleus are in a widely uncoiled state (→ chromosome coiling). Activity of the chromosomes with the genes exerting their effects (→ gene action) is evidently associated with the uncoiled state of the chromosomes.

The **lifetime of interphase nuclei** which undergo further mitosis may be subdivided into subphases if DNA synthesis and chromosome reduplication (which occupy only a part interphase) are used as the criteria for classification. Interphase of the mitotic cell cycle consists of three parts referred to as G_1 (gap 1), S (synthesis), and G_2 (gap 2). During G_1 the chromosomes are as yet unreduplicated; during S DNA synthesis and chromosome reduplication take place; and G_2 is the postsynthetic phase which ends with the beginning of mitotic prophase. A somewhat similar subdivision is possible for those interphases which precede meiosis, but in this case G_2 is usually short or entirely absent.

nulli-haploid (*Riley & Chapman* 1958) — → haploid.

nulliplex (*Belling, Blakeslee & Farnham* 1923) — of a → polyploid in which all chromosomes of one homologous type carry the same recessive → allele of a particular → gene (e.g., aaa in case of triploidy, aaaa in case of tetraploidy, etc.), in contrast to simplex, duplex, triplex, quadruplex, etc., in which the dominant allele is represented one, two, three, four, etc., times (*Darlington & Mather* 1949).

nullisomic (*Blakeslee* 1921; *Sears* 1941) — of an → aneuploid cell, tissue (in the case of → chimeras), or individual with both members of one (or more than one: double nullisomic, etc.) particular pair of homologous chromosomes missing from the chromosome complement. Normally, nullisomics (or nullisomes) cannot survive in diploids, but they are recoverable in analogous forms, e.g., in → allopolyploids (which are functionally diploid) and → autopolyploids. Nullisomy may arise from selfing or intercrossing of → monosomics by fusion of two gametes lacking the same chromosome ("n — 1") of a particular chromosome pair.

O

ochre mutant (*Brenner, Stretton & Kaplan* 1965) — → amber mutant; nonsense mutation.

octoploid — of → auto- or → alloploid cells, tissues, and organisms, respectively, with eight chromosome sets in their nuclei (symbol 8n).

Okazaki piece — any of the short transient nucleotide sequences of DNA formed during discontinuous replication of → deoxyribonucleic acid. They have a short stretch of RNA at their 5′ phosphate ends which is used as a primer for the initiation of new DNA molecules (→ terminal deoxynucleotidyl transferase). Okazaki pieces are subsequently joined by DNA ligase to form a complete daughter strand. The RNA primer is probably removed by nucleases after synthesis of new DNA molecules is initiated but before the completion of the daughter molecule by → DNA polymerases and DNA ligase.

oligogene (*Mather* 1941) — = → major gene.

oligogenic — of → characters controlled by few genes, as opposed to → polygenic characters.

oligolecithal — = → isolecithal.

oligomer (*Changeux* 1964) — a protein composed of identical subunits (polypeptide chains).

omega (*Burgess* 1969) — → RNA polymerase.

oncogene (*Huebner & Todaro* 1969) — → oncogene hypothesis.

oncogene hypothesis (*Huebner & Todaro* 1969) — an hypothesis which proposes that all cells contain in their DNA the information necessary to specify the complete genome of a tumor virus (the virogene). This virus-specific DNA is transmitted vertically from generation to generation, and certain virogene functions may be normally expressed, e.g., in development. The virogene includes oncogenic information (the oncogene) which is responsible for most tumors. This information is normally repressed. A breakdown in the cellular regulation of these viral genes may, however, result in expression of the virogene and oncogene and, therefore, in neoplasia (→ protovirus hypothesis).

oncogenic — of viruses or agents which cause cancer in animals. The process by which a cell acquires new growth properties after virus infection is called → cell transformation. O. viruses may be DNA and RNA viruses.

one gene — one enzyme hypothesis (*Beadle & Tatum* 1941; *Beadle* 1945) — according to this hypothesis (based on studies in biochemical genetics) each → gene controls the synthesis or the activity of but a single protein (enzyme) with catalytic activity. Since its formulation this concept has been verified in principle.

According to current knowledge of the chemical nature and action of the → genetic material, the "one gene — one enzyme" concept has been revised to the "one cistron — one primary function" concept which neither excludes the existence of genes with functions other than specification of the amino acid content and sequence of catalytic proteins, nor excludes cases in which the enzymes have been found to be composed of different polypeptide chains each controlled by one gene.

For those genes (→ structural genes) or → cistrons determining the specificity of enzymes, the "one gene — one enzyme" concept is now modified to the "one cistron — one polypeptide chain" concept, or in the case of several genes representing one → operon, i.e., one unit of genetic transcription and regulation, to the "one operon — one messenger RNA" hypothesis.

one operon — one messenger hypothesis — → operon.

one-plane theory — → chiasmatype theory.

one step meiosis (*Cleveland* 1947) — → meiosis.

ontogenetic — ref. to the development (particularly → embryonic development) of an individual.

ontogeny (*Haeckel* 1886) — the course of development of an individual organism, an organ, or an organelle (→ recapitulation).

oöblastema — a fertilized ovum (= oösperm, oöspore, or zygote).

oöcenter (*Fol* 1891) — the → centrosome of the egg cell (= ovocenter).

oöcyte (*Boveri* 1891) — the egg mother cell ("primary o.") giving rise, in the first meiotic division, to the "secondary o." and the first polar body. At the second meiotic division, the secondary o. gives rise to the ovum and a second polar body (→ oögenesis).

oögamy — the union during → fertilization of a nonmotile female → gamete (egg cell) and a motile male gamete.

oögenesis — the development of the female → germ cell (egg cell or ovum) of animals which takes place in the gonad. The gonad (ovary) contains the primordial germ cells (primary oögonia) which enter a phase of rapid multiplication giving rise to the secondary oögonia. At the end of the multiplication period, the secondary oögonia transform into oöcytes which grow until the end of o. The period of oöcyte development is divisible into three subphases. During the first (premeiotic) phase immediately following the formation of the oöcyte from an oögonium, growth is very slow. At the end of that phase the nucleus starts to swell and the oöcyte enters a second phase of slow growth (mainly due to synthesis of cytoplasm). During the third phase the growth rate increases rapidly and in a relatively short time the oöcyte reaches its final size.

The so-called premeiotic phenomena normally begin immediately after the completion of the last oögonial division. During the first and second meiotic division, oögonia give rise, via primary and secondary oöcytes, to one functional gamete (the ovum or egg cell) and three or (rarely) merely two abortive cells (the polar bodies or polocytes). Meiosis II takes place in the secondary oöcytes and is generally arrested at meta- or anaphase II up to the time of fertilization. After plasmogamy the second meiotic division is completed and the female and male "pronuclei" unite at karyogamy (→ cleavage).

oögonium (*Boveri* 1891) — 1. a primordial → germ cell which gives rise to primary → oöcytes by mitosis (→ oögenesis);
2. a female → gametangium of Thallophyta.

oöplasm — the → cytoplasm of an egg cell. Different regions of the egg cytoplasm, spoken of as "oöplasms", may have specific properties, so that a particular region can only develop (→ embryonic development) in one definite way. Such regions may be precisely localized (sharp boundaries between them), or they may merge into one another (*Waddington* 1956).

oöplasmic segregation — → determination.

oösome (*Sylvestri* 1914) — = → germ line body or germ cell determinant.

oösperm — a fertilized ovum (= oöblastema, oöspore, or zygote).

oösphere — a female gamete before → fertilization.

oöspore — 1. a fertilized ovum (= zygote, oösperm, or oöblastema);
2. an encysted zygote in certain protozoa.

oötid — any of the four meiotic products arising during → oögenesis (= ovotid).

opal mutant (*Brenner* et al. 1967) — a → nonsense mutation giving rise to the chain terminating codon UGA in messenger RNA (→ amber mutant; ochre mutant).

operator (*Jacob* & *Monod* 1959) — according to the → operon model, the recognition site at which negative control of → genetic transcription is mediated through binding of a → repressor. The o. is a stretch of bases (21 bases in the case of the lac o. of *E. coli*) along a DNA molecule, located at the proximal end of an → operon. When the repressor is bound to the o., it prevents the → RNA polymerase from initiating transcription of the operon. Repressor interaction is with the native double-strand of DNA in the o. Cases are known where a single repressor can repress structural genes physically separated but sharing one o.

In vivo, an o. mutation is identified by the cis-dominant constitutive expression of the genes in its operon. *In vitro*, DNA with an altered o. binds less repressor than corresponding wild-type DNA. O. specific mutations (symbol O^x) result in an altered specificity of response to the repressor. O. constitutive mutations (symbol O^c) result in a loss of response to the repressor and interfere with repressor binding to the o. Thus all associated genes of the operon function so as to produce their enzyme products at relatively high unregulated (constitutive) levels. Such mutations are usually deletions of all or part of the o. in an inducible operon. **Operator zero mutations** (symbol O^o) abolish the expression of a whole operon; some mutants have been found to reduce the maximum rate of expression of the operon without suppressing its expression totally. They then render the residual synthesis constitutively.

operator constitutive (*Jacob* & *Monod* 1962) — → operator.

operator zero (*Jacob* & *Monod* 1962) — of mutations which prevent the formation of all the enzymes of an → operon (→ operator).

operon (*Jacob* et al. 1960) — a group of contiguous → structural genes showing coordinate expression and closely associated controlling sites, i.e., elements which determine the → genetic transcription of only those genes to which they are attached (cisdominant effects) (*Epstein* & *Beckwith* 1968). This definition does not presume any particular mechanism of regulation, negative or positive. In the case of negative control of o. expression the controlling sites are so coded that transcription of the o. occurs in the absence of any cytoplasmic regulatory component (repressor) specific for the particular system. In a positive control system the coding at the initiator controlling site prevents expression of the o. in the absence of any specific regulatory product. The o. represents a unit of polarized coordinated genetic transcription in prokaryotes that is initiated and regulated at the → operator site. It consists of the → promoter, → operator, one or more → structural genes coding for enzymes performing related metabolic functions, and the → terminator. Transcription of the o. (eventually resulting in a polycistronic → messenger RNA) starts at the promoter where the → RNA polymerase binds to DNA. Synthesis of mRNA initiates if the operator is free of repressor protein (negative control). Transcription terminates at the end of the o. defined by the terminator site where the mRNA is released. The rate of transcription in a given o. is a function of two parameters: 1. the frequency of transcription initiation, and 2. the rate of transcription propagation.

Some operons seem to have extra regulator loci: in addition to the usual promoter site adjacent to the operator, an extra promoter between structural genes has been found. Since this promoter is not linked to the operator, some transcription of the last few genes always occurs even when the system is switched off.

In the case of negative control an operator can exist in an "open" or "closed" state. In the open state or in the absence (deletion) of the operator, every structural gene (cistron) of the o. to which the operator belongs synthesizes messenger RNA and hence can give rise to the polypeptide for which it codes. In the closed state of the operator, none of the structural genes is able to synthesize messenger RNA and polypeptides. The operator closes whenever it is complexed with a specific cytoplasmic repressor produced by a regulatory gene or regulator. Depending on whether the enzymes are inducible or repressible, at least two kinds of free repressor molecules are known: one type is neutralized by a specific metabolite (often the substrate of the first enzyme specified by the o.), called an → inducer. The inducer prevents the interaction between the operator and the repressor or removes the repressor from the operator and thus induces (turns on), by inactivation of the repressor, the otherwise repressed synthesis of the enzymes specified by all the structural genes of the relevant operon. The second type of repressor must be activated by a specific metabolite, called a → corepressor, before it can interact with its homologous operator, thus turning off the function of all the structural genes of the operon.

The first type of repressor effect is characteristic of operons that are normally nonfunctional and are controlled by a regulatory gene producing a repressor which, in turn, represses the operator and hence the o. (Fig. 71). In addition to inducers, derepression of such operons may be accomplished by mutation of the regulatory gene which changes the repressor, and by mutation of the operator gene which renders it insensitive to the repressor.

The second type of repressor effect is characteristic of operons that are

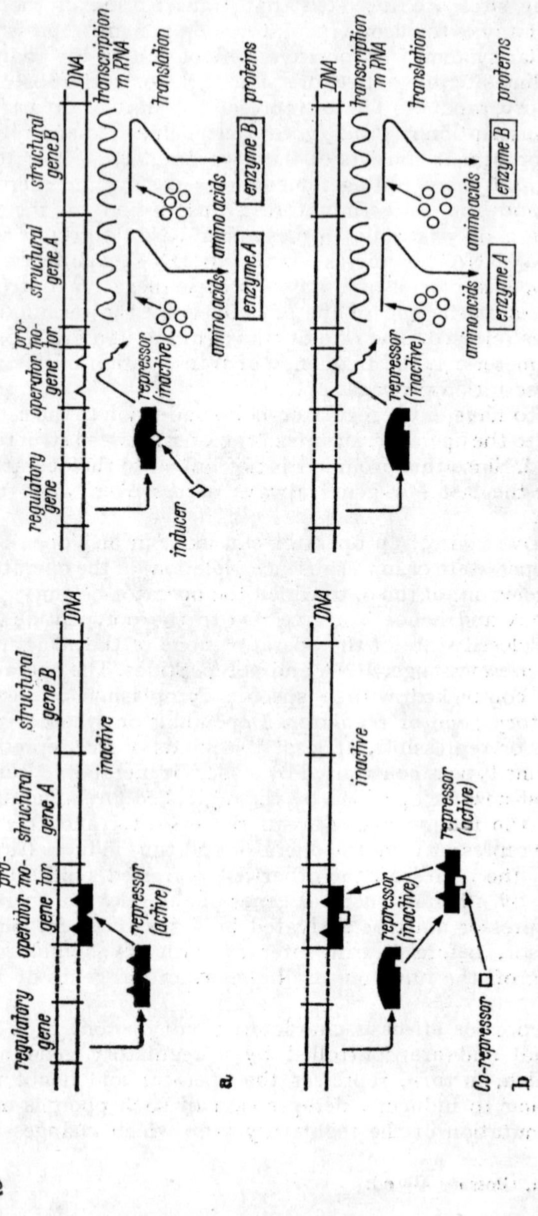

Fig. 71. The "operon concept" for the regulation of (a) inducible and (b) repressible enzyme synthesis.

normally functional and are rendered nonfunctional by repressors (Fig. 71). Repression may then result from mutations in regulatory genes or operator genes which make an operator sensitive to the repressor produced by a foreign regulatory gene.

Comparison of genetic and complementation maps of an o. reveals two major classes of mutants. Some mutants can be arranged into → complementation groups which correspond to site clusters on the genetic map. Other mutants are non-complementing or display polarized asymmetric localization on the genetic map. Suppressors of non-complementing mutants (→ suppressor gene) exhibit the properties of informational suppressors, i.e., are allele and not gene specific. Genes for a number of prokaryotic biosynthetic pathways are not clustered in the form of an operon. In this case, a single species of repressor presumably interacts with several operator sites resulting in a parallel but non-coordinate control of enzyme synthesis. Such a unit of common control is called a → regulon (→ cross-pathway regulation; supraoperon control).

Bacterial operons subject to → catabolite repression require 3′,5′-cyclic AMP and a protein factor → CAP for their trancription. Probably cAMP and CAP are necessary to permit RNA polymerase to initiate transcription at the catabolite sensitive promoters.

If the genes of an o. are differentially derepressed with a higher production for the operator proximal genes and a lower for those located more distally this may be called **semi-coordinate regulation** (*Bauerle* & *Margolin* 1966). This type of regulation may be due to the presence of a second internal → promoter of lower efficiency.

In **eukaryotes**, genes coding for a metabolically related sequence of enzymes are not, in general, closely linked in the form of an operon (→ transcripton).

operon fusion (*Jacob* et al. 1965) — the fusion (by deletion) of two operons such that the genes of one operon are regulated by the controlling elements of a second adjacent operon. One polycistronic mRNA is usually produced (read through) from both operons.

order — a systematic category (below the class) which is composed of one or more families.

organ culture — → cell culture.

organelle — any usually membrane-bound substructure of the eukaryotic cell performing specialized functions, such as the cell → nucleus, the → mitochondria, the → plastids, the → Golgi apparatus, the → lysosomes etc. Some organelles contain DNA called → organelle DNA as opposed to nuclear DNA.

organelle DNA — in eukaryotes, DNA localized in cell organelles other than the cell nucleus with its → nuclear DNA (→ chloroplast DNA; mitochondrial DNA). The interactions between organelle DNA and nuclear genomes are complex.

organism — the unit of a continuous lineage with an individual evolutionary history (*Luria* & *Darnell* 1967).

organization effect (*Levitan* 1954) — any effect (e.g., differential → fitness of individuals which are → heterokaryotypic for two chromosomal structural changes present in either → cis- or trans-configuration) resulting from interaction among adjacent → genetic loci owing to some feature of chromosomal structural organization (→ position effect).

organizer (*Spemann* 1918) — a region ("organization center") of an embryo from which inducing grafts (embryonic → induction) can be obtained, indicating that the group of cells in question may act as an embryonic inductor (or evocator) which evokes and controls the → differentiation of other groups of cells, i.e., determines the fate of cell masses with which it comes into contact.

orientation (*Darlington* 1937) — → centromere orientation.

orthochromatin (*Brink* 1960) — one of the two postulated functional components of the eukaryotic chromosome which comprises the structural genes. → Parachromatin is assumed to comprise the regulatory components that vary in response to ontogenetic signals. Parachromatin may be subdivided into two categories (*Brink* 1972):

1. **Locus-specific parachromatin** which responds in each particular case only to the specific activation signal for the gene with which it is conjoined and collectively represents the regulators of ontogenetical gene action.

2. **Regional parachromatin** which forms a pervasive system of elements throughout the genome, controls general chromosomal processes, and is non-specific in its action upon individual genes.

orthogenesis (*Haacke* 1893) — → evolution of phyletic lines more or less continuous in a single direction (following a rectilinear pathway) over a long interval of time. The term has frequently been used with vitalistic implications in the sense that variation and evolution are determinate and purposive processes assumed to follow a definite course towards a goal.

orthoploid (*Winkler* 1916) — = → euploid.

orthoselection — natural → selection which acts in the same direction for a long period of time and gives rise to what has been called "orthogenesis" (Karyotype o.)

outbreeding — in contrast to → inbreeding, a mating system (= crossbreeding) in which → mating is between individuals less closely related than average pairs chosen at random from the population.

overdominant (*Hull* 1946) — of two → alleles if the heterozygote (Aa) is superior in comparison with the two homozygotes (AA and aa), i.e., lies outside the range of the homozygotes as to the genotypic value of the character under observation (= superdominant).

Real overdominance in the above sense is to be distinguished from apparent or **pseudo-overdominance** which is not a property of a single → genetic locus, but of blocks of loci or of chromosome segments. It is difficult to establish whether a given case of heterozygous advantage is due to real or pseudo-overdominance. Evidence for real overdominance is obtainable only from a genetic locus which has mutated in a highly inbred line, or from a population in which → coupling and repulsion linkages are in equilibrium (*Falconer* 1960).

It is expected that overdominance has a biochemical basis since the physiological versatility of a locus represented by two different alleles is greater than that of a locus represented by a double dose of a single allele. Overdominance might result if

1. one allel of the allele pair is neutral in effect and allele A in single dose (Aa) is nearer to optimal gene dosage than in double dose;

2. both alleles (A and A′) are active and complementary in respect to separate supplementary functions;

3. both alleles (A and A′) are active in respect to different primary functions and give rise to a larger effect if present together;

4. an alternative genetic pathway is available for the heterozygote, thus resulting in the formation of a product different from those manufactured by each allele in the absence of the other.

ovotid — = → oötid.

ovule — the megasporangium of a flowering plant that becomes the seed. It includes the nucellus and the integuments.

oxydosome (*Deley* 1960) — → mitochondrion.

P

P (*Bateson & Saunders* 1902) — a shorthand for the parental generations in crosses. Starting with any individual, P_1 denotes the generation of his parents (the immediate parents of a first filial generation or F_1), P_2 the grandparental, P_3 the great-grandparental generation, etc.

pachynema (*Gregoire* 1907) — = pachytene (→ meiosis).

pachytene (*v. Winiwater* 1900) — → meiosis.

pachytene DNA (*Hotta & Stern* 1971) — DNA synthesized during the pachytene stage of → meiosis as the product of → repair replication DNA synthesis. The synthesis is concomitant with → genetic recombination.

packing factor (*Darlington & Upcott* 1939) — a measure for the degree of → chromosome coiling. It is determined from the diameter of the helix divided by the diameter of the chromosome thread.

paedogamy (*Guillermont* 1910) — a form of automixis (obligatory self-fertilization) characterized by the fusion of iso- or anisogametes produced (in the course of → meiosis) by one and the same individual. Either a parental cell (gamont, gametangium) produces directly copulating gametes, or the individual gives rise to male and female sexual organs (antheridia and oögonia) the products of which fuse. Other types of automixis are → autogamy and → parthenogamy.

paedogenesis (*v. Baer* 1886) — 1. reproduction by parthenogenetic development of egg cells in the larval or embryonic stages of the individual; 2. = → neoteny.

pair-alleles (*Laughnan* 1952) — = → pseudoalleles.

pairing — → chromosome pairing.

pairing block (*Darlington & Mather* 1932) a chromosome segment of definite length (consisting of a definite number of → chromomeres) which acts as a functional unit in meiotic → chromosome pairing. The pairing affinities of the chromosomal pairing blocks and their subunits are satisfied by associations in twos. If there are more than two homologous chromosomes in the meiocyte, there may be changes of partners which may lead to → multivalent formation (→ zygomere).

pairing index (*Pätau* 1941) — → pairing number.

pairing number (*Pätau* 1941) — the number of chromosome arms, in diploids and polyploids, associated in twos by → chromosome pairing which resulted in a minimum of one → chiasma in any of the arm pairs concerned. The proportion between the observed and theoretically expected (i.e., the number of chromosome arms which might be paired in diploids and polyploids) p. n. is referred to as the "pairing index".

pairing segment (*Darlington* 1931) — → interstitial segment.

pair mating — a procedure used to determine the degree of sexual isolation between two groups (A and B) of individuals. Separate tests of mating success are made for the four possible mating combinations, A♀×A♂,

B♀×B♂, (= homogenic matings), A♀×B♂, and B♀×A♂ (heterogenic matings). The mating system is "male multiple choice", if males of one group (A or B) are allowed to mate with females of both groups; it is "female multiple choice", if females of one group are allowed to mate with females from both groups.

palindrome (*Wilson & Thomas* 1974) — any of the hairpin-like structures (folding back of a single linear DNA chain) which result from inverted repetitions of base sequences (e.g., ABCtC′B′A′) located at intervals along the chromatid of eukaryote chromosomes. Since no unpaired single-chain regions in the turn-around region (t) have been found they are called double-chain palindromes which read the same both backwards and forwards. Palindromes may serve as specific recognition sites at the DNA level, possibly for binding of proteins. Besides in DNA, hairpin loops also exist in → transfer RNA, → ribosomal RNA, → heterogeneous nuclear RNA and → messenger RNA and may be responsible for some of the specificity of interaction of these molecules with the appropriate proteins.

palingenesis (*Haeckel* 1874) — recapitulation of ancestral characters during embryonic development.

panmictic — of a randomly interbreeding → population of individuals which mate at random, i.e., each individual is equally likely to mate with any individual of the opposite sex. The population as a whole forms a → deme.

panmictic index — the complement (symbolized by P) of the → inbreeding coefficient, 1 — F.

panmictic unit (*Wright* 1943) — a group of individuals (a local population) all of which have the same chance of mating and of producing progeny, provided they are sexually mature and of opposite sex. Panmictic units represent the smallest → "Mendelian populations" and are integrated into the more or less complex system of populations composing the species.

panmixis (*Weismann* 1895) — a system of → mating (= panmixia) characterized by random choice of mating partners, as opposed to nonrandom mating.

parabasal body — = → kinetoplast.

paracentric — of intrachromosomal structural changes (→ inversion) that do not include the → centromere region of the chromosome involved in the "intrachange", as opposed to pericentric ones which include the centromere.

paracentromeric heterochromatin (*Luyks* 1974) — → heterochromatic regions on both sides of the centromere which contain, in many cases at least, highly repetitions → satellite DNA. In → meiosis the primary role of p. h. may be (*Luyks* 1974): 1. to hold sister chromatids together until the second meiotic division; 2. to determine a particular kind of arrangement of chromatin fibers underlying the centromere itself and determining the orientation and disposition of the sister centromeres on the surface of the chromosomes.

parachromatin (*Brink* 1960) — → orthochromatin.

paragenetic (*Brink* 1962) — of changes within the chromosomes that affect the expression (exemplified by V-type → position effects, unstable loci in plants, mottled expression of sex-linked characters in the female of mammals in consequence of the → dosage compensation mechanism), but not the constitution of the genes conerned. Changes affecting gene constitution are referred to as "genetic".

paraloci (*Dunn* 1954) — closely linked genetic loci with similar or identical activity (= pseudoalleles).

parameiosis (*Battaglia* 1945) → meiosis.

paramutability (*Brink* 1958) — the ability to undergo a directed heritable change in gene expression (→ paramutation). Any factor or influence that results in a change in the expression of a paramutable gene is a paramutagenic factor.

paramutation (*Brink* 1958) — directed heritable alteration of the expression of a (paramutable) allele by heterozygous association with another (paramutagenic) allele. The phenomenon is sometimes also called somatic conversion or conversion-type-inheritance. P. represents a kind of gene instability due to heterochromatization at or near the locus at which metastable repression of gene activity occurs.

paranemic coil (*Sparrow, Huskins & Wilson* 1941) — → chromosome coiling.

paranucleic body — a cap made up of cytoplasmic RNA bound to proteins which surrounds the nuclei of gametes and zoospores of *Allomyces*. The p. b. dissolves at the onset of mitosis or meiosis.

parapatric (*Smith* 1955) — of → allopatric populations whose ranges are in contact and between which gene exchange is geographically possible even without sympatry (→ dichopatric).

paraplasm (*Kupffer* 1875) — a type of cell component (demonstrable by cytological or cytochemical methods) located as transient inclusions in the → cytoplasm of certain cell types or during certain periods of cell activity.

paraselectivity (*Sedlmayr* 1956) — → fertilization.

parasexual (*Pontecorvo* 1954) — of genetic systems that achieve → genetic recombination by means other than the regular alternation of → meiosis and → fertilization; also of the mechanisms providing for nonmeiotic recombination of the → genetic material.

Parasexual processes take place **in viruses** if genetically different mutant strains multiply after mixed infection in the same host. **In bacteria,** three mechanisms leading to parasexual recombination are known and are referred to as bacterial → conjugation, → transduction, and → transformation.

The parasexual cycle **in fungi** is a sequence of events consisting of:

1. the formation of a → heterokaryon in a multinucleate mycelium and the rare, probably accidental, fusion of haploid nuclei differing in → genotype within the heterokaryon;

2. mitotic → crossing-over within the diploid fusion nuclei which multiply side by side with the haploid ones;

3. vegetative → haploidization of the diploid nuclei without the intervention of meiosis via mitotic → non-disjunction. This cycle has the same effect as a regular → sexual cycle in accomplishing genetic recombination and thus increasing available genetic variation.

parasterility (*Brieger* 1928) — sterility which results not from genetically or chromosomally defective gametes but from → incompatibility mechanisms which limit or prevent zygote formation by certain classes of gametes.
parental generation (*Bateson* & *Saunders* 1902) — → filial generation; → P.
parthenapogamy — = diploid → parthenogenesis.
parthenocarpy (*Noll* 1902) — the development either of seedless fruits or of fruits in which the seeds lack embryos (→ apomixis). P. results from a failure of pollination (irrespective of the presence or absence of artificial stimulation), a failure in fertilization (→ incompatibility or gametic sterility), or a failure in embryo development (zygotic sterility).
parthenogamy (*Guilliermond* 1910) — 1. = → parthenogenesis (*Guilliermond*);

2. an extreme case of → automixis: two nuclei fuse within a single undivided cell which has the character of a female gamete or gametangium; the fusion results in the production of a new individual (→ paedogamy);

3. automictic → parthenogenesis: oögenesis involves a normal meiosis and reduction of chromosome number, but the zygotic chromosome number is restored by the fusion of two haploid nuclei, or by the formation of a → restitution nucleus, or by → endomitosis (= meiotic parthenogenesis).

parthenogenesis (*Owen* 1849) — the production of an embryo from a female gamete without the participation of a male gamete, with or without eventual development into an adult. The parthenogenetic organism may be called a "parthenote" or "parthenogenone" (*Beatty* 1957).

P. can be subdivided into several types which in turn may be grouped into several systems. Classification may be based on the mode of reproduction, the mechanism of sex determination, and cytology (*Suomalainen* 1950).

Classification according to the mode of reproduction:

I. Occasional or accidental p. (= tychoparthenogenesis), i.e., unfertilized eggs occasionally develop through p.

II. Normal parthenogenesis.

1. Obligatory p., i.e., the egg develops exclusively by parthenogenesis.

a) Constant or complete p., where all generations are parthenogenetic (= acyclic p.).

b) Cyclic p., where one or more parthenogenetic generations alternate with bisexual (amphigonic) generations (= heterogony).

c) Paedogenesis, where the eggs of individuals at larval stages develop parthenogenetically.

2. Facultative p., where an egg may either be fertilized or develop parthenogenetically.

Classification according to the mechanism of sex determination:

I. Arrhenotoky, where unfertilized eggs develop parthenogenetically into males and fertilized eggs give rise to females.

II. Thelytoky, where unfertilized eggs develop into females.

III. Deuterotoky (= amphitoky), where unfertilized eggs develop into either sex.

Classification based on cytological data:

I. Generative (or haploid) p., where parthenogenetically produced individuals develop from egg cells in which meiotic reduction of chromosome number took place.

II. Somatic p., where parthenogenetically produced individuals have a zygotic (diploid or polyploid) chromosome number.

1. Automictic (= meiotic) p. or parthenogamy, where regular → chromosome pairing and reduction of chromosome number occur but the zygotic number of chromosomes is restored by fusion of two haploid nuclei, the formation of a → restitution nucleus, or → endomitosis.

2. Apomictic (= somatic or ameiotic) p., where neither chromosome reduction nor fusion of nuclei nor any corresponding phenomenon takes place in the egg cell which develops parthenogenetically.

The genetic consequences of p. differ with the cytological features of the parthenogenetic system and may be classified as follows (*Suomalainen* 1950):

1. the progeny are genotypically similar to the mother (as in apomictic p.);

2. heterozygosity is obligatorily replaced by homozygosity of the progeny (as in automictic p. when the somatic (zygotic) chromosome number is restored by fusion of two cleavage nuclei with reduced chromosome number);

3. heterozygosity is replaced by homozygosity if the alleles of heterozygous loci separate prereductionally, i.e. at → meiosis I, but is maintained if they show postreductional disjunction, i.e. separate at meiosis II. This type of genetic behavior occurs in organisms with automictic p. in which the somatic (zygotic) chromosome number is restored by fusion of the second polar nucleus with the egg cell nucleus. It is also characteristic of those cases of automixis in which the zygotic number of chromosomes is reestablished by endomitosis in both anaphase-I groups;

4. heterozygosity is maintained if the alleles of heterozygous loci separate postreductionally. In the case of prereduction for the alleles in question, a mosaic of the two homozygous cell types results;

5. heterozygosity is maintained in the case of prereductional separation of heterozygous allele pairs. If they separate postreductionally, a heterozygous individual results in 50% of the cases, and a mosaic formed by the two homozygous cell types in 50%.

The property of parthenogenesis conferred on amphimictic organisms results from the gradual development of a polygenic system of modifier genes. The advantages connected with p. depend on the system of p. present. Generally, p. is connected with an increase in the reproductive potential since only half as many parents are needed per individual. An additional advantage may be fixation of the genotype and the preservation of extensive heterozygosity as well as → heterosis.

parthenogenone (*Beatty* 1957) — the parthenogenetic organism (→ parthenogenesis).

parthenomixis (*Winkler* 1908) — the fusion of two female nuclei within a multinucleate oögonium. This may replace true fertilization.

parthenote — = → parthenogenone.
patroclinal (*Kerner* 1881) — → matroclinal.
patrogenesis (*Collins & Kempton* 1916) — = → androgenesis.
pattern gene — a gene whose action concerns a particular pattern established in the course of → cytodifferentiation.
pattern modifier (*Waddington* 1939) — any gene modifying the pattern of → differentiation in a quantitative rather than a qualitative way. Pattern modifiers may be grouped into:

1. "disruption genes" which prevent the formation of a particular pattern of differentiation;
2. "reorganization genes" which may lead to disorganization of a pattern of differentiation when present in their normal genotypic milieu but may give rise to new nondeleterious effects in a genotype which has been selected for this activity;
3. genes which, when present, result in a new, functional pattern of differentiation.

pattern of damage (*Hadorn* 1945) — the pattern of action of a → lethal factor. It comprises the sum total of all the locus-specific, abnormal characters resulting in malformation and lethality connected with the presence of a lethal mutation in effective dosage (→ pattern of manifestation) in the genotype.
pattern of manifestation (*Hadorn* 1945) — the sum total of primary and secondary processes and characters which are characteristic products of the action of a particular gene (→ gene action). The p. of m. is defined by comparison of the wild-type and mutant, whereby all statistically significant differences between the two are regarded as direct or indirect consequences of the particular mutation which changed, inactivated, or eliminated the normal allele (→ gene mutation). The p. of m. is thus a differential pattern denoting the differences in phenotype between wild-type and mutant. A so-called "residual pattern" representing those activities which are identical for the normal and mutant allele may remain since it is not accessible to analysis. The total pattern of manifestation is, therefore, composed of the differential and the residual patterns of action. The more drastic the mutational alteration, the clearer the differential pattern becomes, and the smaller is the residual pattern. In the case of → lethal factors, the p. of m. is referred to as the → pattern of damage.
pedigree — a list of ancestry or genealogical register. Pedigree diagrams (Fig. 72) are frequently used in human genetics in the analysis of Mendelian inheritance. Females in such a p. are symbolized by either circles or by the symbol ♀, males by squares or the symbol ♂. Parent symbols are joined by a horizontal "marriage line", and symbols for their offspring are placed in a horizontal row below this line to which they are connected by verticals. The children of a parental pair represent a sibship, and the individual children are referred to as sibs or siblings, irrespective of their sex. They are listed and numbered in a p. from left to right in order of birth. If an individual possesses the character whose inheritance forms the subject of the p. it is designated by a black symbol. An open symbol denotes absence of the character under observation. Twins are represented

by two symbols connected by lines to the same point of the sibship line; a number enclosed in a large symbol shows the number of sibs who are not listed separately.

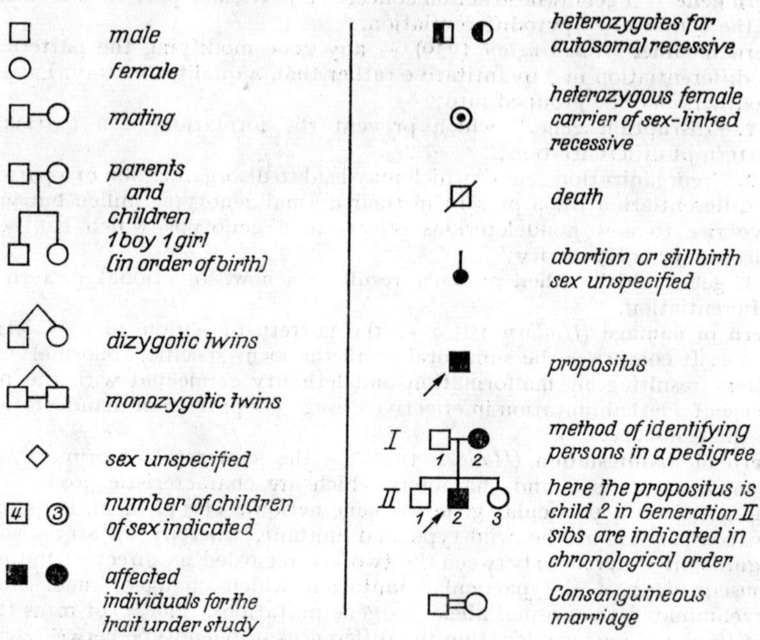

Fig. 72. Symbols commonly used in pedigree diagrams (slightly modified from *Thompson* and *Thompson* 1966).

The pattern of distribution of genetically controlled characters in kindreds, i.e. large groups of persons interconnected by descent and marriage, is called the "pedigree pattern". It provides information on the Mendelian principles of → inheritance. Observation of pedigree patterns is one of the chief methods of genetic study in man since critically informative matings made by design are impossible in human beings.

pegmatypic (*Kalmus* & *Maynard Smith* 1966) — of a → mating system in which imprinting operates during a sensitive period of the individual's youth. Imprinting denotes a special form of sexual selection. It means that different objects to which young animals of certain species are exposed during an early period of their life later become the particular objects towards which they will direct sexual advances when adult. The term **pegmatype** is used for the imprinted characteristics of an individual, or (according to the context) for an individual or class of individuals bearing this acquired characteristic.

penetrance (*Voigt* 1926) — the frequency (in per cent) with which a (dominant or homozygous recessive) gene or gene combination manifests itself in the → phenotype of the carriers. P. (as well as → expressivity) depends both on the genotype and the environment.

P. is complete when all the homozygous recessives show one phenotype, when all of the homozygous dominants show another phenotype, and when all of the heterozygotes are alike. If less than 100% of the carriers of a certain genotype manifest that phenotype characteristic for the class, p. is reduced or incomplete; p. of a gene may be identical or different in either sex of dioecious organisms, or in extreme cases, may be limited to one or the other sex ("sex-limited genes").

pentaploid — of → auto- or alloploid cells, tissues, and organisms, respectively, with five chromosome sets in their nuclei (symbol 5n).

pentasomic — of a → polysomic cell, tissue, or organism with one chromosome in the complement represented five times (e.g., a pentasomic diploid 2n + 3; a pentasomic tetraploid 4n + 1, etc.).

peptide synthetase — = → peptidyl transferase.

peptidyl site — → genetic translation; ribosome.

peptidyl transferase — an enzyme (= peptide synthetase) which catalyzes the formation of peptide bonds (peptide transfer) during → genetic translation. This takes place by transfer of the growing peptidyl group from the CCA terminus of one molecule of → transfer RNA to an aminoacyl group attached to the CCA terminus of a second, incoming molecule of tRNA. The reaction possesses properties of a single group transfer reaction and the p. t. is an integral part of the larger subunit of the → ribosome (→ aminoacyl transferase; translocation).

pericentric — → paracentric.

perichromatin fibril (*Monneron & Bernhard* 1969) — any of the fibrillar ribonucleoprotein components surrounding areas of condensed → chromatin in the cell → nucleus. Quantitative and topological changes of p. fibrils occur after increased or decreased RNA synthesis and correspond to similar changes of → heterogeneous nuclear RNA.

perichromatin granule (*Yamamoto* et al. 1969) — any of the ribonucleoprotein granules surrounding areas of condensed → chromatin in the cell → nucleus.

periclinal (*Baur* 1909) — of a → chimera in which the distinct, idiotypically different tissues are arranged in concentric layers.

For plants in which the vegetative point (the apical meristem) is composed of three tissue layers (L I, L II and L III), four different types of p. chimeras are distinguishable:

1. "haplochlamydeous" p. chimeras in which only the outermost (L I) of the three layers is idiotypically different from the others;
2. "diplochlamydeous" p. chimeras in which only the innermost (L III) of the three layers is idiotypically different from the others;
3. "mesochimeras" in which only L II (the middle layer) is idiotypically different from the other two layers;
4. "trichimeras" in which all three tissue layers are idiotypically different from each other.

perinuclear cisterna (*Watson* 1955) — → nuclear envelope (= perinuclear space).

perinuclear space (*Policard & Bessis* 1956) — → nuclear envelope.

perinucleolar chromatin — → chromatin structures (= nucleolus-associated chromatin) associated with the nucleolus and localized around the nucleolar body as a more or less discontinuous layer of clusters of fibrils usually composed of coiled, fine filaments. The p. c. is connected with the → intranucleolar chromatin by septalike structures or strips which penetrate into the nucleolus.

periplasm (*Rattenbury & Serra* 1952) — that part of the cell plasm which surrounds the → chromosomes and the → spindle after disruption of the → nuclear envelope. It is supposed to be a mixed product arising from the cytoplasm and the nucleoplasm.

perissoploid — → artioploid.

peristromium — a contractile, amoeboid sheath around the chloroplast.

permanent heterozygosity — → heterozygous.

permissive — 1. of a cell that supports lytic infection by a specific virus. Non-p. (or restrictive) cells do not; there are indications that the permissiveness of the cells to virus infection is regulated at the initiation of protein synthesis.

2. → conditional lethal.

peroxisome (*de Duve* 1965) — a special type of cytoplasmic organelle (= microbody) characterized by the association of one or more hydrogen peroxide-generating oxidases with large amounts of catalase. Peroxisomes occur in certain cell types actively involved in the formation of carbohydrate from non-carbohydrate precursors. They are bound by a single membrane, are smaller than → lysosomes (0.3 to 1.5 nm in diameter) and originate from dilated regions of the smooth → endoplasmic reticulum. Frequently, they contain in a finely granular matrix a dense core or nucleoid that may have a regular crystalloid structure. Small peroxisomes (micro-p.) are extensively interconnected with the endoplasmic reticulum.

perpetuation (*Tsuchida, Nonoyama & Ikeda* 1966) — the maintenance and multiplication of a virus genome in (bacterial) host cells without killing them due to a weakening of the virus virulence under appropriate conditions (e.g., incubation at low or high temperature).

persistence — → genetic death.

phage — a bacterial virus (= → bacteriophage).

phage conversion — the acquisition of new characters (e.g., changes in certain antigenic properties, in the production of toxins, and in the sensitivity to infection by unrelated phages) by bacterial cells containing a → prophage ("prophage-mediated conversion" or "lysogenic conversion"), a → temperate phage not yet integrated in the bacterial genome, or virulent mutants of temperate phages. Converted characters disappear and the cells regain their original characters when the converting phage is lost. The phenotypic changes associated with the presence of phages are, therefore, due to the addition of new genetic information rather than substitution of new information for old (→ mutation, → transformation, → transduction, and → conjugation).

phage cross — multiple infection of a single bacterium by bacteriophages that differ at one or more genetic sites. This leads to the production of recombinant progeny phage.

phage exclusion (*Delbrück* 1945; *Streisinger* & *Weigle* 1956) — the partial exclusion of the genes of a bacteriophage from the progeny of crosses with another phage (e.g., partial exclusion of the genome of phage T 2 from crosses with T 4). P. e. is genetically controlled by a specific gene which is expressed as an early function and may show polarity, i.e., occurs with low frequency for genes at the beginning of the chromosome map and with higher frequency for genes localized distally.

phage ghost — the phage protein coat prepared of viable phage by osmotic shock. P. ghosts retain no detectable DNA with genetic functions. They may, however, cause cessation of bacterial synthetic processes.

phage heterozygosity — → heterozygous.

phage restriction — the prevention of phage replication due to a → DNA modification-restriction system in the bacterial host cell.

phagocytosis (*Metchnikoff* 1893) — the ability of cells to ingest ("engulf") solid particles (→ pinocytosis).

phagolysosome — the fusion product of a → lysosome and a → phagosome (= lysophagosome or telolysosome).

phagosome (*Strauss* 1958) — any phagocytic or pinocytic vacuole or vesicle before the fusion with → lysosomes (= endosome). Phagolysosome and lysophagosome are the terms used for the combined granules.

pharmacogenetics (*Vogel* 1958) — the genetics of the metabolism of and response to drugs within one species.

phase specificity (of gene action) — the regular manifestation of a particular gene via → gene action during a certain and limited period of development. P. s. results from differential → gene activation.

phene (*Johannsen* 1909) — a genetically controlled phenotypic → character.

phene therapy (*Siniscalco* 1972) — all methods applicable for correcting the consequences of expression of mutant genes and the manifestation of → genetic diseases (such as dietary measures, avoidance of toxic drugs, replacement of missing gene products or defective tissues, enzyme induction, etc.). (→ eugenics; euphenics).

phenocline (*Huxley* 1939) — → cline.

phenocopy (*Goldschmidt* 1935) — a nonhereditary, phenotypic → modification (caused by special environmental conditions) that mimics a similar phenotype caused by a → gene mutation (→ genocopy).

According to *Landauer* (1958) a true p. is characterized by:

1. an additive effect between the gene action in question and the particular environmental conditions resulting in a p.;

2. a larger effect of the environmental condition in heterozygotes than in homozygotes for the standard allele;

3. an increase of the environmental effect producing a p. in the presence of mutants with low → penetrance;

4. a similar effect of modifiers on both the penetrance and → expressivity of the gene whose effect is phenocopied and on the strength of activity of the phenocopying environmental condition.

It has been shown that certain phenocopies are only inducible when the environmental influence is applied during a certain limited period of development. One and the same influence may mimic different mutations, depending on the time it acts. Various environmental factors (agencies) may be used for the induction of a particular p., that is, there is no specific relationship between the external influence and the induced change which, in the overall effect, copies a mutational event.

In all probability, a p. is most often the result of influences on secondary rather than primary activities of the genes whose activity is changed. Since there are many possibilities of influencing secondary reactions, the nonspecificity of influences leading to a. p. is understandable.

phenocritic (*Haecker* 1918) — of that phase in ontogony when the developmental pathways of two different genotypes, e.g., wild-type and a mutant genotype, begin to differ, either morphologically, physiologically, or biochemically.

phenodeviant (*Lerner* 1954) — a phenotypically abnormal individual that deviates from the populations mean (or norm). The abnormality is due to special gene combinations and normally fails to follow a simple → inheritance pattern.

Phenodeviants are characterized by the following features:

1. they arise sporadically;
2. they occur more frequently in → inbred lines;
3. their frequency and → expressivity may be increased by selection and inbreeding;
4. the genes causally responsible for the production of phenodeviants in different strains are nonalleles;
5. the genes responsible for the production of phenodeviants are difficult to localize;
6. in crosses, phenodeviants behave in a variable and unpredictable manner;
7. in strains in which phenodeviants frequently arise, the reproductive efficiency is decreased.

Phenodeviants (whose origin is poorly understood) have been explained as an excessive proportion of homozygous loci (*Lerner* 1954), as the breakdown of balanced polygonic systems coadapted in heterozygotes (*Dubinin* 1948; *Goldschmidt* 1949), and as defective regulatory systems giving rise to phenodeviants by the erroneous interception by particular genes of information intended for others (*Wallace* 1963).

phenogenesis (*Fischer* 1939) — the development of genetically controlled phenotypic → characters or phenes in the course of ontogenesis. P. is the result of the interaction between → genotype and environmental factors and results in a definite → phenotype (→ phenogenetics).

phenogenetics (*Haecker* 1918) — a branch of genetics involved in the study of the relationships between the genotype and its manifestations (the phenotype). One aspect of p. is developmental genetics.

phenomic lag (*Davis* 1949) — a delay in the phenotypic expression of a → gene mutation owing to the time required for the manifestation of changes in the affected biochemical reactions (→ segregational lag).

phenotype (*Johannsen* 1909) — the observable properties (structural and functional) of an organism, produced by the interaction between the organism's genetic potential (its → genotype) and the environment in which it finds itself. The genotype sets the "norm of reaction" on the environmental influences. The term p. can be applied either to the totality of expressions of the genotype or to only a part, i.e., to particular characters, traits, or phenes.

Phenotypic expression of → genetic information is mediated by → messenger RNA, which carries the transcript of DNA (→ genetic transcription) to the polyribosome, i.e. the ribosomal complex engaged in protein synthesis (→ genetic translation).

Those (adaptational) components of the phenotype which are not directly related to the environment and the survival of the individual may be called **endophenotypic** (*Lewis & John* 1963), as opposed to the (adaptive) **exophenotypic** components which affect an individual's ability to compete and to survive. It is the exophenotype on which → selection acts directly. Both endo- and exophenotype are controlled by the genotype. (→ epiphenotype).

Heredity and environment overlap extensively in the shaping of the phenotype. Phenotypic similarities between individuals may occur, when they posses the same or different genotypes, and phenotypic differences between individuals may or may not be accompanied by genotypic differences. Environmental differences can cause the same genotype to produce a variety of phenotypes. This phenotypic range of reaction of a particular genotype is referred to as its "norm of reaction" or **"reaction norm"**. The capacity of an organism to function in a range of environments which may include plastic and stable responses is called **"phenotypic → flexibility"** (*Thoday* 1953). **"Phenotypic correlations"** between different → characters result from combinations of genetic and environmental correlations. Genetic correlations can arise from → pleiotropy, → linkage, or common introduction of the genes involved into a population. Environmental correlations may be produced by the development of two or more characters in the same individual and their exposure to the same environment.

phenotypic expression — the manifestation via → gene action of a particular gene present in effective dosage in an individual's genotype. Gene expression results in the production of a particular phenotypic character.

phenotypic flexibility (*Thoday* 1953) — → phenotype.

phenotypic intergradation — the occurrence of variability and intermediate forms in areas of geographic overlap between populations which elsewhere are distinct and relatively homogeneous. P. i. often indicates interspecific or interracial hybridization.

phenotypic lag — the delayed phenotypic manifestation of a → gene mutation (e.g., a mutation from prototrophy to auxotrophy) in bacteria. The delay may last several generations and is due (in most cases) either to → phenomic lag or → segregational lag.

phenotypic masking (*Gorini* et al. 1967) — in bacteria, the masking of the true phenotype in drug (antibiotic)-dependent strains. P. m. is due to alterations of the ribosomes by antibiotics.

phenotypic mixing (*Novick & Szilard* 1951; *Hershey, Roesel, Chase & Forman* 1951) — the production in the same host cell of virus particles with discrepant genome and → virion specificities due to the incorporation into the same → capsid or envelope of proteins made under the direction of different viral genomes. The genomes in question produce different structural components of the protein "coats". The components of the coats are drawn, without regard to the genotype of the particle, from a pool of mutant and wild-type protein molecules and, thus, mixed coats may be produced giving rise to p. m. In bacteriophages, phenotypic mixing has been observed for stability at high temperature, serological properties, the properties of particles to attach to different strains of bacteria, as well as other features depending on the protein moiety of mature phages.

phenotypic plasticity — the ability of certain genotypes to respond to varying environments by producing radically different phenotypes.

phenotypic reversion — the non-genetic restoration (repair), partial or complete, of the functional activity of mutated genes (missense and nonsense mutations) without directly substituting the required function itself. P. r. may be due to miscoding or → mistranslation. A number of agents are able to provoke p. r. and result in nonheritable phenotypic changes.

phenotypic stability (*Lewis* 1954) — = developmental → homeostasis.

phenotypic variance — the total → variance observed in a character.

photoenzymatic repair (*Harm* 1968) — the basic mechanism responsible for → photoreactivation, i.e., light-activated enzymatic reaction by which pyrimidine dimers in DNA are monomerized. The photoenzymatically repaired fraction of the total number of lethal UV lesions originally induced is the photoenzymatically repairable sector. In cells completely lacking → dark repair the photoenzymatically repairable sector is expected to be equal to the → photoreactivable sector, otherwise overlap may occur between dark repair and photoenzymatic repair.

photoenzymatically reparable sector (*Harm* 1968) — → photoenzymatic repair.

photomorphosis — the control which may be exerted by light over growth, development, and differentiation of a plant, independently of photosynthesis. Light may be regarded as an elective factor which influences the manner in which those genes present in the particular organ are expressed.

photoprotection (*Latarjet & Gray* 1954) — a decreased sensitivity of a biological system to ultraviolet light caused by pre-irradiation treatment with light (symbol PP). Generally, PP requires larger light doses than → photoreactivation and the PP action spectrum covers a narrower band of wave lengths than does that for photoreactivation. PP is practically temperature independent during illumination, whereas photoreactivation is usually strongly temperature dependent. PP operates by providing more time for intracellular → dark repair to act before DNA replication leads to → "mutation fixation", since the photoprotecting wavelengths (338 nm) induce delayed growth. → Liquid-holding recovery is absent in photoprotected cells (*Rupert & Harm* 1966).

photoreactivation (*Kelner* 1949) — the partial elimination of the lethal and mutagenic effects of ultraviolet light (290–310 nm) by subsequent exposure of irradiated cells to other nonionizing radiation (330–450 nm) leading to a decrease in the effective UV-dose (symbol PR). PR has been found to proceed by at least three distinct mechanisms (*Rupert & Harm* 1966):

1. direct photoenzymatic repair of UV-damage to DNA by splitting the pyrimidine dimers; apparently this is the major mechanism of PR in bacteria and bacteriophage (*Rupert* 1960, 1961);

2. indirect processes (related to → photoprotection) by which light-induced growth delay favors a light independent repair system (*Jagger & Stafford* 1965);

3. direct nonenzymatic photochemical repair as observed in transforming (→ genetic transformation) DNA but not in any organism (*Setlow & Setlow* 1963).

Not every UV induced lesion is photorestorable. The fraction of photoreactivable lesions is called the **"photoreactivable sector"**. PR is to be distinguished from photoprotection and together with → dark repair represents the two basic repair processes with respect to UV effects.

Except for marsupials, mammalian cells lack a photoreactivating enzyme. Cells from non-mammalian vertebrates possess a photoreactivating enzyme.

photoreversal — = → photoreactivation.

phragmoplast (*Errera* 1888) — a defined plasma body representing a modified → spindle region in the equatorial plane of dividing plant cells which is identifiable after → chromosome movement to the poles. The p. is involved in the formation of the → cell plate (→ phragmosome). Initially, the p. is in lateral view a barrel-shaped plasma mass which then flattens out and assumes the form of a biconvex lens. The spindle fibers traversing the p. are simultaneously shortened as it flattens and the two daughter nuclei are drawn to the middle of the dividing cell. They finally come to rest on both sides of the p. The cell plate is formed in the region of the p. which deposits wall lamellae (primary walls) against both sides of the so-called middle lamella, i.e., the cell plate impregnated with pectin.

The p. consists of fibrillar elements (microtubules), which arise at the cell plate and radiate outward, and numerous droplets which are moved to and incorporated into the plate. The microtubules of the p. first appear in the midregion of the cell in early stages of plate formation and assist in establishing the zone along which the cell plate forms.

phragmosome (*Porter & Machado* 1960) — any of the circular to oval structures in dividing plant cells which (in addition to the → endoplasmic reticulum) are associated with the formation of the → cell plate. Such particles (0.2–0.5 μm in diameter) are surrounded by a membrane enclosing their contents which appear homogeneous or finely granular depending on the type of fixation. They are observed in large numbers in the region of cell plate formation. Although they become much less apparent on completion of → cytokinesis, they are also present in interphase cells. Their number increases during pro-, meta-, and anaphase. They are commonly observed in contact with the cisternae of the endoplasmic reticulum and are supposed to arise as blisters from this structure which may be reservoirs of synthetic materials being discharged into the plate region.

phyletic evolution (*Simpson* 1944) — → evolution.

phylogeny (*Haeckel* 1866) — the evolutionary history of an organism or taxonomic group (= phylogenesis).

physiogenesis (*Waddington* 1957) — the → differentiation of a part of the embryo leading to distinctive differences between various regions and to changes within each such region. Frequently the term "histogenesis" which refers to the formation of particular tissues in multicellular embryos is used in the same sense. Since similar progressive changes may concern the regions of a single cell, the more general term p. has been proposed.

pilus (*Brinton, Buzzell & Lauffer* 1954) — any of the numerous outgrowths (= fine threads, bristles, fimbriae, or filaments) from the surface of many bacterial species which have the form of long, thin, straight, thread-like appendages (diameter 25 nm or more, length from 10–20 μm). The number of pili varies from 10 to several thousand per cell. The pili are distinct from the flagellae and are known to occur both in the flagellated and nonflagellated forms.

Piliated strains (symbol P^+) transmit the property of piliation to non-piliated (P^-) forms through genetic recombination, both in intraspecific and interspecific matings.

One variety of pili occurs independently of the sexual nature of the bacterial strain (piliation may occur in F^-, F^+, or Hfr strains; → F plasmid). A second variety of pili was found to be sex specific. Such pili (one to two per cell), called → F-pili, grow exclusively on the surface of male or donor cells (both F^+ and Hfr).

pinocytosis (*Lewis* 1931) — the process of active intake of fluid by the cell. It proceeds by the elevation of thin undulating folds that impound sizeable droplets of fluids, by inpocketing of the surface to form shallow vesicles of submicroscopic size that detach and move into the cytoplasm, or by deeper invagination of the cell membrane to form long channels that bud off small fluid-filled vacuoles at their end (*Fawcett* 1965).

The term "micropinocytosis" is used for the submicroscopic vesiculation of the plasma membrane by invagination of smooth surfaced vesicles or local specialization of both surfaces of the membrane prior to its invagination to form a vesicle (*Odor* 1956).

pistil (*Clusius* 1601; *Tournefort* 1700) — in flowers, the female seed-bearing organ (the gynoecium) consisting of the ovary, style, and stigma. Flowers bearing pistils are termed "pistillate" (→ stamen).

pit — a thin region (in the light microscope, a canal divided into two pit cavities by a pit membrane) of the secondary → cell wall, functioning in the exchange of material between neighbouring cells. According to electron microscopic studies, the primary cell wall is penetrated in those places where pits will later arise by clusters of → plasmodesmata which remain visible in the pit membrane. Cellulose microfibrils are deposited around the future p. and plugs of cytoplasm on both sides of the pit area prevent the apposition of the secondary wall.

In cells which lose their living contents and act as capillaries for water conduction, the simple p. is replaced by so-called "bordered pits". These possess a large pit membrane which is circularly overarched by a border originally laid down as a slender rim of the primary wall. This border is separated from the pit membrane and covered on both the inner and outer sides with a tertiary layer; the space referred to as the pit cavity in simple pits represents a narrow entrance (porus) and a widened pit chamber. The membrane consists of a cushion-like thickening (torus) and a thin marginal zone (margo).

plane (*v. Wettstein* 1928) — = → plasmagene and → plasmid.

planogamete — a motile → gamete, as opposed to a nonmotile aplanogamete.

plaque — a round clear area in a confluent bacterial cell lawn which results from the killing or lysis of contiguous cells by several cycles of → bacteriophage growth.

plasmagene (*Winkler* 1920; *Darlington* 1939) — any extranuclear → hereditary determinant showing non-Mendelian → inheritance. The sum total of plasmagenes represents the → plasmotype which is one component of the → idiotype.

plasmagynogamous (*Michaelis* 1955) — of a → fertilization whereby the cytoplasm of the → zygote is exclusively derived from the female parent, in contrast to "plasmaisogamous" and "plasmaheterogamous" fertilization.

plasmaheterogamous (*Michaelis* 1955) — → plasmagynogamous.

plasmaisogamous (*Michaelis* 1955) — → plasmagynogamous.

plasmalemma (*Seifriz* 1928) — = → cell membrane or plasma membrane.

plasmalemmasome (*Edwards & Stevens* 1963) — = → mesosome.

plasma membrane — = → cell membrane.

plasmatic — pertaining to or mediated by the → cytoplasm (as, e.g., cytoplasmic → inheritance).

plasmatic inheritance — = → cytoplasmic inheritance.

plasmid (*Lederberg* 1952) — originally, an extrachromosomal genetic element; today, any genetic element which is additional to the normal genome of the bacterial cell and, in particular, may be propagated in one or the other of two alternative states: either independently in the cytoplasm or, following insertion, as an integral part of the host's chromosome. Integrated plasmids were formerly called → episomes.

So far all bacterial plasmids have proved to consist of circular DNA (→ genetic circularity) and to represent (in the extrachromosomal state) cytoplasmic → replicons, i.e., genetic elements whose replication and segregation to daughter cells at division is independent of the bacterial chromosome.

Bacterial plasmids may be classified into two major types (*Clowes* 1972):

1. **Infectious** or **self-transmissible plasmids** which control the establishment in the host cell of a donor state. They may be subdivided into two types, F-like and I-like, referring to whether they determine the synthesis of → sex pili resembling those produced by cells harboring the classical → F plasmid (sex factor), or those produced by cells harboring the → colicinogenic factor I. The pili provide a mechanism of bacterial → conjugation

that permits their transfer (→ conjugational DNA synthesis) from the host (donor) cell into a recipient cell (which is thereby itself converted into a donor). Those plasmids that can also promote the transfer of other genetic material are known as → sex factors. In many instances this transfer may result from covalent linkage of p. DNA to the host chromosome.

2. **Non-infectious plasmids** unable to set up a donor state and which require either that a sex factor plasmid be present in the same cell for their conjugal transfer or that transfer be effected by a transducing phage.

Examples of plasmids of enteric bacteria are the F-plasmid, the → colicinogenic factors, and the → drug-resistance factors. The stability of the association of plasmids with the host cell and the small number of plasmids per cell imply that mechanisms exist which coordinate, at the level of initiation of rounds of replication, the frequency of replication with the growth rate of the host and also determine the number of particles per cell.

The exclusion of certain plasmids from cells that might otherwise be hosts is called → entry exclusion. Another phenomenon is restriction, i.e., non-establishment of incoming DNA in certain strains carrying restriction genes. Many plasmids restrict phage DNA and *vice versa*. Curing occurs when the rate of p. replication and/or distribution fails to keep pace with the chromosomal replication and cell division of the host (→ superinfection immunity; plasmid incompatibility).

plasmid incompatibility (*Novick* 1967) — the inability of two distinguishable bacterial plasmids to coexist in the same cell. Incompatibility is exhibited whether the plasmid exists autonomously or is integrated with the host chromosome. The DNA of plasmids belonging to the same incompatibility group (which cannot stably coexist with each other) shows considerable homology. Little DNA homology exists between plasmids of different incompatibility groups.

P. i. may be due to an inhibition of replication and an inhibition of hereditary stability.

plasmodesma (*Strasburger* 1882) — any of the fibrillar plasma connections (in all terrestrial plants) which extend through the cell wall and bridge adjacent cells. Plasmodesmata are unbranched cylinders of cytoplasmic origin which penetrate the cell wall perpendicular to its surface and consist (in cross-section) of two → unit membranes separated from each other by a thin layer of groundplasm and surrounding a thin lumen. Plasmodesmata, collected together in groups within depressions, or narrow regions of the wall, are referred to as → pits. Elements of the → endoplasmic reticulum are assumed to play a role in the origin of plasmodesmata.

plasmodium — = → coenocyte or syncytium.

plasmogamy — the fusion of the cytoplasm of two (or more) cells. P. precedes → karyogamy in the process of fertilization (→ dikaryophase).

plasmon (*v. Wettstein* 1924) — originally (*Wettstein*), the cytoplasm of an individual considered as a single hereditary agent; currently, the sum total of extrachromosomal hereditary determinants (= plasmotype).

plasmon mutation — any mutational change of extrachromosomal hereditary determinants which may give rise to genetically different cytoplasms (= "plasmon alteration" according to *Marquardt* 1952).

plasmon "segregation" — the "segregation" (by dissociation) of qualitatively different extrachromosomal hereditary determinants as manifested by the production of new plasmonic or plasmotypic combinations from a heterogeneous plasmon or plasmotype. P. s. taking place within one individual may give rise to → chimeras.

plasmon-sensitive (*Renner* & *Kupper* 1921) — of genes whose action is confined to a particular cytoplasm (*Goldschmidt* 1955).

plasmosome (*Ogata* 1883) — = → nucleolus.

plasmotropic (*Waddington* 1962) — embryonic → induction.

plasmotype (*Imai* 1936) — the sum total of extrachromosomal hereditary determinants (→ plasmon). P. and → genotype together represent the idiotype or genetic system of the cell.

plast (*Bělař* 1928) — any of the differential, self-reduplicating cell organelles, localized in the cytoplasm, e.g., → plastids, → centrioles, → mitochondria, → kinetosomes.

plasticity — the extent to which the expression of an individual's genotype can be modified by environmental factors (→ flexibility), irrespective of the adaptive value of the (physiological and morphological) changes occurring. The p. of a particular character can be

a) specific for that character,
b) specific in relation to particular environmental influences,
c) specific in direction,
d) under genetic control not necessarily related to heterozygosity,
e) radically altered by selection (*Bradshaw* 1965).

plastid — in plant cells, any of a class of cell organelles which may or may not contain pigments.

Depending on the presence or absence of pigment, p. include **chloroplasts** (containing chlorophyll and active in photosynthesis), **chromoplasts** (containing carotenoids and common in fruits and flowers), and **leucoplasts** (containing no visible pigments and like amyloplasts able to synthesize starch). Leucoplasts can be transformed via chloroplasts into chromoplast which may be regarded as degenerative forms of the plastids.

Chloroplasts of higher plants are lens-shaped, membranous structures (about 2 μm thick and 3 to 5 μm in diameter) surrounded by two smooth membranes separated from each other by a distinct space. Closed, flattened sacs (called "thylakoids") are the structural units of internal organization (Fig. 73). The larger thylakoids make up the so-called stroma lamellae of the internal membrane system which varies considerably from plant to plant and within one plant depending on its physiological state. The smaller thylakoids are stacked together to form the grana lamellae of the membrane system which contain the light absorbing pigments. Each granum represents a column of flattened appressed discs piled one on the other and each of the discs is composed of a pair of membranes.

The inner surface of the two layers forming the lamellae is granular in organization due to oblate spheroids (about 20×10 nm in size, molecular weight about 2×10^6) called "quantasomes" which are probably surrounded by a unit membrane. The quantasome is the smallest subunit capable of carrying on the photochemical reaction and contains about 200 chlorophyll molecules. One cytochrome molecule and one or more acceptor molecules are probably associated with each. This site, directly involved in electron transfer, is referred to as the quantotrope (*Calvin* 1962).

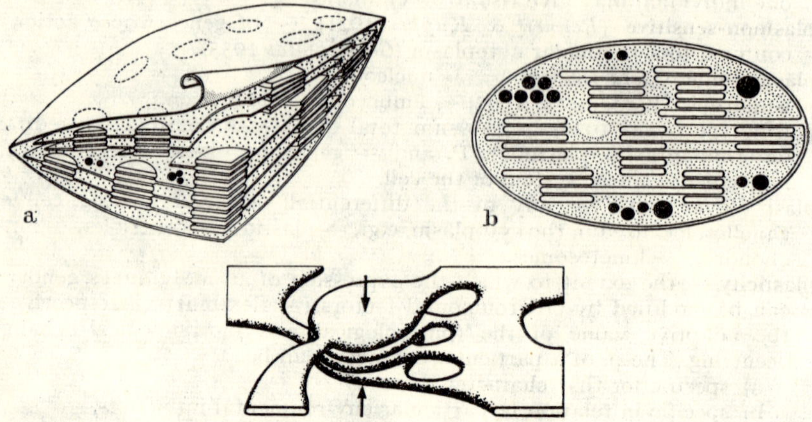

Fig. 73. Diagram showing the ultrastructure of chloroplasts. a) The three dimensional structure (from *Eriksson* et al. 1961); b) the submicroscopic structure in cross-section (black = lipophilic globuli; white ovoid region = starch grain); c) the formation of a granum (arrows indicate the thylakoid column (b and c from *Sitte* 1965).

The thylakoid system is embedded in a proteinaceous "stroma ground substance" or "matrix" which contains the soluble enzymes (the insoluble ones are in the membrane system) taking part in the conversion of photosynthetically reduced CO_2 to sugars and starch. Starch grains are produced inside the matrix which may furthermore contain lipophilic globuli and ribosome-like particles.

All types of plastids arise from so-called "proplastids" which are bounded by a double membrane and filled with a relatively dense ground substance. The inner of the two unit membranes frequently shows tubular invaginations. The proplastids may give rise to leucoplasts which are transformable into chloro- and chromoplasts, or directly to chloroplasts transformable into chromoplasts (Fig. 74).

The plastids contain all the components of a genetic system. Chloroplasts contain DNA (→ chloroplast DNA) and RNA, and are able to synthesize DNA, RNA and protein. DNA functions as a template for RNA synthesis and determines the heritable properties of the plastids (→ plastom). Plastid DNA is a double helix in structure and has a base composition different from that of chromosomal DNA of the same species. The DNA content (1–5% of the total DNA) is about 10^{-11} to 10^{-12} mg per chloroplast (about that of a bacterium) which suffices for coding for amino acids of at least several hundred proteins with a molecular weight of 20000.

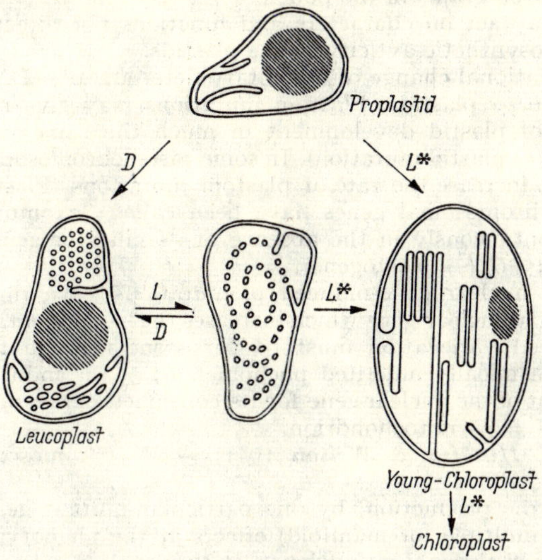

Fig. 74. Diagram showing the development of plastids. D: darkness; L: light; L*: intensive light; hatched areas: starch; upper part of the leucoplast contains a so-called *Heitz-Leyon* crystal; lower part contains a plastid center (from *Sitte* 1965).

plastid DNA — → organelle DNA contained in a → plastid (→ chloroplast DNA).

plastid inheritance — non-Mendelian → inheritance due to hereditary determinants located in the → plastids of plant cells and controlling (generally in cooperation with chromosomal genes) plastid characters.

plastidome (*Dangeard* 1920) — collectively the → plastids of a cell.

plastidotype — that component of the → idiotype of plants which comprises the sum total of the hereditary determinants located in the → plastids (→ plastom).

plastogene (*Imai* 1937) — any one of the hereditary determinants (collectively representing the → plastom of the cell) which are located in the → plastids of plant cells.

plastom (*Renner* 1929, 1934) — the sum total of the extrachromosomal → genetic information (→ plastogene) encoded in → plastid DNA.

The p. is known to determine several characters of the plastids such as:

a) their ability to become green and to have normal photosynthetic activity,

b) their inability to become green in combinations with particular alien genome complexes,

c) their inability to become green due to a loss mutation,

d) their specific multiplication rate, and

e) the shape of the starch grains in the pollen.

In addition, the p. may act on characters and functions not directly connected with the photosynthetic avticity of the plastids.

plastom mutation — a mutational change of hereditary determinants (DNA in nature) located in the → plastids. Plastom mutations may give rise to blocks in all stages of plastid development in much the same way as → gene mutations do (= plastid mutation). In some cases, chromosomal genes have been found to increase the rate of plastom mutations. Plastid mutations induced by chromosomal genes have been called "exomutations", those arising spontaneously in the absence of → mutator genes, "automutations" (*Imai* 1936) (→ plastogene).

Genetic systems with nuclear gene-induced p. mutations have three features: 1. they contain a nuclear gene which increases the → variegation frequency; 2. the induced variegation must be persistent through the sexual cycle as a cytoplasmically inherited phenomenon; 3. the induced variegation is independent of the nuclear gene for its continued expression.

plastosome (*Meves* 1908) — = → mitochondrion.

plectonemic coil (*Sparrow, Huskins & Wilson* 1941) — → chromosome coiling.

pleiotropy (*Plate* 1910) — the production, by one particular mutant gene, of apparently unrelated multiple (or manifold) effects at the phenotypic level (= polypheny). Since there is no evidence at the level of primary → gene action that one gene (one cistron) has more than one primary function (→ one gene — one enzyme hypothesis), pleiotropic gene action should be regarded as a reflection of the highly integrated state of cellular and developmental metabolism. With the remoteness of the phenotype from the primary action of the gene and the number of intervening steps which can be influenced by other genes and by environmental factors, secondary effects may give rise to a "genetic syndrome" consisting of a variety of phenotypic effects.

According to *Goldschmidt* (1955) four types of p. may be distinguished:

1. **Syndromic p.** (= spurious p. of *Grüneberg* 1938): The effect of a particular mutant gene is a syndrome of pathological features which represent a kind of mechanical consequence of primary, early embryonic damage.

2. **Pattern p.:** The mutant gene produces a primary product which affects one process most conspicuously, but simultaneously influences related processes at other places, sometimes in a qualitatively different way.

3. **Dichotomic p.:** The primary activity of the mutant gene (usually an inhibitory one) is of a generalized type affecting the whole of the organism simultaneously according to the reactivity of its parts.

4. **Interference p.:** The nearest approximation to the unlikely type of "genuine" p. whereby one and the same gene affects completely different

processes. A genetically controlled process interferes with other independently determined actions with the result that they shift in a way which might have been brought about by independent mutation.

Another classification of p. has been proposed by *Hadorn* (1954):

1. **Mosaic p.:** Genuine p., due to the differential action of the same gene in different cytoplasms and resulting in a pattern of mutant gene manifestation which consists of a number of independent → autophenes.

2. **Relational p.:** A mutant gene acts directly on only one type of cells but exerts indirect influence on other cells; each of these reacts in its own fashion. The intercellular influence presumably occurs through diffusible substances such as hormones, amino acids, and the like. This is a form of "spurious" p. The pattern of manifestation of the pleiotropic gene consists of autophenes and secondarily arisen allophenes.

Both mosaic and relational p. are based on intercellular mechanisms of gene activity, as opposed to the following types controlled by intracellular mechanisms of gene action:

3. **Cell reactive p.:** A mutant gene exerts identical primary actions in different cell systems but owing to their differential reaction different characters arise.

4. **Induced gene active p.:** The primary action of a mutant gene is not identical in different cell systems because the developmental factors are different in these systems and, therefore, different characters arise in each cell systems.

pleuromitic (*Oguma* 1942) — of chromosomes with diffuse → centromere activity, as opposed to "telomitic" chromosomes with localized centromeres.

ploidy — a term referring to the number of chromosome sets per cell (→ haploid; diploid; polyploid).

poikiloploid (*Levy* 1920) — = → mixoploid.

point error (*La Cour* & *Rutishauser* 1954) — → pseudochiasma.

point mutation (*Bridges* 1923) — a → gene mutation (intragenic mutation) which can recombine with any of a large number of mutants, as opposed to larger alterations where the corresponding mutant does not show → genetic recombination with at least two other mutants each able to recombine with the other. The precise definition of a "point" is dependent on the number and kind of mutants crossed and on the smallest distance in the → genetic map for which recombinants are experimentally, or possibly molecularly, detectable (*Freese* 1961).

point stickiness — localized sticky-effects in chromatids assumed to give rise to a → pseudochiasma.

polar body (*Robin* 1862) — in → oögenesis, the smaller cells produced during meiosis that do not develop into functional egg cells.

polar cap = → polar zone.

polar insertion (*Malamy* et al. 1972) — in bacteria and viruses, any of a class of extreme → polar mutations (insertion mutations) which result from the insertion (→ insertosome) of some "foreign" DNA into an → operon. The insertion process seems to include recognition of a specific sequence of the insertable DNA which serves as the integration site. The insertion may result in complete inactivation of the gene in which DNA insertion occurs and in most cases also of genes within an operon distal to the site of the insertion.

polar mutation (*Franklin* & *Luria* 1961; *Jacob* & *Monod* 1961) — any mutation of the → nonsense, → frameshift, → deletion, or → insertion mutation type which results in → genetic polarity. In the case of a p. m. (= dual effect mutation of *Englesberg* 1961) in an operon 1. the production of the particular enzyme for which the nonmutated gene codes is prevented (ribosomes carrying nascent peptide chains are released as they encounter the mutant site in the mRNA of the mutated gene), and 2. the functioning of the operator-distal genes of the operon is relatively decreased (mRNA distal to the mutation remains free of ribosomes and is highly susceptible to endonucleolytic attack)(→ polarity gradient).

Certain p. mutations have been found to cause an "antipolar effect" (*Ito* & *Crawford* 1965), i.e., the enzyme specified by the gene of the operon immediately preceding the mutated gene is formed in reduced amounts (→ genetic antipolarity; polarity suppressor).

polar rays — fibrillar cytoplasmic differentiations arising from the → centrioles.

polar zone — the functional center for mitosis in cells of higher plants devoid of → centrioles. → Spindle fibers as a rule converge on these cytoplasmic poles (= polar caps) which lack a distinct structure and represent clear cytoplasmic areas.

polarity (*Franklin* & *Luria* 1961; *Jacob* & *Monod* 1961) — 1. a selective deficiency, during steady state production and metabolism of an mRNA species, of operator-distal mRNA portions and their protein products as compared to those of operator-proximal sequences (*Martin* 1969). P. may be due to a specific lesion in the mRNA (nonsense codon; frameshift mutation; insertion) and may then be called → genetic polarity. Another type of p. affects a number of mRNAs and occurs when genetic translation is blocked by antibiotics, starvation for an amino acid, or mutations affecting am → elongation factor or the ribosome.

2. A position-dependent preferential → gene conversion of one of two alleles in an intragenic cross. This type of p. is accounted for by postulating a discontinuous distribution of recombination events (*Lissouba* et al. 1962).

polarity gradient — the quantitative effect a → polar mutation exerts on the expression of genes located more operator-distal in the same operon. The p. g. is a function of the distance between the polar mutation and the next polypeptide chain initiation signal. The greater the distance of the mutation from the end of the gene, the greater the reduction in synthesis of the proteins encoded by the subsequent genes.

polarity mutation — = → polar mutation.

polarity suppressor (*Bertrand* et al. 1975) — a → suppressor which relieves → genetic polarity caused by a → polar mutation of the nonsense mutation type in an operon. Polarity suppressors do not suppress the mutant phenotype itself, i.e., the suppressor does not act on the polar mutation *per se*. These suppressors act by stabilizing mRNA after premature termination of protein synthesis or are deficient in an endonuclease activity and thus mRNA distal to the polar mutation does not become degraded. The efficienc of p. suppressors varies considerably.

polarization — 1. of chromosomes at mitotic telophase and the following interphase, if their proximal parts (adjacent to the → centromere) remain directed to the polar side of the nucleus;

2. of chromosomes at prophase of meiosis I, if their ends are directed towards a definite point on the nuclear surface (→ bouquet stage);

3. of a nonrandom "segregation" (polarized segregation) of the members of a pair of homologous chromosomes, or of a pair of daughter chromosomes (chromatids), with respect to the two poles of the spindle in meiosis (→ assortment);

4. of → genetic recombination via → conversion, if a continuous gradient in the conversion frequency of mutational sites occurs from one end to the other of a → polaron;

5. of replication (→ replicon) and transfer processes (→ episome) of the → genetic material as well as its → genetic transcription and → genetic translation (→ operon) which proceed with a definite polarity.

polaron (*Lissouba & Rizet* 1960) — a segment of the → genetic material (a nucleotide sequence in DNA) within which → genetic recombination is polarized and due primarily to gene → conversion, i.e., mechanisms other than → crossing-over. Polarity of recombination within a p. is evident from the observation that in those cases where one of a group of linearly located alleles undergoes conversion more frequently than another placed to its left, then the right-hand member of any other pair is found to show a similar difference from the left-hand member (*Whitehouse* 1963).

The polarity (gradient) of conversion is interpreted as the formation of a DNA hybrid (heteroduplex) by an initial breaking point outside of the implicated region (→ polaron hybrid DNA model). The often converted site is the one nearest to the breakage point (linkage point) where crossover and noncrossover recombinations are assumed to initiate.

polaron hybrid DNA model (*Whitehouse* 1963) — a model based on the correction of heterozygosity in DNA as a source of intrachromosomal → genetic recombination which can explain both reciprocal (crossing-over) as well as nonreciprocal recombination (conversion).

The entire sequence of events, termed a "dissociation cycle" (Fig. 75), is assumed to proceed as follows (*Whitehouse & Hastings* 1965):

1. Enzymatic breakage (b in Fig. 75) of complementary nucleotide chains, one from each of two DNA molecules, takes place at fixed points (outside or at the ends of the genes), homologous in position at the "linkage points" defining the ends of each chromosomal → polaron (a region between neighboring genes in which recombination is by → conversion only but may give rise to reciprocal as well as non-reciprocal recombination). Breakage is followed by dissociation and uncoiling of these chains on one or both sides of the break.

2. DNA synthesis as a correction process begins from the points of breakage in one or both directions.

3. The newly synthesized polynucleotide chains uncoil.

4. Pairing and coiling of complementary strands, one from each molec-

ule, lead to the reformation of double helices and, thus, the rejoining of breaks.

5. Breakdown of any unpaired chains.

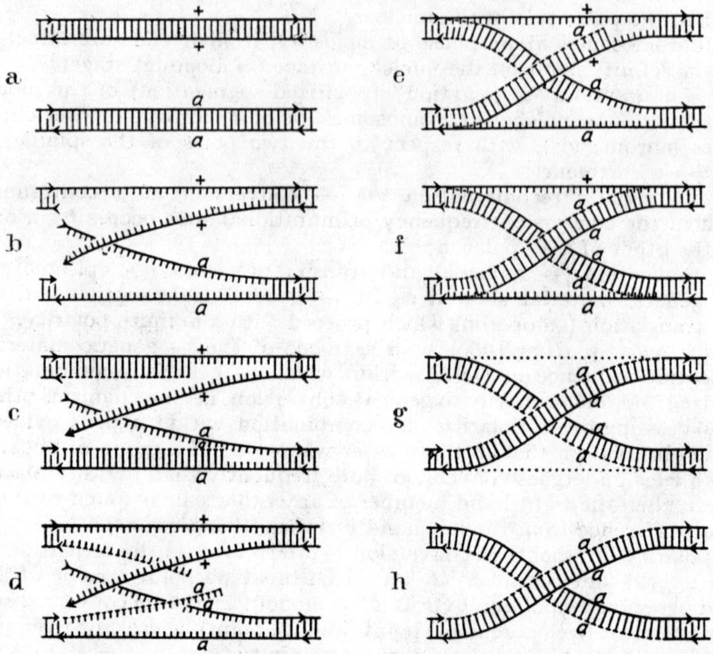

Fig. 75. Diagrammatic representation of the "polaron hybrid DNA model" of reciprocal (crossing-over) and nonreciprocal (conversion) genetic recombination. a) Two homologous chromatids each consisting of a DNA double helix (horizontal lines: nucleotide chains; arrow heads: ends of the polaron and, at the same time, direction of the sugar-phosphate backbones; short vertical lines: hydrogen bonds between the bases of the complementary nucleotide chains). b) Single strand breakage in the nucleotide chains of opposite polarity at one end of the polaron; separation of the broken nucleotide sequences from the complementary chains over the greater part of the length of the polaron. c) Synthesis of new chains (indicated by broken lines). d) Separation of the newly synthesized chains from their templates. e) Pairing of each new chain with its complement from the other molecule. f) Correction of heterozygosity and filling of any gaps with complementary nucleotides. g) Breakdown of unpaired chains (indicated by dotted lines). h) Established crossing-over (from *Whitehouse* 1965).

A crossover results if synthesis has taken place on one side only of the break, while with synthesis on both sides, no overall crossing-over would result. In both cases, lengths of hybrid DNA would arise which provide the opportunity for conversion if they include sites of mutation, i.e., represent heterozygous DNA. The frequency of conversion of any mutant can be related to the frequency with which it falls into the hybrid DNA sequence.

pole — one of the two opposing ends of the → spindle towards which chromosomes move (→ chromosome movement; centriole) at mitotic or meiotic anaphase.

pollen grain — the angiosperm microspore which germinates to form the male → gametophyte. Any of the sporophyte cells which undergo meiosis (in the anther) during → microsporogenesis and give rise to four (haploid) p. grains is called a pollen mother cell or microsporophyte.

pollen grain mitosis — → microsporogenesis.

pollen mother cell — → pollen grain; microsporogenesis.

pollen tube competition — → certation.

pollination — the placing (naturally or artificially) of pollen on the receptive area (stigma in angiosperms, micropyle in gymnosperms) of a female flower.

pollution — an undesirable change in the physical, chemical or biological characteristics of air, land and water that may or will harmfully affect human life or that of relevant species.

polocyte (*Waldeyer* 1888) — = → polar body.

polyadenylic acid (*Edmonds* & *Abrams* 1960) — a sequence of adenylic acid residues (50 to 250) coupled to → heterogeneous nuclear RNA and → messenger RNA in eukaryotes. Poly (A) is attached to the 3′-terminus of the RNA molecules after their trancription from DNA. Histone mRNA seems to lack poly (A) and it is not found in the RNA of prokaryotes. Poly (A) may be required for the processing and transport of mRNA from the nucleus to the cytoplasm and/or for the initiation of → genetic translation of mRNA.

It has been shown that a poly (A) segment is also present in the mRNA of viruses that replicate in the cytoplasm. Cytoplasmic polyadenylation of preexisting mRNA has been shown to occur in the cytoplasm of fertilized sea urchin eggs. The physiological significance of cytoplasmic polyadenylation or cytoplasmic poly (A) elongation is obscure.

poly (ADP-ribose) (*Nishizuka* et al. 1968) — the third polynucleotide which, besides DNA and RNA, exists in eukaryotic nuclei. A chromatin-bound enzyme, poly (ADP-ribose) polymerase, catalyzes the formation of the homopolymer by the successive elongation of ADP-ribose units; these are derived from the substrate NAD. One function of this reaction seems to be the modification (ADP-ribosylation) of histones and other nuclear proteins by the covalent attachment of poly (ADP-ribose) which, in turn, may result in an enhancement of template-primer activity (release of template restriction) of nuclear DNA.

polyandry — the presence of more than one male pronucleus in an egg cell. P. may result from polyspermy and active participation of supernumerary male gametes in fertilization. In animals, p. is an abnormal, but relatively frequent process. In some organisms, p. and → polygyny may involve 25 to 50 percent of (aged) egg cells.

polybasic (*Darlington* & *Janaki Ammal* 1945) — ref. to → alloploids containing chromosome complements derived from two, three, or more species which differ in respect to their → basic number of chromosomes (two different basic numbers = dibasic; three different basic numbers = tribasic, etc.).

polycentric (*Darlington* 1937) — of chromosomes or chromatids (= polykinetic) with more than one localized → centromere: → dicentric (with two centromeres), tricentric (with three centromeres), etc. Polycentrics are in most cases the result of → chromosome mutations. In special cases (→ nuclear differentiation), compound polycentric chromosomes represent a normal feature of chromosome organization. (→ operon).

polycistronic — of a → messenger RNA molecule which contains the information for more than one polypeptide chain (→ operon).

polyembryony (*Braun* 1859) — the production of more than one embryo from a single egg cell, or from other gametophytic or sporophytic cells (in plants).

1. **Simple p.:** several egg cells develop from one → megaspore and each is fertilized by a separate sperm or spore, or their development is parthenogenetic.

2. **Cleavage p.:** more than one embryo results by division (during early cleavage) of the zygote into two or more units, each developing into an embryo. The resultant embryos are "monozygotic" in origin and genetically identical.

3. **Sporophytic p.:** adventitious embryos arise by sporophytic budding from the nucellus and from the integument in flowering plants. The embryos are pseudogamic and usually identical to each other and to the mother plant.

polyenergid (*Sachs* 1892) — 1. of a multinucleate cell able to undergo multiple division (→ cytokinesis);

2. of a → polyploid nucleus (as, e.g., the → macronucleus of most ciliates).

polyergistic (*Waddington* 1955) — of characters controlled by two or more competing systems of alleles.

polygenic (*Plate* 1913; *Mather* 1941) — of traits or → characters (quantitative characters), whose expression is controlled by many genes (polygenes) which individually exert a slight effect on the → phenotype. The mode of inheritance of p. characters is called p. inheritance and its main properties may be summarized as follows (*Lerner* 1968):

1. most metric and meristic characters are affected by a number of genetic loci.

2. The effects of allelic substitution at each of the segregating genes (→ genetic segregation) are usually relatively small and interchangeable in the sense that identical phenotypes may be displayed by a great variety of genotypes.

3. The phenotypic expression of p. characters is subject to considerable modification by environmental influences.

4. P. characters show a continuous rather than a discontinuous distribution.

5. Balanced systems of p. inheritance in a population contain a great deal of potential → genetic variability stored in the heterozygous condition

and released by small increments through → genetic recombination between linked polygenes.

polygenic locus (*Thompson* & *Thoday* 1974) — any individual → locus which is included in the system of genes responsible for the genetic component of variation in a quantitative → character (= polygenic — variation locus). Allelic substitutions contribute to the variance in a specified quantitative character. The p. l. may be either a single or a complex genetic locus in the conventional sense, i.e., either a single gene or a closely linked block of functionally related genes.

polygyny — the union of one male and two or more female pronuclei within the egg cell (→ polyandry).

polyhaploid (*Katayama* 1934) — → haploid.

polyhybrid (*de Vries* 1900) — → monohybrid.

polykaryocyte — → polykaryocytosis.

polykaryotic — of cells with several → nuclei.

polykinetic — = → polycentric.

polylysogenic — of a bacterial strain → lysogenic for more than one → prophage.

polymegaly — the presence, in particular lobes of the testis, of sperm mother cells of different sizes. Such differences may involve not only the cells but also the nuclei, the nucleoli and the cytoplasms separately.

polymerase — any of the enzymes (→ DNA p.; RNA p.) which catalyze the assembly of nucleotides or deoxynucleotides into RNA or DNA on a DNA or RNA (reverse transcriptions) template, processes known as → genetic transcription and DNA replication (→ deoxyribonucleic acid).

polymerase mutant — any alteration of polymerase by → mutation. DNA p. mutants can reduce the fidelity with which a particular base pair is replicated and offer evidence for the role of the enzyme in insuring the specificity of base pairing during DNA replication. Mutants which produce an altered DNA polymerase usually show an increased mutation frequency throughout the genome (→ mutator gene).

polymeric (*Nilsson-Ehle* 1908) — of genes (→ gene interaction) whose individual effects are equivalent, but whose actions intensify each other (→ cumulative genes). Interactions among p. genes which give rise to so-called **"polymery"** (the production of a particular character by cooperation among several p. genes) are of two types: dominance interactions between alleles, and nonallelic interactions. These interactions may be expressed statistically as components of genetic variance (→ variation). For a quantitative character controlled by p. genes, there are thus six components of variance:

V_P = phenotypic variance	V_A = additive variance,
V_E = environmental variance	V_D = dominance variance,
V_G = genetic variance	V_N = nonallelic variance.

polymitosis (*Beadle* 1933) — the intercalation of extra → mitoses at a particular stage of the life cycle in abnormal individuals or varieties of a species (*Darlington* 1965). In plants, p. following → meiosis may take two forms which sterilize the pollen:

1. **Concordant p.** (*Darlington & Thomas* 1941): All mitotic conditions concordantly hastened, the chromosomes being reduplicated and split.

2. **Precocious (discordant) p.** (*Beadle* 1933): The chromosomes and their centromeres are neither reduplicated nor split (→ haplochromosome).

polymorphism — → chromosome polymorphism, genetic polymorphism.

polynemic — 1. of primarily multistranded chromosomes in eukaryotes, as opposed to → binemic chromosomes (→ polytenic);

2. of → cell cycles entailing replication of chromosomes without a sign of M-phase (mitosis), i.e., endoreduplications giving rise to a nuclear condition in which discrete banded chromosomes or chromosomal regions are not seen.

polynuclear — = → multinucleate.

polynucleotide — a linear sequence of → nucleotides in DNA or RNA in which the 3′ position of the sugar of one nucleotide is linked through a phosphate group to the 5′ position on the sugar of the adjacent nucleotide.

polynucleotide ligase (*Weiss & Richardson* 1967) — = → DNA ligase.

polynucleotide phosphorylase — an enzyme widely distributed among bacteria whose activities normally include polymerization of nucleoside phosphates, phosphorolysis of polyribonucleotides, and exchange between inorganic phosphate and the terminal phosphate of nucleoside diphosphates. The physiological role of the enzyme is still unclear. It has been suggested that the function of p. p. is the degradation of → messenger RNA.

polypeptide — a polymeric, covalently bonded arrangement of → amino acids joined together by peptide linkages, i.e., covalent linkage between two amino acids in which the alpha-amino group of one amino acid is linked to the alpha-carboxyl group of the other with the elimination of H_2O (→ protein, elongation factor).

polypeptide chain elongation — the translation (→ genetic translation) of internal → codons by ribonucleoprotein particles (ribosomes) carrying a → peptidyl-tRNA, an N-acylaminoacyl-tRNA, or an aminoacyl-tRNA. Amino acids from aminoacyl-tRNA are added one at a time through a cyclic series of reactions (→ translocation), resulting in the elongation of the polypeptide chain from the N-terminal residue toward the C-terminal amino acid residue (→ aminoacyl transferase; peptidyl transferase). P. c. e. requires the participation of several protein factors, called → elongation factors.

polypeptide-terminating mutation — a nonsense mutation (stop mutation) which results in premature inhibition of → genetic transcription at the site of the mutation (→ nonsense suppression).

polypheny (*Haecker* 1925) — = → pleiotropy.

polyphyletic — → monophyletic.

polyploid (*Strasburger* 1910; *Winkler* 1916) — of somatic cells and tissues, as well as of individuals (polyploids) having three (triploid), four (tetraploid), five (pentaploid), or more complete → chromosome sets instead of two as in diploids. The state of being p. is referred to as **"polyploidy"** and may arise spontaneously or be induced experimentally by → mitotic poisons

(→ C-mitosis; C-meiosis). A polyploid that originated from a zygote containing unlike chromosome sets (usually due to hybridization of two species) is called an → allopolyploid, as opposed to a polyploid containing more than two homologous chromosome sets in its somatic cell which is called an → autopolyploid.

Secondary polyploid (*Darlington & Moffett* 1930) is the term used to denote polyploids with a chromosome complement in which particular chromosomes in the basic set are represented more frequently than others (due to → non-disjunction).

Di-, tri-, or polybasic polyploids (*Darlington & Janaki Ammal* 1945) are allopolyploids derived from the combination of diploid chromosome complements with two, three, or more basic numbers.

In higher plants, polyploidy in the form of → allopolyploidy has been of major significance in → evolution. Among lower plants and animals, polyploidy is quite rare and is, in animals, almost confined to some hermaphroditic groups and to forms reproducing by → parthenogenesis. The main barriers to evolutionary polyploidy in the animal kingdom are (*White* 1973): 1. the almost universal presence of → sex chromosome mechanisms in the bisexual groups; 2. the prevalence of obligatory cross-fertilization mechanisms in the hermaphroditic groups. These facts prevent the establishment of polyploidy because a newly arisen tetraploid will encounter diploid mates and will produce sterile triploid offspring. In higher plants, **"intra-individual somatic polyploidy"** (resulting from → endomitosis and → endoreduplication) is frequently associated with the processes of differentiation and includes conspicuous parts of the plant soma especially those cells and tissues with the most varied degrees of differentiation and of function. Generally, the diploid condition is retained in embryonic cells, apical initials, procambium and some cambial regions, the pericycle, and those cell lines which assure genetic continuity of the species. Intra-individual somatic polyploidy via endomitosis and endoreduplication results when the exact coupling of chromosome duplication and mitosis, controlling diploidy, is relaxed by disturbances or upsets of the spindle function.

polyploidization — the conversion (spontaneous or experimentally induced) of cells or individuals from the haploid or diploid state to a → polyploid state (→ C-mitosis; C-meiosis; endomitosis). Agents capable of inducing polyploidy may be called "polyploidogenic" (→ mitotic poison).

polyribosome (*Warner, Rich & Hall* 1962) — = → polysome.

polysomatic (*Langlet* 1927) — of tissues or individuals containing diploid and polyploid cells side by side as shown by the occurrence of both diploid and endopolyploid mitotic nuclei (→ mixoploid).

polysome (*Warner, Rich & Hall* 1962) — a multiribosomal structure (= ergosome, polyribosome) representing a linear array of → ribosomes held together by → messenger RNA (the single ribosome bound to mRNA is referred to as a monosome by *Haselkorn & Fried* 1964). Probably the number of ribosomes is related to the size of the mRNA chain. Polysomes represent the "active complexes" in cellular protein synthesis and are able to incorporate amino acids into polypeptides both *in vivo* and ***in***

vitro. Such ribosome aggregates are usually 10 to 20 times more active than isolated ribosomes (monosomes) in protein synthesis.

The ribosomes contained in polysomes isolated from → chloroplasts, → mitochondria and the surrounding cytoplasm of eucells can be distinguished from one another by sedimentation analysis. Chloroplast and mitochondrion ribosomes are closely related to bacterial ribosomes.

Cytoplasmic polysomes of eucells (metazoa) occur in at least two and possibly three forms, which are biochemically and morphologically distinguishable. One class of polysomes exists apparently free in the cytoplasm, while a significant fraction of polysomes is closely associated with the → endoplasmic reticulum (especially in cells synthesizing protein for secretion). The fraction of ribosomes occurring on membranes is dependent on the type and state of the cell. One class of polysomes bound to the endoplasmic reticulum, termed "tight", is attached directly to the membrane via the large ribosomal subunit and is assumed to be involved in the production of proteins which are inserted into the cisternae of the rough endoplasmic reticulum. A second class of polysomes with different properties is called "loose"; the properties of this class remain to be determined in detail.

polysomic (*Blakeslee* 1921) — of an otherwise diploid cell or individual with one or more chromosomes represented three (→ trisomic; doubly trisomic) or four (→ tetrasomic) times instead of twice (2n + 1, 2n + 1 + 1, 2n + 2, etc.). **Polysomic inheritance** (*Blakeslee, Belling & Farnham* 1923) is the mode of inheritance which arises when any chromosome (in the case of polysomy or polyploidy) finds more than one partner for → chromosome pairing at meiosis.

polyspermy — the entry of more than one sperm into the egg cell, irrespective of whether the supernumerary sperm are effective or ineffective in fertilization. P. may be a normal or abnormal condition (→ polyandry).

polyspore (*Renner* 1916) — = gonospore (→ germ cell).

polytene chromosome — a special type of → chromosome representing a bundle of numerous fibrils (chromatids) which arise, according to the "polyteny hypothesis" (*Koltzoff* 1934, *Bauer* 1935), by repeated cycles of → endoreduplication of single chromatids. This type of chromosome in its typical form is characteristic for various tissues of Diptera and may be up to 200 (and even more) times the size of corresponding chromosomes at meiosis or in the nuclei of ordinary somatic cells. In Diptera, homologous chromosomes are somatically paired (→ chromosome pairing), and, therefore, the number of p. chromosomes per cell corresponds to half the (diploid) chromosome number of somatic cells. The number of fibrils per p. chromosome is species-specific and may ultimately reach about 2000 in extreme cases. The length and diameter of the fibrils varies between 100—250 μm and 15—25 μm, respectively.

The so-called "banding pattern" (→ band) which results from the intimate association (pairing) at the same level of identical → chromomeres of the constituent fibrils is highly chromosome-specific and may be used for the identification of individual chromosomes as well as for the construction of detailed chromosome maps. Structural modifications of the banding pattern, in the form of reversible "puffs" and "Balbiani rings" are associated with differential → gene activation (→ puffing). In addition to Diptera, p. chromosomes have been found in some angiosperms, protozoa, and in some carcinomas.

In plants, the p. chromosomes are not paired and show a granular structure without distinct bands (*Nagl* 1962).

polyteny (*Koltzoff* 1934; *Darlington* 1937) — a special nuclear differentiation in larval and adult Diptera, Collembola, Protista, and angiosperm ovular nuclei represented by the presence of → polytene chromosomes. P. arises during differentiation of specific cell types which represent end-products of the cell-lineage in question.

polytopic — of a taxonomic category composed of widely separated but phenotypically identical populations (*Mayr* 1963).

polytypic (*Mayr* 1942) — of a taxonomic category (e.g., a species) composed of two or more immediately subordinate categories (e.g., geographic races or subspecies).

population (*Johannsen* 1903) — a community of potentially interbreeding individuals (at a given locality) which share a common → gene pool. Such a → "Mendelian population" (*Dobzhansky* 1935) represents a (dynamic) panmictic unit (→ breeding size). The sum of all factors governing the pattern by which gametes from various individuals unite with each other during fertilization makes up the population structure.

The largest group of potentially interbreeding individuals and thus the largest Mendelian population is the → species. It in turn is composed of numerous local populations, each intercommunicating and intergrading with others.

A p. freely exposed to → gene flow and subject to the input of alien genes, due to immigration, is called an **"open p."**, as opposed to a **"closed p."** with no genetic input other than by → mutation.

Population systems in nature may be classified on the basis of their phenotypic variation, their relative geographical distribution, and their mode of isolation (*Grant* 1963) as follows:

1. The population systems intergrade continuously in their phenotypic characters and are, therefore, judged to interbreed freely:

a) Allopatric = contiguous geographical races.

b) Sympatric = ecological races.

2. The population systems intergrade discontinuously or partially and are, therefore, judged to interbreed on a restricted scale:

a) Allopatric and differentiated morphologically or physiologically to a moderate degree = disjunct geographical races.

b) Allopatric and differentiated morphologically or physiologically to a considerable degree = allopatric semispecies.

c) Sympatric and not isolated reproductively = ecological races.

d) Sympatric and partially isolated reproductively = sympatric semispecies.

3. The population systems are separated by a discontinuity in the pattern of morphological and physiological variations and are evidently not interbreeding:

a) Allopatric and not isolated reproductively = allopatric semispecies.
b) Allopatric and reproductively isolated = allopatric species.
c) Sympatric and reproductively isolated = sympatric species.

population density — the number of individuals of a population per unit of space in the habitat.

population equilibrium — = → genetic equilibrium.

population genetics — a branch of → genetics that describes (in mathematical terms) the consequences of Mendelian → inheritance on the population level. It deals with the frequencies and interactions of genes in so-called "Mendelian → populations" and studies the influences of agencies such as → mutation, natural and artificial → selection, → migration, mixing of races and chance factors which tend to alter → gene frequencies and thus to cause evolutionary changes (evolution).

population monitoring system — any system used to detect an increase in → mutation rates in human or other populations.

population size — = → breeding size.

population system — any of the aggregates of → populations, which form breeding units of large extent and long duration (*Grant* 1963).

population wave — any of the irregular or rhythmic changes of the effective → breeding size of a → population, or any territorial shift ("territorial p. w.") within the area occupied by a population. Such changes may be accompanied by random and relatively rapid alterations of → gene frequency (→ genetic drift) and may thus be a force in evolutionary change (*Ludwig* 1954). Population waves either restrict panmixis or (in case of territorial population waves) confront the population's gene pool with new constellations of environmental influences and thus new conditions of → selection.

pore complex (*Watson* 1959) — → nuclear envelope.

position effect (*Sturtevant* 1925) — a change in the phenotypic effect of one or more genes due to a change in their position with respect to other genes in the → genotype; this may be the result of chromosomal structural changes (→ chromosome mutation) or of → crossing-over. Differences in the phenotypic expression of a gene caused by a change in its spatial relation to other genes on the chromosomes may be brought about by:

a) changes in the production of a given → messenger RNA (→ genetic transcription),

b) changes in the → genetic translation of the messenger RNA in question, or

c) changes in the interactions between the polypeptides resulting from the translation of messenger RNA.

According to *Lewis* (1950), two types of position effects may be distinguished: the "cis-trans type" (or "stable type") and the "variegated type":

1. **The cis-trans position effect** consists (as shown by the → cis-trans test) of different phenotypic expressions of two mutated sites within one gene or → cistron; they are present in the heterozygous state and separable by → genetic recombination. The cis-configuration (both mutated sites

on one chromosome, the normal sites on the other) results in a normal phenotype, the trans-configuration (one normal and one mutated site on each chromosome) in a mutant phenotype. This type of position effect can also occur over an interval greater than one cistron (intergenic cis-trans p. e.), between mutations occurring in the cistrons of one → operon controlled by a single → operator and between mutations of the operator itself.

The intergenic cis-trans p. e. may result if

a) the product of one or both of the wild-type genes involved is unstable (except in combination with the other) and the products decay unless the normal alleles of both genes are translated in the immediate neighborhood of both, or

b) the translation and/or transcription of one or the other of two complementing (→ genetic complementation) wild-type genes is partially suppressed by a → nonsense mutation in the cis-position to it.

2. **The variegated (V-type) position effect** (*Muller* 1930) involves chromosomal structural changes whereby the chromosomal "environment" of the wild-type genes under consideration is changed (by → translocation or → inversion). Classically, genes localized in euchromatin are juxtaposed to → heterochromatin and *vice versa*. V-type position effects are confined to genes present in the heterozygous state and result in a partial repression (by heterochromatinization) of the wild-type gene when it is brought into contact with heterochromatin. This repression is manifested by the expression of the mutant allele and, since the wild-type allele may escape repression in a number of cases, the final phenotype represents a mixture of patches of both the wild-type and the recessive phenotype (variegation or mottling).

The effect of heterochromatin is reversible and nonspecific in the sense that any gene will give rise to a variegated phenotype when present in the heterozygous state and transposed to heterochromatin. For a given V-type p.e., the expression of several adjacent genes may be influenced, i.e., the effect spreads along the chromosome, in which case the amount of variegation is an inverse function of the distance separating the gene from the heterochromatin.

position effect variegation — → variegation due to the suppression of the expression of a wild-type allele when its position on the chromosome is close to a heterochromatic region. P. e. v. is caused by some mechanism that suppresses normal → genetic transcription or → genetic translation (→ position effect). The key characteristics of p. e. v. include (*Cattanach* 1974): 1. almost invariably the variegation is associated with a rearrangement one of whose breakpoints is close to the gene whose expression is affected; 2. one of the breaks is within a → heterochromatic region of a chromosome; 3. the wild-type allele of the affected gene must be carried in the → cis position with the rearrangement and usually with a recessive allele carried on the normal homologous chromosome; 4. the probability of a given locus being affected is dependent upon its proximity to the breakpoint (polarized spreading effect), i.e., the closer the genes are to the breakpoint, the more likely they are to be affected.

position pseudoalleles (*Lewis* 1951) — → pseudoalleles showing a cis-trans → position effect if present in the heterozygous state.

positional information (*Wolpert* 1969) — information specifying the spatial pattern of → differentiation.

positive control — control (→ genetic regulation) by regulating proteins whose presence in the correct conformation is required for gene expression (→ negative control).

postdivision (*Darlington* 1937) — → predivision.

postreduction (*Korschelt & Heider* 1903) — → meiosis.

post-replication repair (*Rupp & Howard-Flanders* 1968) — a type of → dark repair of DNA lesions which does not act on the primary DNA lesions (→ excision repair) but on secondary lesions arising as a consequence of unrepaired primary lesions in replicating DNA. The primary lesions lead to discontinuities (gaps about 1000 to 1600 nucleotides in length) in the daughter strand produced during replication. In bacteria (Fig. 82), the repair of daughter strand gaps involves a recombinational process (= recombinational repair) which operates independently of excision repair and depends on genes also needed for → genetic recombination (rec-dependent repair). After replication of damaged DNA, p.-r. r. reconstructs at least one undamaged copy of the genome of each cell by recombinational filling-in (incorporation of portions of one daughter duplex into the other) of the gaps. P.-r. r. is an error-prone process during which mutations may recur as errors in repair. It is stimulated by UV irradiation and ionizing radiation and inhibited by caffeine.

Secondary lesions repaired by p.-r. r. cannot be repaired by excision repair since they are located opposite a non-coding DNA lesion and are, therefore, not repairable by repolymerization of the missing bases.

In mammalian cells, recombination seems not to be involved in p.-r. r. but rather a process similar to normal replication which is inhibited by hydroxyurea.

postsplit aberration — a chromosomal structural change (→ chromosome mutation) which takes place (in the chromosome segments already duplicated) during or after chromosome duplication, as opposed to a "pre-split aberration" affecting the unreduplicated chromosomes.

post-transcriptional control — control of gene expression by selective translation (→ genetic translation) of → messenger RNA's (→ genetic regulation; transcriptional control).

post-transcriptional processing — in eukaryotes, the stepwise tailoring (trimming) of DNA transcripts) (→ heterogeneous nuclear RNA; pre-messenger RNA; pre-ribosomal RNA) which results in a shortening and modification of the primary nucleotide sequence of the transcript.

post-translocation complex — in → genetic translation, the messenger-ribosome complex with peptidyl-tRNA in the ribosomal P site and a vacant A site (→ pretranslocation complex).

preadaptation (*Davenport* 1903) — the possession of (preadaptive) characters that enable an organism to immediately adapt to an environment to which it has not been previously exposed (→ adaptation). Such characters may be due to the presence of duplicate genes.

P-particle (*Sonneborn* 1959) — → killer.

precocity theory (*Darlington* 1930) — a theory which regards → meiosis as a precocious → mitosis. Meiotic (as compared with mitotic) prophase is assumed to start precociously with chromosomes which have not yet split and this is held responsible for → chromosome pairing, chiasma formation, nondivision of the centromeres and insertion of a second division, together with the reduction of chromosome number and the segregation of genes (*Darlington & Mather* 1949). Since both DNA and protein synthesis have been shown to be completed before meiotic chromosome pairing occurs, the p. t. appears to be no longer tenable.

precursor RNA — the initial RNA products of → genetic transcription which give rise by processing (cleavage) to mature RNAs (→ pre-ribosomal RNA; pre-messenger RNA; pre-transfer RNA).

predetermination (*Kühn* 1927) — the determination of gene-controlled characters by the maternal genotype prior to the fertilization of the egg cell. Hybrid progeny arising from predetermined zygotes are matroclinous with respect to characters showing p. A mode of → inheritance in which each generation (irrespective of the genotype of the zygote) manifests the genotype of the female parent with respect to particular characters is called "delayed inheritance" (*Boycott & Diver* 1923). A similar result may arise if the characters in question are controlled by extrachromosomal hereditary determinants which show matrilinear inheritance.

predivision (*Darlington* 1937) — equational separation of the two chromatids of a → univalent to opposite poles at the first meiotic division, as opposed to postdivision characterized by chromatid separation at the second meiotic division.

preformation — the belief that the egg, sperm, or zygote contains a preformed adult in miniature, which becomes "unfolded" during development, as opposed to "epigenesis" according to which new structures arise in the course of development.

preferential fertilization — = selective → fertilization.

preferential segregation — → genetic segregation.

premature chromosome condensation (*Johnson & Rao* 1970) — the induction of→ chromatin condensation in an interphase nucleus after fusion of mitotic cells with interphase cells (heterophasic → cell fusion). P. c. c. results in the appearance of fragmented chromatin masses or abnormally condensed chromosomes. The morphology of the prematurely condensed chromatin depends on the → interphase period (G 1, S, G 2) of the affected nucleus after cell fusion (→ prophasing). Prematurely condensed G 1 and G 2 chromosomes are usually discrete units with one and two chromatids, respectively. Condensed S-phase chromosomes show a pulverized appearance (→ chromosome pulverization).

pre-messenger RNA (*Georgiev & Mantieva* 1962; *Scherrer & Darnell* 1962) — in eukaryotes, the transcription product of a → transcripton which is the precursor molecule of cytoplasmic and polyribosomal → messenger RNA. Pre-mRNA is associated with specific proteins (→ informofer) and characterized by the following features (*Scherrer* 1973):

1. The coding sequence which represents mRNA (Fig. 76) is read from DNA as a small part of pre-mRNA molecules.

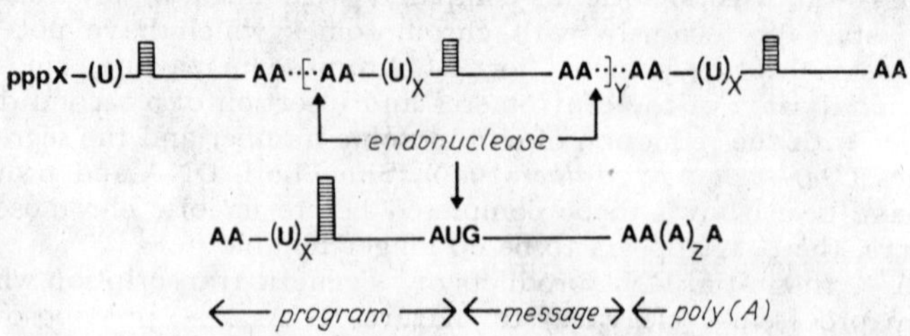

Fig. 76. Schematic representation of pre-mRNA consisting of several units bordered by oligo-(A)-sequences. Each unit is assumed to be composed of one message, programming and nonsense sequences (from *Scherrer* 1973).

2. Oligo(A) and poly(A) regions. The former comprise 5 to 30 adenylic residues, the latter (possibly added after the first cleavage during processing) comprise 100 to 200 adenylic residues.

3. Sequences (of about 30 nucleotides) up to about 80% of uridylate (oligo(U) sequences) which make up about 0.1 to 0.2% of the total pre-mRNA molecule.

4. Double-stranded sequences (about 3% of the molecule) fall into two size classes: one of more than 200000 mol. weight, the other of a 20000 to 50000 mol. weight range. These regions appear to occur ten times per nascent pre-mRNA molecule.

Components 2 to 4 may represent signals in the programming sequence of pre-mRNA, attachment sites for stabilizing proteins, or may expose specific RNA fractions to cleaving enzymes. Programming sequences are probably involved in the mechanisms of post-transcriptional processing and modification of pre-mRNA and may contain signals for transcriptional and/or post-transcriptional regulation, transport and translation.

Probably after the first cleavage of nascent pre-mRNA, intermediate-size pre-mRNA molecules are adenylated post-transcriptionally. The resulting poly (A) sequences are eventually up to 200 nucleotides long; small pre-mRNA molecules originating after further processing contain poly (A) sequences of the same length.

Pre-mRNA molecules are much larger (up to 10^7 daltons) than messenger sized molecules (10^5 to 10^6 mol. weight range).

premetaphase stretch (*White* 1941) — in some animal species, the directed orientation of paired centromeres (→ centromere orientation) towards opposite poles, resulting in a visible stretching of chromosome arms on the spindle. It takes place at prometaphase of first meiotic division. The p. s. does not persist and the centromeres soon approach each other again.

premitosis (*Sagan* 1967) — a type of → mitosis in certain protozoa characterized by an intranuclear → centriole. The → mitotic apparatus is formed within the → nuclear envelope. In flagellates and all higher organisms, the centrioles (or their substitutes) have become extranuclear and division is by eumitosis.

premutation — any DNA lesion (premutational lesion) which may or may not give rise to a → gene mutation or a → chromosome mutation. DNA replication and DNA repair (→ post-replication repair) can lead from premutational changes to permanent changes (→ mutation fixation) in the nucleotide sequence of DNA.

prenatal diagnosis — monitoring of pregnancy through → amniocentesis in the diagnosis of → genetic diseases of metabolism and of changes in normal chromosome number of man.

prepattern — a descriptive term for the general organization of a developing system before a specific and observable pattern of organization is reached.

preprophase inhibitor (*D'Amato* 1954) — → mitotic poison.

prereduction (*Korschelt & Helder* 1903) — → meiosis.

preribosomal particle — any of the ribonucleoprotein particles representing precursors to the cytoplasmic → ribosomes. In eukaryotes, their maturation takes place within the → nucleolus and it seems likely that the ribosome subunits released into the cytoplasm are identical with mature ribosomes apart from their content of a few proteins which must be replaced by those of the ribosome.

pre-ribosomal RNA — in **eukaryotes,** the transcription product (= precursor ribosomal RNA) of → ribosomal RNA which contains the sequences of the 7 S, 18 S, and 28 S species of rRNA and some nonconserved excess or spacer RNA. The fourth rRNA species (5 S) is of independent origin and its trancription takes place on extranucleolar sites. The synthesis and processing of the pre-rRNA molecule (mol. weight 2.3 to 2.9×10^6 in plants, about 2.6×10^6 in Xenopus, and 4.1×10^6 in mammals) proceeds as follows (Fig. 77):

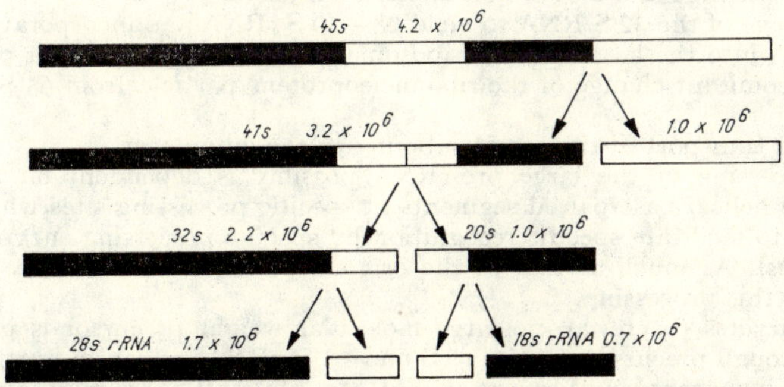

Fig. 77. Processing of pre-rRNA (45 S) in the Hela cell. Black: rRNA regions; white: excess RNA (from *Grierson* et al. 1970).

1. Synthesis of a 35 to 45 S precursor to rRNA (Table 7) by transcription of → ribosomal DNA located in the nucleolar organizer. In mammals, the precursor has a lifetime of several minutes before being cleaved to produce 41 S RNA from 45 S RNA.

Table 7. *rRNA synthesis in various eukaryotic systems*

Fungi (yeast)	
35 S RNA	20 S RNA → 18 S rRNA
	27 S RNA → 25 S rRNA
Insects (Chironomus)	
38 S RNA	18 S rRNA
	30 S RNA → 28 S rRNA
Amphibians (Xenopus)	
40 S RNA	18 S rRNA
	32 S RNA → 28 S rRNA
Mammals (rat, human)	
45 S RNA	23 S RNA → 18 S rRNA
	35 S RNA → 28 S rRNA

2. Methylation (primarily 2′-o-methylation) of the 18 S and 28 S regions of the pre-rRNA occurring during its transcription. The non-conserved portions at the 3′-OH ends contain few, if any, methyl groups and are characterized by a high G + C content.

3. Association of pre-rRNA with protein (including species found on mature ribosomes) to form a complex sedimentating at about 80 S.

4. Processing of pre-rRNA in the nucleolus to yield 16–19 S and 32 S species by cleavage and loss of non-methylated parts.

5. Cleavage of the 32 S RNA to yield 28–30 S rRNA and incorporation of 5 S RNA into the large nascent ribosomal subunit (65 S nucleolar particle). Concomitant change of the ribonucleoprotein particle from 65 S to 60 S.

6. Rapid transport of ribosomal subunits to the cytoplasm.

The processing of the large pre-rRNA possibly is dependent on the presence of helical base-paired segments at specific processing sites which may serve to facilitate specific recognition by specific processing enzymes (convertases). As much as half of the original precursor species may be lost during this processing.

In **prokaryotes** (bacteria), no large molecular weight precursor is produced, although precursors of the 16 S and 23 S rRNAs which are rather close to their mature counterparts in size, etc., do exist. This may imply that cleavage into 16 S, 23 S and 5 S RNA occurs before transcription of the 16 S–23 S–5 S DNA unit is complete.

presplit aberration — → postsplit aberration.

pre-transfer RNA — the product of → genetic transcription of DNA sequences (tDNA) that code for → transfer RNAs. Prokaryotic pre-tRNA is composed of about 120 nucleotides, that of eukaryotes (mammals) of 300 to 400 nucleotides. Its maturation involves post-transcriptional cleavage at the 5′ and 3′ ends (in eukaryotes), modification yielding minor nucleosides in the chain, and perhaps conformational changes. The mammalian pre-tRNA is deficient in modified bases but contains pseudouridine. The secondary structure of pre-tRNA has a "clover-leaf" shape.

pretranslocation complex — in → genetic translation, the messenger-ribosome complex with peptidyl-tRNA in the ribosomal A site and discharged tRNA in the P site (→ post-translocation complex).

prevalence — in human genetics, the frequency with which a medical condition with a genetic basis is found at a particular time in the general population (as, e.g., p. of a → genetic disease). (→ incidence).

primordium — an embryonic area or structure which will ultimately form an adult part or organ.

proband — in human genetics, the "affected" person with whom the study of a particular → character in a family (→ pedigree) usually begins (= propositus if male, proposita if female, or index case). A method of comparing the proportion of progeny in families in which an "index case" or p. shows a specific character with the proportion expected if this character fulfills Mendelian inheritance is referred to as the "proband method".

procentriole (*Gall* 1961) — a short new → centriole arising perpendicular to the old centriole and growing distally as it matures. The maturation and differentiation of a p. involves (*Pitelka* 1969): 1. the synthesis of required polypeptides; 2. the assembly of those into → microtubules, dense material and "cartwheel" substance; 3. the elaboration from these of the polarized, nine-membered cylinder; 4. the synthesis and assembly of the protein of the fibrous appendages at specific times and places.

prochromosome (*Rosenberg* 1904) — any → heterochromatic block in the interphase nucleus that represents a positively → heteropycnotic chromosome segment usually lying on both sides of the → centromere. The number of prochromosomes may be equal to or, if fused, less than the number of chromosomes in the complement.

progamic (*Haecker* 1902) — → sex determination.

progeny — the offspring of a particular → mating, of a particular mate, or of a particular individual in the case of apomictic reproduction (→ apomixis).

progeny selection — the evaluation of an individual's progeny as customarily employed in artificial → selection. Methods of p. s. vary with the types of → matings that are possible with a given animal or plant. Selection may be based on the performance of self-pollinated progeny, of progeny from crosses to an inbred (test crosses), or of progeny from crosses to daughters (backcrosses). When selection of outstanding individuals is based on progeny performance, consideration must be given to the magnitude of phenotypic variance (→ variation), the → heritability of the character being selected, and the degree of → genetic relationship.

progeny test — a method for assessing the → genotype of an individual or of the performance of a parent by a study of its progeny under controlled conditions. This type of test was first used systematically by *G. Mendel.*

progressive double crossing-over (*Weinstein* 1936) — → crossing-over.

prokaryon (*Dougherty* 1957) — the nuclear equivalent (→ nucleus, nucleoid) of → prokaryotic organisms (→ eukaryon) which does not show any "true" → chromosomes, is not bounded by a nuclear envelope, and does not divide by mitosis.

prokaryotic (*Chatton* 1925) — of viruses, bacteria, and blue-green algae ("prokaryota" or "prokaryotes") which lack those features of the eukaryote organization of the genetic material, viz. no membrane-bounded nuclei dividing by → mitosis and → meiosis, no spindle mechanisms and no condensation cycles of the → chromosomes. Nuclear equivalents (→ nucleoid) replace true nuclei to which they are functionally homologous. Pro- and eukaryotes represent different evolutionary stages in the organization of the → genetic material.

prolamellar body (*Hodge, McLean & Mercer* 1956) — an accumulation of vesicles formed by the invagination of the proplastid membrane during etiolation (→ plastid).

The p. b. consists of a central core built of an array of interconnected tubules from which peripheral tubular elements extend into the etioplast matrix. During the greening process of etiolated plants, the p. b. undergoes a series of structural transformations: the tubular elements give rise to flattened vesicles and, finally, to lamellae or → thylakoids which fuse to form the grana system of the → chloroplast. Thus, the p. b. is a pool of readily available precursors of chloroplast lamellae.

prometaphase (*Lawrence* 1931) — → mitosis, → meiosis.

prometaphase movement — → chromosome movement.

promitochondrion — a precursor of a → mitochondrion.

promoter (*Jacob* et al. 1964) — in an → operon, a nucleotide sequence located at the → operator end (adjacent but external to the operator) which contains all the signals for correct initiation of → genetic transcription by the → RNA polymerase holoenzyme and determines the maximal rate of RNA synthesis. Different promoters probably have different rates of initiation with RNA polymerase and some need additional factors for high-level expression of their operons (→ catabolite repression). Genetically, the p. is defined by mutations that coordinately alter the level of expression of all the structural genes in the operon. These mutations lie outside the structural genes and do not affect induction of the operon.

The primary nucleotide sequence of the promoters analysed up to now differs widely indicating that the DNA-dependent RNA polymerase recognizes a tertiary structure in the DNA. The p. sequences contain symmetric elements which may be significant for the attainment of three-dimensional structures.

The p. includes the → RNA polymerase entry (binding) site, a → drift region (which acts as a cyclic AMP binding site to switch on inducible operons) and start sites. The p. may bind several polymerase molecules; their number is characteristic of the individual p. and functionally defines a storage region of the p.

Internal promoters of relatively low efficiency have been found within certain operons at or near the boundary of definite genes in addition to the principal p. in the vicinity of the operator region. Internal promoters may serve as transcriptional initiators (low-level, non repressible synthesis of the products of genes on one side of them). A class of mutations has been found (Ini mutations) which is able to function as internal initiator elements.

P. mutations can decrease or increase ("up-promoter" mutations of *Fraenkel & Parola* 1972) the level of *in vitro* transcription.

promutagen — any non-mutagenic agent convertible to mutagenic metabolites (→ host-mediated assay).

pronucleus (*van Beneden* 1875) — the egg or sperm nucleus from maturation until → karyogamy at which time the pronuclei fuse to produce the zygote nucleus.

propagule — an uni- or multicellular agent of dissemination.

prophage (*Lwoff & Gutman* 1950) — a temperate → bacteriophage integrated into the genome of → lysogenic bacteria which replicates in coordination with the replication of the → genetic material of the host cell. The p. contains the genetic information for phage production (→ induction) and, thus, constitutes a potentially → lethal factor for the lysogenic bacterium. It behaves like a gene or gene group that exercises only a small part of its potential for phenotypic expression.

The presence of a p. in a bacterial cell is recognized by the immunity conferred on the lysogenic cells to further infection by an extrinsic, homologous phage. Furthermore, the p. may endow the host cell with other characters whose phage mediated acquisition is referred to as lysogenic conversion or → phage conversion.

Occasionally lysogenic cells lose their p. and are then said to be "cured". Certain temperate phages have very specific attachment (integration) sites and are capable of effecting only specific (localized) → transduction; the temperate phages responsible for generalized transduction seem to attach at a variety of positions (→ integration system; integrase). Prophages transferred by bacterial → conjugation to a sensitive recipient cell are obliged to enter a lytic cycle (→ zygotic induction).

Defective prophages are those that have undergone a deletion of their genome. A region of the bacterial chromosome has been inserted into the p. genome, probably in substitution for the missing region. Defective prophages can be transmitted lineally in their host strains but cannot produce infective particles (→ prophage complementation).

Cryptic prophages lack the genes necessary for the lytic cycle or contain mutants of the relevant genes.

Phage production by a cell carrying a p. occurs spontaneously with a very low frequency (10^{-2} to 10^{-5} per cell per generation). This frequency can be raised to nearly 100% by certain agents which interfere with DNA replication (prophage → induction).

prophage complementation (*Thomas* 1966) — the phenomenon wherein a defective phage superinfects a bacterial strain → lysogenic for a heteroimmune but closely related phage; the yield of phage progeny is usually higher than following infection of the corresponding, non-lysogenic strain. As a result of heteroimmune superinfection, some of the genes of the prophage are switched on, i.e., there functions are under direct immunity control.

prophage excision — the direct reversal of the integrative recombination (→ integration system) of a temperate phage leading to excision of the prophage DNA and thus prophage → induction.

prophage induction (*Lwoff* 1953) — → induction.

prophage integration — the attachment (prophage insertion) of phage DNA to the bacterial chromosome (→ integration system). Mutants that are defective in p. i. form phages indistinguishable from those of wild-type since they can establish abortive → lysogeny.

prophage interference — a change in the sensitivity of → lysogenic bacteria (containing a prophage) to infection by bacteriophages unrelated to the prophage integrated in the bacterial genome. P. i. is a case of "lysogenic conversion" (→ phage conversion).

prophage substitution — the replacement of a → prophage by a superinfecting phage in a small number of → lysogenic bacterial cells. The frequency of p. s. is a linear function of the multiplicity of superinfection (about 0.01—0.001 substituted cells per superinfecting phage particle).

prophase (*Strasburger* 1884) — → mitosis, meiosis.

prophase poison (*D'Amato* 1948) — → mitotic poison.

prophasing (*Matsui* et al. 1972) — in an interphase-metaphase binucleate cell which originated after Sendai virus-mediated heterophasic → cell fusion, the transformation of the interphase nucleus to a state similar to prophase in a mononucleate cell (→ premature chromosome condensation).

proplastid (*Strugger* 1950) — an immature, partially differentiated → plastid found in the meristematic cells of multicellular algae or higher plants and in etiolated plants (→ prolamellar body) in which chlorophyll synthesis has been suppressed by growth in the dark. Proplastids lack chlorophyll but contain small amounts of protochlorophyllide.

propositus — = → proband.

proreduplication (*Hsu & Moorhead* 1956) — → interreduplication.

prosphase (*Bullough* 1963) — that part of → interphase during which all the syntheses necessary before the cell enters → mitosis are completed. The p. can be divided into at least three main periods: early p., the phase of DNA synthesis (S), and antephase (G_2). (→ apophase).

protamine — any basic polypeptide with a molecular weight between 1000 and 5000, isolated as a component of chromosomes from animal sperm (→ histone).

protandrous (*Hildebrand* 1867) — → dichogamous.

protection — → restoration, photoprotection.

protein (*Berzelius*) — one of a group of nitrogenous organic compounds (molecular weight ranging between 34000 and 200000) of complex shape and composition. Proteins are made up of molecular subunits called "polypeptides" which are themselves composed of smaller building-blocks, called → amino acids, linked together in form of long chains. Such chains may contain 200 or more amino acids of which there are 20 different kinds (Fig. 78).

The composition of a given p. is uniform, i.e., each molecule consists of the same number and kind of polypeptide chains and each of these in turn is composed of the same number, kind and sequence of amino acids.

The **primary structure** of a p. is represented by the amino acid sequence in the polypeptide chain(s) which is determined by one → cistron (or → gene) during → genetic transcription and → genetic translation of its content of → genetic information.

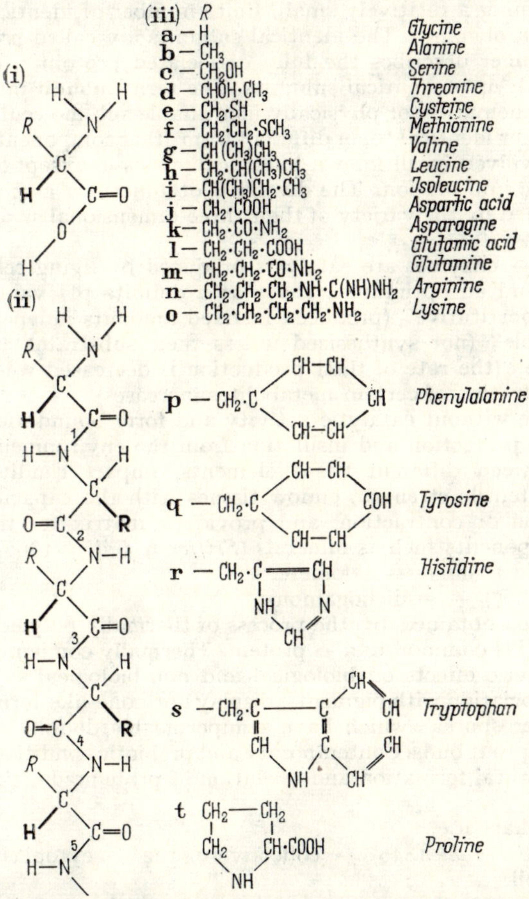

Fig. 78. The structure of amino acids and polypeptide chains (from *Whitehouse* 1965). i) An amino acid; ii) Section of a polypeptide chain with the peptide linkages numbered from the amino group end of the chain,; iii) The formulae of the 20 amino acids normally found in proteins (bold types and italics in i) and ii) indicate projection in front of and behind the plane of the page, respectively).

The **secondary structure** refers to the arrangement of the polypeptide chain in a helical or other regular or irregular manner and the **tertiary structure** to the arrangement of the secondary structures into a three-dimensional (folded and superfolded) structure. The association of more than one polypeptide chain to form a stable unit corresponds to the **quaternary structure.** Interactions between amino acid side-chains largely determine the folding of the polypeptide backbone in proteins and define patterns of specificity and mechanism of action of enzymes.

A **polymeric** p. containing a relatively small, finite number of identical subunits is said to be an oligomer. The identical subunits are called **protomers.** The term **monomer** describes the fully dissociated protomer, or any p. which is not made up of identical subunits. The term **subunit** may be used to refer to any chemically or physically identifiable sub-molecular entity within a p., whether identical to, or different from other components. Proteins are directly involved in all known biological processes except for the storage of → genetic information. The diversity of functions carried out by proteins is matched by the variety of their three-dimensional structures.

Catalytic proteins or → enzymes are catalysts produced by living cells with a specific action and an optimum activity at a definite pH value. They may be either "constitutive" (produced in fixed amounts independent of need), "inducible" (not synthesized unless their substrates are present), or "repressible" (the rate of their production is decreased when the intracellular concentration of certain metabolites increases).

Structural proteins are without catalytic activity and form boundaries, provide containment or protection and insulation from the environment, make connections, between different tissue elements, impart rigidity, hardness, elasticity, or tensile strength, endow tissues with the capacity for movement, extension or contraction, and provide a matrix for the deposition of other components such as minerals (*Seiffer & Gallop* 1966).

protelomere (*Lima-de-Faria* 1958) — → telomere.

protogynous (*Hildebrand* 1867) — → dichogamous.

proteinoid — any substance obtained by the process of thermally polymerizing all of the amino acids common to a → protein. Thermally condensed amino acids show catalytic effects on biological and non-biological substrates, sometimes in association with pigments, display hormone-like activities and can form "microspores" which have semipermeable double outer membranes and can sprout buds. Contemporary and prebiotic conditions seem suitable for the natural formation and evolution of proteinoids (*Fox* 1969).

protokaryotic — = → prokaryotic.

protoplasm (*Purkinje* 1840; *Mohl* 1846) — collectively, the → cytoplasm and → nucleus of the cell.

protoplast (*Hanstein* 1880) — 1. the organized living unit which represents a → cell and consists of the membrane bounded cytoplasm and nucleus (in eukaryotes) or nucleoid (in prokaryotes);

2. a bacterial protoplast is a structure (→ spheroplast) derived from a vegative cell whose entire cell wall has been removed (*McQuillen* 1960).

More or less intact bacterial protoplasts can be isolated by enzymatic digestion of the cell wall (e.g. by lysozyme), or by treatment with antibiotics of the penicillin-basitracin group.

prototrophic (*Ryan & Lederberg* 1946) — of nutritionally independent cells, as opposed to → auxotrophic ones.

protovirus hypothesis (*Temin* 1971) — a hypothesis which postulates that information transfers by transcription and reverse transcription (→ reverse transcriptase; antimessenger) coupled with genetic accidents continually generate oncogenes (→ oncogene hypothesis) and thereby neoplastic events. The p. h. stresses the importance of information flow from DNA to RNA and back to DNA as a process of → gene expression, gene duplication, and gene modification. A protovirus is assumed to arise if genetic information at some stage of this cycle acquires a stable phase distinct from its integrated DNA progenitor. If such protoviruses contain genes relating to cell multiplication, movement, or differentiation, mutation or recombination in the protovirus genome could result in the appearance of genes capable of causing neoplastic transformation.

provirus — the state of a virus in which it integrates into and replicates in coordination with a host cell chromosome and thus is transmitted from one cell generation to another (→ prophage).

2. A DNA copy of RNA tumor viruses produced by RNA-dependent → DNA polymerase (reverse transcriptase). The synthesis of a p. is believed to be an obligatory step in infection and → cell transformation by these viruses.

proximal (*Navashin* 1912) — of a part of a chromosome arm which is nearer to the → centromere than another (distal) part (*Darlington & Mather* 1949).

pseudoallele (*Morgan, Bridges & Sturtevant* 1938; *Lewis* 1948) — any of two (or more) mutations which are allelic (similar or related) in a functional but not in a structural sense. Pseudoalleles map at different positions (→ genetic map) and show a low order of → genetic recombination via crossing-over. Such mutations (m^1 and m^2) are phenotypically mutant in trans-heterozygotes ($m^1 +/+ m^2$), but wild-type in cis-heterozygotes ($m^1m^2/++$), that is, exhibit a cis-trans → position effect ("position pseudoalleles").

Pseudoallelism is evidence that the → gene cannot be defined as a unit of structure **and** function (→ cistron) and may be interpreted in two ways:

1. Each of the separable elements of a (pseudoallelic or complex) locus is a discrete gene (→ operon) with a distinct function and the cis-trans position effect is the result of a gene product acting as the substrate (not able to diffuse from one chromosome to its homologue) for the activity of the adjacent gene.

2. The separable elements represent mutations at different recombinable sites within one and the same gene (→ mutational site). According to this interpretation, the gene functions only when all its parts are assembled together in one sequence, and its function can be impaired by mutation at any of numerous separable points along its length. Today,

pseudoalleles are generally interpreted as closely linked and functionally related genes. A cluster of pseudoalleles is referred to as a "pseudoallelic series", a "complex locus" (*Dunn* 1954), or a "region" (*Benzer* 1957).

pseudoamitosis (*Haecker* 1910) — a type of → karyokinesis which resembles an → amitosis but originates from a mitotic type of nuclear division. P. results from a pycnotic clumping of chromosomes. Pseudoamitoses may be induced by ionizing radiation and various chemicals whose application evokes disturbances of anaphase movement of chromosomes.

pseudoaneuploid (*Rieger* 1963) — of cells or individuals (pseudoaneuploids) characterized by changes in the → basic number of chromosomes within a species due to fragmentation or → centric fusion. This results in chromosome numbers, which deviate from the common one. In contrast to genuine → aneuploids, the genetic material is equivalent to that of individuals with the original chromosome complement, i.e., before the chromosomal structural changes (→ pseudopolyploid) occurred.

pseudoapogamy (*Farmer & Digby* 1907) — the development of the → sporophyte from fused vegetative cells instead of the fertilized egg cell (= pseudomixis of *Winkler* (1908) and somatogamy of *Renner* (1916)).

pseudobivalent (*Levan* 1937) — 1. at → mitosis, a bivalent-like association of two chromosomes due to reciprocal chromatid or subchromatid → translocation.

2. At metaphase of → meiosis, the association of two chromosomes by stickiness rather than by → chiasmata (*Walters* 1954). Associations of this type are called "quasibivalents" by *Östergren* and *Vigfusson* (1953) and may coorientate like true bivalents (→ centromere orientation).

pseudobridge (*Resende* 1943) — a → chromosome bridge due to sticky-adhesions between two chromosomes.

pseudochiasma (*Levan & Tjio* 1948) — a chiasma-like connection between two chromosomes (in meiosis I) or chromatids (in mitosis and meiosis II) due to the adhesion of chromatids (during anaphase separation) at homologous or, rarely, nonhomologous points (= point union, subchromatid error, point error). The points of adhesion may either be terminal or, more frequently, interstitial. P. may result from chromosomal structural changes at the subchromatid level (→ chromosome mutation) which give rise to so-called "two-side arm bridges" (Fig. 79) if they are located interstitially. Under these circumstances the proximal regions become extended during anaphase movement and the distal segments take the form of a pair of arms arranged like a cross. Such configurations, which arise spontaneously and can be induced experimentally by the application of mutagens, have also been interpreted as consequences of localized sticky-effects ("point stickiness") or localized failures of chromosome duplication ("point errors").

pseudocompatibility — the occurrence (= pseudofertility), under exceptional environmental or genotypic conditions, of fertilization that would normally be prevented by → incompatibility mechanisms (*Darlington & Mather* 1949).

pseudodominant — → dominant.

pseudofertility (*East & Park* 1917) — = → pseudocompatibility.

pseudogamy (*Focke* 1881) — apomictic (parthenogenetic) development of a female gamete or cell after stimulation (but not fertilization) by a male gamete or gametophyte. The progeny arising from p. are maternal in character (= gynogenesis).

pseudogonochoristic — = → pseudohermaphroditic.

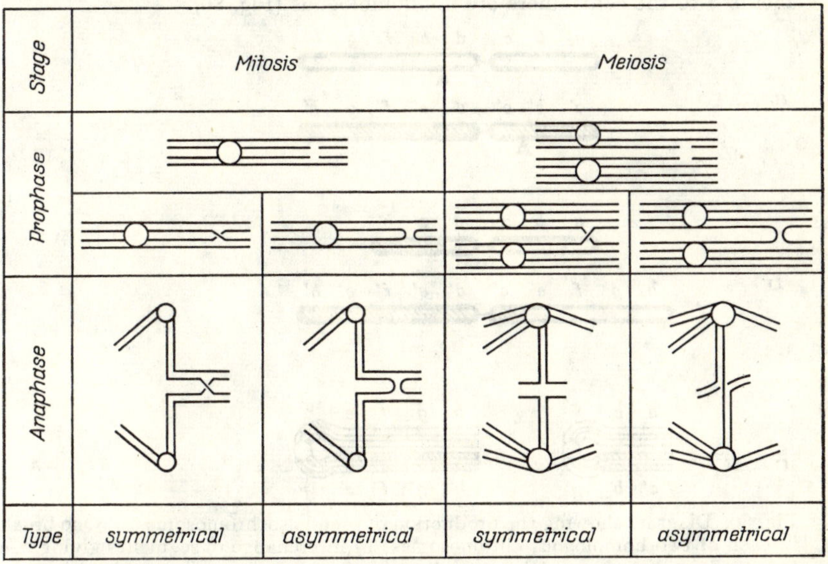

Fig. 79. Diagram illustrating the formation (by subchromatid breakage) of pseudochiasmata in mitosis (left) and meiosis I (right) (from *Lewis* and *John* 1963).

pseudohaploid (*Katayama* 1935; *Ivanov* 1938) — → haploid.

pseudohermaphroditic — of individuals ("pseudohermaphrodites") with gonads of one sex only but possessing reproductive organs with some characteristics of the opposite sex. Mammals are classified as male or female according to the type of gonad present. Male pseudohermaphrodites phenotypically resemble females but possess testes; female pseudohermaphrodites possess ovaries but are phenotypic males as a result of the effects of androgenic influences upon the developing female genital tract. True → hermaphroditic individuals, in contrast to pseudohermaphrodites, possess the gonads of both sexes either as separate ovaries and testes, or combined in the form of an ovotestis.

pseudoheterosis (*Dobzhansky* 1952) — = → luxuriance.

pseudohomeotypic (*Gustafsson* 1935) — of a modified first meiotic division (→ meiosis), characterized by the presence of only unpaired chromosomes (univalents) rather than pairing configurations. The chromatids of each univalent are distributed to opposite spindle poles at anaphase I. Meiosis I is, in fact, a → mitosis and the second meiotic division is omitted.

pseudoisochromosome (*Caldecott* & *Smith* 1952) — a chromosome whose ends are homologous as a result of reciprocal → translocation between end segments of opposite arms of the chromosomes of one pair of homologues. These ends pair (→ chromosome pairing) internally at meiosis like → isochromosomes, but the segments proximal ("interstitial segments") to the centromere are nonhomologous (Fig. 80).

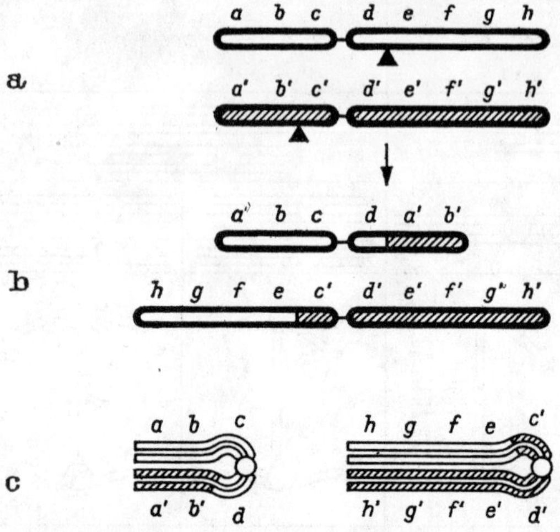

Fig. 80. Diagram showing the production of pseudoisochromosomes. a) One break in each of two homologous chromosomes; b) interchange of segments giving rise to pseudoisochromosomes; c) meiotic chromosome pairing (internal pairing) of pseudoisochromosomes.

pseudolinkage — → affinity.
pseudolysogenic (*Romig* & *Brodetsky* 1961) — of bacteriophages which are capable of establishing a carrier state in bacterial host cells and, in several instances, able to act as generalized transducing agents (→ transduction).
pseudomeiosis (*Battaglia* 1945) — → meiosis.
pseudomixis (*Winkler* 1908) — = → pseudogamy.
pseudomonosomic (*Hiorth* 1948) — → monosomic.
pseudomonothallic (*Ahmad* 1954) — → heteromixis.
pseudomultivalent — an association (→ pseudobivalent) of more than two chromosomes due to sticky adhesions rather than to homologous → chromosome pairing and chiasma formation, in contrast to a true → multivalent.
pseudo-overdominant (*Lerner* 1958) — of a genetic locus whose apparent → overdominance (indicated by heterozygote excess) is in fact due to → linkage disequilibrium of the alleles (= associative overdominance of *Frydenberg* 1963).
pseudopolarity — of effects which mimic → genetic polarity but are not due to factors affecting gene function. The apparent pola reffects may be due

to mutations affecting enzyme activity at the level of protein-protein interaction.

pseudopolyembryony (*Lebegué* 1952) — → polyembryony (in plants) due to fusion of several ovules, the division of the nucellus and the production of several embryo-sacs by the multicellular archespore or by several functional macrospores.

pseudopolyploid (*Battaglia* 1956) — of cells or individuals ("pseudopolyploids") whose chromosome number is doubled or increased to an even higher order without a corresponding increase in the quantity of the → genetic material (→ pseudoaneuploid). The condition ("pseudopolyploidy") arises by one of the following mechanisms:

1. Transverse "breakage" of chromosomes with multiple → centromeres leading to a large number of small chromosomes in some tissues, while the multiple centromere organization is maintained unchanged in others (as, e.g., in *Ascaris*: → nuclear differentiation): **Breakage-pseudopolyploidy.**

2. Transverse breakage of chromosomes with diffuse centromere organization giving rise to fragments which are perpetuated through the normal mitotic cycle. Successive breakage cycles may lead to a corresponding increase in the number of chromosomes which become progressively smaller: **Agmato-pseudopolyploidy.**

3. The converse of agmato-pseudopolyploidy, i.e., fusion of several small chromosomes with centric activity resulting in a few large (polycentric) chromosomes, is known (in *Spirogyra*) and may simulate a polyploid series of chromosome numbers within the systematic category concerned: **Fusion-pseudopolyploidy.**

4. Changes in the lateral multiplicity of chromosomes by longitudinal splitting without prior replication (as in the genus *Thyanta* with diffuse centromeres); this may give rise to numerical polyploidy with the retention of the original DNA quantity if accompanied by reduplication of the individualized longitudinal subunits of the chromosomes: **Pseudopolyploidy due to differential polynemy.**

pseudoreversion — → reversion.

pseudosatellite (*Battaglia* 1956) — → satellite.

pseudoselectivity (*Sedlmayr* 1956) — apparent nonrandom fusion of gametes during zygote formation due to inviability of a number of the zygotes. This may be regarded as a case of selective → fertilization.

pseudo-stemline (*Hughes* 1968) — a → stemline whose modal cells can be characterized only by having the same chromosome number, and not by their karyotype.

pseudotrisomic (*Kush* 1973) — = → doubletelotrisomic.

pseudotype (*Choppin & Compans* 1970) — an enveloped RNA virus which after participation in → phenotypic mixing with a similar virus from a different group shows envelope antigens of one virus and the genome of the other.

pseudovirion (*Michel* et al. 1967) — any viral → capsid containing linear fragments of cellular DNA, instead of the viral genome or fragments thereof. Pseudovirions are not infectious.

pseudowild-type (*Mitchell, Pittenger & Mitchell* 1952) — of a → phenotype arising not by reversion (back mutation) of an original mutational change but rather by nearby suppressor mutations (→ revertant).

psi factor (*Travers* et al. 1970) — any of a class of bacterial positive control elements assumed to act as secondary specificity determinants allowing the → RNA polymerase holoenzyme (not the core enzyme) to transcribe several classes of transcription units. Each element is assumed to be specific for a given class. There are indications that psi factors are regulated by low molecular weight effectors. This provides for a positive control which is readily reversible and would allow the bacterial cell to respond readily to environmental changes.

Psi and → CAP are of approximately the same size, require → sigma function and are functionally involved in transcriptional processes. CAP activity is activated by cyclic AMP, psi is inhibited by guanosine tetraphosphate (pp G pp).

P-site — → ribosome.

puff (*Bridges* 1937) — any of the structurally modified regions in → polytene chromosomes involved in RNA synthesis. Puffs originate from single → bands and involve the despiralization of the deoxyribonucleoprotein (DNP). The despiralized DNP, the accumulated RNA (due to site-specific synthesis), and the accumulation of acidic proteins (synthesized elsewhere) contribute to the morphology of the p. site.

Different tissues and developmental phases are characterized by specific patterns of puff formation (→ puffing), some being common to several tissues, others being tissue-specific.

P. regression is a process that involves the disappearance of both RNA and the proteins from the p. site and the re-establishment of the banded state by the DNP (*Ashburner* 1970). Extremely large puffs are called → Balbiani rings (Fig. 81).

DNA puffs synthesizing excessive amounts of DNA are a characteristic feature of *Sciarid* polytene chromosomes.

puffing — the cell- and stage-specific origination of → puffs from discrete sites of → polytene chromosomes. The first sign of p. is the accumulation at the pre-puff site of an acidic protein followed by an increase in the rate of RNA synthesis at this site. The p. pattern is an expression of differential gene activity and characterized by the time of appearance and regression of individual puffs. The degree of p. of a certain chromosomal site is usually directly correlated with its RNA synthesis rate.

P. phenomena have also been found in polytene plant chromosomes.

pure line (*Johannsen* 1903) — the descendants, through self-fertilization, of a single → homozygous parent, or a highly inbred line of plants or animals obtained by long continued → inbreeding. All members of a p. l. are of the same genotype and are homozygous for all allele pairs within the limits set by the probability of new → gene mutations.

pycnotic — → heteropycnotic.

pyrenoid (*Schmitz* 1882) — a specialized structure occurring within the chloroplasts (→ plastid) and involved in starch synthesis and storage. Pyrenoids are found primarily in certain algae and liverworts. They are composed of a granular, homogeneous or filamentous and fibrillar matrix around which are found starch deposits.

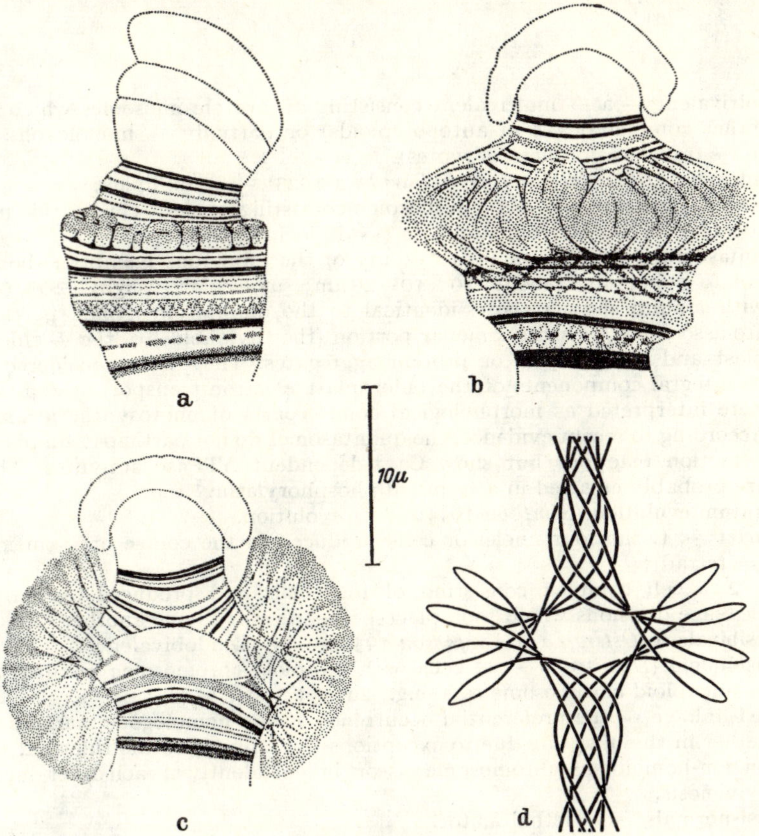

Fig. 81. Different degrees (a to c) of puffing of the Balbiani ring (BR) in the salivary gland chromosome IV of *Chironomus tentans*. The interpretation of the course of the chromosomal fibers in the region of the Balbiani ring is given in d (after *Beermann* 1952 from *Clever* 1965).

pyrimidine cluster (*Szybalski* et al. 1966) — a run of pyrimidines (cytosines) in DNA assumed to serve as a recognition signal for → RNA polymerase, i.e., for initiation of → genetic transcription. The site of transcription initiation in an → operon is called the → promoter.

Q

quadrivalent — a → multivalent consisting of four chromosomes which are either completely (as in autopolyploids) or partially → homologous (as in → translocation heterozygotes).

quadruplex (*Blakeslee, Belling & Farnham* 1923) — → nulliplex.

quadruplochromosome — a chromosome consisting of eight chromatids held together at the centromere region (→ diplochromosome).

quantasome (*Park & Pon* 1961) — any of the numerous membrane-bound oblate spheroids (about $10 \times 10 - 20$ nm; molecular weight 2000000) with subunit characteristics identical to the quantasome. Such particles are associated with the lamellar portion (the thylakoids) of the → chloroplast and are proteins or protein aggregates. They were considered to be integral components of the chloroplast electron transport system and were interpreted as morphological counterparts of photosynthetic units. According to recent evidence, the quantasomes do not participate in photoreduction reactions but show Ca^{++}-dependent ATPase activities. They are probably engaged in a terminal phosphorylating step.

quantum evolution (*Simpson* 1944) — → evolution.

quartet — 1. the four nuclei or cells produced in the course of → meiosis (= tetrad);

2. a cell complex consisting of four cells and produced by two → cleavage divisions which took place perpendicular to one another.

quasibivalent (*Östergren & Vigfusson* 1953) — → pseudobivalent.

quasidiploid (*Hsu* 1957) — of cells with a diploid chromosome number but → aneuploid chromosome sets (e.g., $2n + 1 - 1$).

quasi-linkage — the preferential occurrence of non-homologous markers together in the offspring due to exceptions (→ affinity) to the rule that loci on non-homologous chromosomes assort independently of each other during → meiosis.

quasi-normals — → lethal factor.

R

Rabl-orientation — the maintenance to the next mitotic prophase of the anaphase orientation of the chromosomes (centromeres directed to the spindle pole of the preceding mitosis).

race — an intraspecific category, primarily a population or aggregate of populations, with characteristic gene frequencies or features of chromosome structure that distinguish a particular group of individuals from other groups of the same kind within formally recognizable → subspecies or within → species. Race differences are relative, not absolute. The term subspecies is frequently used in the same sense as race. Any race is able to interbreed freely with any other r. of the same species. Whenever different races of a cross-fertilizing species occupy geographically separate territories, they are said to be allopatric; those occupying the same territory are sympatric. Races may become distinct species (→ speciation) by the production of reproductive → isolation (with respect to the other races of the same species) and thus the formation of isolated → gene pools.

Geographic races are → subspecies occupying a geographic subdivision of the range of a species.

Ecological races are local races owing their most conspicuous attributes to the selective effect of a specific environment (→ ecotype).

Physiological races are races characterized by certain physiological, rather than morphological characters.

Chromosomal races are races differing in respect to features of chromosome structure (cytotypes) or in chromosome number (polyplotypes).

rad equivalent (*Bridges* 1974) — the dose (= roentgen equivalent dose) of ionizing radiation which would produce the same effect as a chemical mutagen at proposed population doses (rad eq units).

radiation chimera — an individual whose blood-forming tissue contains cells that differ genotypically from the individual's own cells. The chimeric state is due to transfusion of hemopoietic cells from a genetically dissimilar donor into an irradiated recipient. Proliferation of the transplanted cells is possible because the recipient's immune system has been depressed by irradiation.

radiation protective — of agents which protect DNA *in vivo* against genetic damage by ionizing radiation. Such agents may protect DNA directly by binding to DNA, or indirectly by preventing radiation-induced release of DNA-destructive agents, or by protection against radiation-induced destruction of → DNA repair systems. Most of r. p. agents are sulfhydryl or disulphide compounds and contain amine or guanine groups which are known to interact with DNA.

radiomimetic (*Dustin* 1947) — of chemical agents (typically represented by alkylating agents) which mimic the most important end effects induced by ionizing radiations in living systems. These include → gene and →

chromosome mutations, carcinostatic and carcinogenic effects and, *in vitro*, depolymerized nuclei acids. In a broader sense, the term r. is used for all chemicals ("radiomimetics") which are capable of inducing gene and chromosome mutations.

random assortment — → assortment.

random fixation (*Wright* 1931) — the complete loss of one of two → alleles in a population, the other allele reaching a frequency of 100% under certain calculable circumstances (→ gene frequency; genetic drift). Whether or not r. f. will occur in a population depends on the interrelationship between the number of breeding individuals, the selective value of the allele(s), the mutation pressure, and the gene flow.

random sample — a random collection from a population, selected in such a way that all items in the population are equally likely to be present in the sample.

rate gene (*Goldschmidt* 1917) — a gene controlling the rate of a certain developmental process.

rate of disintegration — the loss of heterozygosity by random → mating for any generation.

ratio cline — a → cline in the frequency of certain → genotypes (characteristic for systems of → genetic polymorphism showing a gradual change of → genotype frequency).

RC particle (*Nakada, Anderson & Magasanik* 1964) — incomplete ribosomal particles ("relaxed particles") containing 16 S and 23 S ribosomal RNA in their two major subunits, sedimenting at 18 S and 25 S respectively, and protein amounting to about two-thirds of that contained in normal ribosomes. RC particles are produced by a methionine requiring, "relaxed" mutant of *E. coli* during methionine starvation.

R-duction — the transfer in *Enterobacteriaceae* of extrachromosomal drug resistance genes mediated by a → resistance transfer factor.

rDNA — abbreviation for → ribosomal DNA.

rDNA amplification — → ribosomal DNA; gene amplification.

rDNA compensation (*Tartof* 1971) — a disproportionality concerning the rDNA content of *Drosophila* and the number of nucleolus organizers per cell (e.g. the rDNA content of whole XO♂ fly DNA is higher than half the rDNA content of whole XX♀ fly DNA.). rDNA c. appears to be of somatic origin as the extra DNA in the XO♂ is gained or lost in a single generation. The increase in rDNA may be due to a change in the number of tandem rDNA genes within the nucleolus organizer or may take place by independent polytenization of the rDNA (change in the number of lateral copies of the original nucleolus organizer within a polytene nucleus).

rDNA magnification (*Ritossa* 1968) — → ribosomal DNA; gene magnification.

reaction norm (*Woltereck* 1909) — = → norm of reaction.

reactivation — r. or recovery is the term used to describe the phenomenon in which a given physical (radiation) or chemical treatment is rendered nonlethal or nonmutagenic when in the same biological system under different circumstances it would be lethal or mutagenic (→ restoration; photoreactivation; host-cell reactivation; UV-reactivation; catalase reactivation; multiplicity reactivation; cross reactivation).

reactor — a group of cells which react on the activity of an inductor (or evocator) during embryonic → induction.

reading frame — the mechanism which moves the → ribosome three nucleotides at a time during → genetic translation. A shift in the r. f. by any number of nucleotides other than three or a multiple of three will create an entirely new sequence of → codons (→ frameshift mutation).

reading mistake — the incorrect placement of one or more amino acid residues in a polypeptide chain during → genetic translation (→ mistranslation).

read through (*Reznikoff* 1969) — a type of → genetic transcription in which RNA synthesis of one → operon is initiated at the initiation site (promotor) of another operon (→ operon fusion).

rearrangement — a chromosomal structural change (→ chromosome mutation).

recapitulation (*Kielmeyer* 1793) — a theory stating that during ontogeny the individual passes through stages whose form resembles that of adults of its presumptive ancestors; the developmental history of organisms (ontogeny) recapitulates their evolutionary history (phylogeny).

rec-assay (*Kada* et al. 1970) — a screening method for chemical mutagens (→ mutagenicity screening) utilizing → recombination-deficient (rec^-) bacterial stains. These strains are unable to repair diverse chemical damage to DNA through cellular repair functions (→ DNA repair). Agents showing increased lethal activity of rec^- over rec^+ cells are assumed to be able to damage DNA. The r.-a. allows the screening of candidate substances for chemical mutagenesis in a minimum of time (→ spot test).

rec-dependent repair (*Harm* 1968) — = → recombinational repair.

recessive (*Mendel* 1865) — of genetically controlled → characters and the → alleles responsible for their production when the phenotypic expression is associated with alleles in the → homozygous state, as opposed to → dominant characters and alleles. In the first filial generation (F_1) of a cross between two homozygous (purebreeding) strains differing in respect to a dominant and recessive character difference (AA × aa) the recessive character is not evident, but reappears in one quarter of the second filial generation (F_2).

The failure of a gene to express phenotypically its presence in the heterozygous genotype is called "recessivity", as opposed to "dominance".

reciprocal (*Mendel* 1865) — 1. of a second → cross (A♀ × B♂) which is similar to a first one (B♀ × A♂) but with the sexes of the parents (the sources of male and female gametes) interchanged;

2. of two → chiasmata which involve the same two chromatids (= two strand double crossing-over);

3. of inter- or intragenic recombination in which all marked sites involved segregate 2:2, as opposed to nonreciprocal recombination (→ genetic recombination).

4. of → translocations which involve the interchange of segments between two chromosomes, chromatids, or subchromatids.

recombinagenic (*Holliday* 1963) — ref. to agents ("recombinagens") or processes increasing → genetic recombination.

recombinant — any of the individuals or cells arising as a result of interchromosomal and intrachromosomal (via crossing-over or conversion) → genetic recombination. The number of recombinants in relation to the number of offspring in appropriate crosses is referred to as the "recombinant proportion, or fraction".

recombinase (*Kozinski* et al. 1967) — any enzyme involved in → genetic recombination which introduces single-strand breaks in the polynucleotide chain. Enzymes repairing (sealing) single-strand breaks after recombinational events are called repairases.

recombination (*Bridges* & *Morgan* 1923) — any process which gives rise to cells or individuals (recombinants) associating in new ways two or more hereditary determinants (genes) by which their parents differed (→ genetic recombination).

recombination class (*Bridges* & *Morgan* 1923) — any of the types of recombinant gametes produced by crossing-over. If the parents used for crossing differ in respect to two linked pairs of alleles (AB/AB and ab/ab), four classes of gametes are produced: AB, ab, Ab and aB. Ab and aB represent the "recombination classes" resulting from crossing-over between the loci A and B.

recombination-defective — of mutants in eukaryotes that decrease the frequency of meiotic → crossing over, alter the distribution of crossovers, or both.

recombination-deficient (*Clark* & *Margulies* 1965) — of mutants (rec^-) which are defective in recombination functions (→ genetic recombination).

R.-d. mutants fall into two categories: 1. Those modifying recombination in the entire genome. They usually show an increased sensitivity to UV-light, ionizing radiations and alkylating agents and probably correspond to genes directly involved in one of the steps of genetic recombination or in a process which interacts with it. 2. Those affecting recombination only in a small portion of the genome. The corresponding genes seem to be implicated in the regulation of recombination without affecting the overall process itself.

Rec is the general symbol for genes affecting genetic recombination (→ recombination-proficient).

recombination error (*Kondor* et al. 1970) — a mechanism assumed to result in base-substitution mutations due to spontaneous recombination or recombination provoked by DNA damage induced directly (with external agents) or indirectly (secondary damage due to errors in → excision repair of primary DNA damage). After the induction of premutational DNA lesions the realization of mutations requires damage-provoked recombinational events which eventually may result in mutations (→ replication error; post-replication repair; recombination repair).

recombination fraction — → recombination frequency.

recombination frequency — the total number of → recombinants (arising by reciprocal → genetic recombination via crossing-over) divided by the total number of progeny individuals (or gametes) scored. R. f. (= recombination fraction) equals the crossover value (and represents a measure of → map distance) for those chromosomal regions so short that multiple crossovers constitute an insignificant class (complete → interference). As intervals between the → genetic markers under observation increase, this equality no longer holds.

recombination gene (*Smith* 1966) — any gene (allele) that controls the frequency of → genetic recombination (allelic and/or nonallelic). Such genes may determine the structure of an enzyme (which catalyses one of the events leading to genetic recombination) or the structure of a regulatory substance (→ recombinagenic).

recombination index (*Darlington* 1939) — the sum of the gametic (haploid) chromosome number (i.e. the number of bivalents) and the average number of chiasmata per nucleus; it is used as an estimate of the meiotic potential for → genetic recombination via crossing-over in eukaryotes. A high r. i. promotes → flexibility, a low r. i. promotes → fitness.

recombination-proficient — of genotypes capable of complete recombination (symbol rec^+) as opposed to → recombination-deficient (rec^-) mutants.

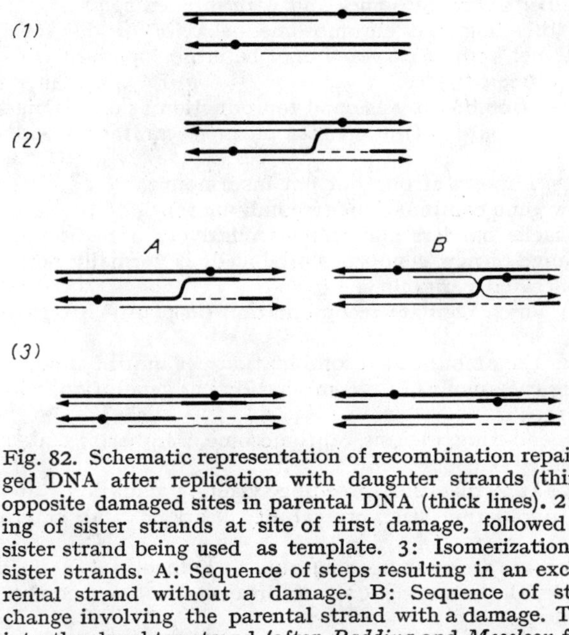

Fig. 82. Schematic representation of recombination repair: 1. A portion of damaged DNA after replication with daughter strands (thin lines) containing gaps opposite damaged sites in parental DNA (thick lines). 2: Crossing-over and pairing of sister strands at site of first damage, followed by gap filling with the sister strand being used as template. 3: Isomerization and exchange between sister strands. A: Sequence of steps resulting in an exchange involving the parental strand without a damage. B: Sequence of steps resulting in an exchange involving the parental strand with a damage. The damage is exchanged into the daughter strand (after *Radding* and *Meselson* from *Ganesan* 1974).

recombination repair (*Howard-Flanders* & *Boyce* 1966) — a → dark repair process by recombination between sister DNA molecules which fills the gaps opposite unrepaired lesions left in the daughter DNA strands after replication of damaged DNA (Fig. 82). R. r. (→ post-replication repair) and → excision repair are complementary mechanisms of DNA repair.

R. r. deals with DNA lesions too close to the replication fork to be dealt with by excision repair which, under such circumstances, might very probably lead to lethal damage equivalent to a double strand break in DNA. R. r. facilitates and mediates mutagenesis, i.e., it is an error-prone process.

Mutants deficient in r. r. are, among other properties, depressed for → genetic recombination (→ recombination-deficient).

recombination system (*Carson* 1957) — the sum total of all factors mediating and controlling the processes of → genetic recombination; the r. s. regulates the generation of genetic variability by restricting the types and frequencies of recombinants produced. In eukaryotes with sexual reproduction, recombination is a function of three main processes: → crossing-over, random distribution of chromosomes and chromatids during meiosis, and random combination of gametes during fertilization. A high level of meiotic recombination is characteristic of organisms with a high frequency of crossing-over, a high number of chromosomes and a high level of outbreeding, as opposed to organisms with a low frequency of crossing-over, with a low number of chromosomes and with inbreeding.

According to the breeding and chromosome behavior of the species under observation, the r. s. of eukaryotes may be either open, relatively restricted, or closed (*Carson* 1957):

1. A closed r. s. is examplified by asexual reproduction as e.g., obligate → apomixis, where the source of new variation is → mutation and → autosegregation.

2. A restricted, r. s. imposes strong but not insurmountable barriers to the realization of new gene combinations (recombinants).

3. An open r. s. lacks barriers and thus is relatively unrestricted as to the number and range of new genotypes produced; it normally permits a free flow of the generated variability.

Among the factors which regulate recombination (in plants) are (*Grant* 1958):

1. those controlling the amount of recombination per unit of time;

2. those controlling the amount of recombination per generation, which may be subdivided into:

a) factors operative during meiosis (chromosome number, frequency of crossing-over, sterility barriers), and

b) factors operative at fertilization (breeding system, pollination system, dispersal potential, population size, crossability barriers, and external isolating mechanisms).

recombinogenic — of agents (recombinogens) able to induce → genetic recombination. Probably, all → mutagenic agents are also effective recombinogens.

recon (*Benzer* 1957) — the smallest unit (corresponding to a single DNA nucleotide) which is exchangeable but not divisible by intragenic recombination (→ genetic recombination) between homologous linkage structures (chromosomes) and which may be detected by the techniques of genetic fine structure analysis of the linear array of → mutational sites within the → genetic material (→ cistron; muton; polaron hybrid DNA hypothesis).

recovery — the ability of an organism to survive in spite of the introduction of lesions (usually in DNA) by some physical or chemical agents as a result of the removal (→ DNA repair) or circumvention of such lesions (→ reinitiation recovery).

recurrence risk — in human genetics, the risk that a woman who has had a child with a certain genetically controlled trait will have another at a later pregnancy.

R. risks will vary considerably. In autosomal dominant disorders, the r. r. after an affected child has been born is 50%, but since not all carriers of a given dominant gene will manifest the disease, the actual risk will often be less than 50%. In recessive disorders, the r. r. is 25%. In X-linked recessive inheritance, there is a 50% risk for future sons to be affected if the mother is a heterozygous carrier, and a 50% risk for future daughters to be carriers.

recurrent parent — in → backcrosses, the parent which is crossed with the first and the subsequent generations, as opposed to the nonrecurrent parent.

reduction (*Weismann* 1887) — 1. **meiotic r.**: the halving of the somatic (zygotic) chromosome number at → meiosis and, by extension, its genetical concomitant, → genetic segregation (*Darlington & Mather* 1949);

2. **somatic (or mitotic) r.**: reductional distribution of chromosomes in tissues other than meiotic ones, which may take place spontaneously or may be induced experimentally (→ reductional grouping).

reductional grouping (*Huskins* 1947) — the separation of mitotic chromosomes prior to anaphase, into two (or more) groups containing the same or different fractions of the somatic chromosome number. Such groups may enter anaphase separately and two groups may give rise to four nuclei at the end of mitosis, each with a number of chromosomes smaller than the original (somatic or mitotic → reduction, as opposed to meiotic reduction of chromosome number). R. g. may take place spontaneously in rare cases (→ genome segregation) and can be induced experimentally.

reductional separation — separation of homologous parts of nonsister chromatids at anaphase I (prereductional s.) or anaphase II (post-reductional s.) of → meiosis, as opposed to equational separation of sister parts of chromatids (*Darlington & Mather* 1949).

reduction division (*Weismann* 1887) — = → meiosis.

redundant DNA — = → repetitious DNA (→ gene redundancy).

refractory (*Hill & Holland* 1967) — of bacterial mutants adsorbing → colicin but resistent to its lethal effect.

regressive (*Weinstein* 1936) — → crossing-over.

regulation — → genetic regulation.

regulation reversal (*Coats & Nester* 1967) — end-product activation in mutants which arose from genotypes with opposite regulatory properties, i.e., show end-product inhibition.

regulator gene (*Jacob & Monod* 1961) —

1. in the broadest sense, any gene which regulates or circumscribes the activity of other genes.

2. A gene which codes for a (allosteric) protein (the → repressor) which

functions alone (inducible systems) or in combination with a → corepressor (repressible system) in regulating the → genetic transcription of the → structural genes in an → operon by binding to the operator. Mutations of r. genes may result either in a complete shutdown in repressor production or may give rise to a repressor whose ability to interact with the operator of the relevant operon is greatly reduced (constitutive enzyme synthesis determined by the structural genes of the operon). A mutationally altered repressor may be unable to bind the corepressor readily or may fail to undergo the proper conformational change (→ allosteric) which activates the repressor after its interaction with the corepressor. Hyperrepressed mutants may act by producing a repressor with very high affinity for the sensitive operator or which more readily assumes the active repressor configuration in the presence of metabolites representing normal cell constituents.

regulatory codon (*Anderson* 1961) — any codon (in mRNA) which is recognized by → transfer RNA species, being present in rate limiting amounts and, thus, exerting → translational control.

regulatory site — any site in an enzyme, different in kind from the active site, which binds → effectors that regulate enzyme activity.

regulon (*Maas & Clark* 1964) — a regulatory system (unit of a common control consisting of scattered genes but subject to → negative control through interaction of a unique repressor with the several operators involved in the expression of each individual gene or group of genes. Regulons as opposed to → operons are not transcribed together as one polycistronic mRNA (→ cross-pathway regulation; supraoperon control).

reinitiation recovery (*Bridges* 1972) — a third type of DNA dark repair which exists in addition to → excision repair and → postreplication repair in bacteria. It is involved in the repair of the damaged replication complex after UV irradiation and in the initiation of a new replication complex at one of the two chromosomal origins. R. r. is inhibited by chloramphenicol and acriflavine and is dependent on both rec^+ and UVr^+ functions.

reinitiation site (*Newton & Zipser* 1967) — = → reinitiator.

reinitiator — a site for the reinitiation of polypeptide synthesis (→ genetic translation) which operates to restore translational activities impaired by polar → nonsense mutations, i.e., reverses the effects of terminator codons (UAG, UAA, UGA in mRNA). The combination of a r. and a terminator codon creates a new cistronlike boundary at the site of the nonsense codon. The further the nonsense codon is from the r., the less efficient is protein synthesis reinitiation. Reinitiation does not behave as a mutation in the absence of a polypeptide chain terminating nonsense codon. They may represent one of the → initiator codons which is not read in the absence of an adjacent terminator codon or may be due to mutations adjacent to an initiation codon which in combination with the initiator codon generate the capacity for reinitiation of polypeptide synthesis.

rejoining — according to the "breakage-reunion hypothesis" the → reunion and restitution of broken ends of chromosomes, chromatids, or subchromatids leading to the production of → chromosome mutations (reunion), or a restoration of the prebreakage structure (restitution).

relational balance (*Mather* 1953) — → coadaptation.
relational coiling (*Darlington* 1935) — → chromosome coiling.
relative sexuality (*Hartmann* 1923) — the ability of a gamete to act either as a male or female when mated to different gametes.
relaxed (*Borek* et al. 1955) — of bacterial mutants (rel) in which, opposed to stringent control in wild-type bacteria, RNA synthesis can occur in the absence of a full complement of charged → transfer RNA species. In the wild-type bacteria, removal of one of the 20 aminoacyl tRNAs results in an immediate cessation of RNA synthesis.
release factor (*Ganoza* 1966; *Capecchi* 1967) — any of the special protein factors (= termination factors) which recognize the → terminator codons UAA, UAG, and UGA in → messenger RNA and stimulate the codon-specific release of polypeptides from ribosomes during → genetic translation. R. f. dependent hydrolysis of the → peptidyl-tRNA ester bond requires location of the peptidyl-tRNA in the ribosomal P-site, NH_4^+ or K^+ ions, and is inhibited by several antibiotics that also inhibit peptide bond formation.

Prokaryotes utilize two r. factors (RF 1 and RF 2), with a third (RF 3) acting to stimulate the reaction.

RF 1 (in *E. coli*) has a mol. weight of 44000, is active as a monomer, and recognizes the terminator codons UAA and UAG.

RF 2 has a mol. weight of 47000 and recognizes UAA and UGA. The binding and release of RF 1 and RF 2 from ribosomes involves a third r. f. (**RF 3** or S) which interacts with GDP and GTP. RF 3 may be identical to the → elongation factor EF-Tu. The fact that both RF 1 and RF 2 recognize UAA (ochre) makes it likely that this codon is the chief terminator with UAG (amber) and UGA (opal) providing accessory fail-safe signals.

Hydrolysis of nascent peptidyl-tRNA requires RF 1 and RF 2 and the ribosome, but not RF 3 or GTP.

R. factors interact with specific terminator codons in three ways (*E. coli*): 1. RF binding to ribosomes; 2. formation of RF-terminator codon-70S ribosome intermediates; 3. release of the nascent polypeptide chain from transfer RNA at the terminator site. The r. f. RF 3 stimulates the release reaction by assisting RF 1 and RF 2 to interact with the ribosome in response to its particular terminator signals. The substrate for the termination reaction is thus a completed polypeptide chain in ester linkage with tRNA and complete with ribosome and messenger RNA. GTP is inhibitory for RF 3 stimulated release.

Codon-specific release of polypeptides in **eukaryotes** (mammals) appears to be fundamentally similar to that in bacteria. One major difference from prokaryotic termination is the requirement for GTP. In mammals, a single protein seems to embody all the requisite functions. It stimulates release with all three terminator codons and is much larger (mol. weight 255000) than the bacterial r. factors.

In bacteria, about 500 RF 1 and 700 RF 2 molecules exist per cell compared with about 30000 ribosomes and an even larger number of → elongation factor molecules per cell.
relic coiling (*Darlington* 1935) — → chromosome coiling.

renaturation — the return by slow conversion of a denatured (→ denaturation) nucleic acid or protein to its "native" configuration. In the case of denatured → deoxyribonucleic acid (DNA), r. is the return to the double helix configuration, usually by annealing at some temperature below the melting temperature (T_m), which brings forth helix-coil transition (*Kohn* et al. 1966). (→ DNA-DNA hybridization).

Renner complex — in → complex heterozygous species (*Oenothera*), a group of chromosomes (and the genes which they contain) which are distributed as one unit from generation to generation.

Renner effect (*Darlington* 1932 — competition (= megaspore competition) among the four genetically different spores formed by one meiosis in regard to which shall form the embryo (*Darlington & Mather* 1949).

repair (*Muller* 1954) — restoration of biological activity of damaged DNA by recovery mechanisms in which physico-chemical changes occur at the actual site of damage (→ repair replication). R. includes both reversion and replacement of an induced lesion. If r. results from the action of some agent applied to the system, two different kinds may be distinguished (*Jagger & Stafford* 1965):

1. **Direct repair** temporally and spatially close to the point of action of this agent;

2. **indirect repair** temporally or spatially distant from the action of the agent that simply helps to provide conditions favoring a r. brought about by some other means.

Direct reconstitution mag occur by → photoreactivation or → dark repair of an undamaged segment of DNA from a molecule containing lesions.

repairase (*Kozinski* et al. 1967) — → recombinase.

repair-deficient — of mutants with defective → DNA repair systems.

repair polymerase — → repair replication.

repair recombination — → recombination repair.

repair replication — a non-conservative synthesis of DNA by repair polymerase which occurs after or in consort with the excision of damaged sites in DNA (= repair synthesis), i.e., the limited insertion of nucleotides into pre-existing, damaged DNA (→ DNA-repair; excision repair). During r. r. there is no net increase in the amount of DNA. The resulting double-stranded DNA molecules are not hybrid but consist of parental strands containing short repaired regions (Fig. 83). R. r. occurs after treatment with ionizing radiation, UV-light and chemical mutagens and has been equated with → unscheduled DNA synthesis in eukaryotes. In bacterial systems, no r.r. has been found after exposure to ionizing radiation. This is probably due to the extensive DNA degradation following treatment with ionizing radiations. The presence of r. r. after treatment of eukaryotic cells with ionizing radiation is indicative of a repair system that does not involve a large scale DNA breakdown as occurs in bacteria.

The amount of r. r. depends on the amount and chemical nature of DNA damage and the manner in which the damage is repaired. The enzyme involved in r. r. is a → DNA polymerase called repair polymerase. It shows the capability of self-correction, i.e., 3′ exonuclease activity enables it to remove incorrectly inserted bases during r. r.

repair synthesis — = → repair replication.

repeat (*Bridges* 1935) — a type of → duplication where the duplicated segments lie adjacent (tandem), inverted or not, in the relation to each other and the rest of the chromosome. Repeats may occur at different structural levels and may involve groups of genes, single genes, or even only parts of a gene.

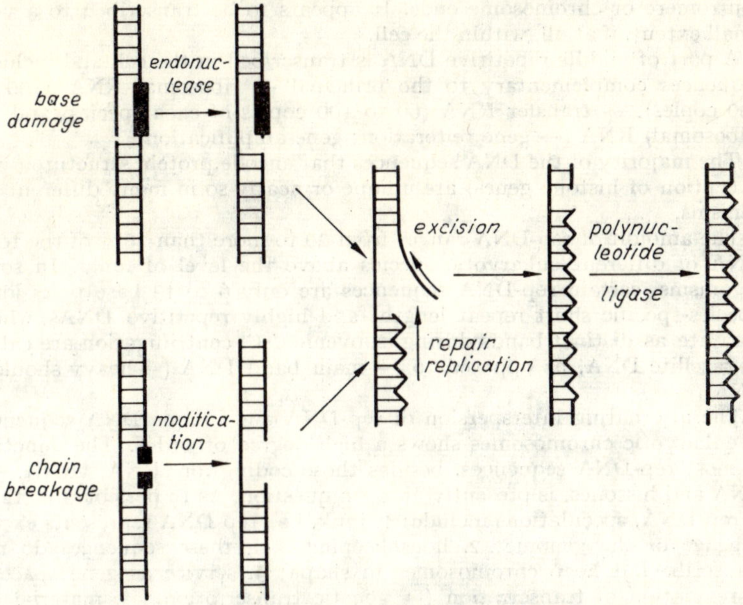

Fig. 83. Schematic representation of excision repair of damaged DNA bases and broken strands involving repair replication (from *Cleaver* 1974).

rep-DNA — abbreviation for → repetitious DNA.

repetitious DNA (*Britten* & *Kohne* 1968) — in eukaryotes, nucleotide sequences (abbreviation rep-DNA) occurring repeatedly in the chromosomal DNA and characterized by 1. the number of nucleotide pairs per sequence, 2. the specific order of base pairs per sequence, and 3. the number of copies of a sequence per genome. A particular family of repeated sequences can exist either dispersed between → unique DNA sequences or as clusters of repeating units limited to one or a few areas (called regional repetitions by *Lee* & *Thomas* 1973) in the genome.

Two categories of rep-DNA may be distinguished on the basis of denaturation-renaturation kinetics (→ Cot value): 1. a **highly repetitive** or "fast fraction" repeated about 10^6 times and consisting of simple-sequence DNA; 2. a **middle repetitive** or intermediate fraction made up of sequences repeated 10^2 to 10^4 times each and 300 ± 200 base pairs long, interspersed in the majority of instances between regions of → unique or non-repetitive DNA. In each of these fractions, several families of repeated sequences are found. The evolution of such families is assumed to involve divergence through mutation of highly repetitive sequences which arose through → saltatory multiplication events (→ Thomas circle).

Highly repetitive or simple-sequence DNA tends to be associated with constitutive → heterochromatin and is often concentrated close to the → centromere or chromosome ends. It appears to be transcribed to a very small extent, if at all, within the cell.

A part of middle repetitive DNA is transcribed in the cell and includes sequences complementary to the principal → ribosomal RNA (100 to 500 copies), → transfer RNA (60 to 100 copies of each species) and 5 S (ribosomal) RNA (→ gene reiteration; gene amplification).

The majority of the DNA sequences that encode protein structure (with exception of histone genes) are unique or nearly so in many different organisms.

The amount of rep-DNA ranges from 20 to more than 80% of the total DNA of different eukaryotic species above the level of fungi. In some organisms, certain rep-DNA sequences are only 6 to 13 base pairs long. Species-specific short repeat lengths and highly repetitive DNAs which separate as distinct bands during isopycnic CsCl centrifugation are called → satellite DNA, as opposed to → main band DNA (→ heavy shoulder DNA).

The alternating interspersion of rep-DNA and unique DNA sequences in eukaryotic chromosomes shows a high degree of order. The function of most rep-DNA sequences, besides those coding for rRNA, tRNA, 5 S RNA and histones, is presently an open question. As to possible functions of rep-DNA, speculations include: 1. junk, i.e., rep-DNA represents excess baggage of the genome; 2. housekeeping, i.e., these sequences do not transcribe but keep chromosomes in shape; 3. service as gene spacers; 4. regulation of transcription (→ genetic transcription); 5. material for creating new genes.

repetitive DNA (*Britten & Kohne* 1968) — = → repetitious DNA.

replicase (*Spiegelman & Hayashi* 1963) — in RNA viruses (where → ribonucleic acid is the sole carrier of → genetic information), any of the RNA-dependent polymerases (= RNA synthetase, RNA polymerase) catalyzing RNA replication. Each of the replicases recognizes the RNA genome of its origin and requires it as a template for normal synthetic activity which gives rise to the formation of complementary RNA chains.

replicating instability (*Auerbach* 1967) — any premutational change in DNA that can replicate in that state so that the locus continues to give rise to mutations in several lines of descent. R. instabilities have been found after treatment with many mutagens.

replication — in genetics, the nondestructive read-out and storage of → genetic information giving rise to a new carrier of information by copying a preexisting unit of the same kind. Subcellular biological systems in which the process of r. occurs are the nucleic acids (in most cases DNA, in some RNA).

replication bubble — → deoxyribonucleic acid.

replication error — any error due to damage to the DNA replication machinery after treatment with mutagens or spontaneously that might lead to base-change mutations (→ recombination error).

replication fork — a Y-shaped region of chromosomes representing the growing point in DNA replication.

replication form — = → replicative form.

replication map (*Strelzoff* & *Ryan* 1962) — a genetic map (= temporal map) based on temporal mapping by sequential mutagenesis during chromosome replication in bacteria. Temporal mapping requires no sexual system and is a means of performing studies on genetically refractory organisms. The r. m. is a map based on the doubling of specific marker mutation frequencies observed when successive aliquotes of synchronous bacterial cultures are treated with a fixed dose of a non-specific mutagen for a specified time. The distance on the map depends on the time necessary for replication of the DNA lying between the markers. This is proportional to the physical distance when the rate of replication, in terms of nucleotides added per second at a given growing point (→ replication fork), is constant. The resolving power is dependent on the degree of induced synchrony of the culture, the rate of DNA synthesis, and the duration of DNA synthesis (→ transcriptional mapping).

replication repair (*Symonds* et al. 1973) — a process for repairing UV damage in bacterial cells infected by bacteriophage T4 distinct from → excision repair and specified by its deficiency in cells infected with the phage mutants T4x and T4y (→ x gene reactivation). R. r. is the main pathway leading to radiation stability; it does not involve the bacterial Kornberg polymerase (→ DNA polymerase) and bears some resemblance to → post-replication repair.

replication section (*Huberman* & *Riggs* 1968) — → replication unit.

replication unit (*Huberman* & *Riggs* 1968) — in eukaryotes, the basic unit of control in the initiation of chromsomal DNA replication (→ chromosome). It is assumed to represent an adjacent pair of tandemly joined, diverging replication sections of DNA, i.e., stretches of DNA each replicated by a single fork-like growing point (Fig. 32). In the r. u., DNA is replicated in opposite directions at adjacent growing points of the two replication sections (bidirectional replication). The maximum rate of DNA synthesis is 2.5 μm per minute or less; the length of each r. section is 7 to 30 μm. Adjacent r. sections share origins or termini of replication. Initiation of replication may be controlled at the level of adjacent pairs of r. sections rather than at the level of indivudal r. sections (→ replicon).

replicative form (*Sinsheimer, Starman, Nagler* & *Guthrie* 1962) — a double-stranded DNA or RNA structure (symbol RF) produced as an intermediate during the replication of naturally occurring viruses with single-stranded DNA or RNA (→ replicative intermediate).

replicative intermediate (*Erikson* et al. 1964) — 1. in the replication of single-stranded DNA and RNA viruses, a structure which consists of several nascent parental nucleic acid molecules associated with a single complementary template strand in a partially double-stranded conformation. R.

intermediates are distinguished from completely double-stranded complexes called → replicative forms which may represent either a very early step in replication or one of its by-products.

2. → Okazaki pieces produced during replication of double-stranded → deoxyribonucleic acid.

replicative synthesis — the semiconservative replication of → deoxyribonucleic acid as distinguished from → repair replication or repair synthesis.

replicator (*Jacob & Brenner* 1963) — → replicon.

replicon (*Jacob & Brenner* 1963) — any genetic element (bacterial chromosome, virus genome, plasmid) that behaves as an autonomous unit of DNA replication (→ deoxyribonucleic acid), i.e., capable of replication under its own control. The r. possesses a functional group which controls its replication and is called the **replicator.** The diffusible product of a particular gene controlling (positive control) the replication of DNA is termed the **initiator** and is assumed to interact specifically with the replicator. Interaction results in the initiation of a new round of replication at specific sites (the origin). Replication ends at the terminus and is bidirectional. In bacteria and DNA viruses, transcription of the replication origin appears to be essential for the initiation of a new round of DNA replication. Newly synthesized RNA may serve as a primer for DNA synthesis and the local unwinding of a double helical structure during the initiation of transcription might facilitate the initiation of DNA replication (*Lark* 1972). Bacterial DNA replication probably begins by the attachment of the chromosome (replicon) to a cell membrane complex (the → mesosome). The "replicons" in eukaryotic chromosomes are called → replication units.

repliconation (*Clark* 1967) — a process that mediates the inheritance of → plasmid DNA as an independent → replicon without integration into the bacterial chromosome.

repressible (*Vogel* 1957) — of enzymes (proteins) normally present in the cell (→ constitutive; inducible) whose production ceases when the intracellular concentration of certain metabolites (usually the end products of the enzyme system) increases (→ operon). R. enzyme systems are usually anabolic.

repression (*Monod & Cohen-Bazire* 1953; *Vogel* 1957) — the alteration in the expression of a gene with the net result that specific enzyme production may fail (→ operon), as opposed to "derepression", when the level of a particular → repressible enzyme increases after removal of a specific metabolite.

R. may be due to → transcriptional control or → translational control superimposed on the former and represents a means of adaptation of enzyme production to the environment (→ induction). Coordinate r. (*Ames & Garry* 1959) means the simultaneous r. of many or all enzymes of a sequence controlled by scattered genes (→ regulon).

repressor (*Pardee* et al. 1959) — any of the specific allosteric protein molecules (products of regulator genes) which bind to the → operator of → operons and prevent → RNA polymerase (which binds at the preceding → promoter) from proceeding into the operon to transcribe → messenger RNA from DNA (negative control). Each set of independently regulated genes is controlled by a different repressor produced by a different regulator gene. For each r.-operator system there exists an effector molecule (the inducer or co-repressor) that interacts with r. affecting its affinity for the operator. The co-repressor (an end product or related compound of anabolic systems) increases the repressor's operator binding activity; the inducer acts in the reverse manner. A fully functional r. must, therefore, contain two sites: one (A) must be specific for the recognition of, and association with, the operator on the DNA; the second (B) must specifically recognize and interact with an effector or antirepressor.

In inducible bacterial operons (→ catabolite repression), the release of the r. is not in itself sufficient to ensure full activity of the structural genes. In this case the intervention of cyclic adenosine monophosphate (cAMP) and of → catabolite activator protein is necessary.

Three genetically defined repressors have been isolated. Two control certain genes of coliphages lambda and 434, the third controls the functioning of the lac operon of *E. coli*. Both phage repressors (mol. weight 28000 and 26000, respectively) bind to their respective operators in the form of oligomers (dimers or tetramers which are in equilibrium with the monomers). The lac r. is a tetramer of identical subunits (mol. weight about 40,000) and the N-terminal regions of the r. subunits (about 50 amino acids) are involved in the specific binding to the lac operator.

Genetic regulation by repressors is only one of several mechanisms for gene control. Specific positive control elements are known to exist; the products of such control genes are required to turn the operon on rather than to turn it off. If the control gene is deleted, the structural genes do not function.

reproduction — the production (self-propagation) of an organism, a cell, or cell organelle by one like itself (→ replication).

Sexual reproduction: the regular alternation (in the life cycle of → haplontic, → diplontic, and → diplohaplontic organisms) of → meiosis and → fertilization (karyogamy) which provides for the production of offspring. The main biological significance of sexual r. lies in the fact that it achieves → genetic recombination. In diploid and polyploid organisms, a large amount of → genetic variability is stored within a framework of phenotypic uniformity. Sexual r. is an effective means of pooling the → genetic information contained in the individuals of a breeding population.

Parasexual reproduction (*Pontecorvo* 1954): all processes of r. and genetic recombination other than a regular alternation of meiosis and fertilization (→ parasexual cycle).

Asexual or agamic reproduction: the development of a new individual from either a single cell (agamogony) or from a group of cells (vegetative r.) in the absence of any sexual process. Asexual r. may be the only mode of r. of a species, or it may take place within the sexual cycle of an organism as an essential or nonessential part of its life cycle. The capacity of an organism for either sexual or asexual r. is referred to as **"versatile r."** (*Crane* & *Thomas* 1941).

Subsexual reproduction (*Darlington* 1937): → parthenogenesis following → ameiosis. There is no reduction of chromosome number but → segregation of genes occurs owing to crossing-over (*Darlington & Mather* 1949).

Differential reproduction: r. in which different invididuals do not contribute to the next generation in proportion to their number.

reproductive cell — any of the → germ cells (usually the gametes) and their immediate predecessors from which they arise by division.

reproductive isolation — → isolation by various genetically controlled mechanisms which prevent → gene exchange between two populations and preserve differences in the → gene pools of populations previously achieved by natural → selection and geogaphic isolation. R. i. is a factor of → evolution.

reproductive potential — an organism's ability to produce surviving offspring. The r. p. includes the individual's capacity to reach the reproductive stage, its fertility during this period, and the viability of its offspring to sexual maturity.

repulsion (*Bateson, Saunders & Punnet* 1905) — = → trans-configuration.

rescue factor (*Ganoza* et al. 1974) — a factor, different from the → release factors, which rescues protein synthesis *in vitro* from temperature-dependent arrest.

residual centric distance (*Southern* 1967) — → chiasma.

residual genotype — → genotype.

residual homology (*Stephens* 1942) — → homoeologous.

residual protein — = → nonhistone chromosomal protein.

residual telomeric distance (*Southern* 1967) — → chiasma.

resistance factor (*Iseki & Sakai* 1953) — any of a class of extrachromosomal elements or → plasmids (R factors) which confer on their bacterial host resistance to one or more antibiotics and (in Enterobacteriaceae) can promote transfer of drug resistance to other bacteria via bacterial → conjugation (transmissible R factors). R factors consist of closed circular DNA. They may be classified on the basis of their ability (fi^+) or inability (fi^-) to repress the fertility properties of an → F plasmid present in the same cell, the type of distinguishable → sex pilus that the r. f. determines, the ability of a r. f. to coexist in a cell specifically with other plasmids, and the presence of genes in the r. f. that specify → DNA modification and restriction systems.

R factors that promote their own transfer (**transmissible r. factors** with → sex factor properties) are of relatively large molecular weight (26 to 76×10^6) and are present in only a limited number of copies (1 to 4) per bacterial chromosome (in *E. coli*).

R factors that promote their own transfer generally promote the transfer of the host chromosome, or in certain cases the transfer of r. plasmids that are non-transmissible.

Transmissible r. factors consist of two components: the transfer unit (RTF: resistance transfer factor) which controls transmission of the plasmid and is linearly linked to drug-resistance determinants (R determinants) which carry genetic information specifying resistance to antimicrobial agents. Both units of at least certain types of r. factors can replicate independently of each other. In certain bacterial species, the RFT and the R determinants remain covalently attached to one another; in others they are able to reversibly dissociate and replicate independently (i.e., both are independent → replicons). Reassociation occurs by a recombinational event between DNA circles.

The mol. weight of the **RTF unit** is about 5×10^6; its major functions comprise:

1. initiation and regulation of plasmid DNA replication;
2. maintenance of plasmids in the cell and their segregation to bacterial daughter cells at division;
3. specification of the gene products needed for bacterial conjugation;
4. transfer of DNA during conjugation.

The resistance determinants (R determinants) specify the molecules that confer the resistance. These are frequently biochemical products that destroy or inactivate the antibiotics; sometimes they prevent the antibacterial agent from reaching its target within the cell.

The RTF units can carry determinants for → colicinogeny as well as drug resistance. Similar agents carry determinants for haemolysis and enterotoxin synthesis. It is possible that such agents transfer a wide range of other characters which await identification.

The frequency of conjugal transfer of drug resistance in enterobacteria is the result of the interaction of a number of influences: 1. the recipient strain; 2. the donor strain; 3. the physiological state of the donor and recipient cells; 4. the rate of transmissibility of the transfer unit alone; 5. the nature of the linkage between the determinants and the transfer unit; 6. the particular determinants and transfer units concerned; 7. a number of unspecific factors.

Bacterial strains carrying r. factors resist superinfection with the same or similar factors. In Gram-positive bacteria, transfer of r. factors occurs by → transduction.

resistance transfer factor (*Watanabe* & *Fukasawa* 1960) — → resistance factor.

respiratory particle — any subcellular particle isolated fom cell homogenates that can carry out the oxidation of suitable substrates. Respiratory particles may be intact or damaged mitochondria, submitochondrial particles (e.g. → electron transport particles) or nonmitochondrial material derived, e.g., from bacteria.

resting stage — = → interphase.

restitution (*Darlington* & *Upcott* 1941) — → reunion.

restitution nucleus (*Rosenberg* 1927) — 1. meiotic r. n.: a single nucleus with unreduced chromosome number, produced in place of two nuclei owing to failure of the first or second meiotic division (→ meiosis);

2. mitotic r. n.: a single nucleus with tetraploid chromosome number, produced in place of two diploid nuclei owing to failure of → mitosis (→ C-mitosis).

restoration — an alleviation of the effects induced by radiation or chemical agents in biological systems brought about by specific treatments following irradiation or the application of a chemical. In contrast to r., similar effects of a treatment prior to irradiation or application of a chemical with damaging activity are called "protection" (*Latarjet* & *Gray* 1954). The terms "recovery" or "reactivation" are frequently used synonymously with r.

restorer gene — a gene that at least superficially, recoups the changes brought about by a cytoplasmic induced sterility.

restriction (*Dussoix* & *Arber* 1962) — 1. the degradation (→ DNA modification-restriction system) of unmodified (foreign) DNA from various sources (bacteria, phages, plasmids) by → restriction endonucleases capable of recognizing particular patterns of specificity (due to DNA modification by methylation or glycosylation). 2. A reduction in the rate of → genetic translation due to mutations in a gene coding for a protein in the 30 S subunit of bacterial ribosomes. These mutations generally reduce (restrict) the efficiency of various → nonsense suppressors.

restriction endonuclease (*Meselson* & *Yuan* 1968) — any of the strain-specific restriction enzymes that enable bacteria to recognize and rapidly degrade DNA introduced from foreign sources (such as bacteria, phages, plasmids) which has not been modified by replication in the appropriate bacterial host cell (→ DNA modification-restriction enzyme).

Such enzymes introduce double-strand scissions at a limited number of specific sites (sequences of base pairs) on the DNA molecule, if this lacks particular modified bases, rendering it susceptible to further degradation by non-specific nucleases (→ cleavage map). There is evidence that at least some r. endonucleases have modification methylase activities.

R. enzymes fit into two groups on the basis of their restriction products: 1. those which cut at unique sites of the DNA yielding a population of specific fragments with unique terminal sequences (r. e. from *Haemophilus* and those from the R I and R II → resistance transfer factors). The terminal sequences imply that the nucleases act at symmetric nucleotide sequences which read the same from both left to right and from right to left, e.g.

5′ .... p G p T p T p A p A p C p ...... 3′.
3′ .... p C p A p A p T p T p G p ...... 5′.

2. Those (from *E. coli* and Phage P 1) which recognize specific sites, but in contrast to group 1 enzymes make a limited number of double strand cuts at apparently random sites, far from the recognition sites. The cuts generate a heterogeneous population of molecules which can be denatured and renatured to reform circular molecules.

R. endonucleases require the cofactors ATP, S-adenosyl methionine, and magnesium ions for their activity (type 1), or only magnesium ions (type 2). The latter are smaller and simpler in subunit composition than type 1 and appear to be more specific in their cleavage sites on heterologous DNA.

Modification of either of the two DNA strands is sufficient to prevent r. endonucleases from action. Usually both strands of DNA are modified. Replication of modified DNA yields DNA of which one strand (the parental) is modified, and the daughter strand is unmodified.

restriction enzyme — = → restriction endonuclease.

restriction gene — → modifier gene.
restriction-modification system — → DNA modification-restriction system.
restrictive — of animal cells (= nonpermissive) which do not allow infecting virus to multiply and produce infectious progenies, although the virus can enter the cell and express some of its gene functions. In permissive cells, the virus can multiply and produce infectious progeny, eventually killing the host (lytic infection).
reticulosome (*Pollak & Shorey* 1964) — a cytoplasmic component (particularly frequent in the → microsomal fraction) which can give rise to a membraneous structure in the presence of phospholipids. It is assumed to be involved in membrane formation.
reunion (*Darlington & Upcott* 1941) — according to the so-called "breakage-reunion hypothesis", the rejoining of broken ends of chromosomes, chromatids, or subchromatids in such a way that chromosomal structural changes (→ chromosome mutations) arise, as opposed to "restitution", i.e., the rejoining of broken ends which restores the prebreakage structure and represents a process for repairing chromosomal damage. The proportion of recognizable breaks in the chromosomes or chromatids of a sample of nuclei which undergo recognizable reunion is referred to as the "empirical coefficient of reunion" (E.C.R.).
reverse mutation — a heritable change in a mutant gene (= → back mutation) which restores the original nucleotide sequence.
reverse pinocytosis (*de Robertis & Vaz Ferreira* 1957) — usually, the secretion of cytoplasmic storage granules from the cell (→ pinocytosis).
reverse transcriptase (*Baltimore* 1970; *Temin & Mizitany* 1970) — RNA-dependent → DNA polymerase (originally isolated from single-stranded RNA viruses but not unique to these, and virus-infected or tumor cells) that transcribes synthetic or natural RNAs into DNA. The enzyme requires Mg^{2+} or Mn^{2+} as a metal activator, deoxyribonucleioside triphosphates, an RNA template, and an RNA primer to initiate the synthesis of a DNA chain, which is covalently bound to the primer RNA. DNA synthesis is initiated by adding a deoxynucleotide to the 3′-hydroxyl group of the primer, i.e., the mechanism of DNA synthesis by r. t. seems to be essentially identical to that of the Kornberg DNA polymerase I of *E. coli*, except that the r. t. can use an RNA primer and an RNA template.

In the case of RNA viruses, a RNA-DNA hybrid molecule and some single-stranded DNA molecules are intermediate in the appearance of double-stranded DNA. A ribonuclease H seems to be responsible for the displacement of single-stranded DNA from RNA-DNA hybrids. Both enzyme activities may reside in the same enzyme molecules. R. t. has been shown to produce faithfully DNA copies of isolated → messenger RNAs, i.e., antimessengers.

The biological role of tumor virus r. t. is the synthesis of a DNA provirus during the establishment of infection, i.e., the enzyme provides a mechanism by which an oncogenic RNA virus can insert stable genetic information into a host chromosome by a multi-step sequence (viral RNA → RNA-DNA hybrid → double-stranded DNA → DNA integration). When "turned on" the virus DNA may transcribe viral RNA molecules.

reverse transcription — RNA-directed DNA synthesis (→ reverse transcriptase).

reversion — a genetic (genotypic r.) or nongenetic (phenotypic r.) **restoration**, partial or complete, of the wild-type phenotype to a **mutant** (→ revertant).

reversion analysis — the analysis of susceptibility of a mutant (→ gene mutation) to reversion induction by specific mutagens as a means of examining the chemical nature of a mutational lesion. Base analogues and proflavin separate mutants into three distinct classes: those responding specifically to one agent, the other or to neither. If a reversion is induced by proflavin, the mutant is considered to contain an addition or a deletion of a small number of base pairs, regardless of whether reversion occurs at the same or at a nearby site. Mutants responding to base analogues are considered to represent a base pair substitution. If the induced reversion can be shown to occur at the same site, then the mutation is identified as a → transition. If reversion frequently occurs by intracistronic → suppression, the originally mutated site may contain either a → transversion or a transition.

revertant — a mutant which has regained, partially or completely, the wild-type phenotype by either genetic or nongenetic mechanisms of → reversion. Genotypic revertants which have regained the enzyme or function that was lost by → gene mutation may arise by several mechanisms and fall into two classes:

1. Revertants repesenting a genuine reversal of the original mutational event, called back or reverse mutation. In this case, the base of a particular nucleotide at a given → mutational site which had been altered by the "forward mutation" is changed either back to the original base or to a third one. These changes result in the incorporation into the corresponding polypeptide (arising via → genetic transcription and → genetic translation) of an amino acid which is identical to or different from that in the wild-type polypeptide.

2. Revertants resulting from second site mutations called → suppressor mutations. They may take place either outside or inside (and then in the same or in another → codon) the → cistron which was altered by the mutation whose effect is reversed. Such mutations give rise to gene activity without changing the nucleotide of the original mutation. Frequently, suppressor mutations do not lead to a full reversal to wild-type but only to a partial reversal. Suppressor mutations can be separated from the original mutation by → genetic recombination.

The different routes which may give rise to a r. within a → structural gene can be distinguished by a combination of genetic fine structure analysis and primary protein structure studies (amino acid sequence analysis). Second site reversions (in contrast to true back mutations) always appear as double mutants separable from each other by either intra-(intracistronic second site reversion) or intergenic (intercistronic second site reversion) recombination.

R factor = → resistance factor.

rho factor (*Roberts* 1969) — a protein factor in *E. coli* (= termination factor) which binds to the → RNA polymerase holoenzyme and catalyzes specific (correct) termination of → genetic transcription and release of RNA *in vitro*. The rho factor (mol. weight about 200,000) is a tetramer or hexamer causing the release of RNA species of discrete sizes, i.e., termination of transcription at DNA sites not recoginized by RNA polymerase alone (→ M factor; psi factor). The rho factor is complementary to the → sigma factor and plays an essential role in the mechanism which results in RNA polymerase recognizing termination signals (terminator codons). Rho is antagonized by the antiterminator proteins synthesized during phage infection of bacterial cells. Rho and sigma determine the specificity of transcription of RNA off DNA templates.

ribonuclease — any of a class of enzymes that inactivate and degrade → ribonucleic acid (RNA).

ribonucleic acid — a polynucleotide (symbol RNA) consisting essentially of chains with a repeating backbone of phsophate and sugar units to which nitrogenous bases are attached. RNA governs all protein manufacture and exists in a variety of different species. Most sequenced RNAs have the potential to form secondary structures by means of local base pairing and loop formation in otherwise single-stranded molecules. These structures (→ pre-messenger RNA; transfer RNA) are evolutionarily the result of requirements for unique functions. RNA metabolism of eukaryotes is more complex than that of prokaryotes and many species of RNA have been found to have functions other than those classically assigned to RNA (→ heterogeneous nuclear RNA; low molecular weight nuclear RNA; chromosomal RNA).

Three main differences distinguish → deoxyribonucleic acid (DNA) and RNA: (Fig. 84).

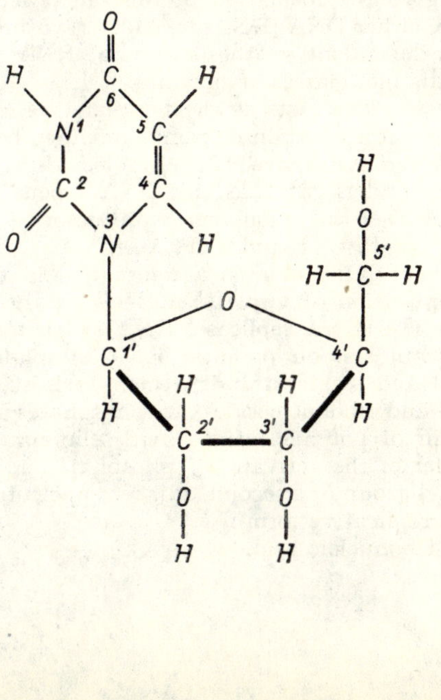

Fig. 84. The chemical structure of uracil ribonucleoside. Ribonucleic acid (RNA) and deoxyribonucleic acid (DNA) differ in that RNA contains uracil instead of thymine, and ribose instead of deoxyribose.

1. The D-ribose sugar of RNA differs from the 2-deoxy-D-ribose of DNA in having a total of four rather than three hydroxyl groups. The extra OH group is at the C 2 position.

2. Both DNA and RNA contain four principal bases but only three of these (adenine, guanine, and cytosine) are common to both nucleic acids. The RNA counterpart of thymine is uracil, the former being the 5-methyl derivative of the latter.

3. The secondary and tertiary structures of RNA show structural and functional heterogeneities which allow one distinguish two main classes of RNA:

3.1 **Genetic RNA:** The carrier of primary genetic information in most plant viruses, some animal viruses and certain bacteriophages (→ reverse transcriptase; replicative form) which may be single- or double-stranded.

3.2 **Non-genetic RNA:** RNA specified by DNA in the course of → genetic transcription. It is heterogeneous and three principal types can be distinguished: ribosomal → RNA (80% of the total cell RNA), → transfer RNA (10 to 15% of the total), and messenger → RNA (5 to 10% of the total). These types of RNA are usually synthesized via the sequential processing of primary transcription units (→ transcripton) which are longer in nucleotide sequence rather than the mature products (precursor molecules).

Regulation of RNA synthesis (transcription) *in vivo* is of two types, generalized and specific. Generalized regulation as a function of growth rate or of the presence of essential amino acids is coordinate, i.e., applies simultaneously to all species of RNA (→ relaxed; stringent). Regulation of the synthesis of specific RNA species occurs at the level of genetic transcription (→ transcriptional control; genetic regulation).

Two types of reactions leading to the formation of functional RNA molecules are known which require either DNA (DNA-dependent synthesis of cellular RNA) or RNA (RNA-dependent synthesis of viral RNA) as the primer and are catalyzed by distinct classes of enzymes.

1. **DNA-dependent RNA synthesis:** DNA acts as the template, i.e., its nucleotide sequence determines by the formation of complementary base pairs the nucleotide sequence of the synthesized RNA molecules. The process is → genetic transcription and is catalyzed by DNA-dependent RNA polymerase which links together the monomeric precursors (the ribonucleoside triphosphates ATP, GTP, UTP and CTP).

2. **RNA-dependent RNA synthesis:** RNA acts as a template and the process corresponds to the → replication of viral RNA. It is catalyzed by RNA-dependent RNA polymerase (RNA replicase, RNA synthetase) capable of forming new RNA strands upon parental RNA templates. When viral RNA enters the cell and single strands attach to host → ribosomes this enzyme is formed and ribonucleoside-5′-triphosphates are polymerized to form a long chain of polynucleotides (with elimination of inorganic phosphate) by transfer of the activated 5′-phosphate end of a nucleotide to the free 3′-hydroxyl group of a second, adjacent nucleotide in the growing polynucleotide (→ replicative form).

ribonucleotide — a nucleotide in → ribonucleic acid.

ribosomal DNA — chromosomal or (in the case of → rDNA amplification) extrachromosomal DNA which codes for → ribosomal RNA (symbol rDNA). In eukaryotes, rDNA coding for the two large species of rRNA (28 S and 18 S) is clustered at or near the → nucleolus organizer and is present as a series of tandemly reiterated genes. For a given species the number of rRNA genes is proportional to the number of nucleolus organizers. Each rDNA transcription unit contains the sequences coding for 28 S and 18 S rRNA, a transcribed spacer, and a non-transcribed spacer. Spacers in different species may differ greatly in their nucleotide sequence, while the regions coding for rRNA molecules appear to be highly conserved during evolution of different species. The rDNA repeats are not interspersed with other kinds of chromosomal DNA sequences. Insects, amphibians, birds and mammals appear to possess from 100 to 1 000 rRNA genes per genome.

Changes in redundancy of rDNA sequences may occur by unequal → crossing over or by **rDNA magnification**, i.e., the stepwise formation of extra copies of rDNA unstably bound to the chromosome in both somatic and germ cells of *Drosophila* (phenotypically bobbed male combinations) when the number of rRNA genes is insufficient to provide for a normal phenotype. Magnification is a mechanism which can adjust rDNA redundancy. Only in germ cells can the extra copies be circularized and integrated into the chromosome. After integration, if the total number of genes in the block is sufficient to provide for a normal phenotype, magnification stops (*Ritossa* et al. 1973).

In some organisms, rDNA becomes amplified (→ gene amplification) in early oögenesis by selective replication which results in many (up to thousands) nucleolus organizer equivalents of rDNA per cell. **rDNA amplification** probably arises by a chromosome copying amplification process which may be of various kinds: 1. nucleolus organizers replicate repeatedly and the released rDNA copies accumulate passively; 2. each detached product of replication may itself serve as a template for further rDNA synthesis; 3. a → reverse transcriptase driven mechanism may produce rDNA from an RNA copy (transcript of the rRNA genes) via an RNA-DNA hybrid. rDNA amplification is associated with extensive synthesis and storage of → ribosomes for later use and permits cells (oöcytes) with extrachromosomal rDNA to uncouple the synthesis of ribosomes from protein synthesis (→ genetic translation), in contrast to somatic cells where these are usually coupled.

Nearly all plants and animals exhibit rDNA with a disproportionately high G + C content (55 to 70%) so that rDNA is easily separated from all other DNA sequences of the genome.

ribosomal peptidyl transferase — an enzyme synthesizing all proteins translated from → messenger RNA. It repeatedly promotes the reaction between a peptidyl ester and an amine to form peptidylamine, i.e., a peptidyl-ester which contains one or more amino acids (→ genetic translation).

ribosomal precursor RNA (*Scherrer* et al. 1963) — = → pre-ribosomal RNA.

ribosomal RNA (*Kurland* 1960) — the RNA components (rRNA) of → ribosomes which make up about 50 to 65% of the mass of these organelles and about 80% of the RNA content of the cell. The amount of DNA (→ ribosomal DNA) that codes for rRNA (on the average about 0.3% of the genome of pro- and eukaryotes) is different in different organisms.

Ribosomes from whatever source contain one molecule of 5 S RNA and one molecule each of 16 to 18 S and 23 to 28 S rRNA. Multiple genes exist for all these RNA species in both pro- and eukaryotes. The number of such genes ranges from 3 to 10 in bacteria; several 100 may be present in eukaryotes. The large rRNA molecules form the structural backbone of ribosomes and some of the ribosomal proteins are attached to a series of specific sites on the rRNA molecules. The rRNA species isolated from pro- and eukaryotic organisms are generally not the immediate products of → genetic transcription. Post-transcriptional alterations of precursor molecules (→ pre-ribosomal RNA) include various modifications of certain nucleotides (up to 5% of nucleotides are modified by 2-o-methylation of the sugar unit), which contribute to qualitative differences in function of rRNAs, as well as cleavage events, which reduce the lengths of the polynucleotide chains of pre-rRNA to those of the mature rRNA molecules. Hairpin loops formed by bonding between pairs of nucleotide bases are structural features of rRNA and → transfer RNAs.

Molecular size of rRNAs as well as rRNA-protein ratios differ in prokaryotic and eukaryotic ribosomes.

1. **Bacterial (prokaryotic) rRNA species:** The mature rRNA components of ribosomes are of the sizes 16 S (mol. weight of about 0.55×10^6), 23 S (mol. weight 1.1×10^6) and 5 S (mol. weight 0.04×10^6). The precursor of mature 23 S rRNA is of about 5% greater molecular length, the precursor to 16 S rRNA is about 10% larger than the mature molecule. In both cases about 160 nucleotides are removed from the precursors. The *in vivo* substrates for the maturation enzymes involved in precursor cleavage are ribonucleoprotein subribosomal particles. The 5 S RNA is a component of the larger ribosomal subunit (50 S). The ratio of 23 S:5 S is always 1:1. 5 S RNA contains 120 nucleotides (no rare bases). Without 5 S RNA, 50 S ribosomal subunits are unable to synthesize proteins.

The cistrons (genes) for the three rRNA species are linked in polycistrons and transcribed (from one → promoter) in the order 16 S–23 S–5 S (there are about 10,10 and 4 copies of each gene, respectively, per genome). 5 S RNA has a G + C : A + U ratio greater than one. All three rRNAs contain a number of adjacent short base-paired regions with a short single-stranded loop at the end (hairpin). Hairpins are interspersed with single-stranded regions. The percentage of bases in the double-stranded regions is about 60 to 70% of the total. The prokaryotic rRNA composition is about 53mole% GC (→ psi factor).

2. **Eukaryotic rRNA species:** In eukaryotes, the genes coding for the two larger species of rRNA (18 and 28 S) are present in multiple copies and arranged as tandem repeats in the → nucleolus organizer. The redundancy varies from around 100 to several 1000 per genome, depending on the organism. Mature rRNA results from rRNA precursor molecules. About 15 to 45% of each precursor molecule is discarded during processing. The DNA coding for 5 S RNA is also redundant (190 to 200 genes per genome

in *Drosophila*, 2000 genes in HeLa cells, 2700 genes in *Xenopus*). 5 S RNA genes are not linked to the tandem repeats coding for large rRNA molecules but are scattered over many chromosomes.

The molecular weight of the smaller rRNA species is identical in almost all eukaryotes (about 0.65 to 0.7×10^6 daltons). The larger rRNA species in plants and most protozoa has a mol. weight of 1.3×10^6 and in lower animals 1.4×10^6 (26 S). In higher animals, this component has evolved slightly in size (mol. weight up to 2.45×10^6 in mammals).

The 28 S rRNA from the large ribosome subunit in mammals consists of a single, uninterrupted polynucleotide chain, hydrogen-bonded to which is a small RNA about 150 nucleotides long (5.8 S). The corresponding bacterial 23 S rRNA is without any associated small RNA. Linked 18 S, 28 S and 5.8 S ribosomal RNA genes are transcribed as a unit into a single polynucleotide precursor (45 S rRNA) which is ultimately cleaved to produce one molecule of each of the three RNA species.

One molecule of 5 S RNA is associated with the larger subunit of biologically active ribosomes. It consists of a single chain of about 120 or 121 ribonucleotides; minor bases are absent. Possibly, double-stranded segments parallel to the long axis of the 5 S RNA are present. There are some indications that 5.8 S RNA, and not eukaryotic 5 S RNA, is the evolutionary homologue of prokaryotic 5 S RNA in ribosomes.

Chloroplast and **mitochondrial rRNAs** in eukaryotes have lower molecular weights than their cytoplasmic counterparts, and different average nucleotide composition. The genes coding for these RNAs are contained in → chloroplast DNA and → mitochondrial DNA, respectively.

ribosome (*Roberts* 1958; *Dintzis* et al. 1958) — an exceedingly complex cell organelle in pro- and eukaryotes which is made up of about 60 unique proteins and three (in prokaryotes) or four (eukaryotes) species of → ribosomal RNA (rRNA) organized into two dissimilar ribosomal subunits (30 S and 50 S in prokaryotes; 40 S and 60 S in eukaryotes). Ribosomes are more or less spherical in shape and measure approximately 20 nm across. Those of bacteria are slightly smaller than those of eukaryotes. A growing bacterial cell contains 10,000 to 30,000 ribosomes, most or perhaps all of which are involved in protein synthesis during the exponential growth phase.

Ribosomes play a decisive role in → genetic translation and specifically bind → messenger RNA (mRNA), → aminoacyl- and peptidyl-transfer RNA (tRNA), GTP, and a number of different translation factors (→ initiation factor; release factor). They are the sites for the reactions of translation initiation (→ initiation complex), amino acid polymerization, and polypeptide chain termination (Fig. 85). Amino acid polymerization involves not only peptide bond formation, but also movement of the r. along the mRNA and GTP splitting. There are two substrate binding sites on the r.: one (the P-site) for holding the peptidyl-tRNA, the other (the A-site) for attachment of the aminoacyl-tRNA.

The larger ribosomal subunit contains one large (23 to 28 S) rRNA molecule and one (5 S in prokaryotes) or two (5 S and 7 S in eukaryotes) small RNA molecules and a variety of (~30) proteins. The smaller subunit contains one 16 to 18 S rRNA molecule and additional (about 30) proteins.

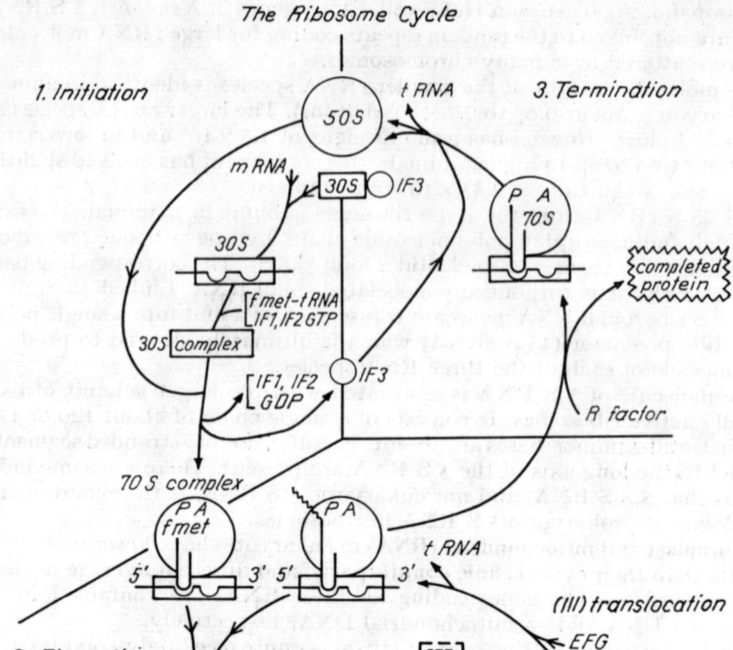

Fig. 85. Schematic representation of the ribosome cycle in genetic translation (from *Clark-Walker* 1973).

Within this framework a number of differences exist between prokaryotic and eukaryotic ribosomes. Structural differences include size, relative RNA to protein content and the ease or difficulty with which the two types of ribosomes can be dissociated into subunits *in vitro*. Both the total number and average molecular weights of eukaryotic (mammalian) ribosomal proteins significantly exceed estimates for the bacterial (*E. coli*) indicating a more complex situation in eukaryotes. Studies of protein components of isolated ribosomal subunits make it evident that the population of ribosomes from even single-celled organisms is heterogeneous. The functional significance of ribosomal heterogeneity is not yet clear. There is also evidence that some ribosomes may be "message-specific".

Some 50 to 63% of the mass of ribosomes consists of rRNA; the rest represents protein. Protein synthesis requires an ordered association of the two ribosomal subunits. Two basic aspects of the cyclic function of ribosomes in protein synthesis are: 1. ribosomes frequently dissociate into their subunits, both in prokaryotic and eukaryotic cells, most likely after each passage of ribosomes over mRNA; 2. during reduced rates of protein synthesis single ribosomes, inactive in protein synthesis and not carrying mRNA, accumulate at the expense of → polyribosomes with the size of the ribosomal subunit pool remaining invariant.

Complete ribosomes from bacteria, blue-green algae, mitochondria, and chloroplasts belong to the 70 S class; cytoplasmic ribosomes from eukaryotes sediment at about 80 S. The S values in the 70 S and 80 S classes show a rather wide range, i.e. from 67 to 73 S and from 77 S to 81 S, respectively. In eukaryotes, the ribosomal proteins are synthesized in the cytoplasm and migrate to the → nucleus where the rRNA is transcribed (→ nucleolus). RNA and protein assemble and finally pass to the cytoplasm as a functional ribosome. Biogenesis of ribosomes in both pro- and eukaryotes is a regulated function, coordinated with cell growth and sensitive to environmental influences. It proceeds through the formation of precursor particles (→ pre-ribosomal RNA), consisting of an incomplete set of ribosomal proteins and of RNA molecules that are longer and less methylated than mature ribosomal RNA. During assembly of the ribosomal subunits, free ribosomal proteins are added in a stepwise manner and in a specific order to RNA and/or intermediate ribonucleoprotein particles. The interaction between ribosomal subunits is affected by a number of macromolecular factors, e.g., aminoacyl-tRNA, peptidyl-tRNA, messenger RNA, and initiation factors.

In many eukaryotic cells, a large proportion of cytoplasmic ribosomes are bound to membranes of the → endoplasmic reticulum. The binding is mediated by the large ribosomal subunits. There is evidence of functional and compositional differences between membrane-bound and free ribosomes. A major function of r.-membrane interaction is to effect a topological segregation of different populations of ribosomes synthesizing different classes of proteins.

There is evidence that free ribosomes synthesize proteins which remain in the cell sap, while membrane-bound ribosomes manufacture products for export from the cell. In the latter, the nascent polypeptide, which grows within a protected region of the large subunit, enters into close relationship with the membrane immediately upon emerging from the r.

There is also evidence that the r. plays an active role in controlling recognition efficiency of → transfer RNA. This is assumed to be due to a ribosomal tRNA recognition site. In *E. coli*, five or six proteins bind specifically to 16 S rRNA, eight specifically to 23 S rRNA, and two to 5 S RNA. Each protein recognizes a specific site on the RNA. The mechanism of action of the r. follows a cycle of conformational changes clearly manifested in interconversions between active and inactive forms.

Mitochondrion-specific ribosomes are the smallest known (about 60 S) and consist of a 45 S and a 35 S subunit. Thus far no RNA equivalent to the 5 S RNA species has been identified. The mitochondrial ribosomes, like those of bacteria and eukaryotes, are associated with mRNA to form polyribosomes. The assembly of mitochondrial ribosomes involves RNA components synthesized inside the organelle and protein components synthesized outside. In the case of chloroplast ribosomes the situation is quite similar.

ribosome cycle — the exchange of the large and small ribosomal subunits which occurs in growing bacteria and in the cytoplasm of eukaryotes. Exchange is due to dissociation of the subunits after each passage over messenger RNA and reformation from a pool of ribosomal subunits that continuously recycle through polyribosomes. Recycling requires an → initiation factor (= ribosome dissociation factor), maintains the polyribosomes, inhibits the association of ribosomal subunits, promotes the quantitative entry of the larger subunit into polysomes, allows the accumulation of ribosomal subunits (instead of single ribosomes), and is required for the binding of messenger RNA to ribosomes (→ initiation complex).

ribosome dissociation (*Kohler* et al. 1968) — → ribosome cycle.

ribosome dissociation factor — → dissociation factor; ribosome cycle

ribosome recognition site — → transfer RNA.

ribosome translocation — the movement of the → ribosome along → messenger RNA by three base increments during → genetic tranlation. R. t. is an active process which requires GTP hydrolysis and the participation of a soluble protein called the → G factor or translocation factor. The ribosomal site(s) for interaction with G and GTP is in the larger ribosomal subunit. For r. t. deacylated tRNA must be expelled from the P site of the ribosome, peptidyl-tRNA must be moved from the A to the P site, and mRNA must be moved across the smaller ribosomal subunit so as to align the next codon in the A site.

ring bivalent — a → bivalent at the first meiotic division consisting of two paired chromosomes (→ chromosome pairing) with terminalized chiasmata in both arms (→ chiasma terminalization; ring multivalent).

ring chromosome — a physically circular chromosome (→ genetic circularity). Ring chromosomes are a normal feature of chromosome organization in some prokaryotes (e.g., *Escherichia coli* and some viruses). In eukaryotes, ring chromosomes arise as a result of chromosomal structural changes (→ chromosome mutation) and are usually meiotically unstable due to the production, by meiotic → crossing-over, of double-sized dicentric chromosomes which normally are eliminated.

During mitosis, a ring chromosome usually gives rise to two daughter rings of equal size which are regularly distributed to daughter nuclei (Fig. 86). Under these circumstances the ring chromosomes are somatically stable.

As a result of "sister strand exchange", ring chromosomes may give rise to "double sized rings" in somatic tissues (Fig. 86b), i.e., continuous dicentric rings of double the original size which may lead to a double

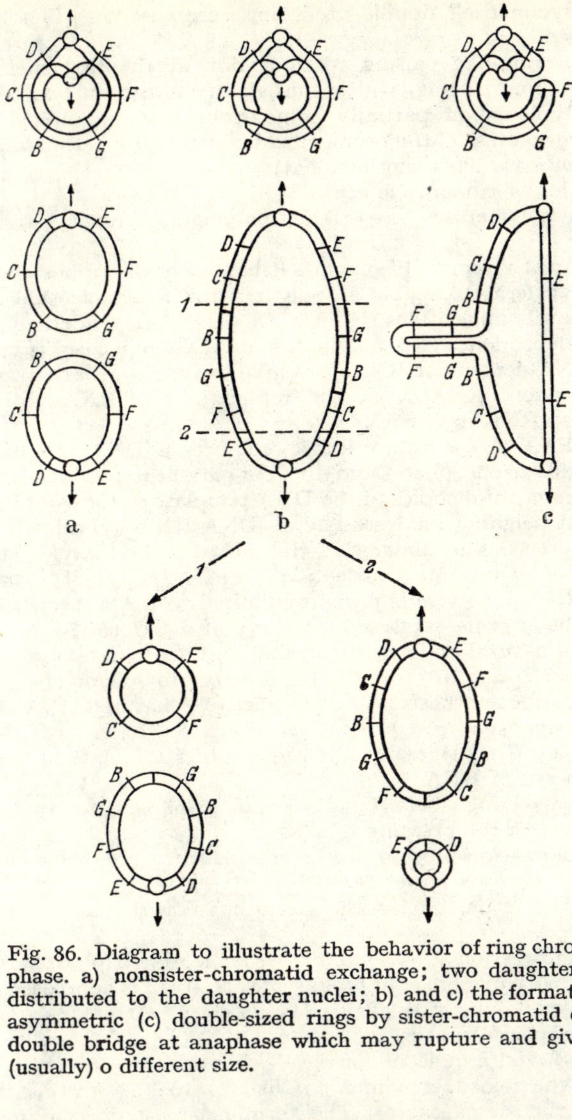

Fig. 86. Diagram to illustrate the behavior of ring chromosomes in mitotic anaphase. a) nonsister-chromatid exchange; two daughter rings of equal size are distributed to the daughter nuclei; b) and c) the formation of symmetric (b) and asymmetric (c) double-sized rings by sister-chromatid exchange; the result is a double bridge at anaphase which may rupture and give rise to two new rings (usually) o different size.

bridge at anaphase (→ chromosome bridge). In the case of bridge rupture, two new rings may be formed which are usually of different size. This process can result in → variegation due to the formation of → duplications and → deletions. Depending on their mode of formation **"symmetric"**

(Fig. 86b) and **"asymmetric"** double-sized ring chromosomes (Fig. 86c) may be distinguished.

ring multivalent — a ring-like pairing configuration at the first meiotic division (→ chromosome pairing) which consists of more than two completely (in autopolyploids) or partially homologous (e.g., in diploid → translocation heterozygotes) chromosomes held together by terminalized chiasmata up to anaphase I (→ ring bivalent).

RNA — a shorthand for → ribonucleic acid.

RNA-dependent DNA polymerase — → DNA polymerase; reverse transcriptase.

RNA polymerase — any of a number of enzymes (ribonucleoside triphosphate: RNA nucleotidyltransferase) capable of polymerizing ribonucleoside triphosphates into polyribonucleotides (RNA). DNA (DNA-dependent RNA p.) or RNA (RNA-dependent RNA polymerase or RNA replicase) is used as the template. DNA-dependent RNA p. is involved in → genetic transcription, RNA-dependent RNA p. in the replications of RNA viruses with double-stranded RNA molecules.

DNA-dependent RNA p. (= transcriptase) requires a DNA template, the four ribonucleotide triphosphates and divalent cations to produce RNA transcripts which are faithful copies of the DNA template in terms of base composition, nearest neighbor analyses and → DNA-RNA hybridization. Chain lengthening proceeds by linkage of the ribose 3′-hydroxyl to the 5′-phosphate group of an incoming nucleotide.

The activity of RNA p. *in vivo* is probably limited to the transcription of DNA. *In vitro,* the enzyme catalyzed a variety of reactions:

1. transcription of natural and synthetic ribo- and deoxyribopolymers;
2. reiterative transcription of short ribo- and deoxyribo-oligomers;
3. homopolynucleotide synthesis in the presence of denatured DNA and a single nucleotide triphosphate;
4. poly(A)- and poly(U)-synthesis in the absence of a template;
5. pyrophosphorolysis of RNA;
6. pyrophosphate exchange between nucleoside triphosphate and inorganic pyrophosphate in the presence of DNA.

The overall polymerization process can be divided into three steps:

1. binding of RNA p. to its template;
2. association of the initial nucleoside triphosphate with the enzyme-template complex;
3. polymerization (direction 5′ → 3′) of further nucleoside triphosphates.

DNA-dependent RNA polymerase in prokaryotes (*E. coli*) is composed of at least four different polypeptide chains ($\alpha_2, \beta\beta'\sigma$): two α chains (mol. weight 39,000), one β chain (mol. weight about 150,000), a β' chain (mol. weight about 160,000) and a molecule of the → sigma factor (mol. weight 90 to 95,000). This is the holoenzyme (mol. weight 7.2×10^5; 21 to 24 S); the core enzyme lacks the sigma factor. The terminator factor rho (mol. weight about 200,000) complements the sigma factor and ensures that transcription is halted at the correct site (prevention of read-through; → psi factor). The core enzyme ($\alpha_2\beta\beta'$) possesses all the enzymatic activities associated with the RNA p., except that it can hardly initiate RNA synthesis on certain templates. Sigma enhances accurate assembly of the $\alpha_2\beta$ complex and the β subunit and confers certain template specificity on the enzyme (recognition of specific → promotors on the DNA).

When an intact double-stranded DNA is used as template, transcription *in vivo* and *in vitro* by RNA p. is asymmetric, i.e., only one of the complementary strands is transcribed.

In eukaryotes, multiple forms of DNA-dependent RNA p. with different intranuclear localization have been shown to exist. This suggests that gene expression in eukaryotic cells is regulated, at least in part, by distinct RNA polymerases with different template specificities. Unlike the prokaryotic RNA p., the eukaryotic enzymes are tightly bound to a complex consisting of DNA, → histones, → nonhistone chromosomal protein, and RNA. The multiple forms of RNA p. are compartmentalized within the nucleus and have different requirements for cations.

RNA p. I (=A) is localized in the → nucleolus and transcribes → ribosomal DNA into → ribosomal RNA (mol. weight 500 to 700,000). The enzyme is composed of smaller subunits and active with both Mn^{2+} and Mg^{2+}.

RNA p. II (= B) is found in the non-nucleolar nucleoplasm and is responsible for the synthesis of → heterogeneous nuclear RNA (→ premessenger RNA) from which → messenger RNA is derived (mol. weight about 700,000). Its activity is much higher with Mn^{2+} than Mg^{2+}. For optimum activity, a relatively high ionic strength is required. The enzyme is specifically inhibited by α-amanitin. Discrete protein components of RNA p. II have molecular weights of about 38,000, 55,000, 150,000, and 190,000, respectively.

RNA p. III (= C) is usually a minor compound found in the non-nucleolar nucleoplasm of a variety of tissues. It is possibly involved in the transcription of transfer RNA genes and 5 S RNA genes.

Mitochondrial RNA p. (RNA p. IV) displays properties different from those of the nuclear polymerases and consists of a single polypeptide chain (mol. weight 64,000). Its activity is unaffected by α-amanitin and inhibited by rifampicin, acriflavin and actinomycin D.

RNA-dependent RNA polymerase is a virus-dependent RNA p. (RNA replicase) necessary for the replication of double-stranded RNA viruses. It requires all four ribonucleotide triphosphates, RNA as a template, and divalent cations. The enzyme is analogous to DNA-dependent RNA p. and transcribes messenger RNA from DNA without affecting the integrity of the template (→ reverse transcriptase).

RNA processing — posttranscriptional tailoring of RNA molecules which represent transcription products of DNA (→ posttranscriptional processing; pre-messenger RNA; pre-ribosomal RNA; pre-transfer RNA).

RNA replicase — → RNA polymerase.

RNA synthetase — → ribonucleic acid.

RNP — abbreviation for RNA-protein complexes.

RNP particle — a ribonucleoprotein particle consisting of eukaryotic RNA associated with proteins (→ informofer; informosome) which is quite distinct from → ribosomes and may function either to transport → messenger RNA from the nucleus to the cytoplasm or to control its expression in the cytoplasm.

Robertsonian — of chromosome structural changes due to → centric fusion or → centric fission.

roentgen equivalent dose (*Crow* 1973) — that dose (symbol RED) of a chemical mutagen which is equivalent to the X ray dose necessary to induce quantitatively and qualitatively the same effect (= radequivalent).

rolling circle model (*Gilbert* & *Dressler* 1968) — a model explaining the replication of circular, double-stranded bacterial DNA by the following steps (Fig. 87): 1. the "negative" strand remains a closed circle; 2. The "positive"

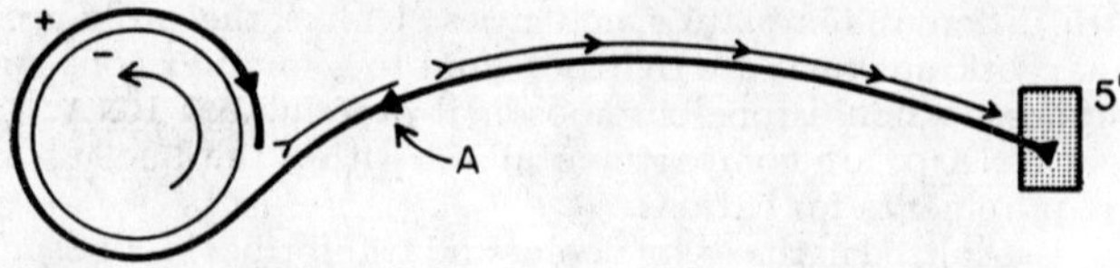

Fig. 87. The rolling circle model of DNA replication. The tail end of the positive strand (5') is fixed to a membrane site (from *Gilbert* and *Dressler* 1968).

strand is nicked and continuously elongated on the circular negative strand template with the newly exposed 3'-hydroxyl end serving as a start for → DNA polymerase; 3. The original positive strand is peeled off from the rolling circle as a single-strand tail; 4. The new negative strand is initiated late and polymerized on the positive single-strand tail; 5. Before the new negative strand has been completed the → replicative intermediate splits into the circular part and the linear tail of the rolling circle. The circular part contains the old circular negative strand and the new linear positive strand. The tail consists of the linear old positive strand and the incomplete negative new strand.

The main difference between the r. c. m. and the other models is the asymmetry of replication: one strand remains circular, the other is cut. The circular strand may be copied several times, while the other parental strand is copied only once.

rotation (*White* 1942) — a change of form in → bivalents between early diplotene and diakinesis of the first meiotic division, consisting in a "rotation" of two chromosome arms relative to the other two in bivalents with one chiasma so that a typically cross-like figure arises. In bivalents with more than one chiasma by r. the loops between successive chiasmata come to lie at right angles to one another (Fig. 62).

S

S_1, S_2, S_3, etc. — shorthands for generations of self-fertilization in plants: S_1 denotes the generation obtained by selfing the parent plants, S_2 the generations obtained by selfing the S_1 plant, etc.

sacculus — a shape-maintaining element in the cellular envelope of bacteria (*E. coli*) which encloses the cell completely and consists exclusively of the polymer murein (a network of polysaccharide chains which are interlinked by short peptide bridges).

S-allele — any → allele of those genes which control → incompatibility in plants. Alleles present in both style and pollen are referred to as "matched S-alleles", whereas those present only in either style or pollen are called "unmatched S-alleles" (*Lewis* 1943). S_F-alleles are alleles belonging to the same series of multiple alleles as S-alleles but which control self-fertility.

saltatory replication — a hypothetical, sudden type of DNA replication held responsible for the production of families of similar nucleotide sequences (→ repetitious DNA).

samesense mutation — a base substitution mutation in the third base of the relevant → codon which is redundant or synonymous (= synonymous mutation) because of the degeneracy of the → genetic code; a substitution of the third base does not necessarily alter the amino acid sequence of a polypeptide chain.

SAT-chromosome (*Heitz* 1931) — a chromosome with a secondary → constriction which separates a → satellite from the rest of the chromosome. SAT-chromosomes whose secondary constriction is associated with the formation of the → nucleolus are referred to as "nucleolar" SAT-chromosomes and the constriction is a "nucleolar" secondary constriction.

satellite (*Navashin* 1912) — a chromosomal segment separated from the main body of the chromosome by one secondary → constriction if located

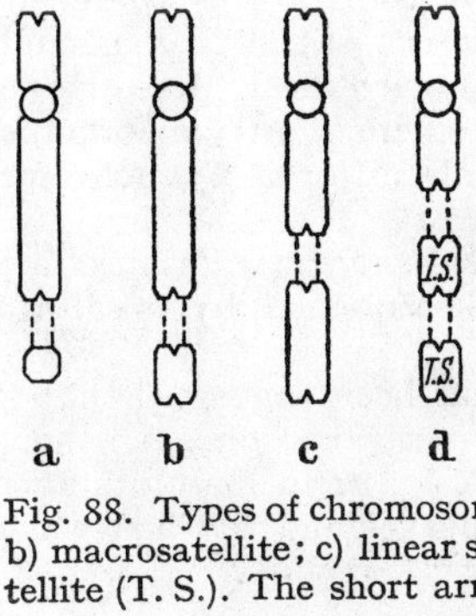

Fig. 88. Types of chromosomal satellites (after *Battaglia* 1955). a) microsatellite; b) macrosatellite; c) linear satellite; d) intercalary satellite (I. S.) and terminal satellite (T. S.). The short arm of the acrocentric chromosome is a pseudosatellite.

terminally (terminal s.), or by two if intercalary (intercalary s.). Together, the s. and the secondary constriction are termed "satellite region".

According to *Battaglia* (1956) the following types of satellites are distinguished (Fig. 88).

1. **Microsatellite:** a relatively small, spheroidal s., i.e., with a diameter equal to or less than one-half the chromosome diameter.

2. **Macrosatellite:** a relatively large, spheroidal s., i.e., with a diameter greater than one-half the chromosome diameter.

3. **Linear s.:** a s. having the shape of a long chromosomal segment.

4. **Pseudosatellite:** a spheroidal part of a chromosome adjacent to the → centromere, e.g., the short arm of a chromosome with a subterminal centromere: terminal pseudosatellite. The presence of a secondary constriction indirectly changes the position of the pseudosatellite from the terminal to an intercalary position: intercalary pseudosatellite.

satellite DNA (*Sueoka* 1961; *Kit* 1961; *Britten* & *Kohne* 1968) — in eukaryotes, any DNA fraction sufficiently different in base composition from the major DNA component to separate as distinct bands during isopycnic CsCl gradient centrifugation. Sat. DNAs may have a buoyant density in neutral CsCl lower or heavier than the → main band DNA. Nuclear and cytoplasmic sat. DNAs may be distinguished: the former are of solely nuclear origin, the latter are composed of DNA from cytoplasmic sources including DNA from → chloroplasts (in plants), → kinetoplasts, and → mitochondria.

Nuclear sat. DNA in eukaryotes is either light (A + T rich) or heavy (G + C rich) as compared with main band DNA (→ heavy-shoulder DNA) on CsCl salt gradients. It is generally highly repetitious (hundred of thousands of clustered sequences per genome) and represents a part of the families of repeated DNA sequences (→ repetitious DNA). In general, sat. DNAs share the properties of sharp melting transitions, extremely rapid renaturation after denaturation (→ Cot value), strand separation in alkaline CsCl, and marked species-specificity.

Sat. DNA may or may not be transcribed and does not seem to code for protein. It has been found in association with constitutive → heterochromatin localized primarily adjacent to the centromeres (pericentric heterochromatin). The repeating unit of different sat. DNAs is highly variable in size (from less than 10 to 300—400 nucleotides). More than one kind of sat. DNA may be localized at the centromere of any one organism. Sat. DNA replicates either as a single functional unit or as a synchronized set of units.

Sat. DNAs are assumed to have arisen independently many times during evolution from unrelated, unrepeated nucleotide sequences by → saltatory replication.

The function of sat. DNA is unknown. It has been suggested that sat. DNA may be involved in processes such as → chromosome pairing, → chromosome movement and chromosome packing, or in maintaining the structural integrity of the chromosomes ("housekeeping"). Clustered repetitious sequences which just happen to have the same average base composition as main band DNA do not form a distinct band by centrifugation but can be analyzed by special techniques. They may be called "hidden satellites".

SAT-zone (*Resende* 1940) — a broad secondary "constriction" in → SAT-chromosomes whose diameter may approximate that of the chromosome.

schizogony — → fission.

Schultz-Redfield-effect (*Schultz* & *Redfield* 1951) — → inversion.

scripton (*Szybalski* et al. 1970) — = → transcripton.

sealase (*Hurwitz* et al. 1967) — → nick.

secondary pairing (*Darlington* 1928) — → chromosome pairing.

second division segregation — reductional ("postreductional") separation and thus → segregation of a → heterozygous pair of alleles during the second division of → meiosis (where the first meiotic division was "equational" for the allele pair in question), as opposed to first division segregation, i.e., reductional ("prereductional") separation of a heterozygous pair of alleles during the first meiotic division (where the second meiotic division is then equational).

second messenger (*Sutherland* et al. 1965) — any agent responsible for the transmission of hormonal stimuli within the cell or within compartments of the cell. 3′, 5′-cyclic adenosine monophosphate (c-AMP) is a s. m. with tissue-specific action and a variety of different effects in different organs and organisms. At the molecular level it activates various protein kinases which in turn are able to regulate the activities of numerous other enzymes. cAMP also influences negatively controlled activities of genes (→ operon; catabolite repression) and, thus, does not act solely via activation mechanisms.

second site reversion (*Yanofsky, Helsinski* & *Maling* 1961) — → suppressor mutation.

secretion granule — in exocrine cells, the intracellular storage compartment for protein destined for export. S. granules are located in the apical cytoplasm of the excorine cells between the Golgi apparatus region, where the granules originate, and the apical cell membrane, where the period of storage is terminated by release of the granule contents into the extracellular space by the process of → exocytosis.

sectorial (*Baur* 1909) — of a → chimera in which the idiotypically distinct tissues are cross-sectionally arranged as sectors of a circle.

seed incompatibility (*Valentine* 1960) — all forms of sterility in plants due to disturbances after fertilization of the development of tissues involved the production of seeds.

segment (*Belling* 1927) — a portion of a chromosome taken as a unit under given circumstances (→ differential s.; interstitial s.; intercentric s.).

segmental allopolyploid (*Stebbins* 1947) — → alloploid.

segregation (*Bateson* & *Saunders* 1902) — = → genetic segregation.

segregational lag (*Witkin* 1951) — a delay in the phenotypic manifestation of a newly arisen → gene mutation in bacterial cells containing more than one → nucleoid. Since there is more than one set of genetic material, phenotypic expression of a mutation in one of the chromosomes must await the partitioning (segregation) of that chromosome into a cell that contains no nonmutant chromosomes (→ phenotypic lag).

segregational load (*Crow* 1958) — → genetic load.

segregational sterility (*Darlington & Mather* 1949) — → sterility resulting from gene segregation in structural and numerical hybrids which gives rise to genetically unbalanced gametes, as opposed to "genotypic sterility" which is due to an unbalance in the zygotic genotype itself.

segregation distortion (*Sandler & Hiraizumi* 1961) — a condition (in *Drosophila* males) due to intrachromosomal changes which results in unequal → genetic segregation (up to 20:1 instead of 1:1) of two alleles present in a heterozygote. S. d. is caused by certain aberrant mechanisms during → meiosis (→ meiotic drive).

segrosome (*Tanaka* 1962) — any cellular granule in which vital stains are segregated.

selection (*Darwin* 1858) — differential and nonrandom reproduction (→ selection coefficient) of different → genotypes representing the most important of the factors that induce evolutionary changes (→ evolution) by affecting the → gene and → genotype frequencies in populations. In the absence of s. and → mutation these frequencies remain constant from generation to generation in large, randomly interbreeding populations (→ Hardy-Weinberg rule).

Biological s., which may result from differential success of various genotypes in → mating, fecundity, viability, longevity and emigration, has both a positive and negative aspect. The preferential reproduction and survival of some genotypes go hand in hand with the preferential elimination of others (*Grant* 1963).

S. may act at the following levels (*Wright* 1956): among genes, cells (→ cell selection), clones, biparental organisms, within populations (demes), between populations, and among species. The effectiveness of s. in producing a response within a previously unselected population depends on (*Mather* 1955) the modes and intensities of selective forces, the action of genes displayed in phenotypes being selected, and the amount and pattern of → genetic variability in the population.

Natural s. acts to preserve in nature favorable variations and ultimately to eliminate those that are "injurious" (*Darwin*). It is a consequence of differences between genotypes (→ fitness) in respect to their ability to produce progeny and represents a process without purpose whose primary form **("Darwinian s.")** takes place between individuals in a population. However, both competition and s. do occur between reproducible biological units of greater and lesser complexity than that of individuals.

Artificial s., in contrast to natural s., is a purposeful process with definite goals set by the breeder and means s. applied under a selected set of environmental conditions. It normally occurs with controlled matings of a few selected genotypes, and its goal is to change specific phenotypic characters of a population. Artificial s. applied to one character almost always leads to changes in others ("correlated response").

The **modes of Darwinian selection** within populations may be classified according to their operation into stabilizing, directional, and disruptive s.

(*Mather* 1953), but in any given situation two or even all three modes may act simultaneously (Fig. 89):

1. **Stabilizing s.** (= centripetal s. of *Simpson* 1953; normalizing s.) favors a single optimum in the population. The main effect of s. is the elimination of peripheral variants which arise by mutation, immigration or recombination. In stable environments and populations which have already achieved a high state of adaptation, a certain range of genotypes of proven fitness is preserved from generation to generation. Stabilizing s. does not result in evolutionary changes, but rather maintains an existing state of adaptation. According to *Waddington* (1953; 1957) two different stabilizing responses may result from s. as parents individuals near the mean: either genotypes which control the development of abnormal pheno-

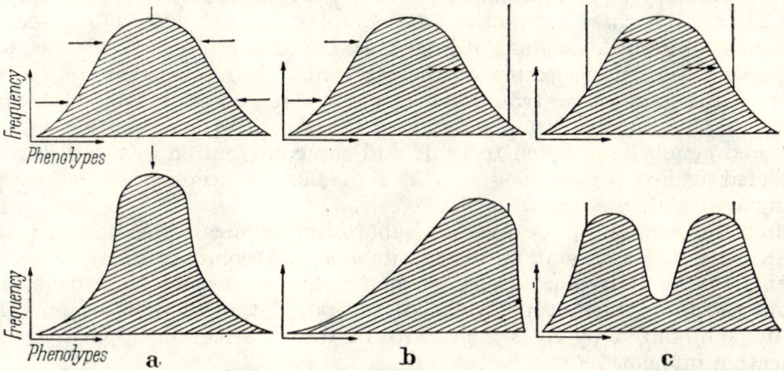

Fig. 89. Modes of selection within populations: a) stabilizing selection; b) directional selection; c) disruptive selection (the horizontal arrows show the direction of selection; the lower curves represent the new pattern of variation after partial response of the population to the selection pressure) (after *Mather* 1948).

types ("normalizing" s.) or which render developmental pathways of the developing individual sensitive to the potentially disturbing effects of environmental stresses ("canalizing s.") will be eliminated.

2. **Directional s.** ("progressive s. of *Schmalhausen* 1949; linear s.; dynamic s.) favors a single optimum like stabilizing s. but results in a systematic shift (in the direction of applied s.) in gene frequencies and in the population mean for the character considered. Directional s. operates in a progressively changing environment and leads to a state of adaptation.

3. **Disruptive s.** (= centrifugal s. of *Simpson* 1953) simultaneously favors more than one optimum in a population occupying a heterogeneous habitat, i.e., two or more different genotypes are at an advantage and intermediate types are at a disadvantage. The favored genotypes are maintained in a state of polymorphic equilibrium by the tendency of each of the several selective pressures to preserve a certain genotypic variant which is optimally adapted to one particular feature of the environment.

Sexual s. is based on male competition or female choice and is responsible for sexual dimorphism, while the term **habitat s.** refers to the capability of dispersing individuals to select an appropriate habitat.

Frequency-dependent s. is a type of s. which favors particular alleles or chromosome rearrangements when rare but not when common.

Apostatic s. (*Clarke* 1962) is a type of s. which favors "apostates", i.e., visually distinct variants from the norm. The hypothesis of a. s. assumes that predators are most likely to form "specific searching images" for common varieties of prey.

In contrast to Darwinian s. within populations, **interpopulation s.** (*Grant* 1963) means the differential reproduction of populations of any size (from local breeding groups to noninterbreeding species).

Without doubt, Darwinian and interpopulation s. interact and balance each other but these interactions are complex and not well understood.

selection coefficient — a quantitative measure of the intensity of selection expressed as **s**, the selection coefficient, which is the proportional reduction in the gametic contribution of a particular genotype compared with a standard genotype, usually the most favored. The contribution of the favored genotype is taken to be **1**, and the contribution of the genotype selected against is then (1 — s). This expresses the fitness of one genotype compared with the other.

selection differential — the average superiority or mean phenotypic value (expressed either in standard deviations or in absolute units of measurement) of the individuals selected as parents, expressed as a deviation from the mean phenotypic value of the population of their origin. The s. d. (symbolized by S) is a measure of the → selection applied, i.e. of selection intensity.

selection limit — a population is said to be at a s. l. when the response to selection has ceased. The s. l. is reached gradually with the response to selection becoming progressively slower. Therefore, it is usually impossible to decide exactly at what point the limit is reached.

selection pressure (*Wright* 1921) — the intensity of natural → selection which is usually measured by the change of → gene frequency per generation due to the influence of selection. The degree of selective advantage of a preferred allele or genotype over the others is one of the fundamental factors determining the rate at which allelic frequencies are changed under s. p. This factor may be measured in terms of → selection coefficients.

selection response — the gain in liability due to selection which is a function of the → heritability of liability, assuming that genotype-environment interactions are absent. If A is the mean liability of affected individuals (those in the tail of the general population distribution), and R the mean of their offspring, then the realized gain in liability is R — A.

selective advantage — an advantage in the competition for survival between one genotype as compared with another showing "selective disadvantage" and producing fewer viable offspring. In **natural selection,** the s. a. of a certain genotype is the property of all its phenotypic expressions and may have different causes, e.g., the production of larger progenies, better resistance of the progeny to environmental factors with deleterious effects, better resistance of the progeny to predators (including parasites), better ability of the progeny to find a suitable habitat, etc. In **artificial selection,** s. a. is conferred by the breeder on some members of a population in preference to others because they or their relatives possess certain phenotypic characters which are visible, measurable, or can be evaluated in some manner by the agency of control over the relative reproductive rate (*Lerner* 1958).

In the absence of other differences, the genotype showing selective disadvantage is gradually replaced by the genotype with s. a. if both are present within one habitat (→ annidation).

selective disadvantage — → selective advantage.

selective fertilization — → fertilization.

selective peak (*Wright* 1956) — a → gene frequency system characterized by homeostatic properties (→ homeostasis). The population occupying a s. p. tends to swing back to that peak after temporary disturbances, provided these are not so drastic as to shift the group of individuals into the sphere of influence and control of another s. p. (→ adaptive peak).

selective value — a value proportional to the probability that an individual (genotype) survives from conception to adulthood (= fitness). The mean **s. v. of a gene** is the mean of the s. values of the genotypes into which the gene enters, weighted by the frequencies of the other genes involved. Mean s. values of genes are functions of the gene frequencies and, therefore, change when the gene frequencies change.

The mean **s. v. of a population** is the mean of the s. values of the various genotypes, weighted by their frequencies at the time of conception. The change of the mean s. v. of a population due to natural selection is proportional to the additive variance (→ genetic variability) of the s. values of the individuals in the population (*Jacquard* 1974).

The "index of the opportunity of selection" (*Crow* 1958) measures the maximum change in the s. v. which the population's mortality and fecundity characteristics can entail.

self-catabolite repression (*Katz & Engelsberg* 1971) — a feedback mechanism involved in → catabolite repression which, in interaction whith induction, provides a means of insuring that optimal levels of enzymes are produced for the efficient metabolism of the → inducer of an operon. This mechanism is probably common to all instances where the inducer of an operon is metabolized. S.-c. r. insures the prevention of enzyme overproduction through the induction process and is due to the production of → effectors which dampen the expression of the relevant operons.

self-compatibility — → self-incompatibility.

self-fertile — of individuals having reproductive organs and gametes which permit zygote formation by fusion of gametes derived from the same parent organism. The phenomenon is referred to as self-fertility.

self-fertilization — the union (= selfing, automixis) of male and female gametes and/or nuclei from the same haploid, diploid, or polyploid organism (→ fertilization; paedogamy; parthenogamy).

self-incompatibility (*Stout* 1917) — the inability of a plant producing functional gametes to set seed when self-pollinated (→ incompatibility).

self-pollination — the transfer of pollen to stigma within the same flower or in the same → clone.

self-regulation (*Smith* & *Magasanik* 1971) — = → autoregulation.

semantide (*Zuckerkandl* & *Pauling* 1965) — any → informational macromolecule.

semiallele (*Komai* 1950) — → = pseudoallele.

semiapospory (*Fagerlind* 1940) — → apomixis following a → pseudohomeotypic division with the gametophyte arising from an unreduced egg cell. Apospory, semiapospory, and displospory blend into one another without clearcut distinctions.

semibivalent (*Battaglia* & *Boyes* 1955) — any of the units of anaphase movement in the case of "postreductional → meiosis", consisting of two chromatids paired side-by-side or end-to-end. With more than two homologues (polyploidy or polysomy) semitrivalents, semiquadrivalents, etc. may arise.

semiconservative (*Delbrück* & *Stent* 1957) — of the normal mode of replication of → deoxyribonucleic acid (Fig. 31) in which each of the two parental strands is conserved in a different daughter molecule (→ nonconservative; DNA polymerase).

semidominant — → dominant.

semigeographic — of a mode of → speciation which consists in the partitioning of species along a line of secondary → intergradation or along lines of strong ecological contrast (*Mayr* 1963).

semiheterotypic (*Rosenberg* 1926) — → nonreduction.

semihomologous — of chromosomes or genomes which are partially → homologous (= homeologous).

semi-isolation — incomplete → isolation between two populations, allowing incidental → gene flow.

semikaryotype (*Battaglia* 1952) — → karyotype.

semilethals (*Muller* & *Altenburg* 1919) — = sublethals (→ lethal factor).

semispecies (*Mayr* 1940) — borderline cases between → species and → subspecies characterized by some, but not all, attributes of species rank. S. represent population systems which exhibit some species-like properties combined with some subspecies-like properties. They may show greater morphological differences and greater discontinuity in their variation pattern than is usual for subspecies (races) without possessing the → isolating mechanisms against interbreeding that are characteristic of true species. Related s. are linked by limited amounts of → gene exchange (interbreeding is not as free as between subspecies, but freer than between species) and represent breeding units of higher order. An assembly of s. is referred to as a **superspecies or syngameon** depending on the geographical relationships between the component semispecies. If the semispecies under observation are allopatric, the assembly represents a superspecies, while a syngameon consists of sympatric semispecies (*Grant* 1963).

semisterility (*Belling* 1914) — sterility in genic and structural hybrids (heterozygous for → chromosome mutations such as reciprocal → translocations) characterized by inviability of approximately half of the male and female gametes.

sense strand — = → codogenic strand.

sensitizing mutation (*Koch* & *Drake* 1970) — a → leaky mutation which is used to enhance the recovery of normally undetectable, → cryptic mutation.

septate junction (*Wood* 1959) — in invertebrate tissues, the most extensive → cell junction which occurs as a zonula, or belt, surrounding the apical or basal regions of cells. In transverse view the s. j. consists of two adjacent cell membranes joined by a periodic arrangement of electron-dense bars, or septa. S. junctions from different sources may show considerable variations in the three-dimensional appearance of the intercellular septa. The s. j. has been proposed as a site of intercellular adhesion, a permeability barrier, and a pathway of low-resistance ionic coupling.

sequence homology map — a physical → chromosome map obtained from → heteroduplex mapping and division of the heteroduplex map into fractions of → map units.

sequential enzyme induction — → induction.

serial genes (*Grant* 1964) — → gene interaction.

service unit (*Lifschytz* 1971) — in eukaryotes, a DNA sequence for which the → chromomere or → band is the cytological counterpart; in the simplest case, a s. u. is assumed to be composed of a structural gene and all sequences needed for its function (such as initiation and termination sites for replication and transcription as well as recognition sequences needed for the regulation of the translation process). In certain cases, more than one structural gene can be associated with one s. u. or, conversely, some service sequences can be duplicated (→ transcripton).

Sewall Wright effect — = → genetic drift.

sex — the contrasting and complementary characteristics exhibited by male and female, + or — (if the differences are only physiological), donor and recipient (in prokaryotes) individuals or clones within the same species, as well as the capacity for performing amphimixis and/or → genetic recombination by such individuals (→ reproduction). The basis for the biological divergence of the sexes may be either inherent in the organism or impressed on it by its environment (→ sex determination).

Sex is able to affect the phenotypic expression and the genetic ratios of chromosomal hereditary determinants in several ways:

1. The → inheritance of genes located in the → sex chromosomes (and showing sex-linked inheritance) follows special rules based on the pattern of homology and transmission of these chromosomes.

2. Dominance of alleles may differ in heterozygotes of the two sexes. This phenomenon is termed "sex-influenced dominance".

3. One sex may be uniform in phenotype with respect to particular characters yet transfer genes which give rise to phenotypic differences in the offspring of the other sex. This is referred to as "sex-limited gene expression".

4. Cumulatively acting genes may be required to exceed different thresholds in males and females before a phenotypic difference becomes manifest and, therefore, phenotypic segregation simply results in two classes, variously reached in the two sexes.

In the case of extrachromosomal hereditary determinants, transfer to the offspring may be exclusively (at least with oögamy) through the female line; this type of → inheritance is called "matrilinear".

sex cell — = → gamete.

sex character — any of the primary sex differences concerning gametes and sexual organs, as well as the secondary characters that do not concern gametes and sexual organs.

sex chromatin (*Barr* & *Bertram* 1949) — one, in special cases more than one, plano-convex, spherical or pyramidal and Feulgen positive intranuclear body (about 0.8×1.1 μm in size), commonly situated at the periphery of the interphase nucleus just inside the nuclear envelope (= Barr body). It represents a single → X-chromosome which is positively → heteropycnotic and was originally called a "nucleolar satellite". Depending on the tissue and technical factors, 20–96% of nuclei of all females of many mammalian species contain s. c., whereas the comparable structure is rare or absent in nuclei of diploid males, with the exception of species with unusual → sex-chromosomes. The s. c. of leukocytes is contained in a specific drumstick-like nuclear appendage which is usually referred to as a → "drumstick".

Depending on the number of X-chromosomes, the maximum number of sex chromatin bodies in any one cell is: none in XY and XO individuals; one in XX, XXY, XXYY; two in XXX, XXXY; and three in XXXX individuals, i.e., one less than the sum of X-chromosomes in a diploid or aneuploid individual. Cells with less than the maximum number have fewer and, accordingly, larger bodies.

Sex chromatin is derived from one of the two X-chromosomes normally present per cell in diploid females and is in a heavily condensed → heteropycnotic state. The main points of the so-called "inactive-X hypothesis" (*Lyon* 1961, 1966) may be summarized as follows:

1. the heteropycnotic X-chromosome, which forms a sex chromatin body, synthesizes its DNA later than the other chromosomes. The single X of males and one X-chromosome of females replicate concurrently with the → autosomes;

2. the heteropycnotic X-chromosome is genetically inactivated. (→ dosage compensation);

3. the inactivation occurs early in embryonic development and remains fixed throughout the further development of each cell line;

4. the inactive X-chromosome can be either maternal or paternal in origin in different cells of the same diploid individual;

5. in → aneuploid individuals (males with two or more X-chromosomes, females with more than two X-chromosomes) all but one X become inactivated and, therefore, sex chromatin may also be present in the male sex, provided the males have more than one X in their chromosome complement. Thus the strict association between presence (female sex) and absence (male sex) of s. c. applies only to diploid individuals with an XX-XY- or XX-XO-system of → sex chromosomes.

sex-chromosome (*Wilson* 1906) — a chromosome or group of chromosomes ("multiple sex chromosomes") in eukaryotic organisms with separate sexes which is represented differentially in the two (male and female) sexes and is causally engaged in the genotypic control of → sex determination.

In organisms that have the sexes combined in a single individual as in hermaphroditic animals and the majority of higher plant species, there are no special s.-chromosomes in the karyotype. The formation of male and female germ cells (microspores and megaspores in plants) is accomplished in such cases by a process of histological differentiation.

In the **basic system of sex-chromosomes,** in one sex a pair of microscopically indistinguishable chromosomes (referred to as the X-chromosomes) is represented in the opposite sex by a pair of visibly different chromosomes. One of these two is identical to the X-chromosome of the opposite sex, the other is structurally and functionally different, restricted to the one sex and termed the Y-chromosome. Regions in which → chromosome pairing and → crossing-over between X- and Y-chromosomes can take place and which are → homologous are called "pairing segments", as opposed to "differential segments" which are nonhomologous and do not pair at meiosis.

Normally, the XX sex is female and the XY sex male. The XX and XY sexes are also called "homo-" and "heterogametic", respectively. The homogametic sex produces (during meiosis) gametes of one kind (all contain one X) while the heterogametic sex gives rise to two types of gametes with either one X or one Y chromosome. The X- and Y-chromosomes of the heterogametic sex regularly disjoin to opposite spindle poles at either the first ("prereduction" of the sex-chromosomes) or the second ("postreduction" of the sex-chromosomes) meiotic division, so that equal numbers of gametes with either an X- or a Y-chromosome are produced. Union of an X-bearing male gamete with an X-bearing female gamete results in a female zygote, while union of a Y-bearing male and an X-bearing female gamete gives rise to a male zygote.

In some organisms, the situation is reversed, i.e., structurally identical sex-chromosomes are present in the (homogametic) male sex and structurally distinctive ones in the female (heterogametic) sex. With female heterogamety the sex chromosomes are often called Z ($\equiv$ X) and W ($\equiv$ Y) chromosomes.

Chromosomes of the complement that are not sex-chromosomes (or allosomes) are referred to as → autosomes. As opposed to the sex-chromosomes, the distribution of autosomes to one or another of the meiotic products has no influence on the sex of the offspring determined by the fusion of one male and one female gamete.

Various types of **derived sex-chromosome systems** differing from the basic XX♀—XY♂- or ZZ♂—ZW♀-systems are known. In some the Y-chromosome has been lost altogether (XX—XO-systems) or the X- and Y-chromosomes have become completely nonhomologous, that is, unable to pair and undergo crossing-over during meiosis. In others, called systems with multiple sex chromosomes, either the X- or Y-chromosome is represented by two or more chromosomes.

The basic and the derived sex-chromosomes systems may be grouped as follows (*John & Lewis* 1965):

1. **Systems with chiasmate meiosis** (pairing and crossing-over) in the heterogametic sex:
a) Basic XX—XY-system;
b) derived XX—XY-systems;
c) multiple systems of the types X_nY, XY_n and X_nY_n.
2. **Systems with achiasmate meiosis** (no pairing and crossing-over) in the heterogametic sex:
a) Absence of pairing partner(s) in the heterogametic sex;
a_1) simple XX—XO-system;
a_2) compound X_nO-systems.
b) Differential partner(s) in the heterogametic sex;
b_1) simple XX—XY-system;
b_2) compound systems of the types X_nY, XY_n and X_nY_n.

Five main mechanisms may give rise to multiple sex-chromosome systems (*Dietz* 1958):

1. Reciprocal → translocations (→ centric fusions) between sex-chromosomes and autosomes may bring autosomes (called neo-X- and neo-Y-chromosomes) into the sex-chromosome system and transform a system of the XX—XO-type into a neo-sex chromosome system of the XX—XY-type, or a system of the XX—XY-type into a X_nY-, XY_n-, or X_nY_n-system.
2. Fragmentation of an X-chromosome with diffuse → centromere organization.
3. Increase due to polyploidy of the number of sex chromosomes.
4. Autonomous transformation of autosomes into X-chromosomes.
5. Increase in the number of Y-chromosomes due to → non-disjunction which gives rise to supernumerary Y-chromosomes.

The behavior in meiotic segregation of multiple sex-chromosomes may be grouped into four basic types (*Cooper* 1946):

1. Systems with a sex-chromosome → multivalent may show a definite orientation and distribution of the constituent chromosomes at meiosis.
2. Systems without physical contact between the segregating chromosomes in which alternative sets of sex-chromosomes take part in distance or "touch-and-go" pairing.
3. Systems (e.g., X_1X_2OO males) characterized by multiple sex-chromosomes whose components are all of the same type and are distributed together to one and the same spindle pole.
4. Systems in which one sex-chromosome bivalent and one univalent, or two and more sex-chromosome bivalents are regularly distributed in such a way that definite groups of sex-chromosomes (not in physical contact with one another) reach opposite spindle poles.

The evolution of X- and Y-chromosomes may be interpreted as a series of progressive, evolutionary modifications of a pair of autosomes according to the following hypothesis (*White* 1950).

1. The origination of a gene pair, whose segregation controls the production of the two sexes.
2. Crossing-over is suppressed in that region of the homologues in which the sex-differentiating genes were located.

3. Mutations of genes and of chromosome structure take place in this region and give rise to a so-called "differential segment" with no meiotic pairing within its limits.

4. Gene mutations make one of the differential segments relatively "inert" genetically while the corresponding segment of the other chromosome gains a new internal balance. The chromosome with the "inert" segment could represent a Y-chromosome, the other an X-chromosome.

5. Further chromosomal structural changes could then result either in a complete elimination of the "inert" segment of the Y-chromosome or include other autosomes into the sex-determining system.

sex composite — = → gynandromorph.

sex-controlled (*Goldschmidt* 1920) — of characters, whose degree of phenotypic expression is controlled by the sexual constitution of the individual. The type of → inheritance in which sex controls or changes the manifestation of characters, but not the chromosomal transmission of the hereditary determinants controlling these characters (→ sex-linkage), is referred to as "sex-controlled inheritance".

sex cycle — the alternation of → karyogamy and → meiosis in eukaryotes, or (by extension) any recurrent series of processes that lead to → genetic recombination (→ parasexual) in prokaryotes.

In eukaryotes, meiosis is necessarily correlated in the s. c. with the complementary process of → fertilization, i.e., the fusion of two gametes of opposite sex which results in a zygote containing the complete chromosome sets introduced by the male and female gametes. The primary function of the sex cycle is the combination and recombination (→ genetic recombination) of genes from different sources.

In prokaryotes (viruses and bacteria), this s. c. is replaced by parasexual processes which lead to genetic recombination. The factors determining "sexual differentiation" in bacteria (→ sex factor) are fundamentally different from those controlling → sex determination in eukaryotes.

sex determination — the determination of sex differentiation in eusexual (→ parasexual) organisms by processes which shift the socalled "bisexual potency") (*Hartmann* 1923) or the balance between male (M) and female (F) determining genes (sex realizators) of cells, organs and individuals in either the male or female direction.

1. **Phenotypic (or environmental) s. d.** (*Burgeff* 1915): sex is mainly (and independent of syngamy or karyogamy) determined by internal and external environmental conditions which decide whether a male or a female is formed. Female and male "realizators" or trigger genes (symbolized by F and M, respectively) are in this case equally balanced and show a norm of reaction that allows a shift in their activity which can be induced by environmental factors. Phenotypic s. d. may take place during haplophase **(haplophenotypic s. d.)** or diplophase **(diplophenotypic s. d.)**.

The term phenotypic s. d. is misleading when contrasted to genotypic s. d.: Phenotypic s. d. is, in fact, genetic s. d. without a → sex chromosome mechanism to shift the balance of F and M sex realizators in either the male or female direction. In the case of phenotypic s. d., this shift results from the influence of various environmental factors that affect the genic action of F and M.

2. **Genotypic s. d.** (*Burgeff* 1915): sex is mainly determined by the genotype of the zygote or spores. If s. d. takes place during haplophase by meiotic segregation of genes, resulting in half male and half female gametes (or spores), this is called **"haplogenotypic s. d."** in contrast to **"diplogenotypic s. d."** where the diplophase (in diplontic organism) is also sexually determined (either male or female).

There are two main cytogenetic mechanisms of diplogenotypic s. d. in which the so-called → sex-chromosomes play a decisive role in the inheritance and determination of sex. The XY-system and its several variations based on the balance between X-chromosomes and autosomes occurs in higher animals and dioecious plants and represents a multigenic mechanism. A given sex is usually the result of the interaction of several, probably many, pairs of genes.

In other XY-systems, the Y-chromosome of the heterogametic sex is actively sex determining and individuals carrying such a chromosome are of a definite sex (male in the case of male heterogamety, female in the case of female heterogamety) irrespective of their autosomal genotype. The sex of the embryo is determined at fertilization by the sex-chromosomes contributed by the sperm and ovum.

With the X-chromosome-autosomes balance system (derived from results with *Drosophila*) the main factor in s. d. (which also takes place at fertilization) is the ratio between X-chromosomes and autosomes. Sex differences are assumed to be due to the action during development of two counteracting sets of genes located both in the X-chromosomes and in the autosomes. The two sets of genes are not equally effective; maleness genes are more effective in the autosomes (individuals with one X and two sets of autosomes are males irrespective of the presence or absence of a Y-chromosome), while femaleness genes are more numerous or more effective in the X-chromosomes (individuals with two X-chromosomes and two sets of autosomes are females; → metafemale). The Y-chromosome has no influence on s. d. In this case, the 2X—1X mechanism of sex chromosomes in diploid individuals acts to shift the balance of the M and F sex realizators in either the male or female direction.

Haplo-diploid s. d. is a mode of s. d. in which the males arise from a haploid egg and the females from a diploid.

Maternal s. d. is a mode of s. d. in which sex is determined by the → idiotype of the mother or egg cell.

Syngamic s. d. is a mode of s. d. in which the sex of the progeny is fixed as a result of gamete fusion and karyogamy.

Progamic s. d. is a mode of s. d. in which the sex of the progeny is determined in the egg prior to fertilization and even prior to meiosis.

Metagamic s. d. is a mode of s. d. in which the sex of the progeny is not fixed by karyogamy and, therefore, is largely subject to environmental influences. Genotypically similar individuals or cell complexes become male or female after fertilization by influences external to the genotype, or a genotypically male individual may be transformed into a phenotypically female individual by external factors and *vice versa*.

sex differentiation — the origin and development of organisms with different (male and female) sex organs and sex characteristics and the production of male and female, or (+) and (—) gametes, by one individual or different individuals. By extension, also the production of donor and recipient strains among bacteria (→ parasexual; sex determination; sex factor).

sex digamety — the capacity of one sex (called heterogametic) to produce male and female determining gametes by means of the → sex chromosome mechanisms. S. d. leads to the differentiation of → unisexual individuals by a wide variety of developmental patterns in the different phyla (*Bacci* 1965) (→ sex monogamety; sex polygamety).

sex duction (*Jacob* & *Wollman* 1960) — the incorporation of bacterial genes by sex factors and their subsequent transfer by → conjugation (→ F-duction). The sex factors act as → conjugons and are → plasmids.

sex factor — in bacteria, transmissible → plasmids that can promote the transfer (by → conjugation) of genetic material of the host chromosome or a coexisting plasmid from the bacterial donor to the recipient besides their own transfer from the host cell into a recipient cell. The fertility properties of most s. factors (some, but not all, of the → colicinogenic factors and → resistance factors) are generally repressed in the majority of cells (the → F-plasmid being an exception to the rule). S. factors determine specific cellular traits (→ sex pilus) that render a male bacterial cell capable of mating with a (female) recipient.

sex fimbrium (*Diguid* & *Anderson* 1967) — = → sex pilus.

sex-influenced — → sex-controlled.

sex-limited (*Morgan* 1910) — of genetically controlled characters which are phenotypically manifest in only one of the two sexes due to sex-limited gene expression. The underlying genes (which may be located in → sex-chromosomes as well as in → autosomes) are normally transmitted in → inheritance by the male and female parents (→sex-controlled).

sex-linkage (*Morgan* 1914) — → linkage of genes located in the → sex-chromosomes (X- and Y-chromosomes) of eukaryotes. Such genes and their mode of → inheritance, as well as the phenotypic characters whose manifestion these genes control, are all referred to as "sex-linked" (→ sex-limited)

1. **Partial (incomplete) s. l.** (*Darlington, Haldane* & *Koller* 1934): The genes concerned have loci on both X and Y and thus are located in pairing segments of the two chromosomes and may be exchanged by → crossing-over between these chromosomes.

2. **Complete s. l.** (*Morgan* 1914): The genes concerned have loci only on the X and thus are in the differential (nonhomologous) segments of the X- and Y-chromosomes. They are not recombinable by crossing-over.

Genes located in the Y-chromosomes are termed → "holandric".

sex monogamety (*Bacci* 1961) — in the case of balanced hermaphrodites, labile → gonochoric and exclusively → parthenogenetic populations, the supposed capacity for producing gametes of only one kind with respect to sex determination (→ sex digamety).

sex mosaic — = → gynandromorph or gynander.

sex piebald — a female with small regions of tissue with male sex constitution distributed over the body.

sex pilus (*Meynell & Lawn* 1967) — any of the specialized filaments extending from the bacterial surface (in *E. coli* and other Enterobacteria) on whose presence in the donor (male) cells → conjugation and gene transfer depend. S. pili are very similar in appearance to the common pili or fimbriae found on many enteric bacteria regardless of whether or not they carry → sex factors.

Two main classes of s. pili may be distinguished: 1. The → F pili determined by the → F plasmid and some → resistance and → colicinogenic factors; 2. I pili determined by colicine I factors or other derepressed factors.

S. pili of each class are involved in the formation of mating pairs between donor and recipient cells and probably also in the transfer of genetic material during mating. They are also adsorption sites for male-specific DNA and RNA phages and responsible for the specific antigenic markers of F^+ bacteria.

F pili are composed of polymerized phosphoglycoprotein subunits (mol. weight 11,800) called pilin. The pilus itself consists of two parallel protein rods (each build by the F-pilin monomer), may be several μm long and has a diameter of 8.5 nm. The structure of I pili is unknown.

Common pili are thought to stabilize bacterial pair formation even though they cannot themselves initiate it.

sex polygamety — the capacity of male and female or hermaphroditic individuals for producing gametes of many different types with respect to sex determination by a polyfactorial mechanism (*Bacci* 1965) (→ sex monogamety; sex digamety).

sex ratio — 1. The number of males per 100 females at zygote formation (primary s. r.), at birth (secondary s. r.), or at sexual maturity (tertiary s. r.).

2. A form of → meiotic drive (in *Drosophila*) characterized by the production of only or mostly female progeny by a male whose → X chromosome carries a certain genetic factor.

sex realizator (*v. Wettstein* 1924) — any of the sex determining or sex deciding genes (symbolized by F for female and M for male determiners) which are assumed either to decide whether the male (A) or female (G) part of the so-called "bisexual potency" (*Hartmann* 1923) will become more active (→ AG complex), or to represent a balanced system with an alternative norm of reaction (*Goldschmidt* 1929/1930) which may be shifted in either the male or female direction (= sex realizers). In phenotypic and genotypic → sex determination, this shift in development to maleness or femaleness is induced by both environmental and genetic factors, which serve as switch mechanisms.

sex realizer — = → sex realizator.

sex reversal — a change (either natural, pathological, or experimentally induced) in the sex of an individual from male to female or *vice versa*.

sexual — of all processes involving both → meiosis and → fertilization by means of which sexual → reproduction and → genetic recombination are achieved in "eusexual" organisms (→ parasexual).

sexual character — any of those traits which define the sex of an individual and are produced, after → sex determination, by the processes of → sexual differentiation. Primary sexual characters are those directly involved in → reproduction (male and female reproductive organs), secondary sexual characters are those expressed differentially in males and females but which do not participate directly in reproduction.

sexual cycle — in eukaryotes, a cycle consisting in its basic form of the pairwise fusion of haploid gametic nuclei to give diploid zygotes. Zygote nuclei, or some of their mitotic products, then undergo → meiosis so that haploid nuclei are reproduced. These, or mitotic descendants, then serve gametic functions and the cycle is complete.

sexual dimorphism — the presence of marked differences (in shape, size, structure, color, etc.) between male and female individuals of the same species.

sexual incompatibility — the inability of genetically similar ("homogenic s. i.") or heterogeneous ("heterogenic s. i.") gametes or nuclei to undergo karyogamy (*Esser* 1956). The term is used for all systems restricting sexual propagation on the basis of their genetic determination (→ incompatibility).

sexual isolation — = reproductive → isolation.

sexual reproduction — → reproduction (= gamogony; gamogenesis).

sexual selection — → selection.

sexuparous (*Caullery* 1913) — producing male and female offspring either by → parthenogenesis or by sexual reproduction (amphigony).

sex vesicle (*Sachs* 1954) — a prominent heteropycnotic body formed by the X/Y chromosome pair of mammals during meiotic prophase.

shift — a chromosomal structural change (also called "internal → translocation") which results in the transposition of a chromosome segment to another place in the same chromosome. Shifts may be homo- or heterobrachial, i.e., take place within one and the same, or between two chromosome arms. Depending on the frequency of meiotic → chromosome pairing and chiasma formation between the transposed segment and its untransposed homologous region in → heterokaryotypes (heterozygosity for the shift under observation), → duplications and → deletions may arise.

sibling(s) — the progeny (= sibs) of the same parents.

sibling species (*Mayr* 1952) — reproductively isolated populations that are morphologically similar or identical and frequently sympatric.

sib mating — brother-sister mating. In many species of plants and animals, specifically those which are self-incompatible or dioecious, s. m. is the closest possible form of → inbreeding.

sib method — in human genetics, a method of deriving the proportions of individuals with and without a character under observation from those members of a sibship other than the → propositus by whom the sibship (→ pedigree) was identified.

side-arm bridge — a type of chromosome structural change inducible only during prophase of mitosis or from pachytene to metaphase I of meiosis and observable at the anaphase immediately following treatment. S.-a. b. also occur spontaneously (→ pseudochiasma).

side-line — → stem line.

sigma factor (*Burgess* et al. 1969) — a positively acting transcription factor (mol. weight 85,000 to 90,000; 4.5 to 5 S) which is the primary determinant for → promotor recognition by bacterial DNA-dependent → RNA-polymerase. Sigma is one of the subunits of the RNA polymerase holoenzyme and restricts the RNA polymerase core enzyme (without s. f.) to initiating transcription at certain genes. Transcription of some bacteriophages during infection is controlled by production of successive phage-coded sigma-like factors which replace the host s. f. and direct the host core RNA polymerase to transcribe specific phage genes (→ psi factor; rho factor; catabolite activating protein; M factor).

Sigma-like factors, each uniquely modifying the range of templates that can be transcribed, are possibly a general feature of the regulation of gene expression in bacteria and phages. They may also play a role in the regulation of transcription in eukaryotes although these possess multiple RNA polymerases.

The bacterial s. f. is released from the RNA polymerase after initiation of transcription (i.e., is free to program another polymerase molecule) and the core enzyme continues to transcribe RNA. Besides its function in transcription initiation (probably by opening local regions of promotor sites), the s. f. plays an essential role in the accurate assembly of the $\alpha_2\beta$ complex of RNA polymerase and the β' subunit.

sign mutation — = → frameshift mutation.

silent mutation (*Sonneborn* 1965) — = → samesense mutation or synonymous mutation.

silent section — any segment of genetic material for which no functions have been found.

simple sequence DNA (*Walker* 1971) — → repetitious DNA.

simplex (*Blakeslee, Belling* & *Farnham* 1923) — → nulliplex.

single-copy DNA — = → unique or non-repeated DNA.

single site mutation — → multisite mutation, gene mutation.

sire — the male parent in mammals (→ dam).

sister chromatid — a → chromatid derived from duplication (replication) of one and the same chromosome during interphase of the → cell cycle, as opposed to non-sister chromatids derived from other homologous (homologous s. c.), or nonhomologous chromosomes of the chromosome complement.

sister chromatid exchange (*Taylor* 1958) — a deviation from the regular semiconservative replication pattern of the chromosomes of eukaryotes consisting in the exchange of segments between → sister chromatids of a chromosome. After incorporation of ^{3}H-thymidine into chromosomal DNA, s. c. e. is evident from segmental switches of radioactive label from one chromatid to the other in the second post-labeling mitosis (→ sister label exchange). S. c. e. may be of two types: **single exchanges** which effect only one of the two pairs of granddaughter chromosomes (prevention of anaphase separation of chromatids results in four homologues per tetraploid cell), and **twin exchanges** in which both chromosomes show an exchange at the same point.

Exceptions to the rule that only one of the two chromatids of second post-labeling division chromosomes is labeled at any point along its length are isolabeling, iso-nonlabeling, and heterolabeling:

1. **Isolabeling** — part of the second division chromosome has both chromatids labeled at second division.

2. **Iso-nonlabeling** — part of the second division chromosome has neither strand labeled.

3. **Heterolabeling** — a labeling pattern possibly due to exchange between labeled and non-labeled half-chromatids.

The phenomenon of s. c. e. (which is stimulated by UV and ionizing radiation, and inhibited by caffeine) may be due to an exchange process analogous to genetic recombination (→ recombination repair).

sister label exchange — after application of ^{3}H-thymidine, reciprocal exchange of labeled segments between the sister chromatids of first anaphase or second metaphase (→ meiosis) chromosomes (= reciprocal switches of label). Non-reciprocal label exchanges between sister chromatids of meiotic chromosomes, which are reciprocal between non-sister chromatids of first anaphase chromosome pairs, are called non-sister label exchanges (= non-reciprocal switches of label).

site (*Demerec* 1956) — the smallest subunit in a → gene capable of independent mutation (mutational site or → muton) and separable by intragenic recombination from such neighboring subunits (→ recon). Two mutants are said to occupy the same site when they fail to recombine with each other to produce wildtype recombinants. → Base-pair substitution mutations clearly identify a site as a single nucleotide in DNA.

SMC — an abbreviation for sperm or spore mother cell.

soluble RNA — = → transfer RNA.

soma (*Weismann*) — the body of an organism apart from the germ line cells.

somatic — of cell types, structures and processes in an individual other than those which are characteristic of, or give rise to, germ cells.

somatic cell fusion — homo- or heterophasic fusion between somatic cells in culture which may give rise to somatic → cell hybridization. As a normal part of growth and development, s. c. f. occurs between genetically identical cells of mammalien skeletal muscle, fungal hyphase and true slime mold plasmodia.

somatic cell hybridization (*Barski* et al. 1960) — → cell hybridization.

somatic crossing-over — = mitotic → crossing-over.

somatic mutation — a → mutation in any cell not destined to become a → germ cell.

somatic pairing (*Metz* 1916) — → chromosome pairing.

somatic reduction — reduction of (normally polyploid) somatic chromosome complements (→ reductional grouping; cell hybridization) by formation of → aneuploid intermediates. S. r. in premeiotic cells suggests a possible origin for some existing diploid and tetraploid taxa from polyploids through a process of depolyploidization. S. r. may give rise to haploid cells (→ haploidization) during the parasexual cycle of certain fungi.

somatic segregation — the → segregation at → mitosis of chromosomally or extrachromosomally inherited genetic determinants (due to gene mutation, somatic recombination, aneuploidy). S. s. may give rise to → variegation or sectoring.

somatogamy (*Renner* 1916) — = → pseudoapogamy.

spacer DNA (*Brown* & *Weber* 1968; *Birnstiel* et al. 1968) — "nongenic" DNA sequences which separate successive genes in eukaryotic chromosomes from one another. S. DNA is assumed to account for a large fraction of the DNA within a chromosome. Spacer gene arrangement of cistrons is generally established for → ribosomal DNA where the spacer separates the blocks of the 28 S and 18 S ribosomal RNA cistrons. In this case, the untranscribed spacer may have an average molecular weight as high as 6.6×10^6 daltons for each repeating unit and seems to vary in length by as much as a factor of 10 even within one species. Its sequence is very different even in two closely related species. Where amplification of ribosomal DNA occurs (→ gene amplification), the spacer takes part in this process.

special segment (*Darlington* & *LaCour* 1938) — a reversible modification of certain (probably → heterochromatic) chromosome segments of definite and species-specific location, in mitotic chromosome complements. It consists of a reduction of staining capacity and (usually) of diameter. Special segments may be induced by growing certain plants and animals at low temperatures. "Differential reactivity" of chromsome segments at low temperature is assumed to result from differential contraction.

speciation (*Simpson* 1944) — the evolutionary formation of → species which, in essence, occurs by the production of barriers to gene exchange (reproductive isolation) between genetically diverging populations. Isolating mechanisms are promoted by natural → selection if gene exchange lowers the → fitness of its products.

The main modes of s. are (*Mayr* 1963):

1. **Sympatric speciation:** s. without geogaphic isolation, i.e., the acquisition of isolating mechanisms within a → deme.

2. **Allopatric speciation:** s. with geographic isolation, i.e., the acquisition of isolating mechanisms by a population during a period of geographic isolation ("geographic s."), or the partitioning of species along lines of secondary → intergradations, or along lines of strong ecological contrast ("semigeographic s.").

3. **Stasipatric speciation** (*White* et al. 1967): s. whereby an ancestral species gives rise to a number of descendent species through chromosomal rearrangements, arising within its range, spreading to occupy limited areas, and acting as incipient isolating mechanisms in narrow zones of hybridization (*White* 1973).

4. **Instaneous or cytocatalytic speciation:** s. by the origination of a single (e.g., → alloploid) individual which is reproductively isolated from the species to which its parents belong. It is reproductively and ecologically capable of establishing a new population. An analogous result may be obtained by multiple structural changes of the karyotype (saltational s. of *Lewis* 1954) which, after hybridization between the ancestral species and the resultant derivatives with structurally changed chromosomes, give rise to a high degree of hybrid sterility.

species (*Ray* 1670) — groups of actually or potentially interbreeding natural → populations which are reproductively isolated from other such groups. S. are the largest and most inclusive reproductive communities of sexual and cross-fertilizing individuals that share in a common → gene pool (*Mayr* 1940; *Dobzhansky* 1950). Species containing only a single → subspecies are called "monotypic", while those containing more than one subspecies are termed "polytypic" (→ semispecies, sibling species, superspecies).

Besides those sexually reproducing s. which are unequivocally defined by 1. morphological and physiological differences, 2. reproductive isolation, and 3. ecological differences, there are many natural populations (especially in the plant kingdom) which reproduce asexually, by apomixis or obligate autogamy and are difficult to characterize by the three aforementioned criteria, or for which there is insufficient knowledge for definition as a species. Provided all three groups of the above-mentioned information are available, the following classification of a group of individuals under observation is proposed by *Mayr* (1948):

The individuals are	not reproductively isolated	reproductively isolated
identical in morphology and sympatric	1. same population	5. sibling species
identical in morphology and allopatric	2. same subspecies	6. sibling species
different in morphology and sympatric	3. individual variants of the same population	7. separate species
different in morphology and allopatric	4. separate subspecies	8. separate species

species group — a group of closely related species (→ superspecies) with (usually) partially overlapping range (*Mayr* 1963).

species hybridization — → hybridization between → species. According to *Mayr* (1963) five kinds may be distinguished:

1. Occasional crossing of sympatric species which gives rise to hybrids that are ecologically or behaviorally isolated and, therefore, unable to backcross with either parental species.

2. Occasional or frequent production of more or less fertile hybrids between sympatric species, some of which backcross to one or both parental species.

3. Formation of a secondary zone of contact and of partial interbreeding between two formerly isolated populations that did not acquire reproductive isolation during the period of geographic isolation.

4. Complete local breakdown of reproductive isolation between sympatric species, resulting in the production of hybrid swarms that may include the whole range of variability of the parental species.

5. Production of a new entity of species rank due to hybridization between individuals from different species and subsequent doubling of the chromosomes in the → alloploid hybrid, giving rise to an allopolyploid (in plants).

S. h. may play a role in → evolution as a way of perfecting isolating mechanisms if the hybrids are rare and inviable, as a source of new species, and as a means of increasing → genetic variability.

specificity (*Timoféeff-Ressovsky* 1926) — the quality of gene action as defined by the location, phenotypic appearance and mode of variation of the corresponding character resulting from the expression of the gene under observation. There is no clearcut distinction between s. and → expressivity.

specific locus test — a test to evaluate mutagenic effects in germ cells (→ mutagenicity screening): Wildtype male or female mice are treated with the test agent and then mated to partners that are homozygous for a number of recessive genes. The progeny are examined for the phenotypic expression of recessive genes. This can result only if one of the dominant alleles in the mating partner treated with the test agent has undergone → gene mutation. Phenotypes usually evaluated concern coat color and other morphological characteristics.

specioid (*Mansfeld* 1950) — a systematic category in cultivated plants which corresponds to a → species.

sperm — the male gamete in animals (= spermatozoon) and in primitive plants (= spermatozoid).

spermateleosis — the differentiation of a spermatid into a spermatozoon (→ spermatogenesis).

spermatid (*La Valette St. George* 1886) — → spermatogenesis.

spermatocyte (*La Valette St. George* 1886) — → spermatogenesis.

spermatogenesis — the development of the male animal → germ cells (spermatozoa, sperm) within the male gonad (the testis).

The testes contain the primordial germ cells **(primary spermatogonia)** which enter into a phase of rapid mitotic multiplication producing the **secondary spermatogonia.** The secondary spermatogonia transform into **primary spermatocytes** (meiocytes). Each of the latter (after increasing in size and a series of complex nuclear changes) undergoes the first meiotic division and gives rise to the two **secondary spermatocytes,** which in turn undergo the second meiotic division. As a result of these two divisions, four **spermatids** with reduced chromosome number arise. The spermatids differentiate into the mature male gametes, the **spermatozoa** or sperm (→ oögenesis; megasporogenesis; microsporogenesis).

spermatogonium (*La Valette St. George* 1886) — any of the immature reproductive cells in male animals after the formation of the testis. By division and differentiation spermatogonia give rise to spermatocytes (→ spermatogenesis).

1. **Predefinitive spermatogonia** are cells that produce (through stem cell divisions) cells like themselves as well as definitive spermatogonia.

2. **Indefinitive spermatogonia** are those generations of spermatogonia occurring between the stem cell divisions and the definitive divisions (= primary spermatogonia).

3. **Definitive spermatogonia** are cells irrevocably determined through (dichotomous) division and differentiation to become the spermatocytes and eventually the spermatozoa.

spermatozoon — → spermatogenesis.

spermiogenesis — = → spermiohistogenesis.

spermiohistogenesis — the production of mature spermatozoa from undifferentiated spermatids (= spermiogenesis).

S phase (*Howard & Pelc* 1953) — in eukaryotes, the period of chromosomal DNA replication during interphase of the → cell cycle.

S phase recovery — chromosome repair during the → S phase in eukaryotes (mammals) after UV treatment. It involves semi-conservative DNA replication. Gaps (about 2,800 nucleotides in size) opposite UV-induced dimers in nascent DNA strands are joined by a process involving *de novo* synthesis without involvement of the parental strand via recombination as in bacterial → recombination repair.

spheroplast (*Tulasne* et al. 1959) — a bacterial cell (stable protoplast) the wall of which is incompletely removed or reduced to a template moiety. Frequently L form is used synonymously with s.

spherosome (*Dangeard* 1919; *Perner* 1952) — in plants, organelles that correspond to the → lysosomes of animal cells.

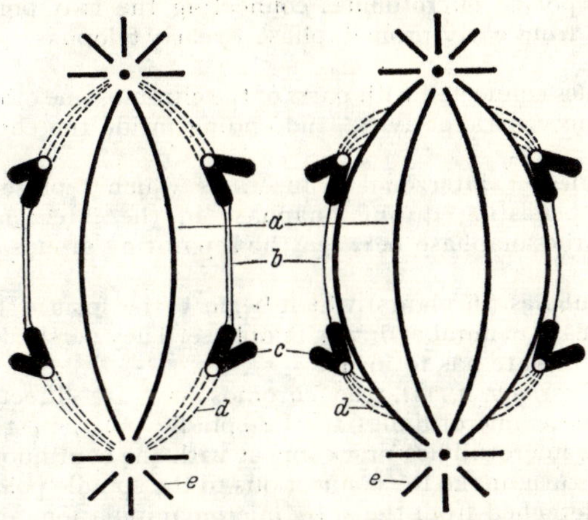

Fig. 90. The main types of spindle organization in anaphase. Left: The direct type of spindle organization; right: the indirect type of spindle organization (a = continuous spindle fibers; b = interzonal connections; c = chromosomes; d = chromosomal spindle fibers; e = spindle poles) (from *Schrader* 1944).

spindle — in eukaryotes, an ellipsoidal (bipolar) collection of fibers (s. fibers) visible during nuclear division (→ mitosis and → meiosis) and → cytokinesis (Fig. 90). Some spindles are long, thin and pointed at the poles, others are much shorter and may appear barrel-shaped. In the meiosis of some animals, each → bivalent has a s. of its own and the → mitotic apparatus as a whole consists of a bundle of these individual spindles. In some protozoa, the → centrioles are intranuclear and the whole mitotic apparatus is formed within the → nuclear envelope (intranuclear s.), which does not break down (premitosis). In flagellates and all higher organisms, the centrioles have become extranuclear and division is by eumitosis. The spindles of higher plants are anastral, i.e., they lack asters and centrioles.

Unipolar spindles are normally present at meiosis I of the → spermatogenesis of certain midges and scale insects. These spindles may have, in fact, two poles: one acuminate and the other diffuse and flared out. Multipolar spindles are characterized by more than two poles and have several → equators which intersect one another.

Chemically the s. contains protein, RNA, lipid, carbohydrate, and zinc; enzymes present include ATPase. Spindles remain intact after being isolated from the cells and *in vivo* can be pushed through the cytoplasm.

Besides → microtubules (4 to 10% of the s.volume), the s. area contains smooth and granular vesicles, membranes, ribosome-like particles, endoplasmic reticulum, amorphous background material, filamentous material associated with the microtubules, small circular bodies, and unstained clear areas around microtubules.

In general, most mitotic spindles are composed of various classes of microtubules or s. fibers (*Bajer* 1968):

1. **Centromere** (or kinetochore) **microtubules** going from the → centromere region towards the s. poles. They are present from early prometaphases of mitosis to early telophase (= chromosomal fibers).

2. **Continuous** (or interpolar) **microtubules** connecting the two polar regions of the s. observed from early prometaphase to early telophase (= continuous s. fibers).

3. **Neocentric microtubules** connected with parts of the chromosome other than the centromere (→ neocentric activity) and ending inside the chromosome.

4. **Interzonal microtubules** or interzonal connections which represent continuous microtubules persisting during anaphase in the interzonal region and found only at the anaphase between the separating groups of chromosomes.

5. **Phragmoplast microtubules** (in plants) which begin to be formed by middle anaphase and increase in number during telophase. They persist for a long time after the → cell plate has formed.

In the **direct s. type** (*Schrader* 1944), the chromosomes are directly connected by the centromere microtubules to the spindle poles, in the **indirect s. type,** centromere microtubules first connect with the continuous microtubules and these, in turn, make the connections to the spindle poles.

A chromosome can be detached from the s. by micromanipulation, and the chromosome reforms an attachment to the s. (→ centromere orientation). There are several reports of aperiodic bridges between s. microtubules.

spindle plaque — an electron-opaque structure associated with the → nuclear envelope at the poles of the intranuclear → spindle in Ascomycetes.

spindle poison (*Dustin* 1938) — any → mitotic poison affecting the formation or function of the → spindle and blocking karyokinesis at metaphase. The prototyp of a s. p. is colchicine (→ C-mitosis; C-meiosis).

spiralization (*Darlington* 1932) — → chromosome coiling.

split spindle — a → spindle type which is split into two parts and gives rise to a double metaphase plate. In certain Hemiptera, the autosomes are contained in one, the sex-chromosomes in the other (smaller) of the two metaphase plates which are rather widely separated from one another.

splitting (*Simpson* 1944) — one of the main modes of → evolution, consisting in the splitting of a population (species) into two or more groups of individuals isolated by geographical or ecological barriers. Within these separated groups, mechanisms of reproductive isolation may evolve and the populations split off from the parental population may eventually become new → species (→ speciation).

sporad (*Webber* 1933) — any group of normally four spores (tetrad) produced by the first and second meiotic divisions from each spore mother cell in higher plants. Instead of a tetrad, sporads with aberrant numbers of cells (monads, dyads, triads, pentads, etc.) may be produced under certain conditions (e.g., in haploids of normally diploid plants) due to irregular → meiosis.

spore — a reproductive cell in plants (→ germ cells). Spores are agametes and may be produced meiotically ("meiospores", "tetraspores") or mitotically ("gonidia"). In higher plants, meiospores (→ megasporogenesis; microsporogenesis) give rise to female and male → gametophytes.

Spory, sporogenesis, and sporogony are the terms used for the process of spore formation. Spory may be subdivided into (*Battaglia* 1955):

1. **Euspory**: spore formation by normal meiotic divisions.
2. **Aneuspory**: spore formation by irregular meiotic divisions.
3. Spore formation without meiosis.

a) **Apomeiotic spory** (= gonial or generative apospory): spores are spore mother cells which develop without the intercalation of meiotic divisions.

b) **Somatic spory** (= apospory or somatic apospory): spores represent somatic cells (e.g., sporophyte cells).

sporocyte — a spore mother cell (→ spore).

sporogenesis — = → spory.

sporogony — → fission.

sporophyte — the (normally) diploid, asexual generation (in → diplo-haplontic plants having an → alternation of generations) which usually produces → spores. In higher plants, the s. bears the haploid → gametophytes; in mosses and liverworts, the s. is smaller than the gametophyte and is borne by the latter.

sporulation — 1. the formation from vegetative cells of bacteria and other organisms of dry, metabolically inactive cells with thick surface coats (spores) which can resist extreme environmental conditions (→ bacterial sporulation);

2. in yeast the induction of → meiosis resulting in four haploid spores.

spory (*Battaglia* 1955) — the formation of → spores (= sporogenesis; sporogony).

spot test (*Ames* 1971) — a genetic system for detecting mutagens and carcinogens. The s. t. involves placing the agent at the center of a petri dish containing minimal medium that is seeded with → auxotrophic bacteria unable to grow (due to a deficiency in a metabolic pathway). Revertants from auxotrophy to prototrophy are identified as colonies in a ring around the area (spot) where the potential mutagen was deposited. By use of appropriate bacterial strains, the s. t. permits the characterization of mutagens inducing → base pair substitutions, → frameshift mutations or large → deletions. (→ rec-assay).

spreading effect (*Lewis* 1950) — → position effect.

stamen — the male organ of a flower (the androecium) consisting of a stalk or filament with an anther containing pollen. Flowers containing only stamens are called "staminate" (→ pistil).

standard type — an arbitrarily specified → genotype used as a basis of comparison for genetic investigations. The genes of the s. t. are usually symbolized by a (+), irrespective of their being → dominant or recessive. All genes deviating from those of the s. t. are designated by names or symbols (→ genetic nomenclature). Where possible, the most frequent → wild-type of the taxonomically typical subspecies of a species of the first thoroughly analyzed wildtype should be used as s. t.

stasigenesis (*Huxley* 1957) — in → evolution, all processes by means of which stability and persistence of types and organizational patterns (from the species up to the phylum) is achieved. S. gives rise to "grades", i.e groups of individuals similar in their level of organization which represent persisting units of → anagenesis.

stathmokinetic — of agents which will arrest cells in mitotic metaphase (→ mitotic poison).

stem cell — → germ cell.

stem cell spermatogonium — any of the → spermatogonia responsible for the maintenance of the → germ line in male animals.

stem-line (*Levan & Hauschka* 1953) — the most frequent karyotype (with respect to chromosome number and/or chromosome structure) of a tumor cell population. Other karyotypes present in at least 10% of the cell population under observation are called side-lines.

stereoisomeric — of molecules ("stereoisomers") that have the same structural formula but different spatial arrangements of dissimilar groups bonded to a common atom. Stereoisomers are identical in respect to many of their physical and chemical properties but show differences in their crystal structure, in the direction of rotation of polarized light, and in their ability to be used in enzyme catalyzed reactions (*Watson* 1965).

sterility — any complete or partial (→ semisterility) failure of an individual to produce functional gametes or viable zygotes ("gametic" and "zygotic s." of *Renner* 1929; "haplontic" and "diplontic s." of *Müntzing* 1930) under a given set of environmental conditions. It may concern both the male and female sexes, or either the male ("male s.") or female ("female s.") sex.

S. may be caused by environmental effects, extrachromosomal hereditary determinants ("cytoplasmic s."), incompatibility between endosperm and embryo in plants ("somatoplastic s." of *Bryant* 1935), genotypic defects, or segregation in hybrids, polyploids and aneuploids ("genic" and "chromosomal s."). Genotypic defects in the gametes may be due to segregation in hybrid parents, and, in zygotes, to unbalance in the hybrid genotype (→ hybrid sterility; segregational sterility).

In contrast to "true s.", parasterility or sexual → incompatibility is the inability of genetically like ("homogenic sexual incompatibility") gametes or nuclei to undergo karyogamy and thus to achieve fertilization. The same gametes may successfully unite with genotypically different partners giving rise to viable zygotes.

stickiness (*Beadle* 1932) — a sort of chromosomal "agglutination" of unknown nature which results in a pycnotic or sticky appearance of chromosomes. S. may give rise to → sticky adhesions between two or more chromosomes and to the formation of "sticky bridges" at anaphase when the chromosomes or chromatids associated in such adhesions move to opposite spindle poles. Sticky effects result from certain gene mutations and may be induced by the application of a wide variety of chemical and physical agents during the condensed state of chromosomes in mitotis and meiosis.

sticky association (*Price* 1956) — any association of two or more chromosomes at mitosis or meiosis due to nonspecific sticky adhesions rather than to homologous → chromosome pairing and chiasma formation (→ pseudobivalent; pseudomultivalent).

sticky end — = → cohesive end.

strain — 1. a group of related individuals distinguished by certain characters from other such groups;

2. the progeny of a single plant irrespective of the mode of pollination.

stringent (*Borek* et al. 1955) — → relaxed.

stroma — the nonlamellar ground substance of the → chloroplast (→ plastid) and → mitochondrion.

structural change (*Darlington* 1929) — any change in the structure of one (intrachromosomal s. c. or intrachange) or more (interchromosomal s. c. or interchange) chromosomes due to → chromosome mutations (deletion, duplication, inversion of a chromosome segment, interchange or reciprocal translocation of segments between chromosomes).

Such changes may be intraradial (= intrabrachial, paracentric) or extraradial (= heterobrachial, pericentric) with respect to chromosome arms; internal (= homosomal), fraternal and external (= heterosomal) with regard to chromosomes; eucentric (= symmetric) or dyscentric (= asymmetric) with respect to possession of a → centromere or to the direction of a segment in relation to the centromere (*Darlington* & *Mather* 1949).

Structural changes that do not give rise to deletions and duplications (inversions, interchanges) are termed euploid (*Muller* 1954) or balanced (*Darlington* 1937) as opposed to aneuploid or unbalanced structural changes connected with the loss (deletion) and duplication of chromosome segments.

"Primary structural changes" of chromosomes occur spontaneously and may be induced experimentally, while "secondary structural changes" (*Darlington* 1932) derive from primary ones by → crossing-over in homologous segments of paired chromosomes which differ in structure on one or both sides of the pairing segments, i.e., in structural hybrids heterozygous for a primary s. c.

structural gene (*Jacob & Monod* 1959) — any → gene (representing a unique segment of genetic material) that determines the primary structure (the amino acid sequence) of a polypeptide by → genetic trancription and → genetic translation of the genetic code contained in its nucleotide sequence. The primary product of a s. g. is → messenger RNA (→ operon; regulatory gene).

structural heterozygosity — heterozygosity for a chromosomal structural change (→ chromosome mutation). Heterokaryotypes characterized by one or more → heterozygous structural changes in the chromosome complement (karyotype) are called "structural heterozygotes" or "structural hybrids". S. h. usually entails a reduction in fertility (→ hybrid sterility) due to the production of genotypically defective gametes and unbalanced zygotes and thus reduces recombination and variation (→ chromosome polymorphism).

"Cryptic structural h." (*Stebbins* 1945) is heterozygosity for structural changes which are presumably so small that they are without influence on meiotic → chromosome pairing. They may give rise to more or less pronounced hybrid sterility. The existence of such cryptic structural changes in the heterozygous form may be inferred from "preferential pairing" (→ differential affinity) of chromosomes after polyploidization of the → heterokaryotype.

structural homozygosity — the structural identity of two chromosome regions, chromosomes, or chromosome sets, in contrast to → structural heterozygosity.

structural hybrid (*Darlington* 1929) — → structural heterozygosity; hybrid.

S-type position effect (*Lewis* 1950) — = cis-trans → position effect.

subandroecious — of primarily male plants with a few hermaphroditic or female flowers, as opposed to → subgynoecious.

subchromatid — any of the fibrillar subunits of a single → chromatid. Subchromatid aberrations are chromosomal structural changes (→ chromosome mutation) in which the subchromatids are the units of change. Like aberrations of chromatids and chromosomes, they may be of the → intra- or → interchange type (→ pseudochiasma).

subdioecious — of plant populations composed in the majority of unisexual plants but containing a few → monoecious individuals (→ dioecious).

subgynoecious — of primarily female plants with a few hermaphroditic or male flowers, as opposed to → subandroecious.

subhaploid (*Kiellander* 1942) — → haploid.

sublethal factor (*Johannsen* 1909) — → lethal factor.

subsexual (*Darlington* 1937) — → reproduction.

subspecies (*v. Wettstein* 1898; *du Rietz* 1930) — an aggregate of (local) breeding populations of a given → species occupying a geographic subdivision of the species range (= geographic → race). A s. usually differs from other similar breeding groups of the same species both taxonomically and with respect to certain → gene pool characteristics (such as the frequency or prevalence of certain genes: → gene frequency). Subspecies like → semispecies and species may be either allo- or sympatric. The small local populations of which a s. consists are called "microgeographic races" or "microsubspecies (*Huxley* 1940). A s. represented by widely separated, phenotypically identical populations is said to be "polytopic".

substitution — → chromosome substitution.

substitution line — a line in which one chromosome pair is transferred from the donor variety into the recipient variety (→ alien addition line; alien substitution line). S. lines are being used in determining the number and dominance relationships of the genes governing a → character, for the genetic analysis of quantitative characters, for determining the number of genes controlling a quantitative character on a single chromosome, and for mapping these genes (*Khush* 1973).

substitutional haploid (*Riley* 1960) — → haploid.

substitutional load — → genetic load.

subvital (*Hadorn* 1948) — of a gene or genotype which, when expressed, reduces the viability of the carrier to a level significantly below that of the → standard type (→ lethal factor).

subvital factor (*Hadorn* 1948) — → lethal factor.

supercoiled — of a DNA configuration in which a double-stranded DNA helix is arranged in larger whirls. S. DNA is characteristic of certain viruses.

superdominant (*Fisher, Immer & Tedin* 1932) — = → overdominant.

superfemale (*Bridges* 1914) — = → metafemale.

supergene (*Darlington & Mather* 1949) — a group of linked genes mechanically held together on a chromosome and usually inherited as a unit. If coadapted, such genes (which are not necessarily functionally related) may cooperatively produce some adaptive characteristic. Supergenes are preserved intact either by heterozygous structural changes in the chromosomes concerned (inversions or reciprocal translocation suppressing or reducing the amount of → genetic recombination via meiotic → crossing-over in the relevant segments) or by localized → chiasmata formation in the orgnism under observation.

superhelix density — the number of superhelical turns per ten base pairs in naturally occurring closed circular DNA (→ chromosome). Such DNA may differ in s. d. within the range from -2×10^{-2} for → mitochondrial DNA (from SV 40 transformed mouse cells in culture) to -5×10^{-2} for the DNA from the bacteriophage PM 2.

superinfection curing — the loss of the → prophage after infection of → lysogenic cells by heteroimmune phages having the same integration site as the resident prophage.

superinfection immunity (*Watanabe* 1963) —

1. in the case of transmissible bacterial → plasmids (→ sex factor), the reduction of the pasmid's ability to bring about its own transfer from one bacterial cell to another by conjugation when both members of a conjugal pair harbor isogenic or closely related plasmids. S. i. may be due to → entry exclusion and → plasmid incompatibility.

2. The inability of two viruses of the same kind to coexist stably in the same cell either due to entry exclusion or immunity proper, where either the resistent or the superinfecting genome is reproduced, but not both.

supermale (*Bridges* 1914) — a phenotypically male, abnormal (sterile) individual with fewer → X-chromosomes than half the sets of autosomes, in organisms (e.g., *Drosophila*) with an X-chromosome-autosomes balance type of → sex-determination. A *Drosophila* s. has three sets of autosomes and one X-chromosome, i.e., a "sex-index" of 0.33 as compared to normal males (two sets of autosomes, one X-chromosome) with a "sex index" of 0.5 (→ metafemale; superfemale).

supernumerary chromosome (*Longley* 1927) — = → B-chromosome.

superspecies (*Mayr* 1942) — a monophyletic group of entirely or essentially allopatric → semispecies that are too distinct to be considered as → subspecies of the same → species and not sufficiently distinct to be regarded as full species (*Grant* 1963) (→ syngameon).

super-suppressor (*Hawthorne & Mortimer* 1963; *Gilmore* 1967) — any allele-specific and locus-nonspecific, partially or completely dominant external → suppressor (in yeast). S.-suppressors result in what is called **s.-suppression,** i.e., a phenomenon in which the phenotypes of specific alleles at a variety of loci are simultaneously reversed to a → pseudo-wildtype. Dominant s.-suppression is expressed as the ability to read → nonsense codons and is caused by mutations in genes which code for → transfer RNA's (→ suppressor tRNA; nonsense suppressor). Super-suppressors are also called multisite suppressors.

supervital — a gene or genotype which, when expressed, is significantly more viable than the → standard type (→ heterosis).

suppression (*Sturtevant* 1920) — the restoration (partial or complete phenotypic repair) of a mutationally altered function of a genotype which may be due to:

1. → suppressor mutations located in a gene different from that involved in the primary mutation **(intergenic s.),** i.e., in a gene coding for some component of the translational apparatus (→ genetic translation), such as → transfer RNA (→ suppressor tRNA) or ribosomal proteins, which exert control over the fidelity of translation.

2. Suppressor mutations within the same gene located at some distance from the site of the primary mutation (intragenic s.) and causing an additional alteration of the mutant protein which compensates for the first. This allows the formation of a functional tertiary protein structure.

3. Non-Mendelian or cytoplasmic suppressors involving changes of → organelle DNA (such as → mitochondrial DNA of fungi). (→ missense suppression; nonsense suppression; frameshift suppression).

suppressor mutation (*Bridges* 1932) — a secondary mutation ("second-site mutation") that totally or partially restores a function lost due to a primary mutation which is located at a → mutational site different from the primary mutation. In contrast to → back mutation, s. m. does not result in a reversal of the original mutation. Frequently suppressor mutations do not lead to a complete → suppression or phenotypic repair in the mutants under observation, but only to a partial reversal toward wild-type activity of the double mutants (→ revertant). Since suppressor mutations occur as parts of double mutants, the s. m. can be separated from the original mutation by → genetic recombination.

Suppressor mutations may be classified as follows (*Gorini* & *Beckwith* 1966):

1. **From the genetic point of view** the s. m. may be located either in the same gene or cistron as the original mutation (intragenic or internal s. m.), within another gene in the same chromosome, or even in a different chromosome (intergenic, extragenic or external s. m.). A s. m. may be strictly allele specific, i.e., it suppresses only mutations which arise at a particular codon within a cistron, or it may be able to suppress equally well several or all mutations occurring at different codons within the same cistron ("cistron specific suppressors"), or mutations of the same codon occurring in different, unrelated cistrons.

Intragenic suppressors restore the wild-type phenotype by allelic interaction between the primary mutation and a compensatory secondary mutation at a site very closely linked to the original mutation. **Intergenic suppressors** restore the wild-type phenotype by genic interaction between linked or unlinked primary mutations and suppressor mutations at a site outside the gene in which the primary mutation took place. Such suppressors usually act on functionally unrelated genes and their mode of action, must involve some common step in the steps from DNA to protein, i.e., the code translating mechanism.

2. **From the point of view of their mode of action** suppressor mutations may be grouped as follows:

a) the second mutation may be in a gene whose mutated product opens an alternative pathway of deficient metabolic synthesis. As a result of a second mutation, an enzyme activity may appear which permits the synthesis of an intermediate whose synthesis, through the usual pathway, was blocked (→ genetic block) by the first mutation (*Davis* & *Woodward* 1962).

b) The second mutation may be in a gene whose mutationally altered product may substitute directly for the function of the product of the first gene. A second enzyme (or protein) may mutate to a broader specificity (or function) encompassing that of the missing first enzyme (or protein).

c) The second mutation provides cytoplasmic conditions which affect the structure of the finished product of the first mutation. The condition may stabilize an enzyme which owing to a structural change is unstable in the wild-type cytoplasm.

d) The second mutation introduces a second alteration within the same peptide chain (a second amino acid substitution) which reintegrates, completely or partially, the protein function, with reappearance of the wild-type or a quasi wild-type phenotype (*Suskind* & *Kurek* 1959).

e) The second mutation reintroduces the correct code meaning, by frameshift (→ frameshift mutation), within the same cistron in which the codon reading-frame was altered by addition or deletion of a base pair.

f) The second mutation concerns one of the factors controlling the mechanism of transfer of genetic information from DNA to protein, i.e., repair of the mutational defect owing to an alteration in → genetic translation of the genetic information (*Champe* & *Benzer* 1962).

Categories a) to c) act indirectly to circumvent, rather than to repair directly the primary genetic lesion **("indirect suppression")**. Categories d) and e) represent intragenic suppressors and are characterized by the fact that (barring the extreme case of intracodon suppression) the primary and suppressor mutations are both transcribed (→ genetic transcription) into one gene product. This gives rise to two amino acid substitutions (or possibly more in case of category e)), in the suppressed mutant protein with respect to the wild-type protein. Suppression according to category f) is due to meaning alterations of a codon and are not themselves transcribed in the suppressed mutant protein as a second amino acid substitution. Suppressors of this type are always intergenic and may be designated as "informational suppressors" (*Gorini* & *Beckwith* 1966).

Several types of informational suppressors are known to affect → transfer RNA structure or ribosomal proteins which control the fidelity of genetic translation (→ mistranslation). → Nonsense and → missense suppressors have been described which involve tRNA species with altered coding specificity. In most cases this is due to changes in the anticodon which allow the → suppressor tRNA to translate a nonsense or missense codon.

After separation of intragenic suppressors from the primary mutation by → genetic recombination, such suppressors, as a rule, appear as nonfunctional mutants. Intergenic suppressors usually only partially compensate for the phenotypic effect of a primary mutation and this may be distinguished from true → back-mutation. Another means of differentiating them from back-mutations is by back-crossing to the wild-type and observing the segregation ratios produced.

suppressor tRNA (*Capecchi* & *Gussin* 1965) — any → transfer RNA species which, due to mutation in a transfer RNA gene, shows altered coding specificity and is able to translate (→ genetic translation) → nonsense or → missense codons. **Nonsense** and **missense suppressor tRNA** is frequently modified in the anticodon loop or in an adjacent region. Some mutant tRNAs with temperature sensitive suppressor activity have been shown to have nucleotide substitutions in variable regions outside the anticodon.

Bacterial suppressor tRNAs are usually codon-specific rather than gene specific.

Frameshift suppressor tRNA compensates for → frameshift mutations by reading non-triplet codons. Such tRNA molecules may carry an additional base in their anticodon, such as CCCC instead of CCC in the wild-type tRNA. This would allow them to translate frameshifts which involve the change of a GGG codon to a GGGG quadruplet. Or an additional base in the tRNA anticodon may serve as a spacer such that the tRNA occupies an unusually large space over the mRNA message and so suppresses +1 frameshifts.

tRNA suppressor mutations are lethal if they result in the inability to translate a normal (non-mutant) codon. Under such circumstances only dispensable tRNA species (such as → isoacceptor tRNAs) can be converted to suppressor tRNA or another, non-mutated tRNA molecule must be synthesized and rescue the function abolished by the suppressor mutation.

Frequently, suppressor tRNAs, besides their major decoding specificity (including → wobble), also possess some minor decoding properties which result in mistranslation.

supraoperon control (*Roth* & *Nester* 1971) — The co-ordinate control of a → gene cluster with multiple → operator and → promotor sites which consists, in essence, of several interacting → operons controlling different biochemical pathways.

surface coat — a layer of material outside the limiting membrane (→ cell membrane) of the animal cell. It assists the cell in either adhering to its substratum or to other cells and in the binding of particles (from ions to viruses) for subsequent → pinocytosis. Furthermore the s. c. plays a role in the determination of the antigenic identity of the cell; the reaction of antibodies with the living cell occurs in the first place at its surface (→ cell wall).

switch gene — a gene which causes the → epigenotype to switch to a different → developmental pathway.

sympatric (*Poulton* 1903; *Mayr* 1942) — of two or more → populations occupying the same geographic area, or of a population existing in a breeding condition within the cruising range of individuals of another population, as opposed to → allopatric. S. populations are not necessarily in physical contact, or at least reproductive contact; they may remain completely separated by different habitat preferences, or by different habits.

S. speciation is species formation in the absence of geographic → isolation, by acquiring mechanisms of reproductive isolation within a → deme (*Mayr* 1963).

symplast (*Hanstein* 1880) — a multinucleate body of protoplasm which according to its origin may be a → plasmodium or a → syncytium.

synapsis (*Moore* 1895) — meiotic → chromosome pairing (= syndesis).

synaptene (*Winiwarter* 1900) — = zygotene (→ meiosis).

synaptic (*Riley* & *Law* 1965) — ref. to the processes and forces of meiotic → chromosome pairing or synapsis. "Synaptic genes" are those which influence the extent of meiotic pairing (→ desynaptic; asynaptic).

synaptic junction (*Palay* 1958) — a specialization of the membrane of two neurons in the region where the neurons make contact. The s. j. consists of highly differentiated parts of the pre- and postsynaptic plasma membranes joined by a synaptic cleft. The presynaptic specialization consists of a hexagonal array of dense projections extending into the cytoplasm. The postsynaptic specialization varies with the type of synapse and most often includes a prominent electron-opaque structure (postsynaptic density) located on the cytoplasmic side of the postsynaptic membrane just within the synaptic area.

synapton (*Whitehouse* 1969) — = → synaptonemal complex.

synaptonemal complex (*Moses* 1958) — as seen in the electron microscope a regularly occurring, coplanar set of proteinaceous parallel strands (usually three), coaxial to chromosomes involved in meiotic → chromosome pairing (Fig. 91).

The presence of the s. c. is restricted to meiotic cells and prerequisite, but not alone sufficient for → chiasma formation (and, hence, → crossing over). The s. c. shows a basic structural similarity in very diverse eukaryotic organisms: two dense **lateral elements** flank a relatively unstructured area (the central space), containing a **central element** of variable prominence. The dimensions and spacing of the various components may vary according to species. The lateral elements vary in diameter from 30 to 65 nm, the central element between 12 and 50 nm, and the central space between 65 and 120 nm. The total width of the s. c. ranges between 160 and 240 nm,

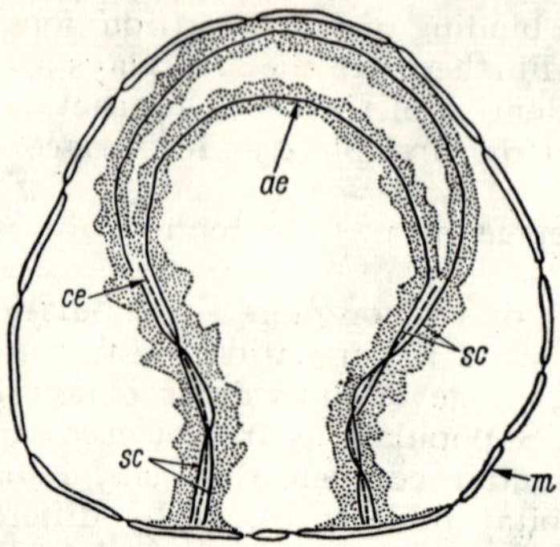

Fig. 91. Schematic representation of a meiotic pairing configuration (prophase of a spermatocyte nucleus of the salamander *Plethodon cinereus*) as observed electron microscopically (ae = axial complex of the homologues; sc = synaptonemal complex; ce = central element; m = nuclear membrane) (after *Moses* 1960 from *Bresch* 1964).

i.e., is less variable than the component parts. The central element of the s. c. is joined to the lateral elements by thin **transverse filaments** that may serve to align and join homologous chromosomes via the lateral elements, and to stabilize them at a fixed separation distance. The transverse filaments are fine (1.5 to 2.0 nm), straight, nonlooping protein fibers that originate in the lateral elements, have a remarkable constant length and interdigate in a complex fashion in the central region, where, together with additional protein material, they form the central element (*Socari* & *Moses* 1973).

The s. complexes of most organisms terminate at the → nuclear envelope. The attachment points may be polarized or may show no discernable pattern of arrangement.

The s. c. is probably formed by self-assembly of specific protein molecules a major component of which is a basic, arginine-rich protein. Current ideas on structure and function of the s. c. propose that the s. c. forms a structural framework around which the chromatin is arranged in such a way that the mass of chromatin, which is not directly involved in → genetic recombination, is sequestered on the outer surface of the lateral elements and only selected stretches of chromatin penetrate the lateral element and enter the central space where they engage in molecular pairing (effective pairing) and genetic exchange.

Usually the s. c. is set free from the chromosome at the diplotene stage of meiosis and either becomes dispersed or aggregated into stacks, socalled **polycomplexes.**

synclone — → caryonide.

syncyte (*Levan* 1942) — a → polyploid or multinucleate cell. The production of syncytes may take place by:

1. Inhibition of → cytokinesis in mitotic divisions. The resulting syncytes may be di- or multinucleate or may contain one polyploid nucleus if a → restitution nucleus is formed.
2. Migration of the nucleus from one cell to another. The phenomenon is called "cytomictic formation of a s." (*Price* 1956).
3. Fusion of cells after dissolution of the cell membranes and cell walls.

syncytium (*Haeckel* 1894) — = → syncyte.

syndesis (*Haecker* 1907) — → synapsis.

syndiploid (*Strasburger* 1907) — of cells which possess twice the diploid number of chromosomes owing to fusion of mitotically produced daughter nuclei or inhibition of the → spindle mechanism. The phenomenon is called "syndiploidy" (= tetraploidy).

syndrome — a group of specific characters (symptoms) characteristic of a certain (abnormal) genetic condition. An "inheritable" or "genetic s." may be transmitted to the offspring as a unit.

synergid — in higher plants, any of the (two) haploid nuclei found with the egg at the micropylar end of the → embryo sac (→ macrosporogenesis).

syngameon (*Lotsy* 1916; *Grant* 1957) — a group of sympatric → semispecies (→ superspecies).

syngamic (*Haecker* 1902) — → sex determination.

syngamodeme (*Gilmour & Heslop-Harrison* 1954) — = → comparium (→ deme).

syngamy — sexual → reproduction: the union (during → fertilization) of male and female gametes, resulting in karyogamy and zygote formation (→ apogamy).

syngen (*Sonneborn* 1957) — a reproductively isolated population representing an evolutionary unit in Ciliates. The term s. formally corresponds to a → species, but since species are defined by their → gene pools and common gene pools are absent in Ciliates, syngens are not species in the true sense.

syngenote (*Morse, Lederberg & Lederberg* 1956) — a bacterial → merozygote which, in addition to its own → genome (referred to as the endogenote), contains an additional genome fragment (referred to as the exogenote). A s. is partially diploid (hyperploid) and may be → heterogenotic or homogenotic.

syngonic — of organisms producing male and female gametes in the same sexual organ.

synizesis (*McClung* 1905) — a clumping (into a more or less dense knot) of the chromosomes in early prophase of first meiotic division. S. may be a "normal" feature of → meiosis in some organisms and can be induced artificially through fixation or other kinds of cellular damage.

synkaryon — the → nucleus of a → zygote, resulting from the fusion (karyogamy) of the nuclei of the female and male gamete during fertilization. A s. consisting of the genomes of two different organisms (interspecific, intergeneric or even interphylar) may be produced by somatic → cell hybridization and is called a heterokaryon (*Harris* 1970). In some fungi, the production of synkaryons characterizes the end of the dikaryophase (→ synkaryotic).

synkaryophyte — = → sporophyte.

synkaryotic (*Buller* 1941) — of diploid cells in Asco- and Basidiomycetes, as opposed to dikaryotic cells (→ dikaryon). S. cells are produced by "synkaryosis" which takes place at the end of the dikaryophase and consists of the pairwise fusion of haploid nuclei of opposite → mating type.

synoecius (*Correns* 1928) — of haploid or diploid organisms producing both male and female gametes. The term includes → monoecious and → hermaphroditic individuals.

synonymous mutation (*Sonneborn* 1965) — a class of → gene mutations (samesense mutations) which represent → base-pair substitutions in DNA that do not result in the substitution of a different amino acid into the encoded protein. This is due to the degeneracy of the → genetic code. The codon produced by mutation is called a synonymous codon or synonym.

syntelic (*Bauer, Dietz & Röbbelen* 1961) — of the → centromeres of the two chromatids of each chromosome if they are oriented to the same spindle pole at first meiotic division. Orientation to different poles is called amphitelic.

syntenic (*Renwick* 1971) — of the two genetic loci presumed to be linked to the same chromosome, whether or not → linkage has been demonstrated, as opposed to asyntenic loci which are linked to different chromosomes.

The term synteny may be used when referring to linkage relationships based upon analysis after somatic → cell hybridization, i.e., determination of linkage relationships by the results of correlated loss or retention of markers in hybrid cells. If concordant segregation of markers occurs, then linkage may be presumed; if discordant segregation occurs, then linkage can be denied.

synthetic lethal (*Dobzhansky* 1946) — → lethal factor.

syntrophic (*Davis* 1950) — of any two bacterial strains, → auxotrophic for different growth factors which are able to cross feed each other and to grow together on minimal medium. The phenomenon is called syntrophism or cross-feeding and is found with many bacterial mutants.

systematic pressure — any of the nonrandom evolutionary pressures; → selection, → mutation, and → migration (→ evolution).

T

tachygenesis — a shortened or accelerated → embryonic development which omits one or more developmental stages.

tachytelic (*Simpson* 1944) — ref. to → evolution at a rate distinctly higher than a standard or → horotelic rate (→ bradytelic). Tachytelic evolutionary rates represent exceptions characteristic of certain relatively short periods of time. The factors causing t. evolution are assumed to be environmental in nature (changes in geological environment, transposition of the evolving group of individuals into a new environment).

tandem duplication — → duplication.

tandem fusion (*White* 1957) — the fusion in the course of chromosome structural changes of two acrocentric chromosomes "end-to-end" with loss of one centromere, so as to give rise to a double-length acrocentric. Alternatively, the fusion of an acrocentric, which lost its centromere, with the end of one arm of a metacentric chromosome. Both types of t. f. played some role in the evolution of the karyotype of certain organisms (→ centric fusion; translocation).

tandem inversion — → inversion.

tandem repeat — = tandem → duplication.

tandem ring (*Novitski* 1954) — → X-chromosome.

tandem satellite (*Taylor* 1926) — two → satellites of the same chromosome, separated from each other by a secondary → constriction.

tautomeric shift — a reversible change in the localization of a proton in a molecule thus changing the chemical properties of the molecule (→ gene mutation) (*Watson* 1965).

taxon (*Rickett* 1958) — a taxonomic group of individuals (population) reconized as a formal unit at any level of a hierarchical classification. The individuals possess certain common characters assumed to be controlled by identical hereditary determinants (genes).

T-chromosome (*Kattermann* 1939) — a monocentric chromosome whose terminal segments (T-ends) show → neocentric activity at meiosis.

tectine (*Mazia & Ruby* 1968) — any of a class of similar proteins which form various eukaryotic structures, such as → microtubules, → microfilaments, and membranes. Tectines show resemblances to each other in, e.g., amino acid composition, and the molecules of each t. specifically aggregate together.

telocentric (*Darlington* 1939) — of chromosomes or chromatids with a terminal → centromere. Such chromosomes may arise by → centromere misdivision or breakage induced within the centromere region. Usually telocentrics are unstable and are either eliminated in the course of a few cell divisions or transformed into so-called → isochromosomes. This instability is assumed to be the reason for the absence or rarity of telocentrics in the chromosome complements of wild species.

telochromomere — any → chromomere localized at the end of a chromosome arm (→ telomere).

telochromosome — a chromosome with a terminally located → centromere.

teloisodisomic (*Kimber & Sears* 1968) — a cell or individual in which one chromosome pair is missing but a telocentric for one arm of the missing pair and an → isochromosome for the same arm are present (→ monotelomonoisosomic).

telolecithal — of egg cells in which the yolk (reserve material) is accumulated towards one end, referred to as the vegetal pole (→ isolecithal).

telomere (*Muller* 1940) — a term used to denote the natural unipolar chromosome ends in eukaryotes. They cannot normally be transposed (by chromosomal structural changes) to intercalary positions in the chromosomes. Their absence (deletion) usually impairs normal chromosome behavior (→ bridge-breakage-fusion-bridge cycle). Any product of a chromosomal structural change usually requires two telomeres and one → centromere in order to survive.

Telomeres have a compound structure, a special cycle of division, and a tendency for nonhomologous association at first meiotic prophase. Like centromeres, telomeres can show active mobility on the spindle and are then referred to as → neo-centromeres. There are indications that homologous telomeres are attached together or in close association on the nuclear envelope. This attachment is assumed to provide a ready basis for the initiation of → chromosome pairing since homologous distal regions of chromosomes would be brought in close proximity and pairing could be initiated in such regions (*Sved* 1966).

According to *Lima-de-Faria* and *Sarvella* (1958) a t. consists of two distinct regions referred to as the protelomere and the eutelomere. The **protelomere** is a terminal deep-staining structure with sharp limits, constituted (usually) of one to three large → chromomeres and deep staining fibrils. The **eutelomere** is a weakly staining subterminal segment adjacent to the protelomere. It is well delimited on the outer but not always so on the inner side and consists of one to two minute chromomeres and transparent fibrils. One complete t. may be composed of as many as eight differentiated segments. Parts of this complex structure may be lost owing to deletion, without impairment of its main function.

Electron microscopic studies demonstrate that the t. contains masses of irregularly folded chromatin fibers. Each varies around 23 nm in diameter. The fibers rarely terminate at the chromosome ends, but loop back into the body of the chromatid.

There is evidence that at least some telomeres contain → repetitious DNA.

telomitic (*Carothers* 1917) — = → telocentric.

telophase (*Heidenhain* 1894) — → mitosis, meiosis.

teloreduplication (*Hsu & Moorhead* 1956) — → interreduplication.

telosome (*Endrizzi & Kohel* 1966) — a → telocentric chromosome.

telotrisomic (*Kimber & Sears* 1968) — an individual deficient in one chromosome but which has a → telocentric chromosome and an isochromosome for the same arm of the missing chromosome (→ diiso-compensating trisomic).

temperate (*Jacob* et al. 1953) — of bacteriophages that can provoke one of two responses upon infection of a host cell: 1. the **lytic response** in which the bacterial cell lyses and releases new virus particles, or 2. the → **lysogenic response** in which the cell survives with the viral DNA associated with the bacterial chromosome and replicates with it in the → prophage state. Whether the lytic or lysogenic pathway is followed is influenced by various phage functions (repression of viral functions required for the lytic response) and the prevailing physiological conditions of the recipient cell (→ coimmune; heteroimmune). A **virulent phage** can provoke only the lytic response.

The relationship between a t. phage and the host bacterium is characterized by three general properties: 1. each prophage has a specific location (attachment site) on the bacterial chromosomes which is different for different t. phages. 2. The prophage confers upon the bacterium immunity against a lytic response upon superinfection by the same phage. 3. A bacterium harboring a prophage still carries the potentiality for producing a lytic response and release of virus particles (→ induction).

Defective t. phages (*Romig* 1968) are those which form particles with all the morphological attributes of normal phage but are incapable of replication.

temperature-conditional (*Davis* 1950) — of temperature sensitive conditional lethal mutations than can be perpetuated at the permissive temperature. The lethal defect may be studied at the non-permissive temperature. This type of conditional lethal mutation is usually due to → missense mutation. → Nonsense and → frameshift mutations are more likely to abolish function at all temperatures and to lead to absolute lethality in the absence of a → nonsense or → frameshift suppressor. The defect in t.-c. mutations is usually attributable to a protein which is non-functional at a high (non-permissive) temperature, but can function normally at lower temperatures.

template — a macromolecular mold for the synthesis of another macromolecule. Through template processes a limited number of building blocks (e.g., four → nucleotides, twenty → amino acids, etc.) are polymerized into macromolecular structures whereby the actual arrangement of the building blocks is uniquely determined by a preexisting one, which is identical, complementary, or otherwise related to the newly synthesized macromolecule. The product of a biosynthetic template process is predetermined by the uniqueness of the template. Replication of → deoxyribonucleic acid (DNA), → genetic transcription of DNA into a number of ribonucleic acid (RNA) species, and → genetic translation of RNA into polypeptides are the best known template processes (*Pontecorvo* 1966).

template RNA — = → messenger RNA.

temporal mapping — → replication map.

T-end (*Prakken* & *Müntzing* 1942) — → T-chromosome.

teratogen — any (teratogenic) agent that produces or increases the → incidence of congenital malformations in a population. The best known types of teratogens in man are viruses, radiation, and drugs. To act as a t., an agent must affect some specific metabolic process in the developing embryo. Most teratogens act during a rather limited period at the time the embryo is undergoing rapid organogenesis. The extent of damage produced varies with the timing of teratogen application.

terminal — ref. to chromosome or chromatid segments terminal in position, as opposed to intercalary segments.

terminal affinity (*Darlington* 1932) — the forces (of unknown nature) by which chromosomes are held together end-to-end from diplotene to metaphase I of → meiosis, due to the inhibition of → chiasma terminalization at the chromosome ends, or are brought together in this way at metaphase in special cases (*Darlington & Mather* 1949).

terminal association — nonspecific → chromosome pairing which results in nonchiasmatic end-to-end association of chromosomes.

terminal deoxynucleotidyl transferase (*Bollum* 1960) — an enzyme able to randomly (without template) add monodeoxyribonucleotides to any 3′-OH end of single-stranded DNA primers and gaps in double-stranded molecules. The enzyme activity is separate from that of → DNA polymerase and is found exclusively in vertebrate thymus, bursa, and leukemic cells.

Recently it has been found that oligoribonucleotides are accepted by t. d. t., giving rise to the synthesis of polydeoxynucleotides covalently linked to the oligoribonucleotide. Possibly, the enzyme may be part of a larger DNA-synthesizing enzyme complex (→ DNA polymerase) which disintegrates during enzyme purification.

terminalization (*Darlington* 1929) — → chiasma terminalization.

terminalization coefficient — → chiasma terminalization.

terminal redundancy (*Streisinger* 1964) — the presence of identical base pair sequences at the two ends of the linear DNA molecule (→ chromosome) of certain bacteriophages. T. r. may be of variable length, ranging from 0.5 to 3% of the phage chromosome.

Terminally redundant chromosomes may also be → circular permutations of each other, i.e., contain the same set of genes arranged in the same order and related to each other as though each had once been a circle and the circles had been opened by randomly distributed breaks. If the terminally redundant sequences contain different alleles of one or more genetic loci, the chromosome is heterozygous for these loci. Such phage heterozygotes are referred to as "terminal redundancy HETs".

terminal riboadenylate transferase — any enzyme (TrT) which catalyzes the transfer of adenylate residues from ATP to the 3′-OH group of certain polyribonucleotides in the presence of Mn^{2+}. Enzymes of this kind may be involved in the processing of → heterogeneous nuclear RNA to → messenger RNA in eukaryotic cells.

terminase system (*Campell* et al. 1969) — an enzymatic maturation system of bacteriophage lambda which produces mature lambda chromosomes by cleavage of multichromosomal lengths of lambda DNA (ter system). Mature chromosomes would be generated by ter cleavage at two cohesive end sites (cos sites) to generate unit length molecules with → cohesive ends (→ integration system; prophage).

termination — in protein biosynthesis (→ genetic translation), the incorporation of the last amino acid into a nascent polypeptide chain and the release of the complete chain from the ribosomes and messenger RNA (→ release factor; terminator codon).

termination factor (*Ganoza* 1966) — = → release factor.

termination signal — in some phages, in which initially early genes are transcribed by a bacterial polymerase, a protein factor which halts polymerase. Subsequently an anti-termination signal (protein) is synthesized to allow transcription beyond these signals.

terminator — the mRNA termination site at the end of each → operon.

terminator codon — in prokaryotes and eukaryotes, any of the specific → codons (UAA, UAG, and UGA in mRNA) which direct polypeptide chain termination during → genetic translation. T. codons are recognized by → release factors. UAA (ochre) is probably the chief t. c., while UAG (amber) and UGA (opal) provide accessory fail-safe signals to ensure that termination never fails. Both UAA and UAG are t. codons that lead to 100% efficient translational termination in bacteria. UGA is seldom more than 98% efficient and in a "leaky" t. c., i.e., a terminator apparently misread as an amino acid codon at a low but significant rate (even in non-suppressing strains of bacteria: → suppressor mutation). This results in polypeptide chain elongation beyond the t. c. (read-through); the amount of misreading appears to vary with the bacterial strain.

Premature polypeptide chain termination may result if a mutation occurs in a gene so that a triplet in the correct phase on the mRNA becomes a t. c. (→ nonsense mutation; nonsense suppression; reinitiator).

test cross (*Bridges* 1934) — a → cross of a double or multiple heterozygote (hybrid) with the corresponding double or multiple recessive (nonhybrid) in order to estimate the → linkage relationships of the gene loci in question.

tetraallelic (*Atwood* 1944) — → monoallelic.

tetrad (*Němec* 1910) — 1. the four chromatids of any → bivalent in the first division of → meiosis (*Němec*).

2. A quartet of cells (each with reduced chromosome number) resulting from a single meiotic event (→ meiosis) which takes place in a → meiocyte or mother cell. In some organisms (certain fungi, bryophytes and algae), all four meiotic products may be recovered and analyzed genetically (→ tetrad analysis). Among such organisms the immediate products of meiosis (referred to as gones) undergo 1 or 2 subsequent mitoses, giving rise to 8 or 16 products which represent exact replicates of the original four.

Tetrads (or their derivatives) may be ordered (in *Neurospora* and some yeasts) or unordered (in most organisms).

1. **Unordered tetrads:** With two segregating gene pairs (A/a and B/b) and the parental combinations AB and ab, three segregation patterns are possible. Tetrads of the type AB/AB/ab/ab (no recombinants) are referred to as **"parental ditype"** (PD), tetrads of the type AB/Ab/aB/ab (two recombinants) are referred to as **"tetratype"** (TT), and tetrads of the type Ab/Ab/aB/aB (four recombinants) are called **"non-parental ditype"** (NPD).

These three types of tetrads arise (by not more than 2 crossovers between the loci A and B) as follows (Fig. 92): PD due to no crossing-

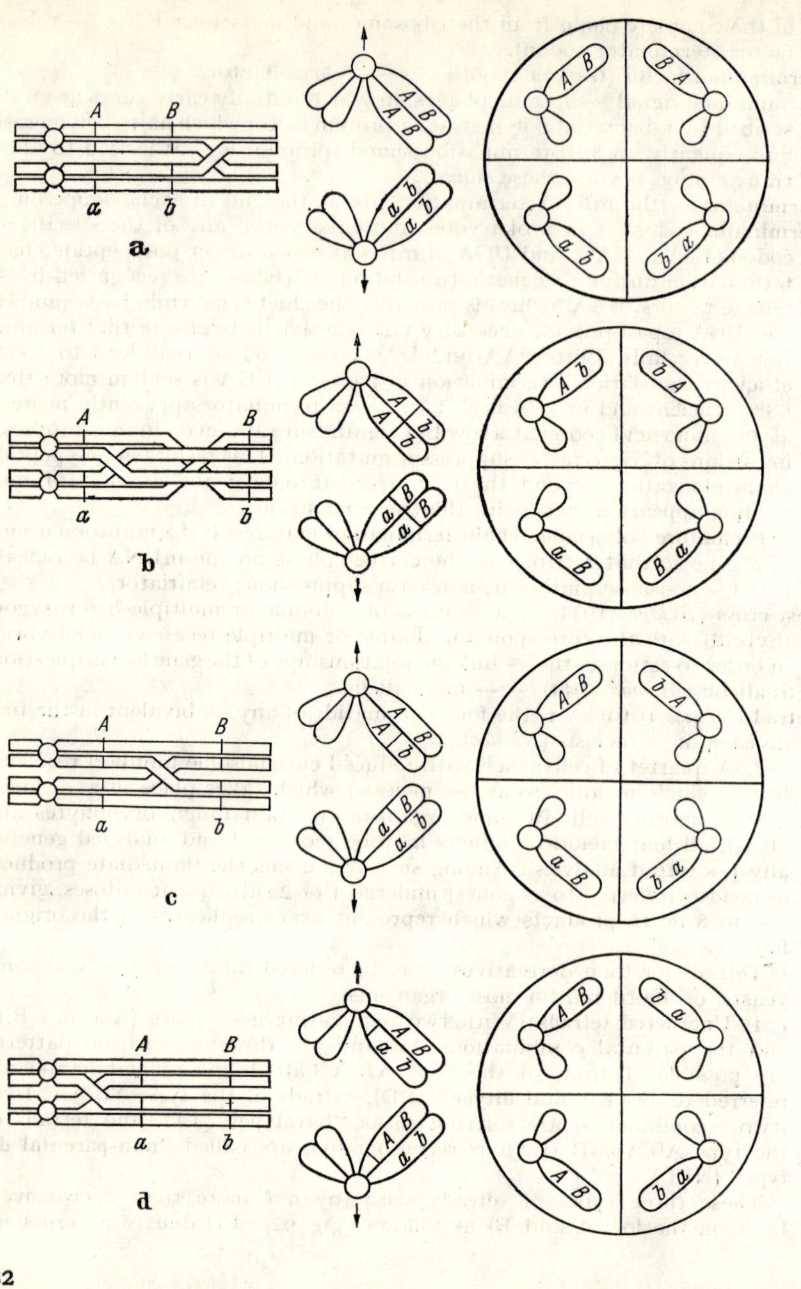
A
B
a
b
a
b
c
d

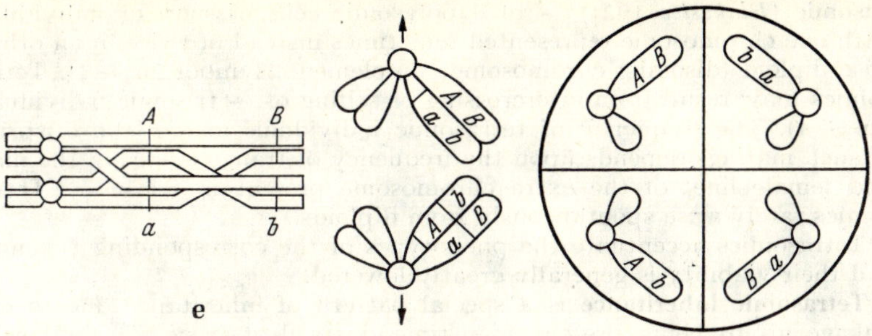

Fig. 92. Different types of tetrads resulting from a single meiotic event which takes place in a meiocyte with two segregating gene pairs (Aa and Bb) belonging to the same linkage group (left: meiotic pairing and crossing-over; middle: anaphase I; right: the resulting quartet of haploid cells). a) "parental ditype" tetrad in the case of prereduction of both gene pairs; b) "non-parental ditype" tetrad in the case of prereduction of both gene pairs (four-strand crossing-over); c) "tetratype" tetrad in the case of prereduction of gene pair Aa and postreduction of gene pair Bb; d) "parental ditype" tetrad in the case of postreduction of both gene pairs; e) "tetratype" tetrad in the case of postreduction of both gene pairs (three-strand crossing-over).

over or two-strand double crossovers between A and B; NPD due to two crossovers of the four-strand type between A and B (this type of t. is rare for linked markers); TT due to either a single crossover or a three-strand double crossover between A and B.

By means of tetrad analysis the occurrence of linkage between two markers may be determined. If there is → linkage, the NPDs will be rare relative to the other two tetrad types. If the markers are unlinked, the NPDs and PDs will be equal in number, since both types result from random → assortment of the chromosomes (with unlinked markers, the TTs result from crossing-over between either A and its → centromere, or B and its centromere). Recombination frequency between A and B equals $^1/_2$TT + NPD divided by the total number of tetrads.

2. **Ordered tetrads:** The meiotic products are ordered and the order indicates the pattern of meiotic events. The analysis of ordered tetrads may be readily used to obtain → recombination frequencies both between different genes and between genes and their centromeres. By classifying the tetrads as PD, TT, and NPD with respect to each gene pair, differentiation among two-, three-, and four-strand double crossing-over can be made.

tetrad analysis — the genetic analysis of all the products of a single meiotic event in organisms with (ordinarily) cytologically observable chromosomes (eukaryotes) and a standard → meiosis. T. a. is possible in those organisms in which the four products (or their derivatives) of a single nucleus which has undergone meiosis are grouped together (→ tetrad).

tetraploid (*Němec* 1910) — of → auto- or → alloploid cells, tissues, and individuals with four chromosome sets in their nuclei (symbol 4n).

tetrasomic (*Blakeslee* 1921) — of→ polysomic cells, tissues, or individuals with one chromosome represented four times instead of twice in an otherwise diploid (disomic) chromosome complement (symbol 2n + 2). Tetrasomics may result from intercrossing or selfing of → trisomic individuals (2n + 1). The frequency of tetrasomic individuals among the offspring of such matings depends upon the frequency of transmission (in the male and female line) of the extra chromosome present in trisomics. Tetrasomics rarely arise spontaneously from diploids.

Tetrasomics accentuate the phenotypes of the corresponding trisomics and their viability is generally greatly lowered.

Tetrasomic inheritance is a special pattern of inheritance due to one linkage group being present four times instead of twice. The following segregation ratios, resulting from a meiotic two-by-two distribution of the four homologous chromosomes, is theoretically expected for a tetrasomic genotype AAaa:

Gametes	1 AA	4 Aa	1 aa
1 AA	1 AAAA	4 AAAa	1 AAaa
4 Aa	4 AAAa	16 AAaa	4 Aaaa
1 aa	1 AAaa	4 Aaaa	1 aaaa

With A completely → dominant over a, a phenotypic segregation ratio of 35A:1a (→ autoploid) is expected.

Compensating nullisome-tetrasome combinations are allotetraploid karyotypes (→ alloploid) in which a certain tetrasome compensates more or less completely for a certain nullisome (→ nullisomic), indicating that the chromosomes under observation are reasonably homologous. Nullisome-tetrasome analysis is a powerful tool for genome analysis in segmental allopolyploids.

tetraspore (*Renner* 1916) — → germ cell.

T factor — = → elongation factor or transfer factor.

theletoky (*v. Siebold* 1871) — → parthenogenesis.

thelygenous — producing exclusively or predominantly female offspring, as opposed to arrhenogenous.

thermal denaturation — → denaturation.

thermal recovery — → liquid-holding recovery.

threshold character (*Dempster & Lerner* 1950) — a term used for those phenotypic → characters whose segregating distributions are phenotypically discontinuous but whose inheritance is multigenic like that of continuously varying (quantitative) characters. The discontinuous segregations of such "quasi-continuous characters" (*Grüneberg* 1952) result from threshold effects, i.e., the characters in question have an underlying continuity with a threshold which imposes a discontinuity on their visible expression. Threshold characters may segregate into many discontinuous phenotypic classes; those that segregate into two classes only are also called "all-or-none characters".

Thomas circle (*Schachat & Hogness* 1973) — a double-stranded DNA fragment from eukaryotic → chromosomes which can form circular structures after treatment with exonucleases (→ nuclease). The formation of T. circles is assumed to be due to the presence of → repetitious DNA sequences in the fragments.

thylakoid (*Menke* 1961) — the morphological unit of the photosynthetic apparatus in → chloroplasts (Fig. 73). The t. consists of lamellae (membranes) which are joined to form closed sacs of variable size in which the photochemical reactions take place. T. membranes are assembled from lipids, apparently synthesized within the chloroplast, and from polypeptides synthesized both inside and outside the → plastid. In higher plants, the thylakoids are formed into dense stacks called → grana. In prokaryotes, the thylakoids are of various shapes and usually free in the cytoplasm.

thymineless death (*Cohen & Barner* 1954) — the loss by a thymine-requiring auxotrophic strain of bacteria of the ability to multiply after a period of thymine starvation. Thymineless death is assumed to result from two processes: 1. the induction of an → episome by thymine starvation and subsequent addition of thymine, and 2. a lethal effect of thymine starvation which is independent of episome induction. There are indirect indications that DNA may be altered under thymineless conditions, including a decrease in messenger RNA synthesis, loss of transforming ability of DNA, difficulties of DNA extraction from thymineless cells and a decrease in priming ability for RNA polymerase by DNA from thymineless cells. The last effect may be due to a heat stable substance produced during thymine starvation, possibly associated with DNA and causing a decrease in the ability of DNA to serve as template for polymerase.

tight junction (*Farquhar & Palade* 1963) — a beltlike region (= zona occludens) completely encircling the cells of epithelia and brain endothelia in which the adjacent cells are so closely apposed that the intercellular space between them is completely or incompletely occluded by an apparent fusion of the outer leaflets of the → cell membranes. The t. j. is a major local permeability barrier but allows to some extent the passage of small solutes between the cell lumen and intercellular space via a completely extracellular pathway (→ gap junction). In addition, the t. j. probably has a role in maintaining intercellular adhesion (→ synaptic junction).

tissue — a spatially ordered assembley of → cells with a particular pattern of interaction among them. A typical **mitotic t.** consists of two main groups of cells, the one involved in → mitosis and the other in ageing and function (→ dichophase; chalone). The balance between cell gain, cell function, and cell loss in any normal t. is determined by a homeostatic mechanism of selection between two alternative gene-controlled programs of syntheses. In a **non-mitotic t.**, composed of cells that are incapable of mitosis, no such balance exists. New cells are never created and although the t. may increase in size any such growth is due solely to an increase in cell size.

The mass of any t. is determined by the relation between the rate of cell production by division and the rate of cell loss.

tissue culture — the growth and maintenance of cells from higher organisms *in vitro*, i.e. outside the tissue of which they are normally a part (→ cell hybridization).

T_m — → melting temperature.

tokozygote (*Renner* 1916) — the → zygote in Thallophyta which is immediately transformed into a gonotokont (= gonotokozygote) and gives rise to haploid → germ cells (gonospores) or vegetative cells (gonocytes).

tonoplast (*de Vries* 1885) — a → unit membrane having the same appearance as the → plasma (cell) membrane and bordering the vacuole of plant cells. The permeability barrier of the t. is generally considerably higher than that of the plasma membrane.

torsion pairing (*Darlington* 1935) — nonspecific pairing between nonhomologous chromosome segments at prophase of the first meiotic division (→ chromosome pairing). T. p. does not result in → crossing-over or → chiasma formation.

touch-and-go pairing (*Wilson* 1925) — a transitory but highly specific (end-to-end), association of chromosomes (especially of → sex-chromosomes) at the second meiotic division in the heterogametic sex of Hemiptera.

trabant (*Navashin* 1912) — = → satellite.

trait — = → character.

trans-configuration (*Haldane* 1942) — → cis-configuration.

transcriptase (*Spiegelman & Hayashi* 1963) — = DNA dependent → RNA polymerase (→ reverse transcriptase) responsible for the transcription of → ribosomal RNA, → transfer RNA, and → messenger RNA from a DNA template.

transcription — → genetic transcription.

transcriptional control — a type of regulation (→ genetic regulation) of protein synthesis which acts at the level of → genetic transcription (→ translational control). It allows RNA synthesis to occur at selected loci of the genetic information and may be negative or positive (→ cascade regulation).

1. **Negative t. c.**: genetic transcription is turned off by → repressor molecules that can bind to an → operator and prevent RNA polymerase from travelling across the operator.

2. **Positive t. c.**: genetic transcription is turned on by a positive controller, which may or may not be a protein (→ sigma factor), and does not involve release of repression.

Generally, the efficiency of genetic transcription is not regulated by changes in the rate of RNA chain elongation but by the spacing of RNA polymerase which, in turn, is probably controlled by factors affecting initiation of transcription.

In eukaryotes selective suppression of genetic transcription may occur by inactivation of 1. whole chromosome sets in certain organisms; 2. individual sex chromosomes in some species, 3. condensed or → heterochromatic regions of interphase chromosomes, and 4. unexpanded → bands of → polytene chromosomes.

transcription bubble — a local opening (destabilization) of the DNA double helix during → genetic transcription in those regions where temporary

DNA-RNA hybrids are formed as intermediates in the transcription of DNA. As transcription proceeds, the RNA chain is gradually displaced from its template DNA chain by the reunion of the two DNA chains to reform the double helix.

transcription error — an error in the transcription of genetic information (→ genetic transcription). If a t. e. occurs in the transcription of a typical → structural gene, all polypeptides translated (→ genetic translation) from the mRNA will be incorrect. The particular error need not be, however, a serious one for the genes if the mRNA output is high and/or under feedback control. In the case of genes whose mRNA output is fixed and of the order of one copy per cell cycle, the t. e. may become a problem if the gene concerned is critical to cell function. If the gene subject to a t. e. codes for one of the components of the translation apparatus, a certain proportion of such errors would cause the affected compound to malfunction which, in turn, may increase its own error frequency.

transcription factor — any of a variety of factors involved in the control of → genetic transcription (→ catabolite activating protein; psi factor; M factor; sigma factor).

transcription mapping (*Bleyman & Woese* 1969) — a → genetic mapping technique that allows genes within an → operon to be mapped (intraoperonal mapping) with DNA-dependent → RNA polymerase as a measuring device. T. m. allows the determination of whether or not specific genes are located in the same → transcripton and, if so, the determination of their order within the transcription unit. The basic procedure is to perturb the transcribing activity of RNA polymerase in a way that alters the extent of expression of different genes within one transcripton. Perturbation may be achieved 1. by creating random blocks to the reading of the polymerases, 2. by preventing initiation of transcription and allowing the transcribing polymerase to read off, and 3. by depleting the genes for polymerases before a synchronous restart of trancription is tiggered.

T. m. is co-linear and co-metric with classical genetic mapping (→ replication mapping; heteroduplex mapping).

transcriptive intermediate (*Dortner & Kingsbury* 1972) — in RNA-virus infected cells, an RNA complex (TI) with properties analogous to the → replicative intermediate. The t. i. is metabolically active, partially double-stranded with base-paired regions and also contains substantial single-stranded structures. T. intermediates are virus-specific and absent in uninfected cells.

transcripton (*Hayashi* et al. 1964; *Szybalski* et al. 1970) — the general unit of → genetic transcription (= scripton) principially defined by the presence of a single autonomous → promotor (i.e., a promotor not under the cis dominant control of another promotor). It represents a cluster of genes and may contain a variety of positive and negative control elements (→ transcriptional control). Its various segments need not be expressed in a coordinate manner and may even become physically separated during certain stages of development. Several → messenger RNA molecules can be synthesized at varying rates by one t., but the transcription of promotor distal regions must always depend on initiation at the single autonomous promoter. Transcriptons may overlap in a divergent manner and, thus, some regions of the (prokaryotic) genome may be part of several transcriptons. The → operon is a special, rather simple kind of t.

In eukaryotes, the t. is assumed to consist of a rather small structural zone (one or few structural genes) and a large acceptor zone (near the promotor), which contains the acceptor sites interacting with structural and regulatory proteins. The whole t., including the non-informative acceptor zone, is assumed to be transcribed into a giant molecule of → pre-messenger RNA (*Georgiev* et al. 1973). During processing of pre-mRNA, most of the acceptor zone is probably degraded, and the informative part is transported into the cytoplasm as mature mRNA (→ informofere).

trans-derepression (*Willard & Echols* 1968) — → derepression of certain bacterial → operons (gal and lac of *E. coli*) by infection with certain bacteriophages (lambda dg and phi 80 d lac, respectively). T.-d. is related to → transduction escape synthesis and suggests that phage-carried bacterial operons effect the removal of cellular → repressors, perhaps by binding of repressor directly to specific operons on DNA.

transdetermination (*Hadorn* 1966) — the process whereby cells determined to form one type of structure change their developmental pathway and develop, under certain conditions, into quite different structures. For any given state of → determination there exists a distinct probability or frequency of t. in a specific direction. T. is based upon activation of new sets of genes.

transdifferentiation — the transition from one differentiated state of a cell to another (→ cytodifferentiation).

transduction (*Zinder & Lederberg* 1952) — The transfer of genes from one cell to another by means of a viral (in the case of bacteria, a bacteriophage) vector or a vector, which, in its external characters at least, is similar to a virus particle (→ pseudovirion). The genes picked up by the virus particles are incorporated at the expense of all or part of the virus genome. This deletion of viral genes leads to defects which vary in kind and extent depending on the nature and the length of the host genome segment. (→ genetic transformation; transfection).

Two types of phage mediated t. of bacterial cells may be distinguished:

1. **generalized t.:** the transducing vector consists of a random fragment of the bacterial chromosome (the → exogenote or → merogenote) and a phage coat after some → phenotypic mixing-like process. The formation of transducing particles (→ pseudovirion) occurs during lytic growth. Upon subsequent infection of a bacterial strain by such a particle the bacterial genes are injected and can recombine with the host chromosome. Generalized t. requires homology between the transduced material and the recipient chromosome as well as the host-recombination system (rec). The recombinants arising by this type of t. are stable.

2. **Specialized or restricted t.:** the transducing genome is a defective → prophage which, as a consequence of inproper prophage excision, carries a specific fragment of the bacterial chromosome incorporated into the phage genome. Lysogeny and prophage → induction are generally prerequisites for the formation of transducing phages. Only bacterial genes located near

the prophage → attachment site are subject to this type of t. Since the transduced genes are part of the phage genome, they can insert into the recipient chromosome by the phage integration systems. Specialized t., therefore, does not require the rec system. Like a prophage, the transducing phage genome can be lost from the host chromosome (curing) and this makes the transductant clones unstable in that the inherited genes are subject to loss. The transducing phage genomes are usually deleted in phage genes on the side opposite to the incorporated bacterial DNA. A generalized transducing phage may be capable of specialized t. when its genome is exceptionally linked to a definite bacterial marker.

Specialized transducing phages carrying a variety of bacterial genes not normally located near the prophage attachment site may be constructed by: 1. translocation of genes of interest to positions near the prophage site; 2. insertion of the prophage at an abnormal site near the genes of interest, and 3. fusion of two F-prime plasmids (→ F-plasmid), one carrying the prophage attachment site and the other carrying the bacterial genes of interest.

Complete t. depends on the attachment or insertion of the transducing DNA into the host chromosome (integration). In this state, the transducing phage genome, like that of a normal prophage or integrated → F plasmid, reproduces in concert with the bacterial chromosome. If integration is not achieved, the transducing DNA, unlike a normal vegetative phage or → sex factor, is unable to replicate, t. is abortive and transmission of the transduced feature is unilinear.

Cotransduction means t. of more than one bacterial marker gene in a single event, indicating that transducing bacteriophages can sometimes transport a segment of the bacterial chromosome large enough for neighboring (closely linked) genes to be transferred together from donor to recipient bacteria. Cotransduction permits the inference of relative → linkage of two genetic markers on the bacterial chromosome from the relative frequency with which they are cotransduced. The greater the frequency of cotransduction, the tighter the linkage. Even very closely linked markers do not invariably occur together in the same transductant bacterium. Such a "transductional segregation" of closely linked markers is interpreted as evidence for the process of → genetic recombination needed for the integration of the donor markers into the genome of the recipient cell, an event which requires a double → crossover for every integration act. Integration of two closely linked markers of the donor.

transduction escape synthesis — → escape synthesis.

transfection (*Földes & Trautner* 1964) — the infection of a cell or protoplast with isolated virus nucleic acid (DNA or RNA) resulting in the production of intact virus particles. T. has been reported for several bacterial species and is distinct from normal bacteriophage infection in that 1. the specific activities of the DNA are low, and 2. with species of phage DNA of molecular weight $>20\times10^6$ the number of infective centers produced varies exponentially rather than linearly with DNA concentration. T. was first reported for the RNA of tobacco mosaic virus. The host range for t. is vastly wider than for infection by the corresponding intact virus. Bacterial cells can even be transfected with some animal virus nucleic acids.

Several systems of bacterial t. are known (Fig. 93) which differ with respect to the manipulations performed in order to facilitate the uptake of phage nucleic acid into bacteria (→ transduction; genetic transformation).

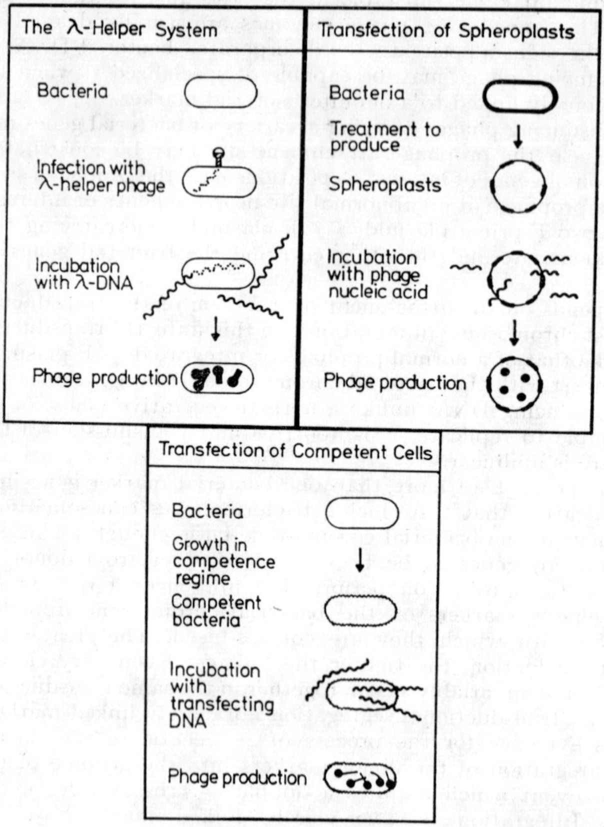

Fig. 93. Schematic representation of different modes of transfection (from *Trautner* et al. 1973).

Genetic recombination between several infective molecules introduced into a bacterial cell must precede replication of virus DNA in order to eliminate DNA lesions introduced into phage DNA by intracellular inactivation events and to reconstruct intact replicons which initiate phage development.

transferase — any of the soluble protein factors (aminoacyl transferases) required, besides GTP and aminoacyl-tRNAs, for the elongation of polypeptide chains on mammalian ribosomes.

Transferase I is the aminoacyl-tRNA binding factor required for the interaction of the substrates with ribosomes. If peptidyl-tRNA is present at the ribosomal peptidyl site, a ribosome-associated peptidyl-transferase activity brings about the formation of a peptide bond between the peptidyl-tRNA and the aminoacyl-tRNA. Transferase I binds to the small ribosomal subunit.

Transferase II is a factor needed to transport peptidyl-tRNA from the acceptor (aminoacyl) site to the donor (peptidyl) site on the mammalian ribosome. The activity of the enzyme depends on GDP and intact sulfhydryl groups. A complex is first formed between t. II and GTP, even in the absence of ribosomes. Binding of the complex requires both ribosomal subunits, reduced sulfhydryl groups on the enzyme and the ribosome, and occurs at the acceptor site.

transfer DNA — any of the cistrons in DNA coding for → transfer RNA (abbreviation tDNA).

transfer element — = → sex factor.

transfer enzyme (*Conway & Lipman* 1964) — any of the enzymes that are responsible for the transfer of amino acids from → transfer RNA to the growing peptide chain on the → ribosome (→ genetic translation).

transfer factor (*Nathans & Lipman* 1962) — 1. = → elongation factor; 2. = → sex factor.

transfer-proficient (*Achtman* et al. 1971) — of bacterial cells (symbol tra^+) carrying a → sex factor and acting as efficient donors in bacterial → conjugation.

transfer RNA (*Hoagland* et al. 1957) — any of a class of ribonucleic acid molecules (= amino acid acceptor RNA; adaptor RNA; soluble RNA. Symbols tRNA or sRNA) each with a molecular weight of about 25,000 (4 S) consisting of 73 to 90 nculeotides (Fig. 94). Transfer RNA molecules are involved in → genetic translation of → messenger RNA into the amino acid sequences of protein molecules (→ aminoacyl tRNA), in → nonsense and → frameshift suppression (→ suppressor tRNA), and (in bacteria) in the regulation of a number of amino acid biosynthetic → operons.

In genetic translation, tRNAs are able to recognize specific amino acids and corresponding genetic code words in → messenger RNA. In doing so, they translate the information encoded in nucleic acid messages into protein products. Multiple steps in this process include amino acid binding (→ aminoacyl-tRNA-synthetase), transport to the → ribosome, binding to one ribosomal site (the A site), peptide bond formation, translocation to another ribosomal site (the P site), and release from the ribosome.

The transcription products of the genes coding for tRNA are → pre-tRNA molecules which undergo a variety of modifications prior to or at specific times during the maturation process.

The → genetic code could be translated by 31 tRNA species (assuming no other restrictions on codon-anticodon recognition than those due to → wobble). However, among prokaryotes 40 to 80 tRNA genes have been found, and for eukaryotic cells the number is even larger: 320 to 1400 depending on the organism. The redundancy (→ gene reiteration) of tRNA genes paralleles that of the major → ribosomal RNA cistrons so that the ratio of tRNA genes: rRNA genes is between 4 and 10 in bacteria (*E. coli*). The genes for some of the tRNA molecules are clustered in groups of 2 or 3 on the chromosome.

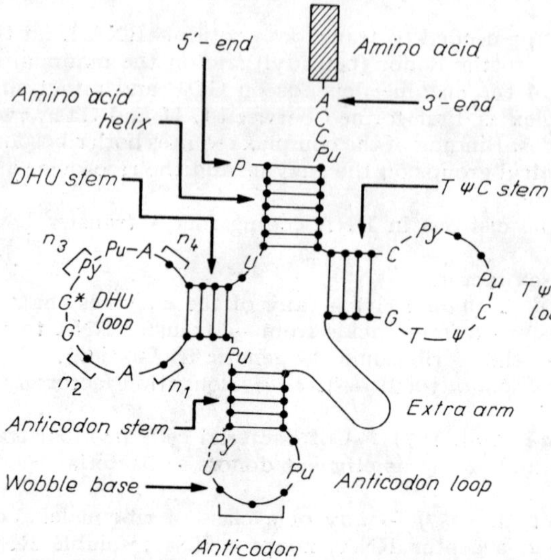

Fig. 94. The cloverleaf structure for transfer RNA. Pu = purine nucleoside; py = pyrimidine nucleoside; G represents either guanosine or 2'-O-methylguanosine; $n_1 = 0 - 1$, $n_2 = 1 - 3$, $n_3 = 1 - 3$ but usually 2, and $n_4 = 0 - 2$ nucleotides. ●—● indicates bases usually paired by hydrogen bonding. The DHU stem is shown with 4 base pairs; in some tRNAs only 3 base pairs are present (from *Letham* 1973).

tRNA makes up about 10—15% of the total RNA content of a cell; its main role in genetic translation is the acceptance of specific amino acids, the binding of aminoacyl tRNA to the messenger RNA-ribosome → initiation complex and the release of the amino acids to the growing polypeptide chain. Every amino acid has its specific tRNA for activation during protein synthesis. tRNA species which accept the same amino acid but differ in their code word recognition patterns are called **isoaccepting tRNAs** (→ isoacceptor tRNA). The amount of redundant tRNA species varies widely. The biological significance of this phenomenon may be seen in the action of isoaccepting tRNA as a safeguard against mutations. They also offer an explanation for the origin of suppressor tRNAs. The terms **heterogeneric** and **homogeneric** tRNAs (*White* et al. 1973) describe tRNA molecules that are transcribed from different DNA cistrons or from the same cistron, respectively. Although being the products of the same gene, homogeneric tRNAs are chromatographically distinct because of different degrees of post-transcriptional modification. The rather high degree of polymorphism is compatible with a molecule whose surface is largely repetitive and densely populated with charged acidic groups.

The rules of pairing between tRNA and mRNA result in the → genetic code. tRNA molecules contain up to some five percent of modified nucleosides. All modifications occur post-transcriptionally and comprise methylation, thiolation, isopentenylation, 2-thiomethyl-6-isopentenyl adenosine formation, pseudouridine formation, and esterification of carboxymethyluridine.

Although the nucleotide chain is small, tRNA has a complicated secondary and tertiary structure. All the tRNAs analyzed so far can be arranged on what is called the "cloverleaf structure". The cloverleaf model postulates helical base-paired regions sometimes referred to as the secondary structure. The modified nucleotides are largely, but not invariably, confined to the single-stranded regions in tRNA.

The cloverleaf structure has double-stranded regions of constant lengths (Fig. 94): seven base pairs near the 5'-end, at least three bases in the dihydrouridine arm, five base pairs in the anticodon loop, and five base pairs in the T ψ arm. The remaining single-stranded regions in the loop show much similarity. There are only a few variable places in the T ψ loop and in the dihydrouridine loop. The finger or extra loop region varies in size between four and 14 bases. The length of the dihydrouridine region may also differ.

The particular positions of 2'-O-methylation of the sugar unit are the middle of the dihydrouridine loop, the turning points from the double-stranded regions to the looped single-stranded regions, and the wobble base position of the anticodons.

The cloverleaf secondary structure of tRNA is further folded to give rise to a tertiary three-dimensional structure characteristic of the tRNA molecule.

This structure essentially consists of three major interlinked substructures orientated to each other in the shape of a T: the long, almost perfect double helix is formed by stacking the acceptor and T ψ C stems. A socalled D system is attached to the long helical segment almost at the middle part and is augmented by several unusual base pairs and postulated triplets. This forms the T junction. The anticodon stem extends outwards from the augmented D helix, with the helices separate and noncolinear. The anticodon helix with its loop out at the far end of the molecule allows for mobility, possibly during genetic translation. It is assumed that bases susceptible to modification are those not involved in tertiary interactions., i.e., double-helical residues are unaffected by modification. The bases reactive in modification are on the surface of the molecule.

The tRNA molecule must be able to recognize its → aminoacyl-tRNA synthetase, the tRNA binding site on the ribosomes, the appropriate ribosomal factors, and its matching codon in the messenger RNA. These recognition functions are fulfilled by four specific sites present in each tRNA molecule:

1. **The amino-acid attachment site** represented by the . . . CCA sequence at one end of the molecule.

2. **A site that recognizes the specific amino-activating enzyme** which is specific for a given amino acid and varies among different tRNA types.

3. The **ribosome-recognition site** necessary for the interaction of tRNA with the → ribosomes.

4. The **coding-recognition site** (anticodon, nodoc) which is specific for a particular tRNA molecule and responsible for the correct recognition (by complementary base pairing) of the corresponding codon in messenger RNA and, thus, for the correct insertion of the amino acid attached to the tRNA molecule into the growing polypeptide (Table 8). Present evidence indicates that a tRNA species can recognize multiple codons

Table 8. *A tentative list of transfer RNA coding triplets (anticodons) complementing with messenger RNA coding triplets (codons) (from Jukes 1966)*

Amino	Triplets		Amino	Triplets	
Acid	Messenger RNA	Transfer RNAs	Acid	Messenger RNA	Transfer RNAs
phe	UUU, UUC	GAA	Ilu	AUU, AUC	GAU
	UUU	AAA		AUU, AUC, AUA	I AU
leu	UUA, UUG	UAA	met	AUG	CAU
	UUG	CAA			
ser	UCU, UCC	GGA	thr	ACU, ACC	GGU
	UCU, UCC, UCA	I GA		ACU, ACC, ACA	I GU
	UCG	CGA		ACG	CGU
	UCA, UCG	UGA		ACA, ACG	UGU
tyr	UAU, UAC	GUA	asN	AAU, AAC	GUU
	UAU	AUA		AAU	AUU
Gap	UAA, UAG	UUA	lys	AAA, AAG	UUU
	UAG	CUA		AAG	CUU
cys	UGU, UGC	GCA	ser	AGU, AGC	GCU
	UGU	ACA		AGU	ACU
try	UGA(?)	UCA(?)	arg	AGA, AGG	UCU
	UGG	CCA		AGG	CCU
leu	CUU, CUC	GAG	val	GUU, GUC	GAC
	CUU, CUC, CUA	I AG		GUU, GUC, GUA	I AC
	CUG	CAG		GUG	CAC
	CUA, CUG	UAG		GUA, GUG	UAC
pro	CCU, CCC	GGG	ala	GCU, GCC	GGC
	CCU, CCC, CCA	I GG		GCU, GCC, GCA	I GC
	CCG	CGG		GCG	CGC
	CCA, CCG	UGG		GCA, GCG	UGC
his	CAU, CAC	GUG	asp	GAU, GAC	GUC
	CAU	AUG		GAU	AUC
gIN	CAA, CAG	UUG	glu	GAA, GAG	UUC
	CAG	CUG		GAG	CUC
arg	CGU, CGC	GCG	gly	GGU, GGC	GCC
	CGU, CGC, CGA	I CG		GGU, GGC, GGA	I CC
	CGG	CCG		GGG	CCC
	CGA, CGG	UCG		GGA, GGG	UCC

differing in the third letter (→ genetic code). The pattern of multiple recognition is consistent with the postulates of the so-called "wobble hypothesis" (*Crick* 1966).

According to the **"wobble hypothesis"**, the standard base pairs may be used rather strictly in the first two positions of the triplet, but there may be some "wobble" in the pairing of the third base. U (uracil; the first base, assuming antiparallel base-pairing) on the anticodon of tRNA can recognize both A (adenine) and G (guanine) in the third place on the messenger RNA codon; the base C (cytosine) on the anticodon can recognize only G in the third place; the base A can only recognize U; the base G can recognize both U and C; and the base hypoxanthine can recognize U, C, and A in the third place.

The tRNA sites with special function, noted above, are assumed to be located at the unpaired nucleotide sequences in the bends or loops of the tRNA structure where the bases are exposed and not in portions of the molecules that are DNA-like in structure.

Mutational alterations of the primary structure of tRNA may result in functional changes of the relevant tRNA species. Such alterations may change codon — anticodon pairing and amino acid recognition. The latter seems to be determined by various components of the tRNA molecule, perhaps including its tertiary structure. In addition, changes have been obtained which lead to the absence of function, temperature-sensitive function and altered recognition by the activating enzyme.

tRNA species which ensure the initiation of genetic translation are called → initiator tRNA.

In certain species of bacteria, aminoacyl tRNA apparently serves multiple synthetic roles. In addition to its function in protein synthesis, it provides for the amino acid residues forming the interpeptide bridge of some bacterial cell walls. Other unusual functions are the role of tRNA in aminoacyl phosphatidylglycerol synthesis and in the terminal addition of amino acid residues to proteins.

transfer RNA methylase — any of the enzymes that post-transcriptionally modify → transfer RNA species by catalyzing the transfer of methyl groups from S-adenosylmethionine to specific sites in the tRNA molecule. These enzymes are unique in being species-specific, organ-specific and base-specific. In biological systems undergoing developmental changes, large alterations occur in the levels of tRNA m. activity. The identification of natural inhibitors of these enzymes has revealed one regulatory mechanism.

transfer RNA recognition — the recognition and interaction of → transfer RNA molecules with the enzyme → aminoacyl-tRNA synthetase resulting in the attachment of the appropriate amino acid to a specific tRNA.

transformation — → genetic transformation; cell transformation.

transgenosis (*Doy* et al. 1973) — the transfer, maintenance, and expression of genes in completely unrelated organisms (as, e.g., the expression of bacterial genes in plant cells). T. is recommended as an appropriate term in those cases where donor and recipient cells are widely separated by evolution. The mechanisms of gene transfer and maintenance are obscure (→ genetic correction; exosome model; insertion model).

transgression — the appearance in a segregating generation (F_2, backcross, etc.) of one or more genotypes (individuals) which fall outside the limits of variation defined by the parents and the F_1 of the cross in respect to one or more characters (*Darlington* & *Mather* 1949).

transition (*Freese* 1959) — the substitution of one pyrimidine for another (e.g., replacement of deoxycytosine by thymidine) or of one purine for another at any base location (site) in DNA (or RNA if this is the primary carrier of genetic information, as in some viruses). At the next replication of the altered region, a similar t. will be induced in the daughter strand, i.e., pyrimidine t. will result in a purine t. in the newly synthesized complementary strand. With this type of → base pair substitution the purine-pyrimidine orientation is preserved. By means of spontaneous or induced transitions → gene mutations (t. mutations) may arise. Any transition mutation can revert spontaneously or by any transitional agent.

The preferred direction of t. mutations induced by different mutagens is as follows (*Freese* & *Yoshida* 1965):

Agent	Base pair change
2-aminopurine	A → G
	T ← C
5-bromodeoxyuridine	G → A
	C ← T
nitrous acid	G → A
	C ← T
hydroxylamine	G/C → A/T
ethylating agents	G → A
	C ← T
low pH	G → A
	C ← T
spontaneous	G → A
	C ← T

translation — → genetic translation.

translation error — → genetic translation; mistranslation.

translational ambiguity (*v. Ehrenstein* 1966) — the fact that a given → codon in mRNA can be translated in more than one way. T. a. may be due to genetic factors (mutations) or external agents (e.g., streptomycin). (→ genetic translation; mistranslation).

translational control — the regulation (→ genetic regulation) of gene expression at the level of → genetic translation, i.e. by controlling the rate at which specific mRNA molecules are translated into polypeptides (→ transcriptional control). This rate may be different for different kinds of messages in the same cell, for the same mRNA in different cells, and for the same mRNA at different times during the life of a cell. Mechanisms of the t. c. are thought to be of particular importance in higher organisms (**eukaryotes**) because of the relative metabolic stability of → messenger RNA in differentiated cells. Such mechanisms may include: 1. the stability of actively translated mRNA; 2. the processing of → pre-messenger RNA; 3. the transport of mRNA into the cytoplasm; 4. "masking" of mRNA (as in some unfertilized eggs); 5. recruitment of previously inactive message etc. Other forms of t. c. (t. c. *sensu strictu*) may involve the efficiency of use of mRNA, i.e., modulations of the rate of peptide chain elongation or non-standard initiation rates of polypeptide chains.

translational reinitiation — the restart of → genetic translation (protein synthesis) after polypeptide chain termination at a → nonsense (terminator) codon (= translational restart). There are indications that internal initiation sites (inactive in the absence of nearby termination) are rather common in bacterial and bacteriophage mRNAs. The codon AUG functions as an initiation signal in protein synthesis, i.e., as an → initiator codon. In internal positions of the mRNA it codes for the insertion of methionine. Under normal conditions, no initiation occurs within a gene. The internal initiation sites of certain cistrons have little or no effect in the wild-type gene product, and function only in the presence of a nearby terminator codon. Then the codon permits the reinitation of the gene product. Once the site is made available by chain termination, initiation presumably occurs by the normal mechanism.

Mutations which cause t. r. are called translation-restart-mutations (*Sarabhai & Brenner* 1967).

translational repression (*McLellan & Vogel* 1970) — a translational regulatory mechanism in bacteria causing a reduction in genetic translation under repressive conditions.

translational repressor (*Jay & Kaempfer* 1974) — = → interference factor.

translation factor — any of a number of factors involved in the regulation of → genetic translation (→ initiation factor; elongation factor; release factor; transferase).

translocation — 1. a chromosomal structural change characterized by the change in position of chromosome segments (and the gene sequences they contain) within the chromosome complement (→ chromosome mutation).

In many organisms, translocations result in specific patterns of → genetic segregation, such as unusual → linkages and reduced meiotic recombination, reduced viability of a fraction of the offspring, familial patterns of trisomy, etc. These can be used for the detection and mapping of translocations in genetically well-analyzed organisms. Generally, in higher organisms cytological investigations are used for their identification or confirmation.

Simple translocations may be classified as follows:

1. **Intrachromosomal (internal) translocations or shifts**: change in position of a chromosome segment within the same chromosome either from one chromosome arm to the other **(extraradial)** or within the same chromosome arm **(intraradial)** (Fig. 95a).

2. **Interchromosomal translocations**: transfer of a chromosome segment from one chromosome into another chromosome **(transposition,** Fig. 95b, c) or reciprocal exchange of segments between chromosomes (**reciprocal t., mutual** t. or **interchange,** Fig. 95d, e). An interchange of segments between

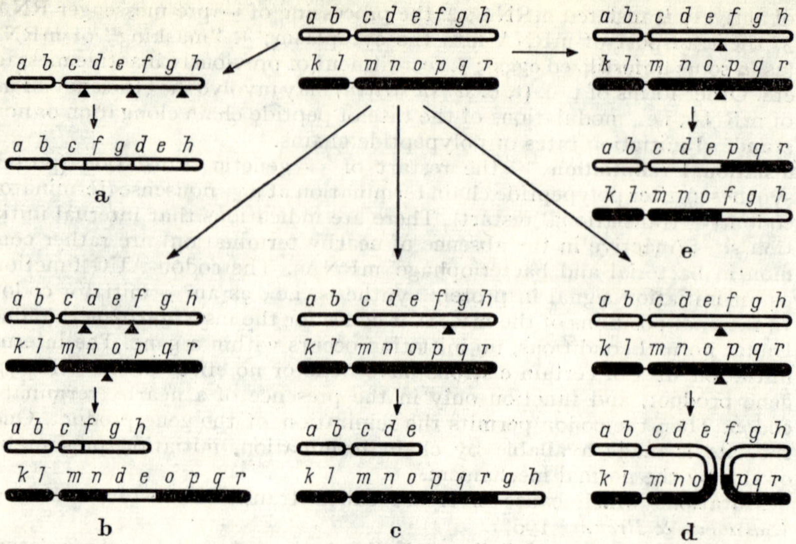

Fig. 95. Diagram showing several types of intrachromosomal and interchromosomal translocations. a) intrachromosomal transposition ("shift") of a chromosome segment; b) interchromosomal transposition of a chromosome segment; the sequence d—e is transferred to a heterologous chromosome; c) "terminal translocation" of the segment g—h to a heterologous chromosome; d) reciprocal (asymmetric) translocation (interchange) of segments f—h and p—r; e) reciprocal (symmetric) translocation of segments f—h and p—r; in contrast to an asymmetric interchange the translocation products are monocentric.

homologous chromosomes is called "fraternal", while one between non-homologous chromosomes is called "external".

In the case of **eucentric** t., the sequence of loci contained in the transposed segments is the same as that in the original segment position with respect to the → centromere, while the sequence is reversed in case of a **"dyscentric"** t.

Intercalary incorporation of a chromosome segment due to transposition is called an **"intercalation"** or **"insertion"** ("insertional translocation"). Fraternal transpositions result in an intrachromosomal → duplication in one, and the corresponding → deletion in the other chromosome of a pair of homologues.

Terminal transpositions of chromosome segments to the natural ends (→ telomere) of another chromosome are not possible.

In those cases where a chromosome segment is laterally transposed to another chromosome, a three-armed configuration arises which is said to be a **"triradial"** due to lateral translocation.

The origin of translocations is interpreted either according to the "breakage-reunion" or the "exchange model" (→ chromosome mutation).

The unit of translocation may be the chromosome **(chromosome t.)**, the single chromatid **(chromatid t.)** or one of the subunits of a chromatid **(sub- or half-chromatid t.)**. The sites of translocation are called **translocation points.**

Reciprocal translocations (interchanges) may be either **asymmetrical** (= aneucentric) or **symmetrical** (= eucentric):

In the former, one → dicentric and one acentric product of t. arise

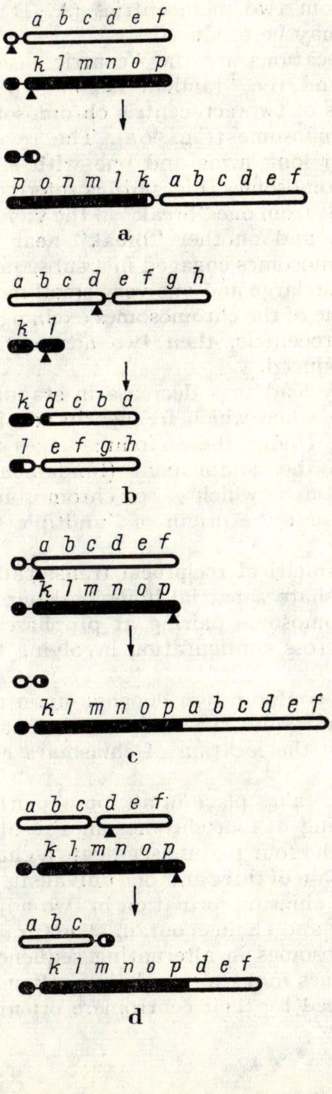

Fig.96. Special types of translocations. a) centric fusion; b) dissociation; c) tandem fusion between two acrocentric chromosomes; d) tandem fusion between one acrocentric and one metacentric chromosome.

(Fig. 95d); in the latter the translocation products are both monocentric (Fig. 95e). Asymmetrical interchanges may lead to the production of a → chromosome bridge at anaphase if the two centromeres of the dicentric product are distributed to opposite spindle poles.

Whole-arm translocations (*Muller* 1940) or whole-arm transfers are translocations in which whole (or nearly whole) arms are transposed or interchanged. By this process one acrocentric (·A) and one metacentric (B · C) chromosome may give rise to one telocentric (·C) and one new metacentric (A · B) chromosome, or from two metacentrics (A · B and C · D) two new ones (A · C and B · D) may be produced.

Special cases of "whole-arm" translocations are the "centric fusion" (*Robertson* 1916), the "dissociation", and the "tandem fusion" (*White* 1957). With **centric fusion** the long arms of two acrocentric chromosomes are united to form a metacentric chromosome (Fig. 96a). The reverse process by which one metacentric with long arms and one with short arms may give rise to two acrocentric chromosomes (Fig. 96b) is referred to as **"dissociation"**. **"Tandem fusions"** result from one "break" in the vicinity of the centromere in one chromosome, and another "break" near the chromosome end in a second. If the chromosomes engaged in a subsequent interchange are two acrocentrics, then one large and one very small metacentric chromosome arise (Fig. 96c). If one of the chromosomes exchanging segments is metacentric, the other acrocentric, then two acrocentrics (one short, the second large) may be produced.

Both centric and tandem fusions may lead to a decrease in standard chromosome number if the small chromosome which frequently consists only of heterochromatin is eliminated. Under these circumstances the number of large chromosome arms in the complement (fundamental number) remains unchanged. Centric fusions in which → sex chromosomes and → autosomes take part may represent the origin of "multiple sex chromosome" systems.

In **individuals heterozygous for a symmetrical reciprocal translocation** (structural hybrids), four chromosomes share a partial homology but no two are identical. Consequently → chromosome pairing at prophase of first division of → meiosis results in a cross configuration involving the four chromosomes (Fig. 97).

The subsequent behavior of this cross-configuration depends upon the frequency and location of the → chiasmata and the mode of → centromere orientation (which is influenced by the location of chiasmata and the morphology of the chromosomes).

If crossing-over and chiasma formation takes place in all four "pairing segments" (→ interstitial segment), a ring of four chromosomes results. If chiasma formation is absent in one of the four pairing segments a chain of four chromosomes results (Fig. 98). A chain of three and one univalent, or two bivalents, result as a consequence of chiasma formation in two adjacent or alternate pairing segments. Rings and chains contain structurally normal and structurally changed chromosomes in alternating sequence.

The distribution of the four chromosomes in a ring or chain-configuration at anaphase I of meiosis is determined by their centromere orientation which is either concordant or discordant (Fig. 98). In the case of → **alternative distribution,** chromosomes located alternately in the pairing configuration are distributed to the same spindle pole, while in → **adjacent distribution** neighboring chromosomes reach the same pole. If neighboring chromosomes with nonhomologous centromeres reach the same pole, this mode of distribution is called "adjacent-I distribution". In the case of "adjacent-II distribution", those neighboring chromosomes in the ring or chain which are distributed to the same pole have homologous centromeres.

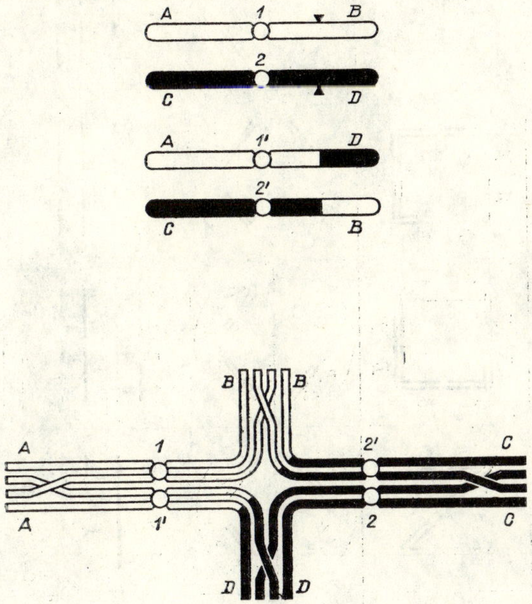

Fig. 97. Diagram illustrating the meiotic pairing configuration (pachytene) in the case of a karyotype heterozygous for a reciprocal translocation of the symmetric type (the heterologous chromosomes AB and CD are involved in the translocation which resulted in the structurally changed chromosomes AD and CB).

In the alternative distribution, a complete genome reaches each spindle pole, but with the two types of adjacent distribution the polar groups are characterized by → duplications and deletions and the resulting gametes are genetically unbalanced. This unbalance may give rise to gametic → sterility in plants and to zygotic sterility in animals. The percent sterility is usually about 50% (between 30 and 70%) and the phenomenon is called "semisterility".

In the case of discordant centromere orientation, gametes with n − 1 and n + 1 chromosomes may arise. Disomic (n + 1) gametes may contain one or both chromosomes with interchanged segments and may

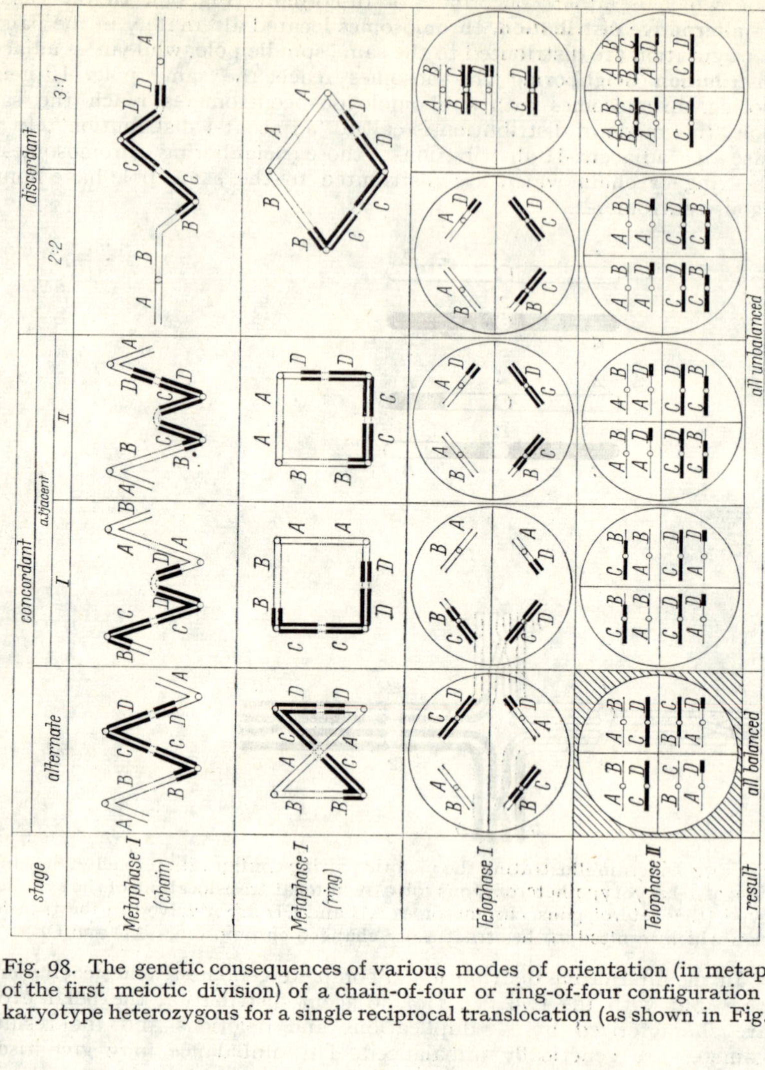

Fig. 98. The genetic consequences of various modes of orientation (in metaphase of the first meiotic division) of a chain-of-four or ring-of-four configuration in a karyotype heterozygous for a single reciprocal translocation (as shown in Fig. 97).

produce → trisomics by the union with normal (n) gametes. These are referred to as tertiary trisomics and translocation trisomics if they contain one or both chromosomes which participated in the interchange.

If crossing-over and chiasma formation occurs in the interstitial segments (located between the centromere and the translocation points) of the pairing configuration, there results either alternative or nonhomol-

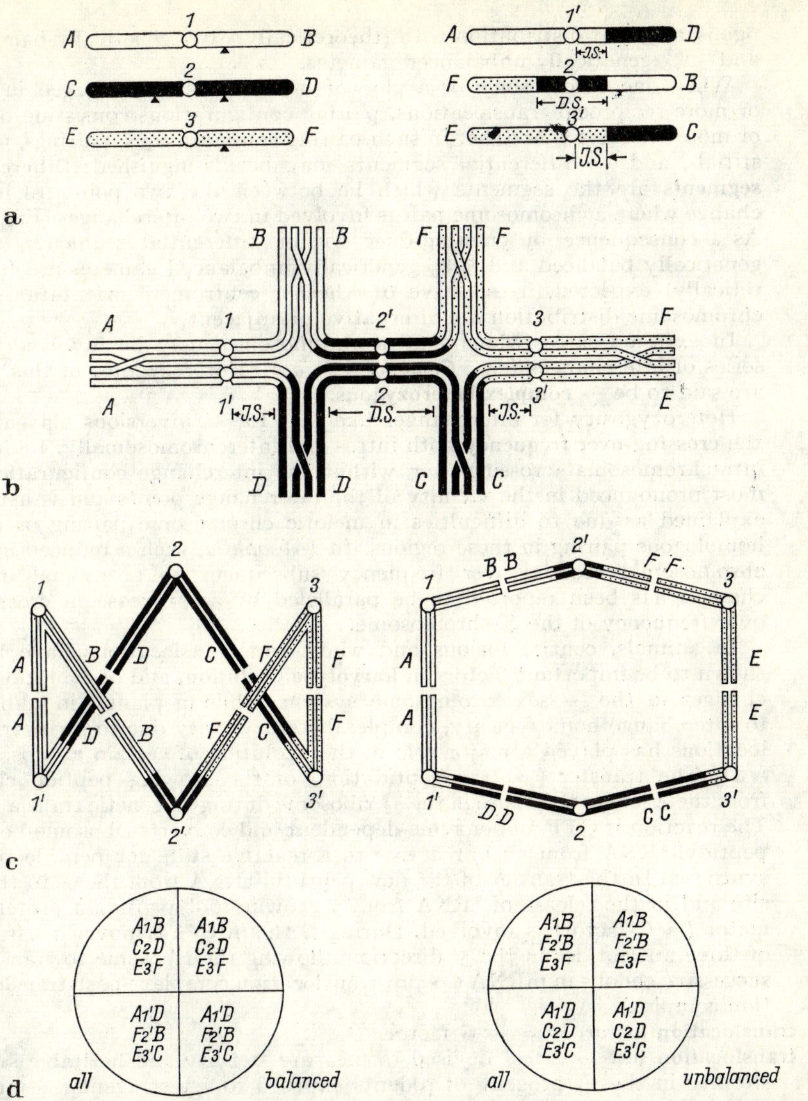

Fig. 99. Diagram showing the meiotic consequences of an interchange association of six chromosomes in a karyotype heterozygous for two reciprocal translocations between heterologous chromosomes. a) Left the normal and right the structurally changed chromosomes (I. S. = interstitial segment; D. S. = differential segment). b) The pachytene pairing configuration with the patterns of crossing-over and chiasma formation; c) double zig-zag and ring-form orientation of the hexavalent at first meiotic metaphase; d) the genetic constitution of the haploid meiotic products (telophase of second meiotic division).

ogous adjacent distribution with (theoretically) 50% genetically balanced and 50% genetically unbalanced gametes.

When one chromosome or a pair of homologues are engaged in two or more reciprocal translocations, pairing configurations consisting of six or more chromosomes arise. In such pairing configurations, pairing-, interstitial-, and → differential segments may be distinguished. **Differential segments** are the segments which lie between the two points of interchange where a chromosome pair is involved in two interchanges (Fig. 99). As a consequence of crossing-over within differential segments, 50% genetically balanced and 50% genetically unbalanced gametes are (theoretically) expected, irrespective of whether centromere orientation and chromosome distribution are alternative or adjacent.

In extreme cases, whole sets of chromosomes may be involved in a series of interchanges (e.g., *Oenothera, Rhoeo*). Heterozygotes of this type are said to be → **complex heterozygous.**

Heterozygosity for interchanges like that for → inversions may affect the **crossing-over frequency** both intra- and interchromosomally. Reduced intrachromosomal crossing-over within the interchange configuration is most pronounced in the vicinity of the interchange points and is usually explained as due to difficulties in meiotic chromosome pairing or nonhomologous pairing in these regions. In *Drosophila,* such a reduced intrachromosomal crossing-over frequency subsequent to autosomal interchanges has been reported to be paralleled by an increase in crossing-over frequency of the X-chromosome.

In animals, centric fusions and whole-arm translocations have been shown to be important factors in **karyotype evolution,** and of evolutionary changes in the → sex-chromosome system, while in plants, in addition to interchange homozygosity, complex heterozygosity due to serial translocations has played a major role in the evolution of certain groups.

II. The transfer (→ transpeptidation) of the growing peptide chain from the A site to the P site in the → ribosome during → genetic translation. The reaction is GTP and enzyme-dependent and converts ribosome-bound peptidyl-tRNA from an unreactive to a reactive state for peptide bond synthesis. In the transfer of the new peptidyl-tRNA from the A to the P site and in the release of tRNA from a growing polypeptide a protein t. factor (→ G-factor) is involved. During t. the mRNA is moved a distance of three nucleotides in the 5′ direction allowing the ribosome to translate successive codons in mRNA (→ pre-translocation complex; post-translocation complex).

translocation factor — = → G factor.

translocation test — a test devised to measure sterility and heritable semisterility in the F_1 progeny of rodents exposed to a test agent (→ mutagenicity testing). Males or females are treated with the test agent, mated with untreated animals, and F_1 individuals (usually males) are selected for further breeding and examination of the litter size produced by the F_1 animals. In general, any animal showing greater than 50% fetal death is considered to be semisterile due to an induced chromosomal translocations. The translocation can then be detected cytogenetically.

transpeptidation — the reaction of peptide bond formation during → genetic translation which is catalyzed by the → ribosome. The peptidyl transferase center is a part of the large (50 S or 60 S) ribosomal subunit. After t., the mechanical transfer (→ translocation) of tRNA residues and mRNA occurs in the ribosome (→ transfer factor).

transposition — transfer of a chromosome segment to another position owing to intra- (→ shift) or interchromosomal structural changes (→ translocation).

transvection effect (*Lewis* 1954) — a → position effect in *Drosophila* which results when → pseudoalleles present in the → trans-configuration are combined with a heterozygous chromosomal structural change disrupting somatic → chromosome pairing. The structural change intensifies the mutant effect associated with pseudoalleles in the trans-configuration, but does not influence the phenotype either of the cis-configuration or the phenotypes of the single mutants in homo- or heterozygous condition.

transversion (*Freese* 1959) — the substitution of a purine for a pyrimidine or of a pyrimidine for a purine at any base location (site) in DNA (or RNA if this is the carrier of primary genetic information). T. results in a distortion of the nucleic acid backbone in addition to base pairing difficulties, since two purines hydrogen-bonded in the helix would increase the diameter of the molecule while hydrogen-bonded pyrimidines would decrease it. Transversions may give rise to → gene mutations (t. mutations) which can be detected by means of amino acid analysis. No chemical or physical mutagen has been shown to induce transversion exclusively (→ transition). Transversions may come about by, e.g., the mistaken pairing of one purine with either another purine or with a gap produced by depurination, or the mistaken pairing of one pyrimidine with another pyrimidine. Transition agents are unable to revert transversion mutants and transversion agents do not revert transition mutants.

The following kinds of base pair substitutions may occur by t.

$$\genfrac{}{}{0pt}{}{A}{T} \rightarrow \genfrac{}{}{0pt}{}{C}{G} \text{ or } \genfrac{}{}{0pt}{}{T}{A}; \genfrac{}{}{0pt}{}{T}{A} \rightarrow \genfrac{}{}{0pt}{}{G}{C} \text{ or } \genfrac{}{}{0pt}{}{A}{T}; \genfrac{}{}{0pt}{}{G}{C} \rightarrow \genfrac{}{}{0pt}{}{A}{T} \text{ or } \genfrac{}{}{0pt}{}{G}{C}; \genfrac{}{}{0pt}{}{G}{C} \rightarrow \genfrac{}{}{0pt}{}{T}{A} \text{ or } \genfrac{}{}{0pt}{}{C}{G}$$

trihybrid — a → hybrid heterozygous for three pairs of alleles (genes).

triisosomic (*Kimber* & *Sears* 1968) — of a cell or individual which lacks one chromosome pair but has three homologous isochromosomes for the same arm of the missing pair (→ monoisosomic).

trimonoecious (*Errera* & *Gevaert* 1878) — of plants with hermaphroditic, male, and female flowers on the same individual (= coenomonoecious).

trioecious — of plants having male, female, and hermaphroditic flowers on different individuals (→ dioecious).

triparental cross (*Hershey* & *Chase* 1951) — in virus genetics, the simultaneous infection of bacteria with three parent bacteriophages which differ from each other in three of their genetically controlled markers (e.g., $ab^+c^+ \times a^+b^+c \times a^+bc^+$). From such triply infected cells a triparental recombinant abc (one genetic marker from each of the three parents) may be recovered. At least two successive genetic exchanges are required for its production: the first between two parents, and the second between the third parent and the biparental recombinant. This course of events indicates that repeated recombinational events occur in the infected cell (→ genetic recombination).

triplex (*Blakeslee, Belling & Farnham* 1923) — → nulliplex.

triploid (*Němec* 1910) — of → auto- or → alloploid cells, tissues, and individuals with three chromosome sets in their nuclei (symbol 3n).

triradial — a chromosome with three arms arising as a result of a "lateral" → translocation.

trisomic (*Blakeslee* 1921) — of → polysomic cells, tissues, or individuals with one or more (doubly t., etc.) extra chromosomes in an otherwise diploid (disomic) chromosome complement. This phenomenon is referred to as trisomy and is symbolized by 2n + 1, 2n + 1 + 1, etc. The extra chromosome is either completely (primary trisomics) or partially homologous (secondary and tertiary trisomics) to certain chromosomes of the standard complement.

1. **Primary trisomy** (*Blakeslee* 1924; *Rhoades* 1933; *Goodspeed & Avery* 1939): The extra chromosome is completely homologous to one pair of homologues in the chromosome complement and, therefore, one chromosome is represented three times in an otherwise diploid complement. Such homology permits trivalent formation at the first meiotic division. The trivalents may assume various configurations (→ multivalent). Instead of one trivalent, a → bivalent plus a rodshaped univalent may be produced (Fig. 100). The number of different primary trisomics possible for a given karyotype corresponds to the number of chromosomes per → chromosome set.

The principal sources of p. trisomics are (*Khush* 1973): 1. normal disomics (spontaneous occurrence or by treatment with physical and chemical agents); 2. asynaptic and desynaptic disomics; 3. polyploids (e.g. haploids, triploids, and tetraploids); 4. translocation heterozygotes; 5. primary trisomics, tetrasomics, multiple trisomics, and secondary, tertiary, and compensating trisomics; 6. monosomics; and 7. other chromosomal abnormalities.

Frequently, primary trisomics differ from normal diploids and from each other in a number of phenotypic characters due to changes in the genic balance brought about by the extra chromosome. Their viability and fertility is usually lower than that of the diploids from which they arose.

2. **Secondary trisomy** (*Blakeslee* 1924; *Rhoades* 1933; *Goodspeed & Avery* 1939): the extra chromosome has two identical arms and is an → isochromosome. For each primary two types of secondaries are possible. Thus, in secondary trisomics, one chromosome segment (arm) is represented four times in an otherwise diploid chromosome complement.

The diagnostic pairing configuration at the first meiotic division is either a ring trivalent (which only rarely occurs in primaries), or a bivalent and one U-shaped ring resulting from the internal pairing of the univalent with identical arms (Fig. 100).

Seondary trisomics may occur among the progeny of normal diploid individuals, but their best source is among the progeny of individuals with one or more univalent chromosomes in which → centromere misdivision takes place.

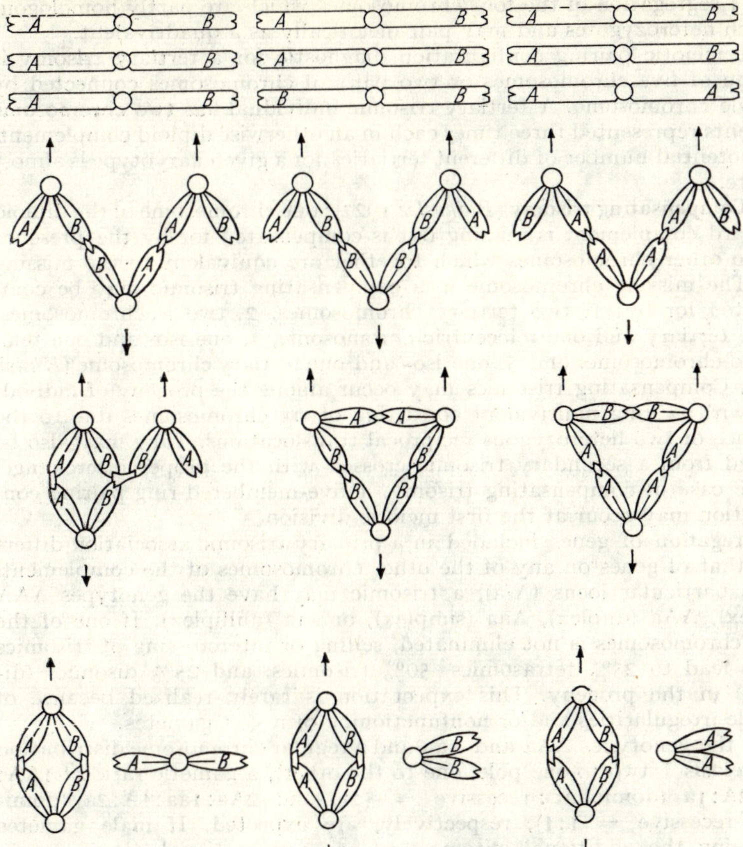

Fig. 100. Primary and secondary trisomics. Left: meiotic pairing configurations (metaphase I) in the case of primary trisomy; middle and right: meiotic pairing configurations in the two secondary trisomics possible for the primary type shown. In the case of secondary trisomy the univalents are U-shaped owing to internal pairing in isochromosomes.

3. **Telosomic trisomy** (*Burnham* 1962): the extra chromosome is a telocentric fragment chromosome homologous to one arm of a chromosome pair in the standard complement (= telotrisomy).

4. **Tertiary trisomy** (*Belling & Blakeslee* 1926; *Rhoades* 1933): the extra chromosome is the result of a reciprocal → translocation (interchange) between two standard chromosomes. Its ends are homologous with the ends of two different chromosomes. Tertiary trisomics occur regularly (and possibly exclusively) in the progenies of translocation heterozygotes

by 3:1 segregation of the four chromosomes which are partly homologous in such heterozygotes and may pair meiotically as a quadrivalent.

The meiotic pairing configuration diagnostic for a tertiary **trisomy** is a chain of five chromosomes or two pairs of chromosomes connected by a single chromosome. A tertiary trisomic individual has two **chromosome** segments represented three times each in an otherwise diploid complement. The potential number of different tertiaries for a given karyotype is almost infinite.

5. **Compensating trisomy** (*Blakeslee* 1927): one chromosome of the diploid standard complement is missing but is compensated for by the presence of two other chromosomes which together are equivalent to the missing one. The missing chromosome in a compensating trisomic may be compensated for by: 1. two tertiary chromosomes, 2. two isochromosomes, 3. one tertiary and one telocentric chromosome, 4. one iso- and one telocentric chromosome, and 5. one iso- and one tertiary chromosome (*Khush* 1973). Compensating trisomics may occur among the progeny of individuals with a ring multivalent consisting of six chromosomes due to the presence of two heterozygous reciprocal translocations. They may also be derived from a secondary trisomic crossed with the proper interchange. In the case of compensating trisomy, a five-membered ring pairing configuration may occur at the first meiotic division.

Segregation of genes included in a primary trisomic association differs from that of genes on any of the other chromosomes of the complement. For a particular locus (A/a), a trisomic may have the genotypes AAA (triplex), AAa (duplex), Aaa (simplex), or aaa (nulliplex). If one of the three chromosomes is not eliminated, selfing or intercrossing of trisomics would lead to 25% tetrasomics, 50% trisomics, and 25% disomics (diploids) in the progeny. This expectation is rarely realized because of meiotic irregularities and/or nonfunctioning of n + 1 gametes.

For the genotypes AAa and Aaa and a regular chromosome distribution (at meiosis I two to one pole, one to the other), a gametic ratio of 1AA: 2Aa:2A:1a (dominant : recessive = 5:1) and 2Aa:1aa:1A:2a (dominant : recessive = 1:1), respectively, are expected. If male gametes possessing the additional chromosome are nonfunctional (as in many plants) and → chromatid segregation is absent, the following genotypic proportions are expected in the progenies of selfed or intercrossed trisomics of the genotypes AAa and Aaa:

Genotype	AAa		Genotype	Aaa	
Gametes	2 A	1 a	Gametes	1 A	2 a
1 AA	2 AAA	1 AAa	2 Aa	2 AAa	4 Aaa
2 Aa	4 AAa	2 Aaa	1 aa	1 Aaa	2 aaa
2 A	4 AA	2 Aa	1 a	1 Aa	2 Aa
1 a	2 Aa	1 aa	2 a	2 Aa	4 aa

The observed genotypic ratio in the progenies may deviate considerably from this theoretical expectation. If instead of → chromosome segregation, chromatid segregation takes place (and the locus in question is so far from the centromere that crossing-over takes place regularly between centromere and locus) then the expected gametic ratio for the genotype AAa will be 6AA:8Aa:1aa:10A:5a instead of 1AA:2Aa:2A:1a.

For → complex-heterozygous species (*Oenothera*) another classification of the trisomic types has been proposed by *Renner* (1941):

1. **Isotrisomy:** the extra chromosome is completely homologous to one pair of chromosomes in the standard form.

2. **Anisotrisomy:** the trisomic type has a maximum of two, and never three, completely homologous chromosomes.

a) **Additive dimorphic trisomy:** one chromosome of the standard complement has been doubled by → non-disjunction. With selfing, the diploid type and the trisomic type are produced side by side.

b) **Compensated monomorphic trisomy:** x chromosomes of one of the two complexes have been replaced by x + 1 chromosomes of the other complex owing to double non-disjunction (e.g., replacement of chromosome 3—6 belonging to one complex by the chromosomes 4—3 and 6—5 of the other results in the addition of chromosome ends 4 and 5). This type of trisomy is stable in the case of selfing or intercrossing identical trisomics.

tritelosomic (*Kimber & Sears* 1968) — of a cell or individual which lacks one chromosome pair but has three homologous telocentric chromosomes (→ ditelosomic).

trivalent — a → multivalent consisting of three meiotically paired chromosomes.

tRNA — symbol for → transfer RNA.

tRNA releasing factor (*Ishitsuka & Kaji* 1970) — a factor (TR) catalyzing the release of → transfer RNA from ribosomes. TR is distinct from the → G factor in that it loses its activity upon heating (50 °C) and does not require GTP for its action. TR may play a role in removing tRNA from the ribosome after the polypeptide chain termination factor has released the completed chain from the ribosome-bound tRNA.

tropokinesis (*Gavaudan* 1943) — the shifting of the axis of a bipolar → spindle. T. is the slightest visible disturbance resulting from treatment with → spindle poisons (→ C-mitosis).

true breeding — of genotypes or certain of their gene loci which do not segregate (→ segregation) because they are → homozygous.

tube nucleus — the → nucleus of the pollen tube which controls the growth and behavior of the pollen tube in higher plants.

tubulin (*Borisy & Taylor* 1967) — the subunit protein of → microtubules. It is a 110,000 dalton dimer composed of 55,000 dalton monomers which are possibly non-identical. The t. dimer is associated with two moles of guanosine nucleotide phosphate per mole of dimer. Both monomers of a dimer are covalently linked to a small number of various sugars.

Two distinguishable guanine nucleotide binding sites exist per molecule of t. One site (the E site) rapidly exchanges GTP with the medium, the other site (the N site) binds GDP or GTP very strongly. By transphosphorylase activity the GDP bound to the N site can be converted to GTP using GTP or ATP as the phosphate donor.

turning point — → intersex.

twin meiosis (*Gutz* 1967) — after copulation of diploid yeast cells, separate meiosis in both nuclei in a common cytoplasm in those cases where karyogamy does not occur immediately after copulation. By t. m. asci with eight haploid spores are formed.

twin method — in human genetics, a method using identical and fraternal twins for studying the influences of heredity and environment on the phenotype.

twins — two embryos developed simultaneously within one uterus, one vitelline membrane, or one testa (similarly triplets, quadruplets, etc.). T. may be fraternal (dizygotic) if they originate from two fertilized eggs, or identical (monozygotic) if they arise from only one egg. Monozygotic t. are genetically identical.

twin species — two → species which are phenotypically very similar but separated from each other by reproductive → isolation (→ sibling species).

twin spot — two adjacent cell groups which differ in genotype **and** phenotype due to mitotic → crossing-over or segregation within the somatic cells of a → heterozygous individual.

twisting (*Cleveland* 1949) — the rotation about one another of the two chromatids of a chromosome ("individual t."), or of two homologous or nonhomologous chromosomes, chromosome arms or chromosome segments of one chromosome ("relational t.") during early meiotic prophase. T. and "relational coiling" (→ chromosome coiling) are assumed to be different phenomena.

two-factor cross — a genetic recombination experiment involving two → genetic markers (e.g., $a^+b^+ \times ab$).

two-plane theory (*Sharp* 1924) — → chiasmatype theory.

two-side arm bridge — → pseudochiasma.

type number (*Harvey* 1917) — the most frequently occurring chromosome number in a particular taxonomic group (= modal number). The term t. n. is often used with the implication of its being the ancestral number of the group in question. Other numbers are then assumed to be derived from the type number.

U

under-replication — slower replication of certain → heterochromatic chromosome regions and of → ribosomal DNA in relation to the rest of the DNA. U.-r. has been observed in certain organisms and is characteristic in → polytenization in Diptera during which the euchromatin is replicated about one thousand-fold while the heterochromatin, located at the → centromere, is not replicated.

undifferentiated — of cells which are descendents of the fertilized ovum and have not yet developed recognizable heterogeneity (→ cytodifferentiation).

undulipodia (*Frey-Wyssling & Mühletahler* 1965) — plasmatic differentiations such as cilia and flagella which differ in their length (a cilium measures about 5—10 μm, a flagellum 150 μm or more) but are alike in their ultrastructure. Each contains a cylindrical shaft (diameter about 0.2 μm) composed of 11 proteinaceous microfibrils, nine of which form a circle around a central pair. These nine peripheral microfibrils are doublets, each consisting of two subfibrils; the central microfibrils are single. All the microfibrils are embedded in a noncontrasting membrane-bound matrix. U. are rooted in a plasmic particle referred to as the "basal granule" (the "kinetosome" of the ciliates and the "blepharoplast" of the flagellates) which is delimited from the extracellular flagellum by the "basal plate".

unilateral inheritance (*Winge* 1927) — → inheritance.

uninemic — of chromatids with only one double helix of DNA, as opposed to → bi- and → polynemic (→ chromosome).

unipolar — of a → spindle with only one spindle pole. Under these circumstances (= monaster) the chromosomes usually lack a polarized movement (→ chromosome movement).

unique DNA — any unique and specifically ordered sequence of nucleotides in DNA (= single-copy DNA; unique sequence DNA) which occurs only once in a genome (chromosome set), as opposed to → repetitious DNA. In the → chromosomes of eukaryotes, a large fraction of the unique and repeated DNA sequences are interspersed. With very few exceptions, u. DNA represents primarily the structural genes coding for polypeptides.

unisexual — of organisms having separate sexes (dioecious, gonochoric) and producing only one kind (male or female) of gametes. Such organisms are thus dimorphic (male and female).

unit character (*Castle* 1905) — a → character difference which is transmitted as a unit in heredity; a character difference controlled by one pair of → alleles.

unit membrane (*Robertson* 1959) — any lipoprotein (about 40% lipid and 60% protein) membrane, irrespective of cellular location (→ cell membrane; endomembrane system; non-unit membrane), which can be resolved as a triple-layered structure (6 to 10 nm wide) consisting of two electron-dense outer layers and a less dense central layer (all about 2.5 nm in diameter).

The u. m. represents a vesicular or tubular system with a continuum being made up of nesting lipoprotein repeating units, each m. possessing its own specific structural unit or repeating particles. A current concept of u. m. structure demands a class of proteins capable of perferential association with a lipid bilayer and called integral proteins. These are intrinsic to the membrane, in distinction to extrinsic cytosol proteins.

Different membranes may show considerable modifications of the basic trilayered structure. Their total thickiness may be greater than 6 to 10 nm, and the two outer dense layers are often of unequal thickness. In some cell types, an additional layer of filamentous material may be present on the outer surface. A wide spectrum of variation also exists with respect to chemical content, enzyme activity, permeability characteristics, metabolism and function.

Cellular and organelle membrane frequently consist of two u. membranes in paired arrangement.

unit of replication — = → replicon; replication unit.

unit of transcription — = → transcripton.

univalent — an unpaired chromosome (→ chromosome pairing) at the first meiotic division which corresponds to a particular chromosome of the mitotic complement (→ bivalent; multivalent). Univalents either lack a homologue or result from → asynapsis (absence of meiotic pairing is due either to asynaptic genes or to the influence of environmental factors) and → desynapsis (premature dissociation of pairing configurations in diplotene due to the absence of → chiasma formation).

Univalent chromosomes usually lie at random on the spindle at metaphase I of meiosis and may follow two main courses at anaphase (*Darlington* 1957):

1. Those lying far from the spindle equator are randomly distributed (without division) to the spindle poles ("postdivision");

2. those lying near the equator move onto the plate where they orientate themselves axially and divide into their two chromatids which in turn pass to opposite poles ("predivision").

Some univalents may not be included in one of the daughter nuclei and are then lost in the cytoplasm (formation of micronuclei). Furthermore, the presence of several univalents in one meiocyte may completely hinder the separation of the chromosomes into two daughter groups and result in a → restitution nucleus. Univalents frequently show → centromere misdivision at first or second meiotic division.

"False univalents" (*Upcott* 1938) is the term proposed for univalents which result from indifferent co-orientation (→ centromere orientation) of the chromosomes contained in multivalents rather than from asynapsis or desynapsis. False univalents behave in the same way as the above-mentioned true univalents in that either the centromeres divide and the chromatids belatedly move to the poles or they remain undivided and then are randomly included into one of the two polar groups of chromosomes.

"Loop univalents" are ring chromatids (= closed univalents; loop chromatids) which may result from crossing-over and chiasma formation proximal to and within a heterozygous paracentric → inversion ("three-strand double crossing-over). If, in addition to a proximal crossover, a four-strand double crossover occurs within the inversion region, two loop univalents may arise. Loop univalents may give rise to univalent bridges.

univalent shift (*Person* 1956) — after crossing disomic (diploid) and → monosomic individuals (which lack a particular chromosome) the occurrence, among the progeny, of monosomic individuals which lack a chromosome other than that which is lacking in the monosomic parent. U. s. has been observed in *Triticum* crosses where the monosomic parent was partially asynaptic.

unscheduled DNA synthesis (*Rasmussen & Painter* 1964) — in eukaryotes, DNA synthesis outside the S phase of the → cell cycle and, in part, nonconservative. U. DNA s. occurs spontaneously and can be induced by UV light, ionizing radiations and radiomimetic agents, i.e., agents that produce damage to DNA and exert lethal and mutagenic effects. The amount of u. DNA synthesis is generally less than 1% of normal S phase synthesis.

U. DNA s. is specific for certain genetic processes, such as → repair of molecular damage to DNA (→ repair replication), → genetic recombination, selective → gene amplification and the formation of → DNA puffs.

untranslated region — any region of → messenger RNA not coding for protein. All the mRNAs that have been purified are known to contain more nucleotides than are required for coding. These extra residues, presumably present on each side of the coding region, and defined as the 5′ and 3′ untranslated regions are of unknown function and probably serve in → translational control, the termination of transcription, the addition of poly(A), the binding of transport proteins and mRNA degradation.

UV-reactivation (*Weigle* 1953) — the → reactivation (increased survival) of UV-inactivated, host cell reactivable bacteriophages adsorbed to their host bacteria, by an additional (slight) UV dose given to the complex (symbol UVR). A similar increase of phage survival is observed when the host cells are pretreated with a slight dose of X-rays or with Mitomycin C.

Two types of UV-r. may be distinguished:

1. **direct UV-r.**: accompanied by a high rate of mutation (UV-mutagenesis) of the reactivated phage due to an error-prone repair process different from → excision and → post-replication repair processes.

2. **Indirect UV-r.**: The introduction of UV-damaged DNA into a host cell which provides a means of promoting UV reactivation of UV-irradiated phage (lambda). This form of UV-r. is also associated with mutagenesis of the restored phage.

UV-inactivation of phages is due to UV-damage to the nucleic acids and UVR is assumed to be a reflection of an increased efficiency of the host cell reactivation repair system owing to the elimination of counteracting processes. Recovery from UV-damage to nucleic acids can proceed by (*Rupert & Harm* 1966):

1. reversion of a damaged site to its original or an equivalent form;
2. replacement of damaged structures with undamaged or newly synthesized substitutes;
3. prevention of damage caused by toxic photoproducts;
4. suppression of an extrasensitivity present in some systems.

V

vacuole (*Dujardin* 1941) — in mature cells, any of the large, customarily transparent, vesicles which are limited by a membrane referred to as the → tonoplast. Vacuoles are filled with a dilute solution called the cell sap. In meristematic tissues vacuoles are often quite small (diameter 2—10 μm) and numerous. Their shape is variable. The → endoplasmic reticulum, the Golgi cisternae, other cell organelles and preexisting provacuoles, all have been implicated in the origin of vacuoles.

vacuome (*Dangeard* 1919) — originally, a system comprising a variety of inclusions of plant cells, from the highly hydrated → tonoplast to the dense aleuron granules, all with the common ability to stain vitally with neutral red. Today, all the membrane-bounded spaces of the cell with the exception of the → mitochondria and → plastids which make up the chondriome (*de Duve* 1969) and the plastome respectively.

variability — the state or quality of being variable, or subject to → variation, i.e., of having the tendency to vary in form, nature, substance, etc. (→ genetic variability; variation).

variance — a numerical calculation describing the extent of dispersion of data around a mean (symbol s^2).

V. component analysis is a method of studying the genetics of continuously distributed variables in segregating populations. The magnitude of, and interrelationships among, components may be interpreted in terms of types of → gene action involved, → genetic relationships among population parents, and the selective potential available to the breeder.

variation — the occurrence of heritable or nonheritable differences in the permanent structures of cells ("intraindividual v."), among individuals of one population ("individual v."), or among populations ("group v.").

The chief sources of v. among the characteristics of related organisms are genic differences (due to → mutation or new assortment of genes) or environmentally induced differences which only cause temporary changes of the phenotype.

Primarily, biological v. may be subdivided into three categories:

1. **phenotypic variation**: The total biological v. of a given → character (symbol V_P). V. without natural discontinuities is called **continuous v.** and characters that exhibit it are quantitative or metric characters, as opposed to **discontinuous v.** and qualitative characters.

The components of continuous and discontinuous phenotypic v. in a segregating population may be grouped into two major classes referred to as genetic (symbol V_G) and nongenetic or environmental (symbol V_E) variation. By definition, phenotypic v. is then $V_P = V_G + V_E$.

2. **Environmental variation**: V. due to all intra- and extracellular factors which influence the expression of the → genotype. This part of v. is expressed in terms of environmental variance (symbol V_E) and

consists of two main components. One is represented by the intangible statistical residue known as "error" and certain genotype-environment interactions; the other is the controllable environment.

3. **Genetic variation:** V. due to the contribution of segregating genes and → gene interactions (symbol V_G) represented as the proportion of the total phenotypic variance that is exclusively genetic. This proportion is called the → heritability (symbol H) and is commonly expressed as a percentage (H = 100% if all the variation is due to genetic causes or there is no environmental variance; as the environmental component of variance increases, H decreases):

$$H = \frac{V_G}{V_P} \quad \text{or} \quad H = \frac{V_G}{V_G + V_E}.$$

The main components of genetic variance are "additive genetic variance" (due to any two alleles that have different quantitative effects contributing to this component), "dominance deviations" (deviations from additive variance due to dominance) and "interaction" or "epistatic deviations" (deviations from additive variance due to → epistasis and other non-allelic interactions).

Genetic v. in a population is controlled by three main factors (*Mayr* 1963): The input of new genetic information due to mutation and → gene flow; the erosion of this variation by selection and sampling errors; and the protection of the stored variability by cytophysiological devices and ecological factors (→ genetic variability).

variegation (*Schultz* 1936) — the occurrence, within one tissue, organ, or individual, of a mosaic phenotype (→ mosaicism) with respect to pigmentation (e.g., diversity in color due to the deprivation of a part of the normally green cells of a plant of their chlorophyll content) or any other easily observable phenotypic character. V. is a widespread phenomenon caused by several (observed or inferred) mechanisms (*Ar-Rushdi* 1957):

1. plastid variation and the behavior of plastid variants in somatic divisions of plants;
2. genic instability and the recurrence of somatic mutations;
3. instability of the phenotypic expression of genes as a result of → position effect;
4. genetically controlled → stickiness of chromosomes;
5. somatic → crossing-over;
6. somatic behavior of → ring chromosomes and → dicentric chromosomes (→ bridge-breakage-fusion-bridge cycle);
7. infection.

variety — in classical taxonomy, a heterogeneous grouping including non-genetic variations of the phenotype, morphs, domestic breeds, and geographic races (*Mayr* 1963).

versatile (*Crane* & *Thomas* 1941) — → reproduction.

vesicle — in eukaryotic cells, any of the closed structures completely surrounded by a → unit membrane (→ lamella). As opposed to vacuoles, the contents of vesicles are not in the liquid state.

vesicular conglomerate (*Yamada, Muta, Montomura & Koga* 1957) — = → multivesicular body.

v-gene reactivation (*Streisinger* 1956) — a phage-specific → reactivation system (symbol vR), due to the presence in the genome of phage T 4 of gene v^+ which makes the phage more than twice as resistant to UV-irradiation as the related phages T 2 and T 6 and the mutants of T 4 carrying the recessive allele v (*Rupert & Harm* 1966) (→ x-gene reactivation).

The product of the v^+ gene is essential for the formation of the single-strand breaks which initiate → excision repair.

viability — the capability for living or continuing to develop (→ lethal factor).

viability polymorphism (*Brues* 1969) — a term proposed to substitute for → genetic load (with the exception of mutational load) which is used with a variety of meanings.

viral transformation — = → cell transformation.

virion (*Lwoff* et al. 1959) — a complete, infectious virus particle, consisting of a nucleic acid core embedded in the protein → capsid (→ pseudovirion).

virogeny (*Koprowski* 1964) — in a nonlysing mammalian cell culture, the induction or active appearance of an infectious virus, the induction of an infectious nucleic acid, the presence of virus antigen, and the resistance of the cell to superinfection.

viroid (*Diener* 1971) — any of a class of plant pathogenic agents consisting of an infectious, single-stranded free RNA molecule with unusually low molecular weight (about 50,000). The RNA has → transfer RNA-like conformation, i.e., the single-strand is folded in such a way that it has a partial double-stranded character.

virogene (*Huebner & Todaro* 1961) — → oncogene hypothesis.

virosome (*Dahl & Kates* 1970) — any of a class of ribonucleoprotein particles (heterogeneous in size and density) which contain (vaccinia) virus DNA and can be found in the cytoplasm of cells infected with this virus. They contain the newly replicated viral DNA and an endogeneous → RNA polymerase activity.

virulent — of bacteriophages which are incapable of establishing the → lysogenic response (→ temperate).

virus — any of the infectious, subcellular and ultramicroscopic particles (usually nucleoprotein, more rarely exclusively DNA or RNA in nature) representing potentially pathogenic agents, whose replication in the cell and transmission by infection results in characteristic reactions of the host cells and host individuals. Viruses are independent genetic systems (endowed with genetic continuity and mutability) which possess an evolutionary history of their own. They use the synthetic machinery of living cells to direct the synthesis of specialized particles, the → virions, which contain the viral genome and transfer it to other cells. Major groups of viruses are bacterial viruses (bacteriophages), animal viruses, and plant viruses.

Nucleic acid molecules from mature → virions can be put into four classes: linear or circular and these either single- or double-stranded (→ chromosome).

virus maturation – a complex process which results in the formation of infective progeny particles inside the cell. It includes the formation of mature viral nucleic acid in the form found in the parent particles, the encoating or packaging of viral nucleic acid within a capsule, and the assambly of packaged nucleic acid with the accessory structures necessary for infectivity.

visibles – mutations recognizable by their phenotypic expression, in contrast to → vitality mutations and → lethal factors.

vitality mutation – any mutation the effect of which, in contrast to → visibles and → lethal factors, is difficult to determine, but which changes the vitality (viability) of the carrier genotype if present in an effective dosage.

Hadorn (1949) classified vitality mutations as follows:

1. **semilethal mutations** (viability of the carrier less than 50% as compared with a standard type).

2. **Subvital mutations** (viability of the carrier higher than 50%, but lower than normal when compared with a standard type).

3. **Supervital mutations** (viability of the carrier higher than that of the standard type).

V-type position effect (*Lewis* 1950) – → position effect.

W

Wallace-effect (*Grant* 1966) — selection for reproductive → isolation in areas of sympatry reinforcing previously existing barriers and thereby reducing genetic wastage, hybridization, and disruptive → gene flow.

W-chromosome — the → sex chromosome which, in cases of female heterogamety, is limited to the female sex. In male heterogamety this is the Y-chromosome (→ Z-chromosome).

whole-arm transfer (*Muller* 1940) — → translocation.

whole-arm transposition — = whole-arm transfer (→ translocation).

wild-type — ref. to a strain, organism, or gene (→ standard type; genetic nomenclature) of the type predominating in the wild population (*Darlington & Mather* 1949).

wobble (*Crick* 1966) — the ability of the third base in the → transfer RNA anticodon (5′ end) to hydrogen bond with any of two or three bases at the 3′ end of the → codon in mRNA, i.e., to undergo unusual base pairing. Due to w. a single tRNA species can recognize several different codons. Although w. allows some limited ambiguity in the codon — anticodon recognition, this is compensated by the degeneracy of the → genetic code, so that no → mistranslation follows. The rules of w. are:

1. If two codons specifying one amino acid begin with different bases, they are recognized by two transfer RNAs.

2. One species of transfer RNA can only recognize three codons if these have the first two bases in common and uracil, cytosine or adenine as the third base (these bases pair with inosine in the anticodon); two codons are recognized by one transfer RNA only if the first two bases are the same and the third is either adenine, guanine, uracil or cytosine, but not if one codon ends in a purine and the other in a pyrimidine.

3. Transfer RNAs specific for only one codon always recognize codons ending in guanine.

X

x — a symbol for the → basic number of chromosomes, i.e., the smallest number of chromosomes in a → chromosome set.

X-chromosome (*Henking* 1891; *McClung* 1902) — in organisms with male heterogamety and diploid sex differentiation ("diplogenotypic → sex determination"), the → sex chromosome which is present in both sexes (→ sex linkage). One sex (the homogametic, female sex) is homozygous (XX) with respect to this chromosome, the other (the heterogametic, male sex) is heterozygous (XO or XY). In organisms with haploid sex differentiation ("haplogenotypic sex determination"), the sex-chromosome of the female.

In **"multiple sex chromosome systems"** the X-chromosome may be represented more than twice in the female and more than once in the male sex (e.g., X_nO, X_nY, X_nY_n).

"Compound X-chromosomes" (*Novitski* 1954) are chromosomes produced by the attachment of two X-chromosomes to a single centromere (as in *Drosophila*); tandem metacentric compound X-chromosomes (metacentric chromosomes composed of two intact, originally acrocentric, X-chromosomes inverted with respect to each other; → attached X-chromosomes); tandem acrocentric compound X-chromosomes (two X-chromosomes attached to each other in tandem and to a single subterminal centromere); tandem compound ring X-chromosomes (monocentric ring chromosomes composed of two complete X-chromosomes in tandem sequence); reversed compound ring X-chromosomes (monocentric ring chromosomes composed of two complete X-chromosomes in reversed sequence).

The **"inactive-X hypothesis"** (*Lyon* 1961) states that in XY-males the single X-chromosome is active in all cells, while in each cell of the early female mammalian embryo (XX), one of the two X-chromosomes becomes inactivated. The choice of X-chromosome to be genetically inactivated in any one cell is assumed to be random. X-inactivation is a means of preventing gene imbalance in mammals with an XX-XY mechanism of sex determination (→ dosage compensation).

xenia (*Focke* 1881) — seeds whose hybrid nature is recognizable by particular phenotypic characters (color, shape, size, etc.) owing to direct influences exerted by the genotype of the pollen on the embryo or on the maternal tissue (endosperm) of the fruit.

xenobiotic (*Mason, North & Vaneste* 1965) — ref. to components of the chemical environment which are foreign to the metabolic network of an organism.

xerograft — = → heterograft.

x-gene reactivation (*Harm* 1963) — a phage-specific → reactivation system which increases the UV-resistance of T-even phages (absent in mutants T 4x and T 4vx) due to the presence of gene v^+ in the genomes (symbol xR).

In combination with gene v^+(T 4v^+x^+), UV-resistance is increased by a factor 4.4 times that of the double mutant (T 4vx). The mutation from x^+ to x is coupled with a decrease in the frequency of genetic recombination in the mutants. The x^+ gene is probably involved in a → post-replication (recombinational) repair system (→ v gene reactivation; UV-reactivation).

***X*-inactivation** — → dosage compensation.

Y

Y-chromosome (*Wilson* 1909) — in organisms with male heterogamety and diploid sex differentiation ("diplogenotypic → sex determination"), the → sex chromosome limited to one sex (the heterogametic, male sex) and which pairs with the → X-chromosome (→ sex linkage). In organisms with haploid sex differentiation ("haplogenotypic sex determination"), the sex chromosome of the male sex.

In **"multiple sex chromosome systems"** the Y-chromosome may be represented more than once (e.g., XY_n, X_nY_n) in the chromosome complement.

y gene reactivation (*Boyle & Symonds* 1969) — a phage T 4 gene that affects radiation sensitivity in a manner similar to → x gene reactivation.

yolk — a complex collection of macromolecules and smaller nutrient molecules with which the oöcyte is preloaded prior to → fertilization. Synthesis and deposition of y. constitute one of the major events of → oögenesis.

yolk nucleus — an intensely osmiophilic body situated in the vicinity of the oöcyte nucleus in invertebrates and vertebrates. The y. n. is a conglomeration of lamellae and vesicles as observed with the electron microscope (= Balbiani body).

Z

Z-chromosome — the → sex chromosome which, in cases of female heterogamety, is present in both sexes, in contrast to the → W-chromosomes.

zona occludens (*Farquhar & Palade* 1963) — = → tight junction.

zygomere (*Sybenga* 1966) — any of the hypothetical units (sites) on each chromosome whose function in eukaryotes is assumed to be the initiation of → chromosome pairing (long-distance attraction). Meiotic pairing is usually initiated at some specific loci (zygomeres) on each chromosome, from where it extends over the chromosome. The activation of zygomeres is obviously limited to specific periods during the development of the cell or organism and results in what is called "zygic activity". Zygomeres are suggested to be discrete units that (dependent upon the organism) may be present in either large numbers along the chromosome, or in lower frequency with concentration in certain regions of a chromosome, or even as a single entity in a large segment or entire chromosome. The z. pattern is held responsible for the determination of the pairing system.

zygonema (*Gregoire* 1907) — = zygotene (→ meiosis).

zygote (*Bateson* 1902) — 1. **Holozygote:** in eukaryotic organisms with sexual → reproduction, a diploid cell formed by the fusion of two haploid gametes during fertilization and (usually) containing two **complete** → genomes (hence holozygote); also the individual derived from it. For those genes which were represented by different → alleles in the genomes of the fusing gametes, the z. is said to be → **heterozygous** or hybrid, while it is **homozygous** with respect to genes represented by identical alleles in the gametes.

2. **Merozygote** (*Wollman, Jacob & Hayes* 1956): in prokaryotic organisms (bacteria), an **incomplete** "zygote" formed by parasexual mechanisms of genetic transfer ("meromixis" due to bacterial → conjugation mediated by → sex factors, → transduction, and → transformation), being diploid only for part of its genetic material and haploid for the remainder. A merozygote contains the complete genome of the recipient cell (the endogenote) and a → merogenote (or exogenote) derived from a donor cell. Merozygotes are usually very instable and may be → **hetero-** or **homogenotic** in respect to their genetic constitution. The potentiality for → genetic recombination in a merozygote is limited to the diploid part of its genetic material.

zygotene (*v. Winiwarter* 1900) — → meiosis.

zygotic induction (*Jacob & Wollman* 1956) — → induction.

zygotic mortality — one of the post-mating mechanisms (in addition to gametic m., hybrid inviability, and hybrid sterility) that prevent the success of an interspecific cross even though mating has taken place. The egg is fertilized but the zygote does not develop into an organism. In the case of **gametic m.** either the male or female gametes are destroyed because of the interspecific cross.

zygotonucleus — the nucleus of a holozygote formed by fusion (karyogamy) of two gametic nuclei.

Bibliography

Abelson, J., and *C. A. Thomas* (1966): The anatomy of the T 5 bacteriophage DNA molecule. J. Mol. Biol. **18**, 262.

Abercrombie, M., and *J. E. M. Heaysman* (1954): Observations on the social behaviour of cells in tissue culture. Exptl. Cell Res. **6**, 293.

Achtmann, M., *N. Willetts*, and *A. J. Clark* (1971): Beginning a genetic analysis of conjugational transfer determined by the F factor in *Escherichia coli* by isolation and characterization of transfer-deficient mutants. J. Bact. **106**, 529.

Adhya, S., *M. Gottesman*, and *B. De Crombrugge* (1974): Release of polarity in *Escherichia coli* by gene N of phage λ: Termination and antitermination of transcription. Proc. nat. Acad. Sci. Wash. **71**, 2534.

Afzelius, B. A. (1969): Ultrastructure of cilia and flagella. In: *A. Lima-de-Faria:* Handbook of Molecular Cytology, North-Holland Publ. Comp., Amsterdam, London, p. 1221.

Agar, W. E. (1911): The spermatogenesis of *Leptosiren paradoxa*. Quartl. J. micr. Sci. **67**, 1.

Ahlström, C. G. (1951): A short review of mitotic poisons. Acta path. microbiol. scand., Suppl. **91**, 52.

Ahmad, M. (1954): A consideration of the terms and mechanisms of heterothallism. Pakist. J. Sci. **5**, 59.

Alberts, B. M., and *L. Frey* (1970): T 4 bacteriophage gene 32: A structural protein in the replication and recombination of DNA. Nature, Lond. **227**, 1313.

Alexeiff, A. (1917): Sur la fonction glycoplastique du kinétoplaste (= kinétonucleus) chez les flagellés. C. R. Soc. Biol., Paris, **80**, 512.

Alfoldi, L., *F. Jacob* et *E. L. Wollman* (1957): Zygose létale dans des croisements entre souches colicinogènes et non colicinogènes d'*Escherichia coli*. C. R. Acad. Sci., Paris, **244**, 2974.

Allen, C. E. (1912): Cell structure, growth and division in the antheridia of *Polytricum juniperinum* Willd. Arch. Zellforsch. **8**, 121.

Allen, J. S. (1967): Cytogenetics of genomic exclusion in *Tetrahymena*. Genetics **55**, 797.

Allen, R. D., and *B. E. Hagström* (1955): Some interrelationships among cortical reaction phenomena in the sea urchin egg. Exp. Cell Res. Suppl. **3**, 1.

Allen, S. L. (1963): Genomic exclusion in *Tetrahymena*: Genetic basis. J. Protozool. **10**, 413.

Alpers, J. B., and *H. Paulus* (1971): Allosteric preconditioning: Role of allosteric ligands in promoting the maturation of enzymes. Nature, Lond., **233**, 478.

Altmann, R. (1889): Über Nukleinsäuren. Arch. Anat. Physiol., Lpz., Physiol. Abt. 524.

Amalric, F., *S. Bernard* and *R. Simard* (1973): Detection of single-stranded DNA in the nucleolus. Nature New Biol. **243**, 38.

Ambrose, E. J., and *D. M. Easty* (1970): Cell Biology. Nelson and sons, London.

Ames, B. N. (1971): The detection of chemical mutagens with enteric bacteria. In: *Hollaender, A.:* "Chemical Mutagens" Vol. 1, p. 267. Plenum Press/New York.

–, and *B. Garry* (1959): Coordinate repression of the synthesis of four histidine biosynthetic enzymes by histidine. Proc. nat. Acad. Sci., Wash., **45**, 1453.

–, and *P. E. Hartman* (1963): The histidine operon. Cold Spr. Harb. Symp. quant. Biol. **28**, 349.

Anderson, E. (1953): Introgressive hybridization. Biol. Rev. **28**, 280.
—, and *L. Hubricht* (1938): Hybridization in *Tradescantia.* I. II. The evidence for introgressive hybridization. Amer. J. Bot. **25**, 396.
Anderson, N. G. (1953): Studies on isolated cell components. V. The effects of various solutions on the nuclear envelope of the isolated rat liver nucleus. Exp. Cell Res. **4**, 306.
— (1953): On the nuclear envelope. Science **117**, 517.
Anderson, W. F. (1969): The effect of tRNA concentration on the rate of protein synthesis. Proc. nat. Acad. Sci., Wash., **62**, 566.
Angelt, T., B. Austen, and *D. G. Catcheside* (1970): Regulation of recombination at the his-3 locus in *Neurospora crassa.* Austral. J. of Biol. Sci. **23**, 1229.
Anker, P., M. Strouth, H. Greppin, and *M. Fredy* (1971): Metabolic DNA in spinach stems in connexion with ageing. Nature, N. B., London, **234**, 184.
Ar-Rushdi, A. H. (1957): The cytogenetics in a species hybridof *Nicotiana.* Genetics **42**, 312.
Ashburner, M. (1970): A prodromus to the genetic analysis of puffing in *Drosophila.* Cold Spring Harbor Symp. quant. Biology **35**, 533.
Astbury, W. T. (1950): Adventures in molecular biology. Harvey Lectures **46**, 3.
Atwood, S. S. (1944): The behavior of oppositional alleles in polyploids of *Trifolium repens.* Proc. nat. Acad. Sci., Wash., **30**, 69.
Auerbach, C. (1967): Heritage from Mendel. Proc. of the Mendel Centennial Symp. 1965. *R. A. Brink* (Ed.) Univ. Wisconsin Press/Madison, Milwaukee, London.
Auerbach, L. (1874): Organologische Studien. Breslau.
Austin, C. R. (1960): Anomalies of fertilization leading to triploidy. J. cell. comp. Physiol. (Symp. on Mammalian Genetics and Reproduction) **56**, (Suppl. 1) 1.
Avery, O. T., C. M. Macleod, and *M. McCarthy* (1944): Studies on the chemical nature of the substance inducing transformation of pneumococcal types. Induction of transformation by a deoxyribonucleic acid fraction isolated from *Pneumococcus* Type III. J. exp. Med. **79**, 137.

Bacci, G. (1950): Alcuni problemi dell'ermafroditismo negli Invertebrati. Boll. Zool. **17**, (Suppl.) 193.
— (1961): Recenti ricerche sulla determinazione polyfattoriale del secco. Boll. Zool. **28**, 469.
— (1965): Sex Determination. Pergamon Press/Oxford.
Bachvarova, R., E. H. Davidson, V. G. Allfrey, and *A. E. Mirsky* (1966): Activation of RNA synthesis associated with gastrulation. Proc. nat. Acad. Sci., Wash., **55**, 359.
Bade, E. G., N. S. Gonzales, and *I. D. Algranati* (1969): Dissociation of 70 S Ribosomes: Some properties of the dissociation factor from *Bacillus stearothermophilus* and *Escherichia coli.* Proc. nat. Acad. Sci., Wash., **64**, 654.
Bainbridge, B. W., and *J. A. Roper* (1966): Observations on the effects of a chromosome duplication in *Aspergillus nidulans.* J. gen. Microbiol. **42**, 417.
Bajer, A. (1968): Behavior and fine structure of spindle fiber during mitosis in endosperm. Chromosoma **25**, 249.
Baker, J. R. (1951): The "Golgi substance". Nature Lond. **168**, 1089.
Balfour, F. M. (1880): Comparative Embryology I.
Ball, O. (1925): Bakteriophage Wirkungen gegen Flexner- und Coli-Bakterien. Wien. klin. Wschr. **34**, 447.
Baltimore, D. (1970): RNA-dependent DNA polymerase in virions of RNA tumour viruses. Nature, Lond., **226**, 1209.
Bangham, A. D., M. M. Staudish, and *J. C. Watkins* (1965): Diffusion of univalent ions across the lamellae of swollen phospholipids. J. Mol. Biol. **13**, 238.

Barber, H. N. (1940): The suppression of meiosis and the origin of diplochromosomes. Proc. roy. Soc. B **128**, 170.
—, and *H. G. Callan* (1943): The effect of cold and colchicine on meiosis in the newt. Proc. roy. Soc. B **131**, 258.
Baricelli, N. A. (1956): A "chromosomic" recombination theory for multiplicity reactivation in phages. Acta biotheor., Leiden, **11**, 107.
Barner, H. D., and *S. S. Cohen* (1954): The induction of thymine synthesis by T 2 infection of a thymine requiring mutant of *Escherichia coli*. J. Bacteriol. **68**, 80.
Barr, M. L., and *E. G. Bertram* (1949): A morphological distinction between neurons of the male and female, and the behaviour of the nucleolar satellite during accelerated nucleoprotein synthesis. Nature, Lond., **163**, 676.
Barski, G., *S. Sorieul* et *F. Cornefert* (1960): Biologie cellulaire. — Production dans des cultures *in vitro* de deux souches cellulaires en association, de cellules de caratére "hybride". Compt. Rend. Acad. Sci., Paris, **251**, 1825.
Bary, A. de (1877): Vergleichende Anatomie der Vegetationsorgane der Phanerogamen und Farne. Leipzig.
Bateman, A. J. (1956): Cryptic self-incompatibility in the wallflower: *Cheiranthus cheiri* L. Heredity **10**, 257.
— (1966): Testing chemicals for mutagenicity in a mammal. Nature, Lond. **210**,205.
Bateson, W. (1894): Materials for the Study of Variation. MacMillan/London.
— (1905): in a letter to *A. Sedgewick*, from *Bateson, W.* (1928) Essays and Adresses, edited by *B. Bateson*, Cambridge Univ. Press/Cambridge.
— (1907): The progress of genetics since the rediscovery of Mendel's paper. Progr. Rei bot. **1**, 368.
—, and *E. R. Saunders* (1902): Experimental studies in the physiology of heredity. Rep. Evolut. Comm. roy. Soc., Rep. I, pp. 160.
— *E. R. Saunders*, and *R. C. Punnett* (1905): Experimental studies in the physiology of heredity. Rep. Evolut. Comm. roy. Soc., Rep. II, 1—55 and 80—99.
—, —, and — (1908): Experimental studies in physiology of heredity. Rep. Evolut. Common. roy. Soc., Rep. IV, pp. 60.
Battaglia, E. (1945): Sulla terminologia dei processi meiotici. Nuovo G. bot. ital. **52**, 42.
— (1947): Sulla terminologia dei processi apomitici. Nuovo G. bot. ital. **54**, 674.
— (1952): Appaiamento cromosomico primario, appaiamento cromosomico secondario ed appaiamento cromatidio secondario nello meiosi. Atti della Soc. Toscana **59**, 166.
— (1955): A consideration of a new type of meiosis (mis-meiosis) in *Juncacea* (*Luzula*) and Hemiptera. Bull. Torrey bot. Cl. **82**, 383.
— (1955): New symbols in cytology. Phytomorphology **5**, 171.
— (1955): The concepts of spore, sporogenesis and apospory. Phytomorphology **5**, 173.
— (1926): The concept of pseudopolyploidy. Caryologia **8**, 214.
—, and *J. W. Boyes* (1955): Post-reductional meiosis: its mechanism and causes. Caryologia **8**, 87.
Bauch, R. (1947): Trypaflavin als Typus der Chromosomengifte. Naturwissenschaften **34**, 346.
Baudhuin, P., *M. Müller*, *B. Poole*, and *C. de Duve* (1965): Non-mitochondrial oxidizing particles (microbodies) in rat liver and kidney and in *Tetrahymena pyriformis*. Biochem. biophys. Res. Commun. **20**, 53.
Bauer, H. (1935): Der Aufbau der Chromosomen aus den Speicheldrüsen von *Chironomus thummi* Kiefer (Untersuchungen an den Riesenchromosomen der Dipteren. I.). Z. Zellforsch. **23**, 280.
— (1943): Chromosomenforschung. Fortschr. Zool. **7**, 256.
— (1952): Die Chromosomen im Soma der Metazoen. Zool. Anz. (Suppl. Bd.) **17**, 252.

Bauer, H. R. Dietz, und *C. Röbbelen* (1961): Die Spermacytenteilungen der Tipuliden. III. Mitteilung. Das Bewegungsverhalten der Chromosomen in Translokationsheterozygoten von *Tipula oleracea*. Chromosoma **12,** 116.

Bauerle, R. H., and *P. Margolin* (1966): The functional organization of the tryptophan gene cluster in *Salmonella typhimurium*. Proc. nat. Acad. Sci., Wash., **56,** 111.

Baur, E. (1909): Das Wesen und die Erblichkeitsverhältnisse der „Varietas albomarginatus hort." von *Pelargonium zonale*. Z. indukt. Abstamm- u. Vererbungslehre **1,** 330.

Beadle, G. W. (1931): A gene in maize for supernumerary cell divisions following meiosis. Cornell. Univ. Agr. Exp. Stat. Mem. **135,** 12.

— (1932): A gene for sticky chromosomes in *Zea mays*. Z. indukt. Abstamm.-Vererbungslehre **63,** 195.

— (1933): Further studies in asynaptic maize. Cytologia, Tokyo **4,** 269.

— (1945): Biochemical genetics. Chem. Rev. **37,** 15.

— (1957): The role of the nucleus in heredity. In: *McElroy, W. D.,* and *B. Glass*: "The Chemical Basis of Heredity". Baltimore/Johns Hopkins Press 3.

—, and *E. L. Tatum* (1941): Genetic control of biochemical reactions in *Neurospora*. Proc. nat. Acad. Sci., Wash., **27,** 499.

Beatty, R. A. (1957): Parthenogenesis and Polyploidy in Mammalian Development. University Press, Cambridge.

Bedford, J. M. (1970): Sperm capacitation and fertilization in mammals. Biol. Reprod. **2** (Suppl. 2), **128.**

Beer, G. R. de (1951): Embryos and Ancestors. Univ. Press, Oxford.

Beermann, W. (1952): Chromomerenkonstanz und spezifische Modifikationen der Chromosomenstruktur in der Entwicklung und Organdifferenzierung von *Chironomus tentans*. Chromosoma **5,** 139.

—, and *G. F. Bahr* (1954): The submicroscopic structure of the Balbiani-Ring. Exp. Cell Res. **6,** 195.

Belar, K. (1928): Die zytologischen Grundlagen der Vererbung. Hdb. Vererbw. I. Bornträger/Berlin.

Belling, J. (1914): A study of semisterility. J. Hered. **5,** 65.

— (1914): The mode of inheritance of semisterility in the offspring of certain hybrid plants. Z. indukt. Abstamm.- u. VererbLehre **12,** 303.

— (1924): Detachment (elimination) of chromosomes in *Cypripedium acaule*. Bot. Gaz. **78,** 458.

— (1925): The origin of chromosomal mutations in *Uvularia*. J. Genet. **15,** 245.

— (1927): The attachments of chromosomes and the reduction division in flowering plants. J. Genet. **18,** 177.

—, and *A. F. Blakeslee* (1926): On the attachment of nonhomologous chromosomes at the reduction division in certain 25 chromosome Daturas. Proc. nat. Acad. Sci., Wash., **12,** 7.

Benda, C. (1898): Über die Spermatogenese der Vertebraten und höherer Evertebraten. II. Teil: Die Histogenese der Spermien. Arch. Anat. Physiol., Physiol. Abtlg., Lpz., 393.

— (1902): Die Mitochondria. Ergebn. Anat. **12,** 743.

Beneden, E. van (1870): Recherches sur la composition et la signification del' oeuf. Mém. cour Acad. R. Belg. **34,** 11.

— (1875): La maturation de l'oeuf, la fecondation et les premières phases du développement embryonnaire des mammifères etc. Bull. Acad. Belg. Cl. Sci. XL. II. Serie, 686.

—, and *E. Bessels* (1868): Mêmoire sur la formation du blastoderme chez les Amphipodes. Mem. cour. Acad. R. Belg. 34.

Bennet, H. S. (1963): Morphological aspects of extracellular polysaccharides. J. Histochem. Cytochem. **11,** 2.

Bennet, H. S. (1969): The cell surface: Components and configurations. In: *Lima-de-Faria, A.* (Ed.): "Handbook of molecular cytology" North-Holland Publ. Comp./Amsterdam, p. 1261.
Benzer, S. (1955): Fine structure of a genetic region in bacteriophage. Proc. nat. Acad. Sci., Wash., **41**, 344.
— (1957): The elementary units of heredity. In: *McElroy, W. D.*, and *B. Glass* "The Chemical Basis of Heredity". Baltimore, Johns Hopkins Press, 70.
— (1961): On the topography of the genetic fine structure. Proc. nat. Acad. Sci., Wash., **47**, 403.
Bernhard, W., Ch. Frayssinet, Ch. Lafarge, and *E. Leberton* (1965): Lésions nucléolaires précoces provoquées par l'aflatoxine dans les cellules hépatiques de Rat. C. R. Acad. Sci., Paris, **261**, 1785.
— *Gautier, A.*, et *C. Rouiller* (1954): La notion de "microsomes" et le problème de la basophilie cytoplasmatique. Arch. d'Anat. microsc. **35**, 236.
—, and *N. Granboulan* (1968): In "The Nucleus" (*A. J. Dalton* and *F. Haguenau*, Eds.), Acad. Press, New York and London, 81.
— *F. Haguenau,* et *U. Oberling* (1952): L'ultrastructure du nucléole de quelques cellules animales révélée par le microscope electronique. Experientia **8**, 58.
Bertani, G., and *J. J. Weigle* (1953): Host-controlled variation in bacterial viruses. J. Bacteriol. **65**, 113.
Bessis, M., J. Breton-Gorius, et *J. P. Thiery* (1958): Centriole, corps de Golgi et aster des leucocytes. Étude au microscope électronique. Rev. Hématol. **13**, 363.
Bhalla, S. C., and *R. R. Sokal* (1964): Competition among genotypes in the housefly at varying densities and proportions (the green strain), Evolution **18**, 312.
Binns, A., and *F. Meins* (1973): Habituation of tobacco pith cells for factors promoting cell division. Proc. nat. Acad. Sci. Wash., **70**, 2660.
Birdsell, J. B. (1950): Some implications of the genetical concept of race in terms of spatial analysis. Cold. Spr. Harb. Symp. quant. Biol. **15**, 259.
Birnstil, M., J. Speirs, J. Purdom, K. Jones, and *U. E. Loening* (1968): Properties and composition of the isolated ribosomal DNA satellite of *Xenopus jaevis.* Nature, London, **219**, 454.
Bischoff, G. W. (1835): Bemerkungen über die Lebermoose, vorzüglich aus den Gruppen der Marchantien und Riccien, nebst Beschreibung mehrerer theils kritischer, theils neuerer Arten. Nova Acta Acad. Caes. Leop. Carol. Nat. Cur. **17**, 911.
Bizzozero, G. (1871): Sulla struttura degli epiteli pavimentosi stratificati. Zentr. Med. Wochenschr. **9**, 482.
Blakeslee, A. F. (1904): Sexual reproduction in the *Mucorineae.* Proc. Amer. Acad. Arts Sci. **40**, 205.
— (1921): Types of mutations and their possible significance in evolution. Amer. Nat. **55**, 254.
Blakeslee, A. F. (1924): Distinction between primary and secondary chromosomal mutants in *Datura.* Proc. Nat. Acad. Sci., Wash., **10**, 109.
— (1927): Nubbin, a compound chromosomal type in *Datura.* Ann. N. Y. Acad. Sci. **30**, 1.
— (1928): Genetics of *Datura.* Z. indukt. Abstamm.- u. VererbLehre Suppl. **1**, 117.
—, *J. Belling,* and *M. Fahrnham* (1932): Inheritance in tetraploid *Datura.* Bot. Gaz. **76**, 329.
Blattner, F. R., J. E. Dahlberg, J. K. Doettiger, M. Fiandt, and *W. Szybalski* (1972): Distance from a promotor mutation to an RNA synthesis startpoint on bacteriophage λ DNA. Nature, New Biol. **237**, 232.
Bleier, H. (1930): Untersuchungen über das Verhalten der verschiedenen Kernkomponenten bei der Reduktionsteilung an Bastarden. Cellule **40**, 85.
Bleyman, M., and *C. Woese* (1969): "Transcriptional mapping", I. Introduction to

the method and the use of Actinomycin D as a transcriptional mapping agent. Proc. nat. Acad. Sci., Wash., **63**, 532.

Bloom, W., and *R. J. Leider* (1962): Optical and electron microscopic changes in ultraviolet-irradiated chromosome segments. J. Cell Biol. **13**, 269.

Boller, K., and *W. Schmid* (1971): Chemische Mutagenese beim Säuger. Das Knochenmark des Chinesischen Hamsters als in vivo-Testsystem. Hämatologische Befunde nach Behandlung mit Trenimon. Humangenetik **11**, 35.

Bollum, F. J. (1960): Oligodeoxyribonucleotide primers for calf thymus polymerase. J. biol. Chem. **235**, PC 18.

— (1963): Primer in DNA polymerase reactions. Progr. Nucleic Acid Res. **1**, 1.

Bonner, J. (1965): The Molecular Biology of Development. Clarendon Press/Oxford.

Böök, J. A. (1945): Cytological studies in *Triton*. Hereditas. Lund, **31**, 177.

Bordet, I. C. (1925): Le problème de l'autolyse microbienne transmissible ou du bacteriophage. Ann. Inst. Pasteur **84**, 273.

Borek, E., and *F. J. Ryan* (1958): The transfer of irradiation-elicited induction in a lysogenic organism. Proc. nat. Acad. Sci., Wash., **44**, 374.

—, *F. J. Ryan*, and *J. Rockenbach* (1955): Nucleic acid metabolism in relation to the lysogenic phenomenon. J. Bacteriol. **69**, 460.

Borisy, G. G., and *E. W. Taylor* (1967): The mechanism of action of colchicine binding of cholchicine-^{3}H to cellular protein. J. Cell. Biol. **34**, 525.

Boveri, Th. (1887): Über die Befruchtung der Eier von *Ascaris megalocephala*. S. B. Ges. Morph. Physiol., Münch., **3**, 71.

— (1888): Zellstudien. II. Die Befruchtung und Teilung des Eies von *Ascaris megalocephala*. Jena. Z. Naturw. **22**, 685.

— (1891): Befruchtung. Anatomische Hefte 1, Abt. II Ergebnisse.

— (1895): Über das Verhalten der Centrosomen bei der Befruchtung des Seeigeleies nebst allgemeinen Bemerkungen über Centrosomen und Verwandtes. Verh. phys.-med. Ges. Würzb., **29**, 1.

— (1899): Die Entwicklung von *Ascaris megalocephala* mit besonderer Rücksicht auf die Kernverhältnisse. Festschr. f. *C. von Kupffer*.

— (1901): Die Polarität von Ovocyte, Ei und Larve des *Strongylocentrotus lividus*. Spengels Zool. Jahrb. **14**, 630.

— (1905): Zellenstudien. V. Über die Abhängigkeit der Kerngröße und Zellenzahl der Seeigellarven von der Chromosomenzahl der Ausgangszellen. Jena. Z. Naturw. **39**, 445.

— (1907): Zellenstudien. VI. Die Entwicklung dispermer Seeigeleier. Ein Beitrag zur Befruchtungslehre und zur Theorie des Kernes. Jena. Z. Naturw. **43**, 1.

Bower, F. O. (1886): On apospory in ferns. J. Linn. Soc. Bot. **21**, 360.

Bowman, W. (1840): On the minute structure and movements of voluntary muscle. Phil. Trans. **130**, 457.

Boyce, R. P., and *P. Howard-Flanders* (1964): Release of ultraviolet light-induced thymine dimers from DNA in *E. coli* K-12. Proc. nat. Acad. Sci., Wash., **51**, 293.

Boycott, A. E., and *C. Diver* (1923): On the inheritance of sinistrality in *Limnea peregra*. Proc. roy. Soc. **195**, 143.

Boyle, J. M., and *N. Symonds* (1969): Radiation-sensitive mutants of T_4D I. T_4y: A new radiation-sensitive mutant; effect of the mutation on radiation survival, growth and recombination. Mutation Res. **8**, 431.

Bradshaw, A. D. (1965): Evolutionary significance of phenotypic plasticity in plants. Advanc. Genet. **13**, 115.

Braun, W. (1965): Bacterial Genetics (2. Ed.). Saunders/Philadelphia.

Bray, R. A., and *R. A. Brink* (1966): Mutation and paramutation at the R locus in maize. Genetics **54**, 137.

Brehme, K. S. (1939): A study of the effect on development of "Minute" mutations in *Drosophila melanogaster*. Genetics **24**, 131.

Brenner, S., and *J. R. Beckwith* (1965): Ochre mutants, a new class of suppressible nonsense mutants. J. Mol. Biol. **13**, 629.
—, *F. Jacob*, and *M. Meselson* (1961): An unstable intermediate carrying information from genes to ribosomes for protein synthesis. Nature, Lond., **190**, 576.
—, *A. O. W. Stretton*, and *S. Kaplan* (1965): Genetic code: "nonsense" triplets for chain termination and their suppression. Nature, Lond., **206**, 994.
—, *L. Barnett, F. H. C. Crick*, and *A. Orgel* (1961): The theory of mutagenesis. J. Mol. Biol. **3**, 121.
—, —, *E. R. Katz*, and *F. H. C. Crick* (1967): UGA: A third nonsense triplet in the genetic code. Nature, Lond., **213**, 449.
Bresch, C. (1962): Replication and recombination in bacteriophage. Z. Vererb-Lehre **93**, 476.
— (1964): Klassische und molekulare Genetik. Springer/Berlin, Göttingen, Heidelberg.
—, und *R. Hausmann* (1972): Klassische und molekulare Genetik (3. Aufl.). Springer-Verlag, Berlin, Heidelberg, New York.
Breswick, E., and *A. Schwartz* (1968): Functional dynamics of the cell. Academic Press/New York and London.
Breuer, M. E., and *C. Pavan* (1955): Behavior of polytene chromosomes of *Rhynchosciara angelae* at different stages of larval development. Chromosoma **7**, 371.
Bridges, B. A. (1972): Evidence for a further dark repair process in bacteria. Nature N. B., Lond., **240**, 52.
— (1973): Some general principles of mutagenicity screening and a possible framework for testing procedures. Environ. Health Perspect. **6**, 221.
— (1974): The three-tier approach to mutagenicity screening and the concept of radiation-equivalent dose. Mut. Res. **26**, 335.
Bridges, C. B. (1913): Non-disjunction of the sex chromosomes of *Drosophila*. J. exp. Zool. **15**, 587.
— (1914): Direct proof through non-disjunction that the sex-linked genes of *Drosophila* are borne by the X-chromosome. Science **40**, 107.
— (1916): Non-disjunction as proof of the chromosome theory of heredity. Genetics **1**, 1 and 107.
— (1917): Deficiency, Genetics **2**. 445.
— (1919): Specific modifiers of eosin eye color in *Drosophila melanogaster*. J. exp. Zool. **28**, 337.
— (1919): Duplications. Anat. Rec. **15**, 357.
— (1922): The origin of variations in sexual and sexlimited chracters. Amer. Nat. **56**, 51.
— (1923): Aberrations in chromosome materials. Scient. Pap. 2nd Int. Congr. Eugenics **1**, 76.
— (1932): The suppressors of purple. Z. indukt. Abstamm.- u. VererbLehre **60**, 207.
— (1932): Recombination and crossing-over. Amer. Nat. **66**, 571.
— (1934): The testcross — a suggested genetic term. J. Hered. **25**, 18.
— (1935): Demonstrations of the first translocation in *Drosophila melanogaster* and of normal repeats in chromosomes. Rec. Genet. Soc. Amer. **4**, and Amer. Nat. **70**, 41 (1936).
— (1935): Salivary chromosome maps. With a key to the banding of the chromosomes of *Drosophila melanogaster*. J. Hered. **26**, 60.
— (1937): Correspondence between linkage maps and salivary chromosome structure, as illustrated in the tip of chromosome 2R of *Drosophila melanogaster*. Cytologia, Tokyo, Fujii Jub. Vol. 745.
—, and *Th. H. Morgan* (1923): The third-chromosome group of mutant characters of *Drosophila melanogaster*. Publ. Carneg. Instn. Nr. 327.
Brieger, F. (1928): Über die Vermehrung der Chromosomenzahl bei dem Bastard

Nicotiana tabacum L. × *N. rubyi* Britt. Z. indukt. Abstamm.- u. VererbLehre **47**, 1.
Brieger, F. (1958): Populationsgenetik. In: *Th. Roemer* und *W. Rudorf* „Handbuch der Pflanzenzüchtung" (2. Aufl.) Parey/Berlin, Bd. **I**, 176.
Brink, R. A. (1932): Are the chromosomes aggregates of groups of physiologically independent genes? Amer. Naturalist **66**, 444.
— (1958): Paramutation at the R locus in maize. Cold Spr. Harb. Symp. quant. Biol. **23**, 379.
— (1960): Paramutation and chromosome organization. Quart. Rev. Biol. **35**, 120.
— (1962): Phase change in higher plants and somatic cell heredity. Quart. Rev. Biol. **37**, 1.
— (1964): Genetic repression in multicellular organisms. Amer. Nat. **98**, 193.
Brinkley, B. R., and *E. Stubblefield* (1966): The fine structure of the kinetochore of a mammalian cell *in vitro*. Chromosoma **19**, 28.
Brinton, C. (1959): Non-flagellar appendages of bacteria. Nature **183**, 782.
— (1972): Epiviruses. In: *G. Raspe* (Ed.) Workshop on mechanisms and prospects of genetic exchange. Adv. in Biosciences **8**, 75.
—, *A. Buzzell*, and *M. A. Lauffer* (1954): Electrophoresis and phage susceptibility studies on a filament producing variant of the *E. coli* B bacterium. Biochim. biophys. Acta **15**, 533.
—, *P. Gemski*, and *J. Carnaham* (1964): A new type of bacterial pilus genetically controlled by the fertility factor of *E. coli* K 12 and its role in chromosome transfer. Proc. nat. Acad. Sci., Wash., **52**, 776.
Britten, R. J., and *E. H. Davidson* (1969): Gene regulation for higher cells.: A theory. Science **165**, 349.
—, and *D. E. Kohne* (1968): Repeated sequences in DNA. Science **161**, 529.
Brody, S., and *C. Yanofsky* (1963): Suppressor gene alteration of protein primary structure. Proc. nat. Acad. Sci., Wash., **50**, 9.
Brongniart, A. (1834): Nouvelles recherches sur la structure de l'épiderme des végétaux. Ann. Sci. Nat. **1**, 65.
Brown, D. D., and *C. S. Weber* (1968): Gene linkage by RNA-DNA hybridization. II. Arrangement of the redundant gene sequences for 28s and 18s ribosomal RNA. J. Mol. Biol. **34**, 681.
—, *P. C. Wensink*, and *E. Jordan* (1972): A comparison of the ribosomal DNA's of *Xenopus laevis* and *Xenopus mulleri*: the evolution of tandem genes. J. Mol. Biol. **63**, 57.
Brown, R. (1831): Observations on the organs and mode of fecundation in *Orchideae* and *Asclepiadeae*. Trans. Linn. Soc., Lond., (Bot.).
Brown, S. W. (1963): The Comstockiella system of chromosome behavior in the armored scale insects (*Coccoidea*: *Diaspididae*). Chromosoma **14**, 360.
— (1964): Automatic frequency response in the evolution of male haploidy and other coccid chromosome systems. Genetics **49**, 797.
— (1966): Heterochromatin. Science **151**, 417.
—, and *C. Cleveland* (1968): Meiosis in the male of *Puto albicans* (*Coccoidea-Homoptera*). Chromosoma **24**, 210.
Brown, W. K., and *E. O. Wilson* (1956): Character displacement. Systemat. Zool. **5**, 49.
Brues, A. M. (1969): Genetic load and its varieties. Science **164**, 1130.
Bryant, L. R. (1935): A study of the factors affecting the development of the embryosac and the embryo in the McIntosh apple. Univ. New Hampsh. agric. Exp. Stat. Bull. **61**, 1.
Bucher, O. (1959): Die Amitose der tierischen und menschlichen Zelle. Protoplasmatologia, Vol. VI E 1, Springer/Wien.
Bütschli, O. (1871): Vorläufige Mitteilung über Bau und Entwicklung der Samenfäden bei Insekten und Crustaceen. Z. wiss. Zool. **21**, 526.

Bütschli, O. (1876): Studien über die ersten Entwicklungsvorgänge der Eizelle, die Zellteilungsvorgänge der Eizelle, die Zellteilung und die Konjugation der Infusorien. Abh. senckenb. naturf. Ges. **10**.

Buller, A. H. R. (1931): The biological significance of conjugate nuclei in the basidiomycetes. Proc. 5th int. bot. Congr. pp. 357.

— (1941): The diploid cell and the diploidization process in plants and animals, with special reference to the higher fungi. I a. II. Bot. Rev. **7**, 335.

Bullough, W. S. (1963): Analysis of the life-cycle in mammalian cells. Nature, Lond., **199**, 859.

— (1965): Mitotic and functional homeostasis. Cancer Res. **25**, 1683.

— (1967): The Evolution of Differentiation. Acad. Press, London and New York.

Burgeff, H. (1915): Untersuchungen über Variabilität, Sexualität und Erblichkeit bei *Phycomyces nitens Kuntze* II. Flora **108**, 353.

Burgess, R. R., A. A. Travers, J. J. Dunn, and *E. K. F. Bautz* (1969): Factor stimulating transcription by RNA polymerase. Nature, London, **221**, 43.

Burnett, I. H. (1956): The mating systems of fungi I. New Phytol. **55**, 50.

Burnham, C. R. (1962): Discussions in Cytogenetics. Burgess/Minneapolis.

Butler, E. G.(1933): The effects of X-radiation on the regeneration of the forelimbs of *Amblystoma larvae*. J. exp. Zool. **65**, 271.

Cahn, R. D. (1969): Factors affecting inheritance and expression of differentiation: Some methods of analysis. In: *Beermann, W.*, et al. (Ed.) Results and Problems in Cell Differentiation, Vol. **1**, 58.

Caldecott, R. S., and *L. Smith* (1952): A study of X-ray induced chromosomal aberrations in barley. Cytologia, Tokyo, **17**, 224.

Calhoun, D. H., and *G. W. Hatfield* (1973): Autoregulation: a role for a biosynthetic enzyme in the control of gene expression. Proc. Nat. Acad. Sci., Wash., **70**, 757.

Callan, H. G., and *L. Lloyd* (1960): Lampbrush chromosomes of crested newts, *Triturus cristatus* (Laurenti). Phil. Trans. Roy. Soc. (London) B. **243**, 135.

—, and *S. G. Tomlin* (1950): Experimental studies on amphibian oocyte nuclei. I. Investigation of the structure of the nuclear membrane by means of the electron microscope. Proc. roy. Soc. B **137**, 367.

Calvin, M. (1962): The path of carbon in photosynthesis. Science **135**, 879.

Camp, W. H., and *C. L. Gilly* (1942): The structure and origin of species. Brittonia,. N. Y. **4**, 323.

Campbell, A. M. (1962): General properties of episomes. Adv. Genetics **11**, **101**.

Cannon, W. B. (1929): Organization for physiological homeostasis. Phys. Rev. **9**, 399.

Cantoni, G. L., H. Ishikura, H. H. Richards, and *K. Tanaka* (1963): Studies on soluble ribonucleic acid. XI. A model for the base sequence of serine s-RNA. Cold Spr. Harb. Symp. quant. Biol. **28**, 123.

Capecchi, M. R. (1967): Polypeptide chain termination in vitro: Isolation of a release factor. Proc. nat. Acad. Sci., Wash., **58**, 1144.

—, and *G. N. Gussin* (1965): Suppression in vitro: Identification of a serine-sRNA as a "nonsense" suppressor. Science **149**, 417.

Carothers, E. E. (1917): The segregation and recombination of homologous chromosomes. J. Morph. **28**, 445.

Carson, H. L. (1957): The species as a field for gene recombination. In: *E. Mayr* (Ed.) "The species problem", Am. Ass. Adv. of Sci. Publ. **50**.

—, *F. E. Clayton,* and *H. D. Stalker* (1967): Karyotypic stability and speciation in Hawaiian *Drosophila*. Proc. Nat. Acad. Sci., Wash., **57**, 1280.

Carter, T. C., and *A. Robertson* (1952): A mathematical treatment of genetical recombination using a four-strand model. Proc. roy. Soc. B. **139**, 410.
Castle, W. (1905): Recent discoveries in heredity and their bearing on animal breeding. Pop. Sci. Mon. **66**, 193.
— (1906): Yellow mice and genetic purity. Science **24**, 275.
Catcheside, D. G. (1958): Introduction. In: "A discussion on the cytoplasm in variation and development". Proc. roy. Soc. B. **148**, 285.
Cattanach, B. M. (1974): Position effect variegation in the mouse. Genet. Research **23**, 291–306.
Caullery, M. (1913): Les Problèms de la Sexualité. Flammarion/Paris.
Chambers, R. (1940): Recent development of the micromanipulative technique and its application. J. Roy. Soc. Microsc. **60**, 113.
Champe, S. P., and *S. Benzer* (1962): Reversal of mutant phenotypes by 5-fluorouracil: an approach to nucleotide sequences in messenger RNA. Proc. nat. Acad. Sci., Wash., **48**, 532.
Changeux, J. P. (1964): Allosteric interactions interpreted in terms of quaternary structure. Brookh. Symp. in Biol. **17**, 232.
Charpak, M., et *R. Dedonder* (1965): Production d'un "facteur de compétence" soluble par *Bacillus subtilis Marburg* ind_{168}^{-}. C. R. Hebd. Seanc. Acad. Sci., Paris, **260**, 5638.
Chase, M., and *A. H. Doermann* (1958): High negative interference over short segments of the genetic structure of bacteriophage T 4. Genetics **43**, 332.
Chatton, E. (1920): Les Péridines parasites: Morphologie, reproduction, éthologie. Arch. Zool. Exp. Gén. **59**, 1.
— (1925): *Pansporella perplexa.* Ann. Sci. natur. Zool. **8**, 5.
Chiarugi, A. (1933): La cariologia nelle sue applicazioni a problemi di botanici. Atti Soc. Ital. par il Prog. della Scienze **3**, 1.
Chouinard, L. A. (1970): Localization of intranucleolar DNA in root meristematic cells of *Allium cepa.* J. Cell. Sci **6**, 73.
Chu, E. H. Y., *H. C. Thuline*, and *D. E. Norby* (1964): Triploid-diploid chimerism in a male tortoiseshell cat. Cytogenetics **3**, 1.
Chun, E. H. L., *M. H. Vaughan*, and *A. Rich* (1963): The isolation and characterization of DNA associated with chloroplast preparations. J. Mol. Biol. **7**, 130.
Clark, A. J. (1967): The beginning of a genetic analysis of recombination proficiency. J. Cell Physiol. **70** (Suppl. 1) 165.
—, and *E. A. Adelberg* (1962): Bacterial conjugation. Annu. Rev. Microbiol. **16**, 289.
—, and *A. D. Margulies* (1965): Isolation and characterisation of recombination-deficient mutants of *Escherichia coli* K 12. Proc. nat. Acad. Sci., Wash., **53**, 451.
Clark-Walker, G. D. (1973): Translocation of messenger RNA. In: *Stewart, P. R.*, and *D. S. Letham* (Eds.): "The Ribonucleic Acids". Springer-Verlag/Berlin, Heidelberg, New York, p. 135.
Clarke, B. (1962): Balanced polymorphism and the diversity of sympatric species. In "Taxonomy and Geography" (*D. Nichols*, Ed.); Systematics Association, Oxford.
Clausen, J., *D. D. Keck*, and *W. M. Hiesey* (1945): Experimental studies on the nature of species. Publ. Carneg. Instn. **564**, pp. 174.
Cleaver, J. E. (1974): Repair processes for photochemical damage in mammalian cells. Adv. in Rad. Biol. **4**, 1.
Cleland, R. E. (1922): The reduction-divisions in the pollen mother cells of *Oenothera Franziskana.* Amer. J. Bot. **9**, 391.
— (1957): Chromosome structure in *Oenothera* and its effect on the evolution of the genus. Cytologia, Tokyo (Proc. Int. Genet. Symp.) 1956, 5.
— (1962): The cytogenetics of *Oenothera.* Advanc. Genet. **11**, 147.
Cleveland, I. R. (1947): The origin and evolution of meiosis. Science **105**, 287.

Cleveland, I. R. (1949): The whole life cycle of chromosomes and their coiling systems. Trans. Amer. phil. Soc. N. S. **39**, Part 1.

Clever, U. (1964): Puffing in giant chromosomes of Diptera and the mechanism of its control. In: *J. Bonner*, and *P. Tso* (Ed.) "The Nucleohistones", Holden-Day/San Francisco.

Clowes, R. C. (1972): Molecular structure of bacterial plasmids. Bact. Rev. **36**, 361.

Clusius, C. (1601): Rariorum plantarum historia. Antwerpen.

Coats, J. H., and *E. N. Nester* (1967): Regulation reversal mutation: Characterization of end-product-activated mutants of *Bacillus subtilis*. J. of biol. Chemistry **242**, 4948.

Cohen, S. N., and *A. C. Y. Chang* (1974): A method for selective cloning of eukaryotic DNA fragments in *E. coli* by repeated transformation. Molec. gen. Genet. **134**, 133.

—, —, *H. W. Boyer*, and *R. B. Helling* (1973): Construction of biologically functional bacterial plasmids in vitro. Proc. nat. Acad. Sci., Wash., **70**, 3240.

Cole, R. D. (1972): zit. nach *S. C. R. Elgin*, and *J. Bonner:* Isolated chromatin in the study of gene expression. In: *I. K. Pollak*, and *J. Wilson Lee*, Eds.: The biochemistry of gene expression in higher organismus.

Collins, G. N., and *J. H. Kempton* (1916): Patrogenesis. J. Hered. **7**, 106.

Comings, D. E. (1972): Heavy shoulder DNA. Exptl. Cell Res. **70**, 259.

—, and *T. A. Okada* (1973): DNA replication and the nuclear membrane. J. Mol. Biol. **75**, 609.

Cone, C. D. (1969): Some theoretical aspects of intercellular bridges as a potential mechanism of cancerous proliferation. J. theoret. Biol. **22**, 365.

Conger, A. D. (1955): Discussion of the last three papers. J. cell. comp. Physiol. **45**, Suppl. 2, 309.

—, and *N. H. Giles* (1950): The cytogenetic effect of slow neutrons. Genetics **35**, 397.

Conway, T. W., and *F. Lipmann* (1964): Characterization of a ribosome-linked guanosine triphosphatase in *Escherichia coli* extracts. Proc. nat. Acad. Sci. Wash., **52**, 1462.

Cook, S. A. (1965): Reproduction, Heredity, and Sexuality. MacMillan/London.

Cooper, K. W. (1941): Bivalent structure in the fly *Melophagus ovium* L. Proc. nat. Acad. Sci., Wash., **27**, 109.

— (1944): Analysis of meiotic pairing in *Olfersia* and consideration of the reciprocal chiasma hypothesis of sex chromosome conjugation in male *Drosophila*. Genetics **29**, 537.

— (1946): The mechanism of non-random segregation of sex chromosomes in male *Drosophila miranda*. Genetics **31**, 181.

— (1959): Cytogenetic analysis of major heterochromatic elements (especially Xh and Y) in *Drosophila melanogaster* and the theory of "heterochromatin". Chromosoma **10**, 535.

Copeland, J. C., and *V. Bryson* (1966): Restriction in matings of *Escherichia coli* strain K-12 with strain B. Genetics **54**, 441.

Correns, C. (1900): Gregor Mendels Regel über das Verhalten der Nachkommenschaft der Rassenbastarde. Ber. dtsch. bot. Ges. **18**, 158.

— (1900): Über Levkojenbastarde. Zur Kenntnis der Grenzen der Mendelschen Regeln. Bot. Zbl. **84**, 97.

— (1901): Über Bastarde zwischen Rassen von *Zea mays*, nebst einer Bemerkung über die „faux hybrids" Milliardet's und die „unechten Bastarde" de Vries'. Ber. dtsch. bot. Ges. **19**, 211.

— (1902): Scheinbare Ausnahmen von der Mendelschen Spaltungsregel für Bastarde. Ber. dtsch. Bot. Ges. **20**, 157.

— (1902): Weitere Beiträge zur Kenntnis der dominierenden Merkmale und der

Mosaikbildung der Bastarde. Die Merkmalspaare beim Studium der Bastarde. Ber. dtsch. bot. Ges. **21**, 195.

Correns, C. (1902): Über den Modus und den Zeitpunkt der Spaltung der Anlagen bei den Bastarden vom Erbsen-Typus. Bot. Ztg. **60**, 65.

— (1904): Experimentelle Untersuchungen über die Gynodioecie. Ber. dtsch. bot. Ges. **22**, 506.

— (1909): Vererbungsversuche mit blaß (gelb) grünen und buntblättrigen Gruppen bei *Mirabilis, Urtica* und *Lunaria*. Z. indukt. Abstamm.- u. VererbLehre **1**, 291.

— (1928): Bestimmung, Vererbung und Verteilung des Geschlechts bei höheren Pflanzen. Hdb. Vererb. Wiss. **2**, Bornträger/Berlin.

Cowdry, E. V. (1916): The mitochondrial constituents of protoplasm. Publ. Carneg. Instn. 271.

Crane, M. B., and *P. T. Thomas* (1949): Reproductive versatility in *Rubus*. III. Raspberry-blackberry hybrids. Heredity **3**, 99.

Crick, F. H. C. (1958): On protein synthesis. Symp. Soc. exp. Biol. **12**, 138.

— (1963): The recent excitement in the coding problem. Progr. Nucleic Acid Res. **1**, 163.

— (1966): Codon-anticodon pairing: The wobble hypothesis. J. Mol. Biol. **19**, 458.

—, and *L. E. Orgel* (1964): The theory of inter-allelic complementation. J. Mol. Biol. **8**, 161.

—, *L. Barnett, S. Brenner,* and *R. J. Watts-Tobin* (1961): General nature of the genetic code for protein. Nature, Lond., **192**, 1227.

Crouch, Y. F., and *M. L. Barr* (1954): Behaviour of the sex chromatin during axon reaction. J. Neuropathol. **13**, 353.

Crow, J. F. (1948): Alternative hypotheses of hybrid vigor. Genetics **33**, 477.

— (1958): Some possibilities for measuring selection intensities in man. Hum. Biol. **30**, 1.

— (1970): Genetic loads and the cost of natural selection. In: *Kojima, K.* (Ed.) Biomathematics, Vol. 1 "Mathematical topics in population genetics". Springer/ Berlin, Heidelberg, New York.

— (1973): Impact of various types of genetic damage and risk assessment. Env. Health Persp. **6**, 1.

—, and *M. Kimura* (1965): The theory of genetic loads. In: Proc. XI. Int. Congr. of Genet., The Hague **3**, 498.

Crowe, J. K. (1964): The evolution of outbreeding in plants. Heredity **19**, 435.

Cunningham, W. P., D. J. Morré, and *H. H. Mollenhauer* (1966): Structure of isolated plant Golgi apparatus revealed by negative staining. J. Cell Biol. **28**, 169.

Curtiss, R. (1969): Bacterial conjugation. Ann. Rev. Microbiol. **23**, 69.

***Dahl**, R.*, and *J. R. Kates* (1970): Intracellular structures containing vaccinia DNA: Isolation and characterization. Virology **42**, 453.

Dahmus, M. E., and *J. Bonner* (1970): Nucleoproteins in regulation of gene function. Fed. Proc. **29**, 1255.

D'Amato, F. (1948): The effect of colchicine and ethylene glycol on sticky chromosomes in *Allium*. Hereditas, Lund, **24**, 83.

— (1954): Osservazioni citoistologiche sulla attiva antimitotica. Caryologia **6**, 160.

Dangeard, P. A. (1919): Sur la distinction du chondriome des auteurs en vacuome, plastidome et sphérome. C. R. Acad. Sci., Paris, **169**, 1005.

— (1920): Plastidome, vacuome et sphérome dans *Sellaginella Kraussiana*. C. R. Acad. Sci., Paris, **170**, 301.

Danielli, J. F. (1956): Cytoplasmic inheritance. Nature, Lond., **178**, 214.

Danser, B. H. (1929): Über die Begriffe Komparium, Komisskuum und Konvivium und über die Entstehungsweise der Konvivien. Genetica **11**, 399.

Darlington, C. D. (1928): Studies in *Prunus* I und II. J. Genet. **19**, 213.
— (1929): Chromosome behaviour and structural hybridity in the *Tradescantiae*. J. Genet. **21**, 207.
— (1929): The significance of chromosome behaviour in polyploids for the theory of meiosis. John Innes Hort. Inst. Conf. on Polyploidy, p. 42.
— (1930): A cytological demonstration of "genetic" crossing-over. Proc. roy. Soc. B. **107**, 50.
— (1931): Chiasma formation and chromosome pairing in *Fritillaria*. Proc. II. Int. Bot. Congr. p. 189.
— (1931): Meiosis. Biol. Rev. **6**, 221.
— (1931): The mechanism of crossing-over. Science, **73**, 561.
— (1932): Recent Advances in Cytology. Churchill/London.
— (1932): The control of the chromosomes by the genotype and its bearing on some evolutionary problems. Amer. Nat. **66**, 25.
— (1935): The time, place and action of crossing-over. J. Genet. **31**, 185.
— (1935): The internal mechanics of the chromosomes. I. The nuclear cycle in *Fritillaria*. Proc. roy. Soc. B. **118**, 33.
— (1935): The internal mechanics fo the chromosomes. II. Prophase pairing at meiosis in *Fritillaria*. Proc. roy. Soc. B. **118**, 59.
— (1935): The internal mechanics of chromosomes. III. Relational coiling and crossing-over in *Fritillaria*. Proc. roy. Soc. B. **118**, 74.
— (1936): Crossing-over and its mechanical relationships in *Chorthippus* and *Stauroderus*. J. Genet. 33, 465.
— (1936): The external mechanics of chromosomes. Proc. roy. Soc. B. **121**, 264.
— (1937): What is a hybrid? J. Hered. **28**, 308.
— (1937): Recent Advances in Cytology. 2nd Ed. Churchill/London.
— (1939): Misdivision and the genetics of the centromere. J. Genet. **37**, 341.
— (1939): The Evolution of Genetic Systems. Cambridge Univ. Press/Cambridge.
— (1940): The origin of iso-chromosomes. J. Genet. **39**, 351.
— (1941): Polyploidy, crossing-over, and heterochromatin in *Paris*. Ann. Bot. **5**, 203.
— (1944): Heredity, development and infection. Nature, Lond., **154**, 164.
— (1957): Messages and movements in the cell. Conference on Chromosomes, Wageningen. Willink/Zwolle.
— (1958): The Evolution of Genetic Systems. 2nd Ed. Oliver & Boyd/Edinburgh.
— (1965): Cytology. Churchill/London.
—, and *A. E. Gairdner* (1937): The variation system in *Campanula persicifolia*. J. Genet. **35**, 97.
—, and *E. K. Janaki-Ammal* (1945): Chromosome Atlas of Cultivated Plants. Allen & Unwin/London.
—, and *L. F. La Cour* (1938): Differential reactivity of the chromosomes. Ann. Bot. **2**, 615.
—, and *K. Mather* (1932): The origin and behaviour of chiasmata. III. Triploid tulipa. Cytologia, Tokyo, **4**, 1.
—, and — (1949): The Elements of Genetics. Allen & Unwin/London.
—, and *A. A. Moffet* (1930): Primary and secondary chromosome balance in *Pyrus*. J. Genet. **22**, 129.
—, and *P. F. Thomas* (1941): Morbid mitosis and the activity of inert chromosomes in *Sorghum*. Proc. roy. Soc. B. **130**, **127**.
—, and *M. B. Upcott* (1939): The measurement of packing and contraction in chromosomes. Chromosoma **1**, 23.
—, and — (1941): The activity of inert chromosomes in *Zea mays*. J. Genet. **41**, 275.
—, and *A. P. Wylie* (1953): A dicentric cycle in *Narcissus*. Heredity **6**, (Suppl. Vol.) 197.

Darlington, C. D., J. B. S. Haldane, and *P. C. Koller* (1934): Possibility of incomplete sex linkage in mammals. Nature, Lond., **133**, 417.

Darlington, P. J. (1972): Non mathematical models for evolution of altruism, and for group selection. Proc. nat. Acad. Sci., Wash., **69**, 293.

Darwin, Ch. (1858/59): On the tendency of species to form varieties, and on the perpetuation of varieties and species by natural means of selection. I. On the variation of organic beings in a state of nature; on the natural means of selection; on the comparison of domestic races and true species. Journ. of the Proc. of the Linnean Soc., Zoology **3**, Nr. 9, 45.

— (1876): The Effects of Cross and Self-fertilization in the Vegetable Kingdom. Murray/London.

— (1877): The different Forms of Flowers on Plants of the same Species. Murray/London.

Datta, P., and *H. Gest* (1964): Alternative patterns of end-product control in biosynthesis of amino-acids of the aspartic family. Nature, Lond., **203**, 1259.

Davidson, W. M., and *D. R. Smith* (1954): A morphological sex difference in the polymorphonuclear neutrophil leucocytes. Brit. med. J. **2**, 6.

Davis, B. D. (1949): Isolation of biochemically deficient mutants of bacteria by means of Penicillin. Proc. nat. Acad. Sci., Wash., **35**, 1.

— (1950): Studies on nutritionally deficient bacterial mutants isolated by means of Penicillin. Experientia **6**, 41.

Davis, H. M. (1908): Spore formation in *Derbesia*. Ann. Bot. **22**, 1.

Davis, R. H., and *V. W. Woodward* (1962): The relationship between gene suppression and aspartate transcarbamylase activity in pyr-3 mutants of *Neurospora*. Genetics **47**, 1075.

Davis, R. W., and *R. W. Hyman* (1971): A study in evolution: The DNA base sequence homology between coliphages T 7 and T 3. J. Mol. Biol. **62**, 287.

Davison, J., L. M. Pilarski, and *H. Echols* (1969): A factor that stimulates RNA synthese by purified RNA polymerase. Proc. nat. Acad. Sci., Wash., **63**, 168.

De Duve, C. (1957): The enzymatic heterogeneity of cell fractions isolated by differential centrifuging. Symp. Soc. exp. Biol. **10**, 50.

— (1963): The lysosome concept. In: *A. V. S. Reuck*, and *M. P. Cameron* (Eds.) "Ciba Foundation Symposium on Lysosomes". Little, Brown and Comp. Boston, Mass. p. 1.

— (1965): Function of microbodies (peroxysomes). J. Cell Biol. **27**, 25.

—, and *P. Baudhuin* (1966): Peroxisomes (microbodies and related particles). Physiol. Rev. **46**, 323.

—, and *E. Wattiaux* (1966): Functions of lysosomes. Ann. Rev. Physiol. **28**, 435.

—, *B. C. Pressman, R. Gianetto, R. Wattiaux*, and *F. Appelmans* (1955): Tissue fractionation studies. 6. Intracellular distribution patterns of enzymes in rat-liver. Biochem. J. **60**, 604.

Deflandre, G. (1934): Existence, sur les flagellus, de filaments latéroux on terminaux (mastigonèmes). C. R. Acad. Sci. (Paris) **198**, 497.

Delage, H. (1899): Études sur la merogonie. Arch. Zool. Exp. **7**, 383.

Delbrück, M. (1945): Interference between bacterial viruses. III. The mutual exclusion effect and the depressor effect. J. Bacteriol. **50**, 151.

— (1963): Die Vererbungschemie. Spektrum **6**, 74.

—, and *G. S. Stent* (1957): On the mechanism of DNA replication. In: *McElroy, W. D.*, and *B. Glass* "The Chemical Basis of Heredity". Johns Hopkins Press/Baltimore.

DeMan, J. C. H., and *N. J. A. Noorduyn* (1969): Ribosomes: properties and function. In: *A. Lima-de-Faria* (Ed.) "Handbook of molecular Cytology" North-Holland Publ. Comp., Amsterdam-London, p. 1079.

Demerec, M. (1937): Frequency of spontaneous mutations in certain stocks of *Drosophila melanogaster*. Genetics **22**, 469.

— (1946): Induced mutations and possible mechanisms of the transmission of heredity in *Escherichia coli*. Proc. nat. Acad. Sci., Wash., **32**, 36.

— (1956): Terminology and nomenclature. In: "Genetic Studies with Bacteria", Publ. Carneg. Instn. **612**, 1.

— (1962): "Selfers" — attributed to unequal crossovers in *Salmonella*. Proc. nat. Acad. Sci., Wash., **48**, 1696.

— (1964): The clustering of functionally related genes in *S. typhimurium*. Proc. nat. Acad. Sci., Wash., **51**, 1057.

—, and *E. Cahn* (1953): Studies of mutability in nutritionally deficient strains of *E. coli*. J. Bacteriol **65**, 27.

—, and *P. E. Hartman* (1959): Complex loci in microorganisms. Annu. Rev. Microbiol. **13**, 377.

—, and *M. E. Hoover* (1936): Three related X-chromosome deficiencies in *Drosophila*. J. Hered. **27**, 207.

—, *E. A. Adelberg*, *A. J. Clark*, and *P. E. Hartman* (1966): A proposal for a uniform nomenclature in bacterial genetics. Genetics **54**, 61.

Dermen, H. (1947): Periclinal cytochimeras and histogenesis of Cranberry. Amer. Bot. **34**, 32.

— (1953): The pattern of tetraploidy. J. Hered. **44**, 31.

DeRobertis, E. D. P., *W. W. Nowinsky*, and *F. A. Saez* (1965): Cell Biology (4th ed.). W. B. Saunders, Philadelphia, London, Toronto.

Dewey, M. M., and *L. Barr* (1962): Intercellular connection between smooth muscle cells: the nexus. Science **137**, 670.

Diamelidis, Th. (1951): Cytologische Studien an einigen *Allium*-Arten aus Nord-Griechenland. Portug. acta biol. (A) **3**, 151.

Dietz, R. (1958): Multiple Geschlechtschromosomen bei den cypriden Ostracoden, ihre Evolution und ihr Teilungsverhalten. Chromosoma **9**, 359.

Digby, L. (1912): The cytology of Primula kewenesis and other related Primula hybrids. Ann. Bot. **26**, 357.

Dintzis, H. M., *H. Borsook*, and *J. Vinograd* (1958): Microsomal particles and protein synthesis (*Roberts, R. B.*, Ed.), Pergamon, N. Y.

Dobzhansky, T. (1935): A critique of the species concept in biology. Phil. Sci. **2**, 344.

— (1937): Genetics and the Origin of Species. (1st Ed.). Columbia Univ. Press/New York.

— (1946): Genetics of natural populations. XIII. Recombination and variability in populations of *Drosophila pseudoobscura*. Genetics **31**, 269.

— (1950): Mendelian populations and their evolution. Amer. Nat. **84**, 401.

— (1951): Mendelian populations and their evolution. In: *L. C. Dunn* (Ed.) "Genetics in the 20th Century" McMillan/New York, 573.

— (1951): Genetics and the Origin of Species (3rd Ed.). Columbia Univ. Press/New York.

— (1952): Nature and origin of heterosis. In: *J. W. Gowen* "Heterosis", Iowa State College Press/Ames, 218.

— (1956): What is an adaptive trait? Amer. Nat. **90**, 337.

— (1962): Mankind Evolving, Yale Univ. Press/New Haven.

— (1965): Genetic diversity and fitness. In: Proc. XI. Int. Congr. Genet., The Hague, **3**, 541.

— (1967): Changing man. Science **155**, 409.

— (1968): On some fundamental concepts of Darwinian biology. Evolutionary Biology **2**, 1.

— (1970): Genetics of the Evolution Prozess. Columbia University Press, New York and London.

Dodge, B. O. (1945): Further remarks on myco-genetic terminology. Mycologia **37**, 629.
Dodge, I. D. (1964): Chromosome structure in the *Dinophyceae*. Arch. Mikrobiol. **48**, 66.
Doermann, A. H. (1953): The vegetative state in the life cycle of bacteriophage: evidence for its occurrence, and its genetic characterization. Cold Spr. Harb. Symp. quant. Biol. **18**, 3.
—, *M. Chase*, and *F. W. Stahl* (1955): Genetic recombination and replication in bacteriophage. J. cell comp.. Physiol. **45**, 51.
Dougherty, E. C. (1957): Neologisms needed for structures of primitive organisms. I. Types of nuclei. J. Protozool. **4**, (Suppl.), 14.
Doy, C. H., P. M. Gresshoff, and *B. G. Rolfe* (1973): The biochemistry of gene expression in higher organisms. Proc. nat. Acad. Sci., Wash., **70**, 723.
Drake, J. W., and *E. F. Allen* (1968): Antimutagenic DNA polymerases of bacteriophage T 4. Cold Spr. Harb. Symp. quant. Biol. **33**, 339.
Druery, C. T. (1886): Observations on a singular mode of development in the Lady-Fern. J. Linn. Soc. Bot. **21**, 354.
Dubinin, N. P. (1948): Experimental investigations concerning the integration of genetic systems during the process of evolution of population (russ.). J. gen. Biol., Moskow (Zhurn. obsc. biol.) **9**, 203.
—, and *B. N. Sidorov* (1935): The position effect of the hairy gene. Biol. Zh. **4**, 555.
Duguid, J. P., and *E. S. Anderson* (1967): Terminology of bacterial fimbriae, or pili, and their types. Nature, Lond., **215**, 89.
—, *I. W. Smith, G. Dempster*, and *P. N. Edwards* (1955): Non-flagellar filamentous appendages (fimbriae) and haemagglutinating activity of *Bacterium coli*. J. Path. Bact. **70**, 335.
Dujardin, F. (1841): Histoire naturelle des Zoophytes. Roret/Paris.
Dunn, L. C. (1954): The study of complex loci. Proc. 9th Int. Congr. Genet. **1**, 156. (Caryologia Suppl. 6).
DuPraw, E. J. (1965): Macromolecular organization of nuclei and chromosomes: a folded fibre model based on wholemount electron microscopy. Nature, Lond., **206**, 338.
— (1966): Evidence for a "folded-fibre" organization in human chromosomes. Nature, Lond., **209**, 577.
DuRietz, (1930): The fundamental units of biological taxonomy. Svensk. bot. Tidskr. **24**, 333.
Dussoix, D., and *W. Arber* (1962): Host specificity of DNA produced by *Escherichia coli*. II. Control over acceptance of DNA from infecting phage λ. J. Mol. Biol. **5**, 37.
Dustin, A. P. (1934): Action de la colchicine sur le sarcome greffé type Crocker, de la souris. Bull. Acad. R. Med. Belg. **14**, 487.
— (1938): L'action des arsénicaux et de la colchicine sur la mitose, la stathmocinèse. C. R. Ass. Anat. **33**.
— (1947): Some new aspects of mitotic poisoning. Nature, Lond., **159**, 794.
Dyer, A., F. K. Jong, and *J. A. Ratter* (1970): Aneuploidy: a redefinition. Notes Roy. Bot. Garden Edinburgh **30**, 177.

East, E. M. (1910): A Mendelian interpretation of variation that is apparently continuous. Amer. Nat. **44**, 65.
—, and *J. B. Park* (1917): Studies on self-sterility. I. The behavior of self-sterile plants. Genetics **2**, 505.
Edmonds, M., and *R. Abrams* (1960): Polynucleotide biosynthesis: Formation of a sequence of adenylate units from adenosine triphosphate by an enzyme from thymus nuclei. J. Biol. Chem. **235**, 1142.

Edström, J.-E. (1956): Chromosomal RNA and other nuclear RNA fractions. In: *M. Locke* (Ed.) "The Role of Chromosomes in Development", Acad. Press/ New York, 137.
— (1968): Masters, slaves and evolution. Nature, Lond., **220**, 1196.
—, and *W. Beermann* (1962): The base composition of nucleic acids in chromosomes, puffs, nucleoli, and cytoplasm of *Chironomus* salivary gland cells. J. Cell Biol. **14**, 371.
Edwards, M. R., and *R. W. Stevens* (1963): Fine structure of *Listeria* monocytogenes. J. Bacteriol. **86**, 414.
Egawa, R., and *Y. Hirota* (1962): Inhibition of fertility by multiple drugresistance factor (R) in *E. coli K 12.* Jap. J. Genet. **37**, 66.
Ehrenstein, v. G. (1966): Translational variations in the amino acid sequence of the α-chain of rabbit hemoglobin. Cold Spr. Harb. Symp. quant. Biol. **31**, 705.
Ehrman, L. (1962): Hybrid sterility as an isolating mechanism in the genus *Drosophila*. Quart. Rev. Biol. **37**, 279.
Eimer, Th. (1897): Die Entstehung der Arten. Jena, Leipzig.
Eisen, G. (1900): The spermatogenesis of *Batrachoseps*. J. Morph. **17**, 1.
Endrizzi, J. E., and *R. J. Kohel* (1966): Use of telosomes in mapping three chromosomes in cotton. Genetics **54**, 535.
Engler, A., und *K. Prantl* (1897): Die natürlichen Pflanzenfamilien. Engelmann/ Leipzig.
Englesberg, E. (1961): Enzymatic characterization of 17 L-arabinose negative mutants of *Escherichia coli*. J. Bacteriol. **71**, 996.
—, and *L. Ingraham* (1957): Meiotrophic mutants of *Pasteurella pestis* and their use in the elucidation of nutritional requirements. Proc. nat. Acad. Sci., Wash., **43**, 369.
Enriques, P. (1922): Hologynic heredity. Genetics **7**, 583.
Ephrussi-Taylor, H. (1951): Genetic mechanisms in bacteria and bacterial viruses. III. Cold Spr. Harb. Symp. quant. Biol. **16**, 445.
Epstein, R. H., A. Bolle, C. M. Steinberg, E. Kellenberger, E. Boy de la Tour, R. Chevalley, R. S. Edgar, M. Susman, G. Denhardt, and *A. Leilausis* (1963): Physiological studies of conditional lethal mutations in bacteriophage T_4D. Cold Spr. Harb. Symp. quant. Biol. **28**, 375.
Epstein, S. S., and *G. Röhrborn* (1971): Recommended procedures for testing genetic hazards from chemicals, based on the induction of dominant lethal mutations in mammals. Nature, Lond., **230**, 459.
Epstein, W., and *J. R. Beckwith* (1968): Regulation of gene expression. Ann. Rev. Biochem. **37**, 411.
Erikson, R. L., M. L. Fenwick, and *R. M. Franklin* (1964): Replication of bacteriophage RNA: Studies on the fate of parental RNA. J. Mol. Biol. **10**, 519.
Eriksson, G., A. Kahn, B. Walles, und *D. v. Wettstein* (1961): Zur makromolekularen Physiologie der Chloroplasten. III. Ber. dtsch. bot. Ges. **74**, 221.
Erlanger, R. V. (1897): Zur Kenntnis der Zell- und Kernteilung. I. Über die Spindelbildung in den Zellen der Cephalopodenkeimscheibe. Biol. Zbl. **17**, 745.
Ernst-Schwarzenbach, M. (1939): Zur Kenntnis des sexuellen Dimorphismus der Laubmoose. Arch. Klaus-Stift. VererbForsch. **14**, 361.
Errera, L. (1888): Über Zellformen und Seifenblasen. Bot. Zbl. **34**, 395.
—, et *G. Gevaert* (1878): Sur la structure et les modes de fécondation des fleurs. Bull. Soc. Bot. Belg. **17**.
Esposito, M. S. (1968): X-ray and meiotic fines tructure mapping of the adenin-8 locus in Saccharomyces cerevisiae. Genetics **58**, 507.
Esser, K. (1956): Die Inkompatibilitätsbeziehungen zwischen geographischen Rassen von *Podospora anserina* (Ces) Rehm. Z. indukt. Abstamm.- u. VererbLehre **87**, 595.

Esser, K. (1971): Breeding systems in fungi and their significance for genetic recombination. Molec. Gen. Genetics **110**, 86–100.
—, und *R. Kuenen* (1965): Genetik der Pilze. Springer-Verlag/Berlin, Heidelberg, New York.
Estable, C., and *J. R. Sotelo* (1950): A new cellular structure: the nucleoloneme. Inst. Ciencias Biológicas Montevideo **1**, 105.
Evans, G. M., *A. Durrant*, and *H. Rees* (1966): Associated nuclear changes in the induction of flax genotrophs. Nature, Lond., **212**, 697.
Evans, H. J. (1962): Chromosome aberrations induced by ionizing radiation. Int. Rev. Cytol. **13**, 221.

Faberge, A. C. (1958): Relation between chromatid-type and chromosome-type breakage-fusion-bridge-cycles in maize endosperm. Genetics **43**, 737.
Fagerlind, F. (1940): Die Terminologie der Apomixisprozesse. Hereditas, Lund, **26**, 1.
— (1944): Is my terminology of the apomictic phenomena of 1940 incorrect and inappropriate? Hereditas, Lund, **30**, 590.
Falconer, D. S. (1960): An Introduction to Population Genetics. Oliver & Boyd/ London.
Farmer, J. B., and *L. Digby* (1907): Studies in apospory and apogamy in Ferns. Ann. Bot. **21**, 161.
—, and *J. E. S. Moore* (1905): On the meiotic phase (reduction-divisions) in animals and plants. Quart. J. micr. Sci. **48**, 489.
Farquhar, M. G., and *G. E. Palade* (1963): Junctional complexes in various epithelia. J. Cell Biol. **17**, 375.
Fawcett, D. W. (1961): Cilia and flagella. In: *J. Brachet*, and *S. E. Mirsky* (Eds.) "The Cell", Vol. **2**, 217, Academic Press/New York.
— (1965): Histochemical society symposium on structure and function at cell surfaces. Surface specialization of absorbing cells. J. Histochem. Cytochem. **13**, 75.
Feodoroff, S. (1967): Proposed usage of animal tissue culture terms. J. nat. Cancer Inst. **38**, 607.
Fernandez-Moran, H., *T. Oda*, *P. V. Blair*, and *D. E. Green* (1964): A macromolecular repeating unit of mitochondrial structure and function. J. Cell Biol. **22**, 63.
Fincham, J. R. S. (1959): On the nature of glutamic dehydrogenase produced by inter-allele complementation at the am locus of *Neurospora crassa*. J. gen. Microbiol. **21**, 600.
— (1966): Genetic Complementation. W. A. Benjamin/New York.
—, and *G. R. K. Sastry* (1974): Controlling elements in maize. Ann. Rev. of Genetics **8**, 15.
Fischer, E. (1939): Versuch einer Phänogenetik der normalen körperlichen Eigenschaften des Menschen. Z. indukt. Abstamm.- u. VererbLehre **76**, 47.
Fisher, R. A. (1928): The possible modification of the response of the wild type to recurrent mutations. Amer. Nat. **62**, 115.
— (1930): The Genetical Theory of Natural Selection. Clarendon Press/Oxford.
—, *F. R. Immer*, and *O. Tedin* (1932): The genetic interpretation of statistics of the third degree in the study of quantitative inheritance. Genetics **17**, 107.
Fitch, W. M., and *J. S. Harris* (1974): Evolutionary trees with minimum nucleotide replacements from amino acid sequences. J. Mol. Evol. **3**, 263.
—, and *E. Markowitz* (1970): An improved method for determining codon variability in a gene and its application to the rate of fixation of mutations in evolution. Biochem. Genet. **4**, 579.
Fitz-James, P. C. (1960): Participation of cytoplasmic membrane in the growth and spore formation of bacilli. J. biophysic. biochem. Cytol. **8**, 507.

Flamm, W. G. (1974): A tier system approach to mutagen testing. Mut. Res. **26**, 329.
Flemming, W. (1879): Beiträge zur Kenntnis der Zelle und ihrer Lebenserscheinungen. I. Arch. mikr. Anat. **16**, 302; II. Arch. mikr. Anat. **18**, 151; III. Arch. micr. Anat. **20**, 1.
— (1882): Zellsubstanz, Kern und Zellteilung. Vogel/Leipzig.
— (1887): Neue Beiträge zur Kenntnis der Zelle. Arch. mikr. Anat. **29**, 389.
Focke, W. O. (1881): Die Pflanzenmischlinge, ein Beitrag zur Biologie der Gewächse. Borntraeger/Berlin.
Földes, J., and *T. A. Trautner* (1964): Infectious DNA from a newly isolated *B. substilis* phage. Z. VererbLehre **95**, 57.
Fol, H. (1871): Die „Centrenquadrille", eine neue Episode aus der Befruchtungsgeschichte. Anat. Anz. **6**, 266.
— (1877): Sur le commencement de l'hénogénie chez divers animaux. Arch. Sci. Nat. et Phys. Genève **58**.
— (1891): (Cited after *C. Pelling* Proc. Roy. Soc. (B), 164 (1966). Le quadrille des centres. Un épisode noveau dans l'histoire de la fécondation. Arch. Sci. Phys. Nat. **25**, 393.
— (1896): Lehrbuch der vergleichenden mikroskopischen Anatomie mit Einschluß der vergleichenden Histologie und Histogenie. Die Zelle. Engelmann/Leipzig.
Ford, E. B. (1940): Polymorphism and taxonomy. In: *J. S. Huxley* (Ed.) "New Systematics", Clarendon Press/Oxford, 493.
Fox, A. S., S. B. Yoon, and *W. M. Gelbart* (1971): DNA-induced transformation in *Drosophila*. Genetic analysis of transformed stocks. Proc. Nat. Acad. Sci., Wash., **68**, 342.
Fox, S. W. (1969): Self-ordered polymers and propagative cell-like systems. Naturwiss. **56**, 1.
Fraenkel, D. G., and *A. Parola* (1972): „Up-promoter" mutations of glucose 6-phosphate dehydrogenase in *Escherichia coli*. J. Mol. Biol. **71**, 107.
Franklin, N. C., and *S. E. Luria* (1961): Transduction by bacteriophage P 1 and the properties of the lac genetic region in *E. coli* and *S. dysenteriae*. Virology **15**, 299.
Fredericq, P., and *M. Netz-Bareau* (1953): Transfert génétique de la propriété de produire un antibiotique. Compt. Rend. Soc. Biol. **147**, 1653.
Freese, E. (1959): The difference between spontaneous and base analogue induced mutations of phage T 4. Proc. nat. Acad. Sci., Wash., **45**, 622.
— (1961): The molecular mechanisms of mutations. Proc. 5. Int. Congr. of Biochem., Moscow, Symp. Nr. 1, Reprint Nr. 163.
—, *H.-J. Rhaese*, and *E. Bautz-Freese* (1969): Chemical DNA alterations causing inactivation and chromosome breaking. Jap. J. Genetics **44**, Suppl. 1, 94.
Freifelder, D. R., and *D. Freifelder* (1968): Studies on *Escherichia coli* sex factors. I. Specific labeling of F'Lac DNA. J. Mol. Biol. **32**, 15.
Frey-Wyssling, A., and *K. Mühlethaler* (1965): Ultrastructural Plant Cytology. Elsevier/Amsterdam.
Frydenberg, D. (1963): Population studies of a lethal mutant in *Drosophila melanogaster*. I. Behavior in populations with discrete generations. Hereditas, Lund, **50**, 89.
Fulton, Ch. (1971): Centrioles. In: *J. Reinert*, and *H. Ursprung* (Eds.) "Origin and continuity of cell organelles". Results and Problems in Cell Differentiation 2, 170. Springer-Verlag/Berlin-Heidelberg-New York.

Gabridge, M. G., and *M. S. Legator* (1969): A host-mediated microbial assay for the detection of mutagenic compounds Proc. Soc. Exp. Biol. Med. **130**, 831.

Gall, J. G. (1961): Centriole replication. A study of spermatogenesis in the snail *Viviparus*. J. Biophys. Biochem. Cytol. **10**, 163.
Galton, F. (1883): Inquiries into Human Faculty and its Development. McMillan/London.
Ganesan, A. K. (1974): Persistence of pyrimidine dimers during postreplication repair in ultraviolet light-irradiated *Escherichia coli* K 12. J. Mol. Biol. **87**, 103.
Ganoza, M. C. (1966): Polypeptide chain termination in cell-free extracts of *E. coli*. Cold Spr. Harb. Symp. quant. Biol. **31**, 273.
—, and *T. Nakamoto* (1966): Studies on the mechanism of polypeptide chain termination in cell-free extracts of *E. coli*. Proc. nat. Acad. Sci., Wash., **55**, 162.
—, *J. Vandermeer, N. Debreceni*, and *S. L. Phillips* (1973): Mechanism of protein chain termination: Further characterization of a mutant defective in a new protein synthesis factor. Proc. nat. Acad. Sci., Wash., **70**, 31.
Garen, A., and *O. Siddiqi* (1962): Suppression of mutations in the alkaline phosphatase structural cistron of *Escherichia coli*. Proc. nat. Acad. Sci., Wash., **48**, 1121.
—, and *N. D. Zinder* (1955): Radiological evidence for partial genetic homology between bacteriophage and host bacteria. Virology **1**, 347.
Garnier, C. (1897): Les filaments baseaux des cellules glandulaires. Bibliogr. anat. **5**, 278.
Garnjobst, L., and *J. F. Wilson* (1956): Heterocaryosis and protoplasmic incompatibility in *Neurospora crassa*. Proc. nat. Acad. Sci., Wash., **42**, 613.
Garrod, A. E. (1902): The incidence of alkaptonuria: a study in chemical individuality. Lancet **2**, 1616.
— (1908): Inborn errors of metabolism. Lectures I—IV. Lancet **175**, **2**, 1, 73, 142, 214.
Gates, R. R. (1911): Pollenformation in *Oenothera gigas*. Ann. Bot. **25**, 909.
Gavaudan, P. (1943): Étude quantitative de l'action mitoinhibridice des substances aromatiques: définition et terminologie des effets cytologiques utilisés comme test. C. R. Soc. Biol. (Paris) **137**, 281.
Gedda, L., and *G. Brenci* (1973): Chronogenetics. Its foundations, scope, and impact. Acta Genet. Med. Gemellol. **22**, 3.
Geitler, L. (1939): Die Entstehung der polyploiden Somakerne der Heteropteren durch Chromosomenteilung ohne Kernteilung. Chromosoma **1**, 1.
— (1953): Endomitose und endomitotische Polyploidisierung. Protoplasmatologia 6C, Springer/Wien.
— (1955): Die Strukturen der Zelle und ihre chemische und physikalische Konstitution. Normale und pathologische Anatomie der Zelle. In: *W. Ruhland* (Ed.) „Handb. d. Pflanzenphysiologie" **1**, 123, Springer/Berlin.
Gellert, M. (1967): Formation of covalent circles of lambda DNA by *E. coli* extracts. Proc. nat. Acad. Sci., Wash., **57**, 148.
Georgiev, G. P. (1961): Ribonucleic acid of nucleolo-chromosomal apparatus. Biokhimia **26**, 1095.
—, and *V. L. Mantieva* (1962): The isolation of DNA-like RNA and rRNA from the nucleolo-chromosomal apparatus of mammalian cells. Biochem. Biophys. Acta **61**, 153.
Giardini, A. (1901): Origine dell'oocite e della cellule nutrici nel *Dytiscus*. Int. Monatsschrift Anat. Physiol. **18**, 418—484.
Gibson, J., and *G. H. Beale* (1961): Genetic basis of the mate-killer trait in *Paramecium aurelia*, stock 540. Genet. Res. **2**, 82.
—, — (1963): The action of ribonuclease and 8-azaguanine on mate killer paramecia. Genet. Res. **4**, 42.
Giese, A. C. (1973): Cell Physiology (4th ed.). W. B. Saunders Co., Philadelphia-London-Toronto.

Gilbert, W., and *D. Dressler* (1968): DNA Replication: The rolling circle model. Cold Spr. Harb. Symp. quant. Biol. **33**, 473.
—, and *B. Müller-Hill* (1966): Isolation of the lac repressor. Proc. nat. Acad. Sci., Wash., **56**, 1891.
Giles, N. H. (1958): Mutations at specific loci in *Neurospora*. Proc. X. Int. Congr. Genet. **2**, 17.
—, *C. W. H. Partridge*, and *N. J. Nelson* (1957): The genetic control of adenylo-succinase in *Neurospora crassa*. Proc. Nat. Acad. Sci., Wash., **43**, 305.
Gilmore, R. A. (1967): Super-suppressors in *Saccharomyces cerevisiae*. Genetics **56**, 641.
Gilmour, J. S. L., and *J. W. Gregor* (1939): A suggested new terminology. Nature, Lond., **144**, 333.
—, and *J. Heslop-Harrison* (1954): The deme terminology and the units of micro-evolutionary change. Genetica **27**, 147.
Gingery, R., and *H. Echols* (1967): Mutants of bacteriophage λ unable to integrate into the host chromosome. Proc. nat. Acad. Sci., Wash., **58**, 1507.
Ginsburg, J. (1954): Certain measures of intergradation and divergence. Zoologica. N. Y. **39**, 31.
Goebel, K. (1880): Beiträge zur vergleichenden Entwicklungsgeschichte der Sporangien. Bot. Ztg. **38**, 545.
Goldberger, R. F. (1974): Autogenous regulation of gene expression. Science **183**, 810.
Goldschmidt, R. (1915): Vorläufige Mitteilung über weitere Versuche zur Vererbung und Bestimmung des Geschlechts. Biol. Zbl. **35**, 565.
— (1917): A further contribution to the theory of sex. J. exp. Zool. **22**, 593.
— (1917): Intersexuality and the endocrine aspect of sex. Endocrinology **1**, 433.
— (1920): Untersuchungen über Intersexualität. Z. indukt. Abstamm.- u. Vererb-Lehre **23**, 1.
— (1929): Geschlechtsbestimmung im Tier- und Pflanzenreich. Biol. Zbl. **49**, 641.
— (1935): Gen und Außeneigenschaft I. (Untersuchungen an *Drosophila*). Z. indukt. Abstamm.- u. VererbLehre **69**, 38.
— (1935): Gen und Außeneigenschaft II. Z. indukt. Abstamm.- u. VererbLehre **69**, 70.
— (1935): Gen und Außencharakter. III. Biol. Zbl. **55**, 534.
— (1937): Does the quantity of chromatin produce a genetic effect? Amer. Nat. **71**, 83.
— (1938): Physiological Genetics. McGraw Hill/New York.
— (1940): The Material Basis of Evolution. Yale Univ. Press/New Haven.
— (1945): The structure of podoptera, a homeotic mutant of *Drosophila melanogaster*. J. Morph. **77**, 71.
— (1949): The interpretation of the triploid intersexes of *Solenobia*. Experientia **5**, 417.
— (1949): Heterochromatic heredity. Proc. VIII. Int. Congr. Genet. Stockholm (Suppl. Vol. Hereditas) 244.
— (1955): Theoretical Genetics. Calif. Univ. Press/Berkely.
Golgi, C. (1898): Sur la structure des cellules nervenses. Arch. ital. Biol. **30**, 60.
Gonzales, A. (1967): Formación y desarrollo de células binucleadas: bimitosis. Genet. Iberica **19**, 1.
Goodspeed, T. H., and *P. Avery* (1939): Trisomic and other types in *Nicotiana sylvestris*. J. Genet. **38**, 381.
Gorini, L., and *J. R. Beckwith* (1966): Suppression. Annu. Rev. Microbiol. **20**, 401.
—, *R. Rosset*, and *R. A. Zimmermann* (1967): Phenotypic masking and streptomycin dependence. Science **157**, 1314.
Granick, S., and *A. Gibor* (1967): The DNA of chloroplasts, mitochondria, and centrioles. Progr. Nucleic Acid Res. and Mol. Biol. **6**, 143.

Grant, V. (1953): The role of hybridization in the evolution of the leafy-stemmed Gilias. Evolution **7**, 51.
— (1957): The plant species in theory and practice. In: *E. Mayr* (Ed.) "The Species Problem", Am. Ass. Adv. of Sci. Publ. **50.**
— (1958): The regulation of recombination in plants. Cold Spr. Harb. Symp. quant. Biol. **23**, 337.
— (1963): The Origin of Adaptations. Columbia Univ. Press/New York.
— (1964): The Architecture of the Germplasm. Wiley/New York.
— (1966): The selective origin of incompatibility barriers in the plant genus *Gilia*. Amer. Nat. **100**, 99.
— (1971): Dynamics of clonal microspecies in *Cholla cactus*. Evolution **25**, 144.
Gratia, A. (1925): Sur un remarquable exemple d'antagonisme entre deux souches de colibacille. C. R. Acad. Sci., Paris, **93**, 1040.
Green, M. M. (1959): Reverse mutation in *Drosophila* and the status of the particulate gene. Genetica **29**, 1.
Gregoire, V. (1905): Les resultats acquis sur le cinésis de maturation dans les deux regnes. I. Cellule **21**, 297.
— (1907): Les fondements cytologiques des theories courantes sur l'hérédité mendélienne. Ann. Soc. zool. Belg. **42**, 267.
— (1907): La formation des gemini hétérotypique dans les végétaux. Cellule **24**, 369.
— (1932): Euchromocentres et chromosomes dans les végétaux. Bull. Acad. Belg. Cl. Sci., V. Ser. **17**, 1435.
Grell, K. G. (1956): Protozoologie. Springer/Berlin.
Grell, R. F. (1962): A new model for secondary nondisjunction: the role of distributive pairing. Genetics **47**, 1737.
— (1962): A new hypothesis on the nature and sequence of meiotic events in the female of *Drosophila melanogaster*. Proc. nat. Acad. Sci., Wash., **48**, 165.
Grierson, D., M. E. Rogers, M. L. Sartirana, and *V. E. Loening* (1970): The synthesis of ribosomal RNA in different organisms: Structure and evolution of the rRNA precursor. Cold Spring Harb. Symp. **35**, 589.
Griffith, F. (1928): Significance of pneumococcal types. J. Hyg. **27**, 113.
Grinnell, A. (1904): The origin and distribution o the chestnutbacked chickadee. Auk **21**, 364.
Grobstein, C. (1959): Differentiation of vertebrate Cells. In: *Brachet, J.*, and *A. E. Mirsky* (Ed.) "The Cell". Academic Press/New York and London, p. 437.
Groner, Y., Y. Pollack, H. Berissi, and *M. Revel* (1972): Cistron specific translation control protein in *Escherichia coli*. Nature, N. B., Lond., **239**, 16.
Grüneberg, H. (1938): An analysis of the "pleiotropic" effects of a new lethal mutation in the rat (*Mus norvegicus*). Proc. roy. Soc. B. **125**, 123.
Grundmann, E. (1964): Allgemeine Cytologie. Thieme/Stuttgart.
Guignard, L. (1899): Sur les antherozoides et la double copulation sexuelle chez les végétaux angiosperms. C. R. Acad. Sci., Paris, **128**, 864.
Guilliermond, A. (1910): La sexualité chez les champignons. Bull. sci. Fr. Belg. **44.**
Gurdon, J. B. (1974): The control of gene expression in animal development. Clarendon Press/Oxford.
Gustafsson, A. (1935): Studies on the mechanism of parthenogenesis. Hereditas, Lund, **21**, 1.
Gutherz, S. (1907): Zur Kenntnis der Heterochromosomen. Arch. mikr. Anat. **69**, 491.
Guttman, B. S. (1971): Biological Principles. W. A. Benjamin, New York.
Gutz, H. (1967): "Twin meiosis" and other ambivalences in the life cycle of *Schizosaccharomyces pombe*. Science **158**, 796.

Haacke, W. (1893): Gestaltung und Vererbung. Weigel/Leipzig.

Hadorn, E. (1945): Zur Pleiotropie der Genwirkung. Arch. Klaus-Stift. Vererb-Forsch. Ergänzungsbd. zu Bd. **20**, 82.
— (1948): Gene action in growth and differentiation of lethal mutants of *Drosophila*. Symp. Soc. exp. Biol. **2**, 179.
— (1949): Begriffe und Termini zur Systematik der Letalfaktoren. Arch. Klaus-Stift. VererbForsch. **24**, 105.
— (1955): Letalfaktoren in ihrer Bedeutung für Erbpathologie und Genphysiologie der Entwicklung. Thieme/Stuttgart.
— (1965): Problems of determination and transdetermination. Brookh. Symp. Biol. **18**, 148.
— (1966): Konstanz, Wechsel und Typus der Determination und Differenzierung in Zellen aus männlichen Genitalanlagen von *Drosophila melanogaster* nach Dauerkultur *in vivo*. Develop. Biol. **13**, 424.
Haeckel, E. (1866): Generelle Morphologie der Organismen. Allgemeine Grundzüge der organischen Formwissenschaft, mechanisch begründet durch die von *Ch. Darwin* reformierte Deszendenztheorie. Reimer/Berlin.
— (1874): Anthropogenie und Entwcklungsgeschichte des Menschen. Leipzig.
— (1891): Anthropogenie und Entwicklungsgeschichte des Menschen. (2. Aufl.) Leipzig.
— (1894): Systematische Phylogenie. Reimer/Berlin.
Haecker, V. (1892): Die heterotypische Kernteilung im Zyklus der generativen Zellen. Ber. naturf. Ges. Freiburg i. Br. **6**, 160.
— (1895): Die Vorstadien der Eireifung. Arch. mikr. Anat. **45**, 200.
— (1895): Über die Selbständigkeit der väter- und mütterlichen Kernsubstanz während der Embryonalentwicklung von Cyclops. Arch. mikr. Anat. **46**, 579.
— (1897): Über weitere Übereinstimmungen zwischen den Fortpflanzungsvorgängen der Tiere und Pflanzen. Die Keim- und Mutterzellen. Biol. Zbl. **17**, 689.
— (1907): Die Chromosomen als angenommene Vererbungsträger. Ergebn. Zool. **1**, 1.
— (1918): Entwicklungsgeschichtliche Eigenschaftsanalyse. (Phänogenetik). Fischer/Jena.
— (1925): Aufgaben und Ergebnisse der Phänogenetik. Bibl. Genet. Lpz. **1**, 93.
Haefner, K. (1965): Zum Inaktivierungskriterium für Einzelzellen unter besonderer Berücksichtigung der Teilungsfähigkeit Röntgen- und UV-bestrahlter Saccharomyceszellen verschiedenen Ploidiegrades. Intern. J. Rad. Biol. **9**, 545.
Hagemann, R. (1958): Somatische Konversion bei *Lycopersicon esculentum Mill.* Z. VererbLehre **89**, 587.
— (1964): Plasmatische Vererbung. Fischer/Jena.
Hakansson, A., and *A. Levan* (1957): Endo-duplicational meiosis in *Allium odorum*. Hereditas, Lund, **43**, 179.
Haldane, J. B. S. (1919): The combination of linkage values and the calculation of distances between the loci of linked factors. J. Genet. **8**, 299.
— (1930): A note on *Fisher's* theory of the origin of dominance. Amer. Nat. **64**, 87.
— (1930): Theoretical genetics of autopolyploids. J. Genet. **22**, 359.
— (1932): The Causes of Evolution. Harper & Brothers/New York.
— (1932): The time of action of genes, and its bearing on some evolutionary problems. Amer. Nat. **66**, 5.
— (1942): New Paths in Genetics. Harper & Brothers/New York.
— (1949): Rates of evolution. Evolution **3**, 51.
— (1957): The cost of natural selection. J. Genet. **55**, 511.
Hansen, H. N., and *R. E. Smith* (1932): The mechanism of variation in *Verticillium alboatrum*. Phytopathology **22**, 953.
Hanstein, J. v. (1880): Einige Züge aus der Biologie des Protoplasmas. Bot. Abh. (Bonn) **4**, H. 2.

Hardy, G. H. (1908): Mendelian proportions in mixed population. Science **28**, 49.
Harm, W. (1963): Mutants of phage T 4 with increased sensitivity to ultraviolet. Virology **19**, 66.
— (1968): Dark repair of photoreparable UV lesions in *Escherichia coli.* Mut. Res. **6**, 25.
Harris, H. (1970): Cell fusion. Harvard Univ. Press/Cambridge, Mass.
— (1963): Nuclear ribonucleic acid. Progr. in Nucleic Acid Res. Vol. **2**, 19.
Hartmann, M. (1904): Die Fortpflanzungsweise der Organismen, Neubenennung und Einteilung derselben, erläutert an Protozoen, Volvovineen und Dicyemiden. Biol. Zbl. **24**, 18.
— (1923): Über sexuelle Differenzierung und relative Sexualität. Studia Mendeliana Brünn **203**.
— (1939): Geschlecht und Geschlechtsbestimmung im Tier- und Pflanzenreich. De Gruyter/Berlin.
Harven, E. de, and *W. Bernhard* (1956): Étude au microscope électronique de l'ultrastructure du centriole chez les vertébrés. Z. Zellforsch. **45**, 378.
Harvey, E. B. (1917): A review of the chromosome number of the Metazoa. Part. I. J. Morphol. **28**, 1.
Haselkorn, R., and *V. A. Fried* (1964): Cell-free protein synthesis: The nature of the active complex. Proc. nat. Acad. Sci., Wash., **51**, 308.
Hashimoto, H., and *Y. Hirota* (1966): Gene recombination and segregation of resistance factors in *E. coli*. J. Bacteriol. **91**, 51.
Haustein, E. (1955): Neuere Arbeiten zur Befruchtungsphysiologie. Z. Bot. **43**, 253.
Hawthorne, D. C., and *R. K. Mortimer* (1963): Super-suppressors in yeast. Genetics **48**, 617.
Hayashi, M., M. N. Hayashi, and *S. Spiegelman* (1964): DNA circularity and the mechanism of strand selection in the generation of genetic messages. Proc. nat. Acad. Sci., Wash., **51**, 351.
Hayes, W. (1964): The Genetics of Bacteria and their Viruses. Wiley/New York.
Hayflick, L., and *P. S. Moorhead* (1961): The serial cultivation of human diploid cell strains. Exp. Cell. Res. **25**, 585.
Heidenhain, M. (1894): Neue Untersuchungen über die Zentralkörper und ihre Beziehungen zum Kern- und Zellenprotoplasma. Arch. mikr. Anat. **43**, 423.
Heilmeyer, L. (1950): Zur Chemotherapie neoplastischer Erkrankungen. Naturwissenschaften **37**, 58.
Heitz, E. (1928): Das Heterochromatin der Moose I. Pringsheims Jb. wiss. Bot. **69**, 762.
— (1929): Heterochromatin, Chromozentren, Chromomeren. Ber. dtsch. bot. Ges. **47**, 274.
— (1931): Die Ursache der gesetzmäßigen Zahl, Lage, Form und Größe pflanzlicher Nukleolen. Planta **12**, 775.
— (1935): Chromosomenstruktur und Gene. Z. indukt. Abstamm.- u. VererbLehre **70**,402.
Henking, H. (1891): Über Spermatogenese bei *Pyrrhocoris apterus.* Z. wiss. Zool. **51**, 685.
Henle, G., F. Deinhardt, and *A. Girardi* (1954): Cytolytic effects of mumps virus in tissue cultures of epithelial cells. Proc. Soc. Exptl. Biol. Med. **87**, 386.
Heribert-Nilsson, N. (1915): Die Spaltungserscheinungen der *Oenothera Lamarckiana.* Lund Univ. Aarskr. N. F. **12**, 131.
Hermann, F. (1891): Beitrag zur Lehre von der Entstehung der karyokinetischen Spindel. Arch. mikr. Anat.
Hershey, A. A. (1952): Reproduction of bacteriophage. Int. Rev. Cytol. **1**, 119.
—, and *M. Chase* (1952): Genetic recombination and heterozygosis in bacteriophage. Cold Spr. Harb. Symp. quant. Biol. **16**, 471.

Hershey, A. A., E. Burgi, and *L. Ingraham* (1963): Cohesion of DNA molecules isolated from phage lambda. Proc. nat. Acad. Sci., Wash., **49**, 748.
—, *C. Roesel, M. Chase,* and *S. Forman* (1951): Growth and inheritance in bacteriophage. Yearb. Carneg. Instn. **50**, 195.
Herskowitz, I. H. (1965): Genetics (2nd Ed.). Little, Brown & Comp./Boston and Toronto.
—, and *H. J. Muller* (1953): Evidence against the healing of X-ray breakages in chromosomes of female *Drosophila melanogaster*. Genetics **38**, 669.
Hertwig, R. (1902): Über Wesen und Bedeutung der Befruchtung. S. B. bayer. Akad. Wiss. **32**, 57.
— (1903): Über Korrelation von Zell- und Kerngröße und ihre Bedeutung für die geschlechtliche Differenzierung und die Teilung der Zelle. Biol. Zbl. **23**, 108.
— (1908): Über neue Probleme der Zellenlehre. Arch. Zellforsch. **1**, 1.
Hess, O. (1966): Funktionelle und strukturelle Organisation der Lampenbürsten-Chromosomen. Naturw. Rdsch. **19**, 176.
Heyn, R. F., A. Rörsch, and *R. A. Schilperoort* (1974): Prospects in genetic engineering of plants. Quart. Rev. of Biophys. **7** 35.
Heywood, St. M. (1970): Specificity of mRNA binding factor in eukaryotes. Proc. nat. Acad. Sci., Wash., **67**, 1782.
Hildebrand, F. (1867): Die Geschlechter-Vertheilung bei den Pflanzen und das Gesetz der vermiedenen und unvortheilhaften stetigen Selbstbefruchtung. Engelmann/Leipzig.
Hill, C., and *I. B. Holland* (1967): The genetic basis of Colicin E sensitivity in *E. coli K 12*. I. Isolation and properties of refractory mutants and the preliminary mapping of their mutations. J. Bact. **99**, 677.
Hill, J. (1967): The environmental induction of heritable changes in *Nicotiana rustica* parental and selection lines. Genetics **55**, 735.
Hiorth, G. (1948): Über das Wesen der monosomen und disomen Anordnung. 3-Kette und Univalent bei *Godetia Whitneyi*. Z. indukt. Abstamm.- u. Vererb-Lehre **82**, 1.
— (1948): Über Hemmungssysteme bei *Godetia Whitneyi* I u. II. Z. indukt. Abstamm.- u. VererbLehre **82**, 12 und 276.
Hirota, Y., and *P. H. Sneath* (1961): F′ and F mediated transduction. in *E. coli* K-12. Jap. J. Genet. **36**, 307.
Hoagland, M. B. (1959): The present status of the adaptor hypothesis. Brookh. Symp. in Biol. **12**, 40.
—, *P. C. Zamecnick,* and *M. L. Stephenson* (1957): Intermediate reactions in protein synthesis. Biochim. biophys. Acta **24**, 215.
Hodge, A. J., J. D. McClean, and *F. V. Mercer* (1956): A possible mechanism for the morphogenesis of lamellar systems in plant cells. J. biophysic. biochem. Cytol. **2**, 597.
—, *E. M. Martin,* and *R. K. Morton* (1957): The structure of some cytoplasmic components of plant cells in relation to the biochemical properties of isolated particles. J. biophysic. biochem. Cytol. **3**, 61.
Höfler, K. (1957): Mikrosomen und Meiosomen. Protoplasma **48**, 167.
Hofmeister, W. (1849): Die Entstehung des Embryo der Phanerogamen. Fr. Hofmeister/Leipzig.
— (1851): Vergleichende Untersuchungen über die Keimung, Entwicklung und Befruchtung der höheren Kryptogamen und über die Befruchtung der Koniferen. Engelmann/Leipzig.
Holley, R. W., J. Apgar, G. A. Everett, J. T. Madison, M. Marquisee, S. H. Merrill, J. R. Penswick, and *A. Zamir* (1965): Structure of a ribonucleic acid. Science **147**, 1462.
Holliday, R. (1963): The genetic effect of mitomycin C. In: *S. J. Geerts*: "Genetics

today", Proc. 11th Int. Congr. Genet. **1**, 13, Pergamon Press/Oxf., Lond., N. Y., Paris.

Holliday, R. (1964): A mechanism for gene conversion in fungi. Genet. Res. **5**, 282.

Hooke, R. (1665): Micrographia, or some physiological descriptions on minute bodies by magnifying glasses. Facsimile ed. publ. by *R. T. Gunther* in: "Early Science in Oxford" XIII: The Life and Works of *R. Hooke*. Oxford 1938.

Hotta, Y., and *H. Stern* (1970): A protein which binds to single-stranded DNA in meiosis. J. Cell Biol. **47**, 91.

—, and — (1971): Meiotic protein in spermatocytes of mammals. Nature, N. B., Lond., **234**, 83.

—, and — (1971): Analysis of DNA synthesis during meiotic prophase in *Lilium*. J. Mol. Biol. **55**, 337.

Howard, A., and *S. R. Pelc* (1953): Synthesis of deoxyribonucleic acid in normal and irradiated cells and its relation to chromosome breakage. Heredity **6**, (Suppl.) 261.

Howard-Flanders, P., and *R. P. Boyce* (1966): DNA repair and genetic recombination: studies on mutants of *E. coli* defective in these processes. Radiat. Res. (Suppl.) **6**, 156.

Hsu, T. C. (1957): Numerical variation of chromosomes in higher animals. In: *D. Rudnick* (Ed.) "Developmental Cytology", Ronald Press/New York, 47.

—, and *P. S. Moorhead* (1956): Chromosome anomalies in human neoplasms with special reference to the mechanisms of polyploidization and aneuploidization in the Hela strain. Ann. N. Y. Acad. Sci. **63**, 1083.

Huberman, J. A., and *A. D. Riggs* (1968): On the mechanism of DNA replication in mammalian chromosomes. J. Mol. Biol. **32**, 327.

Huebner, R. I., and *G. I. Todaro* (1969): Oncogenes of RNA tumor viruses as determinants of cancer. Proc. nat. Acad. Sci., Wash., **64**, 1087.

Hughes, D. T. (1968): Cytogenetical polymorphism and evolution in mammalian somatic cell populations *in vivo* and *in vitro*. Nature **217**, 518.

Hughes-Schrader, S., and *H. Ris* (1941): The diffuse spindle attachment of coccids, verified by the mitotic behavior of induced chromosome fragments. J. exp. Zool. **87**, 429.

Hull, F. H. (1946): Regression analyses of corn yield data (Abstr.). Genetics **31**, 219.

Hurwitz, J., *A. Becker*, *M. L. Gefter*, and *M. Gold* (1967): Enzymatic reactions at termini of DNA. J. Cell. Comp. Physiol. **70**, 181.

Huskins, C. L. (1932): A cytological study of Vilmorin's unfixable dwarf wheat. J. Genet. **25**, 113.

— (1947): The subdivision of the chromosomes and their multiplication in non-dividing tissues: possible interpretations in terms of gene structure and gene action. Amer. Nat. **81**, 401.

Hutt, F. B. (1969): Genetic aspects of infertility. In: *Bernischke, K.* "Comparative mammalian cytogenetics" p. 146.

Huxley, J. S. (1935): Chemical regulation and the hormone concept. Biol. Rev. **10**, 427.

— (1939): Clines: an auxiliary method in taxonomy. Bijdr. Dierk **27**, 491.

— (1940): Towards the new systematics. In: *J. S. Huxley* (Ed.) "New Systematics". Clarendon Press/Oxford.

— (1942): Evolution, the Modern synthesis. Allen & Unwin/London.

— (1948): Evolution, the Modern synthesis (2nd Ed.). Allen & Unwin/London.

— (1955): Morphism and evolution. Heredity **9**, 1.

— (1955): Heterosis and morphism. Proc. roy. Soc. B. **144**, 215.

— (1957): The three types of evolutionary process. Nature, Lond., **180**, 454.

— (1958): Evolutionary processes and taxonomy with special reference to grades. Univ. Arsskr. Uppsala 21.

— (1963): Genes and developmental pathways. Amer. Zoologist. **3**, 33.

Imai, Y. (1936): Geno- and plasmotypes of variegated Pelargoniums. J. Genet. **33**, 169.
— (1936): Recurrent auto- and exomutation of plastids resulting in tricolored variegation of *Hordeum vulgare*. Genetics **21**, 752.
— (1937): The behaviour of the plastid as a hereditary unit: The theory of the plastogene. Cytologia, Tokyo (Fujii Jub. Vol.) 934.
Ingram, V. M. (1965): The Biosynthesis of Macromolecules. Benjamin/New York, Amsterdam.
Inmann, R. B. (1966): A denaturation map of the λ phage DNA molecule determined by electron microscopy. J. Mol. Biol. **18**, 464.
Isaacs, A. (1963): Interferon. Adv. Virus Res. **10**, 1.
—, and *J. Lindemann* (1957): Virus interference. I. Interferon. Proc. roy. Soc. B. **147**, 258.
Iseki, S., and *T. Sakai* (1953): Artificial transformation of O antigens in *Salmonella* E group. I. Transformation by antiserum and bacterial autolysate. Proc. Jap. Acad. **29**, 121.
Ishikawa, H., R. Bischoff, and *H. Holtzer* (1968): Mitosis and intermediate-sized filaments in developing skeletal muscle. J. Cell Biol. **38**, 538.
Ishitsuka, H., and *A. Kaji* (1970): Release of tRNA from ribosomes by a factor other than G factor. Proc. nat. Acad. Sci., Wash., **66**, 168.
Ito, J., and *I. P. Crawford* (1965): Regulation of the enzymes of the tryptophan pathway in *Escherichia coli*. Genetics **52**, 1303.
Ivanov, M. A. (1938): Experimental production of haploids in *Nicotiana rustica* L. Genetika, Leningrad **20**, 295.
Iyengar, N. K. (1945): Cytological investigations in some of the interspecific hybrids of (American × Asiatic) × American cottons and their progenies. Indian J. Genet. **5**, 32.

Jacob, F. (1966): Genetics of the bacterial cell. Science **152**, 1470.
—, et *S. Brenner* (1963): Sur la régulation de la synthèse du DNA chez les bacteries: l'hypothèse du replicon. C. R. Acad. Sci., Paris, **256**, 298.
—, et *J. Monod* (1959): Génes de structure et génes de régulation dans la biosynthése des protéins. C. R. Acad. Sci., Paris, **249**, 1282.
—, and — (1961): Genetic regulatory mechanisms in the synthesis of proteins. J. Mol. Biol. **3**, 318.
—, and — (1961): On the regulation of gene activity. Cold Spr. Harb. Symp. quant. Biol. **26**, 193.
—, et *E. L. Wollman* (1956): Sur les processus de conjugaison et de recombinaison génétique chez *E. coli*: I. L'induction par conjugaison ou induction zygotique. Ann. Inst. Pasteur **91**, 486.
—, et — (1958): Les épisomes, éléments génétique ajoutés. C. R. Acad. Sci., Paris, **247**, 154.
—, et — (1960): Sexduction et analyse fonctionelle chez les bactéries. C. R. Soc. Biol., Paris, **154**, 1960.
—, *S. Brenner,* and *F. Cuzin* (1964): On the regulation of DNA replication in bacteria. Cold Spr. Harb. Symp. quant. Biol. **28**, 329.
—, *P. Schaeffer,* and *E. L. Wollman* (1960): Episomic elements in bacteria. Symp. Soc. gen. Microbiol. **10**, 67.
—, *A. Ullman,* et *J. Monod* (1964): Le promoteur élément génétique nécessaire à l'expression d'un opéron. C. R. Acad. Sci., Paris, **258**, 3125.
—, *A. Ullman,* et *J. Monod* (1965): Délétions fusionant l'opéron lactose et un opéron purine chez *Escherichia coli*. J. Mol. Biol. **13**, 704.
—, *A. Lwoff, L. Simonovitch,* et *E. L. Wollman* (1953): Définition de quelques termes relatifs à la lysogénie. Ann. Inst. Pasteur **84**, 222.

Jacob, F., D. Perrin, C. Sanchez, et *J. Monod* (1960): L'opéron: groupe de génes à expression coordonnée par un opérateur. C. R. Acad. Sci., Paris, **250**, 1727.
Jagger, J., and *R. S. Stafford* (1965): Evidence for two mechanisms of photoreactivation in *E. coli*. B. Biophys. J. **5**, 75.
James, A. P., and *M. W. Werner* (1966): Radiation-induced lethal sectoring in yeast. Rad. Res. **29**, 523.
Janssens, F. A. (1905): Evolution des auxocytes mâles du *Batracoseps attenuatus*. Cellule **22**, 377.
— (1909): La théorie de la chiasmatypie. Nouvelle interprétation des cinèses de maturation. Cellule **25**, 389.
— (1924): La chiasmatypie dans les insectes. Cellule **34**, 135.
Jaquard, A. (1974): The Genetic Structure of Populations. Springer-Verlag/Berlin-Heidelberg-New York.
Jay, G., and *R. Kaempfer* (1974): Host interference with viral gene expression: Mode of action of bacterial factor i. J. Mol. Biol. **82**, 193.
Jinks, J. L. (1965): Somatic segregation of extrachromosomally inherited differences in fungi. In: Proc. XI. Int. Congr. Genet., The Hague, **3**, 567.
Joergensen, C. A. (1928): The experimental formation of heteroploid plants in the genus *Solanum*. J. Genet. **19**, 133.
—, and *M. B. Crane* (1927): Formation and morphology of *Solanum* chimeras. J. Genet. **18**, 247.
Johannsen, W. (1903): Über Erblichkeit in Populationen und reinen Linien. Fischer/Jena.
— (1909): Elemente der exakten Erblichkeitslehre. Fischer/Jena.
— (1926): Elemente der exakten Erblichkeitslehre (3. Aufl.). Fischer/Jena.
John, B., and *K. R. Lewis* (1965): The Meiotic System. Protoplasmatologia **6** (F 1), Springer/Berlin.
Johnson, R. T., and *P. N. Rao* (1970): Mammalian cell fusion: Induction of premature chromosome condensation in interphase nuclei. Nature, Lond., **226**, 717.
Jollos, V. (1921): Experimentelle Protistenstudien I. Arch. Protistenk. **43**, **1**.
— (1939): Grundbegriffe der Vererbungslehre, insbesondere Mutation, Dauermodifikation, Modifikation. In: *E. Baur*, u. *M. Hartmann* (Ed.) „Handbuch der Vererbungswissenschaft" Bd. IV Borntraeger/Berlin.
Jones, K. W., and *D. E. S. Truman* (1964): A hypothesis for deoxyribonucleic acid transcription and messenger ribonucleic acid synthesis *in vivo*. Nature, Lond., **202**, 1264.
Jukes, T. H. (1965): The genetic code. II. Amer. Scient. **53**, 477.
— (1966): Molecules and Evolution. Columbia Univ. Press/New York.

Kada, T., Y. Saoaie, and *K. Tutikawa* (1970): Mutagenic action of phloxine. Ann. Rep. Nat. Inst. Genet. **21**, **72**.
Kalmus, H., and *S. Maynard Smith* (1966): Some evolutionary consequences of pegmatypic mating systems (imprinting). Amer. Nat. **100**, 619.
Kaplan, R. W., and *H. Stoye* (1974): Electiveness of repair of plaque-type mutations induced by hydroxylamine in phage kappa of *Serratia*. Mut. Res. **22**, 95.
Karpechenko, G. D. (1935): Die Theorie der Kreuzung entfernter Formen. In: *N. I. Vavilov* (Ed.) „Theoretische Grundlagen der Pflanzenzüchtung" Bd. I, Kap. 14.
Kasamatsu, H., D. L. Robberson, and *J. Vinograd* (1971): A novel closed-circular mitochondrial DNA with properties of a replicating intermediate. Proc. nat. Acad. Sci., Wash., **68**, 2252.
Katayama, Y. (1934): Haploid formation by X-rays in *Triticum monococcum*. Cytologia, Tokyo, **5**, 235.

Katayama, Y. (1935): Karyological comparisons of haploid plants from octoploid *Aegilotricum* and diploid wheat. Jap. J. Bot. **7**, 349.
Kattermann, G. (1938): Über konstante, halmbehaarte Stämme aus Weizen-Roggen-bastardierung mit 2n = 42 Chromosomen. Ztschr. indukt. Abstamm.- u. Vererbl. **74**, 354.
— (1939): Ein neuer Karyotyp bei Roggen. Chromosoma **1**, 284.
Katz, L., and *E. Engelsberg* (1971): Hyperinducibility as a result of mutation in structural genes and self-catabolite repression in the ara operon. J. Bact. **107**, 34.
Kaufmann, B. P. (1933): Interchange between X- and Y-chromosomes in attached-X females of *Drosophila melanogaster*. Proc. nat. Acad. Sci., Wash., **19**, 830.
Kellenberger, E. (1962): The study of nature and artificial DNA plasms by thin sections. In: "The interpretation of ultrastructure", 1. Symp. Int. Soc. Cell Biol., Acad. Press/London and New York, p. 233.
—, et *L. Huber* (1953): Contribution à l'étude des équivalents les mitochondries dans les bactéries. Experientia **9**, 289.
Kelner, A. (1949): Photoreactivation of ultraviolet-irradiated *Escherichia coli*, with special reference to the dose-reduction principle and to ultraviolet-induced mutations. J. Bacteriol. **58**, 511.
Kempanna, C., and *R. Riley* (1964): Secondary association between genetically equivalent bivalents. Heredity **19**, 289.
Keosian, J. (1965): The Origin of Life. Chapman & Hall/London.
Kerner, A. (1881): Pflanzenleben. Bd. I. Bibl. Inst./Leipzig u. Wien.
Khush, G. S. (1973): Cytogenetics of Aneuploids. Academic Press, New York and London.
—, and *C. M. Rick* (1967): Haplo-triplo-disomics of the tomato: Origin, cytogenetics, and utilization as a source of secondary trisomics. Biol. Zbl. **86**, 257.
—, and — (1967): Novel compensating trisomics of the tomato: Cytogenetics, monosomic analysis, and other applications. Genetics **55**, 297.
—, and — (1968): Tomato telotrisomics: Origin, identification, and use in linkage mapping. Cytologia **33**, 137.
Kiellander, C. L. (1942): A subhaploid *Poa pratensis* L. with 18 chromosomes and its progeny. Svensk bot. Tidskr. **36**, 200.
Kielmeyer, K. F. (1793): Über die Verhältnisse der organischen Kräfte. Stuttgart.
Kihara, H., und *T. Ono* (1926): Chromosomenzahlen und systematische Gruppierung der *Rumex*-Arten. Z. Zellforsch. **4**, 475.
Kimber, F., and *R. Riley* (1963): Haploid angiosperms. Bot. Rev. **29**, 480.
—, and *E. R. Sears* (1968): Nomenclature for the description of aneuploids in the *Triticinae*. Int. Wheat Genet. Symp. Canberra, 468.
Kimura, M. (1956): A model of a genetic system which leads to closer linkage by natural selection. Evolution **10**, 278.
King, J. L., and *T. H. Jukes* (1969): Non-Darwinian Evolution. Science **164**, 788.
Kit, S. (1961): Equilibrium sedimentation in density gradients of DNA preparations from animal tissues. J. Mol. Biol. **3**, 711.
— (1962): Species differences in animal deoxyribonucleic acids as revealed by equilibrium sedimentation in density gradients. Nature, Lond., **193**, 274.
Klima, J. (1967): Cytologie. Fischer/Stuttgart.
Kniep, H. (1928): Die Sexualität der niederen Pflanzen. Fischer/Jena.
Koch, R. E., and *J. W. Drake* (1970): Cryptic mutants of bacteriophage T 4. Genetics **65**, 379.
Koerperich, J. (1930): Étude comparative du noyau, des chromosomes et de leurs relations avec le cytoplasme (*Nothoscordum, Eucomis, Beschorneria*). Cellule **39**, 307.
Kohler, R. E., E. Z. Ron, and *B. D. Davis* (1968): Significance of the free 70s Ribosomes in *Escherichia coli* extracts. J. Mol. Biol. **36**, 71.

Kohn, K. W., C. L. Spears, and *P. Doty* (1966): Inter-strand crosslinking of DNA by nitrogen mustard. J. molecular Biol. **19**, 266.
Koller, P. C. (1935): The internal mechanics of chromosomes. IV. Pairing and coiling in salivary gland nuclei of *Drosophila.* Proc. roy. Soc. B. **118**, 371.
Koltzoff, N. K. (1934): The structure of the chromosomes in the salivary glands of *Drosophila.* Science **80**, 312.
Komai, T. (1950): Semiallelic genes. Amer. Nat. **84**, 381.
Kondo, S., H. Ichikawa, K. Iwo, and *T. Kato* (1970): Base-change mutagenesis and prophage induction in strains of *Escherichia coli* with different DNA repair capacities. Genetics **66**, 187.
Koprowski, H. (1964): Criteria for the proof of virogeny in mammalian cells. In: *R. S. Krooth* (Ed.) "Somatic Cell Genetics", Univ. of Michigan Press/Ann Arbor, 274.
Korschelt, E., und *K. Heider* (1903): Eireifung, Samenreifung und Befruchtung. Lehrbuch d. vergl. Entw.-Gesch. Allg. Teil, Fischer/Jena.
Kossel, A. (1884): Über einen peptonartigen Bestandteil des Zellkerns. Z. physiol. Chemie **8**, 511.
Kostoff, D. (1939): Autosyndesis and structural hybridity in F_1 hybrids *Helianthus tuberosus* L. × *Helianthus annuus* L. and their sequences. Genetics **21**, 285.
Kourilsky, P., and *F. Gros* (1975): In "Regulation of gene expression in cultured cells"; Fogarty Int. Conf. Center Proc. Nr. 22, U. S. Gouvernment Printing Office, Washington D. C.
Kozinski, A. W., P. B. Kozinski, and *R. James* (1967): Molecular recombination in T 4 bacteriophage deoxyribonucleic acid. J. Virol. **1**, 758.
Krichevskaya, A. A., and *G. P. Georgiev* (1969): Further studies in the protein moiety in nuclear DNA-like RNA containing complexes. Biochem. Biophys. Acta **164**, 619.
Krooth, R. S. (1969): Gene action in human diploid cell strains. In: *Bernischke, K.* "Comparative Mammalian Cytogenetics" p. 154.
Kühn, A. (1927): Die Pigmentierung von *Habrobracon juglandis* Ashmed, ihre Prädetermination und ihre Vererbung durch Gene und Plasmon. Nachr. Ges. Wiss. Göttingen, Math.-Phys. Klasse, 407.
— (1961): Grundriß der Vererbungslehre. Quelle & Meyer/Heidelberg.
Küntzel, H., Z. Barath, I. Ali, J. Kind, H. H. Althaus, and *H. C. Blossey* (1973): Expression of the mitochondrial genome in wildtype and in an extranuclear mutant of the *Neurospora crassa.* In: *E. K. F. Bautz, P. Karlson,* and *H. Kersten* (Eds.) "Regulation of Transcription and Translation in Eukaryotes". Springer-Verlag, Berlin-Heidelberg-New York.
Kupffer, C. (1875): Über die Differenzierung des Protoplasmas in den Zellen tierischer Gewebe. Schr. naturw. Ver. Schl.-Holst. **1**.
— (1896): Über Energiden und paraplastische Bildungen. München.
Kurland, C. G. (1960): Molecular characterization of ribonucleic acid from *Escherichia coli* ribosomes. I. Isolation and molecular weights. J. Mol. Biol. **2**, 83.
Kurtén, B. (1958): A differentiation index and a new measure of evolutionary rates. Evolution **12**, 146.
Küster, E. (1935): Die Pflanzenzelle. Fischer/Jena.

Labella, F. S., and *M. E. Krass* (1968): New terminology for cellular ingestion and egestion. Nature, Lond., **220**, 1141.
La Cour, L. F., and *A. Rutishauser* (1953): Chromosome breakage with endosperm: subchromatid breakage. Nature, Lond., **172**, 501.
Lacy, P. E. (1961): Electron microscopy of the beta cell of the pancreas. Amer. J. Med. **31**, 851.

Lambertz, P. (1954): Untersuchungen über das Vorkommen von Plasmodesmen in den Epidermisaußenwänden. Planta **44**, 147.
Landauer, W. (1958): On phenocopies, their developmental physiology and genetic meaning. Amer. Nat. **92**, 201.
Lang, A. (1911): Fortgesetzte Vererbungsstudien. Z. indukt. Abstamm.- u. VerebLehre **5**, 97.
Lange, M. (1952): Species concept in the genus *Coprium*. Dansk bot. Ark. **14**, 6.
Langlet, O. F. (1927): Beiträge zur Cytologie der Ranunculaceen. Svensk bot. Tidskr. **21**, 1.
Lark, K. G. (1966): Regulation of chromosome replication and segregation in bacteria. Bacteriol. Rev. **30**, 3.
— (1972): Evidence for the direct involvement of RNA in the initiation of DNA replication in *Escherichia coli* 15 T^-. J. Mol. Biol. **64**, 47.
Latarjet, R., and *L. H. Gray* (1954): Definition of the terms "protection" and "restoration". Acta Radiol. Stockh. **41**, 61.
Laughnan, J. R. (1952): On the designation of closely linked genes with similar effects. Amer. Nat. **86**, 109.
La Valette St. George (1886): Über die Genese der Samenkörper. V. Mitteilung. Die Spermatogenese bei den Säugetieren und Menschen. Arch. mikr. Anat. **15**, 261.
Lawrence, W. J. (1931): The secondary association of chromosomes. Cytologia Tokyo, **2**, 352.
Lea, D. E. (1947): The induction of chromosome structural changes by radiation: detailed quantitative interpretation. Brit. J. Radiol., Supp., **1**, 75.
—, and *D. G. Catcheside* (1942): The mechanism of the induction by radiation of chromosome aberrations in *Tradescantia*. . Genet. **44**, 216.
Lebégue, A. (1952): La polyembryonie chez les Angiosperms. Bull. Soc. bot. Fr. **99**, 329.
Ledbetter, M. C., and *K. R. Porter* (1963): A "microtubule" in plant cell fine structure. J. Cell Biol. **19**, 239.
Leder, P. (1973): The elongation reactions in protein synthesis. Adv. in Proteinchem. **27**, 213.
—, and *H. Bursztyn* (1966): Initiation of protein synthesis. I. Effect of formylation of methionyl-tRNA on codon recognition. Proc. nat. Acad. Sci., Wash., **56**, 1579.
Lederberg, J. (1952): Cell genetics and hereditary symbiosis. Physiol. Rev. **32**, 403.
— (1954): Phase variation in *Salmonella*. Genetics **39**, 978 (Abstr.).
— (1955): Recombination mechanisms in bacteria. J. cell. comp. Physiol. **45**, Suppl. II, 75.
— (1966): Experimental genetics and human evolution. Amer. Nat. **100**, 519.
— (1966): Remarks. Current Topics in Developm. Biol. **1**, IX.
—, *L. L. Cavalli*, and *E. M. Lederberg* (1952): Sex incompatibility in *Escherichia coli*. Genetics **37**, 720.
Lee, C. S., and *C. A. Thomas* (1973): Formation of rings from *Drosophila* DNA fragments. J. Mol. Biol. **77**, 25.
Lee-Huang, S., and *S. Ochoa* (1972): Specific inhibitors of MS2 and late T4 RNA translation in *E. coli*. Biochem. Biophys. Res. Com. **49**, 371.
Lefevre, G. (1955): Mutation-isoalleles at the yellow locus in *Drosophila melanogaster*. Genetics **40**, 374.
Legator, M. S., and *W. G. Flamm* (1973): Environmental mutagenesis and repair. Ann. Rev. Biochem. **42**, 683.
Lehmann, F. E. (1947): Über die plasmatische Organisation tierischer Eizellen und die Rolle vitaler Strukturelemente der Biosomen. Rev. Suisse Zool. **54**, 246.
Lehninger, A. L. (1965): Bioenergetics. The Molecular Basis of Biological Energy Transformations. W. H. Benjamin/New York.

Leighty, C. E., and *I. W. Taylor* (1924): "Hairy Neck" wheat segregates from wheat-rye hybrids. J. Agr. Res. **28**, 567.
Lenhossek, M. v. (1898): Untersuchungen über Spermatogenese. Arch. mikr. Anat. **51**, 215.
Lenz, F. (1938): Über kombinantes Verhalten alleler Gene. Erbarzt **5**, 83.
Lerner, J. M. (1954): The genotype in Mendelian populations. Caryologia **6**, Suppl. I, 124.
— (1954): Genetic Homeostasis. Oliver & Boyd/London.
— (1958): The Genetic Basis of Selection. Wiley/New York.
— (1968): Heredity, Evolution and Society. Freeman and Comp., San Francisco.
Letham, D. S. (1973): Transfer RNA and cytokinins: In. *Stewart, P. R.*, and *D. S. Letham* (Eds.) "The Ribonucleic Acids". Springer-Verlag/Berlin, Heidelberg, New York. p. 81.
Lettré, H. (1959): Zellen *in vitro*. Studium gen. **12**, 216.
Levan, A. (1937): Cytological studies in the *Allium paniculatum* group. Hereditas, Lund, **23**, 317.
— (1938): The effect of colchicine on root mitosis in *Allium*. Hereditas, Lund, **24**, 471.
— (1939): The effect of colchicine on meiosis in *Allium*. Hereditas, Lund, **25**, 9.
— (1942): Studies on the meiotic mechanism of haploid rye. Hereditas, Lund, **28**, 177.
—, and *T. S. Hauschka* (1953): Endomitotic reduplication mechanisms in ascites tumours of the mouse. J. nat. Cancer Inst. **14**, 1.
—, and *A. Müntzing* (1963): Terminology of chromosome numbers. Portug. acta biol. Ser. A., **7**, 1.
—, and *J. H. Tjio* (1948): Induction of chromosome fragmentation by phenols. Hereditas, Lund, **34**, 453.
—, *K. Fredga*, and *A. A. Sandberg* (1964): Nomenclature for centromeric position on chromosomes. Hereditas, Lund, **52**, 201.
Levi, H. L. (1973): Genetic screening. Adv. in Human Genetics **4**, 1.
Levinthal, C. (1954): Recombination in phage T 2: its relationship to heterozygosis and growth. Genetics **39**, 169.
— (1959): Bacteriophage genetics. In: *F. M. Burnet*, and *W. M. Stanley* (Ed.) "The Viruses", Vol. **2**, 281. Academic Press/New York, London.
—, and *N. Visconti* (1953): Growth and recombination in bacterial viruses. Genetics **38**, 500.
Levitan, M. (1954): Position effects in natural populations. Amer. Nat. **88**, 419.
Levitsky, G. A. (1924): Materielle Grundlagen der Vererbung (russ.). Kiew, Staatsverlag.
— (1925): Die Chondriosomen in der Gonogenese bei *Equisetum palustre* L. Planta **1**, 301.
— (1931): The morphology of chromosomes. Bull. appl. Bot., St. Petersb. **27**, 19.
— (1931): The "karyotype" in systematics. Bull. appl. Bot., St. Petersb. **27**, 187.
Lewis, D. (1943): The physiology of incompatibility in plants. III. Autopolyploids. J. Genet. **45**, 171.
— (1944): Incompatibility in plants: Its genetical and physiological synthesis. Nature, Lond., **153**, 575.
— (1947): Competition and dominance of incompatibility alleles in diploid pollen. Heredity **1**, 85.
— (1954): Comparative incompatibility in angiosperms and fungi. Advanc. Genet. **6**, 235.
— (1954): Gene-environment interaction. Heredity **8**, 333.
Lewis, E. B. (1945): The relation of repeats to position effect in *Drosophila melanogaster*. Genetics **30**, 137.
— (1948): Pseudo-allelism in *Drosophila melanogaster*. Genetics 33, 113 (Abstr.).

Lewis, E. B. (1950): The phenomenon of position effect: Advanc. Genet. **3**, 75.
— (1951): Pseudoallelism and gene evolution. Cold Spr. Harb. Symp. quant. Biol. **16**, 159.
— (1954): The theory and application of a new method of detecting chromosomal rearrangements in *Drosophila melanogaster*. Amer. Nat. **88**, 225.
— (1955): Some aspects of position pseudoallelism. Amer. Nat. **89**, 73.
— (1963): Genes and developmental pathways. Amer. Zoologist **3**, 33.
Lewis, K. R., and *B. John* (1963): Chromosome Marker. Churchill/London.
—, and — (1964): The Matter of Mendelian Heredity. Churchill/London.
—, and — (1970): The Organisation of Heredity. E. Arnold/London.
Lewis, W. H. (1931): Pinocytosis. Johns Hopk. Hosp. Bull. **49**, 17.
Lewontin, R. C. (1955): The effects of population density and composition on viability in *Drosophila*. Evolution **9**, 27.
Li, H. W., *W. K. Pao*, and *C. H. Li* (1945): Desynapsis in common wheat. Am. J. Bot. **32**, 92.
Lieb, M. (1966): Studies of heat-inducible lambda bacteriophage. I. Order of genetic sites and properties of mutant prophages. J. Mol. Biol. **16**, 149.
Lifschytz, E. (1971): Fine-structure analysis of the chromosome recombinational patterns at the base of the X-chromosome of *Drosophila melanogaster*. Mut. Res. **12**, 35.
Lilienfeld, F. A. (1951): *H. Kihara*: Genome-analysis in *Triticum* and *Aegilops*. X. Concluding Review. Cytologia, Tokyo, **16**, 101.
Lima-de-Faria, A. (1949): Genetics, origin, and evolution of kinetochores. Hereditas, Lund, **35**, 422.
— (1949): The structure of the centromere of the chromosomes of rye. Hereditas, Lund, **35**, 77.
— (1952): The chromomere size gradient of the chromosomes of rye. Hereditas, Lund, **38**, 246.
— (1952): Chromomere analysis of the chromosome complement of rye. Chromosoma **5**, 1.
— (1954): Chromosome gradient and chromosome field in *Agapanthus*. Chromosoma **6**, 330.
— (1956): The role of kinetochore in chromosome organization. Hereditas, Lund, **42**, 85.
— (1958): Recent advances in the study of the kinetochore. Int. Rev. Cytol. **7**, 123.
—, and *H. Jaworska* (1972): The relation between the chromosome size gradient and the sequence of DNA replication in rye. Hereditas, Lund, **70**, 39.
—, and *P. Sarvella* (1958): The organization of telomeres in species of *Solanum*, *Salvia*, *Scilla*, *Secale*, *Agapanthus*, and *Ornithogalum*. Hereditas, Lund, **44**, 337.
—, *S. Daskaloff*, and *A. E. Nell* (1973): Amplification of ribosomal DNA in *Acheta*. Hereditas, Lund, **73**, 99.
Lindegren, C. C. (1949): Chromosome maps of *Saccharomyces*. Proc. 8th Int. Congr. Genet., Hereditas (Suppl.), 338.
Lindsley, D. L., and *E. H. Grell* (1968): Genetic variations of *Drosophila melanogaster*. Carnegie Inst. of Wash. Publication Nr. 627.
Lissouba, P., et *G. Rizet* (1960): Sur l'existence d'une génétique polarisée ne subissant que des échanges non réciproques. C. R. Acad. Sci., Paris, **250**, 3408.
—, *J. Mousseau*, *G. Rizet*, and *J. L. Rossignol* (1962): Fine structure of genes in the ascomycete *Ascobolus immersus*. Adv. Genet. **11**, 343.
Litardiére, R. de (1925): Sur l'existence de figures didiploides dans le méristème radiculaire du *Cannabis sativa*. Cellule **35**, 19.
Ljungdahl, H. (1922): Zur Zytologie der Gattung *Papaver*. (Vorläufige Mitteilung) Svensk bot. Tidskr. **16**, 103.

Ljungdahl, H. (1924): Über die Herkunft der in der Meiosis konjugierenden Chromosomen bei *Papaver*-Hybriden. Svensk. bot. Tidskr. **18**, 279.
Löve, A., and *D. Löve* (1949): The geobotanical significance of polyploidy. Portug. acta biol. Goldschmidt Vol. 273.
Loew, E. (1895): Einführung in die Blütenbiologie. Dümmler/Berlin.
Longley, A. E. (1927): Supernumerary chromosomes in *Zea mays*. J. Agr. Res. **35**, 769.
— (1945): Abnormal segregation during megasporogenesis in maize. Genetics **30**, 100.
Lotsy, J. P. (1904): Die Wendung der Dyaden beim Reifen der Tiereier als Stütze für die Bivalenz der Chromosomen nach der numerischen Reduktion. Flora, Jena **93**, 65.
— (1916): Evolution by Means of Hybridization. Nijhoff/The Hague.
Loveless, A., and *S. Revell* (1949): New evidence on the mode of action of "mitotic poisons". Nature, Lond., **164**, 938.
Lucas-Lenard, J., and *F. Lipman* (1966): Separation of three microbial amino acid polymerization factors. Proc. nat. Acad. Sci., Wash., **55**, 1562.
Luck, D. J. L., and *E. Reich* (1964): DNA in mitochondria of *Neurospora crassa*. Proc. Nat. Acad. Sci., Wash., **52**, 931.
Luckhaus, G. (1965): Fortpflanzung im Pflanzen- und Tierreich. Parey/Berlin.
Ludwig, W. (1943): Die Selektionstheorie. In: *G. Heberer* (Ed.) „Die Evolution der Organismen" Fischer/Jena, 479.
— (1948): Darwins Zuchtwahllehre in moderner Fassung. Aufs. senckb. naturf. Ges.
— (1954): Die Selektionstheorie. In: *G. Heberer* (Ed.) „Die Evolution der Organismen", (2. Aufl.), Fischer/Stuttgart, 662.
Lundegårdh, H. (1912): Fixierung, Färbung und Nomenklatur der Kernstrukturen, ein Beitrag zur Theorie der zytologischen Methodik. Arch. mikr. Anat. **80**, 223.
— (1922): Zelle und Zytoplasma. Linsbauers Handb. PflAnat. **1**, Abt. 1, Borntraeger/Berlin.
Luria, S. E. (1945): Mutations of bacterial viruses affecting their host range. Genetics **30**, 84.
— (1947): Reactivation of irradiated bacteriophage by transfer of self-reproducing units. Proc. nat. Acad. Sci., Wash., **33**, 253.
— (1950): An essay on virus reproduction. Science **111**, 507.
— (1963): Molecular and genetic criteria in bacterial classification. Recent Progr. in Microbiol. **8**, 604.
—, and *M. Delbrück* (1943): Interference between bacterial viruses. II: Interference between inactivated bacterial virus and active virus of the same strain and of a different strain. Arch. Biochem. **1**, 207.
—, and — (1943): Mutations of bacteria from virus sensitivity to virus resistance. Genetics **28**, 491.
—, and *M. C. Human* (1952): A non-hereditary, host-induced variation of bacterial viruses. J. Bacteriol. **64**, 557.
Lush, J. L. (1941): Methods of measuring the heritability of individual differences among farm animals. Proc. of 7th Intern. Congr. Genetics 1939. Suppl. Vol. of J. Genetics, p. 199.
Luykx, P. (1974): The organization of meiotic chromosomes. In "The Cell Nucleus", (*H. Busch*, Ed.), Acad. Press New York and London, Vol. **2**, p. 163.
Lwoff, A. (1949): Les organites donés de continuité génétique chez les Protisdes. In: Unités biologiques donés de continuité génétique. Edition du Centre Nat. de la Rech. Sci. Paris, **7**.
— (1953): Lysogeny. Bacteriol. Rev. **17**, 269.
— (1953): L'induction. Ann. Inst. Pasteur **84**, 225.
—, and *A. Gutmann* (1949): Production discontinue de bactériophages par une souche lysogéne de *Bacillus megatherium*. C. R. Acad. Sci., Paris, **229**, 679.

Lwoff, A., and — (1950): Recherches sur un *Bacillus megatherium* lysogène. Ann. Inst. Pasteur **78**, 711.
—, *T. F. Anderson*, et *F. Jacob* (1959): Remarques sur les charactéristiques de la particule virale infectieuse. Ann. Inst. Pasteur **97**, 281.
Lyon, M. F. (1961): Gene action in the X-chromosome of the mouse (*Mus musculus* L.). Nature, Lond., **190**, 372.
— (1962): Sex chromatin and gene action in the mammalian X-chromosome. Amer. J. hum. Genet. **14**, 135.
— (1966): X-chromosome inactivation in mammals. Adv. in Teratology **1**, 25.

Maas, W. K., and *A. J. Clark* (1964): Studies on the mechanism of repression of arginine biosynthesis in *Escherichia coli*. II. Dominance of repressibility in diploids. J. Mol. Biol. **8**, 365.
Magasanik, B. (1961): Catabolite repression. Cold Spr. Harb. Symp. quant. Biol. **26**, 249.
Magni, G. E. (1964): Origin and nature of spontaneous mutations in meiotic organisms. J. Cell. Comp. Physiol. **64**, Suppl. 1, 165.
Magnus, P. v. (1954): Incomplete forms of influenza virus. Adv. Virus-Res. **2**, 59.
Maire, R. (1902): Recherches cytologiques et taxonomiques sur les Basidiomycètes. Bull. Soc. mycol. Fr. **18**, 1.
Malamy, M. H., M. Fiandt, and *W. Szybalski* (1972): Electron microscopy of polar insertions in the lac Operon of *Escherichia coli*. Mol. Gen. Genetics **119**, 207.
Malecot, G. (1948): Les Mathematique de L'hérédité. Masson & Cie/Paris.
Malheiros-Gardé, N. (1951): Agmatoploidia no genero *Luzula* D. C. Genet. iber. **3**, 155.
Malheiros-Gardé, N., and *A. Gardé* (1950): Fragmentation as a possible evolutionary process in the genus *Luzula* D. C. Genet. iber. **2**, 257.
Mansfeld, R. (1950): Das morphologische System der Saatgerste. Züchter **20**, 8.
Marcker, K. A., and *F. Sanger* (1964): N-formyl-methionyl-S-RNA. J. Mol. Biol. **8**, 835.
Margolin, P. (1965): Bipolarity of information transfer from *Salmonella typhimurium* chromosome. Science **147**, 1456.
—, and *R. H. Bauerle* (1966): Determinants for regulation and initiation of expression of tryptophan genes. Cold Spr. Harb. Symp. quant. Biol. **31**, 311.
Mark, E. L. (1881): Maturation, fecundation, and segmentation of *Limax campestris*. Bull. Mus. comp. Zool. Harv. **6**, 173.
Markert, C. L., and *F. Moller* (1959): Multiple forms of enzymes: tissue, ontogenetic, and species-specific patterns. Proc. nat. Acad. Sci., Wash., **45**, 753.
Marquardt, H. (1937): Die Meiosis von *Oenothera* I. Z. Zellforsch. **27**, 159.
— (1952): Die Natur der Erbträger im Zytoplasma. Ber. dtsch. bot. Ges **65**, 197.
Martin, G. M., and *C. A. Sprague* (1973): Life histories of hyperplastoid cell lines from aorta and skin. Exp. Mol. Pathol. **18**, 125.
Martin, G. R. (1969): Control of gene expression. Ann. Review of Genetics **3**, 181.
Mason, H. S., J. C. North, and *M. Vaneste* (1965): Microsomal mixed-function oxidations: The metabolism of xenobiotics. Fed. Proc. **24**, 1172.
Mather, K. (1932): Chromosome variation in *Crocus*, I. J. Genet. **26**, 129.
— (1933): Interlocking as a demonstration of the occurence of genetical crossing-over during chiasma-formation. Amer. Nat. **67**, 476.
— (1933): The relations between chiasmata and crossing-over in diploid and triploid *Drosophila melanogaster*. J. Genet. **27**, 243.
— (1936): Segregation and linkage in autotetraploids. J. Genet. **32**, 287.
— (1936): The determination of position in crossing-over. I. *Drosophila melanogaster*. J. Genet. **33**, 207.

Mather, K. (1936): Competition between bivalents during chiasma formation. Proc. roy. Soc. B. **120**, 208.
— (1941): Variation and selection of polygenetic characters. J. Genet. **41**, 159.
— (1943): Polygenic inheritance and natural selection. Biol. Rev. **18**, 32.
— (1948): Nucleus and cytoplasm in differentiation. Symp. Soc. exp. Biol. **2**, 196.
— (1953): Genetical control of stability in development. Heredity **7**, 297.
— (1955): Polymorphism as an outcome of disruptive selection. Evolution **9**, 52.
—, and *L. Wigan* (1942): The selection of invisible mutations. Proc. roy. Soc. B. **131**, 50.
Matsui, S., H. Weinfeld, and *A. A. Sandberg* (1972): Fate of chromatin of interphase nuclei subjected to "prophasing" in virus fused cells. J. Nat. Cancer Inst. **49**, 1621.
Matsuura, H. (1937): Chromosome studies in *Trillium kamtschaticum*. III. The mode of chromatid disjunction at the first meiotic metaphase of the P. M. C. Cytologia, Tokyo, **8**, 142.
— (1941): Chromosome studies in *Trillium kamtschaticum* Pall. XIII. The structure and behavior of the kinetochore. Cytologia, Tokyo, **11**, 369.
— (1941): Chromosome studies in *Trillium kamtschaticum* Pall. XV. A contribution to the present status of knowledge on the mechanism of chromonema coiling. Cytologia, Tokyo, **11**, 407.
Matthey, R. (1945): L'évolution de la formule chromosomiale chez les vertébrés. Experientia **1**, 50 and 78.
— (1949): Les Chromosomes des Vertébrés. F. Rouge/Lausanne.
Mayr, E. (1940): Speciation phenomena in birds. Amer. Nat. **74**, 249.
— (1942): Systematics and the Origin of Species. Columbia Univ. Press/New York.
— (1948): The bearing of the new systematics on genetical problems. The nature of species. Advanc. Genet. **2**, 205.
— (1963): Animal Species and Evolution. Belknap Press of Havard Univ. Press/ Cambridge, Mass.
Mazia, D. (1956): The life history of the cell. Amer. Scient. **44**, 1.
— (1961): Mitosis and the physiology of cell division. In: *J. Brachet,* and *A. E. Mirsky* (Ed.) "The Cell", **3**, 77, Acad. Press/New York.
—, and *K. Dan* (1952): The isolation and biochemical characterization of the mitotic apparatus of dividing cells. Proc. nat. Acad. Sci., Wash., **38**, 826.
—, and *A. Ruby* (1968): Dissolution of erythrocyte membranes in water and comparison of the membrane protein with other structural proteins. Proc. nat. Acad. Sci., Wash., **61**, 1005.
McCarthy, B. J., R. J. Britten, and *R. B. Roberts* (1962): The synthesis of ribosomes in *E. coli*. III. Synthesis of ribosomal RNA. Biophys. J. **2**, 83.
McClintock, B. (1934): The relation of a particular chromosomal element to the development of the nucleoli in *Zea mays*. Z. Zellforsch. **21**, 294.
— (1944): The relation of homozygous deficiencies to mutations and allelic series in maize. Genetics **29**, 478.
— (1951): Chromosome organization and gene expression. Cold Spr. Harb. Symp. quant. Biol. **16**, 13.
— (1956): Intranuclear systems controlling gene action and mutation. Brookh. Symp. Biol. **8**, 58.
McClung, C. E. (1900): The spermatocyte-divisions of the *Acrididae*. Kans. Univ. Quart. A. **9**, 73.
— (1902): The accessory chromosome — sex determinant. Biol. Bull. **3**, 43.
— (1902): The spermatocyte divisions in the locustidae. Kansas Univ. Sci. Bull. **1**, 185.
— (1905): The chromosome complex of orthopteran spermatocytes. Biol. Bull., Wood's Hole **9**, 304.

McCready, S. J., and *B. S. Cox* (1973): Antisuppressors in Yeast. Molec. Gen. Genetics **124**, 305.

McDaniel, R. G., and *J. V. Sarkissian* (1966): Heterosis: Complementation by mitochondria. Science **152**, 1640.

McElroy, W. D., M. Deluca, and *J. Travis* (1967): Molecular uniformity in biological catalyses. Science **157**, 150.

McLellan, W. L., and *H. J. Vogel* (1970): Translational repression in the arginine system of *Escherichia coli*. Proc. nat. Acad. Sci., Wash., **67**, 1703.

McQuillen, K. (1960): Bacterial protoplasts. In: *J. C. Gunsalus* and *R. Y. Stanier* (Ed.) "The Bacteria", Acad. Press/New York, **1**, 250.

Mendel, J. G. (1875): Versuche über Pflanzenhybriden. Verh. naturf. Ver. Brünn **4**, (Abh.).

Menke, W. (1961): Über die Chloroplasten von *Anthoceros punctatus*. (5. Mitteilg. z. Entwicklungsgesch. d. Plastiden). Z. Naturf. **16b**, 334.

Merrell, D. J. (1950): Measurement of sexual isolation and selective mating. Evolution **4**, 326.

— (1962): Evolution and Genetics. Holt, Rinheart & Winston/New Yorks.

Meselson, M., and *R. Yuan* (1968): DNA restriction enzyme from *E. coli*. Nature, Lond., **217**, 1110.

Metchnikoff, E. (1883): Untersuchungen über die mesodermalen Phagocyten einiger Wirbeltiere. Biol. Centralbl. **3**, 360.

Metz, C. W. (1916): Chromosome studies on the Diptera. II. The paired association of the chromosomes in the Diptera and its significance. J. exp. Zool. **21**, 213.

Metzenberg, R. L. (1972): Genetic regulatory systems in *Neurospora*. Am. Rev. of Genetics **6**, 111.

Meves, F. (1907): Über Mitochondrien, bzw. Chondriokonten in den Zellen junger Embryonen. Anat. Anz. **31**, 399.

— (1908): Die Chondriosomen als Träger erblicher Anlagen. Cytologische Studien an Hühnerembryonen. Arch. mikr. Anat. **72**, 816.

Meyer, A. (1883): Das Chlorophyllkorn in chemischer, morphologischer und biologischer Beziehung. A. Felix/Leipzig.

— (1896): Die Plasmaverbindungen und die Membranen von *Volvox globator, aureus* und *tertius*, mit Rücksicht auf die tierischen Zellen. Bot. Ztg. **54**, 187.

— (1908): Der Zellkern der Bakterien. Fischer/Jena.

Meynell, G. G., and *A. M. Lawn* (1967): Sex pili and common pili in the conjugational transfer of colicin factor Ib by *Salmonella typhimurium*. Genet. Res. **9**, 359.

Michaelis, P. (1955): Plasma-Vererbung. In: *Kappert, H.*, and *W. Rudorf* (Ed.), „Handbuch d. Pflanzenzüchtung", Parey/Berlin, **140**.

— (1957): Genetische, entwicklungsgeschichtliche und cytologische Untersuchungen zur Plasmavererbung. II. Mitteilung: Über eine Plastidenmutation mit intracellulärer Wechselwirkung der Plastiden, zugleich ein Beitrag von Methodik der Plasmonanalyse und zur Entwicklungsgeschichte von *Epilobium*. Planta **50**, 60.

— (1957): Über die Vererbung von Plastidenmerkmalen. (Vorl. Mitteilung.) Protoplasma **48**, 403.

Michel, M. R., B. Hirt, and *R. Weil* (1967): Mouse cellular DNA enclosed in polyoma viral capsidae (Pseudovirions). Proc. nat. Acad. Sci., Wash., **58**, 1381.

Michie, D., and *M. E. Wallace* (1953): Affinity: A new genetic phenomenon in the house mouse. Nature, Lond., **171**, 26.

Milne, A. (1961): Definition of competition among animals. Symp. Soc. exp. Biol. **15**, 40.

Minot, C. S. (1908): The Problems of Age, Growth and Death. A Study of Cytomorphosis. Putman's Sons/New York, London.

Mitchell, M. B., T. H. Pittenger, and *H. K. Mitchell* (1952): Pseudowildtypes in *Neurospora crassa*. Proc. nat. Acad. Sci., Wash., **38**, 442.
Mittwoch, U. (1954): The genetics of *Coprinus lagopus*. I. The formation of monokaryons from dikaryons. J. Genet. **52**, 547.
Mohl, H. v. (1846): Über die Saftbewegungen im Inneren der Zelle. Bot. Ztg. **4**, 73.
Mohr, O. L. (1923): Das Defizienz-Phänomen bei *Drosophila melanogaster*. Z. indukt. Abstamm.- u. VererbLehre **30**, 279.
— (1926): Über die Letalfaktoren, mit Berücksichtigung ihres Verhaltens bei Haustieren und beim Menschen. Z. indukt. Abstamm.- u. VererbLehre **41**, 59.
Monné, L. (1948): Functioning of the cytoplasm. Advanc. Enzymol. **8**, 1.
Monneron, A., and *W. Bernhard* (1969): Fine structural organization of the interphase nuclei in some mammalian cells. J. Ultrastruct. Res. **27**, 266.
Monod, J. (1942): Recherches sur la Croissance des Cultures bacteriénnes. Hermann/ Paris.
—, and *G. Cohen-Bazire* (1953): L'effect d'inhibition spécifique dans la biosynthese de la tryptophane-demase. C. R. Acad. Sci. Paris, **236**, 530–532.
—, and *F. Jacob* (1961): General conclusions: Teleonomic mechanisms in cellular metabolism, growth, and differentiation. Cold Spr. Harb. Symp. quant. Biol. **26**, 389.
—, and *J.-P. Changeux, G. Jacob* (1963): Allosteric Proteins and cellular control systems. J. Mol. Biol. **6**, 306.
—, *G. Cohen-Bazire*, et *M. Cohn* (1951): Sur la biosynthèse de la β-galactosidase (lactase) chez *Escherichia coli*. La spécifité de l'induction. Biochim. biophys. Acta **7**, 585.
—, *A. M. Torriani*, et *M. Jolit* (1949): Biochimie batcérienne. — Sur la réactivation de bactéries stérilisées par le rayonnement UV. C. R. Acad. Sci., Paris, **229**, 557.
—, *J. Wyman*, and *J.-P. Changeux* (1969): On the nature of allosteric transitions: a plausible model. Theoret. Physics and Biol., North-Holland Publ. Co/Amsterdam.
Montgomery, T. H. (1904): Some observations and considerations upon the maturation phenomena of the germ cells. Biol. Bull. **6**, 137.
— (1906): Chromosomes in the spermatogenesis of the *Hemiptera Heteroptera*. Trans. Amer. phil. Soc. **21**, 97.
Moore, J. E. S. (1895): On the structural changes in the reproductive cells during spermatogenesis of Elasmobranchs. Quart. J. micr. Sci. **38**.
— (1895): On the essential similarity of the process of chromosome reduction in animals and plants. Ann. Bot. **9**, 431.
Moore, R. T., and *J. H. McAlear* (1961): Fine structure of mycota. 5. Lomasomes-previously uncharacterized hyphal structures. Mycologia **53**, 194.
Morgan, L. V. (1922): Non criss-cross-inheritance in *Drosophila melanogaster*. Biol. Bull. **42**, 267.
Morgan, R. S., and *B. G. Uzmann* (1966): Nature of the packing of ribosomes within chromatoid bodies. Science **152**, 214.
Morgan, T. H. (1910): Chromosomes and heredity. Amer. Nat. **44**, 449.
— (1911): An attempt to analyze the constitution of the chromosomes on the basis of sex-limited inheritance in *Drosophila*. J. exp. Zool. **11**, 365.
— (1911): Chromosomes and associative inheritance. Science **34**, 636.
— (1914): Heredity and Sex. (2. Ed.). Columbia Univ. Press/New York.
—, and *E. Cattell* (1912): Data for the study of sexlinked inheritance in *Drosophila*. J. exp. Zool. **13**, 79.
—, *C. B. Bridges*, and *A. H. Sturtevant* (1928): The constitution of the germinal material in relation to heredity. Yearb. Carneg. Inst. **27**, 1.
—, *A. H. Sturtevant, H. J. Muller*, and *C. B. Bridges* (1915): The Mechanism of Mendelian Heredity. H. Holt & Comp./New York.

Morré, D. J., H. H. Mollenhauer, and *C. E. Braeth* (1971): Origin and continuity of Golgi apparatus. In: *Reinart, J.*, and *H. Ursprung* (Eds.) "Results and problems of cell differentiation", Vol. **2**, p. 82, Springer/Berlin-Heidelberg-New York.

Morris, D. W., and *J. A. Moses* (1965): Role of aminoacyl-transfer ribonucleic acid in the regulation of ribonucleic acid synthesis in *Escherichia coli*. J. Bacteriol. **90**, 1624.

Morrison, J. H. (1966): Functional Organelles. Reinhold Publ. Corp./New York.

Morse, M. L., E. M. Lederberg, and *J. Lederberg* (1956): Transductional heterogenotes in *E. coli*. Genetics **41**, 758.

Morton, N. E., J. F. Crow, and *H. J. Muller* (1956): An estimate of the mutational damage in man from data on consanguineous marriages. Proc. nat. Acad. Sci., Wash., **42**, 855.

Moscona, A. A. (1968): Cell aggregation: Properties of specific cell-ligands and their role in the formation of multicellular systems. Devel. Biol. **18**, 250.

Moses, M. J. (1958): The relationship between the axial complex of meiotic prophase chromosomes and chromosome pairing in a salamander (*Plethoden cinereus*). J. biophysic. biochem. Cytol. **4**, 633.

— (1960): Patterns of organization in the fine structure of chromosomes. In: *W. Bargmann, D. Peters* and *C. Wolpers* (Ed.), Verhdlg. 4. Int. Kongr. f. Elektronenmikroskopie **2**, 199, Springer/Heidelberg.

Motulsky, A. G. (1957): Drug reactions, enzymes, and biochemical genetics. J. Amer. med. Ass. **165**, 835.

Mühlethaler, K., und *A. Frey-Wyssling* (1959): Entwicklung und Struktur der Proplastiden. J. biophysic. biochem. Cytol. **6**, 507.

Müntzing, A. (1930): Outline to a genetic monograph of the genus *Galeopsis*. Hereditas, Lund, **13**, 185.

— (1953): Ärftlighets Forskning. LTs Förlag/Stockholm.

Muller, H. J. (1916): The mechanism of crossing-over. Amer. Nat. **50**, 193.

— (1917): An *Oenothera*-like case in *Drosophila*. Proc. nat. Acad. Sci., Wash., **3**, 619.

— (1918): Genetic variation, twinhybrids and constant hybrids in a case of balanced lethal factors. Genetics **3**, 422.

— (1927): The problem of genetic modification. V. Int. Kongr. Vererbgsl. 234.

— (1927): Effects of X-radiation on genes and chromosomes. Anat. Rec. **37**, 174 (Abstr.).

— (1930): Types of visible variations induced by X-rays in *Drosophila*. J. Genet. **22**, 299.

— (1930): Radiation and genetics. Amer. Nat. **64**, 220.

— (1932): Some genetic aspects of sex. Amer. Nat. **66**, 118.

— (1932): Further studies on the nature and causes of gene mutations. Proc. 6. Int. Congr. Genet. **1**, 213.

— (1934): The effects of X-rays upon the hereditary material. In: "Science of Radiology", 305, Baillière, Tindall & Cox/London.

— (1940): An analysis of the process of structural change in chromosomes of *Drosophila*. J. Genet. **40**, 1.

— (1941): Induced mutations in *Drosophila*. Cold Spr. Harb. Symp. quant. Biol. **9**, 151.

— (1950): Our load of mutations. Amer. J. hum. Genet. **2**, 111.

— (1954): The manner of production of mutations by radiation. In: *Hollaender, A.*, "Radiation Biology" **1**, 475, McGraw-Hill Book Comp./New York.

— (1954): The manner of dependence of the "permissible dose" of radiation on the amount of genetic damage. Acta radiol., Stockh. **41**, 5.

—, and *E. Altenburg* (1919): The rate of change of hereditary factors in *Drosophila*. Proc. Soc. exptl. Biol. N. Y. **17**, 10.

Muller, H. J., and *T. S. Painter* (1932): The differentiation of the sex chromosome of *Drosophila* into genetically active and inert regions. Z. VererbLehre **62**, 316.
—, *B. B. League*, and *C. A. Offerman* (1931): Effects of dosage changes of sex-linked genes, and the compensatory effect of other gene differences between male and female. Anat. Rec. **51**, (Suppl.), 110.
Murphy, E. A. (1970): The ENSU scoring system in genetic counselling. Ann. of Human Genet. **34**, 73.

Nachtsheim, H. (1975): Mutation und Phänokopie bei Säugetier und Mensch. Ihre theoretische und praktische Bedeutung für Genetik und Eugenik. Experientia **13**, 57.
Naegeli, C. (1884): Mechanisch-physiologische Theorie der Abstammungslehre. R. Oldenbourg/München, Leipzig.
—, and *C. Cramer* (1855): Pflanzenphysiologische Untersuchungen. F. Schulthess/Zürich.
Nagl, W. (1962): 4096-Ploidie und „Riesenchromosomen" im Suspensor von *Phaseolus*. Naturwiss. **49**, 261.
Nakada, D., *J. A. C. Anderson*, and *B. Magasanik* (1964): Fate of the ribosomal RNA produced by a "relaxed" mutant of *Escherichia coli*. J. Mol. Biol. **9**, 472.
Nanney, D. L. (1958): Epigenetic control systems. Proc. nat. Acad. Sci., Wash., **44**, 712.
— (1963): Irregular genetic transmission in *Tetrahymena* crosses. Genetics **48**, 737.
Nath, V. (1956): Cytology of spermatogenesis. Int. Rew. Cytol. **5**, 395.
Nathans, D., and *F. Lipman* (1961): Amino acid transfer from aminoacyl-ribonucleic acids to protein on ribosomes of *Escherichia coli*. Proc. nat. Acad. Sci., Wash., **47**, 497.
Navashin, M. S. (1926): Variabilität des Zellkernes bei *Crepis*-Arten in bezug auf die Artbildung. Z. Zellforsch. **4**, 171.
— (1927): Über die Veränderung von Zahl und Form der Chromosomen infolge der Hybridisation. Z. Zellforsch. **6**, 195.
— (1928): Amphiplasty, a new caryological phenomenon. Z. indukt. Abstamm.- u. Vererbl. Suppl. **2**, 1148.
Navashin, S. G. (1899): Neue Beobachtungen über Befruchtung bei *Fritillaria tenella* und *Lilium martagon*. Bot. Zbl. **77**, 62.
— (1912): Über den Dimorphismus der Kerne in somatischen Zellen von *Galtonia candicans* (russ.). Bull. Acad. Sci., St. Petersb. **22**, 273.
Needham, J., *C. H. Waddington*, and *D. M. Needham* (1934): Physicochemical experiments on the amphibian organizer. Proc. roy. Soc. B. **114**, 393.
Nei, M., and *A. K. Roychoudhury* (1974): Sampling variances of heterozygosity and genetic distance. Genetics **76**, 379.
Němec, B. (1910): Das Problem der Befruchtungsvorgänge und andere zytologische Fragen. Borntraeger/Berlin.
Nemer, M., *M. Graham*, and *L. M. Dubroff* (1974): L'existence of nonhistone messenger RNA species lacking and containing polyadenylic acid in sea urchin embryos. J. Mol. Biol. **89**, 435.
Newcombe, H. B. (1942): The action of X-rays on the cell. J. Genet. **43**, 145.
— (1953): The delayed appearance of radiation-induced genetic change in bacteria. Genetics **38**, 134.
Nichols, W. W., *A. Levan*, *P. Aula*, and *E. Norrby* (1964): Extreme chromosome breakage induced by measles virus in different *in vitro* systems. Preliminary communication. Hereditas, Lund, **51**, 380.
Nicolosi-Roncati, F. (1912): Formazioni mitochondriali negli elementi sessuali maschili dell'*Helleborus foetidus*. Boll. Orto bot. Napoli **2**, 531.

Nilsson-Ehle, H. (1908): Einige Ergebnisse von Kreuzungen bei Hafer und Weizen. Bot. Notiser 257.

Ninio, J. (1973): Recognition in nucleic acids and the anticodon families. Progr. in Nucleic Acid Res. **13**, 337.

Nishizuka, Y., and *F. Lipmann* (1966): Comparison of guanosine triphosphate split and polypeptide synthesis with a purified *E. coli* system. Proc. Nat. Acad. Sci., Wash., **55**, 212.

—, *K. Ueda, T. Honjo,* and *O. Hayaishi* (1968): Enzymic adenosine diphosphate ribosylation of histone and polyadenosine diphosphate ribose synthesis in rat liver nuclei. J. Biol. Chem. **243**, 3765.

Nomura, M., and *C. V. Lowry* (1967): Phage F 2 RNA-directed binding of formylmethionyl-tRNA to ribosomes and the role of 30S ribosomal subunits in initiation of protein synthesis. Proc. nat. Acad. Sci., Wash., **58**, 946.

—, and *C. Witten* (1967): Interaction of colicins with bacterial cells III. Colicin-tolerant mutations in *Escherichia coli*. J. Bact. **94**, 1093.

Novick, A., and *L. Szilard* (1951): Virus strains of identical (same) phenotype but different genotype. Science **113**, 34.

Novick, R. P. (1967): Penicillinase plasmids of *Staphyloccus aureus*. Fed. Proc. **26**, 29.

Novikoff, A. B., E. Essner, and *N. Quintana* (1964): Golgi apparatus and lysosomes. Fed. Proc. **23**, 3, 1010.

Novitski, E. (1954): The compound X-chromosomes in *Drosophila*. Genetics **39**, 127.

Odor, L. (1956): Uptake and transfer of particulate matter from the peritoneal cavity of the rat. J. biophysic. biochem. Cytol. **2**, (Suppl.), 105.

Oehlkers, F. (1956): Das Leben der Gewächse. Vol. I. Die Pflanze als Individuum. Springer/Berlin.

Östergren, G. (1943): Elastic chromosome repulsions. Hereditas, Lund, **29**, 444.

— (1947): Proximal heterochromatin, structure of the centromere, and the mechanism of its misdivision. Bot. Notiser 1947, 176.

— (1950): Considerations of some elementary factors of mitosis. Hereditas, Lund, **36**, 1.

— (1950): Isopycnosis and isopycnotic, two new terms for use in chromosome studies. Hereditas, Lund, **36**, 511.

—, and *E. Vigfusson* (1953): On position correlations of univalents and quasibivalents formed by sticky univalents. Hereditas, Lund, **39**, 33.

—, *R. Morris,* and *T. Wakonig* (1958): A study in *Hyacinthus* on chromosome size and breakability by X-rays. Hereditas, Lund, **44**, 1.

Ogata, M. (1883): Die Veränderung der Pancreaszellen bei der Sekretion. Arch. Anat. Physiol., Lpz. (phys. Abt.), 405.

Ogita, Z. (1968): Genetic control of isozymes in animales. Proc. XII. Int. Genet. Congr. Vol. **2**, p. 77.

Oguma, K. (1942): Considerations on the form and composition of the chromosomes and their arrangement during cell division and proposals for some new terms. Jap. J. Genet. **18**, 205.

Oksala, T. (1952): Chiasma formation and chiasma interference in the Odonata. Hereditas, Lund, **38**, 449.

Olsson, G., and *A. Hagberg* (1955): Investigations on haploid rape. Hereditas, Lund, **41**, 227.

O'Mara, J. G. (1940): Cytogenetic studies on *Triticinae*. I. A method for determining the effects of individual *Secale* chromosomes on *Triticum*. Genetics **25**, 401.

Oppenheim, A. B., Z. Newbauer, and *E. Calef* (1970): The Antirepressor: a new element in the regulation of protein synthesis. Nature, Lond., **226**, 31.

Owen, F. V. (1945): Cytoplasmatically inherited male-sterility in sugar beets. J. agric. Res. **71**, 423.

Ozeki, H., and *S. Howarth* (1961): Colicin factors as fertility factors in bacteria. *Salmonella typhimurium* strain LT 2. Nature, Lond., **190**, 986.

Pätau, K. (1941): Zytologischer Nachweis einer positiven Interferenz über das Centromer. (Der Paarungskoeffizient I). Chromosoma **2**, 36.

Painter, T. S. (1939): The structure of salivary gland chromosomes. Amer. Nat. **73**, 315.

—, and *H. J. Muller* (1929): Parallel cytology and genetics of induced translocations and delections in *Drosophila*. J. Hered. **20**, 287.

Palade, G. E. (1953): The fine structure of mitochondria. An electron microscope study. Histochem. Cytochem. **1**, 188.

—, and *P. Siekevitz* (1955): Correlated structural and chemical analysis of microsomes. Fed. Proc. **14**, 262.

—, and — (1956): Liver microsomes. An integrated morphological and biochemical study. J. biophysic. biochem. Cytol. **2**, 171.

Palay, S. L. (1958): The morphology of synapses in the central nervous system. Exp. Cell Res. **5** (Suppl.), 275.

Pardee, A. B., F. Jacob, and *J. Monod* (1959): The genetic control and cytoplasmic expression of "inducibility" in the sysnthesis of *β*-galactosidase by *E. coli*. J. Mol. Biol. **1**. 165.

Park, R. B., and *N. G. Pon* (1961): Correlation of structure with function in *Spinacea oleracea* chloroplasts. J. Mol. Biol. **3**, 1.

Parry, J. M. (1962): A quantitative analysis of "negative liquid holding", in some UV sensitive mutants of yeast. Mol. Gen. Genetics **118**, 33.

Patrick, M. H., R. H. Haynes, and *R. B. Uretz* (1964): Dark recovery phenomena in yeast. I. Comparative effects with various inactivating agents. Rad. Res. **21**, 144.

Paul, J. (1965): Animal cell culture in cell biology research. In: *C. V. Ramakrishnan* (Ed.) "Tissue Culture", 87, Junk/The Hague.

Pauling, L. (1964): Molecular disease and evolution. Bull. N. Y. Acad. Med. **40**, 334.

—, *H. A. Itano, S. J. Singer*, and *I. C. Wells* (1949): Sickle cell anemia, a molecular disease. Science **110**, 543.

Paulus, H., and *E. Gray* (1964): Multivalent feedback inhibition of aspartokinase in *Bacillus polymyxa*. J. biol. Chem. **239**, PC 4008.

Peacock, W. J. (1963): Chromosome duplication and structure as determined by autoradiography. Proc. nat. Acad. Sci., Wash., **49**, 793.

Pelc, S. R. (1968): Biological implications of DNA-turnover in higher organisms. Acta Histochem. **8** (Suppl.), 41.

Pelling, C. (1966): A replicative and synthetic chromosomal unit — the modern concept of the chromomere. Proc. roy. Soc. B. **164**, 279.

Percival, J. (1932): Cytological studies of some wheat and *Aegilops* hybrids. Ann. Bot. **46**, 479.

Perner, E. S. (1952): Die Vitalfärbung mit Berberinsulfat und ihre physiologische Wirkung auf Zellen höherer Pflanzen. Ber. dtsch. bot. Ges. **65**, 52.

Perroncito, A. (1910): Contribution a l'étude de la biologie cellulaire. Arch. ital. Biol. **54**.

Perry, R. P. (1969): Nucleoli: the cellular sites of ribosome production. In: *A. Lima-de-Faria* "Handbook of molecular Cytology". North Holland Publ. Comp./Amsterdam, London, p. 620.

Person, C. (1956): Some aspects of monosomic wheat breeding. Canad. J. Bot. **34**, 60.

Pfeffer, W. (1877): Osmotische Untersuchungen. Engelmann/Leipzig.

Pfitzner, W. (1883): Beiträge zur Lehre vom Bau des Zellkernes und seinen Teilungserscheinungen. Arch. mikr. Anat. **22**, 616.

Piekarski, G. (1937): Cytologische Untersuchungen an Paratyphus- und Coli-Bakterien. Arch. Mikrobiol. **8**, 428.

Piperno, R., A. Carere, and *G. Sermonti* (1966): Polarized recombination in *Streptomyces coelicolor.* Ann. Ist. Super. Sanità **2**, 393.

Pirie, N. W. (1937): The meaningless of the terms life and living. In: *J. Needham* and *F. Green* (Eds.) "Perspectives in Biochemistry" Univ. Press/Cambridge.

Pitelka, D. R. (1969): Centriole replication. In: *A. Lima-de-Faria* (Ed.) "Handbook of molecular Cytology", North- Holland Publ. Comp./Amsterdam, London.

Plate, L. (1910): Vererbungslehre und Deszendenztheorie. Festschr. f. R. Hertwig, II, Fischer/Jena, 537.

— (1913): Vererbungslehre. Engelmann/Leipzig.

Plowe, J. Q. (1931): Membranes in the plant cell. I. Morphological membranes at protoplasmic surfaces. Protoplasma **12**, 196.

Policard, A., and *M. Bessis* (1956): Sur l'espace périnucléaire. C. R. Acad. Sci. Paris **242**, 2496.

Pollak, K. J., and *C. D. Shorey* (1964): The isolation and characterization of a new cytoplasmic component, the "reticulosome". Biochem. J. **93**, 360.

Pollock, M. R., J. Fleming, and *S. Petrie* (1965): The effects of specific antibodies on the biological activities of wild-type bacterial penicillinases and their mutationally altered analogues. In: *B. Cinader* (Ed.) "Antibodies to Biologically Active Molecules", second Meeting Europ. Fed. Biochem. Soc., Vienna **1**, 139.

Pontecorvo, G. (1952): Genetic formulation of gene structure and gene action. Advanc. Enzymol. **13**, 121.

— (1952): The genetic analysis of cell organization. Symp. Soc. exp. Biol. **6**, 218.

— (1953): The genetics of *Aspergillus nidulans.* Advanc. Genet. **5**, 141.

— (1953): Discussion of the paper of *R. Y. Stanier*: Adaptation in microorganisms. Symp. Soc. exp. Biol. **3**, 16.

— (1954): Mitotic recombination in the genetic systems of filamentous fungi. Proc. 9. Int. Congr. Genet., Caryologia **6**, Suppl. **1**, 192.

— (1955): Gene-structure and action in relation to heterosis. Proc. roy. Soc. B. **144**, 171.

— (1958): Trends in Genetic Analysis. Columbia Univ. Press/New York.

— (1963): The *Leuwenhoek* Lecture. Microbial genetics: retrospect and prospect. Proc. roy. Soc. B. **158**, 1.

— (1966): Template and stepwise processes in heredity. Proc. roy. Soc. B. **164**, 167.

—, *E. T. Gloor,* and *E. Forbes* (1954): Analysis of mitotic recombination in *Aspergillus* nidulans. J. Genetics **52**, 226.

Porter, K. R., and *R. D. Machado* (1960): Studies on the endoplasmic reticulum. IV. Its form and distribution during mitosis in cells of onion root tip. J. biophysic. biochem. Cytol. **7**, 167.

—, *A. Claude,* and *E. F. Fullmann* (1945): A study of tissue culture cells by electron microscopy. Methods and preliminary obsverations. J. exp. Med. **81**, 233.

Portin, P., and *M. Ruohonen* (1972): A new type of allelic interaction at the Ax locus in *Drosophila melanogaster.* Dros. Inf. Service **49**, 70.

Portner, A., and *D. W. Kingsbury* (1972): Identification of transcriptive and replicative intermediates in Sendai virus-infected cells. Virology **47**, 711.

Poulton, E. B. (1903): What is a species? Proc. R. ent. Soc., Lond., 1903, 77.

Prakken, R., and *A. Müntzing* (1942): A meitoic peculiarity in rye, simulating a terminal centromere. Hereditas, Lund., **28**, 441.

Preer, J. R., R. W. Siegel, and *P. S. Stark* (1953): The relationship between kappa and paramecin in *P. aurelia.* Proc. nat. Acad. Sci., Wash., **39**, 1228.

Price, S. (1956): Cytological studies in *Saccharum* and allied Genera. I. Syncytes in certain clones of *Saccharum* and *Erianthus*. Cytologia, Tokyo, **21**, 21.
Ptashne, M. (1967): Specific binding of the λ-phage repressor to lambda-DNA. Nature, Lond., **214**, 232.
Punnet, R. C. (1905): Mendelism. MacMillan/Cambridge.
Purkinje, J. E. (1825): Subjectae sunt symbolae ad ovi avium historiam ante incubationem (Ed. 1). Vratislaviae (Ed. II Lipsiae 1830).
— (1840): Über die Analogien in den Strukturelementen des thierischen und pflanzlichen Organismus. In: Übers. Arb. Veränder. schles. Ges. vaterl. Cultur 1839, 81 (= Opera omina II).

Ramakrishnan, T., and *E. A. Adelberg* (1965): Regulatory mechanisms in the biosynthesis of isoleucine and valine. III. Map order of the structural genes and operator genes. J. Bacteriol. **89**, 661; **90**, 295.
Ramony, Cajal, S. (1908): Histologie du système nerveux de l'homme et des vertébrés. Neudruck; Instituto Ramon y Cajal, Madrid 1952.
Randolph, L. F. (1928): Chromosome numbers in *Zea mays* L. Cornell Univ. Agr. Exp. Mem. **117**.
— (1928): Types of supernumerary chromosomes in maize. Anat. Rec. **41**, 102.
Raper, J. R., and *K. Esser* (1964): The fungi. In: *J. Brachet*, and *A. E. Mirsky* (Ed.) "The Cell", **6**, 139, Acad. Press/New York.
—, and *J. P. SanAntonio* (1954): Heterokaryotic mutagenesis in *Hymenomycetes* I., Am. J. Bot. **41**, 69.
Rasmussen, R. E., and *R. B. Painter* (1964): Evidence for repair of ultra-violet damaged deoxyribonucleic acid in cultured mammalian cells. Nature **203**, 1360.
Rattenburg, J. A., and *J. A. Serra* (1952): Types of nucleolus reconstitution in telophase and the question of the "nucleolar organizer". Portug. acta biol. **3**, 239.
Raven, C. P. (1961): Oogenesis: The Storage of Developmental Information. Pergamon Press/Oxford.
Ravin, A. W. (1963): Experimental approaches to the study of bacterial phylogeny. Amer. Nat. **97**, 307.
Rechler, M. M., and *R. G. Martin* (1970): The intercistronic divide: Translation of an intercistronic region in the histidine operon of *Salmonella typhimurium*. Nature, Lond., **226**, 908.
Reichard, P., R. Eliasson, and *G. Söderman* (1974): Initiator RNA in discontinuous DNA synthesis. Proc. Nat. Acad. Sci., Wash., **71**, 4901.
Reinig, W. F. (1938): Elimination und Selektion. Fischer/Jena.
Renner, O. (1916): Zur Terminologie des pflanzlichen Generationswechsels. Biol. Zbl. **36**, 337.
— (1917): Versuche über die gametische Konstitution der Oenotheren. Z. indukt. Abstamm.- u. VererbLehre **18**, 121.
— (1918): Weitere Vererbungsstudien an Oenotheren. Flora, Jena **111**—**112**, 641.
— (1921): Heterogamie im weiblichen Geschlecht und Embryosackentwicklung bei den Oenotheren. Z. Bot. **131**, 609.
— (1924): Vererbung bei Artbastarden. Z. indukt. Abstamm.- u. VererbLehre **33**, 317.
— (1929): Artbastarde bei Pflanzen. Handb. d. Vererbungswissenschaft Bd. II A), Borntraeger/Berlin.
— (1934): Die pflanzlichen Plastiden als selbständige Elemente der genetischen Konstitution. Ber. Math.-Phys. Kl. Sächs. Acad. d. Wiss. Leipzig **86**, 241.
— (1937): Über *Oenothera atrovirens* Sh. et Bartl. und über somatische Konversion im Erbgang des cruciata-Merkmals der Oenotheren. Z. indukt. Abstamm.- u. VererbLehre **74**, 91.

Renner, O. (1941): Über die Entstehung homozygoter Formen aus komplexheterozygotischen Oenotheren. Flora, Jena **35**, 201.

— (1949): Die 15-chromosomigen Mutanten der *Oenothera Lamarckiana* und ihrer Verwandten. Z. indukt. Abstamm.- u. VererbLehre **83**, 1.

—, und *W. Kupper* (1921): Artkreuzung bei der Gattung *Epilobium*. Ber. dtsch. bot. Ges. **39**, 201.

Rensch, B. (1947): Neuere Probleme der Abstammungslehre. Die transspezifische Evolution. Enke/Stuttgart.

Renwick, J. H. (1971): Assignment and map-positioning of human loci using chromosomal variation. Ann. Human Genet. **35**, 79.

— (1971): Current procedures for analysis and interpretation of linkage data. Ann. Rev. Genet. **5**, 84.

Resch, A. (1952): Untersuchungen über Kerndifferenzierung in peripheren Zellschichten der Sproßachse einiger Blütenpflanzen. Chromosoma **5**, 296.

Resende, F. (1940): Über die Chromosomenstruktur in der Mitose in Wurzelspitzen. Chromosoma **1**, 486.

— (1953): Cytogenetische Folgen des diffusen Kinetochors. Bol. Soc. portug. Sci. nat., Ser. 2, **4**, 252.

Revell, J. P., and *M. I. Karnovsky* (1967): Hexagonal array of subunits in intercellular junctions of the mouse heart and liver. J. Cell Biol. **33**, C 7.

Revell, S. H. (1959): The accurate estimation of chromatid breakage, and its relevance to a new interpretation of chromatid aberrations induced by ionizing radiations. Proc. roy. Soc. B. **150**, 563.

— (1963): Chromatid aberrations — the generalized theory. In: *S. Wolff* (Ed.) "Radiation-induced Chromosome Aberrations", 41, Columbia Univ. Press/New York.

— (1966): Evidence for a dose-squared term in the dose response curve for real chromatid discontinuities induced by X-rays, and some theoretical consequences thereof. Mutation Res. **3**, 34.

Reznikoff, W. S., *J. H. Miller*, *J. G. Scaife*, and *J. R. Beckwith* (1969): A mechanism for repressor action. J. Mol. Biol. **43**, 201.

Rhoades, M. M. (1933): A secondary trisome in maize. Proc. nat. Acad. Sci., Wash., **19**, 1031.

— (1936): The effect of varying gene dosage on aleuron colour in maize. J. Genet. **33**, 347.

— (1938): Effect of the Dt gene on the mutability of the a_1 allele in maize. Genetics **23**, 377.

— (1942): Preferential segregation in maize. Genetics **27**, 395.

— (1952): Preferential segregation in maize. In: *J. W. Gowen* (Ed.) "Heterosis", Iowa State College Press/Ames Iowa, 66.

— (1961): Meiosis. In: *J. Brachet*, and *A. E. Mirsky* (Ed.) "The Cell", **3**, 1, Acad. Press/New York and London.

—, and *W. E. Kerr* (1949): A note on centromere-organization. Proc. nat. Acad. Sci., Wash., **35**, 129.

Rhodin, J. (1954): Correlation of ultrastructural organization and function in normal and experimental changed proximal convoluted tubule cells of the mouse kidney. Doctorale thesis Karolinska Inst., Stockholm.

Rick, C. M. (1943): Cytogenetic consequences of X-ray treatment of pollen in *Petunia*. Bot. Gaz. **104**, 528.

Rickett, H. W. (1958): So what is a taxon? Taxon **7**, 37.

Rieger, R. (1963): Die Genommutationen (Ploidiemutationen). Fischer/Jena.

— (1966): Alternative Interpretationsversuche zur Entstehung chromosomaler Strukturumbauten: Bruch-Reunions- und Austausch-Hypothese. Biol. Zbl. **85**, 29.

Riley, R. (1960): The diploidisation of polyploid wheat. Heredity **15**, 407.

Riley, R., and *V. Chapman* (1958): Genetic control of the cytologically diploid behaviour of hexaploid wheat. Nature, Lond., **182**, 713.

—, and *C. N. Law* (1965): Genetic variation in chromosome pairing. Advanc. Genet. **13**, 57.

Ris, H. (1961): Ultrastructure and molecular organization of genetic systems. Canad. J. Genet. Cytol. **3**, 95.

—, and *B. L. Chandler* (1963): The ulttastructure of genetic systems in prokaryotes and eukaryotes. Cold Spr. Harb. Symp. quant. Biol. **28**, 1.

Ritossa, F. (1968): Unstable redundancy of genes for ribosomal RNA. Proc. Nat. Acad. Sci., Wash., **60**, 509.

—, and *G. Scala* (1969): Equilibrium variations in the redundancy of rDNA in *Drosophila melanogaster*. Genetics (Suppl. 1) **61**, 305.

—, *F. Scalenghe, N. Di. Turi*, and *A. M. Contini* (1973): On the cell stage of X—Y-recombination during rDNA magnification in *Drosophila*. Cold Spr. Harb. Symp. on Quant. Biol. **38**, 483.

Rizet, G., P. Lissouba, et *J. Mousseau* (1961): Sur l'interférence négative au sein d'une série d'allèles chez *Ascobolus immersus*. C. R. Soc. Biol., Paris, **154**, 1967.

Robards, A. W. (1968): A new interpretation of plasmodesmatal ultrastructure. Planta **82**, 200.

Robertis, E. D. P. de, W. W. Nowinski, and *F. A. Saez* (1960): General Cytology (3. Ed.). W. B. Saunders/Philadelphia.

—, —, and — (1965): Cell Biology. W. B. Saunders/Philadelphia.

Roberts, H. S. (1961): Mechanisms of cytokinesis: a critical review. Quart. Rev. Biol. **36**, 155.

Roberts, J.W. (1969): Termination factor for RNA synthesis. Nature, Lond., **224**, 1168.

Roberts, R. B. (1958): Introduction to "Microsomal Particles and Protein Synthesis". Pergamon/New York.

— (1965): The synthesis of ribosomal protein. J. theoret. Biol. **8**, 49.

—, and *E. Aldous* (1949): Recovery from ultraviolet irradiation in *Escherichia coli*. J. Bacteriol. **57**, 363.

—, *R. J Britten*, and *B. J. McCarthy* (1963): Kinetic studies of RNA and ribosomes. In: *J. H. Taylor* (Ed.) "Molecular Genetics", Acad. Press/New York, **1**, 291.

Robertson, J. D. (1959): The ultrastructure of cell membranes and their derivatives. Biochem. Soc. Symp. **16**, 3.

Robertson, W. (1916): Chromosome studies. I. Taxonomic relationships shown in the chromosomes of *Tettigidae* and *Acrididae*. V-shaped chromosomes and their significance in *Acrididae, Locustidae*, and *Gryllidae*: chromosomes and variation. J. Morph. **27**, 179.

Roizman, B. (1962): Polycaryocytosis induced by viruses. Proc. nat. Acad. Sci., Wash., **48**, 228.

Roman, H. L. (1956): Studies of gene mutation in *Saccharomyces*. Cold Spr. Harb. Symp. quant. Biol. **21**, 175.

Romig, W. R. (1968): Infectivity of *Bacillus subtilis* bacteriophage deoxyribonucleic acids extracted from mature particles and from lysogens Bact. Reviews **32**, 349.

—, and *A. M. Brodetsky* (1961): Isolation and preliminary characterization of bacteriophages for *Bacillus subtilis*. J. Bact. **82**, 135.

Rosenberg, O. (1904): Über die Individualität der Chromosomen im Pflanzenreich. Flora, Jena **93**, 251.

— (1917): Die Reduktionsteilung und ihre Degeneration in *Hieracium*. Svensk. Bot. Tidskr. **11**, 145.

— (1926): Über die Verdopplung der Chromosomenzahl nach Bastardierung. Ber. dtsch. bot. Ges. **44**, 455.

— (1926): Die semiheterotypische Teilung und ihre Bedeutung für die Entstehung verdoppelter Chromosomenzahlen. Hereditas, Lund, **8**, 305.

Rosenberg, O. (1927): Homeotpyic division in uninucleate pollen mother cells. Hereditas, Lund, **9**, 285.
Roth, C. W., and *E. W. Nester* (1971): Co-ordinate control of tryptophan, histidine and tyrosine enzyme synthesis in *Bacillus subtilis*. J. Mol. Biol. **62**, 577.
Roux, W. (1905): Die Entwicklungsmechanik, ein neuer Zweig der biologischen Wissenschaft. In: *W. Roux* (Ed.) „Vorträge und Aufsätze über Entwicklungsmechanik der Organismen", **1**, Engelmann/Leipzig.
Rückert, J. (1882): Zur Entwicklungsgeschichte des Ovarialeies bei Selachiern. Anat. Anz. **7**, 107.
Rupert, C. S. (1960): Photoreactivation of transforming DNA by an enzyme from baker's yeast. J. gen. Physiol. **43**, 573.
— (1961): Repair of ultraviolet damage in cellular DNA. J. cell. comp. Physiol. **58**, (Suppl. 1), 57.
—, and *W. Harm* (1966): Reactivation after photobiological damage. Adv. Rad. Biol. **2**, 1.
Rupp, W. D., and *P. Howard-Flanders* (1968): Discontinuities in the DNA synthesized in an excision-defective strain of *E. coli* following ultraviolet radiation. J. Mol. Biol. **31**, 291.
Russell, W. L. (1951): X-ray-induced mutations in mice. Cold Spr. Harb. Symp. quant. Biol. **16**, 327.
Ryan, F. J. (1954): The delayed appearance of mutants in bacterial cultures. Proc. nat. Acad. Sci., Wash., **40**, 178.
—, and *J. Lederberg* (1946): Reverse-mutation and adaptation in leucineless *Neurospora*. Proc. nat. Acad. Sci., Wash., **32**, 163.
Ryter, A., and *F. Jacob* (1964): Étude au microscope électronique de la liaision entre noyau et mésosome chez *Bacillus subtilis*. Ann. Inst. Pasteur **107**, 384.

Sachs, J. v. (1892): Beiträge zur Zellenlehre. Flora, Jena **75**, 57.
Sachs, L. (1954): Sex linkage and sex chromosomes in man. Ann. Eugen. **18**, 255.
Sagan, L. (1967): On the origin of mitosing cell. J. theoret. Biol. **14**, 225.
Sager, R., and *M. R. Ishida* (1963): Chloroplast DNA in *Chlamydomonas*. Proc. nat. Acad. Sci., Wash., **50**, 725.
— (1973): Cytoplasmic Genes and Organelles. Academic Press/London and New York.
Salisbury, E. J. (1940): Ecological aspects of plant taxonomy. In: *J. S. Huxley* "The New Systematics", Univ. Press/Oxford, 329.
Samarina, O. P., *E. M. Lukanidin*, and *G. P. Georgiev* (1967): On the structural organisation of the nuclear complexes containing messenger RNA. Biochem. Biophys. Acta **142**, 561.
—, *E. M. Lukanidin*, *J. Molnar*, and *G. P. Georgiev* (1968): Structural organization of nuclear complexes containing DNA-like RNA. J. Mol.. Biol. **33**, 251.
Sandler, L., and *Y. Hiraizumi* (1961): Meiotic drive in natural populations of *Drosophila melanogaster*. II. A heritable aging effect on the phenomenon of segregation-distortion. Canad. J. Genet. Cytol. **3**, 34.
—, and *E. Novitski* (1957): Meiotic drive as an evolutionary force. Amer. Nat. **91**, 105.
—, *D. L. Lindsley*, *B. Nicoletti*, and *G. Trippa* (1968): Mutants affecting meiosis in natural populations of *Drosophila melanogaster*. Genetics **60**, 525.
Sansome, E. R. (1933): Segmental interchange in *Pisum* II. Cytologia, Tokyo, **5**, 15.
Sarabhai, A., and *S. Brenner* (1967): A mutant which reinitiates the polypeptide chain after chain termination. J. Mol. Biol. **27**, 145.
—, *A. O. W. Stretton*, *S. Brenner*, and *A. Bolle* (1944): Co-linearity of the gene with the polypeptide chain. Nature, Lond., **201**, 13.
Sax, K. (1932): The cytological mechanisms of crossing-over. J. Arnold Abor. **12**, 3.

Sax, K. (1935): Chromosome structure in the meiotic chromosomes of *Rhoeo discolor*. J. Arnold Arbor **16**, 216.
— (1938): Induction by X-rays of chromosome aberrations in *Tradescantia* microspores. Genetics **23**, 494.
— (1941): Types and frequencies of chromosomal aberrations induced by X-rays. Cold Spr. Harb. Symp. quant. Biol. **9**, 93.
Scaife, J. (1963): Conjugation and the sex factor in *Escherichia coli*. Sci. Progr. **51**, 566.
Schachat, F. H., and *D. S. Hogness* (1973): Repetitive sequences in isolated Thomas circles from *Drosophila melanogaster*. Cold Spr. Harb. Symp. quant. Biol. **38**, 371.
Schekman, R., A. Weiner, and *A. Kornberg* (1974): Multienzyme systems of DNA replication. Science **186**, 987.
Scherrer, K. (1973): Messenger RNA in eukaryotic cells: The life history of duck globin messenger RNA. 6th Karolinska Symposium "Protein Synthesis in reproduktive Tissue", Geneva, p. 95.
—, and *J. E. Darnell* (1962): Sedimentation characteristics of rapidly labelled RNA from Hela cells. Biochem. Biophys. Res. Comm. **7**, 486.
—, *H. Latham*, and *J. E. Darnell* (1973): Demonstration of an unstable RNA and of a precursor to ribosomal RNA in Hela cells. Proc. nat. Acad. Sci., Wash., **49**, 240.
—, and *L. Marcaud* (1965): Remarques sur les RNA messenger polycistronique dans les cellules animales. Bull. Soc. Chim. Biol. **47**, 1697.
—, and — (1968): Messenger RNA in avian erythroblasts and the transcriptional and translational levels and the problem of regulation in animal cells. J. Cell Physiol. **72** (Suppl. 1), 181.
Schimper, A. J. W. (1885): Untersuchungen über die Chlorophyllkörner und die ihnen homologen Gebilde. Jahrb. wiss. Bot. **16**, 1.
Schleicher, W. (1878): Die Knorpel-Zellteilung. Arch. mikr. Anat. **16**, 248.
Schlesinger, M. J., and *C. Levinthal* (1963): Hybrid protein formation of *E. coli* alkaline phosphatase leading to *in vitro* complementation. J. Mol. Biol **7**, 1.
Schmalhausen, I. I. (1949): Factors of Evolution. The Theory of stabilizing Selection. Blakiston/Philadelphia.
Schmieger, H. (1972): Phage P22-mutants with increased or decreased transduction abilities. Molec. Gen. Genet. **119**, 75.
Schmitz, F. (1882): Die Chromatophoren der Algen. Bonn.
Schrader, F. (1932): Recent hypotheses on the structure of spindles in the light of certain observations in *Hemiptera*. Z. wiss. Zool. **142**, 520.
— (1936): The kinetochore or spindle fiber locus in *Amphiuma tridactylum*. Biol. Bull. **70**, 484.
— (1939): The structure of the kinetochore at meiosis. Chromosoma **1**, 230.
— (1944): Mitosis. Columbia University Press/New York.
Schultz, J. (1936): Variegation in *Drosophila* and the "inert" chromosome regions. Proc. nat. Acad. Sci., Wash., **22**, 27.
—, and *H. Redfield* (1951): Interchromosomal effects on crossing-over in *Drosophila*. Cold Spr. Harb. Symp. quant. Biol. **16**, 175.
Schwann, Th. (1838): Mikroskopische Untersuchungen über die Übereinstimmung in Struktur und dem Wachsthum der Thiere und Pflanzen. Ostwalds Klassiker Nr. 176. Engelmann/Leipzig.
Scott, F. M. (1950): Perforation of the surface membranes of nuclei and plastids by nucleodesmata and plastodesmata. Bot. Gaz. **111**, 252.
Sears, E. R. (1939): Cytogenetic studies with polyploid species of wheat. I. Chromosomal aberrations in the progeny of a haploid *Triticum vulgare*. Genetics **24**, 509.
— (1941): Nullisomics in *Triticum vulgare*. Genetics **26**, 167.

Sears, E. R., e *A. Camara* (1950): Un chromosoma dicentrico transmissible en trigo. Genet. iber. **2**, 239.
—, and — (1952): A transmissible dicentric chromosome. Genetics **37**, 125.
Sedlmayer, K. (1956): Aktuelle Probleme der Zuckerrübenzüchtung. S. B. dtsch. Akad. Ldwtsch. Wiss. **5**, H. 24, 2.
— (1956): Heterosis bei nicht kastrierbaren, fakultativ allogamen Kulturpflanzen. Wiss. Z. Univ. Leipzig **4**, 258.
Seiffer, S., and *P. M. Gallop* (1966): The structure proteins. In: *H. Neurath* (Ed.) "The Proteins" **4**, 153, Academic Press/New York.
Seifriz, W. (1928): Die physikalischen Eigenschaften des Protoplasmas. Arch. exp. Zellforsch. **6**, 341.
Seiler, J. (1914): Das Verhalten der Geschlechtschromosomen bei Lepidopteren. Arch. Zellforsch. **13**, 159.
—, and *C. B. Haniel* (1921): Das verschiedene Verhalten der Chromosomen in Eireifung und Samenreifung von *Lymantria monacha* L. Z. indukt. Abstamm.- u. VererbLehre **27**, 81.
Sen, K. N. (1952): Isochromosomes in tomato. Genetics **37**, 227.
Sermonti, G., and *A. Carere* (1968): Mechanism for polarized recombination in *Streptomyces*. Molec. Gen. Genetics **103**, 141.
Serra, J. A. (1942): Relations entre la chimie et la morphologie nucléaire. Bol. Soc. Brot. **16**, 83.
— (1948): A note on the nomenclature of allodiploids and alloheteroploids. Portug. acta biol. (A) **2**, 257.
— (1949): A cytophysiological theory of the gene, gene mutation, and position effect. Portug. acta biol. (A) Goldschmidt-Vol. 401.
— (1949): The parallelism between the chemical and the morphological changes in the chromosomes during mitosis and meiosis. Exp. Cell Res. Suppl. **1**, 111.
— (1955): Chemistry of the nucleus. In: *W. Ruhland* „Hdb. d. Pflanzenphysiologie" **1**, 413, Springer/Berlin.
— (1955): Fine structure of the nucleus. In: *W. Ruhland* „Hdb. d. Pflanzenphysiologie" **1**, 445, Springer/Berlin.
— (1955): Physical chemistry of the nucleus. In: *W. Ruhland* „Hdb. d. Pflanzenphysiologie" **1**, 472, Springer/Berlin.
Serres, F. J. de (1963): Studies with purple adenine mutants in *Neurospora crassa*. V. Evidence for allelic complementation among ad-3B mutants. Genetics **48**, 351.
Setlow, R., and *W. Carrier* (1964): The disappearance of thymine dimers from DNA, an error-correcting mechanism. Proc. nat. Acad. Sci., Wash., **51**, 226.
Setlow, J. K., and *R. B. Setlow* (1963): Nature of the photoreactivable ultraviolet lesion in deoxyribonucleic acid. Nature, Lond., **197**, 560.
Seyffert, W. (1960): Theoretische Untesuchungen über die Zusammensetzung tetrasomer Populationen. I. Panmixie. Biometr. Z. **2**, 1.
Shapiro, J. A. (1969): Mutations caused by the insertion of genetic material into the galactose operon of *Escherichia coli*. J. Mol. Biol. **40**, 93.
Sharp, L. W. (1929): Structure of large somatic chromosomes. Bot. Gaz. **88**, 349.
— (1934): Introduction to Cytology. McGraw Hill Comp./New York.
Shaw, M. W. (1970): Human chromosome damage by chemical agents. Ann. Rev. Med. **21**, 409.
Shaw, W. R. (1898): Über die Blepharoplasten bei *Onoclea* und *Marsilia*. Ber. dtsch. bot. Ges. **16**, 177.
Shull, G. H. (1911): Experiments with maize. Bot. Gaz. **52**, 480.
— (1914): Duplicate genes for capsule-form in *Bursa bursa-pastoris*. Z. indukt. Abstamm.- u. VererbLehre **12**, 97.
Sibatani, A., and *S. Hiai* (1964): A model of chromosome duplication in *Escherichia coli*. J. theoret. Biol. **7**, 393.

Siegel, R. W. (1953): A genetic analysis of the mate-killing trait in *Paramecium aurelia,* variety 8. Genetics **38,** 550.
— (1954): Mate-killing in *Paramecium aurelia,* variety 8. Physiol. Zool. **27,** 89.
Siemens, H. W. (1921): Konstitutions- und Vererbungspathologie. Springer/Berlin.
Signal, N., H. Delius, T. Kornberg, M. L. Gefter, and *B. Alberts* (1972): A DNA-unwinding protein isolated from *Escherichia coli*: its interaction with DNA and with DNA polymerases (DNA-proteincomplex/DNA denaturation/DNA polymerase II). Proc. nat. Acad. Sci., Wash., **69,** 3537.
Simpson, G. G. (1944): Tempo and Mode in Evolution. Columbia Univ. Press/New York.
— (1953): The Major Features of Evolution. Columbia Univ. Press/New York.
— (1953): The Baldwin-effect. Evolution **7,** 110.
Singer, M. F., and *P. Leder* (1966): Messenger RNA: an evaluation. Annu. Rev. Biochem. **35** (part 1), 195.
Siniscalco, M. (1972): Correction of genetic defects in cultured mammalian cells. Adv. in Biosciences **8,** 307.
Sinoto, Y., and *D. Sato* (1940): Basicaryotype and its analysis. Cytologia, Tokyo, **10,** 529.
Sinsheimer, R. L., B. Starman, C. Nagler, and *S. Guthrie* (1962): The process of infection with bacteriophage *Φ*×174. I. Evidence for a "replicative form". J. Mol. Biol **4,** 142.
Sitte, P. (1961): Die submikroskopische Organisation der Pflanzenzelle. Ber. dtsch. bot. Ges. **74,** 177.
— (1965): Bau und Feinbau der Pflanzenzelle. Fischer/Jena.
Sjöstrand, F. S. (1955): Étapes du développement de la méthode de préparation des coupes de tissus ultrafines pour la microscopie électronique a haute résolution. C. R. Colloque Intern. "Les Tech. Rec. Microscopie Electronique Corpusc.", Toulouse 151.
— (1956): The ultrastructure of cells as revealed by the electron microscope. Int. Rev. Cytol. **5,** 455.
Slatkin, M. (1974): Cascading speciation. Nature, Lond., **252,** 701.
Slautterback, D. B. (1963): Cytoplasmic microtubules. I. Hydra. J. Cell Biol. **18,** 367.
Slizinski, B. M. (1945): Ectopic pairing and the distribution of heterochromatin in the X-chromosomes of salivary gland nuclei of *Drosophila melanogaster.* Proc. roy. Soc., Edinburgh **62,** 114.
— (1955): Chiasmata in the male mouse. J. Genet. **53,** 597.
— (1955): The sex bivalent of *Mus musculus* L. J. Genet. **53,** 591.
Smith, B. R. (1966): Genetic controls of recombination. I. The recombination —2 gene of *Neurospora crassa.* Heredity **21,** 481.
Smith, G. R., and *B. Magasanik* (1971): Nature and self-regulated synthesis of the repressor of the hut operons in *Salmonella typhimurium.* Proc. Nat. Acad. Sci., Wash., **68,** 1493.
Smith, H. M. (1955): The perspectives of species. Turtox News 33, 74.
— (1965): More evolutionary terms. Systemat. Zool. **14,** 57.
Smith, J. M., and *J. Haigh* (1974): The hitch-hiking effect of a favourable gene. Genet. Res. **23,** 23.
Solari, A. J., and *M. J. Moses* (1973): The structure of the central region in the synaptonemal complexes of hamster and cricket spermatocytes. J. Cell Biol. **56,** 145.
Sonneborn, T. M. (1937): Sex, sex-inheritance, and sex determination in *Paramecium aurelia.* Proc. nat. Acad. Sci., Wash., **23,** 378.
— (1938): Mating types in *P. aurelia*: diverse conditions for mating in different stocks; occurrence, number, and interrelations of the types. Proc. Amer. phil. Soc. **79,** 411.

Sonneborn, T. M. (1943): Gene and cytoplasm. I. The determination and inheritance of the killer character in variety 4 of *Paramecium aurelia*. Proc. nat. Acad. Sci., Wash., **29**, 329.
— (1951): The role of genes in cytoplasmic inheritance. In: *L. C. Dunn* (Ed.) "Genetics in the 20th Century", 291, McMillan/New York.
— (1957): Breeding systems, reproductive methods, and species problems in Protozoa. In: *E. Mayr* (Ed.) "The species problem", Amer. Ass. for the Advancement of Science (AAAS), Washington, D. C. **50**, 155.
— (1959): Kappa and related particles in *Paramecium*. Adv. Virus Res. **6**, 229.
— (1963): Does preformed cell structure play an essential role in cell heredity ? In: *J. M. Allen* (Ed.) "The Nature of Biological Diversity", McGraw Hill/New York, 165.
— (1964): The differentiation of cells. Proc. nat. Acad. Sci., Wash., **51**, 915.
— (1965): The metagon: RNA und cytoplasmic inheritance. Amer. Nat. **99**, 279.
— (1965): Degeneracy of the genetic code. Extent, nature, and genetic implications. In: *Bryson* and *Vogel* (Ed.) "Evolving Genes and Proteins", New York/ Acad. Press, p. 377.
— (1967): The evolutionary integration of the genetic material into genetic systems. In: *R. A. Brink* (Ed.): "Heritage from Mendel", p. 375, The University of Wisconsin Press/Madison.
Sorokin, H. (1929): Idiograms, nucleoli, and satellites of certain *Ranunculaceae*. Amer. J. Bot. **16**, 407.
Sosa, J. M. (1930): Aparato de Golgi y nomenclatura cellular. Act. del Congr. Int. de Biologia de Montevideo, 1955.
Sotelo, J. R., and *K. R. Porter* (1959): An electron microscope study of the rat ovum. J. biophysic. biochem. Cytol. **5**, 327.
—, and *O. Trujillo-Cenóz* (1957): Electron microscope study of the vitelline body of some spider oocytes. J. biophysic. biochem. Cytol. **3**, (1), 301.
Southern, D. J. (1967): Chiasma distribution in Truxaline grasshoppers. Chromosoma **22**, 164.
Sparrow, A. H., *C. L. Huskins*, and *G. B. Wilson* (1941): Studies on the chromosome spiralization cycle in *Trillium*. Canad. J. Res. **19**, 323.
Spemann, H. (1918): Über die Determination der ersten Organanlagen des Amphibienembryo. Roux Arch. f. Entwicklungsmech. **43**, 448.
Spencer, H. (1863): The Principles of Biology. Williams/London.
Spiegelman, S., and *M. Hayashi* (1963): The present status of the transfer of genetic information and its control. Cold Spr. Harb. Symp. quant. Biol. **28**, 161.
Spirin, A. S., *N. V. Belitsina*, and *M. A. Ajtkhozin* (1964): Messenger RNA in early embryogenesis. J. gen. biol., Moscow **25**, 321.
Sprengel, C. K. (1793): Das entdeckte Geheimnis der Natur im Bau und in der Befruchtung der Blumen. Berlin.
Stadler, D., and *B. Kariya* (1973): Marker effects in the genetic transduction of tryptophan mutants of *E. coli*. Genetics **75**, 423.
Stadler, L. J. (1932): On the genetic nature of induced mutations in plants. Proc. VI. Int. Congr. Genet. **1**, 274.
Stahl, F. W., and *N. Murray* (1966): The evolution of gene clusters and genetic circularity in microorganisms. Genetics **53**, 569.
Stanley, W. M., *M. Salas*, *A. J. Wahba*, and *S. Ochoa* (1966): Translation of the genetic message: Factors involved in the initiation of protein synthesis. Proc. nat. Acad. Sci., Wash., **56**, 290.
Stebbins, G. L. (1941): Apomixis in the angiosperms. Bot. Rev. **7**, 507.
— (1945): The cytological analysis of species hybrids. Bot. Rev. **11**, 463.
— (1947): Types of polyploids: their classification and significance. Advanc. Genet. **1**, 403.

Stebbins, G. L. (1950): Variation and Evolution in Plants. Columbia Univ. Press/New York.

Stedman, E., and *E. Stedman* (1943): Chromosomin, a protein constituent of chromosomes. Nature, Lond., **152**, 267.

Stein, G. S., T. C. Spelsberg, and *L. J. Kleinsmith* (1974): Nonhistone chromosomal proteins and gene regulation. Science **183**, 817.

Steiner, R. F. (1965): The Chemical Foundations of Molecular Biology. Van Nostrand Comp./Princeton.

Steitz, J. A. (1969): Polypeptide chain initiation: nucleotide sequences of the three ribosomal binding sites in bacteriophage R 17 RNA. Nature **224**, 957.

Stent, G. S. (1963): Molecular Biology of Bacterial Viruses. Freeman/San Francisco.

— (1964): The operon: On its third anniversary. Modulation of transfer RNA species can provide a workable model of an operatorless operon. Science **144**, 816.

—, and *S. Brenner* (1961): A genetic locus for the regulation of ribonucleic acid synthesis. Proc. nat. Acad. Sci., Wash., **47**, 2005.

Stephens, S. G. (1942): Colchicine-produced polyploids in *Gossypium*. I. An autotetraploid asiatic cotton and certain of its hybrids with wild diploid species. J. Genet. **44**, 272.

Stern, C. (1929): Über die additive Wirkung multipler Allele. Biol. Zbl. **49**, 261.

— (1933): Faktorenkopplung und Faktorenaustausch. In: *H. Baur*, u. *M. Hartmann* (Ed.) „Hdb. d. Vererbungswissensch." **1**, Abt. H.; Borntraeger/Berlin.

— (1958): Selection for subthreshold differences and the origin of pseudoexogenous adaptations. Amer. Nat. **92**, 313.

— (1959): Use of the term "superfemale". The Lancet **12**, 1088.

— (1960): Principles of Human Genetics. Freeman/San Francisco.

—, and *M. Kodani* (1944): Studies on the position effect at the cubitus interruptus locus of *Drosophila melanogaster*. Genetics **40**, 343.

—, and *E. W. Schaeffer* (1943): On wild-type iso-alleles in *Drosophila melanogaster*. Proc. nat. Acad. Sci., Wash., **29**, 361.

Stevens, B. J., and *J. André* (1969): The nuclear envelope. In: *A. Lima-de-Faria* (Ed.) „Handbook of Molecular Cytology". North-Holland Publ. Comp., Amsterdam-London, p. 837.

Stocker, B. A. D., N.D. Zinder, and *J. Lederberg* (1953): Transduction of flagellar characters in *Salmonella*. J. gen. Microbiol. **9**, 410.

Stout, A. B. (1917): Fertility in *Cichorium intybus*. The sporadic appearance of self-fertile plants among the progeny of self-sterile plants. Am. J. Bot. **4**, 395.

— (1918): Fertility in *Cichorium intybus*. Self-compatibility and self-incompatibility among the offspring of self-fertile lines of descent. J. Genet. **8**, 71.

— (1922): Cyclic manifestation of sterility in *Brassica pekinensis* and *B. chinensis*. Bot. Gaz. **73**, 110.

Strandskov, H. H. (1950): The genetics of human populations. Cold Spr. Harb. Symp. quant. Biol. **15**, 1.

Strasburger, E. (1877): Über Befruchtung und Zellteilung. Jena. Z. Naturw. **11**, 435.

— (1878): Über Polyembryonie. Jena Z. Naturw. **12**, 647.

— (1882): Über den Teilungsvorgang der Zellkerne und das Verhältnis der Kernteilung zur Zellteilung. Arch. mikr. Anat. **21**, 476.

— (1884): Neue Untersuchungen über den Befruchtungsvorgang bei den Phanerogamen als Grundlage für eine Theorie der Zeugung. Fischer/Jena.

— (1893): Zu dem jetzigen Stande der Kern- und Zellteilungsfragen. Anat. Anz. **8**, 177.

— (1905): Typische und allotypische Kernteilung. Ergebnisse und Erörterungen. Pringsheims Jb. wiss. Bot. **42**, 1.

— (1907): Apogamie bei *Marsilia*. Flora, Jena **97**, 123.

— (1910): Chromosomenzahl. Flora **100**, 398.

Strauss, W. (1958): Colorimetric analysis with N, N-dimethyl-p-phenylenediamine of the uptake of intravenously injected horseradish peroxidase by various tissues of the rat. J. biophysic. biochem. Cytol. **4**, 541.

Streisinger, G. (1956): The genetic control of ultraviolet sensitivity levels in bacteriophages T 2 and T 4. Virology **2**, 1.

—, and *J. Weigle* (1956): Properties of bacteriophages T 2 and T 4 with unusual inheritance. Proc. nat. Acad. Sci., Wash., **42**, 504.

—, *R. S. Edgar*, and *G. H. Denhardt* (1964): Chromosome structure in phage T 4. I. Circularity of the linkage map. Proc. nat. Acad. Sci., Wash., **51**, 775.

—, *J. Emrich*, and *M. M. Stahl* (1967): Chromosome structure in Phage T 4, III. Terminal redundancy and length determination. Proc. nat. Acad. Sci., Wash., **57**, 292.

Strelzoff, E., and *F. J. Ryan* (1962): The necessary involvement of both complementary strands of DNA in the specification of messenger RNA. Biophys. Biochem. Res. Comm. **7**, 471.

Strugger, S. (1950): Über den Bau der Proplastiden und Chloroplasten. Naturwissenschaften **37**, 166.

Stubbe, H. (1940): Neue Forschungen zur experimentellen Erzeugung von Mutationen. Biol. Zbl. **60**, 113.

Sturtevant, A. H. (1913): The linear arrangement of six sex-linked factors in *Drosophila*, as shown by their mode of association. J. exp. Zool. **14**, 43.

— (1914): The reduplication hypothesis as applied to *Drosophila*. Amer. Nat. **48**, 535.

— (1920): The vermilion gene and gynandromorphism. Proc. Soc. exp. Biol. **17**, 70.

— (1923): Inheritance of direction of coiling in *Limnaea*. Science **58**, 269.

— (1925): The effects of unequal crossing-over at the Bar-locus in *Drosophila*. Genetics **10**, 117.

— (1926): A crossover reducer in *Drosophila melanogaster* due to inversion of a section of the third chromosome. Biol. Zbl. **46**, 697.

—, and *G. W. Beadle* (1936): The relations of inversions in the X-chromosome of *Drosophila melanogaster* to crossing-over and disjunction. Genetics **21**, 554.

—, and *T. Dobzhansky* (1938): Inversions in the chromosomes of *Drosophila pseudoobscura*. Genetics **23**, 28.

Subak-Sharpe, H., *R. R. Bürk*, and *I. D. Pitts* (1969): Metabolic co-operation between biochemically marked mammalian cells in tissue culture. J. Cell Sci. **4**, 353.

Sueoka, N. (1961): Variation and heterogeneity of base composition of deoxyribonucleic acids: A compilation of old and new data. J. Mol. Biol. **3**, 31.

—, and *T. Kano-Sueoka* (1964): A specific modification of leucyl-sRNA of *Escherichia coli* after phage T 2 infection. Proc. nat. Acad. Sci., Wash., **52**, 1535.

Sunshine, M. G., and *B. Kelly* (1961): Extent of host deletions associated with bacteriophage P 2-mediated eduction. J. Bact. **108**, 695.

Suomaleinen, E. (1950): Parthenogenesis in animals. Advanc. Genet. **3**, 193.

— (1954): The significance of the diffuse phase in meiosis with special reference to the spermatogenesis of certain *Neuroptera*. Ann. zool. Soc. zool. bot. fenn. Vanamo **16**, 1.

Suskind, S. R., and *L. J, Kurek* (1959): On a mechanism of suppressor gene regulation of tryptophan synthetase activity in *Neurospora crassa*. Proc. nat. Acad. Sci., Wash., **45**, 193.

Susman, M. (1964): Growth and Development. Foundations of Modern Biology Series; Prentice-Hall/Englewood Cliffs.

— (1965): Developmental phenomena in microorganisms and in higher forms of life. Annu. Rev. Microbiol. **19**, 59.

Sutherland, E. W., *I. Oye*, and *R. W. Butcher* (1965): The action of Epinephrine and the role of the adenyl cyclase system in hormone action. Rec. Progr. in Hormone Res. **21**, 623.

Sutton, W. S. (1903): The chromosomesi n heredity. Biol. Bull. Wood's Hole **4**, 231.
Sved, J. A. (1966): Telomere attachment of chromosomes. Some genetical and cytological consequences. Genetics **53**, 747.
Svedelius, N. (1915): Zytologisch-entwicklungsgeschichtliche Studien über *Scinala furcellata.* Nova Acta Regiae Soc. Scient. Upsalensis **4**, 1.
Swanson, C. P. (1942): The effects of ultra-violet and X-ray treatment on the pollen tube chromosomes of *Tradescantia.* Genetics **27**, 491.
Swift, H. (1950): The deoxyribose nucleic acid content of animal nuclei. Physiol. Zool. **23**, 169.
— (1956): The fine structure of annulate lamellae. J. biophys. biochem. Cytol. **2**, Suppl. 2, 415.
Sybenga, J. (1965): The quantitative analysis of chromosome pairing and chiasma formation based on the relative frequencies of MI configurations. Genetica **36**, 243.
— (1966): The zygomere as hypothetical unit of chromosome pairing initiation. Genetica **37**, 186.
Symonds, N., H. Heindl, and *P. White* (1973): Radiation sensitive mutants of Phage T 4. Mol. Gen. Genetics **120**, 253.
Szybalski, W., K. Bøvre, M. Fiandt, S. Hayes, Z. Hradecna, S. Kumar ,N. A. Lozeron, H. J. J. Nijkamp, and *W. F. Stevens* (1970): Transcriptional units and their controls in *Escherichia coli* phage λ: Operons and scriptons. Cold Spr. Harb. Symp. quant. Biol. **35**, 341.
—, *H. Kubinsky,* and *P. Sheldrick* (1966): Pyrimidine clusters on the transcribing strand of DNA and their possible role in the initiation of RNA synthesis. Cold Spr. Harb. Symp. quant. Biol. **31**, 123.

Täckholm, G. (1922): Zytologische Studien über die Gattung *Rosa.* Acta Hort. Berg. **7**.
Tartof, K. D. (1971): Increasing the multiplicity of ribosomal RNA genes in *Drosophila melanogaster.* Science **171**, 294.
Taschenberg, O. (1892): Historische Entwicklung der Lehre von der Parthenogenesis. Abh. naturf. Ges. Halle **17**, 365.
Taylor, A. C. (1966): Microtubules in the microspikes and cortical cytoplasm of isolated cells. J. Cell Biol. **28**, 155.
Taylor, J. H. (1958): Sister chromatid exchanges in tritium-labeled chromosomes. Genetics **43**, 515.
Taylor, W. R. (1926): Chromosome morphology in *Fritillaria, Alstroemeria, Silphium,* and other genera. Amer. J. Bot. **13**, 179.
Temin, H. M. (1971): The protovirus hypothesis: Speculations on the significance of RNA-directed DNA synthesis for normal development and for carcinogenesis. J. Nat. Cancer Inst. **46**, III.
—, and *S. Mizutani* (1970): RNA-dependent DNA-polymerase in virions of Rous sarcoma virus. Nature **226**, 1211.
Thach, S. S., and *R. E. Thach* (1971): Translocation of messenger RNA and "accomodation" of fMet-tRNA. Proc. nat. Acad. Sci., Wash., **68**, 1791.
Thoday, J. M. (1953): Components of fitness. Symp. Soc. exp. Biol. **7**, 96.
— (1972): Disruptive selection. Proc. Roy. Soc. London B. **182**, 109.
Thomas, C. A. (1966): Recombination of DNA molecules. Progr. in Nucleic Acid Res. and Mol. Biol. **5**, 315.
—, *T. J. Kelly,* and *M. Rhoades* (1968): The intracellular forms of T 7 and P 22 DNA molecules. Cold Spr. Harb. Symp. quant. Biol. **33**, 417.
Thomas, R. (1955): Recherches sur la cinétique des transformation bactériennes. Biochim. biophys. Acta **18**, 467.

Thomas, R. (1966): Control of development in temperate bacteriophages I. Induction of prophage genes following hetero-immune superinfection. J. Mol. Biol. **22**, 79.
— (1971): In "The Bacteriophage Lambda" (A. D. Hershey, ed.), Cold Spr. Harb. Laboratory, Cold Spr. Harb., New York, 211.
Thompson, D. H. (1931): The side chain theory of the structure of the gene. Genetics **16**, 267.
Thompson, J. N., and *J. M. Thoday* (1974): A definition and standard nomenclature for "polygenic loci". Heredity **33**, 430.
Thompson, J. S., and *M. W. Thompson* (1966): Genetics in Medicine. W. B. Saunders Comp./Philadelphia.
Thorell, B. (1944): Behaviour of the nucleolar apparatus during growth and differentiation of cellular proteins. Cold Spr. Harb. Symp. quant. Biol. **12**, 247.
Thornton, R. M., and *K. V. Thimann* (1964): On a crystal-containing body in cells of the oat coleoptile. J. Cell Biol. **20**, 345.
Tilney-Basset, R. A. E. (1963): The structure of periclinal chimeras. Heredity **18**, 265.
Timoféeff-Ressovsky, H. A., und *N. W. Timoféeff-Ressovsky* (1926): Über das phänotypische Manifestieren des Genotyps. II. Über idiosomatische Variationsgruppen bei *Drosophila funebris*. Roux Arch. EntwMech. Organ. **108**, 146.
Tischler, G. (1920): Über die sogenannten „Erbsubstanzen" und ihre Lokalisation in der Pflanzenzelle. Biol. Zbl. **40**, 15.
Togby, H. A. (1943): A cytological study of *Crepis fuliginosa, C. neglecta*, and their F_1-hybrid, and its bearing on the mechanism of phylogenetic reduction in chromosome number. J. Genetics. **45**, 67.
Tobias, P. v. (1953): Trends in the evolution of mammalian chromosomes. S. Afr. J. Sci. **50**, 134.
Tomkins, G. M., *T. D. Gelehrter*, *D. Granner*, *F. Martin*, *H. H. Samuels*, and *B. Thompson* (1969): Control of specific gene expression in higher organisms. Science **166**, 1474.
Tournefort, J. P. (1700): Institutiones Rei Herbarii. Paris.
Travers, A., *R. Kamen*, and *M. Cashel* (1970): The *in vitro* synthesis of ribosomal RNA. Cold Spr. Harb. Symp. quant. Biol. **35**, 415.
—, —, and *R. F. Schleif* (1970): Factor necessary for ribosomal RNA synthesis. Nature, Lond., **228**, 748.
Trow, A. H. (1895): The karyology of *Saprolegnia*. Ann. Bot. **9**, 609.
Tschermak, E. v. (1900): Über künstliche Kreuzung bei *Pisum sativum*. Z. landw. VersWes. Öst. **3**, H. 5; and Ber. dtsch. bot. Ges. **18**, 232.
Tsuchida, N., *M. Nonoyama*, and *Y. Ikeda* (1966): Perpetuation of genome of a temperature-sensitive mutant of RNA bacteriophage β in growing cells of *E. coli* at high temperature. J. Mol. Biol. **20**, 575.
Tulasne, R., *R. Minck*, *A. Kirn*, and *J. Krembel* (1959): Délimitation de la notion de formes L. des Bactéries: Protoplastes, sphéroplastes et formes L. Ann. Inst. Pasteur **99**, 859.
Turesson, G. (1922): The genotypical response of the plant species to the habitat. Hereditas, Lund, **3**, 211.
— (1923): The scope and impact of genecology. Hereditas, Lund, **4**, 171.
— (1926): Studien über *Festuca ovina* L. I. Normalgeschlechtliche, halb- und ganz vivipare Typen nordischer Herkunft. Hereditas, Lund, **8**, 161.
— (1929): Zur Natur und Begrenzung der Arteinheiten. Hereditas, Lund, **12**, 323.

Uexküll-Syllenband, E. v. (1901): Phylogenie der Blütenformen und der Geschlechterverteilung bei den Kompositen. Bibl. bot. **52**, 5.
Umbarger, H. E. (1956): Evidence for a negative feedback mechanism in the biosynthesis of isoleucine. Science **123**, 848.

Unrau, J., C. Person, and *J. Kuspira* (1956): Chromosome substitution in hexaploid wheat. Can. J. Bot. **34**, 629.

Vaarama, A. (1954): Cytological observations on *Pleurozium schreberi*, with special reference to centromere evolution. Ann. bot. Soc. zool.-bot. fenn. Vanamo **28**, 1.
Valentine, D. N. (1960): Seed-incompatibility in *Primula*. Nature, Lond., **185**, 778.
Vandel, A. (1938): Recherches sur la sexualité des Isopodes. III. Bull. biol. **72**, 147.
— (1945): Recherches sur la sexualité des Isopodes. IX. Bull. Biol. **79**, 168.
Vavilov, N. J. (1922): The law of homologous series in variation. J. Genet. **12**, 47.
— (1927): Geographical regularities in the distribution of the genes of cultivated plants. Bull. appl. Bot. Pl.-Breed. **17**, III, 409.
— (1927): Geographische Genzentren unserer Kulturpflanzen. 5. Int. Kongr. Genet. Berlin.
Vejdovsky, F. (1912): Zum Problem der Vererbungsträger. Böhm/Prag.
Verworn, M. (1891): Die physiologische Bedeutung des Zellkernes. Pflüg. Arch. ges. Physiol. **51**, 1.
Visconti, N., and *M. Delbrück* (1953): The mechanism of genetic recombination in phage. Genetics **38**, 5.
Vogel, F. (1958): Bedeutung der Phänogenetik für die innere Medizin. Verh. dtsch. Ges. inn. Med. **64**.
— (1961): Lehrbuch der allgemeinen Humangenetik. Springer/Berlin.
Vogel, H. J. (1957): Repression and induction as control mechanisms of enzyme biogenesis: the adaptive formation of acetylornithinase. In: *W. D. McElroy* and *B. Glass* (Ed.) "The Chemical Basis of Heredity", Johns Hopkins Press/Baltimore, 276.
—, *V. Bryson,* and *J. O. Lampen* (Eds.) (1963): Informational Macromolecules. Acad. Press/New York.
Vogt, O. (1926): Psychiatrisch wichtige Tatsachen der zoologisch-botanischen Systematik. Z. ges. Neurol. Psychiat. **101**, 805.
Vries, H. de (1885): Plasmolytische Studien über die Wand der Vakuolen. Pringsheims Jb. wiss. Bot. **16**, 464.
— (1900): Das Spaltungsgesetz der Bastarde. Ber. dtsch. bot. Ges. **18**, 83.
— (1900): Sur la loi de disjonction des hybrides. C. R. Acad. Sci. **130**, 845.
— (1901): Die Mutationstheorie Bd. I. Veit & Comp./Leipzig.
— (1911): Über doppeltreziproke Bastarde von *Oe. biennis* L. und *Oe. muricata* L. Biol. Zbl. **31**, 97.
— (1917): Halbmutanten und Zwillingsbastarde. Ber. dtsch. bot. Ges. **35**, 128.
— (1929): Über das Auftreten von Mutanten aus *Oenothera Lamarckiana*. Z. indukt. Abstamm.- u. VererbLehre **52**, 121.

Waardenburg, P. J. (1932): Das menschliche Auge und seine Erbanlagen. Bibliogr. genet. 7.
Waddington, C. H. (1932): Experiments on the development of chick and duck embryos cultivated *in vitro*. Phil. Trans. B. **221**, 179.
— (1939): An Introduction to Modern Genetics. Allen & Unwin/London.
— (1940): The genetic control of wing development in *Drosophila*. J. Genet. **41**, 75.
— (1942): Canalization of development and the inheritance of acquired characters. Nature, Lond., **150**, 563.
— (1953): Epigenetics, and evolution. Symp. Soc. exp. Biol. **7**, 186.
— (1953): The evolution of adaptation. Endeavour **12**, 134.
— (1953): The "Baldwin effect", "genetic assimilation ", and "homeostasis". Evolution **7**, 386.
— (1953): The interactions of some morphogenetic genes in *Drosophila*. J. Genet. **51**, 243.

Waddington, C. H. (1955): On a case of quantitative variation on either side of the wild-type. Z. indukt. Abstamm.- u. VererbLehre **87**, 208.
— (1956): Principles of Embryology. Allen & Unwin/London.
— (1957): The Strategy of the Genes. Allen & Unwin/London.
— (1961): Genetic assimilation. Adv. in Genetics **10**, 257.
— (1962): New Patterns in Genetics and Development. Columbia Univ. Press/New York.
—, and *G. A. Schmidt* (1933): Induction by heteroplastic grafts of the primitive streak in birds. Roux Arch. EntwMech. Organ. **128**, 522.
Wahlund, S. (1928): Zusammensetzung von Populationen und Korrelationserscheinungen vom Standpunkt der Vererbungslehre aus betrachtet. Hereditas, Lund, **11**, 65.
Wakker, J. H. (1888): Studien über die Inhaltskörper der Pflanzenzelle. Jb. wiss. Bot. **19**, 423.
Waldeyer, W. (1888): Über Karyokinese und ihre Beziehung zu den Befruchtungsvorgängen. Arch. mikr. Anat. **32**, 1.
Walker, P. M. B. (1971): Repetitive DNA in higher organisms. Progr. Biophys. Mol. Biol. **23**, 145.
Wallace, B. (1963): Genetic diversity, genetic uniformity, and heterosis. Canad. J. Genet. Cytol. **5**, 239.
Walters, M. S. (1954): Pseudobivalents in meiosis of two interspecific hybrids of *Bromus*. Amer. J. Bot. **41**, 160.
— (1963): A nuclear body in meiosis of *Bromus*. Chromosoma **14**, 423.
Warburton, F. E. (1955): Feedback in development and its evolutionary significance. Amer. Nat. **89**, 129.
Warner, J. R., and *R. Soeiro* (1967): Nascent ribosomes from Hela cells. Proc. nat. Acad. Sci., Wash., **59**, 1984.
—, *A. Rich*, and *C. E. Hall* (1962): Electron microscope studies of ribosomal clusters synthesizing hemoglobin. Science **138**, 1399.
Warnke, H. E., and *A. F. Blakeslee* (1939): Induction of simple and multiple polyploidy in *Nicotiana* by colchicine treatment. J. Hered. **30**, 419.
Washington, W. J. (1971): Homoeoallelism in *Triticum aestivum*. Can. J. Genet. Cytol. **13**, 169.
Wassermann, F. (1926): Zur Analyse der mitotischen Kern- und Zellteilung. Z. ges. Anat. I. Z. Anat. EntwGesch. **80**, 344.
Watanabe, T. (1963): Infective heredity of multiple drug resistance in bacteria. Bacteriol. Rev. **27**, 87.
—, and *T. Kukasawa* (1960): Episomic resistance factors in *Enterobacteriaceae*. V. Transfer of resistance factors mediated by a new episome (Jap.). Med. Biol. **56**, 201.
—, *H. Nishida, C. Ogalta, T. Arai*, and *S. Sato* (1964): Episome-mediated transfer of drug resistance in *Enterobacteriaceae*. J. Bacteriol **88**, 716.
Watson, J. D. (1950): The properties of X-ray inactivated bacteriophage. J. Bacteriol. **60**, 697.
— (1963): Involvement of RNA in the synthesis of proteins. Science **140**, 17.
— (1964): The synthesis of proteins upon ribosomes. Bull. Soc. Chim. Biol. **46**, 1399.
— (1965): Molecular Biology of the Gene. Benjamin/New York.
— (1970): Molecular Biology of the Gene (2nd ed.). W. A. Benjamin/New York.
—, and *F. C. H. Crick* (1953): A structure for deoxyribose nucleic acid. Nature, Lond., **171**, 737.
—, and — (1953): Genetical implications of the structure of deoxyribonucleic acid. Nature, Lond., **171**, 964.
Watson, M. L. (1955): The nuclear envelope, its structure and relation to cytoplasmic membranes. J. Biophysic. Biochem. Cytol. **1**, 257.

Watson, M. L. (1959): Further observations on the nuclear envelope of the animal cell. J. Biophys. Biochem. Cytol. **6**, 147.
Webber, H. J. (1897): Notes on the fecundation of *Zamia* and the pollen tube apparatus of *Ginkgo*. Bot. Gaz. **24**, 225.
— (1903): New horticultural and agricultural terms. Science **18**, 501.
Webber, J. M. (1933): Cytological features of *Nicotiana glutinosa* haplonts. J. agric. Res. **47**, 845.
Weigle, J. J. (1953): Induction of mutations in a bacterial virus. Proc. nat. Acad. Sci., Wash., **39**, 628.
Weinberg, W. (1908): Über den Nachweis der Vererbung beim Menschen. Jh. Ver. vaterl. Naturk. Württemberg **64**, 368.
Weinstein, A. (1936): The theory of multiple-strand cross-over. Genetics **21**, 155.
Weismann, A. (1883): Über die Vererbung. In: Aufsätze über Vererbung und verwandte biologische Fragen. Fischer/Jena 1892, 73.
— (1885): Die Kontinuität des Keimplasmas als Grundlage einer Theorie der Vererbung. In: Aufsätze über Vererbung und verwandte biologische Fragen. Fischer/Jena 1892, 191.
— (1887): Über die Zahl der Richtungskörper und ihre Bedeutung für die Vererbung. Fischer/Jena.
— (1891): Amphimixis oder die Vermischung der Individuen. Fischer/Jena.
— (1895): Neue Gedanken zur Vererbungsfrage. Eine Antwort an Herrn Spencer. Fischer/Jena.
Weiss, B., and *C. C. Richardson* (1967): Enzymatic breakage and joining of deoxyribonucleic acid. I. Repair of singlestrand breaks in DNA by an enzyme system from *Escherichia coli* infected with T 4 bacteriophage. Proc. nat. Acad. Sci., Wash., **57**, 1021.
Weiss, P. (1939): The Principles of Development. Holt/New York.
— (1958): Cell contact. Int. Rev. Cytol. **7**, 391.
Went, H. A. (1966): The behavior of centrioles and the structure and formation of the achromatic figure. Protoplasmatologia **6**, G **1**, Springer/Berlin.
Wettstein, F. v. (1928): Morphologie und Physiologie des Formwechsels der Moose auf genetischer Grundlage. II. Bibl. genet. **10**. VI, 216, Borntraeger/Berlin.
— (1932): Bastardpolyploidie als Artbildungsvorgang in Pflanzen. Naturw. **20**, 981.
Wettstein, F. O., T. Staehelin, and *H. Noll* (1963): Ribosomal aggregates engaged in protein synthesis: Characterization of the ergosome. Nature, Lond., **197**, 430.
Wettstein, R. v. (1898): Grundzüge der geographisch-morphologischen Methode der Pflanzensystematik. Fischer/Jena.
White, B. N., and *G. M. Tener* (1973): Activity of a transfer RNA modifying enzyme during the development of *Drosophila* and its relationship to the su(s) locus. J. Mol. Biol. **74**, 635.
White, M. J. D. (1935): The effects of X-rays on mitosis in the spermatogonial divisions of *Locusta migratoria*. Proc. roy. Soc., B. **119**, 61.
— (1935): Eine neue Form von Tetraploidie nach Röntgenbestrahlung. Naturwissenschaften **23**, 390.
— (1941): The evolution of the sex chromosomes. I. The XO and X_1X_2Y mechanisms in praying mantids. J.Genet. **42**, 143.
— (1942): Nucleus, chromosomes, and genes. In: *G. H. Bourne* (Ed.) "Cytology and Cell Physiology", Univ. Press/Oxford.
— (1943): Amount of heterochromatin as a specific character. Nature, Lond., **152**, 536.
— (1945): Animal Cytology and Evolution. Univ. Press/Cambridge.
— (1949): A cytological survey of wild populations of *Trimerotropis* and *Circotettix* (Orthoptera, Acrididae). I. The chromosomes of twelve species. Genetics **34**, 537.
— (1950): The Chromosomes. (4. Ed.) Methuen/London.

White. M. J. D. (1951): Cytogenetics of orthopteroid Insects. Advanc. Genet. **4**, 267.
— (1954): Animal Cytology and Evolution. (2. Ed). Univ. Press/Cambridge.
— (1957: Some general problems of chromosomal evolution and speciation in animals. Surv. biol. Progr. 3, 109.
— (1973): The Chromosomes. 6th Ed. Chapman and Hall/London.
—, *R. E. Blackith, R. M. Blackith,* and *J. Cheney* (1967): Cytogenetics of the viatica group of morabine grasshoppers. I. The coastal species. Austral. J. Zool. **15**, 263.
Whitehouse, H. L. K. (1963): A theory of crossing-over by means of hybrid deoxyribonucleic acid. Nature, Lond., **199**, 1034.
— (1965): Towards an Understanding of the Mechanism of Heredity. E. Arnold/London.
— (1973): Towards an Understanding of the Mechanism of Heredity. 3rd Ed. E. Arnold/London.
—, and *P. J. Hastings* (1965): The analysis of genetic recombination on the polaron hybrid DNA model. Genet. Res., Camb. **6**, 27.
Whiting, P. W., R. J. Greb, and *B. R. Speicher* (1934): A new type of sex-intergrade. Biol. Bull., Wood's Hole **66**, 152.
Whitmann, C. O. (1887): Germ layers in *Clepsine.* J. Morph.
Wichterman, R. (1939): Cytogamy: a new sexual process in joined pairs of *Paramecium caudatum.* Nature, Lond., **144**, 123.
— (1940): Cytogamy: a sexual process occurring in living joined pairs of *Paramecium caudatum* and its relation to other sexual phenomena. J. Morph. **66**, 423.
Williams, E., and *R. A. Brink* (1972): The effect of abnormal chromosome 10 on transposition of modulator from the R locus in maize. Genetics **71**, 97.
Wilson, D. A., and *C. A. Thomas* (1974): Palindromes in chromosomes. J. Mol. Biol. **84**, 115.
Wilson, E. B. (1882): The cell lineage of *Nercis.* J. Morph. **6**.
— (1896): The Cell in Development and Heredity. MacMillan/New York (1 st Ed.).
— (1901): Experimental studies in cytology. Roux Arch. EntwMech. Organ. **13**, 352.
— (1906): Studies on chromosomes III. The sexual difference of the chromosome groups in Hemiptera, with some considerations on the determination and inheritance of sex. J. exp. Zool. **3**, 1.
— (1909): Recent researches on the determination and heredity of sex. Science **29**, 53.
— (1909): Studies on chromosomes IV. The "accessory" chromosome in *Syromastes* and *Phyrrocoris,* with a comparative review of the types of sexual difference of the chromosome groups. J. exp. Zool. **6**, 69.
— (1910): Note on the chromosomes of *Nezara.* A correction and addition. Science, New Series **31**, 788.
— (1911): The sex chromosomes. Arch. mikr. Anat. **77**, 249.
— (1925): The Cell in Development and Heredity. (3rd Ed.). MacMillan/New York.
Wilson, G. B. (1965): The assay of antimitotics. Chromosoma **16**, 133.
Winge, O. (1927): On a Y-linked gene in *Melandrium.* Hereditas, Lund, **9**, 274.
Winiwarter, H. v. (1900): Recherches sur l'ovogénese et l'ovaire des Mammiféres (Lapin et Homme). Arch. Biol., Paris, **17**, 33.
Winkler, H. (1906): Über Parthenogenesis bei *Wikstroemia indica* (L) C. A. Mey. Ann. Jard. bot. Buitenzorg **20**, 208.
— (1907): Über Pfropfbastarde und pflanzliche Chimären. Ber. dtsch. bot. Ges. **25**, 568.
— (1908): Über Parthenogenesis und Apogamie im Pflanzenreich. Fischer/Jena.
— (1916): Über die experimentelle Erzeugung von Pflanzen mit abweichenden Chromosomenzahlen. Z. Bot. **8**, 417.
— (1920): Vererbung und Ursache der Parthenogenese im Pflanzen- und Tierreich. Fischer/Jena.

Winkler, H. (1924): Über die Rolle von Kern und Protoplasma bei der Vererbung. Z. indukt. Abstamm.- u. VererbLehre **33**, 238.
— (1930): Die Konversion der Gene. Fischer/Jena.
Witkin, E. M. (1951): Nuclear segregation and the delayed appearance of induced mutants in *Escherichia coli*. Cold Spr. Harb. Symp. quant. Biol. **16**, 357.
— (1956): Time, temperature, and protein synthesis. A study of ultraviolet-induced mutation in bacteria. Cold Spr. Harb. Symp. quant. Biol. **21**, 123.
— (1969): Ultraviolet-induced mutation and DNA repair. Ann. Rev. of Genet. **3**, 525.
Wolfe, S. L. (1965): The fine structure of isolated metaphase chromosomes. Exp. Cell Res. **37**, 45.
— (1965): The fine structure of isolated chromosomes. J. Ultrastructure Res. **12**, 104.
—, and *B. John* (1965): The organization and ultrastructure of male meiotic chromosomes in *Oncopeltus fasciatus*. Chromosoma **17**, 85.
Wolff, C. F. (1759): Theoria Generationes. Halle.
Wolff, S. (1961): Radiation Genetics. In: *M. Errera*, and *A. Forssberg* (Ed.) "Mechanisms in Radiobiology" **1**, 419, Academic Press/New York.
Wollman, E. L., F. Jacob, and *W. Hayes* (1956): Conjugation and genetic recombination in *Escherichia coli* K 12. Cold Spr. Harb. Symp. quant. Biol. **21**, 141.
Wolpert, L. (1969): Positional information and the apical pattern of cellular differentiation. J. theoret. Biol. **25**, 1.
Woltereck, R. (1909): Weitere experimentelle Untersuchungen über Artveränderung, speziell über das Wesen quantitativer Artunterschiede bei Daphniden. Verh. dtsch. zool. Ges. **19**, 110.
Wood, R. L. (1959): Intercellular attachment in the epithelium of Hydra, as revealed by electron microscopy. J. Biophys. Biochem. Cytol. **6**, 343.
Wright, S. (1921): Systems of mating. Genetics **6**, 111.
— (1929): Fisher's theory of dominance. Amer. Nat. **63**, 247.
— (1931): Evolution in Mendelian populations. Genetics **16**, 97.
— (1932): The roles of mutation, inbreeding, crossbreeding, and selection in evolution. Proc. 6th Int. Congr. Genetics **1**, 356.
— (1943): Isolation by distance. Genetics **28**, 114.
— (1946): Isolation by distance under diverse systems of mating. Genetics **31**, 39.
— (1949): Adaptation and selection. In: *G. L. Jepsen, E. Mayr*, and *G. G. Simpson* "Genetics, Palaeontology, and Evolution", Univ. Press/Princeton, 365.
— (1955): Classification of the factors of evolution. Cold Spr. Harb. Symp. quant. Biol. **20**, 16.
— (1956): Modes of selection. Amer. Nat. **90**, 5.

Yamada, E. (1955): The fine structure of the reanal glomerulus of the mouse. J. Biophys. Biochem. Cytol. **1**, 551.
Yamamoto, Y. (1938): Karyogenetische Untersuchungen bei der Gattung *Rumex*. VI. Geschlechtsbestimmung bei eu- und aneuploiden Pflanzen. Mem. Coll. Agric. Kyoto, **43**.
Yamamoto, H., S. Shiraiwa, T. Ashida, and *Y. Ito* (1969): Perichromatin granules during cell division cycle. J. Electron Microsc. **18**, 57.
Yanofsky, C., and *E. S. Lennox* (1959): Transduction and recombination study of linkage relationships among the genes controlling tryptophan synthesis in *Escherichia coli*. Virology **8**, 425.
—, *E. C. Cox*, and *V. Horn* (1966): The unusual mutagenic specifity of an *E. coli* mutator gene. Proc. nat. Acad. Sci., Wash., **55**, 274.
—, and *J. Ito* (1966): Nonsense codons and polarity in the tryptophan operon. Cold Spr. Harb. Symp. quant. Biol. **31**, 151.

Yanofsky, C., D. Helsinski, and *B. Malling* (1961): The effects of mutation on the composition and properties of the A protein of *Escherichia coli* tryptophan synthetase. Cold Spr. Harb. Symp. quant. Biol. **26**, 11.

Yasuzumi, G. (1951): The micro-structure of chromatin threads in the metabolic stage of the nucleus. Chromosoma **4**, 222.

Yates, R. A., and *A. B. Pardee* (1956): Control of pyrimidine biosynthesis in *Escherichia coli* by a feed-back mechanism. J. Biol. Chem. **221**, 757.

Yoshikawa, H., and *N. Sueoka* (1963): Sequential replication of *Bacillus subtilis* chromosome. I. Comparison of marker frequencies in exponential and stationary growth phases. Proc. nat. Acad. Sci., Wash., **49**, 559.

Yourno, J., D. Barr, and *S. Tanemura* (1959): Externally suppressible frameshift mutants of *Salmonella typhimurium.* J. Bacteriol. **100**, 453.

Zamenhof, S. (1967): Nucleic acids and mutability. Progr. Nucleic Acid Res. and Mol. Biol. **6**, 1.

Zimmermann, W. (1943): Methoden der Phylogenetik. In: *G. Heberer*, ,,Die Evolution der Organismen" (1. Aufl.), 20, Fischer/Jena.

— (1948): Grundfragen der Evolution. Klostermann/Frankfurt.

— (1953): Evolution. Geschichte ihrer Probleme und Erkenntnisse. Alber/Freiburg-München.

Zinder, N. D., and *J. Lederberg* (1953): Genetic exchange in *Salmonella.* J. Bacteriol. **64**, 679.

Zipser, D., and *A. Newton* (1967): The influence of deletion on polarity. J. Mol. Biol. **25**, 567.

Zissler, Y. (1967): Integration-negative (int) mutants of phage lambda. Virology **31**, 189.

Zubay, G., D. Schwartz, and *J. Beckwith* (1970): Mechanism of activation of catabolite sensitive genes: a positive control system .Proc. nat. Acad. Sci., Wash., **66**, 104.

Zuckerkandl, E., and *L. Pauling* (1965): Molecules as documents of evolutionary history. J. theoret. Biol. **8**, 357.

Zukov, N. I. (1941): Alteration of the nature of plants by means of interspecific hybridization in the genus *Nicotiana* (russ.) Vsesojuznyi Instit. tabacn. i. machorocn. Promyslennosh (Krasnodov) **143**, 142.

References added in proof

Achtman, M. A. (1975): Mating aggregates in *Escherichia coli*. J. of Bact. **123**, 505.

Anderson, E. (1968): Oocyte differentiation in the Sea urchin, *Arbacia punctulata*, with particular reference to the origin of corticular granules and their participation in the cortical reaction. J. Cell Biol. **37**, 514.

Anderson, T. F. (1958): Recombination and segregation in *Escherichia coli*. Cold Spr. Harb. Symp. quant. Biol. **23**, 47.

Baccarini, I. (1908): Sulle cinesi vegetative del „*Cynomorium coccineum* L". G. Bot. Ital. **15**, 189.

Bernhard, W. (1966): Ultrastructural aspects of the normal and pathological nucleolus in mammalian cells. Nat. Cancer Inst. Monogr. **23**, 13.

Bertrand, K., L. Korn, F. Lee, T. Platt, C. L. Squires, C. Squires, and *C. Yanofsky* (1975): New features of the regulation of the tryptophan operon. Science **189**, 22.

Bresler, S. E. (1975): Theory of misrepair mutagenesis. Mut. Res. **29**, 467.

Brown, D. D., and *I. B. David* (1968): Specific gene amplification in oocytes. Science **160**, 272.

Cerda-Olmedo, E., P. C. Hanawalt, and *N. Guerola* (1968): Mutagenesis of the replication point by nitrosoguanidine: Map and pattern of replication of the *Escherichia coli* chromosome. J. Mol. Biol. **33**, 705.

Chandra, H. S., and *S. W. Brown* (1975): Chromosome imprinting and the mammalian X chromosome. Nature **253**, 165.

Chetverikov, S. S. (1926): On certain aspects of the evolutionary process from the standpoint of modern genetics. J. Exptl. Biol. (Russian) A 2, 3. Translation Proc. Amer. Phil. Soc. **105**, 167.

Choppin, P. W., and *R. W. Compans* (1970): Phenotypic mixing of envelope proteins of the Parainfluenza virus SV 5 and the vesicular stomatitis virus. J. Virol. **12**, 609.

Coetzee, J. N., N. Datta, and *R. W. Hedges* (1972): R factors from *Proteus rettgeri*. J. gener. Microbiol. **72**, 543.

Crouse, H. V. (1960): The controlling element in sex chromosome behavior in *Sciara*. Genetics **45**, 1429.

Darnell, J. E. (1968): Ribonucleic acids from animal cells. Bact. Reviews **32**, 262.

Decker, P., and *A. Speidel* (1972): Open systems which can mutate between several steady states ("bioids") and a possible prebiological role of the autocatalytic condensation of formaldehyde. Zschr. Naturforschg. **27 b**, 257.

Diener, T. O. (1971): Potato spindle tuber "virus". IV. A replicating, low molecular weight RNA. Virology **45**, 411.

Djordjevic, B., and *L. J. Tolmach* (1967): Responses of synchronous populations of HeLa cells to ultraviolet irradiation at selected stages of the generation cycle. Radiat. Res. **32**, 327.

Donachie, W. D. (1968): Relationship between cell size and time of initiation of DNA replication. Nature **219**, 1077.

Dravid, A. R., N. Pete, and *P. Mandel* (1971): J. Neurochem. **28**, 299.

DuPraw, E. J. (1970): DNA and Chromosomes. Rinehart & Winston, New York.

Durrant, A. (1962): The environmental induction of heritable change in *Linum*. Heredity **17**, 27.

Englund, P. T. (1971): The initial step of *in vitro* synthesis of deoxyribonucleic acid by the T 4 deoxyribonucleic acid polymerase. J. Biol. Chem. **246**, 5684.

Gall, J. G., and *M. L. Pardue* (1969): Formation and detection of RNA-DNA hybrid molecules in cytological preparations. Proc. nat. Acad. Sci. Wash. **63**, 378.

Grant, V. (1971): Plant speciation. Columbia University Press, New York/London.

Green, R. R., and *J. W. Drake* (1974): Misrepair mutagenesis in bacteriophage T 4. Genetics **78**, 81.

Grillo, M. A. (1970): Cytoplasmic inclusions resembling nucleoli in sympathetic neurons of adult rats. J. Cell Biol. **45**, 100.

Gurdon, J. B., C. D. Lane, H. R. Woodland, and *G. Marbaix* (1971): Use of frog eggs and oocytes for the study of messenger RNA and its translation in living cells. Nature **233**, 177.

Gurdon, J. B., J. B. Lingrel, and *G. Marbaix* (1973): Message stability in injected frog oocytes: long life of mammalian α and β globin messages. J. Mol. Biol. **80**, 539.

Hadjivasiliou, A., and *G. B. Blawerman* (1966): Polyadenylic acid in the cytoplasm of rat liver. J. Mol. Biol. **20**, 1.

Hamilton (1964): J. theoret. Biol. **7**, 1.

Hawthorne, E. D. C., and *J. Friis* (1964): Osmotic-remedial mutants. A new classification for nutritional mutants in Yeast. Genetics **50**, 829.

Hedges, R. W., and *A. E. Jacob* (1974): Transposition of ampicillin resistance from RP 4 to other replicons. Molec. Gen. Genet. **132**, 31.

Helmstetter, C., S. Cooper, O. Pierucci, and *E. Revelas* (1968): On the bacterial life sequences. Cold Spr. Harb. Symp. quant. Biol. **33**, 809.

Hershey, A. D. (1970): Persistent heterozygotes in phage T 4. Yearb. Carnegie Inst. **61**, 717.

Heywood, S. M., D. S. Kennedy, and *A. J. Bester* (1974): Separation of specific initiation factors involved in the translation of myosin and myoglobin messenger RNAs and the isolation of a new RNA involved in translation. Proc. nat. Acad. Sci. Wash. **71**, 2428.

John, H. A., M. L. Birnstiel, and *K. W. Jones* (1969): RNA–DNA hybrids at the cytological level. Nature **223**, 582.

References added in proof

Johnson, S. M., A. D. Bangham, M. W. Hill, and *E. D. Korn* (1973): Single bilayer liposomes. Biochem. Biophys. Acta **233**, 820.

Kates, J. (1970): Transcription of the vaccinia virus genome and the occurrence of polyriboadenylic acid sequences in messenger RNA. Cold Spr. Harb. Symp. quant. Biol. **35**, 743.

Kihara, H. (1930): Genomamalyse bei *Triticum* und *Aegilops*. Cytologia **1**, 263.

Kojima, K.-I., and *H. E. Schaffer* (1967): Survival process of linked mutant genes. Evolution **21**, 518.

Kornberg, R. D. (1974): Chromatin structure: a repeating unit of histones and DNA. Science **184**, 868.

Krooth, R. S. (1965): The future of mammalian cell genetics. Birth Defects Original Articles Series **1** (2), 21.

Lane, C. D., G. Marbaix, and *J. B. Gurdon* (1971): Rabbit hemoglobin synthesis in frog cells: the translation of reticulocyte 9 sRNA in frog oocytes. J. Mol. Biol. **61**, 73.

Leblond, C. P. (1964): Classification of cell populations on the basis of their proliferative behavior. J. Nat. Cancer Inst. **14**, 119.

Lehmann, A. R. (1972): Postreplication repair of DNA in ultraviolet-irradiated mammalian cells. J. Mol. Biol. **66**, 319.

Levin, D. H., D. Kyner, and *G. Acs* (1973): Protein synthesis initiation in eukaryotes. Characterization of ribosomal factors from mouse fibroblasts. J. Biol. Chem. **248**, 6416.

Levy, (1974): Amer. J. clin. Path. **62**, 258.

Metzenberg, R. L. (1968): Repair of multiple defects of a regulatory mutant of *Neurospora* by high osmotic pressure and by reversion. Arch. Biochem. Biophys. **125**, 532.

Meynell, E., G. G. Meynell, and *N. Datta* (1968): Phylogenetic relationships of drug-resistance factors and other transmissible bacterial plasmids. Bact. Reviews **32**, 55.

Muller, H. J. (1940): Bearings of the *Drosophila* work on systematics. In: "The New Systematics", Clarendon Press, Oxford, p. 185.

Nass, M. K., and *S. Nass* (1962): Fibrous structures within the matrix of developing chick embryo mitochondria. Exptl. Cell Res. **26**, 424.

Novick, R. P., R. C. Clowes, S. N. Cohen, N. Datta, and *S. Falkow* (1973): Uniform nomenclature for bacterial plasmids: A proposal, Unpublished draft.

Novikoff, A. B., and *P. M. Novikoff* (1973): Microperoxisomes. J. Histochem. Cytochemistry **21**, 963.

Olins, A. L., and *D. E. Olins* (1973): Spheroid chromatin units (nu bodies). J. Cell Biol. **59**, 252a.

Oudet, P., M. Gross-Bellard, and *P. Chambon* (1975): Electronmicroscopic and biochemical evidence that chromatin structure is a repeating unit. Cell **4**, 281.

Paulus, H., and *J. B. Alpers* (1971): Preconditioning: an obligatory step in the biosynthesis of oligomeric enzymes and its promotion by allosteric ligands. Enzyme **12**, 385.

Perry, P., and *H. J. Evans* (1975): Cytological detection of mutage-carcinogen exposure by sister chromatid exchange. Nature **258**, 121.

Pontecorvo, G. (1975): The *Bateson* lecture 1974: "Alternatives to sex"-genetics by means of somatic cells. In: *R. Markham* et al. (Eds.) "Modification of the information content in plant cells"; Second John Innes Symposium, North-Holland/American Elsevier, p. 1.

Radman, M. (1975): Phenomenology of an inducible mutagenic DNA repair pathway in *E. coli*: SOS repair hypothesis. In: *M. Miller* (Ed.) "Molecular and Environmental Aspects of Mutagenesis, C. C. Thomas, Springfield.

Revel, J. P. (1968): Studies on the fine structure of intercellular junctions. Proc. 26th Meeting of the Electron Microscopy Soc. of America. Claitor's Publ. Div./Baton Rouge, p. 40.

Revel, M., *Y. Pollack*, *Y. Groner*, *R. Scheps*, *H. Inouye*, *H. Berissi*, and *H. Zeller* (1973): IF_3-interference factors: protein factors in *E. coli* controlling initiation of mRNA translation. Biochemie **55**, 41.

Sager, R., and *R. Kitchin* (1975): Slective silencing of eukaryotic DNA. Science **189**, 426.

Semancik, J. S., and *L. G. Weathers* (1972): Exocortis disease: evidence for a new species of "infectious" low molecular weight RNA in plants. Nature New Biol. **237**, 242.

Shapiro, A., and *Chargaff, E.* (1964): Studies on the nucleotide arrangement in deoxyribonucleic acids. VIII. A comparison of procedures for the determination of the frequency of pyrimidine nucleotide runs. Biochem. Biophys. Acta **91**, 262.

Solioz, M., *M.-C. Yen*, and *B. Marrs* (1975): Release and uptake of gene transfer agent by *Rhodopseudomonas capsulata*. J. of Bacteriol. **123**, 651.

Taylor, A. L. (1963): Bacteriophage-induced mutation in *Escherichia coli*. Proc. nat. Acad. Sci. Wash. **50**, 1043.

Thoday, J. M. (1975): Non-Darwinian "evolution" and biological progress. Nature **255**, 675.

Trautner, T. A., and *H. Ch. Spatz* (1973): Transfection in *B. subtilis*. Curr. Topics in Microbiology and Immunology **62**, 61.

Wang, J. C. (1971): Interaction between DNA and an *Escherichia coli* protein. J. Mol. Biol. **55**, 523.

Weiss, M. C., and *H. Green* (1967): Human-mouse hybrid cell lines containing partial complements of human chromosomes and functioning human genes. Proc. nat. Acad. Sci. Wash. **58**, 1104.

Weiss, P. (1925): Abhängigkeit der Regeneration entwickelter Amphibienextremitäten vom Nervensystem (Der Begriff des „Gestaltungstonus"). Arch. Mikrosk. Anat. Entwicklungsmech. **104**, 359.

Weissbach, A., *D. Baltimore*, *F. Bollum*, *R. Gallo*, and *D. Korn* (1975): Nomenclature of eukaryotic DNA polymerases. Eur. J. Biochem. **59**, 1.

Whitehouse, H. L. K. (1974): Advances in recombination research. Genetics **78**, 237.

Witkin, E. M., and *D. L. George* (1973): Ultraviolet mutagenesis in POLA and UVRA POLA derivatives of *Escherichia coli* B/R. Evidence for an inducible error-prone repair system. Genetics **73**, Suppl., 91.

Terms added in proof

attenuator (*Bertrand* et al. 1975) — a regulator site (nucleotide sequence) in certain → operons which, in contrast to other regulatory sites at which control is exerted over → RNA polymerase binding, is used to regulate termination of → genetic transcription in a region of the operon preceding the structural genes. Deletion of the a. enhances expression of the operon. A protein factor is thought to interact at the a. site to regulate transcription termination. This protein is assumed to bind at the → promoter and subsequently allows transcription to proceede beyond the a.

autorepressor system — a gene system (operon) one member of which does not code for an enzyme but represents a → regulator gene which encodes a → repressor protein. The repressor binds to the operator of the operon and blocks → genetic transcription (→ autoregulation ; exorepressor system).

basal apparatus (*Wilson* 1925) — the complex of → basal bodies, bridging material, and rootlets in epithelial cells with multiple → cilia. The multiple cilia are interconnected beneath the → cell membrane by amorphous material bridging the adjacent basal bodies; these in turn are anchored into the cytoplasm by means of long, often bifurcating striated rootlets which arise from the proximal end of each basal body.

base pair mismatch — the presence of mismatched bases in a → heteroduplex region of double-stranded DNA. Mismatched bases can occur either because the reading mechanism of → DNA polymerase fails, or because heteroduplex DNA is formed by recombination. Mismatched base pairs may be corrected (→ mismatch repair) with single-stranded breakdown and resynthesis of a section of DNA of fairly constant length.

cell determination — the process by which multipotential cells become committed to a particular developmental pathway (= cellular d.). Classically , c. d. is said to have occurred at the time when cells aquire the ability to differentiate along a particular pathway even if they are removed from their normal environment.

cell fractionation — the homogenization of tissue and the separation of the various components of the cell according to mass and specific gravity. With standard differential centrifugation and increasing centrifugal fields, four main fractions are obtained: 1. nuclear fraction, 2. mitochondrial fraction, 3. microsomal fraction, and 4. soluble fraction. C. f. can be improved through the use of discontinuous or continuous density gradients.

cell surface receptor — any structure at the → cell surface which receives and transduces a variety of environmental signals. The binding of a signalling ligand (= first messenger) induces a change in the c. s. r., initiating the next event(s) by which information is carried across the → cell membrane into the cell interior. The cyclic nucleotides, cAMP and cGMP, which are regulated by specific membrane bound enzymes (cyclases) are "second messengers" in the signalling of cellular activity.

cellular cloning — the production from daughter cells, ultimately derived from a single cell, of large, genetically uniform populations of cells.

chromosome imprinting (*Crouse* 1960) — the process by which one of two genetically homologous chromosomes becomes altered, i.e. predetermined to function differently from its homologue at a subsequent stage of development. C. i. may occur several to many cell generations before the stage at which behavioral differences become obvious (*Chandra* & *Brown* 1975).

chromosome instability syndrome — a class of rare, inherited diseases in man (as, e.g., Fanconi's anemia, Bloom's syndrome, and ataxia telangiectasia) associated with markedly raised frequencies of chromosome aberrations in peripheral blood lymphocytes and cultured fibroblasts. The aberrations are of the chromatid and chromosome type (→ chromosome mutation).

chromosome loss technique — in man-mouse somatic cell hybrids, a technique for detecting synteny and assigning chromosome loci. In this system the human chromosomes are lost one or a few at a time and the rodent chromosomes are retained. Two human chromosome loci are called syntenic if they are either lost or retained together in a series of hybrid clones (*Pontecorvo* 1975). The initial discovery of the loss of human chromosomes in man-mouse somatic cell hybrids was made by *Weiss* and *Green* (1967).

cloning vehicle — a cleaved but still functional → plasmid or phage genome used for the isolation of DNA recombinant molecules after ligation of an enzymatically (→ restriction endonuclease) cleaved DNA fragment and its *in vitro* joining (ligation) to the c. v. (→ molecular cloning; genetic engineering). The recombinant molecules may then be introduced into a suitable host cell (by → transformation) where they replicate autonomously, serving to "clone" the added DNA segments. In this way, DNA from various prokaryotic as well as eukarytoic sources has been replicated and transcribed in *Escherichia coli*. In all cases of molecular cloning, the problem of the potential biohazards must be taken into consideration.

concatemer — a DNA molecule of "multichromosomal length" which represents the precursor from which mature phage chromosomes (e.g., lambda phage) are cut and packaged.

cytochemistry — the identification and localization of chemical constituents at the level of the cell and subcellular structures (→ histochemistry). The cytochemical studies may be based on four main analytical techniques: 1. Separation of cell fractions by conventional biochemical techniques; 2. isolation of minute amounts of tissues, or even single cells, by micro- and ultramicromethods; 3. direct detection of cell components in the cell by staining procedures, and 4. use of measurements of physical parameters.

cytosol — the remaining supernatant (= soluble fraction) after separation of the nuclear, mitochondrial, and microsomal fractions by → cell fractionation. The c. contains the soluble proteins and enzymes found in the cytoplasmic matrix.

deoxyribonucleoprotein fiber — a → chromatin fiber (DNP fiber) which represents the basic unit of the eukaryotic → chromosome; it is composed of DNA, histones, and variable amounts of RNA and nonhistone proteins. Diameter measurements of the DNP f. range from 3 to 50 nm, depending upon circumstances.

dislodgement (*Coetzee* et al. 1972) — a type of interaction between a compatible resident → plasmid and another plasmid transferred by conjugation into a recipient bacterial strain which leads to the elimination of the resident plasmid, loss of some, but not all, of the phenotypic characteristics determined by that plasmid, or recombination between the two plasmids.

DNA silencing (*Sager & Kitchin*) 1975 — in eukaryotes, the selective inactivation or elimination of specific chromosomes, chromosome regions, or DNA molecules in the presence of unaffected homologues (→ chromosome imprinting).

doubling dose — the radiation dose that would double spontaneous mutation rates. It is used for estimates of genetic risk from ionizing radiations.

ecotropic (*Levy* 1974) — → xenotropic.

error correction nuclease (*Englund* 1971) — an error-correcting 3′→5′ exonuclease associated with prokaryotic → DNA polymerase.

evolutionary rate — the speed of evolutionary change. E. rates are highly dissimilar at the species level. Large, populous species with wide distribution show great evolutionary inertia because the co-adapted gene complex is highly resistent to the incorporation of new genes. Founder populations may undergo a genetic evolution within an extremely short time and may possibly replace alleles at 30 to 50 percent of the loci within 5,000 to 10,000 years.

exorepressor system — a gene system (operon) characterized by the positioning of the → regulator gene outside the relevant operon (→ autorepressor system).

field (*Weiss* 1925) — in developmental biology, a dynamic and patterned organizational whole, generated by the collective activity of a group of cells or a group of subunits of a single cell. The pattern is manifested in the form of tissues and organs as each individual cell of the group in turn responds to the f. by differentiating according to its position within it.

forward selection system — any selective system that allows the required mutant to survive.

gene transfer agent (*Solioz* et al. 1975) — any particle released from the cell that contains samples of genetic information representing all parts of the genome. G. t. agents have been found in the nonsulfur purple photosynthetic bacterium *Rhodopseudomonas* and can provide genetic information to other bacteria of the same species, thus effecting genetic transfers. The process resembles genetic → transduction but the particles are much smaller than any known transducing bacteriophage. Cells that express new genetic markers received via g. t. agents are called "transferants".

genetic analysis — the resolution of the genetic material into its component elements.

genetic coherence — a tendency for parental character combinations to remain associated (= character association) in segregating hybrid progenies.

genetic flexibility — the capacity of a population to produce offspring different from their parents. G. f. is promoted by high mutation rates, heterozygosity, and outbreeding.

genetic milieu (*Chetverikov* 1926) — the sum of genetic factors which act from the inside on the manifestation of every gene.

genetic stability — the capacity of individuals to produce adaptive offspring. G. s. is promoted by asexual reproduction, → inbreeding, → homozygosity, and low → mutation rates (*Thoday* 1975).

genome analysis (*Kihara* 1930) — a method designed to determine the diploid ancestors of → allopolyploid species. It consists of the analysis of meiotic → chromosome pairing in hybrids between polyploids and diploids. If the diploid has at one time or another contributed to the formation of the polyploid, chromosome pairing should occur between two sets of homologous chromosomes in the hybrid.

genotypic cohesion — the phenomenon that balanced and superior gene combinations (co-adapted gene complexes or → supergenes) are hold together in the face of the centrifugal forces of → genetic recombination thus reducing the frequency of deleterious recombinants, and with it the → genetic load.

heterozygous advantage — the phenomenon that certain heterozygotes survive or reproduce better than homozygotes in higher plants and animals. H. a. results in the active maintenance of both alleles in the population and may be a mechanism involved in the maintenance of → genetic polymorphism.

histochemistry — the identification and localization of chemical constituents at the level of cells and tissues (→ cytochemistry).

hitch-hiking effect (*Kojima* & *Schaffer* 1967) — an effect, whereby the path of a weakly selected mutant on its way to eventual loss or fixation is influenced by → linkage to a strongly selected locus.

host-controlled restriction — the specific recognition and subsequent degradation of foreign DNA. H.-c. r. is initiated by a class of (bacterial) nucleases, the → restriction enzymes (→ DNA modification-restriction system).

hybrid plasmid — any → plasmid chimera (→ cloning vehicle) containing specific inserted DNA sequences and originating by genetic engineering or → molecular cloning.

initiation mass (*Donachie* 1968; *Helmstetter* et al. 1968) — the cell mass per chromosome origin (initiator) at which bacterial chromosome replication is initiated. The i. m. is constant over a broad range of growth rates. This ensures that the frequency of new rounds of replication and the frequency with which the culture mass doubles will be identical. In a formal sense, this relationship constitutes a mechanism for the control of DNA synthesis.

***in situ* nucleic acid hybridization** (*Gall* & *Pardue* 1969, *John* et al. 1969) — the hybridization of pre-labeled and denatured DNA or RNA fractions (obtained by *in vitro* synthesis from selected DNA fractions) to denatured chromosomes and characterization of the distribution of label by autoradiography. In this way *in situ* DNA/RNA and DNA/DNA hybridization reveals longitudinal differentiation along the chromosomes and allows to localize certain DNA sequences.

interfering protein (*Revel* et al. 1973) — = → interference factor.

intracistronic complementation — the association between differently altered mutant proteins or protein fragments from the same cistron to give biologically functional protein (→ genetic complementation).

kin selection (*Hamilton* 1964) — → selection for altruistic behavior between genetic relatives. K. s. has become the major genetic explanation of many evolutionary phenomena involving social behavior.

mating aggregate (*Achtman* 1975) — in a bacterial mating mixture, two (= mating pair) to twenty bacterial cells physically bound by connections invisible to the light microscope. The quantitative m. a. size distribution and the proportion of "male" cells in a m. a. are dependent on the input ratio of "male" and "female" cells (→ sex factor). Under optimal conditions essentially every "male" cell or every "female" cell in a m. a. can be involved in DNA transfer.

mating pair (*Anderson* 1958) — → mating aggregate.

methionyl-tRNA$_f$ binding factor (*Levin* et al. 1973) — in eukaryotes, a protein factor which forms a ternary complex with Met-tRNA$_f$ (the initiator tRNA) and GTP. Formation of this complex is followed by joining of Met tRNA$_f$ to the 40 S ribosomal subunit. This latter reaction requires additional → initiation factors but does not require added → messenger RNA.

microperoxisome (*Novikoff* & *Novikoff* 1973) — a class of cell organelles which occur in abundance in animal cell types whose major functions are the metabolism, transport, and storage of lipids. Microperoxisomes appear to be dilated areas, or derivatives from, smooth → endoplasmic reticulum.

microsomal mutagenicity assay — an *in vitro* technique which allows tissues to activate promutagens in the presence of a test organism in which induced mutation frequencies are simply determined. Basically, test organisms such as *Salmonella thyphimurium* histidine auxotrophs are incubated at body temperature with mammalian tissue (usually liver microsomes), the promutagen to be tested, and whatever cofactors are necessary for biotransformation of the promutagen to mutagenic metabolites.

The m. m. a. is useful as a screening technique only where a specific type of metabolic activation to a mutagenic product is highly suspected (→ mutagenicity testing).

mismatch repair (*Whitehouse* 1974) — enzymatic → repair of mismatched base pairs in → heteroduplex DNA regions (→ base pair mismatch). M. r. provides a mechanism for → genetic recombination of very closely spaced mutations, enhancing the production of otherwise rare genotypes and also facilitating the loss of deleterious mutations.

misrepair mutagenesis (*Green & Drake* 1974, *Bresler* 1975) — the generation of → gene mutations by a process in which a premutational lesion in DNA triggers the action of an error-prone → repair process. The resulting mutations probably depend primarily upon the error tendencies inherent in the steps of → repair synthesis, and not necessarily upon the chemical nature of the premutational lesion itself. M. m. results in mixtures of → transitions, → transversions, → frameshift mutations, and deletions.

mitochondrial crest — any of the complex infoldings of the inner membrane of the → mitochondrion which project into the matrix. Shape and disposition of m. crests vary in different cells; their number is related to the oxidative activity of the mitochondrion.

mutagen persistence — the stability of an environmental → mutagen in the environment and in the human body as defined by the half-life in the environment and the endogeneous half-life, respectively. The environmental mutagenic efficiency strongly depends on the m. p.

mutator activity — a given mutant is said to possess m. a. when strains carrying that allele show an increased mutation frequency for all genes relative to strains carrying the normal allele of the gene in question. → DNA ligase deficiency has been shown to induce generalized m. a. in *E. coli*.

nematosome (*Grillo* 1970) — in some types of somatic eucells, a structure in the cytoplasm which is suggested to be involved in the storage of long-lived informational RNA. Nematosomes lack a limiting membrane and consist mainly of a tangled reticulum of electron-dense fibrils in a less electron-dense substance (= nucleolus-like bodies).

nicking-closing enzyme (*Wang* 1971) — a class of enzymes which alter the topological winding number of closed circular DNA. The topological winding number is the number of revolutions that one strand of DNA makes about the other if the molecule is constrained to lie in a plane.

novel joint (*Hershey* 1970) — a new base sequence in tandem duplications not present in the parental sequence at a single point.

nu body (*Olins & Olins* 1973) — = → nucleosome.

nucleosome (*Oudet* et al. 1975) — any of the globular nucleoprotein particles (= nu body) along the DNA molecule connected by DNA filaments in → chromatin. The diameter of a n. is about 12.5 nm, it contains equal amounts of → histones H2A, H2B, H3, H4 and about 200 base pairs of DNA. The histones are present as octamers per n. with the composition $H2A_2$, $H2B_2$, $H3_2$ and $H4_2$. The size of the octamer is about 110000 daltons. Adjacent nucleosomes are connected by a strand of free DNA 1.8 to 3 nm long. The histone complex in a n. is probably enveloped by the DNA strand. Histone H1 is irrelevant to n. structure and formation.

occluding junction — = tight junction.

oöcyte assay of mRNA (*Gurdon* et al. 1971) — a sensitive technique for the assay of mRNA involving microinjections in nanogram amounts of polyribosomes or of other preparations containing mRNA into *Xenopus* oöcytes for efficient and long-term translation. Translation of exogeneous mRNAs in the oöcyte hase been shown with minimal estimates to be 100—1 000 times more efficient than in the reticulocyte cell-free system. The oöcyte lacks specificity with respect to the type of mRNA capable of being translated.

osmotic remediability (*Hawthorne* & *Friis* 1964, *Metzenburg* 1968) — a phenomenon demonstrated with a variety of auxotrophs in several organism which show phenotypic repair in growth media of high osmolarities. O. r. is probably the result, in some instances at least, of an effect which prevents the mutant protein from assuming an unstable configuration at the time of synthesis or changes the protein from an unstable to a stable form by some unknown cellular mechanism.

oxygen-mimic — of a variety of electron-affinic radiosensitizers capable of sensitizing hypoxic bacteria and mammalian cell populations to the lethal effects of radiation. Such compounds include various substituted forms of nitrofurans, nitroimidazoles, and nitrobenzenes.

packing ratio (*Du Praw* 1970) — the ratio of the length of a segment of DNA to the length of the → chromatin fiber it is packed in.

phospholipid vesicle (*Huang* 1969, *Johnson* et al. 1973) — a unilammellar → liposome consisting of a single, closed spherical bilayer of phospholipid, of outside diameter approximately 30 nm, which defines an internal aqueous space about 20 nm in diameter. Water-soluble molecules can readily be incorporated into the vesicle's interior space, and lipophilic molecules can be incorporated into the phospholipid bilayer.

phylogenetic tree — a graphic representation showing the descent relations of different organism (→ evolution).

pinocytotic vesicle — in plant → protoplasts, any of a class of vesicles smaller than → plasmolytic vesicles which arise both during removal of the cell wall and following isolation of the protoplast.

plasmid engineering — → genetic engineering utilizing plasmids to introduce foreign genes (→ molecular cloning) or composite DNA molecules. P. e. allows the reaction of a wide variety of novel genetic combinations in microorganisms. The procedure of p. e. consists of several distinct biochemical and biological manipulations which essentially consist of 1) a method of cleaving and joining DNA molecules derived from different sources; 2) a suitable gene carrier (plasmid) that can replicate both itself and a foreign DNA segment linked to 2); 3) a means of introducing the composite DNA molecules into a functional (bacterial) cell, and 4) a method of selecting from a large population of cells a clone of recipient cells that has acquired the molecular chimera.

plasmid integration — the integration of a plasmid into the bacterial chromosome. P. i. is sporadic, involves preferred sites, and is usually not mutagenic but may sometimes suppress the replication defect of certain mutants.

plasmid superinfection immunity (*Meynell* et al. 1968) — the reduction in recipient ability for one plasmid caused by a second plasmid (= superinfection inhibition). The phenomenon involves at least two operationally distinguishable processes which have been referred to as "surface exclusion" or → "entry exclusion" and intracellular incompatibility.

plasmolytic vesicle — in plant → protoplasts, any of a class of single-membraned vesicles, several nm across and originating as gross infoldings of the plasmalemma into the cytoplasm as the protoplast contracts during plasmolysis (→ pinocytotic vesicle).

polyadenylation — the post-transcriptional addition of poly(A) tracts (50–200 nucleotides long) to the 3'-OH terminus of → heterogeneous nuclear RNA and → messenger RNA of eukaryotic cells, as well as viral and viral-directed RNAs.

poly(A) polymerase (*Dravid* et al. 1971): any of a class of enzymes which catalyze the non-transcriptive addition of AMP residues to → heterogeneous nuclear RNA (hn RNA). Nuclear polyadenylation may proceed by initiation of poly(A) chains on a specific acceptor sequence of hnRNA or elongation of the pre-existing poly(A) tract.

preinitiation stretch (*Schäfer* et al. 1973) — an untranscribed stretch of nucleotides in the → promoter region of an → operon which is assumed to serve as a storage site for RNA polymerase molecules.

protoplast fusion — in plant cell culture, the fusion of protoplasts, i.e., cells isolated from cell cultures and from plant tissues by enzymatic removal of the → cell walls. A prerequisite for effective fusion is that extensive membrane contact and activation must be established. The process usually requires an aggregating agent. Polyethylene glycol has been found to effectively agglutinate and to facilitate fusion of protoplasts of different genera.

rate-doubling concentration — the concentration of a chemical that produces as much genetic damage as occurs spontaneously in the same period of time in a particular system used for → mutagenicity testing (→ doubling dose; rem-equivalent-chemical).

relative biological effectiveness — the inverse ratio of the absorbed dose from one radiation type to that of a reference radiation required to produce the same degree of a stipulated biological effect (symbol RBE).

rem-equivalent-chemical — that dose (REC) or product of concentration multiplied by time which produces an amount of genetic damage equal to that produced by 1 rem (= one radiation-equivalent-man) of chronic irradiation (→ rate-doubling concentration).

secretory vesicle — a membrane-bounded container in secretory eucells (= s. granule). In the s. v. secretory products accumulate. It is generally located at the apex of the cell.

sequential mutagenesis (*Cerda-Olmedo* et al. 1968) — a technique which exploits the increased susceptibility of the replication point of the bacterial → chromosme to → mutagenesis by N-methyl-N'-nitro-N-nitrosoguanidine. The mutagen treatment of successive samples from a synchronized bacterial culture and the scoring of treated cells for mutants results in a maximum number of mutants for particular genes occurring at the time of gene replication.

somatic cell genetics — a branch of genetics which includes two distinct, though complementary and overlapping approaches: 1) the use of somatic cells in culture (very much like the use of cultures of microorganisms) for the study of general problems as e.g., mutation, replication of DNA, differentiation, metabolism of macromolecules, selection. 2) The use of somatic cells, mainly in culture, as an alternative to germ cells for the genetic analysis of the individual from which the somatic cells were taken (*Pontecorvo* 1975). (→ somatic cell fusion).

SOS repair (*Witkin & George* 1973; *Radman* 1975) — an error-prone inducible variant of bacterial → post-replication or recombinational repair (= SOS replication of *Radman*). DNA damage signals the induction of this type of repair components of which are required for → UV reactivation, UV mutagenesis, and → prophage induction.

S phase recovery (*Lehmann* 1972) — a → repair system active during normal DNA replication of eukaryotes which enables DNA to be replicated on damaged templates.

superinduction — the increase of the activity and/or the concentration of numerous proteins in eukaryotic cells by actinomycin D, an inhibitor of RNA synthesis. The phenomenon is paradoxical because one might expect the drug to inhibit protein synthesis through its inhibition of RNA synthesis.

tc RNA (*Heywood* et al. 1974) — abbreviation for → translational control RNA.

teratology — the study of the causes and development of → congenital malformations.

terminal deoxynucleotid yltransferase — an unusual type of → DNA polymerase exhibiting a unique distribution in lymphoid cells. It catalyzes the synthesis of a deoxynucleotide sequence without template direction. Its normal presence in thymus and, to a lesser extent, in bone marrow suggests a possible role in immunological diversification.

transconjugant (*Novick* et al. 1973) — a bacterial cell which has received genetic material from another bacterium by → conjugation.

transferant (*Solioz* et al. 1975) — → gene transfer agent.

transfer replication — the synthesis of DNA associated with DNA transfer during bacterial → conjugation (mating).

translational control RNA (*Heywood* et al. 1974) — in eukaryotes, RNA molecules (symbol tc RNA) involved in the utilization of long-lived (masked, or stored) → messenger RNAs and in the the maintenance of these molecules in the inactive form. tc RNA molecules are effective in the regulation of protein synthesis within the cell from which they are derived.

transposon (*Hedges & Jacob* 1974) — in bacteria, a discrete translocateable element capable of → translocation between chromosomal, phage, and → plasmid DNA. Transposons are apparently able to insert into and excise from DNA molecules in the absence of a functional bacterial recombination (recA) system. They have not been found to exist autonomously, i.e., they need to reside in a functional replicon. Possibly, a t. has on both ends inverted complementary DNA sequences (inverted repetitions) of between 100 and 1,500 nucleotide pairs. Insertions of transposons are mutagenic when occurring in structural genes and polar when occurring within operons (→ insertion mutation). Generally, insertion of a t. is a relatively precise process which does not damage the informational capacity of host DNA. Excision, however, frequently results in permanent loss of host genetic information.

xenotropic (*Levy* 1974) — of viruses which, in contrast to ecotropic viruses, can propagate infectiously in their host species of origin and appear to be transmitted in the natural hosts as Mendelian traits, probably representing integrated DNA provirus forms of the genome. When activated to form mature virus particles, x. viruses can only infect and replicate in host cells of certain foreign species.

yolk platelet (*Anderson* 1968) — in mature unfertilized animal eggs, any of a distinct species of subcellular organelles which are ensembled from derivatives of the → Golgi apparatus during → oögenesis. Yolk platelets are storage granules for metabolites required during embryogenesis. They and the → cortical granules exhibit many structural and functional features usually associated with → lysosomes in adult tissues and have been termed "lysosome-like" organelles. Mucopolysaccharides and other glycoproteins as well as acid phosphates are biochemical constituents of y. platelets.

Other Books of Interest

T. C. Hsu, K. Benirschke:
An Atlas of Mammalian Chromosomes
Vol. 1: 50 plates with 212 figs. X, 200 pages. 1867
Vol. 2: 50 plates. XX, 200 pages. 1968
Vol. 3: 50 plates. XX, 200 pages. 1969
Vol. 4: 50 plates. XX, 200 pages. 1970
Vol. 5: 50 plates. XXVII, 200 pages. 1971
Vol. 6: 50 plates. XLIII, 200 pages. 1971
Vol. 7: 50 plates. XVIII, 264 pages. 1973
Vol. 8: 50 plates. XXXIX, 211 pages. 1974
Vol. 9: 51 plates. XVI, 261 pages. 1975
All volumes are loose-leaf boxed.

Related Journals

Chromosoma
Managing Editor: H. Bauer
1976, Vols. 54 – 57 (4 issues each)

Human Genetics
Editorial Board: W. Lenz, A. G. Motulsky, U. W. Schnyder, F. Vogel, G. G. Wendt
1976, Vols. 31 – 35 (3 issues each)

Immunogenetics
Editor-in-Chief: G. D. Snell
1976, Vol. 3 (6 issues)

MGG – Molecular & General Genetics
An International Journal
Managing Editors: G. Melchers, H. Stubbe
1976, Vols. 143 – 147 (3 issues each)

Theoretical and Applied Genetics – TAG
Internationale Zeitschrift für Theoretische und Angewandte Genetik
Managing Editor: H. Stubbe
1976, Vols. 47 – 48 (6 issues each)

Springer-Verlag Berlin Heidelberg New York